BRIEF CONTENTS

T0200452

Book contents

Online contents

iii

CONTENTS

Book contents

STATISTICS
FOR BUSINESS AND ECONOMICS

SWEENEY
WILLIAMS
CAMM
COCHRAN
FRY
OHLMANN
FREEMAN
SHOESMITH

Australia • Brazil • Canada • Mexico • South Africa • Singapore • United Kingdom • United States

SIXTH EDITION

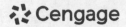

Statistics for Business and Economics,
6th Edition
David R. Anderson, Dennis J. Sweeney,
Thomas A. Williams, Jeffrey D. Camm,
James J. Cochran, Michael J. Fry and
Jeffrey W. Ohlmann
James Freeman and Eddie Shoesmith

Publisher: Annabel Ainscow

List Manager: Birgit Gruber

Marketing Manager: Louise Corless

Associate Content Project Manager:
Narmada Kaushal

Manufacturing Manager: Eyvett Davis

Manufacturing Buyer: Elaine Bevan

Typesetter: Lumina Datamatics, Inc.

Text Design: Lumina Datamatics, Inc.

Cover Design: Simon Levy

Cover and Interior Heading Banner
Image(s): © gremlin/iStock

For product information and technology assistance, contact us at
emea.info@cengage.com

For permission to use material from this text or product and for permission queries, email **emea.permissions@cengage.com**

British Library Cataloguing-in-Publication Data

A catalogue record for this book is available from the British Library.

ISBN: 978-1-4737-9135-0

Cengage Learning, EMEA
Cheriton House, North Way
Andover, Hampshire, SP10 5BE
United Kingdom

Cengage Learning is a leading provider of customized learning solutions with employees residing in nearly 40 different countries and sales in more than 125 countries around the world. Find your local representative at: **cengage.uk.**

To learn more about Cengage platforms and services, register or access your online learning solution, or purchase materials for your course, visit **cengage.uk.**

Printed in the United Kingdom by CPI Antony Rowe
Print Number: 01   Print Year: 2024

DEDICATION

'To the memory of my grandparents, Lizzie and Halsey'

JAMES FREEMAN

'To all my family, past, present and future'

EDDIE SHOESMITH

PREFACE

The purpose of *Statistics for Business and Economics* is to give students, primarily those in the fields of business, management and economics, a conceptual introduction to the field of statistics and its many applications. The text is applications oriented and written with the needs of the non-mathematician in mind. The mathematical prerequisite is knowledge of algebra.

Applications of data analysis and statistical methodology are an integral part of the organization and presentation of the material in the text. The discussion and development of each technique are presented in an application setting, with the statistical results providing insights to problem solution and decision-making.

Although the book is applications oriented, care has been taken to provide sound methodological development and to use notation that is generally accepted for the topic being covered. Hence, students will find that this text provides good preparation for the study of more advanced statistical material. A revised and updated bibliography to guide further study is included as an appendix.

The online platform introduces the student to the software packages IBM SPSS Statistics 29, Minitab 21, R 4.3 and Microsoft® Office Excel 2021, and emphasizes the role of computer software in the application of statistical analysis. Minitab and SPSS are illustrated as they are two of the leading commercial statistical software packages for both education and statistical practice. R is a free software environment and programming language for statistical computing, supported by the R Foundation for Statistical Computing, which is being increasingly used in education as well as research. Excel is not a statistical software package, but the wide availability and use of Excel makes it important for students to understand the statistical capabilities of this package. SPSS, Minitab, R and Excel procedures are provided on the dedicated online platform so that instructors have the flexibility of using as much computer emphasis as desired for the course.

THE EMEA EDITION

This is the 6th EMEA edition of *Statistics for Business and Economics*. It is based on the 5th EMEA edition and the 14th US edition. The US editions have a distinguished history and deservedly high reputation for clarity and soundness of approach. We have aimed to maintain the style and readability of those editions, while recognizing that the EMEA edition is intended for students in many parts of the world, many of whom do not have English as a first language. We have replaced most of the US-based examples, case studies and exercises with equally interesting and appropriate ones sourced from a wider geographical base, particularly the UK, Ireland, continental Europe, South Africa and the Middle East. Other notable changes in this 6th EMEA edition are summarized here.

CHANGES IN THE 6TH EMEA EDITION

- **Content revisions** The following content revisions appear in the new edition:
  - The section on data mining in Chapter 1 has been expanded to include a discussion of ethical guidelines for statistical practice.

- Sections on data visualization and data dashboards have been added to Chapters 2 and 3.
- Chapters 7, 8, 9, 14 and 15 have been expanded to include practical advice on big data and the implications for statistical inference.
- A number of case problems has been added or updated. These are in the chapters on Descriptive statistics: tabular and graphical presentations, Descriptive statistics: numerical measures, Continuous probability distributions, Sampling and sampling distributions, Hypothesis tests, Inferences about population variances and Experimental design and analysis of variance. These case problems provide students with the opportunity to analyze somewhat larger data sets and prepare managerial reports based on the results of the analysis.
- Each chapter begins with a Statistics in Practice article that describes an application of the statistical methodology to be covered in the chapter. The Statistics in Practice articles for Chapters 10, 11, 16 and 17 are new, and most other articles have been substantially updated and revised for this new edition.
- New examples and exercises have been added throughout the book, based on real data and recent reference sources of statistical information. We believe that the use of real data helps generate more student interest in the material and enables the student to learn about both the statistical methodology and its application.
- As with the previous EMEA edition, to accompany the exercises and examples, data files are available on the online platform in only one format, CSV (comma separated values). This format is accessible to Minitab, R, Excel and SPSS. Data set icons are used in the text to identify the data sets that are available on the online platform. Data sets for all case problems as well as data sets for larger exercises are included.
- Appendix D, part of the online resources for students, has solutions for all odd numbered exercises.
- **Software sections** In the 6th EMEA edition, we have updated the software sections (on the online platform) to provide step-by-step instructions for the following software packages: SPSS 29, Minitab 21, R 4.3 and Microsoft® Office Excel 2021.

Online Learning Platform **WebAssign** is available for this title, providing flexible and customizable courses with learning resources, assignments and secure testing.

DEDICATED INSTRUCTOR WEBSITE RESOURCES

This includes the following resources for lecturers:
- Solutions' Manual: all exercises in the book, plus extra exercises, with solutions
- Cognero test bank with approximately 500 extra questions and answers
- PowerPoint slides
- Case studies (internationally focused) with solutions
- Downloadable figures and tables from the book

ADDITIONAL ONLINE RESOURCES

Also available are appendices, all data sets, four online chapters (Chapter 19: Index Numbers; Chapter 20: Statistical Methods for Quality Control; Chapter 21: Decision Analysis; Chapter 22: Sample Surveys), sample papers and software sections: SPSS 29, Minitab 21, R 4.3 and Microsoft® Office Excel 2021.

ACKNOWLEDGEMENTS

The authors and publisher acknowledge the contribution of the following reviewers throughout the six editions of this textbook:

- Mona Fouad AlWakel – Dar Al Uloom University (Saudi Arabia)
- Alban Asllani – Coventry University (UK)
- John R. Calvert – Loughborough University (UK)
- Deliang Dai – Linnaeus University (Sweden)
- Naomi Feldman – Ben-Gurion University of the Negev (Israel)
- Luc Hens – Vesalius College (Belgium)
- Theo Jansen – Hotel Management School Maastricht (Netherlands)
- Martyn Jarvis – University of Glamorgan (UK)
- Khalid M. Kisswani – Gulf University for Science & Technology (Kuwait)
- Tim Low – University of Cape Town (South Africa)
- Issam Malki – University of Westminster (UK)
- Alan Matthews – Trinity College Dublin (Ireland)
- Suzanne McCallum – Glasgow University (UK)
- Chris Muller – University of Stellenbosch (South Africa)
- Surette Oosthuizen – University of Stellenbosch (South Africa)
- Zhan Pang – Lancaster University (UK)
- Karim Sadrieh – Otto von Guericke University Magdeburg (Germany)
- Tulonga Shaalukeni – University of Namibia (Namibia)
- Mark Stevenson – Lancaster University (UK)
- Nicholas Vasilakos – University of East Anglia (UK)
- Dave Worthington – Lancaster University (UK)

ABOUT THE AUTHORS

James Freeman is former Senior Lecturer in Statistics and Operational Research at Alliance Manchester Business School (AMBS), UK. He was born in Tewkesbury, Gloucestershire. After taking a first degree in Pure Mathematics at UCW Aberystwyth, he went on to receive MSc and PhD degrees in Applied Statistics from Bath and Salford universities respectively. In 1992/3 he was Visiting Professor at the University of Alberta. Before joining AMBS, he was Statistician at the Distributive Industries Training Board – and prior to that – the Universities Central Council on Admissions. He has taught undergraduate and postgraduate courses in business statistics and operational research courses to students from a wide range of management and engineering backgrounds. Until 2017 he taught the statistical core course on AMBS's Business Analytics masters programme – since rated top in Europe and sixth in the world. For many years he was also responsible for providing introductory statistics courses to staff and research students at the University of Manchester's Staff Teaching Workshop. Through his gaming and simulation interests he has been involved in a significant number of external consultancy and grant-aided projects. More recently he received significant government ('KTP') funding for research in the area of risk management.

Between July 2008 and December 2014 he was Editor of the Operational Research Society's *OR Insight* journal and between 2017 and 2021 was Editor of the Tewkesbury Historical Society Bulletin.

In November 2012 he received the Outstanding Achievement Award at the *Decision Sciences Institutes 43rd Annual Meeting* in San Francisco. In 2018 he was awarded an Honorary Fellowship by the University of Manchester.

Eddie Shoesmith is a Fellow of the University of Buckingham, UK, where he was formerly Senior Lecturer in Statistics and Programme Director for undergraduate business and management programmes in the School of Business. He was born in Barnsley, Yorkshire. He was awarded an MA (Natural Sciences) at the University of Cambridge, and a BPhil (Economics and Statistics) at the University of York. Prior to taking an academic post at Buckingham, he worked for the UK Government Statistical Service (now the UK Office for National Statistics), in the Cabinet Office, for the London Borough of Hammersmith and for the London Borough of Haringey. At Buckingham, before joining the School of Business, he held posts as Dean of Sciences and Head of Psychology. He has taught introductory and intermediate-level applied statistics courses to undergraduate and postgraduate student groups in a wide range of disciplines: business and management, economics, accounting, psychology, biology and social sciences. He has also taught statistics to social and political sciences undergraduates at the University of Cambridge.

David R. Anderson is Professor Emeritus of Quantitative Analysis in the College of Business Administration at the University of Cincinnati, USA. Born in Grand Forks, North Dakota, he earned his BS, MS and PhD degrees from Purdue University, USA. Professor Anderson has served as Head of the Department of Quantitative Analysis and Operations Management and as Associate Dean of the College of Business Administration at the University of Cincinnati. In addition, he was the coordinator of the college's first executive programme.

At the University of Cincinnati, Professor Anderson has taught introductory statistics for business students as well as graduate-level courses in regression analysis, multivariate analysis and management science. He also has taught statistical courses at the Department of Labor in Washington, DC. Professor Anderson has been honoured with nominations and awards for excellence in teaching and excellence in service to student organizations. Professor Anderson has co-authored ten textbooks in the areas of statistics, management science, linear programming, and production and operations management. He is an active consultant in the field of sampling and statistical methods.

Dennis J. Sweeney is Professor Emeritus of Quantitative Analysis and founder of the Center for Productivity Improvement at the University of Cincinnati, USA. Born in Des Moines, Iowa, he earned a BSBA degree from Drake University, and his MBA and DBA degrees from Indiana University, where he was an NDEA Fellow. Professor Sweeney has worked in the management science group at Procter & Gamble and spent a year as a visiting professor at Duke University, USA. Professor Sweeney served as Head of the Department of Quantitative Analysis and as Associate Dean of the College of Business Administration at the University of Cincinnati.

Professor Sweeney has published more than 30 articles and monographs in the area of management science and statistics. The National Science Foundation, IBM, Procter & Gamble, Federated Department Stores, Kroger and Cincinnati Gas & Electric have funded his research, which has been published in *Management Science, Operations Research, Mathematical Programming, Decision Sciences* and other journals.

Professor Sweeney has co-authored ten textbooks in the areas of statistics, management science, linear programming, and production and operations management.

Thomas A. Williams is Professor Emeritus of Management Science in the College of Business at Rochester Institute of Technology (RIT), USA. Born in Elmira, New York, he earned his BS degree at Clarkson University, USA. He did his graduate work at Rensselaer Polytechnic Institute, USA, where he received his MS and PhD degrees.

Before joining the College of Business at RIT, Professor Williams served for seven years as a faculty member in the College of Business Administration at the University of Cincinnati, USA, where he developed the first undergraduate programme in Information Systems and then served as its coordinator. At RIT he was the first chair of the Decision Sciences Department. He teaches courses in management science and statistics, as well as graduate courses in regression and decision analysis.

Professor Williams is the co-author of 11 textbooks in the areas of management science, statistics, production and operations management, and mathematics. He has been a consultant for numerous *Fortune* 500 companies and has worked on projects ranging from the use of elementary data analysis to the development of large-scale regression models.

Jeffrey D. Camm is the Inmar Presidential Chair and Associate Dean of Analytics in the School of Business at Wake Forest University, USA. Born in Cincinnati, Ohio, he holds a BS from Xavier University (Ohio) and a PhD from Clemson University, both in the USA. Prior to joining the faculty at Wake Forest, he was on the faculty of the University of Cincinnati. He has also been a visiting scholar at Stanford University and a visiting professor of business administration at the Tuck School of Business at Dartmouth College, both in the USA.

Dr Camm has published over 45 papers in the general area of optimization applied to problems in operations management and marketing. He has published his research in *Science, Management Science, Operations Research, Interfaces* and other professional journals. Dr Camm was named the Dornoff Fellow of Teaching Excellence at the University of Cincinnati and he was the recipient of the 2006 INFORMS Prize for the Teaching of Operations Research Practice. A firm believer in practising what he preaches, he has served as an operations research consultant to numerous companies and government agencies. From 2005 to 2010 he served as editor-in-chief of *Interfaces*. In 2017, he was named an INFORMS Fellow. From 2005 to 2010, he served as editor-in-chief of *INFORMS Journal of Applied Analytics* (formerly *Interfaces*). In 2017, he was named an INFORMS Fellow.

James J. Cochran is Associate Dean for Faculty and Research, Professor of Applied Statistics and the Rogers-Spivey Faculty Fellow at the University of Alabama, USA. Born in Dayton, Ohio, he earned his BS, MS and MBA degrees from Wright State University and a PhD from the University of Cincinnati, both in the USA. He has been at the University of Alabama since 2014 and has been a visiting scholar at Stanford University, USA; Universidad de Talca, Chile; the University of South Africa; and Pole Universitaire Leonard de Vinci, France.

Professor Cochran has published over 45 papers in the development and application of operations research and statistical methods. He has published his research *in Management Science, The American Statistician, Communications in Statistics – Theory and Methods, Annals of Operations Research, European Journal of Operational Research, Journal of Combinatorial Optimization, INFORMS Journal of Applied Analytics, Statistics and Probability Letters* and other professional journals. He was the recipient of the 2008 INFORMS Prize for the Teaching of Operations Research Practice and the 2010 recipient of the Mu Sigma Rho Statistical Education Award.

Professor Cochran was elected to the International Statistics Institute in 2005 and named a Fellow of the American Statistical Association in 2011. He received the Founders Award in 2014 and the Karl E. Peace Award in 2015 from the American Statistical Association. In 2017 he received the American Statistical Association's Waller Distinguished Teaching Career Award and was named a Fellow of INFORMS, and in 2018 he received the INFORMS President's Award.

A strong advocate for effective statistics and operations research education as a means of improving the quality of applications to real problems, Professor Cochran has organized and chaired teaching effectiveness workshops in Montevideo, Uruguay; Cape Town, South Africa; Cartagena, Colombia; Jaipur, India; Buenos Aires, Argentina; Nairobi, Kenya; Buea, Cameroon; Kathmandu, Nepal; Osijek, Croatia; Havana, Cuba; Ulaanbaatar, Mongolia; Chisinău, Moldova; Dar es Salaam, Tanzania; Sozopol, Bulgaria; Tunis, Tunisia; and Saint George's, Grenada. He has served as an operations research consultant to numerous companies and not-for-profit organizations. He served as editor-in-chief of *INFORMS Transactions on Education* from 2006 to 2012 and is on the editorial board of *INFORMS Journal of Applied Analytics* (formerly *Interfaces*), *International Transactions in Operational Research, and Significance.*

Michael J. Fry is Professor of Operations, Business Analytics and Information Systems and Academic Director of the Center for Business Analytics in the Carl H. Lindner College of Business at the University of Cincinnati, USA. Born in Killeen, Texas, he earned a BS from Texas A&M University and MSE and PhD degrees from the University of Michigan, USA. He has been at the University of Cincinnati since 2002, where he was previously Department Head and has been named a Lindner Research Fellow. He has also been a visiting professor at the Samuel Curtis Johnson Graduate School of Management at Cornell University, USA, and the Sauder School of Business at the University of British Columbia, Canada.

Professor Fry has published more than 25 research papers in journals such as *Operations Research, M&SOM, Transportation Science, Naval Research Logistics, IISE Transactions, Critical Care Medicine* and *INFORMS Journal of Applied Analytics* (formerly *Interfaces*). His research interests are in applying quantitative management methods to the areas of supply chain analytics, sports analytics and public-policy operations.

He has worked with many different organizations for his research, including Dell, Inc., Starbucks Coffee Company, Great American Insurance Group, the Cincinnati Fire Department, the State of Ohio Election Commission, the Cincinnati Bengals and the Cincinnati Zoo & Botanical Garden. He was named a finalist for the Daniel H. Wagner Prize for Excellence in Operations Research Practice, and he has been recognized for his research as well as teaching excellence at the University of Cincinnati.

Jeffrey W. Ohlmann is Associate Professor of Management Sciences and Huneke Research Fellow in the Tippie College of Business at the University of Iowa, USA. Born in Valentine, Nebraska, he earned a BS from the University of Nebraska, USA, and MS and PhD degrees from the University of Michigan, USA. He has been at the University of Iowa since 2003.

Professor Ohlmann's research on the modelling and solution of decision-making problems has produced more than two dozen research papers in journals such as *Operations Research, Mathematics of Operations Research, INFORMS Journal on Computing, Transportation Science, European Journal of Operational Research* and *INFORMS Journal of Applied Analytics* (formerly *Interfaces*). He has collaborated with companies such as Transfreight, LeanCor, Cargill, the Hamilton County Board of Elections and three National Football League franchises. Because of the relevance of his work to industry, he was bestowed the George B. Dantzig Dissertation Award and was recognized as a finalist for the Daniel H. Wagner Prize for Excellence in Operations Research Practice.

WebAssign

Develop Confident, Independent Learners

Built by educators, *WebAssign* provides you with flexible settings at every step to customize your course with online learning resources, assignments and secure testing. Students can access rich content and study resources designed to fuel deeper understanding, plus a dynamic, intuitive eTextbook, in one place. Proven to help hone problem-solving skills, *WebAssign* fosters learning in any course delivery model.

Save time on grading and administrative work

Customize your course and content

Accelerate student progress with analytics

Connect with a dedicated support team— for you and your students

cengage.com/webassign

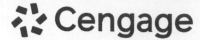

Teaching & Learning Support Resources

Cengage's peer-reviewed content for higher and further education courses is accompanied by a range of digital teaching and learning support resources. The resources are carefully tailored to the specific needs of the instructor, student and the course. Examples of the kind of resources provided include:

A password-protected area for instructors with, for example, a test bank, PowerPoint slides and an instructor's manual.

Lecturers: to discover the dedicated teaching digital support resources accompanying this textbook, please register here for access: **account.cengage.com/login**

Students: to discover the dedicated learning digital support resources accompanying this textbook, contact your instructor or Cengage via our website

1
Data and Statistics

CHAPTER CONTENTS

Statistics in Practice *The Economist*

1.1 Applications in business and economics
1.2 Data
1.3 Data sources
1.4 Descriptive statistics
1.5 Statistical inference
1.6 Analytics
1.7 Big data and data mining
1.8 Computers and statistical analysis
1.9 Ethical guidelines for statistical practice

LEARNING OBJECTIVES After reading this chapter and doing the exercises, you should be able to:

1 Appreciate the breadth of statistical applications in business and economics.

2 Understand the meaning of the terms elements, variables and observations, as they are used in statistics.

3 Understand the difference between qualitative, quantitative, cross-sectional and time series data.

4 Find out about data sources available for statistical analysis both internal and external to the firm.

5 Appreciate how errors can arise in data.

6 Understand the meaning of descriptive statistics and statistical inference.

7 Distinguish between a population and a sample.

8 Understand the role a sample plays in making statistical inferences about the population.

Frequently, we see the following kinds of statements in newspaper and magazine articles:

- The IFO Business Climate Index dipped to 84.3 points in October 2022 from 84.4 points the previous month. (*IFO Institute*, 25 October 2022)
- The IMF projected global growth to slow from an estimated 6.1 per cent in 2021 to 3.6 per cent for 2023. (*IMF*, 28 January 2023)
- According to renowned financial services provider, AJ Bell, the FTSE 100 will reach 8,250 by the end of 2023. The previous high for the index was 7,877 in June 2019. (*The Motley Fool*, 30 December 2022)
- China's GDP is likely to have grown just 2.8 per cent in 2022 as COVID lockdowns weighed on activity and confidence, according to the median forecasts of 49 economists polled by Reuters. (*Reuters*, 16 January 2023)
- After anaemic sales in 2022, the German auto industry association (VDA) said it forecast 2 per cent growth in new car sales in Germany this year, to 2.7 million vehicles – 'a quarter fewer than in 2019' before the economic crisis unleashed by the coronavirus pandemic. (*The Economic Times*, 11 January 2023)
- The *IDC Smartphone Tracker* reported a below par record of worldwide smartphone shipments for Q4 2022 as the market contracted by 18.3 per cent to 300.3 million shipments. (*IDC*, 23 January 2023)

The numerical facts in the preceding statements (84.3 points, 3.6 per cent, 8,250, 2.8 per cent, 2.7m, 300.3m) are called statistics. Thus, in everyday usage, the term *statistics* refers to numerical facts. However, the field, or subject, of statistics involves much more than numerical facts. In a broad sense, statistics is the art and science of collecting, analyzing, presenting and interpreting data. Particularly in business and economics, the information provided by collecting, analyzing, presenting and interpreting data gives managers and decision-makers a better understanding of the business and economic environment and thus enables them to make more informed and better decisions. In this text, we emphasize the use of statistics for business and economic decision-making.

STATISTICS IN PRACTICE
The Economist

The *Economist* is published by the Economist Group – an international company employing 1,325 staff (2021 average) worldwide – with offices in Beijing, Birmingham, Berlin, Chicago, Dubai, Frankfurt, Geneva, Gurugram, Hong Kong, Johannesburg, London, New York, Paris, San Francisco, Sao Paolo, Shanghai, Singapore, Sydney, Tokyo and Washington DC.

Between July and December 2019, the magazine's global circulation was in excess of 1.6m, made up of 909,476 (print) and 748,459 (digital) editions per week. By June 2021, some 53 per cent of its circulation was in North America compared to 17 per cent in the UK.

Impressively, its digital presence has soared in recent years – as reflected by such statistics as:

13.1m LinkedIn followers
281m YouTube views
6m Instagram followers
2.9m subscribers (52 per cent of them digital)

Complementing *The Economist* magazine division within the Economist Group family are the Economist Intelligence division – renowned for its custom research and analysis services – and the Economist Education division – provider of tailored executive courses – and client solutions division, Economist Impact.

Source: pressgazette.co.uk/news/economist-profit-2022-results/.

© golibo/iStock

Chapter 1 begins with some illustrations of the applications of statistics in business and economics. In Section 1.2 we define the term *data* and introduce the concept of a data set. This section also introduces key terms such as *variables* and *observations*, discusses the difference between quantitative and categorical data, and illustrates the uses of cross-sectional and time series data. Section 1.3 discusses how data can be obtained from existing sources or through survey and experimental studies designed to obtain new data. The important role that the internet now plays in obtaining data is also highlighted. The use of data in developing descriptive statistics and in making statistical inferences is described in Sections 1.4 and 1.5. The next three sections of Chapter 1 outline respectively the role of analytics and computers in statistical analysis and introduce the relatively new field of data mining. The final section is concerned with ethical guidelines for statistical practice.

1.1 APPLICATIONS IN BUSINESS AND ECONOMICS

In today's global business and economic environment, anyone can access vast amounts of statistical information. The most successful managers and decision-makers understand the information and know how to use it effectively. In this section we provide examples that illustrate some of the uses of statistics in business and economics.

Accounting

Public accounting firms use statistical sampling procedures when conducting audits for their clients. For instance, suppose an accounting firm wants to determine whether the amount of accounts receivable shown on a client's balance sheet fairly represents the actual amount of accounts receivable. Usually the large number of individual accounts receivable makes reviewing and validating every account too time-consuming and expensive. As common practice in such situations, the audit staff selects a subset of the accounts called a sample. After reviewing the accuracy of the sampled accounts, the auditors draw a conclusion as to whether the accounts receivable amount shown on the client's balance sheet is acceptable.

Finance

Financial analysts use a variety of statistical information to guide their investment recommendations. In the case of stocks, the analysts review a variety of financial data including price/earnings ratios and dividend yields. By comparing the information for an individual stock with information about the stock market averages, a financial analyst can begin to draw a conclusion as to whether an individual stock is over- or under-priced. Similarly, historical trends in stock prices can provide a helpful indication on when investors might consider entering (or re-entering) the market.

Marketing

Electronic scanners at retail checkout counters collect data for a variety of marketing research applications. For example, data suppliers such as ACNielsen purchase point-of-sale scanner data from grocery stores, process the data and then sell statistical summaries of the data to manufacturers. Manufacturers spend vast amounts per product category to obtain this type of scanner data. Manufacturers also purchase data and statistical summaries on promotional activities such as special pricing and the use of in-store displays. Brand managers can review the scanner statistics and the promotional activity statistics to gain a better understanding of the relationship between promotional activities and sales. Such analyses often prove helpful in establishing future marketing strategies for the various products.

Production

Today's emphasis on quality makes quality control an important application of statistics in production. A variety of statistical quality control charts are used to monitor the output of a production process. In particular, an x-bar chart can be used to monitor the average output. Suppose, for example, that a machine fills containers with 330g of a soft drink. Periodically, a production worker selects a sample of containers and computes the average number of grams in the sample. This average, or x-bar value, is plotted on an x-bar chart. A plotted value above the chart's upper control limit indicates overfilling, and a plotted value below the chart's lower control limit indicates underfilling. The process is termed 'in control' and is allowed to continue as long as the plotted x-bar values fall between the chart's upper and lower control limits. Properly interpreted, an x-bar chart can help determine when adjustments are necessary to correct a production process.

Economics

Economists frequently provide forecasts about the future of the economy or some aspect of it. They use a variety of statistical information in making such forecasts. For instance, in forecasting inflation rates, economists use statistical information on such indicators as the Producer Price Index (PPI), the unemployment rate and manufacturing capacity utilization. Often these statistical indicators are entered into computerized forecasting models that predict inflation rates.

Applications of statistics such as those described in this section are an integral part of this text. Such examples provide an overview of the breadth of statistical applications. To supplement these examples, chapter-opening Statistics in Practice articles obtained from a variety of topical sources are used to introduce the material covered in each chapter. These articles show the importance of statistics in a wide variety of business and economic situations.

1.2 DATA

Data are the facts and figures collected, analyzed and summarized for presentation and interpretation. All the data collected in a particular study are referred to as the **data set** for the study. Table 1.1 shows a data set containing information for 60 nations that participate in the World Trade Organization (WTO). The WTO encourages the free flow of international trade and provides a forum for resolving trade disputes.

Elements, variables and observations

Elements are the entities on which data are collected. Each nation listed in Table 1.1 is an element with the nation or element name shown in the first column. With 60 nations, the data set contains 60 elements.

A **variable** is a characteristic of interest for the elements. The data set in Table 1.1 includes the following four variables:

- *WTO Status*: The nation's membership status in the WTO; this can be either as a member or an observer.
- *Per Capita Gross Domestic Product (GDP) ($)*: The total market value ($) of all goods and services produced by the nation divided by the number of people in the nation; this is commonly used to compare economic productivity of the nations.
- *Rating*: The nation's sovereign credit rating as appraised by the credit ratings provided by one of the three main independent rating agencies; the credit ratings range from a high of AA to a low of F and can be modified by + or −.
- *Outlook*: An indication of the direction the credit rating is likely to move over the upcoming two years; the outlook can be negative, stable or positive.

Measurements collected on each variable for every element in a study provide the data. The set of measurements obtained for a particular element is called an **observation**. Referring to Table 1.1, we identify that the first observation (Armenia) contains the following measurements: Member, 4,267, B1 and Stable. The second observation (Australia) contains the following measurements: Member, 51,812, AAA and Negative and so on. A data set with 60 elements contains 60 observations.

TABLE 1.1 Data set for 60 nations in the World Trade Organization

Nation	WTO status	Per capita GDP ($)	Rating	Outlook
Armenia	Member	4,267	B+	Stable
Australia	Member	51,812	AAA	Negative
Austria	Member	48,328	AA+	Stable
Azerbaijan	Observer	4,214	BB+	Stable
Bahrain	Member	28,608	B+	Stable
Belgium	Member	44,594	AA−	Negative
Brazil	Member	6,797	BB−	Negative
Bulgaria	Member	9,976	BBB	Positive
Canada	Member	43,258	AA+	Stable
Cape Verde	Member	43,258	B−	Stable
Chile	Member	13,232	A−	Stable
China	Member	10,500	A+	Stable
Colombia	Member	5,333	BBB−	Negative
Costa Rica	Member	12,077	B	Negative
Croatia	Member	13,828	BBB−	Stable
Cyprus	Member	26,624	BBB−	Stable
Czech Republic	Member	22,932	AA−	Stable
Denmark	Member	61,063	AAA	Stable
Ecuador	Member	5,600	B−	Stable
Egypt	Member	5,600	B+	Stable
El Salvador	Member	3,799	B−	Negative
Estonia	Member	23,027	AA−	Stable
France	Member	39,030	AA	Negative
Georgia	Member	4,279	BB	Negative
Germany	Member	46,208	AAA	Stable
Hungary	Member	15,899	BBB	Stable
Iceland	Member	59,270	A	Negative
Ireland	Member	85,268	A+	Stable
Israel	Member	43,611	A+	Stable
Italy	Member	31,676	BBB−	Stable
Japan	Member	39,539	A	Negative
Kazakhstan	Member	9,056	BBB	Stable
Kenya	Member	1,838	B+	Negative
Latvia	Member	17,620	A−	Stable
Iraq	Observer	4,157	B−	Stable
Lithuania	Member	19,998	A	Stable
Malaysia	Member	10,402	BBB+	Stable
Mexico	Member	8,347	BBB−	Stable
Peru	Member	6,127	BBB+	Stable
Philippines	Member	3,299	BBB	Negative
Poland	Member	15,656	A−	Stable
Portugal	Member	22,437	BBB	Positive
South Korea	Member	1,805	AA−	Stable
Romania	Member	12,896	BBB−	Negative
Russia	Member	12,896	BBB	Stable
Rwanda	Member	798	B+	Stable
Serbia	Observer	7,666	BB+	Stable
Singapore	Member	59,798	AAA	Stable
Slovakia	Member	19,157	A	Negative

(Continued)

TABLE 1.1   (*Continued*)

Nation	WTO status	Per capita GDP ($)	Rating	Outlook
Slovenia	Member	25,517	A	Stable
South Africa	Member	5,091	BB−	Negative
Spain	Member	27,063	A−	Stable
Sweden	Member	52,259	AAA	Stable
Switzerland	Member	87,097	AAA	Stable
Thailand	Member	7,189	BBB+	Stable
Turkey	Member	8,538	BB−	Stable
United Kingdom	Member	40,285	AA−	Negative
Uruguay	Member	15,438	BBB−	Negative
United States	Member	63,544	AAA	Negative
Vietnam	Member	2,786	BB	Positive

Scales of measurement

Data collection requires one of the following scales of measurement: nominal, ordinal, interval or ratio. The scale of measurement determines the amount of information contained in the data and indicates the most appropriate data summarization and statistical analyses.

When the data for a variable consist of labels or names used to identify an attribute of the element, the scale of measurement is considered a **nominal scale**. For example, referring to the data in Table 1.1, we see that the scale of measurement for the WTO Status variable is nominal because the data 'member' and 'observer' are labels used to identify the status category for the nation. In cases where the scale of measurement is nominal, a numeric code as well as non-numerical labels may be used. For example, to facilitate data collection and to prepare the data for entry into a computer database, we might use a numerical code for the WTO Status variable by letting 1 denote a member nation in the WTO and 2 denote an observer nation. The scale of measurement is nominal even though the data appear as numeric values.

The scale of measurement for a variable is called an **ordinal scale** if the data exhibit the properties of nominal data and, in addition, the order or rank of the data is meaningful. For example, referring to the data in Table 1.1, the scale of measurement for the rating is ordinal because the rating labels, which range from AAA to F, can be rank ordered from best credit rating (AAA) to poorest credit rating (F). Thus, ordinal data can also be recorded by a non-numerical code.

The scale of measurement for a variable becomes an **interval scale** if the data show the properties of ordinal data and the interval between values is expressed in terms of a fixed unit of measure. Interval data are always numeric. Graduate Management Admission Test (GMAT) scores are an example of interval-scaled data. For example, three students with GMAT scores of 620, 550 and 470 can be ranked or ordered in terms of best performance to poorest performance. In addition, the differences between the scores are meaningful. For instance, student one scored 620 − 550 = 70 points more than student two, while student two scored 550 − 470 = 80 points more than student three.

The scale of measurement for a variable is a **ratio scale** if the data have all the properties of interval data and the ratio of two values is meaningful. Variables such as distance, height, weight and time use the ratio scale of measurement. This scale requires that a zero value be included to indicate that nothing exists for the variable at the zero point. For example, consider the cost of a car. A zero value for the cost would indicate that the car has no cost and is free. In addition, if we compare the cost of €30,000 for one car to the cost of €15,000 for a second car, the ratio property shows that the first car is €30,000/€15,000 = two times, or twice, the cost of the second car.

Categorical and quantitative data

Data can be further classified as either categorical or quantitative. **Categorical data** include labels or names used to identify an attribute of each element. Categorical data use either the nominal or ordinal scale of measurement and may be non-numeric or numeric. **Quantitative data** require numeric values that indicate how much or how many. Quantitative data are obtained using either the interval or ratio scale of measurement.

A **categorical variable** is a variable with categorical data, and a **quantitative variable** is a variable with quantitative data. The statistical analysis appropriate for a particular variable depends upon whether the variable is categorical or quantitative. If the variable is categorical, the statistical analysis is rather limited. We can summarize categorical data by counting the number of observations in each category or by computing the proportion of the observations in each category. However, even when the categorical data use a numeric code, arithmetic operations such as addition, subtraction, multiplication and division do not provide meaningful results. Section 2.1 discusses ways to summarize categorical data.

On the other hand, arithmetic operations often provide meaningful results for a quantitative variable. For example, for a quantitative variable, the data may be added and then divided by the number of observations to compute the average value. This average is usually meaningful and easily interpreted. In general, more alternatives for statistical analysis are possible when the data are quantitative. Section 2.2 and Chapter 3 provide ways of summarizing quantitative data.

Cross-sectional and time series data

For purposes of statistical analysis, distinguishing between cross-sectional data and time series data is important. **Cross-sectional data** are data collected at the same or approximately the same point in time. The data in Table 1.1 are cross-sectional because they describe the five variables for the 60 WTO nations at the same point in time. **Time series data** are data collected over several time periods. For example, Figure 1.1 provides a graph of the diesel and crude oil prices per gallon from 2018 to 2024. It shows that soon after the start of 2018 the monthly retail price of diesel rose slightly above $3 per gallon.

Most of the statistical methods presented in this text apply to cross-sectional rather than time series data.

Quantitative data that measure how many are discrete. Quantitative data that measure how much are continuous because no separation occurs between the possible data values.

FIGURE 1.1
Monthly US retail fuel prices (Jan 2018–Dec 2024)

Source: EIA. Available at www.eia.gov

1.3 DATA SOURCES

Data can be obtained from existing sources or from surveys and experimental studies designed to collect new data.

Existing sources

In some cases, data needed for a particular application already exist. Companies maintain a variety of databases about their employees, customers and business operations. Data on employee salaries, ages and years of experience can usually be obtained from internal personnel records. Other internal records contain data on sales, advertising expenditures, distribution costs, inventory levels and production quantities. Most companies also maintain detailed data about their customers. Table 1.2 shows some of the data commonly available from internal company records.

TABLE 1.2 Examples of data available from internal company records

Source	Some of the data typically available
Employee records	Name, address, social security number, salary, number of vacation days, number of sick days and bonus
Production records	Part or product number, quantity produced, direct labour cost and materials cost
Inventory records	Part or product number, number of units on hand, reorder level, economic order quantity and discount schedule
Sales records	Product number, sales volume, sales volume by region and sales volume by customer type
Credit records	Customer name, address, telephone number, credit limit and accounts receivable balance
Customer profile	Age, gender, income level, household size, address and preferences

Organizations that specialize in collecting and maintaining data make available substantial amounts of business and economic data. Companies access these external data sources through leasing arrangements or by purchase. Dun & Bradstreet, Bloomberg and the Economist Intelligence Unit are three sources that provide extensive business database services to clients. ACNielsen has built successful businesses collecting and processing data that they sell to advertisers and product manufacturers.

Data are also available from a variety of industry associations and special interest organizations. The European Tour Operators Association and European Travel Commission provide information on tourist trends and travel expenditures by visitors to and from countries in Europe. Such data would be of interest to firms and individuals in the travel industry. The Graduate Management Admission Council maintains data on test scores, student characteristics and graduate management education programmes. Most of the data from these types of source are available to qualified users at a modest cost.

The internet continues to grow as an important source of data and statistical information. Almost all companies maintain websites that provide general information about the company as well as data on sales, number of employees, number of products, product prices and product specifications. In addition, a number of companies now specialize in making information available over the internet. As a result, it is possible to obtain access to stock quotes, meal prices at restaurants, salary data and an almost infinite variety of information. Government agencies are another important source of existing data. For example, Eurostat maintains considerable data on employment rates, wage rates, size of the labour force and union membership. Table 1.3 lists selected governmental agencies and some of the data they provide. Most government agencies that collect and process data also make the results available through a website. For instance, Eurostat has a wealth of data on its website, www.ec.europa.eu/eurostat. Figure 1.2 shows the home page for Eurostat.

TABLE 1.3 Examples of data available from selected European sources

Source	Some of the data available
Eurostat (www.ec.europa.eu/eurostat/)	Education and training, labour market, living conditions and welfare, population and social conditions
European Central Bank (www.ecb.europa.eu/stats/)	Monetary, financial markets, interest rate and balance of payments statistics, unit labour costs, compensation per employee, labour productivity, consumer prices, construction prices, travel, VAT (value added tax), euro exchange employment

FIGURE 1.2
Eurostat home page
(www.ec.europa.eu
/eurostat)

Statistical studies

Sometimes the data needed for a particular application are not available through existing sources. In such cases, the data can often be obtained by conducting a statistical study. Statistical studies can be classified as either *experimental* or *observational*.

Experimental study

In an experimental study, a variable of interest is first identified. Then one or more other variables are identified and controlled so that data can be obtained about how they influence the variable of interest. For example, a pharmaceutical firm might be interested in conducting an experiment to learn about how a new drug affects blood pressure. Blood pressure is the variable of interest in the study. The dosage level of the new drug is another variable that is hoped to have a causal effect on blood pressure. To obtain data about the effect of the new drug, researchers select a sample of individuals. The dosage level of the new drug is controlled, as different groups of individuals are given different dosage levels. Before and after data on blood pressure are collected for each group. Statistical analysis of the experimental data can help determine how the new drug affects blood pressure.

Observational study

Non-experimental, or observational, statistical studies make no attempt to control the variables of interest. A survey is perhaps the most common type of observational study. For instance, in a personal interview survey, research questions are first identified. Then a questionnaire is designed and administered to a sample of individuals. Some restaurants use observational studies to obtain data about their customers' opinions of the quality of food, service, atmosphere and so on. A questionnaire used by the Lobster Pot Restaurant in Limerick City, Ireland, is shown in Figure 1.3. Note that the customers completing the questionnaire are asked to provide ratings for five variables: food quality, friendliness of service, promptness of service, cleanliness and management. The response categories of excellent, good, satisfactory and unsatisfactory provide ordinal data that enable Lobster Pot's managers to assess the quality of the restaurant's operation.

Managers wanting to use data and statistical analyses as an aid to decision-making must be aware of the time and cost required to obtain the data. The use of existing data sources is desirable when data must be obtained in a relatively short period of time.

If important data are not readily available from an existing source, the additional time and cost involved in obtaining the data must be taken into account. In all cases, the decision-maker should consider the contribution of the statistical analysis to the decision-making process. The cost of data acquisition and the subsequent statistical analysis should not exceed the savings generated by using the information to make a better decision.

FIGURE 1.3
Customer opinion
questionnaire used
by the Lobster Pot
Restaurant, Limerick
City, Ireland

The
LOBSTER
Pot
RESTAURANT

We are happy you stopped by the Lobster Pot Restaurant and want to make sure you will come back. So, if you have a little time, we will really appreciate it if you will fill out this card. Your comments and suggestions are extremely important to us. Thank you!

Server's Name _____

	Excellent	Good	Satisfactory	Unsatisfactory
Food Quality	❑	❑	❑	❑
Friendly Service	❑	❑	❑	❑
Prompt Service	❑	❑	❑	❑
Cleanliness	❑	❑	❑	❑
Management	❑	❑	❑	❑

Comments _____

What prompted your visit to us?_____
_____

Please drop in suggestion box at entrance. Thank you.

Data acquisition errors

Managers should always be aware of the possibility of data errors in statistical studies. Using erroneous data can be worse than not using any data at all. An error in data acquisition occurs whenever the data value obtained is not equal to the true or actual value that would be obtained with a correct procedure. Such errors can occur in a number of ways. For example, an interviewer might make a recording error, such as a transposition in writing the age of a 24-year-old person as 42, or the person answering an interview question might misinterpret the question and provide an incorrect response.

Experienced data analysts take great care in collecting and recording data to ensure that errors are not made. Special procedures can be used to check for internal consistency of the data. For instance, such procedures would indicate that the analyst should review the accuracy of data for a respondent shown to be 22 years of age but reporting 20 years of work experience. Data analysts also review data with unusually large and small values, called outliers, which are candidates for possible data errors. In Chapter 3 we present some of the methods statisticians use to identify outliers.

Errors often occur during data acquisition. Blindly using any data that happen to be available or using data that were acquired with little care can result in misleading information and bad decisions. Thus, taking steps to acquire accurate data can help ensure reliable and valuable decision-making information.

1.4 DESCRIPTIVE STATISTICS

Most of the statistical information in newspapers, magazines, company reports and other publications consists of data that are summarized and presented in a form that is easy for the reader to understand. Such summaries of data, which may be tabular, graphical or numerical, are referred to as descriptive statistics.

Refer to the data set in Table 1.1 showing data for 60 nations that participate in the WTO. Methods of descriptive statistics can be used to summarize these data. For example, consider the variable outlook, which indicates the direction the nation's credit rating is likely to move over the next two years. The outlook is recorded as being negative, stable or positive. A tabular summary of the data showing the number of nations with each of the outlook ratings is shown in Table 1.4. A graphical summary of the same data, called a bar chart, is shown in Figure 1.4. These types of summary make the data easier to interpret.

TABLE 1.4 Frequencies and per cent frequencies for the credit rating outlook of 60 nations

Outlook	Frequency	Per cent frequency (%)
Positive	3	5.0
Stable	39	65.0
Negative	18	30.0

FIGURE 1.4
Bar chart for the credit rating outlook for 60 nations

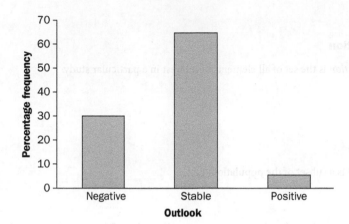

Referring to Table 1.4 and Figure 1.4, we can observe that the majority of outlook credit ratings are stable, with 65 per cent of the nations having this rating. More nations have a negative outlook (30 per cent) than a positive outlook (5 per cent). A graphical summary of the data for the quantitative variable per capita GDP in Table 1.1, called a histogram, is provided in Figure 1.5. Using the histogram, it is easy to see that per capita GDP for the 60 nations ranges from $0 to $90,000, with the highest concentration between $0 and $10,000. There are no countries with per capita GDP in the range of $70,000 to $79,999 and two countries with per capita GDP over $80,000. In addition to tabular and graphical displays, numerical descriptive statistics are used to summarize data. The most common numerical measure is the average, or mean. Using the data on per capita GDP for the 60 nations in Table 1.1, we can compute the average by adding per capita GDP for all 60 nations and dividing the total by 60. Doing so provides an average per capita GDP of $23,704. This average provides a measure of the central tendency or central location of the data. There is a great deal of interest in effective methods for developing and presenting descriptive statistics.

FIGURE 1.5
Histogram of per capita GDP for 60 nations

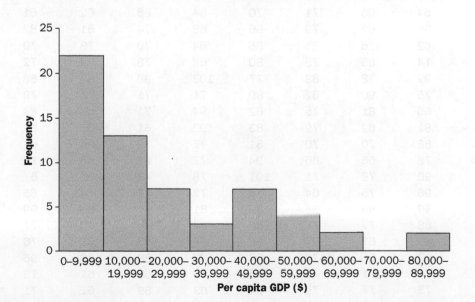

1.5 STATISTICAL INFERENCE

Many situations require data for a large group of elements (individuals, companies, voters, households, products, customers and so on). Because of time, cost and other considerations, data can be collected from only a small portion of the group. The larger group of elements in a particular study is called the population, and the smaller group is called the sample. Formally, we use the following definitions:

Population

A *population* is the set of all elements of interest in a particular study.

Sample

A *sample* is a subset of the population.

The process of conducting a survey to collect data for the entire population is called a census. The process of conducting a survey to collect data for a sample is called a sample survey. As one of its major contributions, statistics uses data from a sample to make estimates and test hypotheses about the characteristics of a population through a process referred to as statistical inference.

As an example of statistical inference, let us consider the study conducted by Electronica Nieves. Nieves manufactures a high-intensity light bulb used in a variety of electrical products. In an attempt to increase the useful life of the light bulb, the product design group developed a new light bulb filament. In this case, the population is defined as all light bulbs that could be produced with the new filament. To evaluate the advantages of the new filament, 200 bulbs with the new filament were manufactured and tested. Data collected from this sample showed the number of hours each light bulb operated before the filament burned out or the bulb failed, refer to Table 1.5.

TABLE 1.5 Hours until failure for a sample of 200 light bulbs for the Electronica Nieves example

NIEVES

107	73	68	97	76	79	94	59	98	57
54	65	71	70	84	88	62	61	79	98
66	62	79	86	68	74	61	82	65	98
62	116	65	88	64	79	78	79	77	86
74	85	73	80	68	78	89	72	58	69
92	78	88	77	103	88	63	68	88	81
75	90	62	89	71	71	74	70	74	70
65	81	75	62	94	71	85	84	83	63
81	62	79	83	93	61	65	62	92	65
83	70	70	81	77	72	84	67	59	58
78	66	66	94	77	63	66	75	68	76
90	78	71	101	78	43	59	67	61	71
96	75	64	76	72	77	74	65	82	86
66	86	96	89	81	71	85	99	59	92
68	72	77	60	87	84	75	77	51	45
85	67	87	80	84	93	69	76	89	75
83	68	72	67	92	89	82	96	77	102
74	91	76	83	66	68	61	73	72	76
73	77	79	94	63	59	62	71	81	65
73	63	63	89	82	64	85	92	64	73

Suppose Nieves wants to use the sample data to make an inference about the average hours of useful life for the population of all light bulbs that could be produced with the new filament. Adding the 200 values in Table 1.5 and dividing the total by 200 provides the sample average lifetime for the light bulbs: 76 hours. We can use this sample result to estimate that the average lifetime for the light bulbs in the population is 76 hours. Figure 1.6 provides a graphical summary of the statistical inference process for Electronica Nieves.

FIGURE 1.6

The process of statistical inference for the Electronica Nieves example

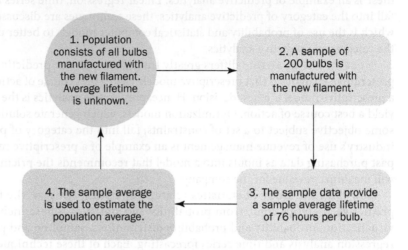

Whenever statisticians use a sample to estimate a population characteristic of interest, they usually provide a statement of the quality, or precision, associated with the estimate. For the Nieves example, the statistician might state that the point estimate of the average lifetime for the population of new light bulbs is 76 hours with a margin of error of ± four hours. Thus, an interval estimate of the average lifetime for all light bulbs produced with the new filament is 72 hours to 80 hours. The statistician can also state how confident they are that the interval from 72 hours to 80 hours contains the population average.

1.6 ANALYTICS

Because of the dramatic increase in available data, more cost-effective data storage, faster computer processing, and recognition by managers that data can be extremely valuable for understanding customers and business operations, there has been a dramatic increase in data-driven decision-making. The broad range of techniques that may be used to support data-driven decisions comprise what has become known as analytics.

Analytics is the scientific process of transforming data into insight for making better decisions. Analytics is used for data-driven or fact-based decision-making, which is often seen as more objective than alternative approaches to decision-making. The tools of analytics can aid decision-making by creating insights from data, improving our ability to more accurately forecast for planning, helping us quantify risk and yielding better alternatives through analysis.

Analytics can involve a variety of techniques from simple reports to the most advanced optimization techniques (algorithms for finding the best course of action). Analytics is now generally thought to comprise three broad categories of techniques. These categories are descriptive analytics, predictive analytics and prescriptive analytics.

Descriptive analytics encompasses the set of analytical techniques that describe what has happened in the past. Examples of these types of techniques are data queries, reports, descriptive statistics, data visualization, data dash boards and basic what-if spreadsheet models.

Predictive analytics consists of analytical techniques that use models constructed from past data to predict the future or to assess the impact of one variable on another. For example, past data on sales of a product may be used to construct a mathematical model that predicts future sales. Such a model can account for factors such as the growth trajectory and seasonality of the product's sales based on past growth and seasonal patterns. Point-of-sale scanner data from retail outlets may be used by a packaged food manufacturer to help estimate the lift in unit sales associated with coupons or sales events. Survey data and past purchase behaviour may be used to help predict the market share of a new product. Each of these is an example of predictive analytics. Linear regression, time series analysis and forecasting models fall into the category of predictive analytics; these techniques are discussed later in this text. Simulation, which is the use of probability and statistical computer models to better understand risk, also falls under the category of predictive analytics.

Prescriptive analytics differs greatly from descriptive or predictive analytics. What distinguishes prescriptive analytics is that prescriptive models yield a best course of action to take. That is, the output of a prescriptive model is a best decision. Hence, prescriptive analytics is the set of analytical techniques that yield a best course of action. Optimization models, which generate solutions that maximize or minimize some objective subject to a set of constraints, fall into the category of prescriptive models. The airline industry's use of revenue management is an example of a prescriptive model. The airline industry uses past purchasing data as inputs into a model that recommends the pricing strategy across all flights that will maximize revenue for the company.

How does the study of statistics relate to analytics? Most of the techniques in descriptive and predictive analytics come from probability and statistics. These include descriptive statistics, data visualization, probability and probability distributions, sampling and predictive modelling, including regression analysis and time series forecasting. Each of these techniques is discussed in this text. The increased use of analytics for data-driven decision-making makes it more important than ever for analysts and managers to understand statistics and data analysis. Companies are increasingly seeking data savvy managers who know how to use descriptive and predictive models to make data-driven decisions.

At the beginning of this section, we mentioned the increased availability of data as one of the drivers of the interest in analytics. In the next section we discuss this explosion in available data and how it relates to the study of statistics.

1.7 BIG DATA AND DATA MINING

With the aid of magnetic card readers, bar code scanners and point-of-sale terminals, most organizations obtain large amounts of data on a daily basis. And, even for a small local restaurant that uses touch screen monitors to enter orders and handle billing, the amount of data collected can be substantial. For large retail companies, the sheer volume of data collected is hard to conceptualize, and figuring out how to effectively use these data to improve profitability is a challenge. Mass retailers such as Walmart and Amazon capture data on 20 to 30 million transactions every day, telecommunication companies such as Orange S.A. and AT&T generate over 300 million call records per day, and Visa processes 6,800 payment transactions per second or approximately 600 million transactions per day. In addition to the sheer volume and speed with which companies now collect data, more complicated types of data are now available and are proving to be of great value to businesses. Text data are collected by monitoring what is being said about a company's products or services on social media such as Twitter. Audio data are collected from service calls (on a service call, you will often hear 'this call may be monitored for quality control'). Video data are collected by in-store video cameras to analyze shopping behaviour. Analyzing information generated by these non-traditional sources is more complicated because of the complex process of transforming the information into data that can be analyzed.

Larger and more complex data sets are now often referred to as **big data**. Although there does not seem to be a universally accepted definition of *big data*, many think of it as a set of data that cannot be managed, processed or analyzed with commonly available software in a reasonable amount of time. Many data analysts define *big data* by referring to the three v's of data: volume, velocity and variety.

Volume refers to the amount of available data (the typical unit of measure is now a terabyte, which is 10^{12} bytes); *velocity* refers to the speed at which data are collected and processed; and *variety* refers to the different data types.

The term *data warehousing* is used to refer to the process of capturing, storing and maintaining the data. Computing power and data collection tools have reached the point where it is now feasible to store and retrieve extremely large quantities of data in seconds. Analysis of the data in the warehouse may result in decisions that will lead to new strategies and higher profits for the organization. For example, General Electric (GE) captures a large amount of data from sensors on its aircraft engines each time a plane takes off or lands. Capturing these data allows GE to offer an important service to its customers; GE monitors the engine performance and can alert its customers when a service is needed or a problem is likely to occur.

The subject of data mining deals with methods for developing useful decision-making information from large databases. Using a combination of procedures from statistics, mathematics and computer science, analysts 'mine the data' in the warehouse to convert it into useful information, hence the name *data mining*. Dr Kurt Thearling, a leading practitioner in the field, defines data mining as 'the automated extraction of predictive information from (large) databases'. The two key words in Dr Thearling's definition are 'automated' and 'predictive'. Data mining systems that are the most effective use automated procedures to extract information from the data using only the most general or even vague queries by the user. And data mining software automates the process of uncovering hidden predictive information that in the past required hands-on analysis.

The major applications of data mining have been made by companies with a strong consumer focus, such as retail businesses, financial organizations and communication companies. Data mining has been successfully used to help retailers such as Amazon determine one or more related products that customers who have already purchased a specific product are also likely to purchase. Then, when a customer logs on to the company's website and purchases a product, the website uses pop-ups to alert the customer about additional products that the customer is likely to purchase. In another application, data mining may be used to identify customers who are likely to spend more than $20 on a particular shopping trip. These customers may then be identified as the ones to receive special email or regular mail discount offers to encourage them to make their next shopping trip before the discount termination date.

Data mining is a technology that relies heavily on statistical methodology such as multiple regression, logistic regression and correlation. But it takes a creative integration of all these methods and computer science technologies involving artificial intelligence and machine learning to make data mining effective. A substantial investment in time and money is required to implement commercial data mining software packages developed by firms such as Oracle, Teradata and SAS. The statistical concepts introduced in this text will be helpful in understanding the statistical methodology used by data mining software packages and enable you to better understand the statistical information that is developed.

Because statistical models play an important role in developing predictive models in data mining, many of the concerns that statisticians deal with in developing statistical models are also applicable. For instance, a concern in any statistical study involves the issue of model reliability. Finding a statistical model that works well for a particular sample of data does not necessarily mean that it can be reliably applied to other data. One of the common statistical approaches to evaluating model reliability is to divide the sample data set into two parts: a training data set and a test data set. If the model developed using the training data is able to accurately predict values in the test data, we say that the model is reliable. One advantage that data mining has over classical statistics is that the enormous amount of data available allows the data mining software to partition the data set so that a model developed for the training data set may be tested for reliability on other data. In this sense, the partitioning of the data set allows data mining to develop models and relationships, and then quickly observe if they are repeatable and valid with new and different data. On the other hand, a warning for data mining applications is that with so much data available, there is a danger of overfitting the model to the point that misleading associations and cause/effect conclusions appear to exist. Careful interpretation of data mining results and additional testing will help avoid this pitfall.

1.8 COMPUTERS AND STATISTICAL ANALYSIS

Statisticians use computer software to perform statistical computations and analyses. For example, computing the average hours until failure for the 200 lightbulbs in the Electronica Nieves example (refer to Table 1.5) would be quite tedious without a computer. Online resources cover the step-by-step procedures for using Microsoft Excel and the statistical packages R, SPSS and Minitab to implement the statistical techniques presented in the chapter.

Special data manipulation and analysis tools are needed for big data, which was described in the previous section. Open-source software for distributed processing of large data sets such as Hadoop, open-source programming languages such as R and Python, and commercially available packages such as SAS and IBM SPSS Statistics (SPSS) are used in practice for big data.

EXERCISES

1. Discuss the differences between statistics as numerical facts and statistics as a discipline or field of study.

2. Every year surveys of subscribers are conducted to determine the best new places to stay throughout the world. Table 1.6 shows the ten hotels that were highly favoured in one such survey in 2019.
   a. How many elements are in this data set?
   b. How many variables are in this data set?
   c. Which variables are categorical and which variables are quantitative?
   d. What type of measurement scale is used for each of the variables?

3. Refer to Table 1.6:
   a. What is the average number of rooms/suites/villas for the ten hotels?
   b. Compute the average room rate.
   c. What is the percentage of hotels located in Italy?
   d. What is the percentage of hotels with 100 rooms/suites/villas or more?

HOTELS
2019

TABLE 1.6   Ten of the best new hotels in 2019

Name of the property	Country	Capacity	Room rate (£)
Savoy Palace	Madeira	352 rooms	124
Anantara Quy Nhon Villas	Vietnam	26 villas	329
Caer Rhun Hall	North Wales	76 rooms	140
Faarufushi Maldives, Raa Atoll	Maldives	80 villas	673
Hotel de la Ville, Rome	Italy	15 suites	573
Masseria Torre Maizza, Puglia	Italy	40 rooms	460
Calilo, Ios	Greece	30 rooms	600
Paradisus Los Cayos	Cuba	802 rooms	243
COMO Uma Canggu	Bali	119 Rooms	208
W Brisbane	Australia	312 rooms	152

Source: *Evening Standard*/CNN Travel

4. Micro hi-fi systems typically have a wireless bluetooth facility, DAB/FM reception and a CD player. The data in Table 1.7 show the product rating and retail price range for a popular selection of systems. Note that the code Y is used to confirm when a facility is included in the system, N when it is not. Output power (watts) details are also provided (Tektouch, 2019).
   a. How many elements does this data set contain?
   b. What is the population?
   c. Compute the average output power for the sample.

5. Consider the data set for the sample of ten micro hi-fi systems in Table 1.7.
   a. How many variables are in the data set?
   b. Which of the variables are quantitative and which are categorical?
   c. What percentage of the hi-fi systems has a four star rating or higher?
   d. What percentage of the hi-fi systems includes a CD player?

TABLE 1.7 A sample of ten micro hi-fi systems

	Brand and model	Rating (no. of stars)	Price (£)	Wireless Bluetooth	DAB/ FM	CD	Output (W)
1	Sony CMT-S20B	5	115		Y	Y	10
2	Philips BTM1360	4	125	Y	Y	Y	30
3	Pioneer X-CM	5	184	Y	Y		30
4	Trevi BT	5	94	Y	Y		40
5	Pioneer Wi-Fi XW	4	69	Y	Y	Y	20
6	LG CM4350	5	79			Y	260
7	Neon	5	149		Y	Y	90
8	Denon D-M41DAB	5	229	Y	Y	Y	30
9	Panasonic	5	144		Y		40
10	Duronic RCD144	5	54			Y	3

Source: www.tektouch.net/Music/10-Best-Micro-Hi-Fi-Systems.php

HIFI SYSTEMS

6. State whether each of the following variables is categorical or quantitative and indicate its measurement scale.
   a. Annual sales.
   b. Soft drink size (small, medium, large).
   c. Occupational classification (SOC 2000).
   d. Earnings per share.
   e. Method of payment (cash, cheque, credit card).

7. In Spring 2015 the first ever SPA/Police Scotland opinion survey was commissioned to determine officer and staff opinion. All 23,438 police officers, staff and special constables were invited to participate in the survey. Response categories were: strongly agree, agree, neither agree nor disagree, disagree and strongly disagree, don't know/no opinion, no reply.
   a. What was the population size for this survey?
   b. Are the data categorical or quantitative?
   c. Would it make more sense to use averages or percentages as a summary of the data for this question?
   d. Of those surveyed, 50.4 per cent responded. How many individuals provided this response?

8. State whether each of the following variables is categorical or quantitative and indicate its measurement scale.
   a. Age.
   b. Gender.
   c. Class rank.
   d. Make of car.
   e. Number of people favouring closer European integration.

9. Figure 1.7 provides a bar chart showing the number of Netflix subscribers from 2011 to 2022. (www.demandsage.com/netflix-subscribers/)
   a. What is the variable of interest?
   b. Are the data categorical or quantitative?
   c. Are the data time series or cross-sectional?
   d. Comment on the trend in Netflix subscribers over time.

FIGURE 1.7
Netflix subscribers (millions)

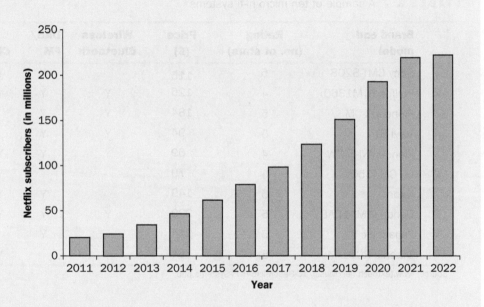

10. The Hawaii Visitors' Bureau collects data on visitors to Hawaii. The following questions were among 16 asked in a questionnaire handed out to passengers during incoming airline flights:

   • This trip to Hawaii is my: 1st, 2nd, 3rd, 4th, etc.
   • The primary reason for this trip is: (ten categories including vacation, convention, honeymoon).
   • Where I plan to stay: (11 categories including hotel, apartment, relatives, camping).
   • Total days in Hawaii.

   a. What is the population being studied?
   b. Is the use of a questionnaire a good way to reach the population of passengers on incoming airline flights?
   c. Comment on each of the four questions in terms of whether it will provide categorical or quantitative data.

11. The International Federation of Robotics estimates the worldwide supply of industrial robots each year. Figure 1.8 shows estimates of the worldwide supply of industrial robots for the years 2015 to 2021.
    a. What is the variable of interest?
    b. Are the data quantitative or categorical?
    c. Are the data cross-sectional or time series?

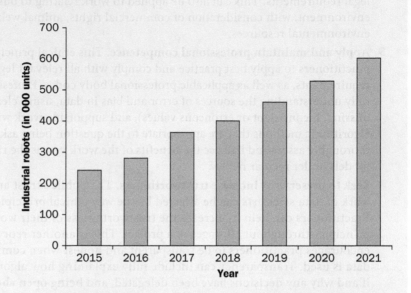

FIGURE 1.8

Estimated industrial robot supply (000 units)

12. In a recent study of causes of death in men 60 years of age and older, a sample of 120 men indicated that 48 died as a result of some form of heart disease.
    a. Develop a descriptive statistic that can be used as an estimate of the percentage of men 60 years of age or older who die from some form of heart disease.
    b. Are the data on cause of death categorical or quantitative?
    c. Discuss the role of statistical inference in this type of medical research.

1.9 ETHICAL GUIDELINES FOR STATISTICAL PRACTICE

Ethical behaviour is characterized by honesty, fairness and equity in interpersonal, professional and academic relationships and in research and scholarly activities. Ethical behaviour respects the dignity, diversity and rights of individuals and groups of people.

Though statistics practice in government is bound by official codes of conduct, the discipline of data ethics more generally is still very much in development (Spiegelhalter, 2020). Aimed at addressing the ethical and professional challenges of working in a data science setting, the IFA and RSS produced a Guide in 2019 that focuses attention on five ethical themes associated with data science and AI:

1 **Seek to enhance the value of data science for society.** As the impact that data science can have on society could be significant, an important ethical consideration is what the potential implications could be on society as a whole. A common theme within ethical frameworks discussing data science and AI is for practitioners to attempt to seek outcomes within their work which support the improvement of public wellbeing. This could involve practitioners seeking to share the benefits of data science and balancing this with the wellbeing of potentially affected individuals.

2 **Avoid harm.** Data science has the potential to cause harm and this ethical consideration therefore focuses on how practitioners can avoid this by working in a manner that respects the privacy, equality and autonomy of individuals and groups, and speaking up about potential harm or ethical violations. Practitioners may be subject to legal and regulatory obligations in relation to the privacy of individuals, relevant to the jurisdiction in which they are working, as well as regulatory obligations to speak up about harm or violations of legal requirements. This can also be applied to work relating to businesses, animals or the environment, with consideration of commercial rights, animal welfare and the protection of environmental resources.

3 **Apply and maintain professional competence.** This ethical principle expects data science practitioners to apply best practice and comply with all relevant legal and regulatory requirements, as well as applicable professional body codes. Professional competence involves fully understanding the sources of error and bias in data, using 'clean' data (e.g. edited for missing, inconsistent or erroneous values), and supporting work with robust statistical and algorithmic methods that are appropriate to the question being asked. Practitioners can also thoroughly assess and balance the benefits of the work versus the risks posed by it and keep models under regular review.

4 **Seek to preserve or increase trustworthiness.** The public's trust and confidence in the work of data scientists can be affected by the way ethical principles are applied. Practitioners can help to increase the trustworthiness of their work by considering ethical principles throughout all stages of a project. This is another reoccurring theme that encourages practitioners to be transparent and honest when communicating about the way data is used. Transparency can include fully explaining how algorithms are being used, if and why any decisions have been delegated, and being open about the risks and biases. Engaging widely with a diverse range of stakeholders and considering public perceptions from the outset and throughout projects, can help to build trustworthiness and ensure all potential biases are understood.

5 **Maintain accountability and oversight.** Another key issue in data ethics around automation and AI is the question of how practitioners maintain human accountability and oversight within their work. Being accountable can include being mindful of how and when to delegate any decision-making to systems, and having governance in place to ensure systems deliver the intended objectives. When deciding to delegate any decision-making, it would be useful to fully understand and explain the potential implications of doing so, as the work could lead to introducing advanced AI systems which do not have adequate governance. Practitioners should note that delegating any decisions to these systems does not remove any of their individual responsibilities.

Within each of these themes are examples of corresponding working practices which aim to help data scientists consider relevant ethical issues. A particularly helpful feature at the end of the Guide is a checklist covering the different project implementation phases to assist with the embedding of ethics in the resulting data science work:

- Project planning
- Data management
- Analysis and development
- Implementation and delivery
- Communication.

In summary, the IFA and RSS's Guide provides a valuable and wide-ranging outline on data ethics. In contrast, the no less insightful ethics self-assessment tool recently developed by the UK Statistics Authority (UKSA, 2022) is very much directed at researchers in the Government Statistical Service.

ETHICAL GUIDELINES FOR STATISTICAL PRACTICE

ONLINE RESOURCES

For the data files, additional questions and answers, and the software section for Chapter 1, go to the online platform.

SUMMARY

Statistics is the art and science of collecting, analyzing, presenting and interpreting data. Nearly every college student majoring in business or economics is required to take a course in statistics. We began the chapter by describing typical statistical applications for business and economics.

Data consist of the facts and figures that are collected and analyzed. A set of measurements obtained for a particular element is an observation. Four scales of measurement used to obtain data on a particular variable include nominal, ordinal, interval and ratio. The scale of measurement for a variable is nominal when the data use labels or names to identify an attribute of an element. The scale is ordinal if the data demonstrate the properties of nominal data and the order or rank of the data is meaningful. The scale is interval if the data demonstrate the properties of ordinal data and the interval between values is expressed in terms of a fixed unit of measure. Finally, the scale of measurement is ratio if the data show all the properties of interval data and the ratio of two values is meaningful.

For purposes of statistical analysis, data can be classified as categorical or quantitative. Categorical data use labels or names to identify an attribute of each element. Categorical data use either the nominal or ordinal scale of measurement and may be non-numeric or numeric. Quantitative data are numeric values that indicate how much or how many. Quantitative data use either the interval or ratio scale of measurement. Ordinary arithmetic operations are meaningful only if the data are quantitative. Therefore, statistical computations used for quantitative data are not always appropriate for categorical data.

In Sections 1.4 and 1.5 we introduced the topics of descriptive statistics and statistical inference. Definitions of the population and sample were provided and different types of descriptive statistics – tabular, graphical and numerical – used to summarize data. The process of statistical inference uses data obtained from a sample to make estimates or test hypotheses about the characteristics of a population.

The next three sections of the chapter provide information on analytics, the role of computers in statistical analysis and a brief overview of the relatively new field of data mining. The last section is concerned with ethical considerations in statistics practice.

KEY TERMS

Analytics
Big data
Categorical data
Categorical variable
Census
Cross-sectional data
Data
Data mining
Data set
Descriptive analytics
Descriptive statistics
Elements
Ethical behaviour
Interval scale
Nominal scale

Observation
Ordinal scale
Population
Predictive analytics
Prescriptive analytics
Quantitative data
Quantitative variable
Ratio scale
Sample
Sample survey
Statistical inference
Statistics
Time series data
Variable

CASE PROBLEM

Customer satisfaction in the large hotel chains sector

In 2016 the Which? company surveyed customer satisfaction in the large UK hotel sector with results as follows:

Hotel chain	Sample size	Average price paid (£)	Cleanliness	Quality of bedrooms	Quality of bathrooms	Value for money	Customer score
Premier Inn	1,462	70	★★★★	★★★★★	★★★★★	★★★★	83%
Hampton by Hilton	51	83	★★★★	★★★★★	★★★★★	★★★★	76%
Novotel	76	98	★★★★	★★★★★	★★★★	★★★★	75%
Hilton	154	126	★★★★★	★★★★★	★★★★★	★★★	72%
Double Tree by Hilton	64	113	★★★★★	★★★★★	★★★★★	★★★	71%
Best Western	296	96	★★★★	★★★★	★★★★	★★★★	70%
Marriot	109	125	★★★★	★★★★★	★★★★★	★★★	70%
Holiday Inn Express	230	76	★★★★	★★★★	★★★★	★★★	69%
Radisson Blu	64	120	★★★★★	★★★★★	★★★★★	★★★	69%
Crowne Plaza	75	114	★★★★	★★★★★	★★★★★	★★★	68%
Jury's Inn	56	87	★★★★	★★★★	★★★★	★★★	67%
Holiday Inn	252	96	★★★★	★★★★	★★★★	★★★	66%
Ibis	101	76	★★★★	★★★★	★★★★	★★★	66%
Ibis Styles	34	80	★★★★	★★★	★★★★	★★★★	66%
Park Inn by Radisson	30	85	★★★★	★★★★	★★★★★		66%
Ibis Budget	38	57	★★★★	★★★	★★★	★★★★	65%
Old English Inns	38	70	★★★	★★★	★★★	★★★★	65%
Travelodge	483	58	★★★	★★★	★★★	★★★★	65%
Days Inn	44	60	★★★★	★★★	★★★	★★★★	63%
MacDonald	80	139	★★★★	★★★★★	★★★★	★★	63%
Mercure	128	107	★★★★	★★★★	★★★★	★★★	63%
Copthorne	31	113	★★★★	★★★	★★★★	★★★	60%
Ramada	32	77	★★★★	★★★	★★★	★★★	55%
Britannia	50	80	★★★	★★	★★	★★	44%

The methodology used for obtaining these results – from the relevant Which? Methodology and Technical report – is summarized below:

Survey form: A web-based survey form was posted on www.which.co.uk in July 2016.

Invitations: Email invitations were sent to all members of Which? Connect (Which?'s online community)

Response: From circa 39,000 panelists contacted, 4,283 Which? Connect members were reported to have completed the survey.

Questions and Categories: Data were collected in two forms:

To obtain the **customer score** (overall evaluation), responses were collected for the key **Satisfaction** and **Endorsement** questions – more later.

In addition, customers were asked to rate the following **categories**:

Cleanliness;
Quality of bedrooms;
Quality of bathrooms;
Bed comfort;
Hotel description;
Value for money using a five star scale (explained later).

Thresholds: For the **customer score** evaluation, both **Satisfaction** and **Endorsement** questions had to be answered.

For the **category** data, each brand (chain) had to have at least 30 valid responses.

Analysis: For the **customer score**:

Responses to the **Satisfaction** and **Endorsement** questions were first weighted (explained below). Weights were then averaged and adjusted to fit with a scale from zero to 100 per cent.

For the **category** data:
Analysis involved averaging responses across all hotels and then, using an undisclosed statistical procedure, individual brand averages were converted back into a one to five star format.

CUSTOMER
SATISFACTION

Results Available from 'Best and Worst UK hotel chains' Which? magazine (15 December 2016 edition). Also refer to www.which.co.uk/reviews/uk-hotel-chains/article/best-and-worst-uk-hotel-chains-aaVVF4u1jZpe.

The scale used for category items was as follows:

Very poor
Poor
Fair
Good
Excellent

Correspondingly, those for the Satisfaction and Endorsement questions (including details of the weights used for generating the customer score) are given below:

Satisfaction ('Overall satisfaction with the brand')

Very dissatisfied	1
Fairly dissatisfied	2
Neither satisfied or dissatisfied	4
Fairly satisfied	8
Very satisfied	16

Endorsement ('Likelihood to recommend the brand to a friend/family member')

Definitely will not recommend	1
Probably will not recommend	2
Not sure	4
Probably will recommend	8
Definitely will recommend	16

Managerial report

1. Critically assess the Which? summary results, particularly in relation to their consistency and completeness, and the sample sizes realized for each hotel chain.

2. How fair do you think the scheme used for calculating the customers' scores is?

3. Was it a good idea for Which? to rely only on its own membership for the research?

Source: www.which.co.uk

2
Descriptive Statistics: Tabular and Graphical Presentations

CHAPTER CONTENTS

Statistics in Practice OFCOM: *Market Communications Report*

2.1 Summarizing categorical data
2.2 Summarizing quantitative data
2.3 Summarizing relationships between two categorical variables
2.4 Summarizing relationships between two quantitative variables
2.5 Data visualization: best practices in creating effective graphical displays

LEARNING OBJECTIVES   After reading this chapter and doing the exercises, you should be able to:

1 Construct and interpret several types of tabular and graphical data summaries.

2 For single categorical variables construct frequency, relative frequency and percentage frequency distributions; bar charts and pie charts.

3 For single quantitative variables construct frequency, relative frequency and percentage frequency distributions; cumulative frequency, cumulative relative frequency and cumulative percentage frequency distributions; dot plots, stem-and-leaf plots and histograms.

4 For pairs of categorical variables construct: cross tabulations, with row and column percentages.

5 For pairs of categorical variables construct: clustered bar charts and stacked bar charts.

6 For pairs of quantitative variables construct: scatter diagrams.

7 You should be able to give an example of Simpson's paradox and explain the relevance of this paradox to the cross tabulation of variables.

As explained in Chapter 1, data can be classified as either categorical or quantitative. Categorical data use labels or names to identify categories of like items. We shall also use the term qualitative data. Quantitative data are numerical values that indicate how much or how many.

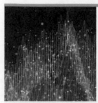

STATISTICS IN PRACTICE
OFCOM: *Market Communications Report*

We are exposed to statistical charts daily: on social media, on TV, in newspapers and magazines, in market research reports and in company annual reports.

The UK Office of Communications, more commonly referred to as Ofcom, is a body set up by central government to regulate the fixed-line telecoms, mobile telecoms, radio, TV, video-on-demand and postal sectors. It is responsible for (among other things) promoting competition in these sectors of the economy, for licensing UK commercial radio and TV services, for deciding and enforcing the regulations under which phone and broadband companies operate, and for looking after consumers' interests in these areas. Each year Ofcom produces a *Communications Market Report*, based on their own tracker and audience research, as well as on

data collated from stakeholders and third parties. The *Communications Market Report* is now presented as an interactive portal on the Ofcom website (www. ofcom.org.uk).

The charts below are reproduced from the *Market Communications Report 2022* (www.ofcom.org.uk/ research-and-data/multi-sector-research/cmr/the-communications-market-2022/communications-market-report-2022-interactive-data, accessed April 2023). The first panel contains a simple bar chart showing the top 20 websites in the UK by audience number. The second panel contains a clustered bar chart comparing domestic postal rates in 10 countries including the UK. In the interactive version of the charts, further details for each country can be displayed by hovering the cursor over the appropriate cluster of bars.

In this chapter, you will learn about tabular and graphical methods of descriptive statistics such as frequency distributions, bar charts, histograms, stem-and-leaf displays, cross tabulations and others. The goal of these methods is to summarize data so they can be easily understood and interpreted.

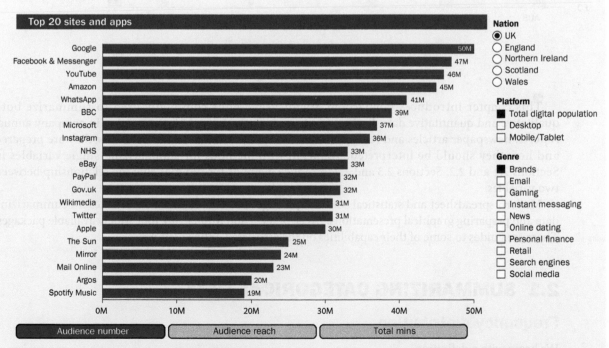

Source: Ipsos iris Online Audience Measurement Service, Ranking reports, 1st September–30th September 2021, age: 15+, UK.
Note: Avg. mins pp and Total mins (mm) are not available for YouTube.
Erroneous figures for WhatsApp when the 'Social Media' genre was selected have been removed in July 2022. The correct figures for WhatsApp can be found in the 'Instant messaging' and 'Brands' genres.

(*Continued*)

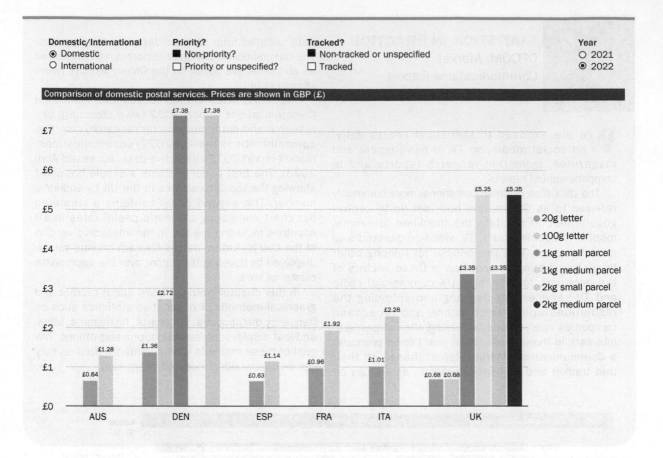

This chapter introduces tabular and graphical methods commonly used to summarize both qualitative and quantitative data. Everyone is exposed to these types of presentation in company annual reports, newspaper articles and research studies. It is important to understand how they are prepared and how they should be interpreted. We begin with methods for summarizing single variables in Sections 2.1 and 2.2. Sections 2.3 and 2.4 introduce methods for summarizing the relationship between two variables.

Modern spreadsheet and statistical software packages provide extensive capabilities for summarizing data and preparing graphical presentations. Excel, SPSS, R and Minitab are four widely available packages. There are guides to some of their capabilities on the associated online platform.

2.1 SUMMARIZING CATEGORICAL DATA

Frequency distribution

We begin with a definition.

Frequency distribution

A frequency distribution is a tabular data summary showing the number (frequency) of items in each of several non-overlapping classes.

The following example demonstrates the construction and interpretation of a **frequency distribution** for qualitative data. Audi, BMW, Mercedes, Opel and VW are five popular brands of car in Germany. The data in Table 2.1 are for a sample of 50 new car purchases of these five brands.

TABLE 2.1 Data from a sample of 50 new car purchases

VW	BMW	Mercedes	Audi	VW
VW	Mercedes	Audi	VW	Audi
VW	VW	VW	Audi	Mercedes
VW	VW	Opel	Opel	BMW
VW	Audi	Mercedes	Audi	Mercedes
VW	Mercedes	Mercedes	VW	Mercedes
VW	VW	Mercedes	Opel	Mercedes
Mercedes	BMW	VW	VW	VW
BMW	Opel	Audi	Opel	Mercedes
VW	Mercedes	BMW	VW	Audi

To construct a frequency distribution, we count the number of times each brand appears in Table 2.1. The frequency distribution in Table 2.2 summarizes these counts: there were 19 VW purchases, 13 Mercedes purchases and so on. The summary offers more immediate insight than the original data. We see that VW is the leader, Mercedes is second, Audi is third. Opel and BMW are tied for fourth.

TABLE 2.2
Frequency distribution of new car purchases

Brand	Frequency
Audi	8
BMW	5
Mercedes	13
Opel	5
VW	19
Total	**50**

CAR
BRANDS

Relative frequency and percentage frequency distributions

A frequency distribution shows the number (frequency) of items in each of several non-overlapping classes. We are often interested in the proportion, or percentage, of items in each class. The *relative frequency* of a class is the fraction or proportion of items belonging to a class. For a data set with n observations, the relative frequency of each class is:

Relative frequency

$$\text{Relative frequency of a class} = \frac{\text{Frequency of the class}}{n} \qquad (2.1)$$

The *percentage frequency* of a class is the relative frequency multiplied by 100. A **relative frequency distribution** is a tabular summary showing the relative frequency for each class. A **percentage frequency distribution** summarizes the percentage frequency for each class. Table 2.3 shows these distributions for the car purchase data. The relative frequency for VW is 19/50 = 0.38, the relative frequency for Mercedes is 13/50 = 0.26 and so on. From the percentage frequency distribution, we see that 38 per cent of the purchases were VW, 26 per cent were Mercedes, etc.

TABLE 2.3 Relative and percentage frequency distributions of new car purchases

Brand	Relative frequency	Percentage frequency
Audi	0.16	16
BMW	0.10	10
Mercedes	0.26	26
Opel	0.10	10
VW	0.38	38
Total	**1.00**	**100**

Bar charts and pie charts

A bar chart, or bar graph, is a graphical representation of a frequency, relative frequency or percentage frequency distribution. On one axis of the chart, we specify the labels for the data classes (categories). A frequency, relative frequency or percentage frequency scale is used for the other axis. Then, using a bar of fixed width drawn for each class, we make the length of the bar equal to the frequency, relative frequency or percentage frequency of the class. For qualitative data, the bars are separated to emphasize that each class is distinct.

Figure 2.1(a) shows an SPSS bar chart of the frequency distribution for the 50 new car purchases. Here the categories are on the horizontal axis and the frequencies on the vertical axis. This is the more common default bar chart orientation for statistical software programs. Figure 2.1(b) shows an Excel percentage frequency distribution for the same data set, with the categories on the vertical axis and the percentage frequencies on the horizontal axis. Excel refers to this orientation as a bar chart, and to the orientation of Figure 2.1(a) as a column chart. Note that in both charts of Figure 2.1, the categories have been arranged in frequency order. This makes comparisons between categories easier.

In quality control applications, bar charts are used to summarize the most important causes of problems. When the bars are arranged in descending order of height from left to right with the most frequently occurring cause appearing first, the bar chart is called a *Pareto diagram*, named after its founder, Vilfredo Pareto, an Italian economist.

A pie chart is another way of presenting relative frequency and percentage frequency distributions. A circle (the pie) represents all the data. The circle is divided into sectors or segments that correspond to the relative frequency for each class. For example, because a circle contains 360 degrees and VW shows a relative frequency of 0.38, the sector of the pie chart labelled VW consists of 0.38(360) = 136.8 degrees. The sector of the pie chart labelled Mercedes consists of 0.26(360) = 93.6 degrees. Similar calculations for the other classes give the pie chart in Figure 2.2. The numerical values shown for each sector can be frequencies, relative frequencies or percentage frequencies.

We have described and illustrated pie charts here because you will see them so frequently in newspapers, company reports and so on. However, research has shown that relative frequency information is more effectively conveyed by a bar chart than by a pie chart, because the human eye is more adept at making comparisons of lengths than of angles.

Often the number of classes in a frequency distribution is the same as the number of categories in the data, as for the car purchase data. Data that included all car brands would require many categories, most of which would have a small number of purchases. Classes with smaller frequencies can be grouped into an aggregate class labelled 'other'. Classes with frequencies of 5 per cent or less would most often be treated in this way.

FIGURE 2.1
Bar chart of new
car purchases:
(a) SPSS chart,
(b) Excel chart

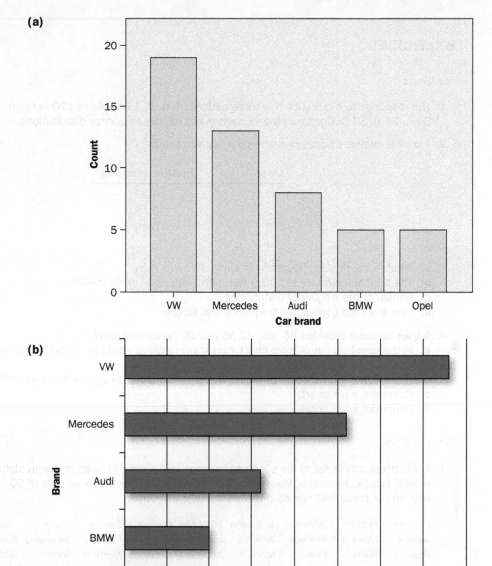

FIGURE 2.2
Pie chart of new car purchases

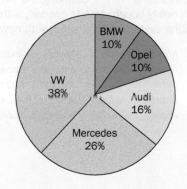

EXERCISES

Methods

1. The response to a question has three options: A, B, C. A sample of 120 responses provides 60 A, 24 B, 36 C. Construct the frequency and relative frequency distributions.

2. A partial relative frequency distribution is given below.

Class	Relative frequency
1	0.22
2	0.18
3	0.40
4	

   a. What is the relative frequency of class 4?
   b. The total sample size is 200. What is the frequency of class 4?
   c. Construct the frequency distribution.
   d. Construct the percentage frequency distribution.

3. A questionnaire provides 58 Yes, 42 No and 20 No-opinion answers.
   a. In the construction of a pie chart, how many degrees would be in the sector of the pie showing the Yes answers?
   b. How many degrees would be in the sector of the pie showing the No answers?
   c. Construct a pie chart.
   d. Construct a bar chart.

Applications

BELGIAN NAMES

4. A Wikipedia article listed the six most common last names in Belgium as (in alphabetical order): Jacobs, Janssens, Maes, Mertens, Peeters and Willems. A sample of 50 individuals with one of these last names provided the following data.

Peeters	Peeters	Willems	Janssens	Janssens	Peeters	Jacobs	Maes	Janssens	Mertens
Jacobs	Maes	Peeters	Willems	Jacobs	Maes	Peeters	Janssens	Maes	Maes
Peeters	Maes	Peeters	Maes	Janssens	Janssens	Mertens	Jacobs	Jacobs	Peeters
Mertens	Maes	Peeters	Janssens	Willems	Willems	Peeters	Janssens	Willems	Mertens
Jacobs	Willems	Peeters	Janssens	Mertens	Janssens	Peeters	Mertens	Mertens	Janssens

   Summarize the data by constructing the following:
   a. Relative and percentage frequency distributions.
   b. A bar chart.
   c. A pie chart.
   d. Based on these data, what are the three most common last names?

MOBILE OS

5. In many European countries, including France, Germany, the UK and Italy, Android is the most common operating system in newly bought mobile phones, with Apple iOS second most common. A sample of operating systems for 50 users with newly bought phones follows.

Android	Android	Android	Android	Apple	Apple	Android	Apple	Apple	Android
Android	Android	Android	Apple	Windows	Android	Apple	Apple	Apple	Windows
Windows	Apple	Android	Apple	Windows	Android	Android	Windows	Android	Apple
Apple	Windows	Other	Android	Android	Android	Apple	Android	Android	Android
Android	Android	Android	Apple	Android	Android	Apple	Android	Android	Other

    a. Are these data qualitative or quantitative?
    b. Construct frequency and percentage frequency distributions.
    c. Construct a bar chart.
    d. On the basis of the sample, which mobile operating system was the most popular? Which one was second?

6. A YouGov survey asked respondents about their use of an alarm to wake them in the morning. Those who used an alarm were asked how many times they typically pressed the snooze button before getting up. The following are the answers from a sample of 20 respondents.

0	3	4	2	1	1	3	3	1	0
3	3	2	2	1	1	3	0	4	3	2

    Summarize the data by constructing the following:
    a. A frequency distribution.
    b. A percentage frequency distribution.
    c. A bar chart.
    d. What do the summaries tell you about respondents' typical response to their wake-up alarm?

7. A customer satisfaction survey carried out by an electricity supply company asked customers to indicate how satisfied they were with the member of the customer service team they had spoken to on the telephone. Customer responses were coded 1 to 7, with 1 indicating 'not at all satisfied' and 7 indicating 'extremely satisfied'. The following data are from a sample of 60 responses for a particular member of the customer service team.

5	7	6	6	7	5	5	7	3	6
7	7	6	6	6	5	5	6	7	6
6	6	4	4	7	6	7	6	7	6
5	7	5	7	6	4	7	5	7	6
6	5	3	7	7	6	6	6	6	5
5	6	6	7	7	5	6	4	6	6

    a. Construct a frequency distribution and a relative frequency distribution for the data.
    b. Construct a bar chart.
    c. On the basis of your summaries, comment on the customers' overall evaluation of the customer service team member.

2.2 SUMMARIZING QUANTITATIVE DATA

Frequency distribution

As defined in Section 2.1, a frequency distribution is a tabular data summary showing the number (frequency) of items in each of several non-overlapping classes. This definition holds for quantitative as well as qualitative data. However, with quantitative data there is usually more work involved in defining the non-overlapping classes.

Consider the quantitative data in Table 2.4. These data show the time, in days, required to complete year-end audits for a sample of 20 clients of Asad & Shahid, a small accounting firm. The data are rounded to the nearest day. The three steps necessary to define the classes for a frequency distribution with quantitative data are:

1 Determine the number of non-overlapping classes.
2 Determine the width of each class.
3 Determine the class limits.

We demonstrate these steps using the audit time data.

TABLE 2.4 Year-end audit times (in days)

AUDIT

| 12 | 14 | 19 | 18 | 15 | 15 | 18 | 17 | 20 | 27 |
| 22 | 23 | 22 | 21 | 33 | 28 | 14 | 18 | 16 | 13 |

Number of classes

Classes are created by specifying ranges to group the data. As a general guideline, we recommend using between 5 and 20 classes. For a small sample of data, you may use as few as five or six classes to summarize the data. For larger samples, more classes are usually required. The aim is to use enough classes to show the pattern of variation in the data, but not so many classes that some contain very few data points. Because the sample in Table 2.4 is relatively small ($n = 20$), we chose to construct a frequency distribution with five classes.

Width of the classes

The second step is to choose a width for the classes. As a general guideline, we recommend using the same width for each class. This reduces the chance of inappropriate interpretations. The choices for the number and the width of classes are not independent decisions. More classes imply a smaller class width and vice versa. To determine an approximate class width, we identify the largest and smallest data values. Then we can use the following expression to determine the approximate class width:

Approximate class width

$$\frac{\text{Largest data value} - \text{Smallest data value}}{\text{Number of classes}} \qquad \textbf{(2.2)}$$

You can round the approximate width given by equation (2.2) to a more convenient value. For example, an approximate class width of 9.28 might be rounded to 10. For the year-end audit times, the largest value is 33 and the smallest value is 12. We decided to summarize the data with five classes, so equation (2.2) provides an approximate class width of $(33 - 12)/5 = 4.2$. We decided to round up and use a class width of 5 days.

In practice, the analyst may try out several combinations of number of classes and class width, to see which provides a good summary of the data. Different people may construct different, but equally acceptable, frequency distributions. The goal is to reveal the natural grouping and variation in the data.

For the audit time data, after deciding to use five classes, each with a width of five days, the next task is to specify the class limits for each of the classes.

Class limits

Class limits must be chosen so that each data item belongs to one and only one class. The *lower class limit* identifies the smallest possible data value assigned to the class. The *upper class limit* identifies the largest possible data value assigned to the class. In constructing frequency distributions for qualitative data, we did not need to specify class limits because each data item naturally fell into a distinct class (category). But with quantitative data, class limits are necessary to determine where each data value belongs.

Using the audit time data, we selected 10 days as the lower class limit and 14 days as the upper class limit for the first class. This class is denoted 10–14 in Table 2.5. The smallest data value, 12, is included in the 10–14 class. We then selected 15 days as the lower class limit and 19 days as the upper class limit of the next class. We continued defining the lower and upper class limits to obtain a total of five classes: 10–14, 15–19, 20–24, 25–29 and 30–34. The largest data value, 33, is included in the 30–34 class. The difference between the lower class limits of adjacent classes is the class width. Using the first two lower class limits of 10 and 15, we see that the class width is $15 - 10 = 5$.

A frequency distribution can now be constructed by counting the number of data values belonging to each class. For example, the data in Table 2.5 show that four values (12, 14, 14 and 13) belong to the 10–14 class. The frequency for the 10–14 class is 4. Continuing this counting process for the 15–19, 20–24, 25–29 and 30–34 classes provides the frequency distribution in Table 2.5. Using this frequency distribution, we can observe that:

1 The most frequently occurring audit times are in the class 15–19 days. Eight of the 20 audit times belong to this class.

2 Only one audit required 30 or more days.

Other comments are possible, depending on the interests of the person viewing the frequency distribution. The value of a frequency distribution is that it provides insights about the data not easily obtained from the data in their original unorganized form.

TABLE 2.5
Frequency distribution for the audit time data

Audit time (days)	Frequency
10–14	4
15–19	8
20–24	5
25–29	2
30–34	1
Total	**20**

The appropriate values for the class limits with quantitative data depend on the level of accuracy of the data. For instance, with the audit time data, the limits used were integer values because the data were rounded to the nearest day. If the data were rounded to the nearest one-tenth of a day (e.g. 12.3, 14.4), the limits would be stated in tenths of days. For example, the first class would be 10.0–14.9. If the data were rounded to the nearest one-hundredth of a day (e.g. 12.34, 14.45), the limits would be stated in hundredths of days, e.g. the first class would be 10.00–14.99.

An *open-ended* class requires only a lower class limit or an upper class limit. For example, in the audit time data, suppose two of the audits had taken 58 and 65 days. Rather than continuing with (empty) classes 35–39, 40–44, 45–49 and so on, we could simplify the frequency distribution to show an open-ended class of '35 or more'. This class would have a frequency count of 2. Most often the open-ended class appears at the upper end of the distribution. Sometimes an open-ended class appears at the lower end of the distribution and occasionally such classes appear at both ends.

Class midpoint

In some applications, we want to know the midpoints of the classes in a frequency distribution for quantitative data. The **class midpoint** is the value halfway between the lower and upper class limits. For the audit time data, the five class midpoints are 12, 17, 22, 27 and 32.

Relative frequency and percentage frequency distributions

We define the relative frequency and percentage frequency distributions for quantitative data in the same way as for qualitative data. The relative frequency is simply the proportion of the observations belonging to a class. With n observations:

$$\text{Relative frequency of a class} = \frac{\text{Frequency of the class}}{n}$$

The percentage frequency of a class is the relative frequency multiplied by 100.

Based on the class frequencies in Table 2.5 and with $n = 20$, Table 2.6 shows the relative frequency and percentage frequency distributions for the audit time data. Note that 0.40 of the audits, or 40 per cent, required from 15 to 19 days. Only 0.05 of the audits, or 5 per cent, required 30 or more days. Again, additional interpretations and insights can be obtained by using Table 2.6.

TABLE 2.6 Relative and percentage frequency distributions for the audit time data

Audit time (days)	Relative frequency	Percentage frequency
10–14	0.20	20
15–19	0.40	40
20–24	0.25	25
25–29	0.10	10
30–34	0.05	5
Total	**1.00**	**100**

Histogram

A histogram is a chart showing quantitative data previously summarized in a frequency distribution (i.e. absolute, relative or percentage). The variable of interest is placed on the horizontal axis and the frequency on the vertical axis. A rectangle is drawn to depict the frequency of each class (absolute, relative or percentage). The base of the rectangle is determined by the class limits on the horizontal axis. The frequencies are represented by the areas of the rectangles: so, if the classes are all of equal width, the heights of the rectangles correspond to the frequencies.

Figure 2.3 is a histogram for the audit time data. The class with the greatest frequency is shown by the rectangle above the class 15–19 days. The height of the rectangle shows that the frequency of this class is 8. A histogram for the relative or percentage frequency distribution would look the same as the histogram in Figure 2.3 except that the vertical axis would be labelled with relative or percentage frequency values.

FIGURE 2.3
Histogram for the audit time data

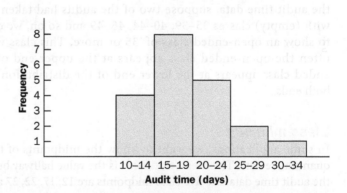

As Figure 2.3 shows, the adjacent rectangles of a histogram touch one another. This is the usual convention for a histogram, unlike a bar chart. The classes for the audit time data are stated as 10–14, 15–19, 20–24 and so on. This would seem to imply that one-unit spaces are needed between 14 and 15, 19 and 20, etc. Eliminating the spaces in the histogram helps show that, taking into account the rounding of the data to the nearest full day, all values between the lower limit of the first class and the upper limit of the last class are possible.

One of the most important uses of a histogram is to provide information about the shape, or form, of a distribution. Figure 2.4 contains four histograms constructed from relative frequency distributions.

FIGURE 2.4
Histograms showing differing levels of skewness

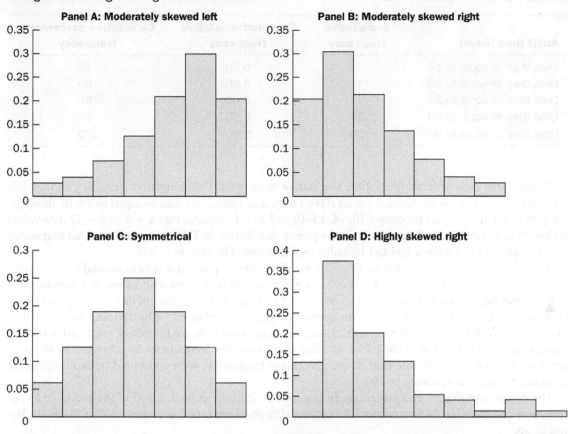

Panel A shows the histogram for a set of data moderately skewed to the left. A histogram is skewed to the left, or negatively skewed, if its tail extends further to the left. A histogram like this might be seen for scores from a relatively simple test, in which only a few really low scores occur, most of the scores are above 70 per cent and scores cannot be greater than 100 per cent. Panel B shows the histogram for a set of data moderately skewed to the right. A histogram is skewed to the right, or positively skewed, if its tail extends further to the right. An example of this type of histogram would be for data such as house values. A relatively small number of expensive homes create the skewness in the right tail.

Panel C shows a symmetrical histogram. In a symmetrical histogram, the left tail mirrors the shape of the right tail. Histograms for real data are never perfectly symmetrical, but may be roughly symmetrical. Data for IQ scores, heights and weights of people and so on, lead to histograms that are roughly symmetrical. Panel D shows a histogram highly skewed to the right (positively skewed). This histogram was constructed from data on the amount of customer purchases over one day at a women's clothing store. Data from applications in business and economics often lead to histograms that are skewed to the right: for instance, data on wealth, salaries, purchase amounts and so on.

Cumulative distributions

A variation of the frequency distribution that provides another tabular summary of quantitative data is the **cumulative frequency distribution**. The cumulative frequency distribution uses the number of classes, class widths and class limits adopted for the frequency distribution. However, rather than showing the frequency of each class, the cumulative frequency distribution shows the number of data items with values *less than or equal to the upper class limit* of each class. The first two columns of Table 2.7 show the cumulative frequency distribution for the audit time data.

TABLE 2.7 Cumulative frequency, cumulative relative frequency and cumulative percentage frequency distributions for the audit time data

Audit time (days)	Cumulative frequency	Cumulative relative frequency	Cumulative percentage frequency
Less than or equal to 14	4	0.20	20
Less than or equal to 19	12	0.60	60
Less than or equal to 24	17	0.85	85
Less than or equal to 29	19	0.95	95
Less than or equal to 34	20	1.00	100

Consider the class with the description 'less than or equal to 24'. The cumulative frequency for this class is simply the sum of the frequencies for all classes with data values less than or equal to 24. In Table 2.5, the sum of the frequencies for classes 10–14, 15–19 and 20–24 indicates that $4 + 8 + 5 = 17$ data values are less than or equal to 24. The cumulative frequency distribution in Table 2.7 also shows that four audits were completed in 14 days or less and 19 audits were completed in 29 days or less.

A **cumulative relative frequency distribution** shows the proportion of data items and a **cumulative percentage frequency distribution** shows the percentage of data items with values less than or equal to the upper limit of each class. The cumulative relative frequency distribution can be computed either by summing the relative frequencies in the relative frequency distribution, or by dividing the cumulative frequencies by the total number of items. Using the latter approach, we found the cumulative relative frequencies in column 3 of Table 2.7 by dividing the cumulative frequencies in column 2 by the total number of items ($n = 20$). The cumulative percentage frequencies were computed by multiplying the cumulative relative frequencies by 100.

The cumulative relative and percentage frequency distributions show that 0.85 of the audits, or 85 per cent, were completed in 24 days or less; 0.95 of the audits, or 95 per cent, were completed in 29 days or less and so on.

The last entry in a cumulative frequency distribution always equals the total number of observations. The last entry in a cumulative relative frequency distribution always equals 1.00 and the last entry in a cumulative percentage frequency distribution always equals 100.

Dot plots and stem-and-leaf displays

One of the simplest graphical summaries of data is a **dot plot**. A horizontal axis shows the range of values for the observations. Each data value is represented by a dot placed above the axis. Figure 2.5 is a dot plot produced in Minitab for the audit time data in Table 2.4. The three dots located above 18 on the horizontal axis indicate that three audit times of 18 days occurred. Dot plots show the details of the data and are useful for comparing data distributions for two or more samples. An example is shown in the Minitab section of the online software guide for this chapter.

FIGURE 2.5
Dot plot for the audit time data

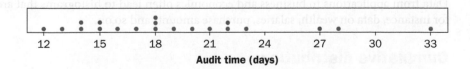

Audit time (days)

A **stem-and-leaf display** shows both the rank order and shape of a data set simultaneously. To illustrate the stem-and-leaf display, consider the data in Table 2.8. These came from a 150-question aptitude test given to 50 individuals recently interviewed for a position at Hawkins Manufacturing. The data indicate the number of questions answered correctly.

TABLE 2.8 Number of questions answered correctly on an aptitude test

112	72	69	97	107	73	92	76	86	73
126	128	118	127	124	82	104	132	134	83
92	108	96	100	92	115	76	91	102	81
95	141	81	80	106	84	119	113	98	75
68	98	115	106	95	100	85	94	106	119

APTITUDE

To construct a stem-and-leaf display, we first arrange the leading digits of each data value to the left of a vertical line. To the right of the vertical line, on the row corresponding to the appropriate first digit, we record the last digit for each data value as we pass through the observations in the order they were recorded. The results are shown in Figure 2.6(a). Sorting the digits on each line into rank order is now relatively simple. This leads to the stem-and-leaf display shown in Figure 2.6(b).

FIGURE 2.6
Construction of a stem-and-leaf plot

```
 6 | 9 8                         6 | 8 9
 7 | 2 3 6 3 6 5                 7 | 2 3 3 5 6 6
 8 | 6 2 3 1 1 0 4 5             8 | 0 1 1 2 3 4 5 6
 9 | 7 2 2 6 2 1 5 8 8 5 4       9 | 1 2 2 2 4 5 5 6 7 8 8
10 | 7 4 8 0 2 6 6 0 6          10 | 0 0 2 4 6 6 6 7 8
11 | 2 8 5 9 3 5 9             11 | 2 3 5 5 8 9 9
12 | 6 8 7 4                   12 | 4 6 7 8
13 | 2 4                       13 | 2 4
14 | 1                         14 | 1
```

(a) unordered leaves   (b) ordered leaves

The numbers to the left of the vertical line (6, 7, …, 14) form the *stem*, and each digit to the right of the vertical line is a *leaf*. For example, the first row has a stem value of 6 and leaves of 8 and 9. It indicates that two data values have a first digit of 6. The leaves show that the data values are 68 and 69. Similarly, the second row indicates that six data values have a first digit of 7. The leaves show that the data values are 72, 73, 73, 75, 76 and 76. Rotating the page counter clockwise onto its side provides a picture of the data that is similar to a histogram with classes of 60–69, 70–79, 80–89 and so on.

The stem-and-leaf display has two potential advantages over a histogram. First, it is easier to construct by hand for small data sets. Second, the stem-and-leaf display provides more information than the histogram because the stem-and-leaf shows the actual data within each class interval.

Just as a frequency distribution or histogram has no absolutely 'correct' number of classes, neither does a stem-and-leaf display have an absolutely correct number of rows or stems. If we believe that our original stem-and-leaf display condensed the data too much, we can stretch the display by using two stems for each leading digit. Using two stems for each leading digit, we would place all data values ending in 0, 1, 2, 3 and 4 in one row and all values ending in 5, 6, 7, 8 and 9 in a second row. This stretched stem-and-leaf display is similar to a frequency distribution with intervals of 65–69, 70–74, 75–79 and so on. Using five stems for each leading digit is also a possibility (for data values ending 0 or 1 / 2 or 3 / 4 or 5 / 6 or 7 / 8 or 9). Examples are shown in the SPSS and Minitab software guides on the online platform.

Stem-and-leaf displays for data with more than two or three digits are possible by scaling the leaf unit to a value different from 1, e.g. 10, 100, etc. Again, an example is shown in the SPSS and Minitab software guides for this chapter on the online platform.

EXERCISES

Methods

8. Consider the following data.

14	21	23	21	16	19	22	25	16	16
24	24	25	19	16	19	18	19	21	12
16	17	18	23	25	20	23	16	20	19
24	26	15	22	24	20	22	24	22	20

   a. Construct a frequency distribution using classes of 12–14, 15–17, 18–20, 21–23 and 24–26.

   b. Construct a relative frequency distribution and a percentage frequency distribution using the classes in (a).

9. Consider the following frequency distribution. Construct a cumulative frequency distribution and a cumulative relative frequency distribution.

Class	Frequency
10–19	10
20–29	14
30–39	17
40–49	7
50–59	2

10. Construct a dot plot using the following data.

| 8.9 | 10.2 | 11.5 | 7.8 | 10.0 | 12.2 | 13.5 | 14.1 | 10.0 | 12.2 |
| 6.8 | 9.5 | 11.5 | 11.2 | 14.9 | 7.5 | 10.0 | 6.0 | 15.8 | 11.5 |

11. Construct a frequency distribution and a percentage frequency distribution using the data from Exercise 10.

12. Construct a stem-and-leaf display for the following data.

| 70 | 72 | 75 | 64 | 58 | 83 | 80 | 82 | 76 | 75 | 68 | 65 | 57 | 78 | 85 | 72 |

13. Construct a stem-and-leaf display for the following data.

| 11.3 | 9.6 | 10.4 | 7.5 | 8.3 | 10.5 | 10.0 | 9.3 | 8.1 | 7.7 | 7.5 | 8.4 | 6.3 | 8.8 |

Applications

14. The London School of Economics and the Harvard Business School have conducted studies of how chief executive officers (CEOs) spend their time. These studies have found that CEOs spend many hours per week in meetings that include conference calls, business meals and public events. Suppose that the data below show the time spent per week in meetings (hours) for a sample of 25 CEOs.

| 14 | 15 | 18 | 23 | 15 | 19 | 20 | 13 | 15 | 23 | 23 | 21 | 15 |
| 20 | 21 | 16 | 15 | 18 | 18 | 19 | 19 | 22 | 23 | 21 | 12 |

   a. What is the least amount of time spent per week in meetings? The highest?

   b. Use a class width of two hours to prepare a frequency distribution and a percentage frequency distribution for the data.

   c. Prepare a histogram and comment on the shape of the distribution.

15. Data for the numbers of landing-page hits on an online retailer's website during a sample of 20 one-minute periods are shown here.

160	170	181	156	176	148	198	179	162	150
162	156	179	178	151	157	154	179	148	156

Summarize the data by constructing the following:
a. A frequency distribution.
b. A relative frequency distribution.
c. A cumulative frequency distribution.
d. A cumulative relative frequency distribution.

16. The closing prices of 40 company shares (in Kuwaiti dinar) follow.

SHARES

44.00	0.80	69.00	226.00	68.00	51.00	265.00	130.00
172.00	202.00	52.00	134.00	81.00	50.00	550.00	28.50
13.00	435.00	218.00	270.00	52.00	108.00	248.00	0.45
188.00	800.00	59.00	65.00	355.00	410.00	102.00	174.00
136.00	34.00	64.00	660.00	122.00	62.00	290.00	90.00

a. Construct frequency and relative frequency distributions.
b. Construct cumulative frequency and cumulative relative frequency distributions.
c. Construct a histogram.
d. Using your summaries, make comments and observations about the price of shares.

17. The table below shows the estimated 2021 mid-year population of the UK, by age group, rounded to the nearest thousand (ONS, release date 21 December 2022).

Age group	Population (000)
0–9	7,514
10–19	7,829
20–29	8,402
30–39	9,133
40–49	8,441
50–59	9,214
60–69	7,310
70–79	5,835
80+	3,348

a. Construct a percentage frequency distribution.
b. Construct a cumulative percentage frequency distribution.
c. Using the cumulative distribution from (b), estimate the age that divides the population into halves (you will learn in Chapter 3 that this is called the *median*).

18. The *Nielsen Home Technology Report* provided information about home technology and its usage by individuals aged 12 and older. The following data are the hours of personal computer usage during one week for a sample of 50 individuals.

COMPUTER

4.1	1.5	5.9	3.4	5.7	1.6	6.1	3.0	3.7	3.1
4.8	2.0	3.3	11.1	3.5	4.1	4.1	8.8	5.6	4.3
7.1	10.3	6.2	7.6	10.8	0.7	4.0	9.2	4.4	5.7
7.2	6.1	5.7	5.9	4.7	3.9	3.7	3.1	12.1	14.8
5.4	4.2	3.9	4.1	2.8	9.5	12.9	6.1	3.1	10.4

Summarize the data by constructing the following:
a. A frequency distribution (use a class width of three hours).
b. A relative frequency distribution.
c. A histogram.
d. Comment on what the data indicate about personal computer usage at home.

19. Below are high and low temperatures (in degrees Celsius) for 20 cities on one particular day in January 2023.

City	High	Low	City	High	Low
Athens	17	10	Munich	7	3
Brussels	8	5	Nairobi	26	16
Cairo	18	13	Paris	8	4
Copenhagen	6	4	Riyadh	18	13
Dublin	8	4	Rome	16	12
Durban	27	17	Seoul	7	−3
Hong Kong	20	18	Singapore	32	25
Johannesburg	25	15	Tel Aviv	18	11
London	9	5	Tokyo	12	11
Marrakech	22	7	Vienna	9	1

a. Prepare a stem-and-leaf display for the high temperatures.
b. Prepare a stem-and-leaf display for the low temperatures.
c. Compare the stem-and-leaf displays from parts (a) and (b), and comment on the differences between daily high and low temperatures.
d. Provide frequency distributions for both high and low temperature data.

2.3 SUMMARIZING RELATIONSHIPS BETWEEN TWO CATEGORICAL VARIABLES

So far in this chapter, we have focused on methods for summarizing *one variable at a time*. Often a manager, decision-maker or researcher requires tabular and graphical methods to assist in the understanding of the *relationship between two variables*. Cross tabulation, clustered bar charts and stacked bar charts are appropriate methods for qualitative/categorical variables.

Cross tabulation

RESTAURANT

A **cross tabulation** is a tabular summary of data for two variables. Consider the following data from a consumer review, based on a sample of 300 restaurants in a large European city. Table 2.9 shows the data for the first five restaurants: the quality rating of the restaurant and typical meal price. Quality rating is a qualitative variable with categories 'disappointing', 'good' and 'excellent'. Meal price is a quantitative variable that ranges from €10 to €49.

TABLE 2.9 Quality rating and meal price for the first five of 300 restaurants

Restaurant	Quality rating	Meal price (€)
1	Disappointing	18
2	Good	22
3	Disappointing	28
4	Excellent	38
5	Good	33

A cross tabulation of the data is shown in Table 2.10. The left and top margin labels define the classes for the two variables. In the left margin, the row labels (disappointing, good and excellent) show the three classes of the quality rating variable. In the top margin, the column labels (€10–19, €20–29, €30–39 and €40–49) correspond to the four classes of the meal price variable. Each restaurant in the sample provides a quality rating and a meal price, and so is associated with a cell in one of the rows and one of the columns of the cross tabulation. For example, restaurant 5 has a good quality rating and a meal price of €33. This restaurant belongs to the cell in row 2 and column 3 of Table 2.10. In constructing a cross tabulation, we simply count the number of restaurants that belong to each of the cells in the cross tabulation.

TABLE 2.10 Cross tabulation of quality rating and meal price for 300 restaurants

| Quality rating | Meal price | | | | |
	€10–19	€20–29	€30–39	€40–49	Total
Disappointing	42	40	2	0	84
Good	34	64	46	6	150
Excellent	2	14	28	22	66
Total	**78**	**118**	**76**	**28**	**300**

We see that the highest cell frequency (64) relates to restaurants with a 'good' rating and a meal price in the €20–29 range. Only two restaurants have an 'excellent' rating and a meal price in the €10–19 range.

Note that the right and bottom margins of the cross tabulation provide the frequency distributions for quality rating and meal price separately. From the frequency distribution in the right margin, we see the quality rating data showing 84 disappointing restaurants, 150 good restaurants and 66 excellent restaurants.

Dividing the totals in the right margin by the total for that column provides relative and percentage frequency distributions for the quality rating variable.

Quality rating	Relative frequency	Percentage frequency
Disappointing	0.28	28
Good	0.50	50
Excellent	0.22	22
Total	1.00	100

We see that 28 per cent of the restaurants were rated disappointing, 50 per cent were rated good and 22 per cent were rated excellent.

Dividing the totals in the bottom row of the cross tabulation by the total for that row provides relative and percentage frequency distributions for the meal price variable. In this case the values do not add exactly to 100, because the values being summed are rounded. From the percentage frequency distribution, we quickly see that 26 per cent of the meal prices are in the lowest price class (€10–19), 39 per cent are in the next highest class and so on.

Meal price	Relative frequency	Percentage frequency
€10–19	0.26	26
€20–29	0.39	39
€30–39	0.25	25
€40–49	0.09	9
Total	1.00	100

The frequency and relative frequency distributions constructed from the margins of a cross tabulation provide information about each of the variables individually, but they do not shed any light on the relationship between the variables. The primary value of a cross tabulation lies in the insight it offers about this relationship. Converting the entries in a cross tabulation into row percentages or column percentages can provide the insight.

For row percentages, the results of dividing each frequency in Table 2.10 by its corresponding row total are shown in Table 2.11. Each row of Table 2.11 is a percentage frequency distribution of meal price for one of the quality rating categories. Of the restaurants with the lowest quality rating (disappointing), we see that the greatest percentages are for the less expensive restaurants (50.0 per cent have €10–19 meal prices and 47.6 per cent have €20–29 meal prices). Of the restaurants with the highest quality rating (excellent), we see that the greatest percentages are for the more expensive restaurants (42.4 per cent have €30–39 meal prices and 33.4 per cent have €40–49 meal prices). Hence, the cross tabulation reveals that higher meal prices are associated with the higher quality restaurants, and the lower meal prices are associated with the lower quality restaurants.

TABLE 2.11 Row percentages for each quality rating category

Quality rating	Meal price				
	€10–19	€20–29	€30–39	€40–49	Total
Disappointing	50.0	47.6	2.4	0.0	100
Good	22.7	42.7	30.6	4.0	100
Excellent	3.0	21.2	42.4	33.4	100

Cross tabulation is widely used for examining the relationship between two variables. The final reports for many statistical studies include a large number of cross tabulations. In the restaurant survey, the cross tabulation is based on one qualitative variable (quality rating) and one quantitative variable (meal price). Cross tabulations can also be constructed when both variables are qualitative and when both variables are quantitative. When quantitative variables are used, we must first create classes for the values of the variable. For instance, in the restaurant example we grouped the meal prices into four classes (€10–19, €20–29, €30–39 and €40–49).

Simpson's paradox

In many cases, a summary cross tabulation showing how two variables are related has been aggregated across a third variable (or across more than one variable). If so, we must be careful in drawing conclusions about the relationship between the two variables in the cross tabulation. In some circumstances the conclusions based upon the aggregated cross tabulation can be completely reversed if we look at the non-aggregated data, something known as Simpson's paradox. To provide an illustration, we consider an example involving the analysis of sales success for two sales executives in a mobile phone company.

The two sales executives are Aaron and Theo. They handle renewal enquiries for two types of mobile phone agreement: pre-pay contracts and pay-as-you-go (PAYG) agreements. The cross tabulation below shows the outcomes for 200 enquiries each for Aaron and Theo, aggregated across the two types of agreement. The cross tabulation involves two variables: outcome (sale or no sale) and sales executive (Aaron or Theo). It shows the number of sales and the number of no-sales for each executive, along with the column percentages in parentheses next to each value.

	Sales executive		
	Aaron	*Theo*	*Total*
Sales	82 (41%)	102 (51%)	184
No-sales	118 (59%)	98 (49%)	216
Total	200 (100%)	200 (100%)	400

The column percentages indicate that Aaron's overall sales success rate was 41 per cent, compared with Theo's 51 per cent success rate, suggesting that Theo has the better sales performance. However, there is a problem with this conclusion. The following cross tabulations show the enquiries handled by Aaron and Theo for the two types of agreement separately.

	Pre-pay				*PAYG*		
	Aaron	*Theo*	*Total*		*Aaron*	*Theo*	*Total*
Sales	56 (35%)	18 (30%)	74	*Sales*	26 (65%)	84 (60%)	110
No-sales	104 (65%)	42 (70%)	146	*No-sales*	14 (35%)	56 (40%)	70
Total	160 (100%)	60 (100%)	220	*Total*	40 (100%)	140 (100%)	180

We see that Aaron achieved a 35 per cent success rate for pre-pay contracts and 65 per cent for PAYG agreements. Theo had a 30 per cent success rate for pre-pay and 60 per cent for PAYG. This comparison suggests that Aaron has a better success rate than Theo for both types of agreement, a result that contradicts the conclusion reached when the data were aggregated across the two types of agreement. This example illustrates Simpson's paradox.

Note that for both sales executives the sales success rate was considerably higher for PAYG than for pre-pay contracts. Because Theo handled a much higher proportion of PAYG enquiries than Aaron, the aggregated data favoured Theo. When we look at the cross tabulations for the two types of agreement separately, however, Aaron shows the better record. Hence, for the original cross tabulation, we see that the *type of agreement* is a hidden variable that should not be ignored when evaluating the records of the sales executives.

Because of Simpson's paradox, we need to be especially careful when drawing conclusions using aggregated data. Before drawing any conclusions about the relationship between two variables shown for a cross tabulation – or, indeed, any type of display involving two variables (like the bar charts and scatter diagrams illustrated in the remainder of the chapter) – you should consider whether any hidden variable or variables could affect the results.

Clustered and stacked bar charts

In Section 2.1 we saw how a frequency distribution for a categorical (qualitative) variable can be displayed graphically using a simple bar chart. Similarly, more elaborate forms of bar chart, in particular clustered bar charts and stacked bar charts, can be used to show the joint distribution of two categorical variables, and to help understand the relationship between them.

We take the restaurant data in Table 2.10 as an example. Figure 2.7 shows an Excel clustered column chart (recall, Excel refers to charts with vertical bars as column charts). Here each bar represents one of the 12 cell counts in the three rows by four columns cross tabulation of Table 2.10. The bars are arranged in four groups or clusters corresponding to the four meal price categories. The different colours of the bars in each cluster identify the three different quality ratings within each meal price category. It is evident that the shapes and colour mix of the clusters vary as we move from the lowest meal price category, where blue ('disappointing') predominates, to the highest meal price category, where green ('excellent') is the longest bar.

Clustered bar charts, like Figure 2.7, are sometimes referred to as multiple bar charts or side-by-side bar charts. An alternative form of presentation is a stacked bar chart. Figure 2.8 shows an SPSS stacked bar chart for the restaurant data, with the data presented in percentages rather than absolute frequencies. Each bar represents one of the meal price categories in Table 2.10, scaled to 100 per cent. The bars are divided into segments, with a different colour for each quality rating category, and the lengths of the segments representing column percentages calculated from the data in Table 2.10. In other words, these segments show us how the restaurants in each meal price category fall, in relative terms, into the three quality rating categories: disappointing, good, excellent. The bar chart shows that, as we move from the lowest meal price category to the highest, the percentage of restaurants with a 'disappointing' quality rating decreases, and the percentage with an 'excellent' quality rating increases.

FIGURE 2.7
Clustered bar chart for
the restaurant data

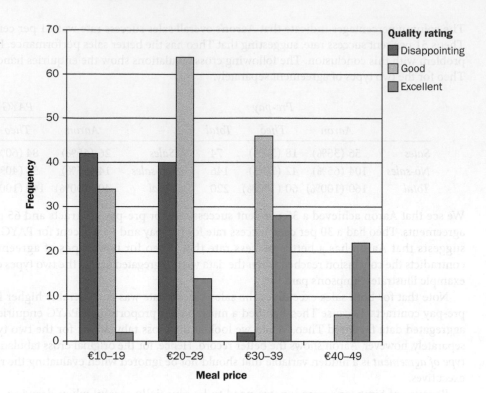

FIGURE 2.8
Stacked bar chart for
the restaurant data

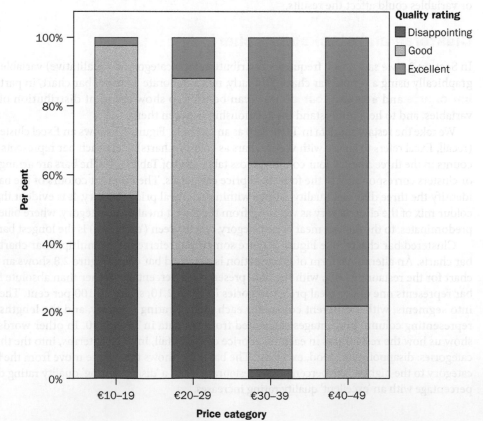

EXERCISES

Methods

20. The following data are for 30 observations involving two qualitative variables, X and Y. The categories for X are A, B and C; the categories for Y are 1 and 2.

CROSSTAB

Observation	X	Y	Observation	X	Y
1	A	1	16	B	2
2	B	1	17	C	1
3	B	1	18	B	1
4	C	2	19	C	1
5	B	1	20	B	1
6	C	2	21	C	2
7	B	1	22	B	1
8	C	2	23	C	2
9	A	1	24	A	1
10	B	1	25	B	1
11	A	1	26	C	2
12	B	1	27	C	2
13	C	2	28	A	1
14	C	2	29	B	1
15	C	2	30	B	2

a. Construct a cross tabulation for the data, with X as the row variable and Y as the column variable.
b. Calculate the row percentages.
c. Calculate the column percentages.
d. Construct a clustered bar chart showing absolute frequencies, and with the Y categories as the 'clusters'.
e. What is the relationship, if any, between X and Y?

Applications

21. The file 'House Prices' on the online platform contains data for a sample of 50 houses advertised for sale in a regional UK newspaper. The first four rows of data are shown for illustration below.

HOUSE
PRICES

Price	Location	House type	Bedrooms	Reception rooms	Bedrooms + receptions	Garage capacity
234,995	Town	Detached	4	2	6	1
319,000	Town	Detached	4	2	6	1
154,995	Town	Semi-detached	2	1	3	0
349,950	Village	Detached	4	2	6	2

a. Prepare a cross tabulation using sale price (rows) and house type (columns). Use classes of 100,000–199,999, 200,000–299,999, etc. for sale price.
b. Compute row percentages for the cross tabulation in (a).
c. Construct a stacked bar chart showing the percentage distribution of house types within each sale price category.
d. Comment on any relationship between the variables.

22. Recently, management at Oak Tree Golf Course received a few complaints about the condition of the greens. Several players complained that the greens are too fast. Rather than react to the comments of just a few, the Golf Association conducted a survey of 100 male and 100 female golfers. The survey results are summarized here.

	Male golfers Greens condition			Female golfers Greens condition	
Handicap	Too fast	Fine	Handicap	Too fast	Fine
Under 15	10	40	Under 15	1	9
15 or more	25	25	15 or more	39	51

   a. Combine these two cross tabulations into one with 'male' and 'female' as the row labels and the column labels 'too fast' and 'fine'. Which group shows the highest percentage saying that the greens are too fast?
   b. Refer to the initial cross tabulations. For those players with low handicaps (better players), which group (male or female) shows the highest percentage saying the greens are too fast?
   c. Refer to the initial cross tabulations. For those players with higher handicaps, which group (male or female) shows the highest percentage saying the greens are too fast?
   d. What conclusions can you draw about the preferences of male players and female players concerning the speed of the greens? Are the conclusions you draw from part (a) as compared with parts (b) and (c) consistent? Explain any apparent inconsistencies.

23. Refer to the data in Exercise 21.
   a. Prepare a cross tabulation using number of bedrooms and house type.
   b. Calculate the percentage distribution for numbers of bedrooms within each house type.
   c. Construct a clustered bar chart with house types as the 'clusters'.
   d. Comment on the relationship between number of bedrooms and house type.

2.4 SUMMARIZING RELATIONSHIPS BETWEEN TWO QUANTITATIVE VARIABLES

A scatter diagram is a graphical presentation of the relationship between two quantitative variables, and a trend line is a line that provides an approximation of the relationship. Consider the advertising/sales relationship for a digital equipment online store. On ten occasions during the past three months, the store used weekend Facebook ads to promote sales. The managers want to investigate whether a relationship exists between the number of Facebook promotions and online sales during the following week. Sample data for the ten weeks with sales in thousands of euros (€000) are shown in Table 2.12.

TABLE 2.12 Sample data for the digital equipment store

FACEBOOK

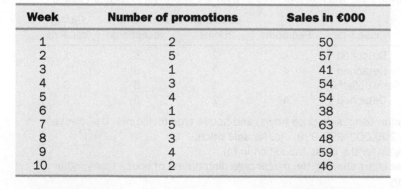

Week	Number of promotions	Sales in €000
1	2	50
2	5	57
3	1	41
4	3	54
5	4	54
6	1	38
7	5	63
8	3	48
9	4	59
10	2	46

Figure 2.9 is an Excel scatter diagram that includes the trend line* for the data in Table 2.12. The number of promotions (x) is shown on the horizontal axis and the sales (y) are shown on the vertical axis. For week 1, $x = 2$ and $y = 50$. A point with those coordinates is plotted on the scatter diagram. Similar points are plotted for the other nine weeks. Note that during two of the weeks there was one promotion, during two of the weeks there were two promotions and so on.

---

* The equation of the trend line is $y = 4.95x + 36.15$. The slope of the trend line is 4.95 and the y-intercept (the point where the line intersects the y axis) is 36.15. We will discuss in detail the interpretation of the slope and y-intercept for a linear trend line in Chapter 14 when we study simple linear regression.

FIGURE 2.9
Scatter diagram and trend line
for the digital equipment store

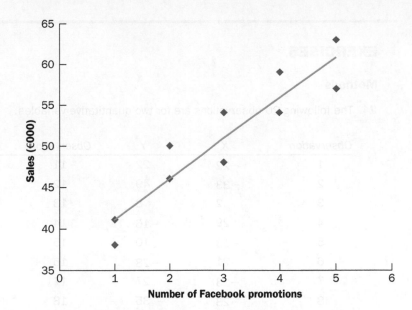

The completed scatter diagram in Figure 2.9 indicates a positive relationship between the number of promotions and sales. Higher sales are associated with a higher number of promotions. The relationship is not perfect in that all points are not on a straight line. However, the general pattern of the points and the trend line suggest that the overall relationship is positive.

Some general scatter diagram patterns and the types of relationships they suggest are shown in Figure 2.10. The top left panel depicts a positive relationship similar to the one for the number of promotions and sales example. In the top right panel, the scatter diagram shows no apparent relationship between the variables. The bottom panel depicts a negative relationship where *y* tends to decrease as *x* increases.

FIGURE 2.10
Types of relationships
depicted by scatter
diagrams

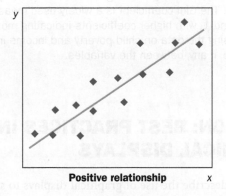

Positive relationship   *x*

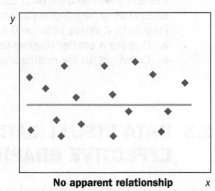

No apparent relationship   *x*

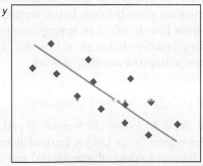

Negative relationship   *x*

EXERCISES

Methods

SCATTER

24. The following 20 observations are for two quantitative variables.

Observation	X	Y	Observation	X	Y
1	−22	22	11	−37	48
2	−33	49	12	34	−29
3	2	8	13	9	−18
4	29	−16	14	−33	31
5	−13	10	15	20	−16
6	21	−28	16	−3	14
7	−13	27	17	−15	18
8	−23	35	18	12	17
9	14	−5	19	−20	−11
10	3	−3	20	−7	−22

a. Construct a scatter diagram for the relationship between X and Y.
b. What is the relationship, if any, between X and Y?

Applications

OECD 2020

25. The file 'OECD 2020' on the online platform contains data for 13 countries taken from the website of the Organisation for Economic Co-operation and Development. The two variables are the Gini coefficient for each country and the percentage of children (0–17 years) in the country estimated to be living in poverty. The Gini coefficient is a widely used measure of income inequality. It varies between 0 and 1, with higher coefficients indicating more inequality.
a. Prepare a scatter diagram using the data on child poverty and income inequality.
b. Comment on the relationship, if any, between the variables.

2.5 DATA VISUALIZATION: BEST PRACTICES IN CREATING EFFECTIVE GRAPHICAL DISPLAYS

Data visualization is a term used to describe the use of graphical displays to summarize and present information about a data set. The goal of data visualization is to communicate as effectively and clearly as possible, the key information about the data. In this section, we provide guidelines for creating an effective graphical display, discuss how to select an appropriate type of display given the purpose of the study, illustrate the use of data dashboards and show how the Cincinnati Zoo and Botanical Garden uses data visualization techniques to improve decision-making.

Creating effective graphical displays

The data presented in Table 2.13 show the forecast or planned value of sales (£000) and the actual value of sales (£000) by sales region in the UK for Purity Chemical for the past year. Note that there are two quantitative variables (planned sales and actual sales) and one categorical variable (sales region). Suppose we would like to construct a graphical display that would enable management of Purity Chemical to visualize how each sales region performed relative to planned sales and simultaneously enable management to visualize sales performance across regions.

TABLE 2.13 Planned and actual sales by sales region (£000)

Sales region	Planned sales (£000)	Actual sales (£000)
Northeast	540	447
Northwest	420	447
Southeast	575	556
Southwest	360	341

Figure 2.11 shows a clustered bar chart of the planned versus actual sales data. Note how this bar chart makes it very easy to compare the planned versus actual sales in a region, as well as across regions. This graphical display is simple, contains a title, is well labelled, and uses colour to differentiate the two types of sales. Here, different shades of blue are used to prevent the chart from being distracting to the reader. Note also that the scale of the vertical axis begins at zero. The four sales regions are separated by space so that it is clear that they are distinct, whereas the planned versus actual sales values are side-by-side for easy comparison within each region. The clustered bar chart in Figure 2.11 makes it easy to see that the Southwest region is the lowest in both planned and actual sales and that the Northwest region slightly exceeded its planned sales.

FIGURE 2.11

Clustered bar chart of planned versus actual sales

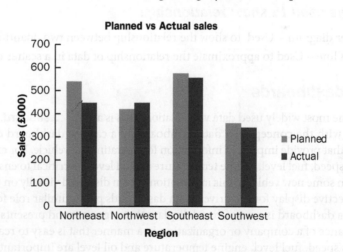

Creating an effective graphical display is as much art as it is science. By following the general guidelines listed below you can increase the likelihood that your display will effectively convey the key information in the data.

- Give the display a clear and concise title.
- Keep the display simple. Do not use three dimensions when two dimensions are sufficient.
- Clearly label each axis and provide the units of measure.
- If colour is used to distinguish categories, make sure the colours are distinct but not distracting.
- If multiple colours or line types are used, use a key to define how they are used and place the key close to the representation of the data.

Choosing the type of graphical display

In this chapter, we discussed a variety of graphical displays, including bar charts, pie charts, dot plots, histograms, stem-and-leaf plots, scatter diagrams, side-by-side bar charts and stacked bar charts. Each of these types of displays was developed for a specific purpose. To provide guidelines for choosing the appropriate type of graphical display, we now provide a summary of the types of graphical displays categorized by their purpose. We note that some types of graphical displays may be used effectively for multiple purposes.

Displays used to show the distribution of data

- Bar chart – Used to show the frequency distribution and relative frequency distribution for categorical data
- Pie chart – Used to show the relative frequency and percentage frequency for categorical data; generally not preferred to the use of a bar chart
- Dot plot – Used to show the distribution for quantitative data over the entire range of the data
- Histogram – Used to show the frequency distribution for quantitative data over a set of class intervals
- Stem-and-leaf display – Used to show both the rank order and shape of the distribution for quantitative data

Displays used to make comparisons

- Clustered bar chart – Used to compare two variables
- Stacked bar charts – Used to compare the relative frequency or percentage frequency of two categorical variables

Displays used to show relationships

- Scatter diagram – Used to show the relationship between two quantitative variables
- Trend line – Used to approximate the relationship of data in a scatter diagram

Data dashboards

One of the most widely used data visualization tools is a **data dashboard**. If you drive a car, you are already familiar with the concept of a data dashboard. In a car, the dashboard contains gauges and other visual displays that provide important information for operating the vehicle. For example, the gauges used to display the car's speed, fuel level, engine temperature and oil level are critical to ensure safe and efficient operation of the car. In some new vehicles, this information is even displayed visually on the windscreen to provide an even more effective display for the driver. Data dashboards play a similar role for managerial decision-making.

A data dashboard is a set of visual displays that organizes and presents information used to monitor the performance of a company or organization in a manner that is easy to read, understand and interpret. Just as a car's speed, fuel level, engine temperature and oil level are important information to monitor in a car, every business has key performance indicators (KPIs) that need to be monitored to assess how a company is performing. Examples of KPIs are inventory on hand, daily sales, percentage of on-time deliveries and sales revenue per quarter. A data dashboard should provide timely summary information (potentially from various sources) on KPIs that is important to the user, and it should do so in a manner that informs rather than overwhelms its user.

To illustrate the use of a data dashboard in decision-making, we shall discuss an application involving the Grogan Oil Company. Grogan has offices located in three cities in Texas: Austin (its headquarters), Houston and Dallas. Grogan's Information Technology (IT) call centre, located in the Austin office, handles calls from employees regarding computer-related problems involving software, internet and email issues. For example, if a Grogan employee in Dallas has a computer software problem, the employee can call the IT call centre for assistance.

The data dashboard shown in Figure 2.12 was developed to monitor the performance of the call centre. This data dashboard combines several displays to monitor the call centre's KPIs. The data presented are for the current shift, which started at 8:00 a.m. The stacked bar chart in the upper left-hand corner shows the call volume for each type of problem (software, internet or email) over time. This chart shows that call volume is heavier during the first few hours of the shift, calls concerning email issues appear to decrease over time, and volume of calls regarding software issues are highest mid-morning.

FIGURE 2.12

Grogan Oil information technology call centre data dashboard

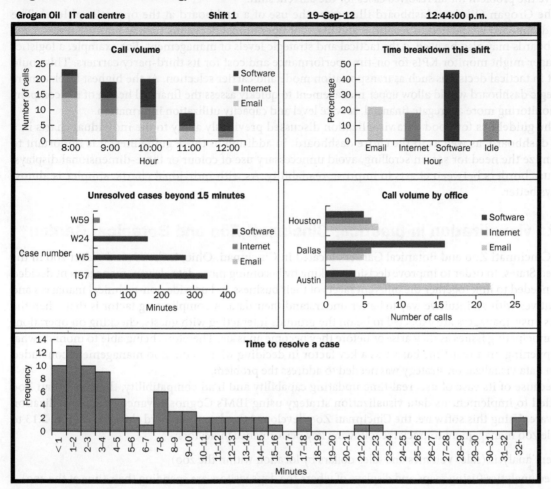

The bar chart in the upper right-hand corner of the dashboard shows the percentage of time that call centre employees spent on each type of problem or were idle (not working on a call). These top two charts are important displays in determining optimal staffing levels. For instance, knowing the call mix and how stressed the system is, as measured by percentage of idle time, can help the IT manager make sure that enough call centre employees are available with the right level of expertise.

The clustered bar chart titled 'Call volume by office' shows the call volume by type of problem for each of Grogan's offices. This allows the IT manager to quickly identify if there is a particular type of problem by location. For example, it appears that the office in Austin is reporting a relatively high number of issues with email. If the source of the problem can be identified quickly, then the problem for many might be resolved quickly. Also, note that a relatively high number of software problems are coming from the Dallas office. The higher call volume in this case was simply due to the fact that the Dallas office is currently installing new software, and this has resulted in more calls to the IT call centre. Because the IT manager was alerted to this by the Dallas office last week, the IT manager knew there would be an increase in calls coming from the Dallas office and was able to increase staffing levels to handle the expected increase in calls.

For each unresolved case that was received more than 15 minutes ago, the bar chart shown in the middle, left-hand side displays the length of time that each of these cases has been unresolved. This chart enables Grogan to quickly monitor the key problem cases and decide whether additional resources may be needed to resolve them. The worst case, T57, has been unresolved for over 300 minutes and is actually left over from the previous shift. Finally, the histogram at the bottom shows the distribution of the time to resolve the problem for all resolved cases for the current shift.

The Grogan Oil data dashboard illustrates the use of a dashboard at the operational level. The data dashboard is updated in real time and used for operational decisions such as staffing levels. Data dashboards may also be used at the tactical and strategic levels of management. For example, a logistics manager might monitor KPIs for on-time performance and cost for its third-party carriers. This could assist in tactical decisions such as transportation mode and carrier selection. At the highest level, a more strategic dashboard would allow upper management to quickly assess the financial health of the company by monitoring more aggregate financial, service level and capacity utilization information.

The guidelines for good data visualization discussed previously apply to the individual charts in a data dashboard, as well as to the entire dashboard. In addition to those guidelines, it is important to minimize the need for screen scrolling, avoid unnecessary use of colour or three-dimensional displays, and use borders between charts to improve readability. As with individual charts, simpler is almost always better.

Data visualization in practice: Cincinnati Zoo and Botanical Garden*

The Cincinnati Zoo and Botanical Garden, located in Cincinnati, Ohio, is the second oldest zoo in the United States. In order to improve decision-making by becoming more data-driven, management decided they needed to link together the different facets of their business and provide non-technical managers and executives with an intuitive way to better understand their data. A complicating factor is that when the zoo is busy, managers are expected to be on the grounds interacting with guests, checking on operations and anticipating issues as they arise or before they become an issue. Therefore, being able to monitor what is happening on a real-time basis was a key factor in deciding what to do. Zoo management concluded that a data visualization strategy was needed to address the problem.

Because of its ease of use, real-time updating capability and iPad compatibility, the Cincinnati Zoo decided to implement its data visualization strategy using IBM's Cognos advanced data visualization software. Using this software, the Cincinnati Zoo developed the data dashboard shown in Figure 2.13 to enable zoo management to track the following key performance indicators:

- Item analysis (sales volumes and sales dollars by location within the zoo)
- Geo analytics (using maps and displays of where the day's visitors are spending their time at the zoo)
- Customer spending
- Cashier sales performance
- Sales and attendance data versus weather patterns
- Performance of the zoo's loyalty rewards programme.

An iPad mobile application was also developed to enable the zoo's managers to be out on the grounds and still see and anticipate what was occurring on a real-time basis. The Cincinnati Zoo's iPad data dashboard, shown in Figure 2.14, provides managers with access to the following information:

- Real-time attendance data, including what 'types' of guests are coming to the zoo
- Real-time analysis showing which items are selling the fastest inside the zoo
- Real-time geographical representation of where the zoo's visitors live.

Having access to the data shown in Figures 2.13 and 2.14 allows the zoo managers to make better decisions on staffing levels within the zoo, which items to stock based upon weather and other conditions, and how to better target its advertising based on geodemographics.

The impact that data visualization has had on the zoo has been significant. Within the first year of use, the system has been directly responsible for revenue growth of over $500,000, increased visits to the zoo, enhanced customer service and reduced marketing costs.

---

* The authors are indebted to John Lucas, formerly of the Cincinnati Zoo and Botanical Garden, for providing this application.

FIGURE 2.13

Data dashboard for the Cincinnati Zoo

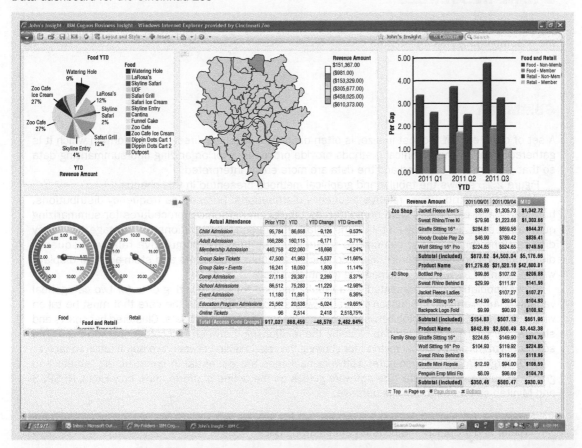

FIGURE 2.14

The Cincinnati Zoo
iPad data dashboard

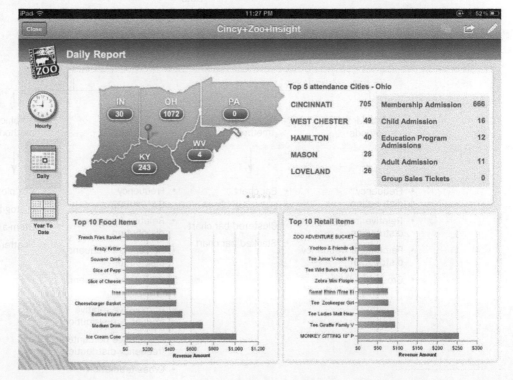

ONLINE RESOURCES

For the data files, additional questions and answers, and the software section for Chapter 2, go to the accompanying online platform.

SUMMARY

A set of data, even if modest in size, is often difficult to interpret directly in the form in which it is gathered. Tabular and graphical methods provide procedures for organizing and summarizing data so that patterns are revealed and the data are more easily interpreted.

Figure 2.15 shows the tabular and graphical methods presented in this chapter.

Frequency distributions, relative frequency distributions, percentage frequency distributions, bar charts and pie charts were presented as tabular and graphical procedures for summarizing qualitative data. Frequency distributions, relative frequency distributions, percentage frequency distributions, histograms, cumulative frequency distributions, cumulative relative frequency distributions, cumulative percentage frequency distributions, dot plots and stem-and-leaf displays were presented as ways of summarizing quantitative data.

Cross tabulation was presented as a tabular method for summarizing data for two categorical variables. An example of Simpson's paradox was set out to illustrate the care that must be taken when interpreting relationships between two variables using aggregated data. Clustered bar charts and stacked bar charts can be used to examine the relationship between two categorical variables. The scatter diagram is a graphical method for showing the relationship between two quantitative variables.

With large data sets, computer software packages are essential in constructing tabular and graphical summaries of data. The software guides on the online platform show how Excel, R, SPSS and Minitab can be used for this purpose.

FIGURE 2.15
Tabular and graphical methods for summarizing data

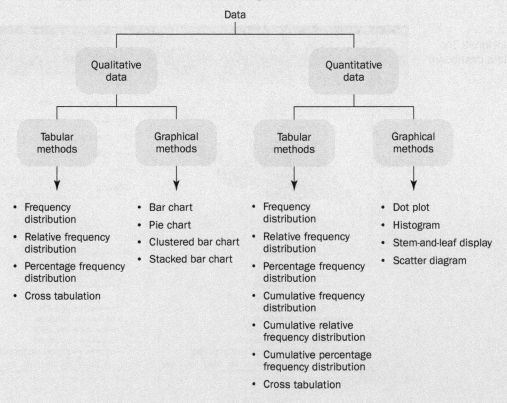

KEY TERMS

Bar chart	Frequency distribution
Bar graph	Histogram
Class midpoint	Percentage frequency distribution
Clustered bar chart	Pie chart
Cross tabulation	Qualitative data
Cumulative frequency distribution	Relative frequency distribution
Cumulative percentage frequency distribution	Scatter diagram
Cumulative relative frequency distribution	Simpson's paradox
Data dashboard	Stacked bar chart
Data visualization	Stem-and-leaf display
Dot plot	Trend line

KEY FORMULAE

Relative frequency

$$\text{Relative frequency of a class} = \frac{\text{Frequency of the class}}{n} \qquad (2.1)$$

Approximate class width

$$\frac{\text{Largest data value} - \text{Smallest data value}}{\text{Number of classes}} \qquad (2.2)$$

CASE PROBLEM 1

© ljubaphoto/iStock

In The Mode fashion stores

In The Mode is a chain of fashion stores. The chain recently ran a promotion in which discount coupons were sent to customers. Data collected for a sample of 100 in-store credit card transactions during a single day following the promotion are contained in the file 'Mode' on the online platform. A portion of the data set has been reproduced and is shown in the following table. A non-zero amount for the discount variable indicates that the customer brought in the promotional coupons and used them. For a very few customers, the discount amount is actually greater than the sales amount (for example, customer 4). In The Mode's management would like to use this sample data to learn about its customer base and to evaluate the promotion involving discount coupons.

Managerial report

Use tables and charts to help management develop a customer profile and evaluate the promotional campaign. At a minimum, your report should include the following.

1. Percentage frequency distributions for key variables.

2. A bar chart showing the percentage of customer purchases possibly attributable to the promotional campaign.

MODE

(Continued)

3. A cross tabulation of type of customer (regular or promotional) versus sales. Comment on any similarities or differences present.

4. A scatter diagram of sales versus discount for those customers responding to the promotion (i.e. omit customers who did not use discount coupons). Comment on any relationship apparent between sales and discount.

5. A scatter diagram to explore the relationship between sales and customer age. Comment on any apparent relationship.

Customer	Payment_Method	Items	Discount	Sales	Sex	Marital_Status	Age
1	Switch	1	0	39.5	Male	Married	32
2	Store Card	1	26	102.4	Female	Married	36
3	Store Card	1	0	22.5	Female	Married	32
4	Store Card	5	121	100.4	Female	Married	28
5	Mastercard	2	0	54.0	Female	Married	34
6	Mastercard	1	0	44.5	Female	Married	44
7	Store Card	2	20	78.0	Female	Married	30
8	Visa	1	0	22.5	Female	Married	40
9	Store Card	2	22	56.5	Female	Married	46
10	Store Card	1	0	44.5	Female	Married	36
11	Store Card	1	0	29.5	Female	Married	48
12	Store Card	1	8	31.6	Female	Married	40

CASE PROBLEM 2

Development of human capital in Egypt

Globalization has been a dominant feature of business and economic development over recent decades. The data in this case problem were collected in a survey of over 2,000 people of working age in Cairo, Egypt, as part of a study of globalization and its effects on the development of human capital and women's economic empowerment.*

The data set on the online platform is a subset of the Cairo survey data, comprising survey respondents who were in work and had at least some university education (though not necessarily to the level of a Bachelor's degree). The variables in the data file are gender (male/female), age group, level of education, length of experience in the current company, length of experience before joining the current company, salary level (in Egyptian pounds per month). All variables are categorized. The first few rows of data have been reproduced (from R) in the following table.

The focus of the case problem is the comparison between salary levels for men and for women.

* The authors are grateful to Dr Mohga Bassim, University of Buckingham, UK, for permission to use the data she collected for her study.

CAIRO SURVEY

	Gender	Age	Education	Exp_with_Co	Exp_Outside	Income
1	Female	26–30	University Education	0–5	0–5	less than 1 thousand
2	Female	21–25	University Education	0–5	0–5	less than 1 thousand
3	Female	26–30	University Education	0–5	0–5	less than 1 thousand
4	Female	31–35	University Education	6–10	0–5	less than 1 thousand
5	Female	26–30	University Education	6–10	0–5	less than 1 thousand
6	Female	26–30	University Education	6–10	0–5	less than 1 thousand
7	Female	31–35	University Degree and Diploma	11–15	6–10	less than 1 thousand
8	Male	36–40	University Education	6–10	0–5	less than 1 thousand
9	Male	31–35	University Education	0–5	0–5	1–2 thousand
10	Male	31–35	University Education	6–10	0–5	less than 1 thousand

Analyst's report

1. Produce graphical displays showing the distribution of each of the variables in the data file. Comment briefly on your results.

2. Construct a cross tabulation of salary level by gender (male/female), using salary level as the row variable. Include column percentages in your cross tabulation.

3. Produce a clustered bar chart to illustrate the cross tabulation in (2). What conclusions do you draw from your cross tabulation and bar chart about the comparison between salary levels for men and women?

4. Produce stacked bar charts to illustrate the relationships between gender and age, and between gender and level of education.

5. Do your results from (4) suggest that your conclusions in (3) might be modified if age and educational level differences between men and women were taken into account?

© Oleksii Hlembotskyi/iStock

CASE PROBLEM 3

CUTRATE

Cut-Rate Machining Ltd

Jon Weideman, manager of Cut-Rate Machining Ltd, is deciding on a vendor from whom to purchase a drilling machine. Jon narrows the alternatives to four vendors: The Hole-Maker Ltd (HM), Shafts & Slips Ltd (SS), Judge's Jigs (JJ) and Drill-for-Bits (DB). Each vendor is offering machines of similar capabilities at similar prices, so the effectiveness of the machines is the only selection criterion Jon can use. Jon invites each vendor to ship one machine to his manufacturing facility for a test. Jon starts all four machines at 8:00 a.m. and lets them warm up for two hours before starting to use any of the machines. After the warmup, one of Jon's employees uses each of the machines for two hours to drill holes 3 cm in diameter in stainless-steel sheets that are 25 cm thick. The exact diameters of the holes drilled with each machine are then measured and recorded.

The results of Jon's data collection are in the file 'CutRate'. The file contains details of the time period when the drilling took place, the employee who did the drilling, the machine vendor and the diameter of the hole. Based on these results, from which vendor would you suggest Jon purchase the new machine?

Managerial report

Use graphical methods to investigate the effectiveness of the machines from each vendor. Include the following in your report:

1. Relevant plots of the measured diameters of each hole (cm).

2. Based on the plots, a discussion of the effectiveness of each vendor and under which conditions (if any) that vendor would be acceptable.

3. A discussion of possible sources of error in the approach taken to assess these vendors.

© photosour/iStock

3

Descriptive Statistics: Numerical Measures

CHAPTER CONTENTS

Statistics in Practice   TV audience measurement

3.1 Measures of location
3.2 Measures of variability
3.3 Measures of distributional shape, relative location and detecting outliers
3.4 Exploratory data analysis
3.5 Measures of association between two variables
3.6 Data dashboards: adding numerical measures to improve effectiveness

LEARNING OBJECTIVES   After reading this chapter and doing the exercises, you should be able to:

1 Calculate and interpret the following statistical measures that describe the central location, variability and shape of data sets:

   1.1   Arithmetic mean, weighted mean, geometric mean, median and mode.

   1.2   Percentiles (including quartiles), range, interquartile range, variance, standard deviation and coefficient of variation.

   1.3   Five-number summaries and box plots.

2 Understand the concept of distributional skewness.

3 Calculate z-scores and understand their role in identifying data outliers.

4 Understand the role of Chebyshev's theorem and of the empirical rule in estimating the spread of data sets.

5 Calculate and interpret covariance and Pearson's correlation coefficient in describing the relationship between two variables.

In Chapter 2 we discussed tabular and graphical data summaries. In this chapter, we present numerical measures for summarizing data. We start with summary measures for a single variable. In Section 3.5 we examine measures of the relationship between two variables.

STATISTICS IN PRACTICE
TV audience measurement

© cyano66/iStock

TV audience numbers and TV audience share are important issues for advertisers, for sponsors and, in the case of public service broadcasting, for governments. In recent years, the number of available TV channels and streaming options has increased substantially in many countries because of the introduction of digital, satellite, cable and video-on-demand (VOD) services. Recording technology and the existence of catch-up and VOD services allow viewers to time-shift their viewing – what broadcasters refer to as 'non-linear' viewing. Accurate audience measurement consequently becomes a more difficult task.

TV audience figures in the UK are produced by the Broadcasters' Audience Research Board (BARB), a not-for-profit company set up in 1981. The BARB website (www.barb.co.uk) gives access to a wealth of viewing figures: for example, for individual shows, both live and time-shifted, for different genres and for a multitude of channels. Averages, such as 'average daily reach' and 'average weekly viewing', feature in many of the data tables.

In December 2022, for example, BARB figures showed that UK viewers were watching 2 hours 57 minutes of TV per day on average. About 1 hour 58 minutes of this total were shared between the five main 'terrestrial' channels BBC1, BBC2, ITV, Channel 4 and Channel 5. In addition, viewers watched an average of between 43 and 44 minutes per day via VOD services, among which Netflix was the biggest player.

In this chapter, you will learn how to compute and interpret some of the statistical measures used in reports such as those presented by BARB. You will learn about the mean, median and mode, and about other descriptive statistics such as the range, variance, standard deviation, percentiles and correlation. These numerical measures will assist in the understanding and interpretation of data.

We introduce numerical measures of location, dispersion, shape and association. When computed for sample data, they are called **sample statistics**. If they are computed for a whole population, they are called **population parameters**. In statistical inference, a sample statistic is often used as a **point estimator** of the corresponding population parameter. In Chapter 7 we shall discuss in more detail the process of point estimation. In the guides on the associated online platform, we show how Excel, R, SPSS and Minitab can be used to compute many of the numerical measures described in the chapter.

3.1 MEASURES OF LOCATION

Mean

The most commonly used measure of central location for a data set is the arithmetic mean, usually referred to simply as the **mean**. When people refer to the 'average' value, they are usually referring to the arithmetic mean. If the data are from a sample, we denote the mean by putting a bar over the data symbol, e.g. $\bar{x}$ (spoken as 'ex bar'). If the data refer to a whole population, the Greek letter μ (mu) is frequently used to denote the mean.

As we shall discuss in Chapters 5 and 6, a common convention in statistics is to *name* variables using capital letters, e.g. X, but to refer to specific values of those variables using small letters, e.g. x. In statistical formulae, it is customary to denote the value of variable X for the first sample observation by x_1, for the second sample observation by x_2 and so on. In general, the value of variable X for the ith observation is denoted by x_i. For a sample with n observations, the formula for the sample mean is as follows:

Sample mean

$$\bar{x} = \frac{\Sigma x_i}{n} \qquad (3.1)$$

In equation (3.1), the numerator is the sum of the values of the n observations. That is:

$$\Sigma x_i = x_1 + x_2 + \ldots + x_n$$

The Greek letter Σ (upper case sigma) is the summation sign.

To illustrate the computation of a sample mean, consider the following class size data for a sample of five university classes:

$$46 \quad 54 \quad 42 \quad 46 \quad 32$$

We use the notation x_1, x_2, x_3, x_4, x_5 to represent the number of students in each of the five classes:

$$x_1 = 46 \quad x_2 = 54 \quad x_3 = 42 \quad x_4 = 46 \quad x_5 = 32$$

To compute the sample mean, we can write:

$$\bar{x} = \frac{\Sigma x_i}{n} = \frac{x_1 + x_2 + x_3 + x_4 + x_5}{n} = \frac{46 + 54 + 42 + 46 + 32}{5} = 44$$

The sample mean class size is 44 students.

Here is a second illustration. Suppose a university careers office has sent a questionnaire to a small sample of business school graduates requesting information on monthly starting salaries. Table 3.1 shows the data collected. The mean monthly starting salary for the sample of 12 business school graduates is computed as:

SALARY

$$\bar{x} = \frac{\Sigma x_i}{n} = \frac{x_1 + x_2 + \ldots + x_{12}}{12} = \frac{2{,}020 + 2{,}075 + \ldots + 2{,}040}{12} = \frac{24{,}840}{12} = 2{,}070$$

TABLE 3.1 Monthly starting salaries for a sample of 12 business school graduates

Graduate	Monthly starting salary (€)	Graduate	Monthly starting salary (€)
1	2,020	7	2,050
2	2,075	8	2,165
3	2,125	9	2,070
4	2,040	10	2,260
5	1,980	11	2,060
6	1,955	12	2,040

Equation (3.1) shows how the mean is computed for a sample with n observations. The formula for computing the mean of a population remains the same, but we use different notation to indicate that we are working with the entire population. We denote the number of observations in a population by N, and the population mean as μ.

Population mean

$$\mu = \frac{\Sigma x_i}{N} \qquad (3.2)$$

Median

The **median** is another measure of central location for a variable. The median is the value in the middle when the data are arranged in ascending order (smallest value to largest value).

> **Median**
>
> Arrange the data in ascending order
>
> **1.** For an odd number of observations, the median is the middle value.
> **2.** For even number of observations, the median is the mean of the two middle values.

Consider the median class size for the sample of five university classes. We first arrange the data in ascending order:

$$32 \quad 42 \quad 46 \quad 46 \quad 54$$

Because $n = 5$ is odd, the median is the middle value. This data set contains two observations with values of 46 (the 3rd and 4th ordered observations). Each observation is treated separately when we arrange the data in ascending order. The median class size is 46 students (the 3rd ordered observation).

Now consider the median starting salary for the 12 business school graduates in Table 3.1. We first arrange the data in ascending order:

1,955   1,980   2,020   2,040   2,040   <u>2,050   2,060</u>   2,070   2,075   2,125   2,165   2,260

Middle two values

Because $n = 12$ is even, we identify the middle two values: 2,050 and 2,060. The median is the mean of these values:

$$\text{Median} = \frac{2,050 + 2,060}{2} = 2,055$$

Although the arithmetic mean is the more frequently used measure of central location, in some situations the median is preferred. The mean is influenced by extremely small and large data values. For example, suppose one of the graduates (refer to Table 3.1) had a starting salary of €5,000 per month. If we change the highest monthly starting salary in Table 3.1 from €2,260 to €5,000, the sample mean changes from €2,070 to €2,298. The median of €2,055, however, is unchanged, because €2,050 and €2,060 are still the middle two values. The median provides a more robust measure of central location than the mean in the presence of extreme values.

Mode

A third measure of location is the **mode**, defined as follows:

> **Mode**
>
> The mode is the value that occurs with the greatest frequency.

To illustrate the identification of the mode, consider the sample of five class sizes. The only value that occurs more than once is 46. This value occurs twice and consequently is the mode. In the sample of starting salaries for the business school graduates, the only monthly starting salary that occurs more than once is €2,040, and therefore this value is the mode for that data set.

The highest frequency could occur at two or more different data values. In these instances more than one mode exists. If the data contain exactly two modes, we say that the data are *bimodal*. If data contain more than two modes, we say that the data are *multimodal*. In multimodal cases the modes are almost never reported, because listing three or more modes would not be particularly helpful in describing a central location for the data.

The mode is an important measure of location for qualitative data. For example, the qualitative data set in Table 2.2 in Chapter 2 resulted in the following frequency distribution for new car purchases:

Car brand	Frequency
Audi	8
BMW	5
Mercedes	13
Opel	5
VW	19
Total	50

The mode, or most frequently purchased car brand, is VW. For this type of data it obviously makes no sense to speak of the mean or median. The mode provides the information of interest, the most frequently purchased car brand.

Weighted mean

In equation (3.1) defining the arithmetic mean of a sample with n observations, each x_i is given equal weight. Although this practice is most common, in some instances the mean is computed by giving each observation a weight that reflects its importance. A mean computed like this is referred to as a **weighted mean**. The weighted mean is computed as follows:

Weighted mean

$$\bar{x} = \frac{\Sigma w_i x_i}{\Sigma w_i}$$ (3.3)

where:

x_i = value of observation i
w_i = weight for observation i

For sample data, equation (3.3) provides the weighted sample mean. For population data, μ replaces $\bar{x}$ and equation (3.3) provides the weighted population mean.

As an example of the need for a weighted mean, consider the following sample of five purchases of a raw material over a recent period. The cost per kilogram has varied from €2.80 to €3.40 and the quantity purchased has varied from 500 to 2,750 kilograms.

Purchase	Cost per kilogram (€)	Number of kilograms
1	3.00	1,200
2	3.40	500
3	2.80	2,750
4	2.90	1,000
5	3.25	800

Suppose a manager asked for information about the mean cost per kilogram of the raw material. Because the quantities ordered vary, we should use the formula for a weighted mean. The five cost-per-kilogram values are $x_1 = 3.00$, $x_2 = 3.40$, ... etc. The weighted mean cost per kilogram is found by weighting each cost by its corresponding quantity. The weights are $w_1 = 1,200$, $w_2 = 500$, ... etc. Using equation (3.3), the weighted mean is calculated as follows:

$$\bar{x} = \frac{\Sigma w_i x_i}{\Sigma w_i} = \frac{1,200(3.00) + 500(3.40) + 2,750(2.80) + 1,000(2.90) + 800(3.25)}{1,200 + 500 + 2,750 + 1,000 + 800}$$

$$= \frac{18,500}{6,250} = 2.96$$

The weighted mean computation shows that the mean cost per kilogram for the raw material is €2.96. Using equation (3.1) rather than the weighted mean formula would have given misleading results. In this case, the mean of the five cost-per-kilogram values is $(3.00 + 3.40 + 2.80 + 2.90 + 3.25)/5 = 15.35/5 = $ €3.07, which overstates the actual mean cost per kilogram purchased.

When observations vary in importance, the analyst must choose the weight that best reflects the importance of each observation in the determination of the mean, in the context of the particular application.

Geometric mean

As described, the arithmetic mean involves totalling the data items and then dividing the total by the number of data items. There is also a mean value, the geometric mean, calculated by finding the product of all n data items, then taking the nth root of this product.

Geometric mean

$$\bar{x}_g = \sqrt[n]{x_1 \times x_2 \times \cdots \times x_n} \tag{3.4}$$

It can be shown that the geometric mean of a data set is always less than the arithmetic mean, unless the data items are all equal to each other, in which case the geometric mean equals the arithmetic mean. Using equation (3.4), you can verify that the geometric mean for the class size data set is 43.3 (arithmetic mean 44.0):

$$\bar{x}_g = \sqrt[5]{46 \times 54 \times 42 \times 46 \times 32} = 43.3$$

and for the salary data set is 2,068.5 (arithmetic mean 2,070):

$$\bar{x}_g = \sqrt[12]{2,020 \times 2,075 \times \cdots \times 2,040} = 2,068.5$$

The geometric mean is most appropriate in situations where the data items to be summarized result from a ratio-type calculation, such as with growth rates or index numbers (index numbers are covered in Chapter 19, on the online platform). Consider the following five data values:

1.11   1.35   0.80   1.40   1.05

These represent the share price of a company at the beginning of five successive years, relative to the price at the start of the previous year. Values calculated like this are known as price relatives. For example, 1.11 is the share price at the start of year 1 divided by the share price at the start of year 0, reflecting a rise of 11 per cent over the year. The fifth data value 1.05 is the share price at the start of year 5 divided by the share price at the start of year 4, representing a rise of 5 per cent. The third data value, 0.80, reflects a fall of 20 per cent in share price from the start of year 2 to the start of year 3. The geometric mean share price relative is:

$$\bar{x}_g = \sqrt[5]{1.11 \times 1.35 \times 0.80 \times 1.40 \times 1.05} = 1.120$$

This figure suggests that the average percentage rise year on year was 12.0 per cent. The product of these five data values is 1.76, which represents the share price at the beginning of year 5 relative to the share price at the start of year 0. The geometric mean, which is the fifth root of this product, is the ratio that would convert the initial share price to the final share price if applied for five years in succession. The arithmetic mean share price relative of 1.142 (implying an annual 14.2 per cent increase) overestimates the average annual change.

Percentiles

A percentile provides information about how the data are spread over the interval from the smallest value to the largest value. For data that do not contain numerous repeated values, the pth percentile divides the data into two parts: approximately p per cent of the observations have values less than the pth percentile; approximately $(100 - p)$ per cent of the observations have values greater than the pth percentile. The pth percentile is formally defined as follows:

> ### Percentile
>
> The pth percentile is a value such that at least p per cent of the observations are less than or equal to this value and at least $(100 - p)$ per cent of the observations are greater than or equal to this value.

Colleges and universities sometimes report admission test scores in terms of percentiles. For instance, suppose an applicant gets a raw score of 54 on an admission test. It may not be readily apparent how this student performed in relation to other students taking the same test. However, if the raw score of 54 corresponds to the 70th percentile, we know that approximately 70 per cent of the students scored lower than this individual and approximately 30 per cent of the students scored higher than this individual.

The following procedure can be used to compute the pth percentile:

> ### Calculating the pth percentile
>
> **1.** Arrange the data in ascending order (smallest value to largest value).
> **2.** Compute a 'positional marker' i
>
> $$i = \left(\frac{p}{100}\right)n$$
>
> where p is the percentile of interest and n is the number of observations.
>
> **3. a.** If i is not an integer, *round up*. The next integer *greater* than i denotes the position of the pth percentile.
> **b.** If i is an integer, the pth percentile is the mean of the values in positions i and $i + 1$.

As an illustration, consider the 85th percentile for the starting salary data in Table 3.1.

1 Arrange the data in ascending order:

1,955 1,980 2,020 2,040 2,040 2,050 2,060 2,070 2,075 2,125 2,165 2,260

2 $i = \left(\frac{p}{100}\right)n = \left(\frac{85}{100}\right)12 = 10.2$

3 Because i is not an integer, *round up*. The position of the 85th percentile is the next integer greater than 10.2: the 11th position.

Returning to the data, we see that the 85th percentile is the data value in the 11th position: 2,165.

As another illustration of this procedure, consider the calculation of the 50th percentile for the starting salary data. Applying step 2, we obtain:

$$i = \left(\frac{p}{100}\right)n = \left(\frac{50}{100}\right)12 = 6$$

Because i is an integer, step 3(b) states that the 50th percentile is the mean of the sixth and seventh data values; that is $(2,050 + 2,060)/2 = 2,055$. Note that the 50th percentile is also the median.

Quartiles

When describing data distribution, it is often useful to consider the values that divide the data set into four parts, with each part containing approximately one-quarter (25 per cent) of the observations. Figure 3.1 shows a data distribution divided into four parts. The division points are referred to as the quartiles and are defined as:

Q_1 = first quartile, or 25th percentile
Q_2 = second quartile, or 50th percentile (also the median)
Q_3 = third quartile, or 75th percentile

FIGURE 3.1
Location of the quartiles

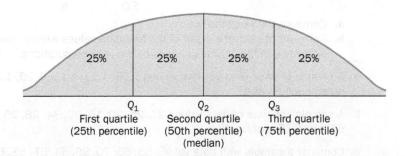

For the starting salary data, we have already identified Q_2, the second quartile (median), as 2,055. The computations of quartiles Q_1 and Q_3 use the rule for finding the 25th and 75th percentiles.

For Q_1:

$$i = \left(\frac{p}{100}\right)n = \left(\frac{25}{100}\right)12 = 3$$

Because i is an integer, step 3(b) indicates that the first quartile, or 25th percentile, is the mean of the third and fourth data values; hence:

$$Q_1 = (2,020 + 2,040)/2 = 2,030$$

For Q_3:

$$i = \left(\frac{p}{100}\right)n = \left(\frac{75}{100}\right)12 = 9$$

Again, because i is an integer, step 3(b) indicates that the third quartile, or 75th percentile, is the mean of the ninth and tenth data values; hence:

$$Q_3 = (2,075 + 2,125)/2 = 2,100$$

1,955 1,980 2,020 | 2,040 2,040 2,050 | 2,060 2,070 2,075 | 2,125 2,165 2,260
$\quad\quad\quad\quad Q_1 = 2,030 \quad\quad\quad\quad\quad Q_2 = 2,055 \quad\quad\quad\quad Q_3 = 2,100$
$\quad\quad\quad\quad\quad\quad\quad\quad\quad\quad\quad\quad$ (Median)

We defined the quartiles as the 25th, 50th and 75th percentiles. Hence we computed the quartiles in the same way as percentiles. However, other conventions are sometimes used to compute quartiles. The actual values reported for quartiles may vary slightly depending on the convention used. Nevertheless, the objective in all cases is to divide the data into four approximately equal parts.

EXERCISES

Methods

1. A sample consists of data values 10, 20, 12, 17, 16. Compute the mean and median.

2. A sample consists of data values 10, 20, 21, 17, 16, 12. Compute the mean and median.

3. Consider the following data and corresponding weights.

x_i	Weight
3.2	6
2.0	3
2.5	2
5.0	8

   a. Compute the weighted mean.
   b. Compute the sample mean of the four data values without weighting. Comment on the difference in the results provided by the two computations.

4. A sample of price relatives has values 1.05, 1.11, 1.13, 1.20, 1.25. Calculate the geometric mean price relative.

5. A sample consists of data values 27, 25, 20, 15, 30, 34, 28, 25. Compute the 20th, 25th, 65th and 75th percentiles.

6. Consider a sample with data values 53, 55, 70, 58, 64, 57, 53, 69, 57, 68 and 53. Compute the mean, median and mode.

Applications

7. A sample of 30 engineering graduates had the following starting salaries. Data are in thousands of euros.

ENGSAL

36.8	34.9	35.2	37.2	36.2	35.8	36.8	36.1	36.7	36.6
37.3	38.2	36.3	36.4	39.0	38.3	36.0	35.0	36.7	37.9
38.3	36.4	36.5	38.4	39.4	38.8	35.4	36.4	37.0	36.4

   a. What is the mean starting salary?
   b. What is the median starting salary?
   c. What is the mode?
   d. What is the first quartile?
   e. What is the third quartile?

8. The following data were obtained for the number of minutes spent listening to recorded music for a sample of 30 individuals on one particular day.

MUSIC

88.3	4.3	4.6	7.0	9.2	0.0	99.2	34.9	81.7	0.0
85.4	0.0	17.5	45.0	53.3	29.1	28.8	0.0	98.9	64.5
4.4	67.9	94.2	7.6	56.6	52.9	145.6	70.4	65.1	63.6

   a. Compute the mean.
   b. Compute the median.

c. Compute the first and third quartiles.

d. Compute and interpret the 40th percentile.

9. OutdoorGearLab is an organization that tests outdoor gear used for climbing, camping, mountaineering and backpacking. Suppose the following data show the ratings of hardshell jackets based on the breathability, durability, versatility, features, mobility and weight of each jacket. The ratings range from 0 (worst) to 100 (best).

42	66	67	71	78	62	61	76	71	67
61	64	61	54	83	63	68	69	81	53

JACKET RATINGS

a. Compute the mean, median and mode.

b. Compute the first and third quartiles.

c. Compute and interpret the 90th percentile.

10. The following is a sample of age data for individuals working from home by 'telecommuting'.

18	54	20	46	25	48	53	27	26	37
40	36	42	25	27	33	28	40	45	25

a. Compute the mean and the mode.

b. Suppose the median age of the population of all adults is 35.5 years. Use the median age of the preceding data to comment on whether the at-home workers tend to be younger or older than the population of all adults.

c. Compute the first and third quartiles.

d. Compute and interpret the 32nd percentile.

11. The assessment for a statistics module comprises a multiple-choice test, a data analysis project, an Excel test and a written examination. Scores for Jil and Ricardo on the four components are shown below.

Assessment	Jil (%)	Ricardo (%)
Multiple-choice test	80	48
Data analysis project	60	78
Excel test	62	60
Written examination	57	53

a. Calculate weighted mean scores (%) for Jil and Ricardo assuming the respective weightings for the four components are 20, 20, 30, 30.

b. Calculate weighted mean scores (%) for Jil and Ricardo assuming the respective weightings for the four components are 10, 25, 15, 50.

12. **a.** The current value of a company is €25 million. If the value of the company six years ago was €10 million, what is the company's mean annual growth rate over the past six years?

b. If an asset declines in value from €5,000 to €3,500 over nine years, what is the mean annual growth rate in the asset's value over these nine years?

3.2 MEASURES OF VARIABILITY

As well as measures of location, it is often desirable to consider measures of variability or dispersion. For example, suppose you are a purchaser for a large manufacturer and you regularly place orders with two different suppliers. After some time, you find that the mean number of days required to fill orders is 10 days for both suppliers. Histograms summarizing the order fulfilment times from the suppliers are shown in Figure 3.2. Although the mean number of days is 10 for both suppliers, do the two suppliers demonstrate the same degree of reliability? Note the dispersion, or variability, in delivery times indicated by the histograms. Which supplier would you prefer?

FIGURE 3.2
Historical data showing the number of days required to fulfil orders

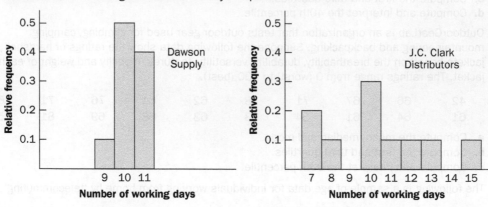

For most firms, receiving materials and supplies on schedule is important. The 7- or 8-day deliveries shown for J.C. Clark Distributors might be favourably viewed. However, a few of the slow 13- to 15-day deliveries could be disastrous in terms of keeping a workforce busy and production on schedule. This example illustrates a situation in which the variability in the delivery times may be an overriding consideration in selecting a supplier. For most purchasers, the lower variability for Dawson Supply would make Dawson the preferred supplier.

We turn now to a discussion of some commonly used measures of variability.

Range

The simplest measure of variability is the **range**.

> **Range**
>
> $$\text{Range} = \text{Largest value} - \text{Smallest value}$$

Refer back to the data on starting salaries for business school graduates in Table 3.1. The largest starting salary is 2,260 and the smallest is 1,955. The range is $2,260 - 1,955 = 305$.

Although the range is easy to compute, it is seldom used as the only measure of variability. Because it is based solely on the minimum and maximum values, it is by definition highly influenced by extremes. If one of the graduates had a starting salary of €5,000 per month, the range would be $5,000 - 1,955 = 3,045$, rather than 305. This would not be especially descriptive of the overall variability in the data because 11 of the 12 starting salaries are relatively closely grouped between 1,955 and 2,165.

Interquartile range

A measure of variability that overcomes the dependency on extreme values is the **interquartile range** (IQR). The IQR is simply the difference between the third quartile, Q_3, and the first quartile, Q_1. In other words, the interquartile range is the range for the middle 50 per cent of the data.

> **Interquartile range**
>
> $$IQR = Q_3 - Q_1 \qquad (3.5)$$

For the data on monthly starting salaries, the quartiles are $Q_3 = 2,100$ and $Q_1 = 2,030$. The interquartile range is $2,100 - 2,030 = 70$.

Variance

The **variance** is a measure of variability that uses all the data values, and is based on the difference between each data value and the mean. This difference is called a *deviation about the mean*: written $(x_i - \bar{x})$ for sample data and $(x_i - \mu)$ for a population. In the computation of the variance, the deviations about the mean are *squared*.

If the data are for a population, the average of the squared deviations is called the *population variance*. It is denoted by the Greek symbol σ^2 (sigma squared). For a population of N observations with mean value μ, the definition of the population variance is:

Population variance

$$\sigma^2 = \frac{\Sigma(x_i - \mu)^2}{N}$$

(3.6)

In most statistical applications, the data under analysis are for a sample. When we compute a sample variance, we are often interested in using it to estimate the population variance σ^2. Although a detailed explanation is beyond the scope of this text, it can be shown that if the sum of the squared deviations about the sample mean is divided by $n - 1$, not by n, the resulting sample variance provides an unbiased estimate of the population variance (a formal definition of unbiasedness is given in Chapter 7).

For this reason, the *sample variance*, denoted by s^2, is defined as follows:

Sample variance

$$s^2 = \frac{\Sigma(x_i - \bar{x})^2}{n - 1}$$

(3.7)

Consider the data on class size for the sample of five university classes (Section 3.1). A summary of the data, including the computation of the deviations about the mean and the squared deviations about the mean, is shown in Table 3.2. The sum of squared deviations about the mean is $\Sigma(x_i - \bar{x})^2 = 256$. Hence, with $n - 1 = 4$, the sample variance is:

$$s^2 = \frac{\Sigma(x_i - \bar{x})^2}{n - 1} = \frac{256}{4} = 64$$

TABLE 3.2 Computation of deviations and squared deviations about the mean for the class size data

Number of students in class (x_i)	Mean class size $(\bar{x})$	Deviation about the mean $(x_i - \bar{x})$	Squared deviation about the mean $(x_i - \bar{x})^2$
46	44	2	4
54	44	10	100
42	44	−2	4
46	44	2	4
32	44	−12	144
Totals		**0**	**256**
		$\Sigma(x_i - \bar{x})$	$\Sigma(x_i - \bar{x})^2$

The units associated with the sample variance can cause confusion. Because the values summed in the variance calculation, $(x_i - \bar{x})^2$, are squared, the units associated with the sample variance are also squared. For instance, the sample variance for the class size data is $s^2 = 64$ (students)^2. The squared units make it difficult to obtain an intuitive understanding and interpretation of the variance. We recommend that you think of the variance as a measure useful in comparing the amount of variability for two (or more) variables measured in the same units. The one with the larger variance shows the greater variability.

As another illustration, consider the starting salaries in Table 3.1 for the 12 business school graduates. In Section 3.1, we showed that the sample mean starting salary was 2,070. The computation of the sample variance ($s^2 = 6,754.5$) is shown in Table 3.3.

TABLE 3.3 Computation of the sample variance for the starting salary data

Monthly salary (x_i)	Sample mean ($\bar{x}$)	Deviation about the mean ($x_i - \bar{x}$)	Squared deviations ($x_i - \bar{x}$)^2
2,020	2,070	−50	2,500
2,075	2,070	5	25
2,125	2,070	55	3,025
2,040	2,070	−30	900
1,980	2,070	−90	8,100
1,955	2,070	−115	13,225
2,050	2,070	−20	400
2,165	2,070	95	9,025
2,070	2,070	0	0
2,260	2,070	190	36,100
2,060	2,070	−10	100
2,040	2,070	−30	900
Totals		**0**	**74,300**

Using equation (3.7)

$$s^2 = \frac{\Sigma(x_i - \bar{x})^2}{n - 1} = \frac{74,300}{11} = 6,754.5$$

In Tables 3.2 and 3.3 we show both the sum of the deviations about the mean and the sum of the squared deviations about the mean. Note that in both tables, $\Sigma(x_i - \bar{x}) = 0$. The positive deviations and negative deviations cancel each other. For any data set, the sum of the deviations about the mean will *always equal zero*.

An alternative formula for the computation of the sample variance is:

$$s^2 = \frac{\Sigma x_i^2 - n\bar{x}^2}{n - 1}$$

where:

$$\Sigma x_i^2 = x_1^2 + x_2^2 + \ldots + x_n^2$$

Standard deviation

The **standard deviation** is defined as the positive square root of the variance. Following the notation we adopted for sample variance and population variance, we use s to denote the sample standard deviation and σ to denote the population standard deviation. The standard deviation is derived from the variance as shown in equations (3.8) and (3.9).

Standard deviation

$$\text{Population standard deviation} = \sigma = \sqrt{\sigma^2} \tag{3.8}$$

$$\text{Sample standard deviation} = s = \sqrt{s^2} \tag{3.9}$$

We calculated the sample variance for the sample of class sizes in five university classes as $s^2 = 64$. Hence the sample standard deviation is:

$$s = \sqrt{64} = 8$$

For the data on starting salaries, the sample standard deviation is:

$$s = \sqrt{6,754.5} = 82.2$$

Recall that the units associated with the variance are squared. For example, the sample variance for the starting salary data of business school graduates is $s^2 = 6,754.5$ (€^{2}). Because the standard deviation is the square root of the variance, the units are euros for the standard deviation, $s = $ €82.2. In other words, the standard deviation has the advantage that it is measured in the same units as the original data. It is therefore more easily compared to the mean and other statistics measured in the same units as the original data.

Coefficient of variation

In some situations we may be interested in a descriptive statistic that indicates how large the standard deviation is relative to the mean. A suitable measure is the **coefficient of variation**, which is usually expressed as a percentage.

Coefficient of variation

$$\left(\frac{\text{Standard deviation}}{\text{Mean}} \times 100\right)\%$$

                                                                                (3.10)

For the class size data, we found $\bar{x} = 44$ and $s = 8$. The coefficient of variation is $(8/44) \times 100\% = 18.2\%$. The coefficient of variation tells us that the sample standard deviation is 18.2 per cent of the value of the sample mean. For the starting salary data with $\bar{x} = 2,070$ and $s = 82.2$, the coefficient of variation, $(82.2/2,070) \times 100\% = 4.0\%$, tells us the sample standard deviation is only 4.0 per cent of the value of the sample mean. In general, the coefficient of variation is a useful statistic for comparing the variability of variables that have different standard deviations and different means.

EXERCISES

Methods

13. A sample has data values 10, 20, 12, 17, 16. Calculate the range and interquartile range.

14. Consider a sample with data values of 10, 20, 12, 17, 16. Calculate the variance and standard deviation.

15. A sample has data values 27, 25, 20, 15, 30, 34, 28, 25. Calculate the range, interquartile range, variance and standard deviation.

Applications

16. The goals scored in six handball matches were 41, 34, 42, 45, 35, 37. Using these data as a sample, compute the following descriptive statistics.
    a. Range.
    b. Variance.
    c. Standard deviation.
    d. Coefficient of variation.

17. The following data were used to construct the histograms of the number of days required to fill orders for Dawson Supply and for J.C. Clark Distributors (refer to Figure 3.2).

Dawson Supply days for delivery: 11   10   9   10   11   11   10   11   10   10
Clark Distributors days for delivery: 8   10   13   7   10   11   10   7   15   12

Use the range and standard deviation to support the previous observation that Dawson Supply provides the more consistent and reliable delivery times.

CRIME

18. Police records show the following numbers of daily crime reports for a sample of days during the winter months and a sample of days during the summer months.

Winter:	18	20	15	16	21	20	12	16	19	20
Summer:	28	18	24	32	18	29	23	38	28	18

a. Compute the range and interquartile range for each period.
b. Compute the variance and standard deviation for each period.
c. Compute the coefficient of variation for each period.
d. Compare the variability of the two periods.

19. A production department uses a sampling procedure to test the quality of newly produced items. The department applies the following decision rule at an inspection station: if a sample of 14 items has a variance of more than 0.005, the production line must be shut down for repairs. Suppose the following data have just been collected:

3.43	3.45	3.43	3.48	3.52	3.50	3.39
3.48	3.41	3.38	3.49	3.45	3.51	3.50

Should the production line be shut down? Why or why not?

3.3 MEASURES OF DISTRIBUTIONAL SHAPE, RELATIVE LOCATION AND DETECTING OUTLIERS

We described several measures of location and variability for data distributions. It is often important to also have a measure of the shape of a distribution. In Chapter 2 we noted that a histogram offers an excellent graphical display showing distributional shape. An important numerical measure of the shape of a distribution is skewness.

Distributional shape

Figure 3.3 shows four histograms representing relative frequency distributions. The histograms in Panels A and B are moderately skewed. The one in Panel A is skewed to the left: its skewness is −0.85 (negative skewness). The histogram in Panel B is skewed to the right: its skewness is +0.85 (positive skewness). The histogram in Panel C is symmetrical: its skewness is zero. The histogram in Panel D is highly skewed to the right: its skewness is 1.62. The formula used to compute skewness is somewhat complex.* However, the skewness can be easily computed using statistical software (refer to the software guides on the online platform).

For a symmetrical distribution, the mean and the median are equal. When the data are positively skewed, the mean will usually be greater than the median. When the data are negatively skewed, the mean will usually be less than the median. The data used to construct the histogram in Panel D are customer

---

* The formula for the skewness of sample data is:

$$\text{Skewness} = \frac{n}{(n-1)(n-2)} \sum \left(\frac{x_i - \bar{x}}{s} \right)^3$$

purchases at a fashion store. The mean purchase amount is €77.60 and the median purchase amount is €59.70. A relatively small number of large purchase amounts pull up the mean, but the median remains unaffected. The median provides a better measure of typical values when the data are highly skewed.

FIGURE 3.3
Histograms showing the skewness for four distributions

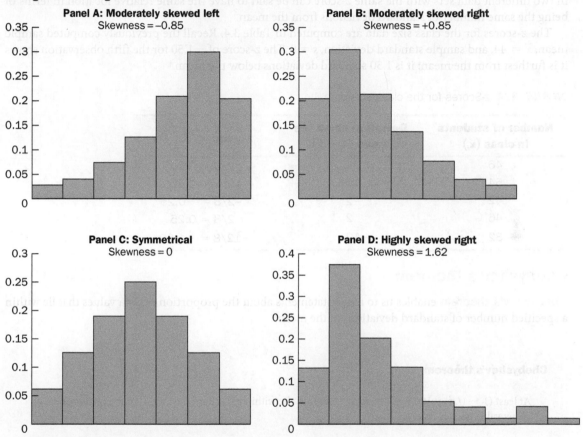

z-Scores

We are sometimes interested in the relative location of data items within a data set. Such measures can help us determine whether a particular item is close to the centre of a data set or far out in one of the tails. By using both the mean and standard deviation, we can determine the relative location of any observation. Suppose we have a sample of n observations, with values denoted by $x_1, x_2, ..., x_n$. Assume the sample mean $\bar{x}$ and the sample standard deviation s are already computed. Associated with each value x_i is a value called its **z-score**. Equation (3.11) shows how the z-score is computed for each x_i.

z-score

$$z_i = \frac{x_i - \bar{x}}{s}$$

(3.11)

where z_i = the z-score for x_i, $\bar{x}$ = the sample mean, s = the sample standard deviation.

The z-score, z_i, is often called the *standardized value* or the *standard score*. It represents the *number of standard deviations x_i is from the mean $\bar{x}$*. Data values above the mean have a z-score greater than zero. Data values below the mean have a z-score less than zero. For example, $z_1 = 1.2$ would indicate that x_1 is 1.2 standard deviations higher than the sample mean. Similarly, $z_2 = -0.5$ would indicate that x_2 is 0.5, or 1/2, standard deviation lower than the sample mean. A z-score of zero indicates that the data value is equal to the sample mean.

The z-score is a measure of the relative location of the observation in a data set. Hence, observations in two different data sets with the same z-score can be said to have the same relative location in terms of being the same number of standard deviations from the mean.

The z-scores for the class size data are computed in Table 3.4. Recall the previously computed sample mean, $\bar{x} = 44$, and sample standard deviation, $s = 8$. The z-score of -1.50 for the fifth observation shows it is furthest from the mean: it is 1.50 standard deviations below the mean.

TABLE 3.4 z-Scores for the class size data

Number of students in class (x_i)	Deviation about the mean ($x_i - \bar{x}$)	z-score = $\dfrac{x_i - \bar{x}}{s}$
46	2	2/8 = 0.25
54	10	10/8 = 1.25
42	-2	-2/8 = -0.25
46	2	2/8 = 0.25
32	-12	-12/8 = -1.50

Chebyshev's theorem

Chebyshev's theorem enables us to make statements about the proportion of data values that lie within a specified number of standard deviations of the mean.

> **Chebyshev's theorem**
>
> At least $(1 - 1/z^2) \times 100\%$ of the data values must be within z standard deviations of the mean, where z is any value greater than 1.

Some of the implications of this theorem are:

- At least 75 per cent of the data values must be within $z = 2$ standard deviations of the mean.
- At least 89 per cent of the data values must be within $z = 3$ standard deviations of the mean.
- At least 94 per cent of the data values must be within $z = 4$ standard deviations of the mean.

Suppose that the mid-term test scores for 100 students in a university business statistics course had a mean of 70 and a standard deviation of 5. How many students had test scores between 60 and 80? How many students had test scores between 58 and 82?

We note that 60 is two standard deviations below the mean and 80 is two standard deviations above the mean. Using Chebyshev's theorem, we see that at least 75 per cent of the students must have scored between 60 and 80.

For the test scores between 58 and 82, we see that $(58 - 70)/5 = -2.4$, i.e. 58 is 2.4 standard deviations below the mean. Similarly, $(82 - 70)/5 = +2.4$, so 82 is 2.4 standard deviations above the mean. Applying Chebyshev's theorem with $z = 2.4$, we have:

$$\left(1 - \frac{1}{z^2}\right) = \left(1 - \frac{1}{(2.4)^2}\right) = 0.826$$

At least 82.6 per cent of the students must have test scores between 58 and 82.

Empirical rule

Chebyshev's theorem applies to any data set, regardless of distributional shape. It could be used, for example, with any of the skewed distributions in Figure 3.3. In many practical applications, however, data sets exhibit a symmetrical mound-shaped or bell-shaped distribution like the one shown in Figure 3.4. When the data are believed to approximate this distribution, the empirical rule can be used to determine the approximate percentage of data values that lie within a specified number of standard deviations of the mean. The empirical rule is based on the normal probability distribution, which will be discussed in Chapter 6. The normal distribution is used extensively throughout this book.

FIGURE 3.4
A symmetrical mound-shaped or bell-shaped distribution

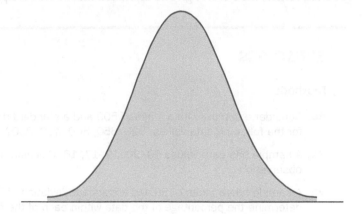

Empirical rule

For data with a bell-shaped distribution:

- Approximately 68 per cent of the data values lie within one standard deviation of the mean.
- Approximately 95 per cent of the data values lie within two standard deviations of the mean.
- Almost all of the data values lie within three standard deviations of the mean.

For example, the empirical rule allows us to say that *approximately* 95 per cent of the data values will be within two standard deviations of the mean (Chebyshev's theorem allows us to conclude only that at least 75 per cent of the data values will be in that interval).

Consider bottles of liquid detergent being filled automatically on a production line. Filling volumes frequently have a bell-shaped distribution. If the mean filling volume is 510 ml and the standard deviation is 7 ml, we can use the empirical rule to draw the following conclusions:

- Approximately 68 per cent of the filled bottles will contain volumes between 503 and 517 ml (that is, within one standard deviation of the mean).
- Approximately 95 per cent of the filled bottles will contain volumes between 496 and 524 ml (that is, within two standard deviations of the mean).
- Almost all filled bottles will contain volumes between 489 and 531 ml (that is, within three standard deviations of the mean).

Detecting outliers

Sometimes a data set will have one or more observations with unusually large or unusually small values. An extreme value is called an outlier. Experienced statisticians take steps to identify outliers and review each one carefully. An outlier may be a data value that has been incorrectly recorded. If so, it can be corrected before further analysis. An outlier may be from an observation that was incorrectly included in the data set. If so, it can be removed. An outlier may be an unusual data value that has been recorded correctly and belongs in the data set. In such cases it should remain.

Standardized values (z-scores) can be used to identify outliers. The empirical rule allows us to conclude that, for data with a bell-shaped distribution, almost all the data values will be within three standard deviations of the mean. Hence, we recommend treating any data value with a z-score less than -3 or greater than $+3$ as an outlier, if the sample is small or moderately sized. Such data values can then be reviewed for accuracy and to determine whether they belong in the data set.

Refer to the z-scores for the class size data in Table 3.4. The z-score of -1.50 shows the fifth class size is furthest from the mean. However, this standardized value is well within the -3 to $+3$ guideline for outliers. Hence, the z-scores give no indication that outliers are present in the class size data.

EXERCISES

Methods

20. Consider a sample with a mean of 500 and a standard deviation of 100. What are the z-scores for the following data values: 520, 650, 500, 450, 280?

21. A sample has data values 10, 20, 12, 17, 16. Calculate the z-score for each of the five observations.

22. A sample has a mean of 30 and a standard deviation of 5. Use Chebyshev's theorem to determine the percentage of the data within each of the following ranges.
    a. 20 to 40
    b. 15 to 45
    c. 22 to 38
    d. 18 to 42
    e. 12 to 48

23. Suppose a data set has a bell-shaped distribution with a mean of 30 and a standard deviation of 5. Use the empirical rule to determine the percentage of data within each of the following ranges.
    a. 20 to 40
    b. 15 to 45
    c. 25 to 35

Applications

24. Suppose that IQ scores have a bell-shaped distribution with a mean of 100 and a standard deviation of 15.
    a. What percentage of people have an IQ score between 85 and 115?
    b. What percentage of people have an IQ score between 70 and 130?
    c. What percentage of people have an IQ score of more than 130?
    d. A person with an IQ score greater than 145 is considered a genius. Does the empirical rule support this statement? Explain.

25. The results of a survey of 1,154 adults showed that, on average, adults sleep 6.9 hours per day during the working week. Suppose that the standard deviation is 1.2 hours.
    a. Use Chebyshev's theorem to calculate the percentage of individuals who sleep between 4.5 and 9.3 hours per day.
    b. Use Chebyshev's theorem to calculate the percentage of individuals who sleep between 3.9 and 9.9 hours per day.
    c. Assume that the number of hours of sleep follows a bell-shaped distribution. Use the empirical rule to calculate the percentage of individuals who sleep between 4.5 and 9.3 hours per day. How does this result compare to the value that you obtained using Chebyshev's theorem in part (a)?

26. Suppose the average charge for a 7-day hire of an economy-class car in Kuwait City is KWD (Kuwaiti dinar) 75.00, and the standard deviation is KWD 20.00.
  a. What is the z-score for a 7-day hire charge of KWD 56.00?
  b. What is the z-score for a 7-day hire charge of KWD 153.00?
  c. Interpret the z-scores in parts (a) and (b). Comment on whether either should be considered an outlier.

27. *Consumer Review* posts reviews and ratings of a variety of products on the internet. The following is a sample of 20 speaker systems and their ratings, on a scale of 1 to 5, with 5 being best.

SPEAKERS

Speaker	Rating	Speaker	Rating
Infinity Kappa 6.1	4.00	ACI Sapphire III	4.67
Allison One	4.12	Bose 501 Series	2.14
Cambridge Ensemble II	3.82	DCM KX-212	4.09
Dynaudio Contour 1.3	4.00	Eosone RSF1000	4.17
Hsu Rsch. HRSW12V	4.56	Joseph Audio RM7si	4.88
Legacy Audio Focus	4.32	Martin Logan Aerius	4.26
26 Mission 73li	4.33	Omni Audio SA 12.3	2.32
PSB 400i	4.50	Polk Audio RT12	4.50
Snell Acoustics D IV	4.64	Sunfire True Subwoofer	4.17
Thiel CS1.5	4.20	Yamaha NS-A636	2.17

  a. Compute the mean and the median.
  b. Compute the first and third quartiles.
  c. Compute the standard deviation.
  d. The skewness of this data is −1.67. Comment on the shape of the distribution.
  e. What are the z-scores associated with Allison One and Omni Audio?
  f. Do the data contain any outliers? Explain.

3.4 EXPLORATORY DATA ANALYSIS

In Chapter 2 we introduced the stem-and-leaf display, an exploratory data analysis technique. In this section we consider five-number summaries and box plots.

Five-number summary

In a **five-number summary** the following five numbers are used to summarize the data.

  1 Smallest value (minimum)
  2 First quartile (Q_1)
  3 Median (Q_2)
  4 Third quartile (Q_3)
  5 Largest value (maximum).

The easiest way to construct a five-number summary is to first place the data in ascending order. Then it is easy to identify the smallest value, the three quartiles and the largest value. The monthly starting salaries shown in Table 3.1 for a sample of 12 business school graduates are repeated here in ascending order:

1,955   1,980   2,020 | 2,040   2,040   2,050 | 2,060   2,070   2,075 | 2,125   2,165   2,260
$\quad\quad\quad\quad\quad$ $Q_1 = 2,030$ $\quad\quad\quad\quad\quad\quad$ $Q_2 = 2,055$ $\quad\quad\quad\quad\quad$ $Q_3 = 2,100$
$\quad\quad\quad\quad\quad\quad\quad\quad\quad\quad\quad\quad\quad$ (Median)

The median of 2,055 and the quartiles $Q_1 = 2,030$ and $Q_3 = 2,100$ were computed in Section 3.1. The smallest value is 1,955 and the largest value is 2,260. Hence the five-number summary for the salary data is 1,955; 2,030; 2,055; 2,100; 2,260. Approximately one-quarter, or 25 per cent, of the observations are between adjacent numbers in a five-number summary.

Box plot

A box plot is a slightly elaborated, graphical version of the five-number summary. Figure 3.5 shows the construction of a box plot for the monthly starting salary data.

1 A box is drawn with the box ends located at the first and third quartiles. For the salary data, $Q_1 = 2,030$ and $Q_3 = 2,100$. This box contains the middle 50 per cent of the data.

2 A line is drawn across the box at the location of the median (2,055 for the salary data).

3 By using the interquartile range, IQR $= Q_3 - Q_1$, *limits* are located. The limits for the box plot are 1.5(IQR) below Q_1 and 1.5(IQR) above Q_3. For the salary data, IQR $= Q_3 - Q_1 = 2,100 - 2,030 = 70$. Hence, the limits are $2,030 - 1.5(70) = 1,925$ and $2,100 + 1.5(70) = 2,205$. Data outside these limits are considered *outliers*.

4 The dashed lines in Figure 3.5 are called *whiskers*. The whiskers are drawn from the ends of the box to the smallest and largest values *inside the limits* computed in step 3. Hence the whiskers end at salary values of 1,955 and 2,165.

5 Finally, the location of each outlier is shown with a symbol, often *. In Figure 3.5 we see one outlier, 2,260. (Note that box plots do not necessarily identify the same outliers as using z-scores less than -3 or greater than $+3$.)

FIGURE 3.5
Box plot of the starting salary data with lines showing the lower and upper limits

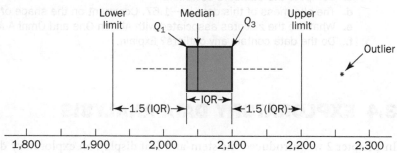

Figure 3.5 includes the upper and lower limits, to show how these limits are computed and where they are located for the salary data. Although the limits are always computed, they are not generally drawn on the box plots.

The box plots in Figure 3.6 illustrate the usual appearance, and demonstrate that box plots provide an excellent graphical tool for making comparisons among two or more groups. Figure 3.6 compares monthly starting salaries for a sample of 111 graduates, by major discipline. The major is shown on the horizontal axis and each box plot is arranged vertically above the relevant major label. The box plots indicate that, for example:

- The highest median salary is in Accounting, the lowest in Management
- Accounting salaries show the highest variation
- There are high salary outliers for Accounting, Finance and Marketing.

FIGURE 3.6
Box plot of monthly
salary

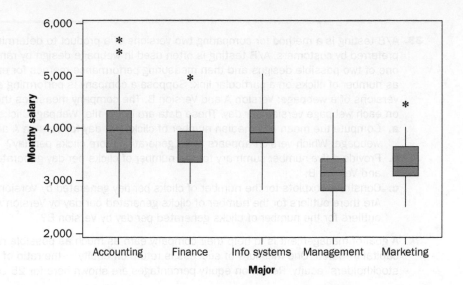

EXERCISES

Methods

28. Consider a sample with data values of 27, 25, 20, 15, 30, 34, 28, 25. Provide the five-number summary for the data.

29. Construct a box plot for the data in Exercise 28.

30. Prepare the five-number summary and the box plot for the following data: 5, 15, 18, 10, 8, 12, 16, 10, 6.

31. A data set has a first quartile of 42 and a third quartile of 50. Compute the lower and upper limits for the corresponding box plot. Should a data value of 65 be considered an outlier?

Applications

32. Annual sales, in millions of euros, for 21 pharmaceutical companies follow:

8,408	1,374	1,872	8,879	2,459	11,413	608
14,138	6,452	1,850	2,818	1,356	10,498	7,478
4,019	4,341	739	2,127	3,653	5,794	8,305

a. Provide a five-number summary.
b. Compute the lower and upper limits for the box plot.
c. Do the data contain any outliers?
d. Johnson & Johnson's sales are the largest on the list at €14,138 million. Suppose a data entry error (a transposition) had been made and the sales had been entered as €41,138m. Would the method of detecting outliers in part (c) identify this problem and allow for correction of the data entry error?
e. Construct a box plot.

WEBPAGE
CLICKS

33. A/B testing is a method for comparing two versions of a product to determine which version is preferred by customers. A/B testing is often used in webpage design by randomly presenting one of two possible designs and then measuring performance metrics for each webpage such as number of clicks on a particular link. Suppose a company is performing an A/B test for two versions of a webpage: Version A and Version B. The company measures the number of clicks on each webpage version per day. These data are in the file 'WebpageClicks'.
   a. Compute the mean and median number of clicks per day for Version A and Version B of the webpage. Which version appears to be generating more clicks per day?
   b. Provide a five-number summary for the number of clicks per day generated by Version A and Version B.
   c. Construct boxplots for the number of clicks per day generated by Version A and Version B. Are there outliers for the number of clicks generated per day by Version A? Are there outliers for the number of clicks generated per day by Version B?

EQUITY

34. A goal of management is to help their company earn as much as possible relative to the capital invested. One measure of success is return on equity — the ratio of net income to stockholders' equity. Return on equity percentages are shown here for 25 companies.

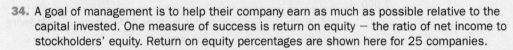

9.0	19.6	22.9	41.6	11.4	15.8	52.7	17.3	12.3	5.1
17.3	31.1	9.6	8.6	11.2	12.8	12.2	14.5	9.2	16.6
5.0	30.3	14.7	19.2	6.2					

   a. Provide a five-number summary.
   b. Compute the lower and upper limits for the box plot.
   c. Do the data contain any outliers? How would this information be helpful to a financial analyst?
   d. Construct a box plot.

3.5 MEASURES OF ASSOCIATION BETWEEN TWO VARIABLES

We have examined methods used to summarize *one variable at a time*. Often a manager or decision-maker is interested in the *relationship between two variables*. In this section we present covariance and correlation as descriptive measures of the relationship between two variables.

We begin by reconsidering the digital equipment online store discussed in Section 2.4. The store manager wants to determine the relationship between the number of Facebook promotions shown and the online sales during the following week. Sample data with sales expressed in €000 were given in Table 2.12, and are repeated here in the first three columns of Table 3.5. It shows ten pairs of observations ($n = 10$), one for each week.

The scatter diagram in Figure 3.7 shows a positive relationship, with higher sales (vertical axis) associated with a greater number of promotions (horizontal axis). The scatter diagram suggests that a straight line could be used as an approximation of the relationship. In the following discussion, we introduce **covariance** as a descriptive measure of the linear association between two variables.

TABLE 3.5 Calculations for the sample covariance

Week	Number of promotions x_i	Sales volume (€000) y_i	$x_i - \bar{x}$	$y_i - \bar{y}$	$(x_i - \bar{x})(y_i - \bar{y})$
1	2	50	-1	-1	1
2	5	57	2	6	12
3	1	41	-2	-10	20
4	3	54	0	3	0
5	4	54	1	3	3
6	1	38	-2	-13	26
7	5	63	2	12	24
8	3	48	0	-3	0
9	4	59	1	8	8
10	2	46	-1	-5	5
Totals	30	510	0	0	99

FIGURE 3.7
Scatter diagram for the digital equipment online store

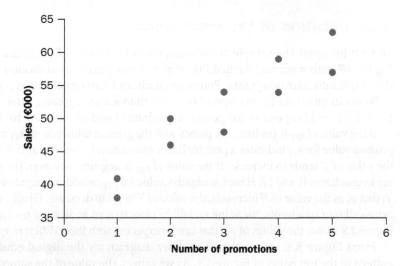

Covariance

For a sample of size n with the observations (x_1, y_1), (x_2, y_2), … , the sample covariance is defined as follows:

Sample covariance

$$s_{XY} = \frac{\Sigma(x_i - \bar{x})(y_i - \bar{y})}{n - 1}$$

(3.12)

This formula pairs each x_i with a corresponding y_i. We then sum the products obtained by multiplying the deviation of each x_i from the sample mean $\bar{x}$ by the deviation of the corresponding y_i from the sample mean $\bar{y}$. This sum is then divided by $n - 1$.

To measure the strength of the linear relationship between the number of promotions X and the sales volume Y in the digital equipment online store problem, we use equation (3.12) to compute the sample covariance. The calculations in Table 3.5 show the computation of $\Sigma(x_i - \bar{x})(y_i - \bar{y})$. Note that $\bar{x} = 30/10 = 3$ and $\bar{y} = 510/10 = 51$. Using equation (3.12), we obtain a sample covariance of:

$$s_{XY} = \frac{\Sigma(x_i - \bar{x})(y_i - \bar{y})}{n - 1} = \frac{99}{10 - 1} = 11$$

The formula for computing the covariance σ_{XY} in a population of size N is similar to equation (3.13), but we use different notation to indicate that we are working with the entire population: μ_X for the population mean of X and μ_Y for the population mean of Y.

Population covariance

$$\sigma_{XY} = \frac{\Sigma(x_i - \mu_X)(y_i - \mu_Y)}{N} \tag{3.13}$$

Interpretation of the covariance

To help interpret the sample covariance, consider Figure 3.8. It is the same as the scatter diagram of Figure 3.7 with a vertical dashed line at $\bar{x} = 3$ and a horizontal dashed line at $\bar{y} = 51$. The lines divide the graph into four quadrants. Points in quadrant I correspond to x_i greater than $\bar{x}$ and y_i greater than $\bar{y}$. Points in quadrant II correspond to x_i less than $\bar{x}$ and y_i greater than $\bar{y}$ and so on. Hence, the value of $(x_i - \bar{x})(y_i - \bar{y})$ is positive for points in quadrants I and III, negative for points in quadrants II and IV.

If the value of s_{XY} is positive, the points with the greatest influence on s_{XY} are in quadrants I and III. Hence, a positive value for s_{XY} indicates a positive linear association between X and Y; that is, as the value of X increases, the value of Y tends to increase. If the value of s_{XY} is negative, however, the points with the greatest influence are in quadrants II and IV. Hence, a negative value for s_{XY} indicates a negative linear association between X and Y; that is, as the value of X increases, the value of Y tends to decrease. Finally, if the points are evenly distributed across all four quadrants, the value s_{XY} will be close to zero, indicating no linear association between X and Y. Figure 3.9 shows the values of s_{XY} that can be expected with three different types of scatter diagram.

From Figure 3.8, we see that the scatter diagram for the digital equipment online store follows the pattern in the top panel of Figure 3.9. As we expect, the value of the sample covariance indicates a positive linear relationship with $s_{XY} = 11$.

FIGURE 3.8

Partitioned scatter diagram for the digital equipment online store

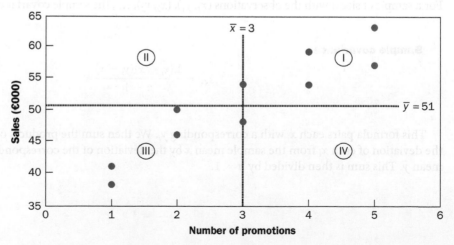

FIGURE 3.9
Interpretation of sample covariance

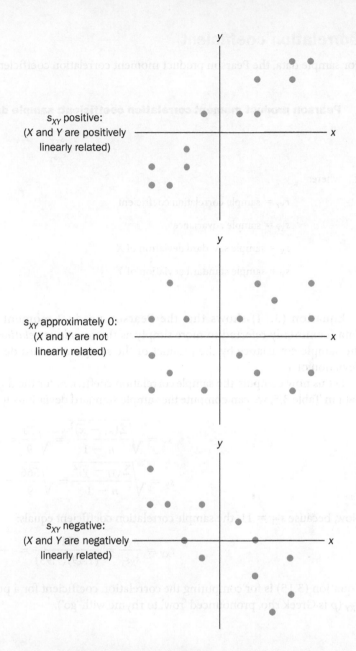

There is a problem with using covariance as a measure of the strength of the linear relationship: namely, its value depends on the units of measurement for X and Y. For example, suppose we are interested in the relationship between height X and weight Y for individuals. Clearly the strength of the relationship should be the same whether we measure height in metres or in centimetres (or in feet). Measuring the height in centimetres, however, gives much larger numerical values for $(x_i - \bar{x})$ than when we measure height in metres. Hence, we would obtain a larger value for the numerator $\Sigma(x_i - \bar{x})(y_i - \bar{y})$ in equation (3.13) — and hence a larger covariance — when in fact the relationship does not change. The **correlation coefficient** is a measure of the relationship between two variables that is not affected by the units of measurement for X and Y.

Correlation coefficient

For sample data, the Pearson product moment correlation coefficient is defined as follows:

Pearson product moment correlation coefficient: sample data

$$r_{XY} = \frac{s_{XY}}{s_X s_Y} \qquad (3.14)$$

where:

r_{XY} = sample correlation coefficient

s_{XY} = sample covariance

s_X = sample standard deviation of X

s_Y = sample standard deviation of Y

Equation (3.14) shows that the Pearson product moment correlation coefficient for sample data (commonly referred to more simply as the *sample correlation coefficient*) is computed by dividing the sample covariance by the product of the sample standard deviation of X and the sample standard deviation of Y.

Let us now compute the sample correlation coefficient for the digital equipment online store. Using the data in Table 3.5, we can compute the sample standard deviations for the two variables:

$$s_X = \sqrt{\frac{\Sigma(x_i - \bar{x})^2}{n-1}} = \sqrt{\frac{20}{9}} = 1.49$$

$$s_Y = \sqrt{\frac{\Sigma(y_i - \bar{y})^2}{n-1}} = \sqrt{\frac{566}{9}} = 7.93$$

Now, because $s_{XY} = 11$, the sample correlation coefficient equals:

$$r_{XY} = \frac{s_{XY}}{s_X s_Y} = \frac{11}{(1.49)(7.93)} = +0.93$$

Equation (3.15) is for computing the correlation coefficient for a population, denoted by the Greek letter ρ_{XY} (ρ is Greek rho, pronounced 'row', to rhyme with 'go').

Pearson product moment correlation coefficient: population data

$$\rho_{XY} = \frac{\sigma_{XY}}{\sigma_X \sigma_Y} \qquad (3.15)$$

where:

ρ_{XY} = population correlation coefficient

σ_{XY} = population covariance

σ_X = population standard deviation for X

σ_Y = population standard deviation for Y

The sample correlation coefficient r_{XY} provides an estimate of the population correlation coefficient ρ_{XY}.

Interpretation of the correlation coefficient

First let us consider a simple example that illustrates the concept of a perfect positive linear relationship. The scatter diagram in Figure 3.10 depicts the relationship between X and Y based on the following sample data:

x_i	y_i
5	10
10	30
15	50

FIGURE 3.10
Scatter diagram depicting a perfect positive linear relationship

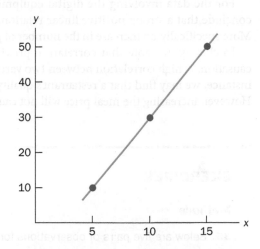

The straight line drawn through the three points shows a perfect linear relationship between X and Y. In order to apply equation (3.14) to compute the sample correlation we must first compute s_{XY}, s_X and s_Y. Some of the computations are shown in Table 3.6.

TABLE 3.6 Computations used in calculating the sample correlation coefficient

	x_i	y_i	$x_i - \bar{x}$	$(x_i - \bar{x})^2$	$y_i - \bar{y}$	$(y_i - \bar{y})^2$	$(x_i - \bar{x})(y_i - \bar{y})$
	5	10	−5	25	−20	400	100
	10	30	0	0	0	0	0
	15	50	5	25	20	400	100
Totals	**30**	**90**	**0**	**50**	**0**	**800**	**200**
	$\bar{x} = 10$	$\bar{y} = 30$					

Using the results in Table 3.6, we find:

$$s_{XY} = \frac{\Sigma(x_i - x)(y_i - y)}{n - 1} = \frac{200}{2} = 100$$

$$s_X = \sqrt{\frac{\Sigma(x_i - \bar{x})^2}{n - 1}} = \sqrt{\frac{50}{2}} = 5$$

$$s_Y = \sqrt{\frac{\Sigma(y_i - \bar{y})^2}{n - 1}} = \sqrt{\frac{800}{2}} = 20$$

$$r_{XY} = \frac{s_{XY}}{s_X s_Y} = \frac{100}{5 \times 20} = +1$$

We see that the value of the sample correlation coefficient is +1. In general, it can be shown that if all the points in a data set fall on a positively sloping straight line, the value of the sample correlation coefficient is +1. A sample correlation coefficient of +1 corresponds to a perfect positive linear relationship between X and Y. If the points in the data set fall on a straight line with a negative slope, the value of the sample correlation coefficient is −1. A sample correlation coefficient of −1 corresponds to a perfect negative linear relationship between X and Y.

Suppose that a data set indicates a positive linear relationship between X and Y but that the relationship is not perfect. The value of r_{XY} will be less than 1, indicating that the points in the scatter diagram are not all on a straight line. As the points deviate more and more from a perfect positive linear relationship, the value of r_{XY} becomes closer and closer to zero. A value of r_{XY} equal to zero indicates no linear relationship between X and Y, and values of r_{XY} near zero indicate a weak linear relationship.

For the data involving the digital equipment online store, recall that $r_{XY} = +0.93$. Therefore, we conclude that a strong positive linear relationship exists between the number of promotions and sales. More specifically, an increase in the number of promotions tends to be associated with an increase in sales.

In closing, we note that correlation provides a measure of linear association and not necessarily causation. A high correlation between two variables does not mean that one variable causes the other. For instance, we may find that a restaurant's quality rating and its typical meal price are positively correlated. However, increasing the meal price will not cause quality to increase.

EXERCISES

Methods

35. Below are five pairs of observations for two variables X and Y.

x_i	6	11	15	21	27
y_i	6	9	6	17	12

  a. Construct a scatter diagram for these data.
  b. What does the scatter diagram indicate about a relationship between X and Y?
  c. Compute and interpret the sample covariance.
  d. Compute and interpret the sample correlation coefficient.

36. Below are five pairs of observations for two variables X and Y.

x_i	4	6	11	3	16
y_i	50	50	40	60	30

  a. Construct a scatter diagram with the x_i values on the horizontal axis.
  b. What does the scatter diagram from part (a) indicate about the relationship between the two variables?
  c. Compute and interpret the sample covariance.
  d. Compute and interpret the sample correlation coefficient.

Applications

37. A department of transport study on driving speed and litres per 100km for family-size cars resulted in the following data:

Speed (kilometres per hour)	50	85	70	95	50	40	100	40	85	95
Litres per 100km	7.1	8.0	8.0	8.7	6.6	6.2	9.5	5.7	7.7	8.0

  a. Construct a scatter diagram with driving speed on the horizontal axis.
  b. Is there any relationship between the two sets of measurements? Explain.

c. Compute and interpret the sample covariance.

d. Compute and interpret the sample correlation coefficient. What does the sample correlation coefficient tell you about the relationship between the two variables?

38. The Nasdaq Composite is a stock market index based on the stock prices of nearly 3,500 companies that include many technology-based companies. The Dow Jones Industrial Average is based on 30 large companies. The file 'NasdaqDow' gives the daily percentage returns for each of these stock indexes in 2021.

NASDAQDOW

a. Plot the percentage returns of the Nasdaq against the percentage returns of the Dow Jones using a scatter plot.

b. Compute the sample mean and standard deviation for each index.

c. Compute the sample correlation.

d. Discuss similarities and differences in these two indexes.

39. The file 'UKUniversities' contains data taken from the *The Guardian University Guide 2023* (www.theguardian.com/uk). The data in the file details two measures given in *The Guardian's* University League Table, which ranks UK universities. One is a measure of 'value added' (scored out of 10, and based on an analysis of entry qualification levels and class of degree achieved). The other is a measure of 'career after 15 months' (scored out of 100, and based on a national survey called the Graduate Outcomes Survey).

UK
UNIVERSITIES

a. Plot a scatter diagram with the value-added measure on the horizontal axis. What does your plot suggest about the relationship between the two measures?

b. Calculate the sample correlation coefficient between the value-added measure and the career after 15 months measure. Interpret your result.

3.6 DATA DASHBOARDS: ADDING NUMERICAL MEASURES TO IMPROVE EFFECTIVENESS

In Section 2.5, we gave an introduction to data visualization, a term used to describe the use of graphical displays to summarize and present information about a data set. The goal of data visualization is to communicate key information about the data as effectively and clearly as possible. One of the most widely used data visualization tools is a data dashboard, a set of visual displays that organizes and presents information used to monitor the performance of a company or organization in a manner that is easy to read, understand and interpret. In this section we extend the discussion of data dashboards to show how the addition of numerical measures can improve the overall effectiveness of the display.

The addition of numerical measures to a data dashboard, such as the mean and standard deviation of key performance indicators (KPIs), is critical because numerical measures often provide benchmarks or goals by which KPIs are evaluated. The purpose of a data dashboard is to provide information on the KPIs in a manner that is easy to read, understand and interpret. Adding numerical measures and graphs that utilize numerical measures can help us accomplish these objectives.

To illustrate the use of numerical measures in a data dashboard, recall the Grogan Oil Company application that we used in Section 2.5 to introduce the concept of a data dashboard. Grogan Oil has offices located in three Texas cities: Austin (its headquarters), Houston and Dallas. Grogan's Information Technology (IT) call centre, located in the Austin office, handles calls regarding computer-related problems (software, internet and email) from employees in the three offices. Figure 3.11 shows the data dashboard Grogan developed to monitor the performance of the call centre. The key components of this dashboard are as follows.

- The stacked bar chart in the upper left corner of the dashboard shows the call volume for each type of problem (software, internet or email) over time.

- The bar chart in the upper right-hand corner of the dashboard shows the percentage of time that call centre employees spent on each type of problem or were idle (not working on a call).

- For each unresolved case received more than 15 minutes ago, the bar chart shown in the middle left portion of the dashboard shows the length of time each of these cases has been unresolved.
- The bar chart in the middle right portion of the dashboard shows the call volume by office (Houston, Dallas and Austin) for each type of problem.
- The histogram at the bottom of the dashboard shows the distribution of the time to resolve a case for all resolved cases for the current shift.

FIGURE 3.11

Initial Grogan Oil information technology call centre data dashboard

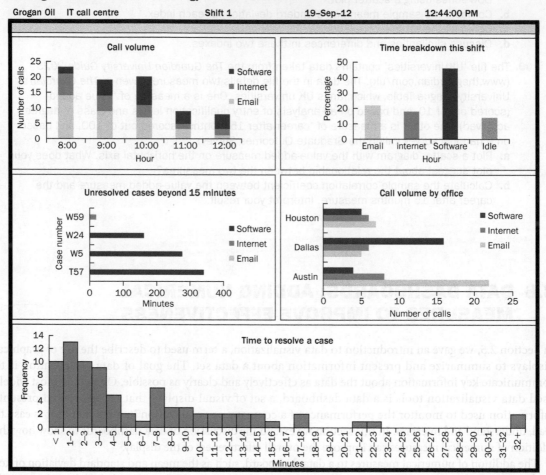

To gain additional insight into the performance of the call centre, Grogan's IT manager has decided to expand the current dashboard by adding boxplots for the time required to resolve calls received for each type of problem (email, internet and software). In addition, a graph showing the time to resolve individual cases has been added in the lower left portion of the dashboard. Finally, the IT manager added a display of summary statistics for each type of problem and summary statistics for each of the first few hours of the shift. The updated dashboard is shown in Figure 3.12.

The IT call centre has set a target performance level or benchmark of 10 minutes for the mean time to resolve a case. Furthermore, the centre has decided it is undesirable for the time to resolve a case to exceed 15 minutes. To reflect these benchmarks, a black horizontal line at the mean target value of 10 minutes and a red horizontal line at the maximum acceptable level of 15 minutes have been added to both the graph showing the time to resolve cases and the boxplots of the time required to resolve calls received for each type of problem.

FIGURE 3.12

Updated Grogan Oil information technology call centre data dashboard

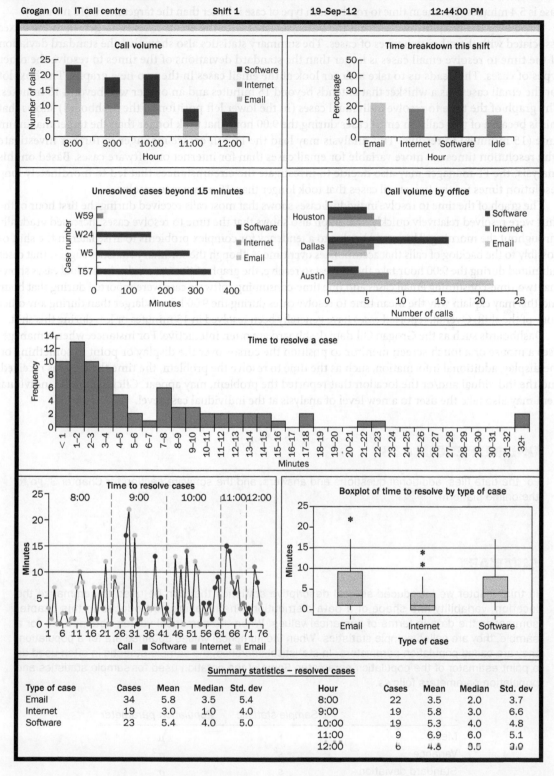

The summary statistics in the dashboard in Figure 3.12 show that the mean time to resolve an email case is 5.8 minutes, the mean time to resolve an internet case is 3.0 minutes, and the mean time to resolve a software case is 5.4 minutes. The mean time to resolve each type of case is better than the target mean (10 minutes).

Reviewing the boxplots, we see that the box associated with the email cases is 'larger' than the boxes associated with the other two types of cases. The summary statistics also show that the standard deviation of the time to resolve email cases is larger than the standard deviations of the times to resolve the other types of cases. This leads us to take a closer look at the email cases in the two new graphs. The boxplot for the email cases has a whisker that extends beyond 15 minutes and an outlier well beyond 15 minutes. The graph of the time to resolve individual cases (in the lower left position of the dashboard) shows that this is because of two calls on email cases during the 9:00 hour that took longer than the target maximum time (15 minutes) to resolve. This analysis may lead the IT call centre manager to further investigate why resolution times are more variable for email cases than for internet or software cases. Based on this analysis, the IT manager may also decide to investigate the circumstances that led to inordinately long resolution times for the two email cases that took longer than 15 minutes to resolve.

The graph of the time to resolve individual cases shows that most calls received during the first hour of the shift were resolved relatively quickly; the graph also shows that the time to resolve cases increased gradually throughout the morning. This could be due to a tendency for complex problems to arise later in the shift or possibly to the backlog of calls that accumulates over time. Although the summary statistics suggest that cases submitted during the 9:00 hour take the longest to resolve, the graph of time to resolve individual cases shows that two time-consuming email cases and one time-consuming software case were reported during that hour, and this may explain why the mean time to resolve cases during the 9:00 hour is larger than during any other hour of the shift. Overall, reported cases have generally been resolved in 15 minutes or less during this shift.

Dashboards such as the Grogan Oil data dashboard are often interactive. For instance, when a manager uses a mouse or a touch screen monitor to position the cursor over the display or point to something on the display, additional information, such as the time to resolve the problem, the time the call was received, and the individual and/or the location that reported the problem, may appear. Clicking on the individual item may also take the user to a new level of analysis at the individual case level.

ONLINE RESOURCES

For the data files, additional questions and answers, and the software section for Chapter 3, go to the online platform.

SUMMARY

In this chapter we introduced several descriptive statistics that can be used to summarize the location, variability and shape of a data distribution. The measures introduced in this chapter summarize the data in terms of numerical values. When the numerical values obtained are for a sample, they are called sample statistics. When the numerical values obtained are for a population, they are called population parameters. In statistical inference, the sample statistic is often used as a point estimator of the population parameter. Some of the notation used for sample statistics and population parameters follows.

	Sample statistic	Population parameter
Mean	$\bar{x}$	μ
Variance	s^2	σ^2
Standard deviation	s	σ
Covariance	s_{XY}	σ_{XY}
Correlation	r_{XY}	ρ_{XY}

As measures of central location, we defined the arithmetic mean, the geometric mean and median. The mode and the percentiles were used to describe other locations in the data set. Next, we presented the range, interquartile range, variance, standard deviation and coefficient of variation as measures of variability or dispersion. Our primary measure of the shape of a data distribution was the skewness. Negative values indicate a data distribution skewed to the left. Positive values indicate a data distribution skewed to the right. We showed how to calculate z-scores, and indicated how they can be used to identify outlying observations. We then described how the mean and standard deviation could be used, applying Chebyshev's theorem and the empirical rule, to provide more information about the distribution of data and to identify outliers.

In Section 3.4 we showed how to construct a five-number summary and a box plot to provide simultaneous information about the location, variability and shape of the distribution.

Section 3.5 introduced covariance and the correlation coefficient as measures of association between two variables.

KEY TERMS

Box plot
Chebyshev's theorem
Coefficient of variation
Correlation coefficient
Covariance
Empirical rule
Five-number summary
Interquartile range (IQR)
Mean
Median
Mode
Outlier

Percentile
Point estimator
Population parameter
Quartiles
Range
Sample statistic
Skewness
Standard deviation
Variance
Weighted mean
z-score

KEY FORMULAE

Sample mean

$$\bar{x} = \frac{\sum x_i}{n} \tag{3.1}$$

Population mean

$$\mu = \frac{\sum x_i}{N} \tag{3.2}$$

Weighted mean

$$\bar{x} = \frac{\sum w_i x_i}{\sum w_i} \tag{3.3}$$

Geometric mean

$$\bar{x}_g = \sqrt[n]{x_1 \times x_2 \times \cdots \times x_n} \tag{3.4}$$

Interquartile range

$$IQR = Q_3 - Q_1 \tag{3.5}$$

Population variance

$$\sigma^2 = \frac{\Sigma(x_i - \mu)^2}{N} \tag{3.6}$$

Sample variance

$$s^2 = \frac{\Sigma(x_i - \bar{x})^2}{n - 1} \tag{3.7}$$

Standard deviation

$$\text{Population standard deviation} = \sigma = \sqrt{\sigma^2} \tag{3.8}$$

$$\text{Sample standard deviation} = s = \sqrt{s^2} \tag{3.9}$$

Coefficient of variation

$$\left(\frac{\text{Standard deviation}}{\text{Mean}} \times 100\right)\% \tag{3.10}$$

z-score

$$z_i = \frac{x_i - \bar{x}}{s} \tag{3.11}$$

Sample covariance

$$s_{XY} = \frac{\Sigma(x_i - \bar{x})(y_i - \bar{y})}{n - 1} \tag{3.12}$$

Population covariance

$$\sigma_{XY} = \frac{\Sigma(x_i - \mu_X)(y_i - \mu_Y)}{N} \tag{3.13}$$

Pearson product moment correlation coefficient: sample data

$$r_{XY} = \frac{s_{XY}}{s_X s_Y} \tag{3.14}$$

Pearson product moment correlation coefficient: population data

$$\rho_{XY} = \frac{\sigma_{XY}}{\sigma_X \sigma_Y} \tag{3.15}$$

CASE PROBLEM 1

Chocolate Perfection website transactions

Chocolate Perfection manufactures and sells quality chocolate products in Dubai. Two years ago the company developed a website and began selling its products over the internet. Website sales have exceeded the company's expectations, and management is now considering strategies to increase sales even further. To learn more about the website customers, a sample of 50 Chocolate Perfection transactions was selected from the previous month's sales. Data showing the day of the week each transaction was made, the type of browser the customer used, the time spent on the website, the number of website pages viewed and the amount spent by each of the 50 customers are contained in the online file named 'Shoppers'. Amount spent is in United Arab Emirates dirham (AED). (One euro is around five AED.) A portion of the data has been reproduced and is shown on the adjacent page.

© ALEAIMAGE/iStock

Chocolate Perfection would like to use the sample data to determine if online shoppers who spend more time and view more pages also spend more money during their visit to the website. The company would also like to investigate the effect that the day of the week and the type of browser has on sales.

Managerial report

Use the methods of descriptive statistics to learn about the customers who visit the Chocolate Perfection website. Include the following in your report:

1. Graphical and numerical summaries for the length of time the shopper spends on the website, the number of pages viewed and the mean amount spent per transaction. Discuss what you learn about Chocolate Perfection's online shoppers from these numerical summaries.

2. Summarize the frequency, the total amount spent and the mean amount spent per transaction for each day of the week. What observations can you make about Chocolate Perfection's business based on the day of the week? Discuss.

3. Summarize the frequency, the total amount spent and the mean amount spent per transaction for each type of browser. What observations can you make about Chocolate Perfection's business, based on the type of browser? Discuss.

4. Construct a scatter diagram and compute the sample correlation coefficient to explore the relationship between the time spent on the website and the amount spent. Use the horizontal axis for the time spent on the website. Discuss.

5. Construct a scatter diagram and compute the sample correlation coefficient to explore the relationship between the number of website pages viewed and the amount spent. Use the horizontal axis for the number of website pages viewed. Discuss.

6. Construct a scatter diagram and compute the sample correlation coefficient to explore the relationship between the time spent on the website and the number of pages viewed. Use the horizontal axis to represent the number of pages viewed. Discuss.

SHOPPERS

Customer	Day	Browser	Time (min)	Pages Viewed	Amount Spent (AED)
1	Mon	Internet Explorer	12.0	4	200.09
2	Wed	Other	19.5	6	348.28
3	Mon	Internet Explorer	8.5	4	97.92
4	Tue	Firefox	11.4	2	164.16
5	Wed	Internet Explorer	11.3	4	243.21
6	Sat	Firefox	10.5	6	248.83
7	Sun	Internet Explorer	11.4	2	132.27
8	Fri	Firefox	4.3	6	205.37
9	Wed	Firefox	12.7	3	260.35
10	Tue	Internet Explorer	24.7	7	252.24

CASE PROBLEM 2

Business schools of Asia-Pacific

The pursuit of a higher education degree in business is now international. A survey shows that more and more Asians choose the Master of Business Administration (MBA) degree route to corporate success. As a result, the number of applicants for MBA courses at Asia-Pacific schools continues to increase. MBA programmes are notoriously tough and include economics, banking, marketing, behavioural sciences, labour relations, decision making, strategic thinking, business law and more. The data set 'AsiaMBA' contains some of the characteristics of the leading Asia-Pacific business schools: full-time enrolment, students per faculty, local tuition fees, foreign tuition fees, student age, % foreign students, GMAT test, English test, work experience and starting salary.

ASIAMBA

Managerial report

Use the methods of descriptive statistics to summarize the data in the file 'AsiaMBA'. Discuss your findings.

1. Include a summary for each variable in the data set. Make comments and interpretations based on maximums and minimums, as well as the appropriate means and proportions. What new insights do these descriptive statistics provide concerning Asia-Pacific business schools?

2. Summarize the data to compare the following:

   a. Any difference between local and foreign tuition costs.

   b. Any difference between mean starting salaries for schools requiring and not requiring work experience.

   c. Any difference between starting salaries for schools requiring and not requiring English tests.

3. Do starting salaries appear to be related to tuition costs?

4. Present any additional graphical and numerical summaries that will be beneficial in communicating the data in 'AsiaMBA' to others.

© cnythzl/iStock

4

Introduction to Probability

CHAPTER CONTENTS

Statistics in Practice   Combatting junk email

4.1 Experiments, counting rules and assigning probabilities
4.2 Events and their probabilities
4.3 Some basic relationships of probability
4.4 Conditional probability
4.5 Bayes' theorem

LEARNING OBJECTIVES   After reading this chapter and doing the exercises, you should be able to:

1 Appreciate the role probability information plays in the decision-making process.

2 Understand probability as a numerical measure of the likelihood of occurrence.

3 Appreciate the three methods commonly used for assigning probabilities and understand when they should be used.

4 Use the laws that are available for computing the probabilities of events.

5 Understand how new information can be used to revise initial (prior) probability estimates using Bayes' theorem.

Managers often base their decisions on an analysis of uncertainties such as the following:

1 What are the chances that sales will decrease if we increase prices?
2 What is the likelihood a new assembly method will increase productivity?
3 How likely is it that the project will be finished on time?
4 What is the chance that a new investment will be profitable?

STATISTICS IN PRACTICE
Combatting junk email

$$P(\text{spam} \mid \text{message}) = \frac{P(\text{message} \mid \text{spam})\, P(\text{spam})}{P(\text{message})}$$

where $P(\text{message}) = \begin{array}{l} P(\text{message} \mid \text{spam})\, P(\text{spam}) \\ + P(\text{message} \mid \text{good})\, P(\text{good}) \end{array}$

Despite a marked decline in levels over the last ten years, in December 2021 spam still accounted for 47.3 per cent of email traffic worldwide – with nearly a quarter (24.77 per cent) found to be generated by Russia alone.

Various initiatives have been undertaken to help counter the problem. However, determining which messages are 'good' ('ham') and which are 'spam' is difficult to establish even with the most sophisticated spam filters (spam-busters). One of the earliest and most effective techniques for dealing with spam is the adaptive Naïve Bayes' method which exploits the probability relationship:

© anyaberkut/iStock

Here:

$P(\text{spam})$ is the prior probability a message is spam based on past experience;

$P(\text{message} \mid \text{spam})$ is estimated from a training corpus (a set of messages known to be good or spam) on the (naïve) assumption that every word in the message is independent of every other so that:

$P(\text{message} \mid \text{spam}) = P(\text{first word} \mid \text{spam})$
$P(\text{second word} \mid \text{spam}) \ldots$
$P(\text{last word} \mid \text{spam})$

Similarly:

$P(\text{message} \mid \text{good}) = P(\text{first word} \mid \text{good})$
$P(\text{second word} \mid \text{good}) \ldots$
$P(\text{last word} \mid \text{good})$

Advantages of Naïve Bayes are its simplicity and ease of implementation. Indeed it is often found to be very effective – even compared to methods based on more complex modelling procedures.

Sources:
www.statista.com/statistics/420391/spam-email-traffic-share/.
The importance of a quality training dataset cannot be overemphasized. Available at iopscience.iop.org/article/10.1088/1757-899X/226/1/012091/meta.

Probability is a numerical measure of the likelihood that an event will occur. Hence, probabilities can be used as measures of the degree of uncertainty associated with events such as the four listed earlier. If probabilities are available, we can determine the likelihood of each event occurring.

Probability values are always assigned on a scale from 0 to 1. A probability of near zero indicates an event is unlikely to occur; a probability near 1 indicates an event is almost certain to occur. Other probabilities between 0 and 1 represent degrees of likelihood that an event will occur. For example, if we consider the event of 'rain tomorrow', we understand that when the weather report indicates 'a near-zero probability of rain', it means almost no chance of rain. However, if a 0.90 probability of rain is reported, we know that rain is very likely to occur. A 0.50 probability indicates that rain is just as likely to occur as not. Figure 4.1 depicts the view of probability as a numerical measure of the likelihood of an event occurring.

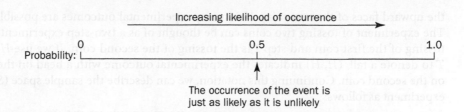

FIGURE 4.1
Probability as a
numerical measure
of the likelihood of an
event occurring

4.1 EXPERIMENTS, COUNTING RULES AND ASSIGNING PROBABILITIES

We define an **experiment** as a process that generates well-defined outcomes. On any single repetition of an experiment, one and only one of the possible outcomes will occur. Several examples of experiments and their associated outcomes follow.

Experiment	Experimental outcomes
Toss a coin	Head, tail
Select a part for inspection	Defective, non-defective
Conduct a sales call	Purchase, no purchase
Roll a die	1, 2, 3, 4, 5, 6
Play a football game	Win, lose, draw

By specifying all possible experimental outcomes, we identify the **sample space** for an experiment.

Sample space

The sample space for an experiment is the set of all experimental outcomes.

An experimental outcome is also called a **sample point** to identify it as an element of the sample space. Consider the first experiment in the preceding table – tossing a coin. The upward face of the coin – a head or a tail – determines the experimental outcomes (sample points). If we let S denote the sample space, we can use the following notation to describe the sample space:

$$S = \{\text{Head, Tail}\}$$

The sample space for the second experiment in the table – selecting a part for inspection – can be described as follows:

$$S = \{\text{Defective, Non-defective}\}$$

Both of the experiments just described have two experimental outcomes (sample points). However, suppose we consider the fourth experiment listed in the table – rolling a die. The possible experimental outcomes, defined as the number of dots appearing on the upward face of the die, are the six points in the sample space for this experiment:

$$S = \{1, 2, 3, 4, 5, 6\}$$

Counting rules, combinations and permutations

Being able to identify and count the experimental outcomes is a necessary step in assigning probabilities. We now discuss three useful counting rules.

Multiple-step experiments

The first counting rule applies to multiple-step experiments. Consider the experiment of tossing two coins. Let the experimental outcomes be defined in terms of the pattern of heads and tails appearing on the upward faces of the two coins. How many experimental outcomes are possible for this experiment? The experiment of tossing two coins can be thought of as a two-step experiment in which step 1 is the tossing of the first coin and step 2 is the tossing of the second coin. If we use H to denote a head and T to denote a tail, (H, H) indicates the experimental outcome with a head on the first coin and a head on the second coin. Continuing this notation, we can describe the sample space (S) for this coin-tossing experiment as follows:

$$S = \{(H, H), (H, T), (T, H), (T, T)\}$$

Hence, we see that four experimental outcomes are possible. In this case, we can easily list all of the experimental outcomes.

The counting rule for multiple-step experiments makes it possible to determine the number of experimental outcomes without listing them.

A counting rule for multiple-step experiments

If an experiment can be described as a sequence of k steps with n_1 possible outcomes on the first step, n_2 possible outcomes on the second step and so on, then the total number of experimental outcomes is given by:

$$n_1 \times n_2 \times \ldots \times n_k$$

Viewing the experiment of tossing two coins as a sequence of first tossing one coin ($n_1 = 2$) and then tossing the other coin ($n_2 = 2$), we can see from the counting rule that there are $2 \times 2 = 4$ distinct experimental outcomes. They are $S = \{(H, H), (H, T), (T, H), (T, T)\}$. The number of experimental outcomes in an experiment involving tossing six coins is $2 \times 2 \times 2 \times 2 \times 2 \times 2 = 64$.

A **tree diagram** is a graphical representation that helps in visualizing a multiple-step experiment. Figure 4.2 shows a tree diagram for the experiment of tossing two coins. The sequence of steps moves from left to right through the tree. Step 1 corresponds to tossing the first coin, and step 2 corresponds to tossing the second coin. For each step, the two possible outcomes are head or tail. Note that for each possible outcome at step 1, two branches correspond to the two possible outcomes at step 2. Each of the points on the right end of the tree corresponds to an experimental outcome. Each path through the tree from the leftmost node to one of the nodes at the right side of the tree corresponds to a unique sequence of outcomes.

FIGURE 4.2

Tree diagram for the experiment of tossing two coins

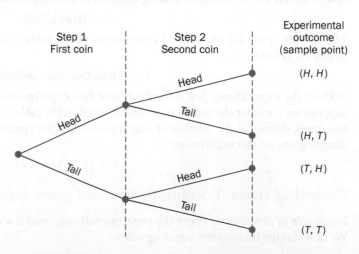

Step 1 First coin	Step 2 Second coin	Experimental outcome (sample point)
	Head	(H, H)
Head	Tail	(H, T)
Tail	Head	(T, H)
	Tail	(T, T)

Let us now see how the counting rule for multiple-step experiments can be used in the analysis of a capacity expansion project for Kristof Projects Limited (KPL). KPL is starting a project designed to increase the generating capacity of one of its plants in southern Norway. The project is divided into two sequential stages or steps: stage 1 (design) and stage 2 (construction). Even though each stage will be scheduled and controlled as closely as possible, management cannot predict beforehand the exact time required to complete each stage of the project. An analysis of similar construction projects revealed possible completion times for the design stage of two, three or four months and possible completion times for the construction stage of six, seven or eight months.

In addition, because of the critical need for additional electrical power, management set a goal of ten months for the completion of the entire project.

Because this project has three possible completion times for the design stage (step 1) and three possible completion times for the construction stage (step 2), the counting rule for multiple-step experiments can be applied here to determine a total of $3 \times 3 = 9$ experimental outcomes. To describe the experimental outcomes, we use a two-number notation; for instance, $(2, 6)$ indicates that the design stage is completed in two months and the construction stage is completed in six months. This experimental outcome results in a total of $2 + 6 = 8$ months to complete the entire project. Table 4.1 summarizes the nine experimental outcomes for the KPL problem. The tree diagram in Figure 4.3 shows how the nine outcomes (sample points) occur.

The counting rule and tree diagram help the project manager identify the experimental outcomes and determine the possible project completion times. We see that the project will be completed in 8 to 12 months, with six of the nine experimental outcomes providing the desired completion time of ten months or less. Even though identifying the experimental outcomes may be helpful, we need to consider how probability values can be assigned to the experimental outcomes before making an assessment of the probability that the project will be completed within the desired ten months.

Combinations

A second useful counting rule allows us to count the number of experimental outcomes when the experiment involves selecting n objects from a (usually larger) set of N objects. It is called the counting rule for combinations.

Counting rule for combinations

The number of combinations of N objects taken n at a time is:

$$^{N}C_{n} = \binom{N}{n} = \frac{N!}{n!(N-n)!}$$

(4.1)

where:

$$N! = N \times (N-1) \times (N-2) \times \ldots \times (2) \times (1)$$

$$n! = n \times (n-1) \times (n-2) \times \ldots \times (2) \times (1)$$

and, by definition:

$$0! = 1$$

The notation ! means *factorial*: for example, 5 factorial is $5! = 5 \times 4 \times 3 \times 2 \times 1 = 120$.

Consider a quality control procedure in which an inspector randomly selects two of five parts to test for defects. In a group of five parts, how many combinations of two parts can be selected? The counting rule in equation (4.1) shows that with $N = 5$ and $n = 2$, we have:

$$^{5}C_{2} = \binom{5}{2} = \frac{5 \times 4 \times 3 \times 2 \times 1}{(2 \times 1) \times (3 \times 2 \times 1)} = \frac{120}{12} = 10$$

Hence, ten outcomes are possible for the experiment of randomly selecting two parts from a group of five. If we label the five parts as A, B, C, D and E, the ten combinations or experimental outcomes can be identified as AB, AC, AD, AE, BC, BD, BE, CD, CE and DE.

TABLE 4.1 Experimental outcomes (sample points) for the KPL project

Completion time (months)			
Stage 1 Design	**Stage 2 Construction**	**Notation for experimental outcome**	**Total project completion time (months)**
2	6	(2, 6)	8
2	7	(2, 7)	9
2	8	(2, 8)	10
3	6	(3, 6)	9
3	7	(3, 7)	10
3	8	(3, 8)	11
4	6	(4, 6)	10
4	7	(4, 7)	11
4	8	(4, 8)	12

FIGURE 4.3
Tree diagram for the
KPL project

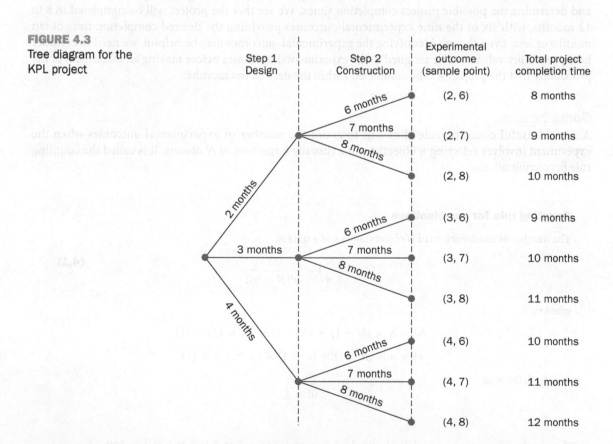

Step 1 Design	Step 2 Construction	Experimental outcome (sample point)	Total project completion time
	6 months	(2, 6)	8 months
	7 months	(2, 7)	9 months
	8 months	(2, 8)	10 months
	6 months	(3, 6)	9 months
	7 months	(3, 7)	10 months
	8 months	(3, 8)	11 months
	6 months	(4, 6)	10 months
	7 months	(4, 7)	11 months
	8 months	(4, 8)	12 months

As another example, consider that the Spanish Lotto 6–49 system uses the random selection of six integers from a group of 49 to determine the weekly lottery winner. The counting rule for combinations, equation (4.1), can be used to determine the number of ways six different integers can be selected from a group of 49:

$$\binom{49}{6} = \frac{49!}{6!(49-6)!} = \frac{49!}{6!43!} = \frac{49 \times 48 \times 47 \times 46 \times 45 \times 44}{6 \times 5 \times 4 \times 3 \times 2 \times 1} = 13{,}983{,}816$$

The counting rule for combinations tells us that more than 13 million experimental outcomes are possible in the lottery draw. An individual who buys a lottery ticket has one chance in 13,983,816 of winning.

Permutations

A third counting rule that is sometimes useful is the counting rule for permutations. It allows us to compute the number of experimental outcomes when n objects are to be selected from a set of N objects where the order of selection is important. The same n objects selected in a different order are considered a different experimental outcome.

Counting rule for permutations

The number of permutations of N objects taken at n is given by:

$$^N P_n = n!\binom{N}{n} = \frac{N!}{(N-n)!} \tag{4.2}$$

The counting rule for permutations closely relates to the one for combinations; however, an experiment results in more permutations than combinations for the same number of objects because every selection of n objects can be ordered in $n!$ different ways.

As an example, consider again the quality control process in which an inspector selects two of five parts to inspect for defects. How many permutations may be selected? The counting rule in equation (4.2) shows that with $N = 5$ and $n = 2$, we have:

$$^5 P_2 = \frac{5!}{(5-2)!} = \frac{5!}{3!} = \frac{5 \times 4 \times 3 \times 2 \times 1}{3 \times 2 \times 1} = \frac{120}{6} = 20$$

Hence, 20 outcomes are possible for the experiment of randomly selecting two parts from a group of five when the order of selection must be taken into account. If we label the parts A, B, C, D and E, the 20 permutations are AB, BA, AC, CA, AD, DA, AE, EA, BC, CB, BD, DB, BE, EB, CD, DC, CE, EC, DE and ED.

Assigning probabilities

Now let us see how probabilities can be assigned to experimental outcomes. The three approaches most frequently used are the classical, relative frequency and subjective methods. Regardless of the method used, two **basic requirements for assigning probabilities** must be met.

Basic requirements for assigning probabilities

1. The probability assigned to each experimental outcome must be between 0 and 1, inclusively. If we let E_i denote the ith experimental outcome and $P(E_i)$ its probability, then this requirement can be written as:

$$0 \leq P(E_i) \leq 1 \text{ for all } i \tag{4.3}$$

2. The sum of the probabilities for all the experimental outcomes must equal 1.0. For n experimental outcomes, this requirement can be written as:

$$P(E_1) + P(E_2) + \ldots + P(E_n) = 1 \tag{4.4}$$

The classical method of assigning probabilities is appropriate when all the experimental outcomes are equally likely. If n experimental outcomes are possible, a probability of $1/n$ is assigned to each experimental outcome. When using this approach, the two basic requirements for assigning probabilities are automatically satisfied.

For example, consider the experiment of tossing a fair coin: the two experimental outcomes – head and tail – are equally likely. Because one of the two equally likely outcomes is a head, the probability of observing a head is 1/2 or 0.50. Similarly, the probability of observing a tail is also 1/2 or 0.50.

As another example, consider the experiment of rolling a die. It would seem reasonable to conclude that the six possible outcomes are equally likely, and hence each outcome is assigned a probability of 1/6. If $P(1)$ denotes the probability that one dot appears on the upward face of the die, then $P(1) = 1/6$. Similarly, $P(2) = 1/6$, $P(3) = 1/6$, $P(4) = 1/6$, $P(5) = 1/6$ and $P(6) = 1/6$. Note that these probabilities satisfy the two basic requirements of equations (4.3) and (4.4) because each of the probabilities is greater than or equal to zero and they sum to 1.0.

The relative frequency method of assigning probabilities is appropriate when data are available to estimate the proportion of the time the experimental outcome will occur if the experiment is repeated a large number of times. As an example, consider a study of waiting times in the X-ray department for a local hospital. A clerk recorded the number of patients waiting for service at 9:00 a.m. on 20 successive days, and obtained the following results:

Number waiting	Number of days outcome occurred
0	2
1	5
2	6
3	4
4	3
	Total = 20

These data show that on 2 of the 20 days, zero patients were waiting for service; on five of the days, one patient was waiting for service and so on. Using the relative frequency method, we would assign a probability of 2/20 = 0.10 to the experimental outcome of zero patients waiting for service, 5/20 = 0.25 to the experimental outcome of one patient waiting, 6/20 = 0.30 to two patients waiting, 4/20 = 0.20 to three patients waiting and 3/20 = 0.15 to four patients waiting. As with the classical method, using the relative frequency method automatically satisfies the two basic requirements of equations (4.3) and (4.4).

The subjective method of assigning probabilities is most appropriate when we cannot realistically assume that the experimental outcomes are equally likely and when little relevant data are available. When the subjective method is used to assign probabilities to the experimental outcomes, we may use any information available, such as our experience or intuition. After considering all available information, a probability value that expresses our *degree of belief* (on a scale from 0 to 1) that the experimental outcome will occur, is specified. Because subjective probability expresses a person's degree of belief, it is personal. Using the subjective method, different people can be expected to assign different probabilities to the same experimental outcome.

The subjective method requires extra care to ensure that the two basic requirements of equations (4.3) and (4.4) are satisfied. Regardless of a person's degree of belief, the probability value assigned to each experimental outcome must be between 0 and 1, inclusive, and the sum of all the probabilities for the experimental outcomes must equal 1.0.

Consider the case in which Tomas and Margit Elsbernd make an offer to purchase a house. Two outcomes are possible:

$$E_1 = \textit{their offer is accepted}$$
$$E_2 = \textit{their offer is rejected}$$

Margit believes that the probability their offer will be accepted is 0.8; hence, Margit would set $P(E_1) = 0.8$ and $P(E_2) = 0.2$. Tomas, however, believes that the probability that their offer will be accepted is 0.6; hence, Tomas would set $P(E_1) = 0.6$ and $P(E_2) = 0.4$. Note that Tomas' probability estimate for E_1 reflects a greater pessimism that their offer will be accepted.

Both Margit and Tomas assigned probabilities that satisfy the two basic requirements. The fact that their probability estimates are different emphasizes the personal nature of the subjective method.

Even in business situations where either the classical or the relative frequency approach can be applied, managers may want to provide subjective probability estimates. In such cases, the best probability estimates often are obtained by combining the estimates from the classical or relative frequency approach with subjective probability estimates.

Probabilities for the KPL project

To perform further analysis on the KPL project, we must develop probabilities for each of the nine experimental outcomes listed in Table 4.1. On the basis of experience and judgement, management concluded that the experimental outcomes were not equally likely. Hence, the classical method of assigning probabilities could not be used. Management then decided to conduct a study of the completion times for similar projects undertaken by KPL over the past three years. The results of a study of 40 similar projects are summarized in Table 4.2.

TABLE 4.2 Completion results for 40 KPL projects

Completion times (months)			
Stage 1 Design	Stage 2 Construction	Sample point	Number of past projects having these completion times
2	6	(2, 6)	6
2	7	(2, 7)	6
2	8	(2, 8)	2
3	6	(3, 6)	4
3	7	(3, 7)	8
3	8	(3, 8)	2
4	6	(4, 6)	2
4	7	(4, 7)	4
4	8	(4, 8)	6
			Total = 40

After reviewing the results of the study, management decided to employ the relative frequency method of assigning probabilities. Management could have provided subjective probability estimates, but felt that the current project was quite similar to the 40 previous projects. Hence, the relative frequency method was judged best.

In using the data in Table 4.2 to compute probabilities, we note that outcome (2, 6) – stage 1 completed in two months and stage 2 completed in six months – occurred 6 times in the 40 projects. We can use the relative frequency method to assign a probability of 6/40 = 0.15 to this outcome. Similarly, outcome (2, 7) also occurred in 6 of the 40 projects, providing a 6/40 = 0.15 probability. Continuing in this manner, we obtain the probability assignments for the sample points of the KPL project shown in Table 4.3.

TABLE 4.3 Probability assignments for the KPL project based on the relative frequency method

Sample point	Project completion time	Probability of sample point
(2, 6)	8 months	P(2, 6) = 6/40 = 0.15
(2, 7)	9 months	P(2, 7) = 6/40 = 0.15
(2, 8)	10 months	P(2, 8) = 2/40 = 0.05
(3, 6)	9 months	P(3, 6) = 4/40 = 0.10
(3, 7)	10 months	P(3, 7) = 8/40 = 0.20
(3, 8)	11 months	P(3, 8) = 2/40 = 0.05
(4, 6)	10 months	P(4, 6) = 2/40 = 0.05
(4, 7)	11 months	P(4, 7) = 4/40 = 0.10
(4, 8)	12 months	P(4, 8) = 6/40 = 0.15
		Total 1.00

Note that $P(2, 6)$ represents the probability of the sample point $(2, 6)$, $P(2, 7)$ represents the probability of the sample point $(2, 7)$ and so on.

EXERCISES

Methods

1. An experiment has three steps with three outcomes possible for the first step, two outcomes possible for the second step and four outcomes possible for the third step. How many experimental outcomes exist for the entire experiment?

2. How many ways can three items be selected from a group of six items? Use the letters A, B, C, D, E and F to identify the items, and list each of the different combinations of three items.

3. How many permutations of three items can be selected from a group of six? Use the letters A, B, C, D, E and F to identify the items, and list each of the permutations of items B, D and F.

4. Consider the experiment of tossing a coin three times.
   a. Develop a tree diagram for the experiment.
   b. List the experimental outcomes.
   c. What is the probability for each experimental outcome?

5. Suppose an experiment has five equally likely outcomes: E_1, E_2, E_3, E_4, E_5. Assign probabilities to each outcome and show that the requirements in equations (4.3) and (4.4) are satisfied. What method did you use?

6. An experiment with three outcomes has been repeated 50 times, and it was learned that E_1 occurred 20 times, E_2 occurred 13 times and E_3 occurred 17 times. Assign probabilities to the outcomes. What method did you use?

7. A decision-maker subjectively assigned the following probabilities to the four outcomes of an experiment: $P(E_1) = 0.10$, $P(E_2) = 0.15$, $P(E_3) = 0.40$ and $P(E_4) = 0.20$. Are these probability assignments valid? Explain.

8. Let X1 and X2 denote the interest rates (%) that will be paid on one-year certificates of deposit that are issued on the first day of next year (year 1) and the following year (year 2) respectively. X1 takes the values 2, 3 and 4 as does X2.
   a. How many sample points are there for this experiment? List the sample points.
   b. Construct a tree diagram for the experiment.

9. Code Churn is a common metric used to measure the efficiency and productivity of software engineers and computer programmers. It is usually measured as the percentage of a programmer's code that must be edited over a short period of time. Programmers with higher rates of code churn must rewrite code more often because of errors and inefficient programming techniques. The following table displays sample information for ten computer programmers.

Programmer	Total lines of code written	Number of lines of code requiring edits
Liwei	23,789	4,589
Andrew	17,962	2,780
Jaime	31,025	12,080
Sherae	26,050	3,780

Binny	19,586	1,890
Roger	24,786	4,005
Dong-Gil	24,030	5,785
Alex	14,780	1,052
Jay	30,875	3,872
Vivek	21,546	4,125

a. Use the data in the table above and the relative frequency method to determine probabilities that a randomly selected line of code will need to be edited for each programmer.

b. If you randomly select a line of code from Liwei, what is the probability that the line of code will require editing?

c. If you randomly select a line of code from Sherae, what is the probability that the line of code will *not* require editing?

d. Which programmer has the lowest probability of a randomly selected line of code requiring editing? Which programmer has the highest probability of a randomly selected line of code requiring editing?

10. A company that manufactures toothpaste is studying five different package designs. Assuming that one design is just as likely to be selected by a consumer as any other design, what selection probability would you assign to each of the package designs? In an actual experiment, 100 consumers were asked to pick the design they preferred. The following data were obtained. Do the data confirm the belief that one design is just as likely to be selected as another? Explain.

Design	Number of times preferred
1	5
2	15
3	30
4	40
5	10

a. Define the experiment being conducted. How many times was it repeated?

b. Prior to conducting the experiment, it is reasonable to assume preferences for the four blends are equal. What probabilities would you assign to the experimental outcomes prior to conducting the taste test? What method did you use?

c. After conducting the taste test, what probabilities would you assign to the experimental outcomes? What method did you use?

11. Refer to Exercise 8.

		Year 2 X2		
		2	3	4
Year 1	2	0.1	0.1	0
X1	3	0.1	0.3	0.2
	4	0	0.1	0.1

Given the probabilities tabulated above (where for example 0.3 = P(X1,X2) = P(3,3)), what is the probability of:

a. $X1 < X2$?

b. $X1 = X2$?

c. $X1 \geq X2$?

4.2 EVENTS AND THEIR PROBABILITIES

In the introduction to this chapter we used the term *event* much as it would be used in everyday language. Then, in Section 4.1 we introduced the concept of an experiment and its associated experimental outcomes or sample points. Sample points and events provide the foundation for the study of probability. We must now introduce the formal definition of an event as it relates to sample points. Doing so will provide the basis for determining the probability of an event.

> **Event**
>
> An event is a collection of sample points.

For example, let us return to the KPL project and assume that the project manager is interested in the event that the entire project can be completed in ten months or less. Referring to Table 4.3, we see that six sample points – (2, 6), (2, 7), (2, 8), (3, 6), (3, 7) and (4, 6) – provide a project completion time of ten months or less. Let C denote the event that the project is completed in ten months or less; we write:

$$C = \{(2, 6), (2, 7), (2, 8), (3, 6), (3, 7), (4, 6)\}$$

Event C is said to occur if *any one* of these six sample points appears as the experimental outcome. Other events that might be of interest to KPL management include the following:

$$L = \text{The event that the project is completed in } less \text{ than ten months}$$
$$M = \text{The event that the project is completed in } more \text{ than ten months}$$

Using the information in Table 4.3, we see that these events consist of the following sample points:

$$L = \{(2, 6), (2, 7), (3, 6)\}$$
$$M = \{(3, 8), (4, 7), (4, 8)\}$$

A variety of additional events can be defined for the KPL project, but in each case the event must be identified as a collection of sample points for the experiment.

Given the probabilities of the sample points shown in Table 4.3, we can use the following definition to compute the probability of any event that KPL management might want to consider:

> **Probability of an event**
>
> The probability of any event is equal to the sum of the probabilities of the sample points for the event.

Using this definition, we calculate the probability of a particular event by adding the probabilities of the sample points (experimental outcomes) that make up the event. We can now compute the probability that the project will take ten months or less to complete. Because this event is given by $C = \{(2, 6), (2, 7), (2, 8), (3, 6), (3, 7), (4, 6)\}$, the probability of event C, denoted $P(C)$, is given by:

$$P(C) = P(2, 6) + P(2, 7) + P(2, 8) + P(3, 6) + P(3, 7) + P(4, 6)$$
$$= 0.15 + 0.15 + 0.05 + 0.10 + 0.20 + 0.05 = 0.70$$

Similarly, because the event that the project is completed in less than ten months is given by $L = \{(2, 6), (2, 7), (3, 6)\}$, the probability of this event is given by:

$$P(L) = P(2, 6) + P(2, 7) + P(3, 6)$$
$$= 0.15 + 0.15 + 0.10 = 0.40$$

Finally, for the event that the project is completed in more than ten months, we have $M = \{(3, 8), (4, 7), (4, 8)\}$ and thus:

$$P(M) = P(3, 8) + P(4, 7) + P(4, 8)$$
$$= 0.05 + 0.10 + 0.15 = 0.30$$

Using these probability results, we can now tell KPL management that there is a 0.70 probability that the project will be completed in ten months or less, a 0.40 probability that the project will be completed in less than ten months, and a 0.30 probability that the project will be completed in more than ten months. This procedure of computing event probabilities can be repeated for any event of interest to the KPL management.

EXERCISES

Methods

12. An experiment has four equally likely outcomes: E_1, E_2, E_3 and E_4.
   a. What is the probability that E_2 occurs?
   b. What is the probability that any two of the outcomes occur (e.g. E_1 or E_3)?
   c. What is the probability that any three of the outcomes occur (e.g. E_1 or E_2 or E_4)?

13. Consider the experiment of selecting a playing card from a deck of 52 playing cards. Each card corresponds to a sample point with a 1/52 probability.
   a. List the sample points in the event an ace is selected.
   b. List the sample points in the event a club is selected.
   c. List the sample points in the event a face card (jack, queen or king) is selected.
   d. Find the probabilities associated with each of the events in parts (a), (b) and (c).

14. Consider the experiment of rolling a pair of dice. Suppose that we are interested in the sum of the face values showing on the dice.
   a. How many sample points are possible? (*Hint:* Use the counting rule for multiple-step experiments.)
   b. List the sample points.
   c. What is the probability of obtaining a value of 7?
   d. What is the probability of obtaining a value of 9 or greater?
   e. Because each roll has six possible even values (2, 4, 6, 8, 10 and 12) and only five possible odd values (3, 5, 7, 9 and 11), the dice should show even values more often than odd values. Do you agree with this statement? Explain.
   f. What method did you use to assign the probabilities requested?

Applications

15. Refer to the KPL sample points and sample point probabilities in Tables 4.2 and 4.3.
   a. The design stage (stage 1) will run over budget if it takes four months to complete. List the sample points in the event the design stage is over budget.
   b. What is the probability that the design stage is over budget?
   c. The construction stage (stage 2) will run over budget if it takes eight months to complete. List the sample points in the event the construction stage is over budget.
   d. What is the probability that the construction stage is over budget?
   e. What is the probability that both stages are over budget?

16. Suppose that a manager of a large apartment complex provides the following subjective probability estimates about the number of vacancies that will exist next month.

Vacancies	Probability
0	0.10
1	0.15
2	0.30
3	0.20
4	0.15
5	0.10

Provide the probability of each of the following events.
a. No vacancies.
b. At least four vacancies.
c. Two or fewer vacancies.

17. When three sharpshooters take part in a shooting contest, their chances of hitting the target are 1/2, 1/3 and 1/4 respectively. If all three sharpshooters fire at it simultaneously:
a. What is the chance that one and only one bullet will hit the target?
b. What is the chance that two sharpshooters will hit the target (and therefore one will not)?
c. What is the chance that all three sharpshooters will hit the target?

Any time that we can identify all the sample points of an experiment and assign probabilities to each, we can compute the probability of an event using the definition. However, in many experiments the large number of sample points makes the identification of the sample points, as well as the determination of their associated probabilities, extremely cumbersome, if not impossible. In the remaining sections of this chapter, we present some basic probability relationships that can be used to compute the probability of an event without knowledge of all the sample point probabilities.

4.3 SOME BASIC RELATIONSHIPS OF PROBABILITY

Complement of an event

Given an event A, the **complement of A** is defined to be the event consisting of all sample points that are *not* in A. The complement of A is denoted by $\overline{A}$. Figure 4.4 is a diagram, known as a **Venn diagram**, which illustrates the concept of a complement. The rectangular area represents the sample space for the experiment and as such contains all possible sample points. The circle represents event A and contains only the sample points that belong to A. The shaded region of the rectangle contains all sample points not covered by event A, and is by definition the complement of A.

FIGURE 4.4
Complement of event
A is shaded

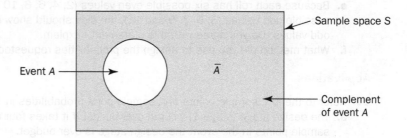

Event A $\overline{A}$ Sample space S   Complement of event A

In any probability application, either event A or its complement $\overline{A}$ must occur. Therefore, we have:

$$P(A) + P(\overline{A}) = 1$$

Solving for $P(A)$, we obtain the following result:

Computing probability using the complement

$$P(A) = 1 - P(\overline{A}) \tag{4.5}$$

Equation (4.5) shows that the probability of an event A can be computed easily if the probability of its complement, $P(\overline{A})$, is known.

As an example, consider the case of a sales manager who, after reviewing sales reports, states that 80 per cent of new customer contacts result in no sale. By allowing A to denote the event of a sale and $\overline{A}$ to denote the event of no sale, the manager is stating that $P(\overline{A}) = 0.80$. Using equation (4.5), we observe that:

$$P(A) = 1 - P(\overline{A}) = 1 - 0.80 = 0.20$$

We can conclude that a new customer contact has a 0.20 probability of resulting in a sale.

In another example, a purchasing agent states a 0.90 probability that a supplier will send a shipment that is free of defective parts. Using the complement, we can conclude that there is a $1 - 0.90 = 0.10$ probability that the shipment will contain defective parts.

Addition law

The addition law is helpful when we are interested in knowing the probability that at least one of two events occurs. That is, with events A and B we are interested in knowing the probability that event A or event B or both occur.

Before we present the addition law, we need to discuss two concepts related to the combination of events: the *union* of events and the *intersection* of events. Given two events A and B, the **union of A and B** is defined as follows:

Union of two events

The *union* of A and B is the event containing *all* sample points belonging to A *or* B *or both*. The union is denoted by $A \cup B$.

The Venn diagram in Figure 4.5 depicts the union of events A and B. Note that the two circles contain all the sample points in event A as well as all the sample points in event B.

FIGURE 4.5
Union of events A and B
is shaded

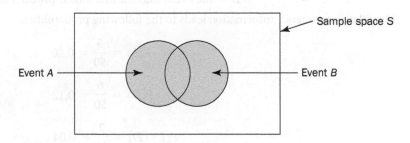

Event A ────→   ←──── Event B

Sample space S

The fact that the circles overlap indicates that some sample points are contained in both A and B. The definition of the **intersection of A and B** follows.

Intersection of two events

Given two events A and B, the *intersection* of A and B is the event containing the sample points belonging to *both A and B*. The intersection is denoted by $A \cap B$.

The Venn diagram depicting the intersection of events A and B is shown in Figure 4.6. The area where the two circles overlap is the intersection; it contains the sample points that are in both A and B.

The **addition law** provides a way to compute the probability that event A or event B or both occur. In other words, the addition law is used to compute the probability of the union of two events. The addition law is written as follows in equation (4.6).

Addition law

$$P(A \cup B) = P(A) + P(B) - P(A \cap B) \qquad \textbf{(4.6)}$$

FIGURE 4.6
Intersection of events
A and B is shaded

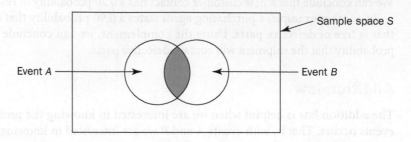

To understand the addition law intuitively, note that the first two terms in the addition law, $P(A) + P(B)$, account for all the sample points in $A \cup B$. However, because the sample points in the intersection $A \cap B$ are in both A and B, when we compute $P(A) + P(B)$, we are in effect counting each of the sample points in $A \cap B$ twice. We correct for this over-counting by subtracting $P(A \cap B)$.

As an example of an application of the addition law, consider the case of a small assembly plant with 50 employees. Each worker is expected to complete work assignments on time and in such a way that the assembled product will pass a final inspection. On occasion, some of the workers fail to meet the performance standards by completing work late or assembling a defective product. At the end of a performance evaluation period, the production manager found that 5 of the 50 workers completed work late, 6 of the 50 workers assembled a defective product and 2 of the 50 workers both completed work late *and* assembled a defective product. Let:

$$L = \text{the event that the work is completed late}$$

$$D = \text{the event that the assembled product is defective}$$

The relative frequency information leads to the following probabilities:

$$P(L) = \frac{5}{50} = 0.10$$

$$P(D) = \frac{6}{50} = 0.12$$

$$P(L \cap D) = \frac{2}{50} = 0.04$$

After reviewing the performance data, the production manager decided to assign a poor performance rating to any employee whose work was either late or defective; hence the event of interest is $L \cup D$. What is the probability that the production manager assigned an employee a poor performance rating? Using equation (4.6), we have:

$$P(L \cup D) = P(L) + P(D) - P(L \cap D)$$
$$= 0.10 + 0.12 - 0.04 = 0.18$$

This calculation tells us that there is a 0.18 probability that a randomly selected employee received a poor performance rating.

As another example of the addition law, consider a recent study conducted by the personnel manager of a major computer software company. The study showed that 30 per cent of the employees who left the firm within two years did so primarily because they were dissatisfied with their salary, 20 per cent left because they were dissatisfied with their work assignments and 12 per cent of the former employees indicated dissatisfaction with *both* their salary and their work assignments. What is the probability that an employee who leaves within two years does so because of dissatisfaction with salary, dissatisfaction with the work assignment or both? Let:

S = the event that the employee leaves because of salary

W = the event that the employee leaves because of work assignment

We have $P(S) = 0.30$, $P(W) = 0.20$ and $P(S \cap W) = 0.12$

Using equation (4.6) we have:

$$P(S) + P(W) - P(S \cap W) = 0.30 + 0.20 - 0.12 = 0.38$$

We find a 0.38 probability that an employee leaves for salary or work assignment reasons.

Before we conclude our discussion of the addition law, let us consider a special case that arises for **mutually exclusive events**.

Mutually exclusive events

Two events are said to be mutually exclusive if the events have no sample points in common.

Events A and B are mutually exclusive if, when one event occurs, the other cannot occur. Hence, a requirement for A and B to be mutually exclusive is that their intersection must contain no sample points. The Venn diagram depicting two mutually exclusive events A and B is shown in Figure 4.7. In this case $P(A \cap B) = 0$ and the addition law can be written as follows:

Addition law for mutually exclusive events

$$P(A \cup B) = P(A) + P(B)$$

FIGURE 4.7
Mutually exclusive events

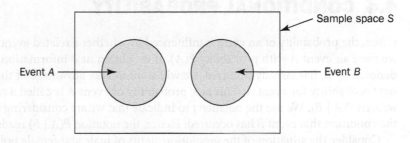

Sample space S

Event A

Event B

EXERCISES

Methods

18. Suppose that we have a sample space with five equally likely experimental outcomes: F_1, F_2, F_3, F_4, F_5. Let:

$$A = \{E_1, E_2\}$$
$$B = \{E_3, E_4\}$$
$$C = \{E_2, E_3, E_5\}$$

a. Find $P(A)$, $P(B)$ and $P(C)$.
b. Find $P(A \cup B)$. Are A and B mutually exclusive?
c. Find $\overline{A}$, $\overline{C}$, $P(\overline{A})$ and $P(\overline{C})$.
d. Find $A \cup \overline{B}$ and $P(A \cup \overline{B})$.
e. Find $P(B \cup C)$.

19. Suppose that we have a sample space $S = \{E_1, E_2, E_3, E_4, E_5, E_6, E_7\}$, where $E_1, E_2, \ldots, E_7$ denote the sample points. The following probability assignments apply:

$P(E_1) = 0.05$, $P(E_2) = 0.20$, $P(E_3) = 0.20$, $P(E_4) = 0.25$, $P(E_5) = 0.15$, $P(E_6) = 0.10$ and $P(E_7) = 0.05$. Let:

$$A = \{E_1, E_2\}$$
$$B = \{E_3, E_4\}$$
$$C = \{E_2, E_3, E_5\}$$

a. Find $P(A)$, $P(B)$, and $P(C)$.
b. Find $A \cup B$ and $P(A \cup B)$.
c. Find $A \cap B$ and $P(A \cap B)$.
d. Are events A and C mutually exclusive?
e. Find $\overline{B}$ and $P(\overline{B})$.

Applications

20. A survey of magazine subscribers showed that 45.8 per cent rented a car during the past 12 months for business reasons, 54 per cent rented a car during the past 12 months for personal reasons, and 30 per cent rented a car during the past 12 months for business as well as personal reasons.
a. What is the probability that a subscriber rented a car during the past 12 months for business or personal reasons?
b. What is the probability that a subscriber did not rent a car during the past 12 months for either business or personal reasons?

4.4 CONDITIONAL PROBABILITY

Often, the probability of an event is influenced by whether a related event has already occurred. Suppose we have an event A with probability $P(A)$. If we obtain new information and learn that a related event, denoted by B, has already occurred, we will want to take advantage of this information by calculating a new probability for event A. This new probability of event A is called a **conditional probability** and is written $P(A \mid B)$. We use the notation $\mid$ to indicate that we are considering the probability of event A *given* the condition that event B has occurred. Hence, the notation $P(A \mid B)$ reads 'the probability of A given B'.

Consider the situation of the promotion status of male and female police officers of a regional police force in France. The police force consists of 1,200 officers: 960 men and 240 women. Over the past two years, 324 officers on the police force received promotions. The specific breakdown of promotions for male and female officers is shown in Table 4.4.

TABLE 4.4 Promotion status of police officers over the past two years

	Men	**Women**	**Total**
Promoted	288	36	324
Not promoted	672	204	876
Totals	**960**	**240**	**1,200**

After reviewing the promotion record, a committee of female officers raised a discrimination case on the basis that 288 male officers had received promotions but only 36 female officers had received promotions.

The police administration argued that the relatively low number of promotions for female officers was not due to discrimination, but to the fact that relatively few women are members of the police force. Let us show how conditional probability could be used to analyze the discrimination charge.
Let:

$$M = \text{event an officer is a man}$$
$$W = \text{event an officer is a woman}$$
$$A = \text{event an officer is promoted}$$
$$\overline{A} = \text{event an officer is not promoted}$$

Dividing the data values in Table 4.4 by the total of 1,200 officers enables us to summarize the available information with the following probability values:

$P(M \cap A) = 288/1,200 = 0.24 =$ probability that a randomly selected officer is a man *and* is promoted

$P(M \cap \overline{A}) = 672/1,200 = 0.56 =$ probability that a randomly selected officer is a man *and* is not promoted

$P(W \cap A) = 36/1,200 = 0.03 =$ probability that a randomly selected officer is a woman *and* is promoted

$P(W \cap \overline{A}) = 204/1,200 = 0.17 =$ probability that a randomly selected officer is a woman *and* is not promoted

Because each of these values gives the probability of the intersection of two events, the probabilities are called **joint probabilities**. Table 4.5 is referred to as a *joint probability table*.

TABLE 4.5 Joint probability table for promotions

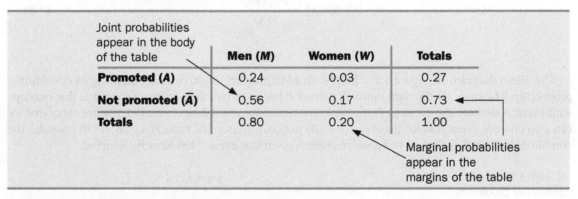

Joint probabilities appear in the body of the table	Men (M)	Women (W)	Totals
Promoted (A)	0.24	0.03	0.27
Not promoted (Ā)	0.56	0.17	0.73
Totals	0.80	0.20	1.00

Marginal probabilities appear in the margins of the table

The values in the margins of the joint probability table provide the probabilities of each event separately. That is, $P(M) = 0.80$, $P(W) = 0.20$, $P(A) = 0.27$ and $P(\overline{A}) = 0.73$. These probabilities are referred to as **marginal probabilities** because of their location in the margins of the joint probability table. We note that the marginal probabilities are found by summing the joint probabilities in the corresponding row or column of the joint probability table. For instance, the marginal probability of being promoted is $P(A) = P(M \cap A) + P(W \cap A) = 0.24 + 0.03 = 0.27$. From the marginal probabilities, we see that 80 per cent of the force is male, 20 per cent of the force is female, 27 per cent of all officers received promotions and 73 per cent were not promoted.

Consider the probability that an officer is promoted given that the officer is a man. In conditional probability notation, we are attempting to determine $P(A \mid M)$. By definition, $P(A \mid M)$ tells us that we are concerned only with the promotion status of the 960 male officers. Because 288 of the 960 male officers received promotions, the probability of being promoted given that the officer is a man is $288/960 = 0.30$. In other words, given that an officer is a man, that officer had a 30 per cent chance of receiving a promotion over the past two years.

This procedure was easy to apply because the values in Table 4.4 show the number of officers in each category. We now want to demonstrate how conditional probabilities such as $P(A \mid M)$ can be computed directly from related event probabilities rather than from the frequency data of Table 4.4.

We have shown that $P(A \mid M) = 288/960 = 0.30$. Let us now divide both the numerator and denominator of this fraction by 1,200, the total number of officers in the study:

$$P(A|M) = \frac{288}{960} = \frac{288/1{,}200}{960/1{,}200} = \frac{0.24}{0.80} = 0.30$$

We now see that the conditional probability $P(A \mid M)$ can be computed as 0.24/0.80. Refer to the joint probability table (Table 4.5). Note in particular that 0.24 is the joint probability of A and M; that is, $P(A \cap M) = 0.24$. Also note that 0.80 is the marginal probability that a randomly selected officer is a man; that is, $P(M) = 0.80$. Hence, the conditional probability $P(A \mid M)$ can be computed as the ratio of the joint probability $P(A \cap M)$ to the marginal probability $P(M)$:

$$P(A|M) = \frac{P(A \cap M)}{P(M)} = \frac{0.24}{0.80} = 0.30$$

The fact that conditional probabilities can be computed as the ratio of a joint probability to a marginal probability provides the following general formula (equations (4.7) and (4.8)) for conditional probability calculations for two events A and B.

Conditional probability

$$P(A|B) = \frac{P(A \cap B)}{P(B)} \qquad \textbf{(4.7)}$$

or

$$P(B|A) = \frac{P(A \cap B)}{P(A)} \qquad \textbf{(4.8)}$$

The Venn diagram in Figure 4.8 is helpful in obtaining an intuitive understanding of conditional probability. The circle on the right shows that event B has occurred; the portion of the circle that overlaps with event A denotes the event $(A \cap B)$. We know that, once event B has occurred, the only way that we can also observe event A is for the event $(A \cap B)$ to occur. Hence, the ratio $P(A \cap B)/P(B)$ provides the conditional probability that we will observe event A given that event B has already occurred.

FIGURE 4.8
Conditional probability

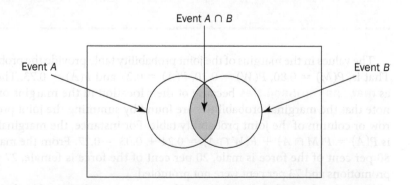

Event $A \cap B$

Event A

Event B

Let us return to the issue of discrimination against the female officers. The marginal probability in row 1 of Table 4.5 shows that the probability of promotion of an officer is $P(A) = 0.27$ (regardless of whether that officer is male or female). However, the critical issue in the discrimination case involves the two conditional probabilities $P(A \mid M)$ and $P(A \mid W)$. That is, what is the probability of a promotion *given* that the officer is a man, and what is the probability of a promotion *given* that the officer is a woman? If these two probabilities are equal, a discrimination argument has no basis because the chances of a promotion are the same for male and female officers. However, a difference in the two conditional probabilities will support the position that male and female officers are treated differently in promotion decisions.

We already determined that $P(A \mid M) = 0.30$. Let us now use the probability values in Table 4.5 and the basic relationship of conditional probability in equation (4.7) to compute the probability that an officer is promoted given that the officer is a woman; that is, $P(A \mid W)$. Using equation (4.7), with W replacing B, we obtain:

$$P(A \mid W) = \frac{P(A \cap W)}{P(W)} = \frac{0.03}{0.20} = 0.15$$

What conclusion do you draw? The probability of a promotion given that the officer is a man is 0.30, twice the 0.15 probability of a promotion given that the officer is a woman. Although the use of conditional probability does not in itself prove that discrimination exists in this case, the conditional probability values support the argument presented by the female officers.

Independent events

In the preceding illustration, $P(A) = 0.27$, $P(A \mid M) = 0.30$ and $P(A \mid W) = 0.15$. We see that the probability of a promotion (event A) is affected or influenced by whether the officer is a man or a woman. Particularly, because $P(A \mid M) \neq P(A)$, we would say that events A and M are dependent events. That is, the probability of event A (promotion) is altered or affected by knowing that event M (the officer is a man) exists. Similarly, with $P(A \mid W) \neq P(A)$, we would say that events A and W are *dependent events*. However, if the probability of event A is not changed by the existence of event M – that is, $P(A \mid M) = P(A)$ – we would say that events A and M are **independent events**. More generally, two independent events would satisfy equations (4.9) and (4.10) as follows:

Independent events

Two events A and B are independent if

$$P(A \mid B) = P(A) \tag{4.9}$$

or

$$P(B \mid A) = P(B) \tag{4.10}$$

Multiplication law

Whereas the addition law of probability is used to compute the probability of a union of two events, the multiplication law is used to compute the probability of the intersection of two events. The multiplication law is based on the definition of conditional probability. Using equations (4.7) and (4.8) and solving for $P(A \cap B)$, we obtain the **multiplication law**, as in equations (4.11) and (4.12).

Multiplication law

$$P(A \cap B) = P(A)P(B \mid A) \tag{4.11}$$

or

$$P(A \cap B) = P(B)P(A \mid B) \tag{4.12}$$

To illustrate the use of the multiplication law, consider a newspaper circulation department where it is known that 84 per cent of the households in a particular neighbourhood subscribe to the daily edition of the paper. If we let D denote the event that a household subscribes to the daily edition, then $P(D) = 0.84$. In addition, it is known that the probability that a household that already holds a daily subscription also subscribes to the Sunday edition (event S) is 0.75; that is, $P(S \mid D) = 0.75$.

What is the probability that a household subscribes to both the Sunday and daily editions of the newspaper? Using the multiplication law, we compute the desired $P(S \cap D)$ as:

$$P(S \cap D) = P(D) \, P(S \mid D) = 0.84 \times 0.75 = 0.63$$

We now know that 63 per cent of the households subscribe to both the Sunday and daily editions.

Before concluding this section, let us consider the special case of the multiplication law when the events involved are independent. Recall that events A and B are independent whenever $P(A \mid B) = P(A)$ or $P(B \mid A) = P(B)$. Hence, using equations (4.11) and (4.12) for the special case of independent events, we obtain the following multiplication law (equation (4.13)):

> ### Multiplication law for independent events
>
> $$P(A \cap B) = P(A) \, P(B) \qquad\qquad \textbf{(4.13)}$$

To compute the probability of the intersection of two independent events, we simply multiply the corresponding probabilities. Note that the multiplication law for independent events provides another way to determine whether A and B are independent. That is, if $P(A \cap B) = P(A)P(B)$, then A and B are independent; if $P(A \cap B) \neq P(A)P(B)$, then A and B are dependent.

As an application of the multiplication law for independent events, consider the situation of a service station manager who knows from past experience that 80 per cent of the customers use a credit card when they purchase petrol. What is the probability that the next two customers purchasing petrol will each use a credit card? If we let:

A = the event that the first customer uses a credit card

B = the event that the second customer uses a credit card

then the event of interest is $A \cap B$. Given no other information, we can reasonably assume that A and B are independent events. Hence:

$$P(A \cap B) = P(A)P(B) = 0.80 \times 0.80 = 0.64$$

To summarize this section, we note that our interest in conditional probability is motivated by the fact that events are often related. In such cases, we say the events are dependent and the conditional probability formulae in equations (4.7) and (4.8) must be used to compute the event probabilities. If two events are not related, they are independent; in this case neither event's probability is affected by whether the other event occurred.

EXERCISES

Methods

21. Suppose that we have two events, A and B, with $P(A) = 0.50$, $P(B) = 0.60$ and $P(A \cap B) = 0.40$.
    a. Find $P(A \mid B)$.
    b. Find $P(B \mid A)$.
    c. Are A and B independent? Why or why not?

22. Assume that we have two events, A and B, that are mutually exclusive. Assume further that we know P(A) = 0.30 and P(B) = 0.40.
  a. What is P(A ∩ B)?
  b. What is P(A | B)?
  c. A student in statistics argues that the concepts of mutually exclusive events and independent events are really the same, and that if events are mutually exclusive they must be independent. Do you agree with this statement? Use the probability information provided for this problem to support your answer.
  d. What general conclusion would you make about mutually exclusive and independent events given the results of this problem?

Applications

23. Consider the following initial survey results of 18- to 34-year-olds, in response to the question 'Are you currently living with your family?'

	Yes	No	Totals
Men	106	141	247
Women	92	161	253
Totals	198	302	500

Develop the joint probability table for these data and use it to answer the following questions.
  a. What are the marginal probabilities?
  b. What is the probability of living with family given you are an 18- to 34-year-old man?
  c. What is the probability of living with family given you are an 18- to 34-year-old woman?
  d. What is the probability of an 18- to 34-year-old living with family?
  e. If 49.4 per cent of 18- to 34-year-olds are men, do you consider this a good representative sample? Why?

24. Test marketing involves the duplication of a planned national marketing campaign in a limited geographical area. The test market of a new home sewing machine was conducted among 75 retail outlets, 50 of which were stores affiliated with a major department store chain. In 40 of the retail outlets, the retailer offered customers a free one-year service contract with the purchase of a new sewing machine. Twenty-five of the 50 department stores offered customers a free one-year servicing contract. Suppose a customer buys one of the new sewing machines from a participating test-market retailer:
  a. What is the probability that the customer did not receive a free one-year service contract with the purchase of the sewing machine?
  b. What is the probability that the customer did receive a free one-year service contract but did not buy the sewing machine from a department store?
  c. If it is known that the customer did receive a free one-year service contract with the purchase of the sewing machine, what is the probability that the customer purchased the machine from a department store?

25. A sample of convictions and compensation orders issued at a number of Manx courts was followed up to see whether the offender had paid the compensation to the victim. Details by sex of offender are as follows:

	Payment outcome		
Offender sex	Paid in full	Part paid	Nothing paid
Male	754	62	61
Female	157	7	8

  a. What is the probability that no compensation was paid?
  b. What is the probability that the offender was not male given that compensation was part paid?

26. A purchasing agent in Haifa placed rush orders for a particular raw material with two different suppliers, *A* and *B*. If neither order arrives in four days, the production process must be shut down until at least one of the orders arrives. The probability that supplier *A* can deliver the material in four days is 0.55. The probability that supplier *B* can deliver the material in four days is 0.35.

    a. What is the probability that both suppliers will deliver the material in four days? (Assume supplier deliveries are independent.)

    b. What is the probability that at least one supplier will deliver the material in four days?

    c. What is the probability that the production process will be shut down in four days because of a shortage of raw material (that is, both orders are late)?

4.5 BAYES' THEOREM

In the discussion of conditional probability, we indicated that revising probabilities when new information is obtained is an important phase of probability analysis. Often, we begin the analysis with initial or **prior probability** estimates for specific events of interest. Then, from sources such as a sample, a special report or a product test, we obtain additional information about the events. Given this new information, we update the prior probability values by calculating revised probabilities, referred to as **posterior probabilities**. **Bayes' theorem** provides a means for making these probability calculations. The steps in this probability revision process are shown in Figure 4.9.

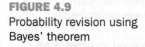

FIGURE 4.9
Probability revision using
Bayes' theorem

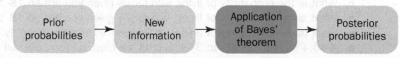

As an application of Bayes' theorem, consider a manufacturing firm that receives shipments of parts from two different suppliers. Let A_1 denote the event that a part is from supplier 1, and A_2 denote the event that a part is from supplier 2. Currently, 65 per cent of the parts purchased by the company are from supplier 1 and the remaining 35 per cent are from supplier 2. Hence, if a part is selected at random, we would assign the prior probabilities $P(A_1) = 0.65$ and $P(A_2) = 0.35$.

The quality of the purchased parts varies with the source of supply. Historical data suggest that the quality ratings of the two suppliers are as shown in Table 4.6.

TABLE 4.6 Historical quality levels of two suppliers

	Percentage good parts	Percentage bad parts
Supplier 1	98	2
Supplier 2	95	5

If we let *G* denote the event that a part is good and *B* denote the event that a part is bad, the information in Table 4.6 provides the following conditional probability values:

$$P(G\,|\,A_1) = 0.98 \quad P(B\,|\,A_1) = 0.02$$

$$P(G\,|\,A_2) = 0.95 \quad P(B\,|\,A_2) = 0.05$$

The tree diagram in Figure 4.10 depicts the process of the firm receiving a part from one of the two suppliers and then discovering that the part is good or bad as a two-step experiment. We see that four experimental outcomes are possible: two correspond to the part being good and two correspond to the part being bad.

FIGURE 4.10

Tree diagram for two-supplier example

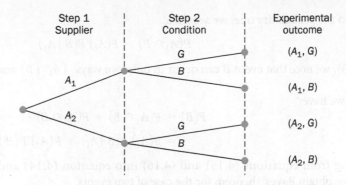

| Step 1 Supplier | Step 2 Condition | Experimental outcome |

Note: Step 1 shows that the part comes from one of two suppliers, and step 2 shows whether the part is good or bad.

Each of the experimental outcomes is the intersection of two events, so we can use the multiplication rule to compute the probabilities. For instance:

$$P(A_1, G) = P(A_1 \cap G) = P(A_1) P(G \mid A_1) = (0.65)(0.98) = 0.6370$$

The process of computing these joint probabilities can be depicted in what is called a probability tree (refer to Figure 4.11). From left to right through the tree, the probabilities for each branch at step 1 are prior probabilities and the probabilities for each branch at step 2 are conditional probabilities. To find the probabilities of each experimental outcome, we simply multiply the probabilities on the branches leading to the outcome. Each of these joint probabilities is shown in Figure 4.11 along with the known probabilities for each branch.

FIGURE 4.11

Probability tree for two-supplier example

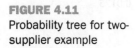
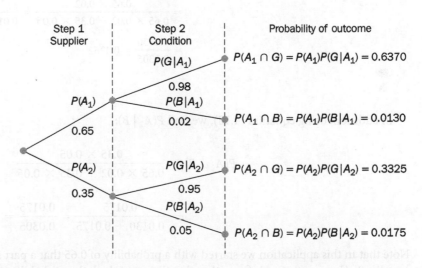

Suppose now that the parts from the two suppliers are used in the firm's manufacturing process and that a machine breaks down because it attempts to process a bad part. Given the information that the part is bad, what is the probability that it came from supplier 1 and what is the probability that it came from supplier 2? With the information in the probability tree (Figure 4.11), Bayes' theorem can be used to answer these questions.

Letting B denote the event that the part is bad, we are looking for the posterior probabilities $P(A_1 \mid B)$ and $P(A_2 \mid B)$. From the law of conditional probability, we know that:

$$P(A_1 \mid B) = \frac{P(A_1 \cap B)}{P(B)}$$

(4.14)

Referring to the probability tree, we see that:

$$P(A_1 \cap B) = P(A_1)\, P(B \,|\, A_1) \tag{4.15}$$

To find $P(B)$, we note that event B can occur in only two ways: $(A_1 \cap B)$ and $(A_2 \cap B)$.

Therefore, we have:

$$P(B) = P(A_1 \cap B) + P(A_2 \cap B)$$
$$= P(A_1)\, P(B \,|\, A_1) + P(A_2)\, P(B \,|\, A_2) \tag{4.16}$$

Substituting from equations (4.15) and (4.16) into equation (4.14) and writing a similar result for $P(A_2 \,|\, B)$, we obtain Bayes' theorem for the case of two events.

Bayes' theorem (two-event case)

$$P(A_1 \,|\, B) = \frac{P(A_1)\, P(B \,|\, A_1)}{P(A_1)\, P(B \,|\, A_1) + P(A_2)\, P(B \,|\, A_2)} \tag{4.17}$$

$$P(A_2 \,|\, B) = \frac{P(A_2)\, P(B \,|\, A_1)}{P(A_1)\, P(B \,|\, A_1) + P(A_2)\, P(B \,|\, A_2)} \tag{4.18}$$

Using equation (4.17) and the probability values provided in the example, we have:

$$P(A_1 \,|\, B) = \frac{P(A_1)\, P(B \,|\, A_1)}{P(A_1)\, P(B \,|\, A_1) + P(A_2)\, P(B \,|\, A_2)}$$

$$= \frac{0.65 \times 0.02}{0.65 \times 0.02 + 0.35 \times 0.05} = \frac{0.0130}{0.0130 + 0.0175}$$

$$= \frac{0.0130}{0.0305} = 0.4262$$

In addition, using equation (4.18), we find $P(A_2 \,|\, B)$:

$$P(A_2 \,|\, B) = \frac{0.35 \times 0.05}{0.65 \times 0.02 + 0.35 \times 0.05}$$

$$= \frac{0.0175}{0.0130 + 0.0175} = \frac{0.0175}{0.0305} = 0.5738$$

Note that in this application we started with a probability of 0.65 that a part selected at random was from supplier 1. However, given information that the part is bad, the probability that the part is from supplier 1 drops to 0.4262. In fact, if the part is bad, it has better than a 50–50 chance that it came from supplier 2; that is, $P(A_2 \,|\, B) = 0.5738$.

Bayes' theorem is applicable when the events for which we want to compute posterior probabilities are mutually exclusive and their union is the entire sample space.* For the case of n mutually exclusive events $A_1, A_2, \ldots, A_n$, whose union is the entire sample space, Bayes' theorem can be used to compute any posterior probability $P(A_i \,|\, B)$ as shown in equation (4.19).

---

* If the union of events is the entire sample space, the events are said to be collectively exhaustive.

Bayes' theorem

$$P(A_i \mid B) = \frac{P(A_i)\, P(B \mid A_i)}{P(A_1)\, P(B \mid A_1) + P(A_2)\, P(B \mid A_2) + \ldots + P(A_n)\, P(B \mid A_n)} \qquad (4.19)$$

With prior probabilities $P(A_1)$, $P(A_2)$, ... , $P(A_n)$ and the appropriate conditional probabilities $P(B \mid A_1)$, $P(B \mid A_2)$, ... , $P(B \mid A_n)$, equation (4.19) can be used to compute the posterior probability of the events A_1, A_2, ... , A_n.

Tabular approach

A tabular approach is helpful in conducting the Bayes' theorem calculations. Such an approach is shown in Table 4.7 for the parts supplier problem. The computations shown there are conducted as follows:

Step 1 Prepare the following three columns:
  Column 1 – The mutually exclusive events A_i for which posterior probabilities are desired.
  Column 2 – The prior probabilities $P(A_i)$ for the events.
  Column 3 – The conditional probabilities $P(B \mid A_i)$ of the new information B given each event.

Step 2 In column 4, compute the joint probabilities $P(A_i \cap B)$ for each event and the new information B by using the multiplication law. These joint probabilities are found by multiplying the prior probabilities in column 2 by the corresponding conditional probabilities in column 3: that is, $P(A_i \cap B) = P(A_i)P(B \mid A_i)$.

Step 3 Sum the joint probabilities in column 4. The sum is the probability of the new information, $P(B)$. Hence we see in Table 4.7 that there is a 0.0130 probability that the part came from supplier 1 and is bad and a 0.0175 probability that the part came from supplier 2 and is bad. Because these are the only two ways in which a bad part can be obtained, the sum $0.0130 + 0.0175$ shows an overall probability of 0.0305 of finding a bad part from the combined shipments of the two suppliers.

Step 4 In column 5, compute the posterior probabilities using the basic relationship of conditional probability:

$$P(A_i \mid B) = \frac{P(A_i \cap B)}{P(B)}$$

Note that the joint probabilities $P(A_i \cap B)$ are in column (4) and the probability $P(B)$ is the sum of column (4).

TABLE 4.7 Tabular approach to Bayes' theorem calculations for the two-supplier problem

(1) Events A_i	(2) Prior probabilities $P(A_i)$	(3) Conditional probabilities $P(B \mid A_i)$	(4) Joint probabilities $P(A_i \cap B)$	(5) Posterior probabilities $P(A_i \mid B)$
A_1	0.65	0.02	0.0130	0.0130/0.0305 = 0.4262
A_2	0.35	0.05	0.0175	0.0175/0.0305 = 0.5738
			$P(B) = 0.0305$	1.0000

EXERCISES

Methods

27. The prior probabilities for events A_1 and A_2 are $P(A_1) = 0.40$ and $P(A_2) = 0.60$. It is also known that $P(A_1 \cap A_2) = 0$. Suppose $P(B \mid A_1) = 0.20$ and $P(B \mid A_2) = 0.05$.
   a. Are A_1 and A_2 mutually exclusive? Explain.
   b. Compute $P(A_1 \cap B)$ and $P(A_2 \cap B)$.
   c. Compute $P(B)$.
   d. Apply Bayes' theorem to compute $P(A_1 \mid B)$ and $P(A_2 \mid B)$.

28. The prior probabilities for events A_1, A_2 and A_3 are $P(A_1) = 0.20$, $P(A_2) = 0.50$ and $P(A_3) = 0.30$. The conditional probabilities of event B given A_1, A_2 and A_3 are $P(B \mid A_1) = 0.50$, $P(B \mid A_2) = 0.40$ and $P(B \mid A_3) = 0.30$.
   a. Compute $P(B \cap A_1)$, $P(B \cap A_2)$ and $P(B \cap A_3)$.
   b. Apply Bayes' theorem, equation (4.19), to compute the posterior probability $P(A_2 \mid B)$.
   c. Use the tabular approach to applying Bayes' theorem to compute $P(A_1 \mid B)$, $P(A_2 \mid B)$ and $P(A_3 \mid B)$.

Applications

29. A company is about to sell to a new client. It knows from past experience that there is a real possibility that the client may default on payment. As a precaution, the company checks with a consultant on the likelihood of the client defaulting in this case and is given an estimate of 20 per cent. Sometimes the consultant gets it wrong. Your own experience of the consultant is that he is correct 70 per cent of the time when he predicts that the client will default but that 20 per cent of clients whom he believes will not default actually do.
   What is the probability that the new client will not default?

30. A consulting firm submitted a bid for a large research project. The firm's management initially felt they had a 50:50 chance of winning the contract. However, the agency to which the bid was submitted subsequently requested additional information on the bid. Past experience indicates that for 75 per cent of the successful bids and 40 per cent of the unsuccessful bids the agency requested additional information.
   a. What is the prior probability of the bid being successful (that is, prior to the request for additional information)?
   b. What is the conditional probability of a request for additional information given that the bid will ultimately be successful?
   c. Compute the posterior probability that the bid will be successful given a request for additional information.

31. A bank reviewed its credit card policy with the intention of recalling some of its credit cards. In the past approximately 5 per cent of cardholders defaulted, leaving the bank unable to collect the outstanding balance. Hence, management established a prior probability of 0.05 that any particular cardholder will default. The bank also found that the probability of missing a monthly payment is 0.20 for customers who do not default. Of course, the probability of missing a monthly payment for those who default is 1.
   a. Given that a customer missed one or more monthly payments, compute the posterior probability that the customer will default.
   b. The bank would like to recall its card if the probability that a customer will default is greater than 0.20. Should the bank recall its card if the customer misses a monthly payment? Why or why not?

32. An electronic component is produced by four production lines in a manufacturing operation. The components are costly, are quite reliable and are shipped to suppliers in 50-component lots. Because testing is destructive, most buyers of the components test only a small number before deciding to accept or reject lots of incoming components. All four production lines usually only produce 1 per cent defective components, which are randomly dispersed in the output. Unfortunately, production line 1 suffered a mechanical difficulty and produced 10 per cent defectives during the month of April. This situation became known to the manufacturer after the components had been shipped. A customer received a lot in April and tested five components. Two failed. What is the probability that this lot came from production line 1?

ONLINE RESOURCES

For the data files, additional questions and answers, and the software section for Chapter 4, go to the online platform.

SUMMARY

In this chapter we introduced basic probability concepts and illustrated how probability analysis can be used to provide helpful information for decision making. We described how probability can be interpreted as a numerical measure of the likelihood that an event will occur and reviewed classical, relative frequency and subjective methods for deriving probabilities. In addition, we saw that the probability of an event can be computed either by summing the probabilities of the experimental outcomes (sample points) comprising the event or by using the relationships established by the addition, conditional probability, and multiplication laws of probability. For cases in which new information is available, we showed how Bayes' theorem can be used to obtain revised or posterior probabilities.

KEY TERMS

Addition law	Multiplication law
Basic requirements for assigning probabilities	Mutually exclusive events
Bayes' theorem	Permutation
Classical method	Posterior probabilities
Combination	Prior probability
Complement of A	Probability
Conditional probability	Relative frequency method
Event	Sample point
Experiment	Sample space
Independent events	Subjective method
Intersection of A and B	Tree diagram
Joint probability	Union of A and B
Marginal probability	Venn diagram

KEY FORMULAE

Counting rule for combinations

$$^{N}C_{n} = \binom{N}{n} = \frac{N!}{n!(N-n)!} \tag{4.1}$$

Counting rule for permutations

$$^N P_n = n! \binom{N}{n} = \frac{N!}{(N-n)!} \qquad (4.2)$$

Computing probability using the complement

$$P(A) = 1 - P(\overline{A}) \qquad (4.5)$$

Addition law

$$P(A \cup B) = P(A) + P(B) - P(A \cap B) \qquad (4.6)$$

Conditional probability

$$P(A \mid B) = \frac{P(A \cap B)}{P(B)} \qquad (4.7)$$

$$P(B \mid A) = \frac{P(A \cap B)}{P(A)} \qquad (4.8)$$

Multiplication law

$$P(A \cap B) = P(B)\, P(A \mid B) \qquad (4.11)$$

$$P(A \cap B) = P(A)\, P(B \mid A) \qquad (4.12)$$

Multiplication law for independent events

$$P(A \cap B) = P(A)\, P(B) \qquad (4.13)$$

Bayes' theorem

$$P(A_i \mid B) = \frac{P(A_i)\, P(B \mid A_i)}{P(A_1)\, P(B \mid A_1) + P(A_2)\, P(B \mid A_2) + \ldots + P(A_n)\, P(B \mid A_n)} \qquad (4.19)$$

CASE PROBLEM

BAC and the alcohol test

In 2019, 7.7 per cent (32 of the 416) road deaths recorded in Austria were attributed to alcohol. The police in Wachau, Austria, a region which is famous for its wine production, is interested in buying equipment to test drivers' blood alcohol levels. The law in Austria requires that the driver's licence be withdrawn if the driver is found to have more than 0.05 per cent BAC (blood alcohol concentration).

Due to the large number of factors that come into play regarding the consumption and reduction (burn-off) rates of different people, there is no blood alcohol calculator that is 100 per cent accurate. Factors include the sex of the drinker, differing metabolism rates, various health issues and the combination of medications being taken, drinking frequency, amount of food in the stomach and small intestine and when it was eaten, elapsed time and many others. The best that can be done is a rough estimate of the BAC level based on known inputs.

There are three types of equipment available with the following conditions:

1. The Saliva Screen is a disposable strip which can be used once – this is the cheapest method.

2. The Alcometer™ is an instrument attached to a container into which the driver breathes, with the Alcometer™ then measuring the BAC concentration through an analysis of the driver's breath. The drawback to the Alcometer™ is that it can only detect the alcohol level correctly if it is used within two hours of alcohol consumption. It is less effective if used beyond this two-hour period.

3. The Intoximeter is the most expensive of the three and it works through a blood sample of the driver. The advantage for this is that it can test the BAC up to 12 hours after alcohol consumption. False positive is the situation where the test indicates a high BAC level in a driver that actually does not have such a level. The false negative is when the test indicates a low level of BAC when the driver is actually highly intoxicated.

Type	False positive	False negative
Saliva Screen	0.020	0.03
Alcometer™	0.015	0.02
Intoximeter	0.020	0.01

Police records show that the percentage of drivers (late night) that drink heavily and drive, ranges between 6 per cent on weekdays and 10 per cent at the weekend.

Managerial report

Carry out an appropriate probability analysis of this information on behalf of the police and advise them accordingly. (Note that it would be particularly helpful if you could assess the effectiveness of the different equipment types separately for weekdays and the weekend.)

Case problem provided by Dr Ibrahim Wazir, Webster University, Vienna

© SashaFoxWalters/iStock

Source: etsc.eu/issues/drink-driving/drink-driving-in-austria/

5

Discrete Probability Distributions

CHAPTER CONTENTS

Statistics in Practice Improving the performance reliability of combat aircraft

LEARNING OBJECTIVES After reading this chapter and doing the exercises, you should be able to:

1 Understand the concepts of a random variable and a probability distribution.

2 Distinguish between discrete and continuous random variables.

3 Compute and interpret the expected value, variance and standard deviation for a discrete random variable.

4 Compute and work with probabilities, covariance and correlation for a bivariate probability distribution.

5 Compute and work with probabilities involving a binomial probability distribution.

6 Compute and work with probabilities involving Poisson probability distribution.

7 Know when and how to use the hypergeometric probability distribution.

In this chapter we continue the study of probability by introducing the concepts of random variables and probability distributions. The focus of this chapter is discrete probability distributions. Three special discrete probability distributions – the binomial, Poisson and hypergeometric – are covered.

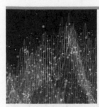

STATISTICS IN PRACTICE
Improving the performance
reliability of combat aircraft

Modern combat aircraft are expensive to acquire and maintain. For example, in 2021 the Eurofighter Typhoon and Lockheed Martin F35-B – successors to the recently decommissioned RAF Tornado fleet – were ranked the two most expensive fighter jets in the world. From a national perspective, the emphasis is therefore on deploying as few aircraft as are required and for these to be made to perform as reliably as possible in conflict and peace-keeping situations. Different strategies have been considered by manufacturers for improving the performance reliability of aircraft. One strategy is to build 'redundancy' into the design. In practice this would involve the aircraft carrying additional engines which would only come into use if one of the operational engines failed. To determine the number of additional engines required, designers have relied on the Poisson distribution. Calculations based on this distribution show that an aircraft with two engines

would need at least four redundant engines to achieve a target maintenance-free operating period (MFOP) of 150 hours. Given that each engine weighs over a tonne, occupies a space of at least 2m³ and costs some €3m, clearly this has enormous implications for those wishing to pursue this solution further.

© Luke Webster/iStock

Sources:
www.aerotime.aero/articles/27553-Top-10-most-expensive-fighter-jets-in-2021.
Kumar, U. D., Knezevic, J. and Crocker, J. (1999) 'Maintenance-free operating period: An alternative measure to MTBF and failure rate for specifying reliability'. *Reliability Engineering & System Safety*, 64: 127–131.
Shaalane, A. A. and Vlok, P. J. (2013) 'Application of the aviation derived maintenance free operating period concept in the South African mining industry'. *South African Journal of Industrial Engineering*, 24(3): 150–165.

5.1 RANDOM VARIABLES

In Chapter 4 we defined the concept of an experiment and its associated experimental outcomes. A **random variable** provides a means for describing experimental outcomes using numerical values. Random variables must assume numerical values.

> **Random variable**
>
> A random variable is a numerical description of the outcome of an experiment.

In effect, a random variable associates a numerical value with each possible experimental outcome. The particular numerical value of the random variable depends on the outcome of the experiment. A random variable can be classified as being either *discrete* or *continuous* depending on the numerical values it assumes.

Discrete random variables

A random variable that may assume either a finite number of values or an infinite – but countable – sequence of values such as 0, 1, 2, … is referred to as a **discrete random variable**. For example, consider the experiment of an accountant taking the chartered accountancy (CA) examination.

The examination has four parts. We can define a random variable as $X =$ the number of parts of the CA examination passed. It is a discrete random variable because it may assume the finite number of values 0, 1, 2, 3 or 4.

As another example of a discrete random variable, consider the experiment of cars arriving at a tollbooth. The random variable of interest is $X =$ the number of cars arriving during a one-day period.

The possible values for X come from the sequence of integers 0, 1, 2 and so on. Hence, X is a discrete random variable assuming one of the values in this infinite sequence. Although the outcomes of many experiments can naturally be described by numerical values, others cannot. For example, a survey question might ask an individual to recall the message in a recent online commercial. This experiment would have two possible outcomes: the individual cannot recall the message and the individual can recall the message.

We can still describe these experimental outcomes numerically by defining the discrete random variable X as follows: let $X = 0$ if the individual cannot recall the message and $X = 1$ if the individual can recall the message. The numerical values for this random variable are arbitrary (we could use 5 and 10), but they are acceptable in terms of the definition of a random variable – namely, X is a random variable because it provides a numerical description of the outcome of the experiment.

Table 5.1 provides some additional examples of discrete random variables. Note that in each example the discrete random variable assumes a finite number of values or an infinite sequence of values such as 0, 1, 2, ... These types of discrete random variable are discussed in detail in this chapter.

TABLE 5.1 Examples of discrete random variables

Experiment	Random variable (X)	Possible values for the random variable
Contact five customers	Number of customers who place an order	0, 1, 2, 3, 4, 5
Inspect a shipment of 50 radios	Number of defective radios	0, 1, 2, ..., 49, 50
Operate a restaurant for one day	Number of customers	0, 1, 2, 3, ...
Sell a car	Sex of the customer	0 if male; 1 if female

Continuous random variables

A random variable that may assume any numerical value in an interval or collection of intervals is called a **continuous random variable**. Experimental outcomes based on measurement scales such as time, weight, distance and temperature can be described by continuous random variables. For example, consider an experiment of monitoring incoming telephone calls to the claims office of a major insurance company. Suppose the random variable of interest is $X =$ the time between consecutive incoming calls in minutes. This random variable may assume any value in the interval $X \geq 0$. Actually, an infinite number of values are possible for X, including values such as 1.26 minutes, 2.751 minutes, 4.3333 minutes and so on. As another example, consider a 90-kilometre section of the A8 Autobahn in Germany.

For an emergency ambulance service located in Stuttgart, we might define the random variable as $X =$ number of kilometres to the location of the next traffic accident along this section of the A8. In this case, X would be a continuous random variable assuming any value in the interval $0 \leq X \leq 90$. Additional examples of continuous random variables are listed in Table 5.2. Note that each example describes a random variable that may assume any value in an interval of values. Continuous random variables and their probability distributions will be the topic of Chapter 6.

TABLE 5.2 Examples of continuous random variables

Experiment	Random variable (X)	Possible values for the random variable
Operate a bank	Time between customer arrivals	$X \geq 0$ in minutes
Fill a soft drink can (max = 350g)	Number of grams	$0 \leq X \leq 350$
Construct a new library	Percentage of construction complete after six months	$0 \leq X \leq 100$
Test a new chemical process	Temperature when the desired reaction achieved	$65 \leq X \leq 100$

EXERCISES

Methods

1. Consider the experiment of tossing a coin twice.
   a. List the experimental outcomes.
   b. Define a random variable that represents the number of heads occurring on the two tosses.
   c. Show what value the random variable would assume for each of the experimental outcomes.
   d. Is this random variable discrete or continuous?

2. Consider the experiment of a worker assembling a product.
   a. Define a random variable that represents the time in minutes required to assemble the product.
   b. What values may the random variable assume?
   c. Is the random variable discrete or continuous?

Applications

3. Three students have interviews scheduled for summer employment. In each case the interview results in either an offer for a position or no offer. Experimental outcomes are defined in terms of the results of the three interviews.
   a. List the experimental outcomes.
   b. Define a random variable that represents the number of offers made. Is the random variable continuous?
   c. Show the value of the random variable for each of the experimental outcomes.

4. Suppose we know home mortgage rates for 12 Danish lending institutions. Assume that the random variable of interest is the number of lending institutions in this group that offers a long-term rate of 1.38 per cent or less. What values may this random variable assume?

5. To perform a certain type of blood analysis, lab technicians must perform two procedures. The first procedure requires either one or two separate steps, and the second procedure requires either one, two or three steps.
   a. List the experimental outcomes associated with performing the blood analysis.
   b. If the random variable of interest is the total number of steps required to do the complete analysis (both procedures), show what value the random variable will assume for each of the experimental outcomes.

6. Listed is a series of experiments and associated random variables. In each case, identify the values that the random variable can assume and state whether the random variable is discrete or continuous.

Experiment	Random variable (X)
a. Take a 20-question examination.	Number of questions answered correctly.
b. Observe cars arriving at a tollbooth for one hour.	Number of cars arriving at tollbooth.
c. Audit 50 tax returns.	Number of returns containing errors.
d. Observe an employee's work.	Number of non-productive hours in an eight-hour workday.
e. Weigh a shipment of goods.	Number of kilograms.

5.2 DISCRETE PROBABILITY DISTRIBUTIONS

The probability distribution for a random variable describes how probabilities are distributed over the values of the random variable. For a discrete random variable X, the probability distribution is defined by a probability function, denoted by $p(x) = p(X = x)$ for all possible values, x. The probability function provides the probability for each value of the random variable. Consider the sales of cars at DiCarlo Motors in Sienna, Italy. Over the past 300 days of operation, sales data show 54 days with no cars sold, 117 days with 1 car sold, 72 days with 2 cars sold, 42 days with 3 cars sold, 12 days with 4 cars sold and 3 days with 5 cars sold. Suppose we consider the experiment of selecting a day of operation at DiCarlo Motors and define the random variable of interest as X = the number of cars sold during a day. Using the relative frequencies to assign probabilities to the values of the random variable X, we can develop the probability distribution for X. From historical data, we know X is a discrete random variable that can assume the values 0, 1, 2, 3, 4 or 5. In probability function notation, $p(0)$ provides the probability of 0 cars sold, $p(1)$ provides the probability of one car sold and so on. Because historical data show 54 of 300 days with no cars sold, we assign the relative frequency 54/300 = 0.18 to $p(0)$, indicating that the probability of no cars being sold during a day is 0.18. Similarly, because 117 of 300 days had one car sold, we assign the relative frequency 117/300 = 0.39 to $p(1)$, indicating that the probability of exactly one car being sold during a day is 0.39. Continuing in this way for the other values of the random variable, we compute the values for $p(2)$, $p(3)$, $p(4)$ and $p(5)$ as shown in Table 5.3, the probability distribution for the number of cars sold during a day at DiCarlo Motors. Because Table 5.3 was derived using the relative frequency method it can be described as an **empirical discrete distribution**.

TABLE 5.3
Probability distribution for the number of cars sold during a day at DiCarlo Motors

x	p (x)
0	0.18
1	0.39
2	0.24
3	0.14
4	0.04
5	0.01
	Total 1.00

A primary advantage of defining a random variable and its probability distribution is that once the probability distribution is known, it is relatively easy to determine the probability of a variety of events that may be of interest to a decision-maker. For example, using the probability distribution for DiCarlo Motors, as shown in Table 5.3, we see that the most probable number of cars sold during a day is one with a probability of $p(1) = 0.39$. In addition, there is a $p(3) + p(4) + p(5) = 0.14 + 0.04 + 0.01 = 0.19$ probability of selling three or more cars during a day. These probabilities, plus others the decision-maker may ask about, provide information that can help the decision-maker understand the process of selling cars at DiCarlo Motors.

In the development of a probability function for any discrete random variable, the following two conditions must be satisfied.

$$p(x) \geq 0 \tag{5.1}$$

$$\Sigma p(x) = 1 \tag{5.2}$$

Table 5.3 shows that the probabilities for the random variable X satisfy equation (5.1); $p(x)$ is greater than or equal to 0 for all values of x. In addition, because the probabilities sum to 1, equation (5.2) is satisfied. Hence, the DiCarlo Motors probability function is a valid discrete probability function.

We can also present probability distributions graphically. In Figure 5.1 the values of the random variable X for DiCarlo Motors are shown on the horizontal axis and the probability associated with these values is shown on the vertical axis. In addition to tables and graphs, a formula that gives the probability function, $p(x)$, for every value of $X = x$ is often used to describe probability distributions. The simplest example of a discrete probability distribution given by a formula is the **discrete uniform probability distribution**. Its probability function is defined by equation (5.3).

FIGURE 5.1

Graphical representation of the
probability distribution for the number
of cars sold during a day at
DiCarlo Motors

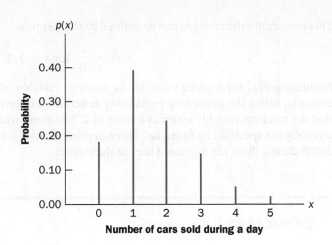

Discrete uniform probability function

$$p(x) = 1/n \qquad\qquad (5.3)$$

where:

$n = $ the number of values the random variable may assume

For example, suppose that for the experiment of rolling a die we define the random variable X to be the number of dots on the upward face. There are $n = 6$ possible values for the random variable: $X = 1, 2, 3, 4, 5, 6$. Hence, the probability function for this discrete uniform random variable is:

$$p(x) = 1/6 \quad x = 1, 2, 3, 4, 5, 6$$

The possible values of the random variable and the associated probabilities are as follows:

x	$p(x)$
1	1/6
2	1/6
3	1/6
4	1/6
5	1/6
6	1/6

As another example, consider the random variable X with the following discrete probability distribution:

x	$p(x)$
1	1/10
2	2/10
3	3/10
4	4/10

This probability distribution can be defined by the formula:

$$p(x) = \frac{x}{10} \quad \text{for } x = 1, 2, 3 \text{ or } 4$$

Evaluating $p(x)$ for a given value of the random variable will provide the associated probability. For example, using the preceding probability function, we see that $p(2) = 2/10$ provides the probability that the random variable assumes a value of 2. The more widely used discrete probability distributions generally are specified by formulae. Three important cases are the binomial, Poisson and hypergeometric distributions; these are discussed later in the chapter.

EXERCISES

Methods

7. The probability distribution for the random variable X follows.

X	$p(x)$
20	0.20
25	0.15
30	0.25
35	0.40

   a. Is this probability distribution valid? Explain.
   b. What is the probability that $X = 30$?
   c. What is the probability that X is less than or equal to 25?
   d. What is the probability that X is greater than 30?

Applications

8. The following data were collected by counting the number of operating rooms in use at a general hospital over a 20-day period. On three of the days only one operating room was used, on five of the days two were used, on eight of the days three were used and on four days all four of the hospital's operating rooms were used.
   a. Use the relative frequency approach to construct a probability distribution for the number of operating rooms in use on any given day.
   b. Draw a graph of the probability distribution.
   c. Show that your probability distribution satisfies the required conditions for a valid discrete probability distribution.

9. In a monthly charity competition, CAD\$10,000 is awarded for every 2,666,667 bonds (each costing CAD\$1) in the competition. The prize money is divided into 295 prizes with the following values:

CAD\$	Number of prizes
1,000	1
500	1
250	2
100	3
50	20
25	268

a. What is the probability of winning a prize with one bond in each draw?
b. If a prize is won, what is the probability that it is one for CAD$100 or more?
c. What is the expected gain each month for each bond?
d. What annual rate of interest does this correspond to?

10. A technician services magnetic resonance imaging (MRI) machines at hospitals in the Trieste area. Depending on the type of malfunction, the service call can take 1, 2, 3 or 4 hours. The different types of malfunctions of the MRI machines occur at about the same frequency.
   a. Develop a probability distribution for the duration of a service call.
   b. Draw a graph of the probability distribution.
   c. Show that your probability distribution satisfies the conditions required for a discrete probability function.
   d. What is the probability a service call for an MRI machine will take three hours?
   e. A service call for an MRI machine has just come in, but the type of malfunction is unknown. It is 3.00 p.m. and service technicians usually leave work at 5.00 p.m. What is the probability the service technician will have to work overtime to fix the machine today?

11. Let X be a random variable indicating the number of sessions required by a psychologist to gain a patient's trust. Then it has been determined that X takes the value of either 1, 2 or 3. The following probability function has been proposed for X:

$$p(x) = \frac{x}{6} \quad \text{for } x = 1, 2 \text{ or } 3$$

   a. Is this probability function valid? Explain.
   b. What is the probability that $X = 2$?
   c. What is the probability that at least two sessions are required to gain the patient's trust?

12. The following table represents an incomplete probability distribution for the MRA Company's projected profits (X = profit in €000s) for the first year of operation (the negative value denotes a loss).

x	$p(x)$
−100	0.10
0	0.20
50	0.30
100	0.25
150	0.10
200	

   a. What is the proper value for $p(200)$? What is your interpretation of this value?
   b. What is the probability that MRA will be profitable?
   c. What is the probability that MRA will make at least €100,000?

5.3 EXPECTED VALUE AND VARIANCE

Expected value

The expected value, or mean, of a random variable is a measure of the central location for the random variable. The formula for the expected value of a discrete random variable X follows in equation (5.4).

Expected value of a discrete random variable

$$E(X) = \mu = \Sigma x p(x) \tag{5.4}$$

Both the notations $E(X)$ and μ are used to denote the expected value of a random variable. Equation (5.4) shows that to compute the expected value of a discrete random variable, we must multiply each value of the random variable by the corresponding probability $p(x)$ and then add the resulting products. Using the DiCarlo Motors car sales example from Section 5.2, we show the calculation of the expected value for the number of cars sold during a day in Table 5.4. The sum of the entries in the $xp(x)$ column shows that the expected value is 1.50 cars per day. We therefore know that although sales of 0, 1, 2, 3, 4 or 5 cars are possible on any one day, over time DiCarlo can anticipate selling an average of 1.50 cars per day. Assuming 30 days of operation during a month, we can use the expected value of 1.50 to forecast average monthly sales of $30(1.50) = 45$ cars.

TABLE 5.4 Calculation of the expected value for the number of cars sold during a day at DiCarlo Motors

x	$p(x)$	$xp(x)$
0	0.18	0 (0.18) = 0.00
1	0.39	1 (0.39) = 0.39
2	0.24	2 (0.24) = 0.48
3	0.14	3 (0.14) = 0.42
4	0.04	4 (0.04) = 0.16
5	0.01	5 (0.01) = 0.05
		1.50

$$E(X) = \mu = \Sigma xp(x)$$

Variance

Even though the expected value provides the mean value for the random variable, we often need a measure of variability, or dispersion. Just as we used variance in Chapter 3 to summarize the variability in data, we now use variance to summarize the variability in the values of a random variable. The formula for the variance of a discrete random variable follows in equation (5.5).

Variance of a discrete random variable

$$\text{Var}(X) = \sigma^2 = \Sigma(x - \mu)^2 p(x) \tag{5.5}$$

As equation (5.5) shows, an essential part of the variance formula is the deviation, $x - \mu$, which measures how far a particular value of the random variable is from the expected value, or mean, μ. In computing the variance of a random variable, the deviations are squared and then weighted by the corresponding value of the probability function. The sum of these weighted squared deviations for all values of the random variable is referred to as the *variance*. The notations $\text{Var}(X)$ and σ^2 are both used to denote the variance of a random variable.

The calculation of the variance for the probability distribution of the number of cars sold during a day at DiCarlo Motors is summarized in Table 5.5. We see that the variance is 1.25. The standard deviation, σ, is the positive square root of the variance. Hence, the standard deviation for the number of cars sold during a day is:

$$\sigma = \sqrt{1.25} = 1.118$$

The standard deviation is measured in the same units as the random variable ($\sigma = 1.118$ cars) and therefore is often preferred in describing the variability of a random variable. The variance σ^2 is measured in squared units and is thus more difficult to interpret.

TABLE 5.5 Calculation of the variance for the number of cars sold during a day at DiCarlo Motors

x	x − μ	(x − μ)²	p(x)	(x − μ)²p(x)
0	0 − 1.50 = −1.50	2.25	0.18	2.25 × 0.18 = 0.4050
1	1 − 1.50 = −0.50	0.25	0.39	0.25 × 0.39 = 0.0975
2	2 − 1.50 = 0.50	0.25	0.24	0.25 × 0.24 = 0.0600
3	3 − 1.50 = 1.50	2.25	0.14	2.25 × 0.14 = 0.3150
4	4 − 1.50 = 2.50	6.25	0.04	6.25 × 0.04 = 0.2500
5	5 − 1.50 = 3.50	12.25	0.01	12.25 × 0.01 = 0.1225
				1.2500

$$\sigma^2 = \Sigma(x - \mu)^2 p(x)$$

EXERCISES

Methods

13. The following table provides a probability distribution for the random variable X.

x	p(x)
3	0.25
6	0.50
9	0.25

a. Compute $E(X)$, the expected value of X.
b. Compute σ^2, the variance of X.
c. Compute σ, the standard deviation of X.

14. The following table provides a probability distribution for the random variable Y.

y	p(y)
2	0.20
4	0.30
7	0.40
8	0.10

a. Compute $E(Y)$.
b. Compute $Var(Y)$ and σ.

Applications

15. The following data have been collected on the number of times owner-occupied and renter-occupied units suffered a serious water supply stoppage in the past three months.

	Number of units (000)	
Number of times	Owner occupied	Renter occupied
0	439	394
1	1,100	760
2	249	221
3	98	92
4+	120	111

a. Define a random variable X = number of times that owner-occupied units experienced a serious water supply stoppage in the past three months and develop a probability distribution for the random variable. (Let $X = 4$ represent 4+ times.)

b. Compute the expected value and variance for X.

c. Define a random variable Y = number of times that renter-occupied units had a serious water supply stoppage in the past three months and develop a probability distribution for the random variable. (Let $Y = 4$ represent 4+ times.)

d. Compute the expected value and variance for Y.

e. What observations can you make from a comparison of the number of water supply stoppages reported by owner-occupied units versus renter-occupied units?

16. Odds in horse race betting are defined as follows: 3/1 (three to one against) means a horse is expected to win once for every three times it loses; 3/2 means two wins out of five races; 4/5 (five to four on) means five wins for every four defeats, etc.

a. Translate the above odds into 'probabilities' of victory.

b. In the 2.45 race at L'Arc de Triomphe the odds for the five runners were:

Phillipe Bois	1/1
Gallante Effort	5/2
Satin Noir	11/2
Victoire Antheme	9/1
Comme Rambleur	16/1

Calculate the 'probabilities' and their sum.

c. How much would a bookmaker expect to profit in the long run at such odds if it is assumed each horse is backed equally? (Hint: Assume the true probabilities are proportional to the 'probabilities' just calculated and consider the payouts corresponding to a notional €1 wager being placed on each horse.)

d. What would the bookmaker's expected profit have been if horses had been backed in line with the true probabilities?

17. A machinist works an eight-hour shift. An efficiency expert wants to assess the value of this machinist where value is defined as value added minus the machinist's labour cost. The value added for the work the machinist does is €30 per item and the machinist earns €16 per hour. From past records, the machinist's output per shift is known to have the following probability distribution:

Output/shift	Probability
5	0.2
6	0.4
7	0.3
8	0.1

a. What is the expected monetary value of the machinist to the company per shift?

b. What is the corresponding variance value?

18. A company is contracted to finish a €100,000 project by 31 December. If it does not complete on time, a penalty of €8,000 per month (or part of a month) is incurred. The company estimates that if it continues alone there will be a 40 per cent chance of completing on time and that the project may be one, two, three or four months late with equal probability.

Subcontractors can be hired by the firm at a cost of €18,000. If the subcontractors are hired then the probability that the company completes on time is doubled. If the project is still late it will now be only one or two months late with equal probability.

a. Determine the expected profit when:
  i. subcontractors are not used
  ii. subcontractors are used
b. Which is the better option for the company?

19. A typical slot machine has 3 dials, each with 20 symbols (cherries, plums, lemons, oranges, bells and bars). A set of dials is configured as follows:

	Dial		
	1	2	3
Cherries	7	7	0
Oranges	3	7	6
Lemons	3	0	4
Plums	4	1	6
Bells	2	2	3
Bars	1	3	1
	20	20	20

According to this table, of the 20 slots on dial 1, 7 are cherries, 3 oranges, etc.

Payoffs (€) for a €1 bet are given below:

	Dial		
1	2	3	Payoff (€)
Bar	Bar	Bar	60
Bell	Bell	Bell	20
Bell	Bell	Bar	18
Plum	Plum	Plum	14
Orange	Orange	Orange	10
Orange	Orange	Bar	8
Cherry	Cherry	Anything	4
Cherry	No cherry	Anything	2
	Anything else		−1

Compute the player's expected winnings on a single play of the slot machine. (Assume that each dial acts independently.)

20. The following probability distributions of job satisfaction scores for a sample of information systems (IS) senior executives and middle managers range from a low of 1 (very dissatisfied) to a high of 5 (very satisfied).

	Probability	
Job satisfaction score	IS senior executives	IS middle managers
1	0.05	0.04
2	0.09	0.10
3	0.03	0.12
4	0.42	0.46
5	0.41	0.28

a. What is the expected value of the job satisfaction score for senior executives?
b. What is the expected value of the job satisfaction score for middle managers?
c. Compute the variance of job satisfaction scores for executives and middle managers.
d. Compute the standard deviation of job satisfaction scores for both probability distributions.
e. Compare the overall job satisfaction of senior executives and middle managers.

5.4 BIVARIATE DISTRIBUTIONS, COVARIANCE AND FINANCIAL PORTFOLIOS

A probability distribution involving two random variables is called a **bivariate probability distribution**. In discussing bivariate probability distributions, it is useful to think of a bivariate experiment. Each outcome for a bivariate experiment consists of two values, one for each random variable. For example, consider the bivariate experiment of rolling a pair of dice. The outcome consists of two values, the number obtained with the first die and the number obtained with the second die. As another example, consider the experiment of observing the financial markets for a year and recording the percentage gain for a stock fund and a bond fund. Again, the experimental outcome provides a value for two random variables, the per cent gain in the stock fund and the per cent gain in the bond fund. When dealing with bivariate probability distributions, we are often interested in the relationship between the random variables. In this section, we introduce bivariate distributions and show how the covariance and correlation coefficient can be used as a measure of linear association between the random variables. We shall also see how bivariate probability distributions can be used to construct and analyze financial portfolios.

A bivariate empirical discrete probability distribution

Recall that in Section 5.2 we developed an empirical discrete distribution for daily sales at the DiCarlo Motors automobile dealership in Sienna. DiCarlo has another dealership in Geneva, Switzerland. Table 5.6 shows the number of cars sold at each of the dealerships over a 300-day period. The numbers in the bottom (total) row are the frequencies we used to develop an empirical probability distribution for daily sales at DiCarlo's Sienna dealership in Section 5.2. The numbers in the right-most (total) column are the frequencies of daily sales for the Geneva dealership. Entries in the body of the table give the number of days the Geneva dealership had a level of sales indicated by the row, when the Sienna dealership had the level of sales indicated by the column. For example, the entry of 33 in the Geneva dealership row labelled 1 and the Sienna column labelled 2 indicates that for 33 days out of the 300, the Geneva dealership sold 1 car and the Sienna dealership sold 2 cars.

TABLE 5.6 Number of cars sold at DiCarlo's Sienna and Geneva dealerships over 300 days

Geneva dealership	Sienna dealership						
	0	**1**	**2**	**3**	**4**	**5**	**Total**
0	21	30	24	9	2	0	86
1	21	36	33	18	2	1	111
2	9	42	9	12	3	2	77
3	3	9	6	3	5	0	26
Total	**54**	**117**	**72**	**42**	**12**	**3**	**300**

Suppose we consider the bivariate experiment of observing a day of operations at DiCarlo Motors and recording the number of cars sold. Let us define X = number of cars sold at the Geneva dealership and Y = the number of cars sold at the Sienna dealership. We can now divide all of the frequencies in Table 5.6 by the number of observations (300) to develop a bivariate empirical discrete probability distribution for cars sales at the two DiCarlo dealerships. Table 5.7 shows this bivariate discrete probability distribution. The probabilities in the lower margin provide the marginal distribution for the DiCarlo Motors Sienna dealership. The probabilities in the right margin provide the marginal distribution for the DiCarlo Motors Geneva dealership.

TABLE 5.7 Bivariate empirical discrete probability distribution for daily sales at DiCarlo dealerships in Sienna and Geneva

Geneva dealership	Sienna dealership						
	0	**1**	**2**	**3**	**4**	**5**	**Total**
0	0.0700	0.1000	0.0800	0.0300	0.0067	0.0000	0.2867
1	0.0700	0.1200	0.1100	0.0600	0.0067	0.0033	0.3700
2	0.0300	0.1400	0.0300	0.0400	0.0100	0.0067	0.2567
3	0.0100	0.0300	0.0200	0.0100	0.0167	0.0000	0.0867
Total	**0.18**	**0.39**	**0.24**	**0.14**	**0.04**	**0.01**	**1.0000**

The probabilities in the body of the table provide the bivariate probability distribution for sales at both dealerships. Bivariate probabilities are often called joint probabilities. We see that the joint probability of selling 0 cars at Geneva and 1 car at Sienna on a typical day is p(0, 1) = 0.1000, the joint probability of selling 1 car at Geneva and 4 cars at Sienna on a typical day is 0.0067, and so on. Note that there is one bivariate probability for each experimental outcome. With 4 possible values for X and 6 possible values for Y, there are 24 experimental outcomes and bivariate probabilities.

Suppose we would like to know the probability distribution for total sales at both DiCarlo dealerships and the expected value and variance of total sales. We can define S = X + Y as total sales for DiCarlo Motors. Working with the bivariate probabilities in Table 5.7, we see that $p(s = 0) = 0.0700$, $p(s = 1) = 0.0700 + 0.1000 = 0.1700$, $p(s = 2) = 0.0300 + 0.1200 + 0.0800 = 0.2300$, and so on. We show the complete probability distribution for S = X + Y along with the computation of the expected value and variance in Table 5.8. The expected value is $E(S) = 2.6433$ and the variance is $Var(S) = 2.3895$.

TABLE 5.8 Calculation of the expected value and variance for total daily sales at DiCarlo Motors

s	p(s)	sp(s)	s − E(S)	(s − E(S))²	(s − E(S))² p(s)
0	0.0700	0.0000	−2.6433	6.9872	0.4891
1	0.1700	0.1700	−1.6433	2.7005	0.4591
2	0.2300	0.4600	−0.6433	0.4139	0.0952
3	0.2900	0.8700	0.3567	0.1272	0.0369
4	0.1267	0.5067	1.3567	1.8405	0.2331
5	0.0667	0.3333	2.3567	5.5539	0.3703
6	0.0233	0.1400	3.3567	11.2672	0.2629
7	0.0233	0.1633	4.3567	18.9805	0.4429
8	0.0000	0.0000	5.3567	28.6939	0.0000
		E(S) = 2.6433			**Var(S) = 2.3895**

With bivariate probability distributions, we often want to know the relationship between the two random variables. The covariance and/or correlation coefficient are good measures of association between two random variables. We saw in Chapter 3 how to compute the covariance and correlation coefficient for sample data. The equation we will use for computing the covariance between two random variables X and Y is given in equation (5.6).

> **Covariance of random variables X and Y**
>
> $$Cov(X, Y) = \sigma_{XY} = [Var(X + Y) - Var(X) - Var(Y)]/2 \qquad \textbf{(5.6)}$$

We have already computed $Var(S) = Var(X + Y)$ and, in Section 5.3, we computed $Var(Y)$. Now we need to compute $Var(X)$ before we can use equation (5.6) to compute the covariance of X and Y. Using the probability distribution for X (the right margin of Table 5.7), we compute $E(X)$ and $Var(X)$ in Table 5.9.

TABLE 5.9   Calculation of the expected value and variance of daily car sales at DiCarlo Motors' Geneva dealership

x	p(x)	xp(x)	x − E(x)	[(x − E(x)]²	[x − E(x)]² p(x)
0	0.2867	0.0000	−1.1435	1.3076	0.3749
1	0.3700	0.3700	−0.1435	0.0206	0.0076
2	0.2567	0.5134	0.8565	0.7336	0.1883
3	0.0867	0.2601	1.8565	3.4466	0.2988
		E(X) = 1.1435			**Var(X) = 0.8696**

We can now use equation (5.6) to compute the covariance of the random variables X and Y.

$$\sigma_{XY} = [Var(X + Y) - Var(X) - Var(Y)]/2 = (2.3895 - 0.8696 - 1.25)/2 = 0.1350$$

A covariance of 0.1350 indicates that daily sales at DiCarlo's two dealerships have a positive relationship. To get a better sense of the strength of the relationship we can compute the correlation coefficient. The correlation coefficient for the two random variables x and y is given by equation (5.7).

> **Correlation between random variables X and Y**
>
> $$\rho_{XY} = \frac{\sigma_{XY}}{\sigma_X \sigma_Y} \qquad \textbf{(5.7)}$$

From equation (5.7), we see that the correlation coefficient for two random variables is the covariance divided by the product of the standard deviations for the two random variables.

To compute the correlation coefficient between daily sales at the two DiCarlo dealerships we first compute the standard deviations for sales at the Sienna and Geneva dealerships by taking the square root of the variance.

$$\sigma_X = \sqrt{0.8696} = 0.9325$$

$$\sigma_Y = \sqrt{1.25} = 1.1180$$

Following on, the correlation coefficient is given by:

$$\rho_{XY} = \frac{\sigma_{XY}}{\sigma_X \sigma_Y} = \frac{0.1350}{(0.9325)(1.1180)} = 0.1295$$

In Chapter 3 we defined the correlation coefficient as a measure of the linear association between two variables. Values near +1 indicate a strong positive linear relationship; values near −1 indicate a strong negative linear relationship; and values near zero indicate a lack of a linear relationship. This interpretation is also valid for random variables. The correlation coefficient of 0.1295 indicates there is a weak positive relationship between the random variables representing daily sales at the two DiCarlo dealerships. If the correlation coefficient had equalled zero, we would have concluded that daily sales at the two dealerships were uncorrelated.

Financial applications

Let us now see how what we have learned can be useful in constructing financial portfolios that provide a good balance of risk and return. A financial advisor is considering four possible economic scenarios for the coming year and has developed a probability distribution showing the per cent return, $X = x$, for investing in a large-cap stock fund and the per cent return, $Y = y$, for investing in a long-term government bond fund given each of the scenarios. The bivariate probability distribution function for X and Y is shown in Table 5.10. Table 5.10 is simply a list with a separate row for each experimental outcome (economic scenario). Each row contains the joint probability for the experimental outcome and a value for each random variable. Since there are only four joint probabilities, the tabular form used in Table 5.10 is simpler than the one we used for DiCarlo Motors where there were $(4)(6) = 24$ joint probabilities.

TABLE 5.10 Probability distribution of per cent returns for investing in a large-cap stock fund, X, and investing in a long-term government bond fund, Y.

Economic scenario	Probability $p(x, y)$	Large-cap stock fund (x)	Long-term government bond fund (y)
Recession	0.10	−40	30
Weak growth	0.25	5	5
Stable growth	0.50	15	4
Strong growth	0.15	30	2

Using the formula in Section 5.3 for computing the expected value of a single random variable, we can compute the expected per cent return for investing in the stock fund, $E(X)$, and the expected per cent return for investing in the bond fund, $E(Y)$.

$$E(X) = 0.10(-40) + 0.25(5) + 0.5(15) + 0.15(30) = 9.25$$
$$E(Y) = 0.10(30) + 0.25(5) + 0.5(4) + 0.15(2) = 6.55$$

Using this information, we might conclude that investing in the stock fund is a better investment. It has a higher expected return, 9.25 per cent. But financial analysts recommend that investors also consider the risk associated with an investment. The standard deviation of per cent return is often used as a measure of risk. To compute the standard deviation, we must first compute the variance. Using the formula in Section 5.3 for computing the variance of a single random variable, we can compute the variance of the per cent returns for the stock and bond fund investments.

$$Var(X) = 0.1(-40 - 9.25)^2 + 0.25(5 - 9.25)^2 + 0.50(15 - 9.25)^2 + 0.15(30 - 9.25)^2 = 328.1875$$

$$Var(Y) = 0.1(30 - 6.55)^2 + 0.25(5 - 6.55)^2 + 0.50(4 - 6.55)^2 + 0.15(2 - 6.55)^2 = 61.9475$$

The standard deviation of the return from an investment in the stock fund is $\sigma_X = \sqrt{328.1875} = 18.1159\%$ and the standard deviation of the return from an investment in the bond fund is $\sigma_Y = \sqrt{61.9475} = 7.8707\%$.

So, we can conclude that investing in the bond fund is less risky. It has the smaller standard deviation. We have already seen that the stock fund offers a greater expected return, so if we want to choose between investing in either the stock fund or the bond fund it depends on our attitude towards risk and return. An aggressive investor might choose the stock fund because of the higher expected return; a conservative investor might choose the bond fund because of the lower risk. But there are other options. What about the possibility of investing in a portfolio consisting of both an investment in the stock fund and an investment in the bond fund?

Suppose we would like to consider three alternatives: investing solely in the large-cap stock fund, investing solely in the long-term government bond fund, and splitting our funds equally between the stock fund and the bond fund (i.e. half in each). We have already computed the expected value and standard deviation for investing solely in the stock fund or the bond fund. Let us now evaluate the third alternative: constructing a portfolio by investing equal amounts in the large-cap stock fund and in the long-term government bond fund.

To evaluate this portfolio, we start by computing its expected return. We have previously defined X as the per cent return from an investment in the stock fund and Y as the per cent return from an investment in the bond fund so the per cent return for our portfolio is R = 0.5X + 0.5Y. To find the expected return for a portfolio with half invested in the stock fund and half invested in the bond fund, we want to compute $E(R) = E(0.5X + 0.5Y)$. The expression 0.5X + 0.5Y is called a linear combination of the random variables X and Y. Equation (5.8) provides an easy method for computing the expected value of a linear combination of the random variables X and Y when we already know $E(X)$ and $E(Y)$. In equation (5.8), a represents the coefficient of X and b represents the coefficient of Y in the linear combination.

Expected value of a linear combination of random variables X and Y

$$E(aX + bY) = aE(X) + bE(Y)$$

(5.8)

Since we have already computed $E(X) = 9.25$ and $E(Y) = 6.55$, we can use equation (5.8) to compute the expected value of our portfolio.

$$E(0.5X + 0.5Y) = 0.5E(X) + 0.5E(Y) = 0.5(9.25) + 0.5(6.55) = 7.9$$

We see that the expected return for investing in the portfolio is 7.9 per cent. With $100 invested, we would expect a return of $100(0.079) = $7.90; with $1,000 invested we would expect a return of $1,000(0.079) = $79.00; and so on. But, what about the risk? As mentioned previously, financial analysts often use the standard deviation as a measure of risk.

Our portfolio is a linear combination of two random variables, so we need to be able to compute the variance and standard deviation of this in order to assess the portfolio risk. This is possible using the formula in equation (5.9) where a represents the coefficient of X and b represents the coefficient of Y in the linear combination.

Variance of a linear combination of two random variables

$$Var(aX + bY) = a^2Var(X) + b^2Var(Y) + 2abCov(X,Y)$$

(5.9)

where $Cov(X, Y) = \sigma_{XY}$ is the covariance of X and Y.

From equation (5.9), we see that both the variance of each random variable individually and the covariance between the random variables are needed to compute the variance of a linear combination of two random variables and hence the variance of our portfolio.

We have already computed the variance of each random variable individually: $Var(X) = 328.1875$ and $Var(Y) = 61.9475$. Also, it can be shown that $Var(X + Y) = 119.46$. So, using equation (5.6), the covariance of the random variables X and Y is:

$$\sigma_{XY} = [Var(X + Y) - Var(X) - Var(Y)]/2 = [119.46 - 328.1875 - 61.9475]/2 = -135.3375$$

A negative covariance between X and Y, such as this, means that when X tends to be above its mean, Y tends to be below its mean and vice versa.

We can now use equation (5.9) to compute the variance of return for our portfolio.

$$Var(0.5X + 0.5Y) = 0.5^2(328.1875) + 0.5^2(61.9475) + 2(0.5)(0.5)(-135.3375) = 29.865$$

The standard deviation of our portfolio is then given by $\sigma_{0.5X+0.5Y} = \sqrt{29.865} = 5.4650\%$. This is our measure of risk for the portfolio consisting of investing 50 per cent in the stock fund and 50 per cent in the bond fund.

Table 5.11 shows the expected returns, variances and standard deviations for each of the three alternatives.

TABLE 5.11   Expected values, variances and standard deviations for three investment alternatives

Investment alternative	Expected return (%)	Variance of return	Standard deviation of return (%)
100% in stock fund	9.25	328.1875	18.1159
100% in bond fund	6.55	61.9475	7.8707
Portfolio (50% in stock fund, 50% in bond fund)	7.90	29.8650	5.4650

Which of these alternatives would you prefer? The expected return is highest for investing 100 per cent in the stock fund, but the risk is also highest. The standard deviation is 18.1159 per cent. Investing 100 per cent in the bond fund has a lower expected return, but a significantly smaller risk. Investing 50 per cent in the stock fund and 50 per cent in the bond fund (the portfolio) has an expected return that is halfway between that of the stock fund alone and the bond fund alone. But note that it has less risk than investing 100 per cent in either of the individual funds. Indeed, it has both a higher return and less risk (smaller standard deviation) than investing solely in the bond fund. So we would say that investing in the portfolio dominates the choice of investing solely in the bond fund.

Whether you would choose to invest in the stock fund or the portfolio depends on your attitude towards risk. The stock fund has a higher expected return. But the portfolio has significantly less risk and also provides a fairly good return. Many would choose it. It is the negative covariance between the stock and bond funds that has caused the portfolio risk to be so much smaller than the risk of investing solely in either of the individual funds.

The portfolio analysis we just performed was for investing 50 per cent in the stock fund and the other 50 per cent in the bond fund. How would you calculate the expected return and the variance for other portfolios? Equations (5.8) and (5.9) can be used to make these calculations easily.

Suppose we wish to create a portfolio by investing 25 per cent in the stock fund and 75 per cent in the bond fund? What are the expected value and variance of this portfolio? The per cent return for this portfolio is $R = 0.25X + 0.75Y$, so we can use equation (5.8) to get the expected value of this portfolio:

$$E(0.25X + 0.75Y) = 0.25X + 0.75Y = 0.25(9.25) + 0.75(6.55) = 7.225$$

Likewise, we may calculate the variance of the portfolio using equation (5.9):

$$Var(0.25X + 0.75Y) = (0.25)^2 Var(X) + (0.75)^2 Var(Y) + 2(0.25)(0.75)\sigma_{XY}$$
$$= 0.0625(328.1875) + (0.5625)(61.9475) + (0.375)(-135.3375)$$
$$= 4.6056$$

The standard deviation of the new portfolio is $\sigma_{0.25X+0.75Y} = \sqrt{4.6056} = 2.1461$.

EXERCISES

Methods

21. Given below is a bivariate distribution for the random variables X and Y taking the values x and y respectively.

p(x, y)	x	y
0.2	50	80
0.5	30	50
0.3	40	60

   a. Compute the expected value and the variance for X and Y.
   b. Develop a probability distribution for $X + Y$.
   c. Using the result of part (b), compute $E(X + Y)$ and $Var(X + Y)$.
   d. Compute the covariance and correlation for X and Y. Are X and Y positively correlated, negatively correlated, or uncorrelated?
   e. Is the variance of the sum of X and Y bigger, smaller, or the same as the sum of the individual variances? Why?

22. A person is interested in constructing a portfolio. Two stocks are being considered. Let X = per cent return for an investment in stock 1, and Y = per cent return for an investment in stock 2. The expected return and variance for stock 1 are $E(X) = 8.45\%$ and $Var(X) = 25$. The expected return and variance for stock 2 are $E(Y) = 3.20\%$ and $Var(Y) = 1$. The covariance between the returns is $\sigma_{XY} = -3$.
   a. What is the standard deviation for an investment in stock 1 and for an investment in stock 2? Using the standard deviation as a measure of risk, which of these stocks is the riskier investment?
   b. What is the expected return and standard deviation, in dinars, for a person who invests 500 dinars in stock 1?
   c. What is the expected per cent return and standard deviation for a person who constructs a portfolio by investing 50% in each stock?
   d. What is the expected per cent return and standard deviation for a person who constructs a portfolio by investing 70% in stock 1 and 30% in stock 2?
   e. Compute the correlation coefficient for X and Y and comment on the relationship between the returns for the two stocks.

23. The Chamber of Commerce in a Canadian city has conducted an evaluation of 300 restaurants in its metropolitan area. Each restaurant received a rating on a 3-point scale on typical meal price (1 least expensive to 3 most expensive) and quality (1 lowest quality to 3 greatest quality). A cross tabulation of the rating data is shown below. Forty-two of the restaurants received a rating of 1 on quality and 1 on meal price, 39 of the restaurants received a rating of 1 on quality and 2 on meal price, and so on. Forty-eight of the restaurants received the highest rating of 3 on both quality and meal price.

	Meal price (Y)			
Quality (X)	1	2	3	Total
1	42	39	3	84
2	33	63	54	150
3	3	15	48	66
Total	78	117	105	300

    a. Develop a bivariate probability distribution for quality and meal price of a randomly selected restaurant in this Canadian city. Let X = quality rating and Y = meal price.

    b. Compute the expected value and variance for quality rating, X.

    c. Compute the expected value and variance for meal price, Y.

    d. The $Var(X + Y) = 1.6691$. Compute the covariance of X and Y. What can you say about the relationship between quality and meal price? Is this what you would expect?

    e. Compute the correlation coefficient between quality and meal price. What is the strength of the relationship? Do you suppose it is likely that a low-cost restaurant in this city that is also high quality can be found? Why or why not?

24. PortaCom has developed a design for a high-quality portable printer. The two key components of manufacturing cost are direct labour and parts. During a testing period, the company has developed prototypes and conducted extensive product tests with the new printer. PortaCom's engineers have developed the bivariate probability distribution shown below for the manufacturing costs. Parts cost (in euros) per printer is represented by the random variable X and direct labour cost (in euros) per printer is represented by the random variable Y. Management would like to use this probability distribution to estimate manufacturing costs.

	Direct labour (Y)			
Parts (X)	43	45	48	Total
85	0.05	0.2	0.2	0.45
95	0.25	0.2	0.1	0.55
Total	0.30	0.4	0.3	1.00

    a. Show the marginal distribution of direct labour cost and compute its expected value, variance and standard deviation.

    b. Show the marginal distribution of parts cost and compute its expected value, variance and standard deviation.

    c. Total manufacturing cost per unit is the sum of direct labour cost and parts cost. Show the probability distribution for total manufacturing cost per unit.

    d. Compute the expected value, variance and standard deviation of total manufacturing cost per unit.

    e. Are direct labour and parts costs independent? Why or why not? If you conclude that they are not, what is the relationship between direct labour and parts cost?

    f. PortaCom produced 1,500 printers for its product introduction. The total manufacturing cost was €198,350. Is that about what you would expect? If it is higher or lower, what do you think may have caused the difference?

5.5 BINOMIAL PROBABILITY DISTRIBUTION

The binomial probability distribution is a discrete probability distribution that has many applications. It is associated with a multiple-step experiment that we call the binomial experiment.

A binomial experiment

A **binomial experiment** exhibits the following four properties:

Properties of a binomial experiment

1. The experiment consists of a sequence of n identical trials.

2. Two outcomes are possible on each trial. We refer to one outcome as a *success* and the other outcome as a *failure*.

3. The probability of a success, denoted by π, does not change from trial to trial. Consequently, the probability of a failure, denoted by $1 - \pi$, does not change from trial to trial.

4. The trials are independent.

If properties 2, 3 and 4 are present, we say the trials are generated by a Bernoulli process. If, in addition, property 1 is present, we say we have a binomial experiment. Figure 5.2 depicts one possible sequence of successes and failures for a binomial experiment involving eight trials.

FIGURE 5.2
One possible sequence of successes and failures for an eight-day trial binomial experiment

Property 1: The experiment consists of $n = 8$ identical trials.

Property 2: Each trial results in either success (S) or failure (F).

Trials $\longrightarrow$	1	2	3	4	5	6	7	8
Outcomes $\longrightarrow$	S	F	F	S	S	F	S	S

In a binomial experiment, our interest is in the *number of successes occurring in the* n *trials*. If we let X denote the number of successes occurring in the n trials, we see that X can assume the values of 0, 1, 2, 3, ..., n. Because the number of values is finite, X is a *discrete* random variable. The probability distribution associated with this random variable is called the **binomial probability distribution**. For example, consider the experiment of tossing a coin five times and on each toss observing whether the coin lands with a head or a tail on its upward face. Suppose we want to count the number of heads appearing over the five tosses. Does this experiment show the properties of a binomial experiment? What is the random variable of interest? Note that:

1 The experiment consists of five identical trials; each trial involves the tossing of one coin.

2 Two outcomes are possible for each trial: a head or a tail. We can designate head a success and tail a failure.

3 The probability of a head and the probability of a tail are the same for each trial, with $\pi = 0.5$ and $1 - \pi = 0.5$.

4 The trials or tosses are independent because the outcome on any one trial is not affected by what happens on other trials or tosses.

Hence, the properties of a binomial experiment are satisfied. The random variable of interest is $X =$ the number of heads appearing in the five trials. In this case, X can assume the values of 0, 1, 2, 3, 4 or 5.

As another example, consider an insurance salesperson who visits ten randomly selected families. The outcome associated with each visit is classified as a success if the family purchases an insurance policy and a failure if the family does not. From past experience, the salesperson knows the probability that a randomly selected family will purchase an insurance policy is 0.10. Checking the properties of a binomial experiment, we observe that:

1 The experiment consists of ten identical trials; each trial involves contacting one family.
2 Two outcomes are possible on each trial: the family purchases a policy (success) or the family does not purchase a policy (failure).
3 The probabilities of a purchase and a non-purchase are assumed to be the same for each sales call, with $\pi = 0.10$ and $1 - \pi = 0.90$.
4 The trials are independent because the families are randomly selected.

Because the four assumptions are satisfied, this example is a binomial experiment. The random variable of interest is the number of sales obtained in contacting the ten families. In this case, X can assume the values of 0, 1, 2, 3, 4, 5, 6, 7, 8, 9 and 10.

Property 3 of the binomial experiment is called the *stationarity assumption* and is sometimes confused with property 4, independence of trials. To see how they differ, consider again the case of the salesperson calling on families to sell insurance policies. If, as the day wore on, the salesperson got tired and lost enthusiasm, the probability of success (selling a policy) might drop to 0.05, for example, by the tenth call. In such a case, property 3 (stationarity) would not be satisfied, and we would not have a binomial experiment. Even if property 4 held – that is, the purchase decisions of each family were made independently – it would not be a binomial experiment if property 3 was not satisfied.

In applications involving binomial experiments, a special mathematical formula, called the *binomial probability function*, can be used to compute the probability of x successes in the n trials. We will show in the context of an illustrative problem how the formula can be developed.

Marrine Clothing Store problem

Let us consider the purchase decisions of the next three customers who enter the Marrine Clothing Store. On the basis of past experience, the store manager estimates the probability that any one customer will make a purchase is 0.30. What is the probability that two of the next three customers will make a purchase?

Using a tree diagram (Figure 5.3), we see that the experiment of observing the three customers each making a purchase decision has eight possible outcomes. Using S to denote success (a purchase) and F to denote failure (no purchase), we are interested in experimental outcomes involving two successes in the three trials (purchase decisions). Next, let us verify that the experiment involving the sequence of three purchase decisions can be viewed as a binomial experiment. Checking the four requirements for a binomial experiment, we note that:

1 The experiment can be described as a sequence of three identical trials, one trial for each of the three customers who will enter the store.
2 Two outcomes – the customer makes a purchase (success) or the customer does not make a purchase (failure) – are possible for each trial.
3 The probability that the customer will make a purchase (0.30) or will not make a purchase (0.70) is assumed to be the same for all customers.
4 The purchase decision of each customer is independent of the decisions of the other customers.

Hence, the properties of a binomial experiment are present.

The number of experimental outcomes resulting in exactly x successes in n trials can be computed using equation (5.10).*

---

* Equation (5.10), introduced in Chapter 4, determines the number of combinations of n objects selected x at a time. For the binomial experiment, this combinatorial formula provides the number of experimental outcomes (sequences of n trials) resulting in x successes.

FIGURE 5.3
Tree diagram for the
Marrine Clothing Store
problem

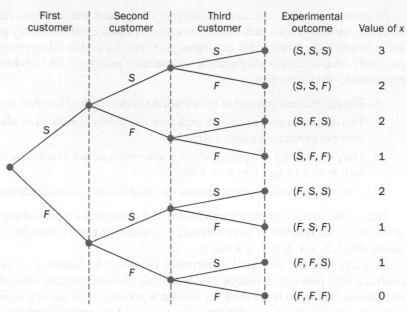

S = Purchase
F = No purchase
x = Number of customers making a purchase

Number of experimental outcomes providing exactly x successes in n trials

$$\binom{n}{x} = \frac{n!}{x!(n-x)!} \tag{5.10}$$

where:

$$n! = n \times (n-1) \times (n-2) \times \ldots \times (2) \times (1)$$

and, by definition,

$$0! = 1$$

Now let us return to the Marrine Clothing Store experiment involving three customer purchase decisions. Equation (5.10) can be used to determine the number of experimental outcomes involving two purchases; that is, the number of ways of obtaining $X = 2$ successes in the $n = 3$ trials. From equation (5.10) we have:

$$\binom{n}{x} = \binom{3}{2} = \frac{3!}{2! \times (3-2)!} = \frac{3 \times 2 \times 1}{(2 \times 1) \times (1)} = \frac{6}{2} = 3$$

Equation (5.10) shows that three of the experimental outcomes yield two successes. From Figure 5.3 we see these three outcomes are denoted by (S, S, F), (S, F, S) and (F, S, S). Using equation (5.10) to determine how many experimental outcomes have three successes (purchases) in the three trials, we obtain:

$$\binom{n}{x} = \binom{3}{3} = \frac{3!}{3! \times (3-2)!} = \frac{3 \times 2 \times 1}{(3 \times 2 \times 1) \times (1)} = \frac{6}{6} = 1$$

From Figure 5.3 we see that the one experimental outcome with three successes is identified by (S, S, S).

We know that equation (5.10) can be used to determine the number of experimental outcomes that result in X successes. If we are to determine the probability of x successes in n trials, however, we must also know the probability associated with each of these experimental outcomes. Because the trials of a binomial experiment are independent, we can simply multiply the probabilities associated with each trial outcome to find the probability of a particular sequence of successes and failures.

The probability of purchases by the first two customers and no purchase by the third customer, denoted (S, S, F), is given by:

$$\pi\pi(1 - \pi)$$

With a 0.30 probability of a purchase on any one trial, the probability of a purchase on the first two trials and no purchase on the third is given by:

$$0.30 \times 0.30 \times 0.70 = 0.30^2 \times 0.70 = 0.063$$

Two other experimental outcomes also result in two successes and one failure. The probabilities for all three experimental outcomes involving two successes are as follows:

Trial outcomes			Experimental	Probability of
1st customer	2nd customer	3rd customer	outcome	experimental outcome
Purchase	Purchase	No purchase	(S, S, F)	$\pi\pi (1 - \pi) = \pi^2(1 - \pi)$ $= (0.30)^2(0.70) = 0.063$
Purchase	No purchase	Purchase	(S, F, S)	$\pi(1 - \pi)\pi = \pi^2(1 - \pi)$ $= (0.30)^2(0.70) = 0.063$
No purchase	Purchase	Purchase	(F, S, S)	$(1 - \pi)\pi\pi = \pi^2(1 - \pi)$ $= (0.30)^2(0.70) = 0.063$

Observe that all three experimental outcomes with two successes have exactly the same probability. This observation holds in general. In any binomial experiment, all sequences of trial outcomes yielding x successes in n trials have the *same probability* of occurrence.

The probability of each sequence of trials yielding x successes in n trials is shown in equation (5.11).

Probability of a particular sequence of trial outcomes $= \pi^x(1 - \pi)^{(n-x)}$ **(5.11)**
with x successes in n trials

For the Marrine Clothing Store, equation (5.11) shows that any experimental outcome with two successes has a probability of $\pi^2(1 - \pi)^{(3-2)} = \pi^2(1 - \pi)^1 = (0.30)^2 (0.70)^1 = 0.063$. Combining equations (5.10) and (5.11) we obtain the following **binomial probability function**:

Binomial probability function

$$p(x) = \binom{n}{x}\pi^x(1 - \pi)^{(n-x)} \tag{5.12}$$

where $p(x) =$ the probability of x successes in n trials

$n =$ the number of trials

$$\binom{n}{x} = \frac{n!}{x!(n - x)!}$$

$\pi =$ the probability of a success on any one trial

$1 - \pi =$ the probability of a failure on any one trial

In the Marrine Clothing Store example, we can use this function to compute the probability that no customer makes a purchase, exactly one customer makes a purchase, exactly two customers make a purchase and all three customers make a purchase. The calculations are summarized in Table 5.12, which gives the probability distribution of the number of customers making a purchase. Figure 5.4 is a graph of this probability distribution.

FIGURE 5.4
Graphical representation of the probability distribution for the number of customers making a purchase

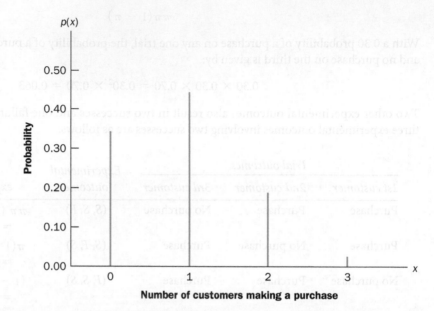

The binomial probability function can be applied to *any* binomial experiment. If we are satisfied that a situation demonstrates the properties of a binomial experiment, and if we know the values of n and π, we can use equation (5.12) to compute the probability of x successes in the n trials.

If we consider variations of the Marrine experiment, such as ten customers rather than three entering the store, the binomial probability function given by equation (5.12) is still applicable.

TABLE 5.12
Probability distribution for the number of customers making a purchase

x	p(x)
0	$\frac{3!}{0!3!}(0.30)^0(0.70)^3 = 0.343$
1	$\frac{3!}{1!2!}(0.30)^1(0.70)^2 = 0.441$
2	$\frac{3!}{2!1!}(0.30)^2(0.70)^1 = 0.189$
3	$\frac{3!}{3!0!}(0.30)^3(0.70)^0 = \underline{0.027}$
	1.000

Suppose we have a binomial experiment with $n = 10$, $x = 4$ and $\pi = 0.30$. The probability of making exactly four sales to ten customers entering the store is:

$$p(4) = \frac{10!}{4!6!}(0.30)^4(0.70)^6 = 0.2001$$

Historically tables of binomial probabilities have been available with many statistics textbooks – including the first three editions of this text. Even though the tables of binomial probabilities are relatively easy to use, it is impossible to have tables that show all possible values of n and π that might be encountered in a binomial experiment. However, with today's calculators, using equation (5.12) to calculate the desired probability is not difficult, especially if the number of trials is not large. In the exercises, you should practise using equation (5.12) to compute the binomial probabilities unless the problem specifically requests that you use a binomial probability table.

Statistical software packages and spreadsheet packages such as Excel also provide the capability to compute binomial probabilities. Consider the Marrine Clothing Store example with $n = 10$ and $\pi = 0.30$. Figure 5.5 shows the binomial probabilities generated by computer for all possible values of x.

FIGURE 5.5
Computer output showing binomial probabilities for the Marrine Clothing Store problem

x	P (X = x)
0	0. 028248
1	0. 121061
2	0. 233474
3	0. 266828
4	0. 200121
5	0. 102919
6	0. 036757
7	0. 009002
8	0. 001447
9	0. 000138
10	0. 000006

Expected value and variance for the binomial distribution

In Section 5.3 we provided formulae for computing the expected value and variance of a discrete random variable. In the special case where the random variable has a binomial distribution with a known number of trials n and a known probability of success π, the general formulae for the expected value and variance can be simplified. The results are as follows:

Expected value and variance for the binomial distribution

$$E(X) = \mu = n\pi \qquad (5.13)$$

$$\text{Var}\,(X) = \sigma^2 = n\pi(1 - \pi) \qquad (5.14)$$

For the Marrine Clothing Store problem with three customers, we can use equation (5.13) to compute the expected number of customers who will make a purchase.

$$E(X) = n\pi = 3 \times 0.30 = 0.9$$

Suppose that for the next month the Marrine Clothing Store forecasts 1,000 customers will enter the store. What is the expected number of customers who will make a purchase? The answer is $\mu = n\pi = 1,000 \times 0.3 = 300$. Hence, to increase the expected number of purchases, Marrine must induce more customers to enter the store and/or somehow increase the probability that any individual customer will make a purchase after entering.

For the Marrine Clothing Store problem with three customers, we see that the variance and standard deviation for the number of customers who will make a purchase are:

$$\sigma^2 = n\pi(1 - \pi) = 3 \times 0.3 \times 0.7 = 0.63$$
$$\sigma = \sqrt{0.063} = 0.79$$

For the next 1,000 customers entering the store, the variance and standard deviation for the number of customers who will make a purchase are:

$$\sigma^2 = n\pi(1 - \pi) = 1,000 \times 0.3 \times 0.7 = 210$$
$$\sigma = \sqrt{210} = 14.49$$

EXERCISES

Methods

25. Consider a binomial experiment with two trials and $\pi = 0.4$.
  a. Draw a tree diagram for this experiment (refer to Figure 5.3).
  b. Compute the probability of one success, $p(1)$.
  c. Compute $p(0)$.
  d. Compute $p(2)$.
  e. Compute the probability of at least one success.
  f. Compute the expected value, variance and standard deviation.

26. Consider a binomial experiment with $n = 10$ and $\pi = 0.10$.
  a. Compute $p(0)$.
  b. Compute $p(2)$.
  c. Compute $P(X \leq 2)$.
  d. Compute $P(X \geq 1)$.
  e. Compute $E(X)$.
  f. Compute $Var(X)$ and σ.

27. Consider a binomial experiment with $n = 20$ and $\pi = 0.70$.
  a. Compute $p(12)$.
  b. Compute $p(16)$.
  c. Compute $P(X \geq 16)$.
  d. Compute $P(X \leq 15)$.
  e. Compute $E(X)$.
  f. Compute $Var(X)$ and σ.

Applications

28. A smart speaker is an internet-connected speaker with which the user can interact with the speaker through voice commands. A survey by Reuters/Oxford University in 2018 estimates that the BBC accounts for 64 per cent of all news update usage by smart speakers in the UK. Suppose eight smart speaker users are randomly selected to be interviewed.
  a. What is the probability that none of the eight smart speaker users listens to the news through their smart speaker?
  b. What is the probability that all eight smart speaker users listen to the news through their smart speaker?
  c. What is the probability that five of the eight smart speaker users listen to the news through their smart speaker?
  d. What is the probability that six or more of the eight smart speaker users listen to the news through their smart speaker?

29. It takes at least nine votes from a 12-member jury to convict a defendant. Suppose that the probability that a juror votes a guilty person innocent is 0.2 whereas the probability that the juror votes an innocent person guilty is 0.1.
  a. If each juror acts independently and 65 per cent of defendants are guilty, what is the probability that the jury renders a correct decision?
  b. What percentage of defendants is convicted?

30. A firm bills its accounts at a 1 per cent discount for payment within ten days and the full amount is due after ten days. In the past, 30 per cent of all invoices have been paid within ten days. If the firm sends out eight invoices during the first week of January, what is the probability that:
  a. No one receives the discount?
  b. Everyone receives the discount?
  c. No more than three receive the discount?
  d. At least two receive the discount?

31. In a game of 'Chuck a luck' a player bets on one of the numbers 1 to 6. Three dice are then rolled and if the number bet by the player appears i times ($i = 1, 2, 3$) the player then wins i units. On the other hand, if the number bet by the player does not appear on any of the dice the player loses 1 unit. If X is the player's winnings in the game, what is the expected value of X?

5.6 POISSON PROBABILITY DISTRIBUTION

In this section we consider a discrete random variable that is often useful in estimating the number of occurrences over a specified interval of time or space. For example, the random variable of interest might be the number of arrivals at a car wash in one hour, the number of repairs needed in ten kilometres of highway, or the number of leaks in 100 kilometres of pipeline.

If the following two properties are satisfied, the number of occurrences is a random variable described by the Poisson probability distribution.

Properties of a Poisson experiment

1. The probability of an occurrence is the same for any two intervals of equal length.
2. The occurrence or non-occurrence in any interval is independent of the occurrence or non-occurrence in any other interval.

The Poisson probability function is defined by equation (5.15).

Poisson probability function

$$p(x) = \frac{\mu^x e^{-\mu}}{x!} \qquad (5.15)$$

where:

$p(x) = $ the probability of x occurrences in an interval

$\mu = $ expected value or mean number of occurrences in an interval

$e = 2.71828$

Before we consider a specific example to see how the Poisson distribution can be applied, note that the number of occurrences, x, has no upper limit. It is a discrete random variable that may assume an infinite sequence of values ($x = 0, 1, 2, \ldots$).

An example involving time intervals

Suppose that we are interested in the number of arrivals at the payment kiosk of a car park during a 15-minute period on weekday mornings. If we can assume that the probability of a car arriving is the same for any two time periods of equal length and that the arrival or non-arrival of a car in any time period is independent of the arrival or non-arrival in any other time period, the Poisson probability function is applicable. Suppose these assumptions are satisfied and an analysis of historical data shows that the average number of cars arriving in a 15-minute period of time is ten; in this case, the following probability function applies:

$$p(x) = \frac{10^x e^{-10}}{x!}$$

The random variable here is X = number of cars arriving in any 15-minute period.

If management wanted to know the probability of exactly 5 arrivals in 15 minutes, we would set $X = 5$ and thus obtain:

$$\text{Probability of exactly 5 arrivals in 15 minutes} = p(5) = \frac{10^5 e^{-10}}{5!} = 0.0378$$

In the preceding example, the mean of the Poisson distribution is $\mu = 10$ arrivals per 15-minute period. A property of the Poisson distribution is that the mean of the distribution and the variance of the distribution are *equal*. Thus, the variance for the number of arrivals during 15-minute periods is $\sigma^2 = 10$. The standard deviation is:

$$\sigma = \sqrt{10} = 3.16$$

Our illustration involves a 15-minute period, but other time periods can be used. Suppose we want to compute the probability of one arrival in a three-minute period. Because ten is the expected number of arrivals in a 15-minute period, we see that $10/15 = 2/3$ is the expected number of arrivals in a one-minute period and that $2/3 \times 3$ minutes $= 2$ is the expected number of arrivals in a three-minute period. Hence, the probability of x arrivals in a three-minute time period with $\mu = 2$ is given by the following Poisson probability function:

$$p(x) = \frac{2^x e^{-2}}{x!}$$

The probability of one arrival in a three-minute period is calculated as follows:

$$\text{Probability of exactly one arrival in three minutes} = P(1) = \frac{2^1 e^{-2}}{1!} = 0.2707$$

Earlier we computed the probability of five arrivals in a 15-minute period; it was 0.0378. Note that the probability of one arrival in a three-minute period (0.2707) is not the same. When computing a Poisson probability for a different time interval, we must first convert the mean arrival rate to the time period of interest and then compute the probability.

An example involving length or distance intervals

Consider an application not involving time intervals in which the Poisson distribution is useful. Suppose we are concerned with the occurrence of major defects in a highway, one month after resurfacing. We will assume that the probability of a defect is the same for any two highway intervals of equal length and that the occurrence or non-occurrence of a defect in any one interval is independent of the occurrence or non-occurrence of a defect in any other interval. Hence, the Poisson distribution can be applied.

Suppose that major defects one month after resurfacing occur at the average rate of two per kilometre. Let us find the probability of no major defects in a particular three-kilometre section of the highway. Because we are interested in an interval with a length of three kilometres, $\mu = 2$ defects/kilometre $\times 3$ kilometres $= 6$ represents the expected number of major defects over the three-kilometre section of highway. Using equation (5.15), the probability of no major defects is $p(0) = 6^0 e^{-6}/0! = 0.0025$. Hence, it is unlikely that no major defects will occur in the three-kilometre section. Equivalently there is a $1 - 0.0025 = 0.9975$ probability of at least one major defect in the three-kilometre highway section.

EXERCISES

Methods

32. Consider a Poisson distribution with $\mu = 3$.
    a. Write the appropriate Poisson probability function.
    b. Compute $p(2)$.
    c. Compute $p(1)$.
    d. Compute $P(X \geq 2)$.

33. Consider a Poisson distribution with a mean of two occurrences per time period.
    a. Write the appropriate Poisson probability function.
    b. What is the expected number of occurrences in three time periods?
    c. Write the appropriate Poisson probability function to determine the probability of x occurrences in three time periods.
    d. Compute the probability of two occurrences in one time period.
    e. Compute the probability of six occurrences in three time periods.
    f. Compute the probability of five occurrences in two time periods.

Applications

34. A local company employs 150 people and operates on three eight-hour shifts a day. A re-arranged shift rota resulted in 1 per cent of all employees arriving late for work on any given day. Using the Poisson distribution find the probability that on any given day:
    a. No-one was late for work.
    b. Three people were late for work.
    c. More than two but fewer than six people were late for work.

35. It is known that the injection of a vaccine to counter the latest COVID-19 virus causes an adverse reaction in one patient in 100,000. In a national immunization campaign 30,000 people were injected with the vaccine. What is the probability that:
    a. No-one suffered an adverse reaction?
    b. At least two people suffered an adverse reaction?
    c. More than two people suffered an adverse reaction?

36. According to a 2019 survey – conducted by the Radicati Group – office workers receive an average of 121 emails per day (*Campaign Monitor* magazine website). Assume the number of emails received per hour follows a Poisson distribution and that the average number of emails received per hour is five.
    a. What is the probability of receiving no emails during an hour?
    b. What is the probability of receiving at least three emails during an hour?
    c. What is the expected number of emails received during 15 minutes?
    d. What is the probability that no emails are received during 15 minutes?

5.7 HYPERGEOMETRIC PROBABILITY DISTRIBUTION

The hypergeometric probability distribution is closely related to the binomial distribution. The two probability distributions differ in two key ways. With the hypergeometric distribution, the trials are not independent, and the probability of success changes from trial to trial.

In the usual notation for the hypergeometric distribution, r denotes the number of elements in the population of size N labelled success, and $N - r$ denotes the number of elements in the population labelled failure. The hypergeometric probability function is used to compute the probability that in a random selection of n elements, selected without replacement, we obtain x elements labelled success and $n - x$ elements labelled failure. For this outcome to occur, we must obtain x successes from the r successes in the population and $n - x$ failures from the $N - r$ failures. The hypergeometric probability function in equation (5.16) provides $p(x)$, probability of obtaining x successes in a sample of size n.

Hypergeometric probability function

$$p(x) = \frac{\binom{r}{x}\binom{N-r}{n-x}}{\binom{N}{n}} \tag{5.16}$$

where:

$p(x) = $ probability of x successes in n trials

$n = $ number of trials

$N = $ number of elements in the population

$r = $ number of elements in the population labelled success

Note that $\binom{N}{n}$ represents the number of ways a sample of size n can be selected from a population of size N; $\binom{r}{x}$ represents the number of ways that x successes can be selected from a total of r successes in the population; and $\binom{N-r}{n-x}$ represents the number of ways that $n-x$ failures can be selected from a total of $N-r$ failures in the population.

To illustrate the computations involved in using equation (5.16), consider the following quality control application. Electric fuses produced by Warsaw Electric are packaged in boxes of 12 units each. Suppose an inspector randomly selects 3 of the 12 fuses in a box for testing. If the box contains exactly five defective fuses, what is the probability that the inspector will find exactly one of the three defective fuses? In this application, $n = 3$ and $N = 12$. With $r = 5$ defective fuses in the box the probability of finding $x = 1$ defective fuse is:

$$p(1) = \frac{\binom{5}{1}\binom{7}{2}}{\binom{12}{3}} = \frac{\frac{5!}{1!4!}\frac{7!}{2!5!}}{\frac{12!}{3!9!}} = \frac{5 \times 21}{220} = 0.4773$$

Now suppose that we wanted to know the probability of finding *at least* one defective fuse. The easiest way to answer this question is to first compute the probability that the inspector does not find any defective fuses. The probability of $x = 0$ is:

$$p(0) = \frac{\binom{5}{0}\binom{7}{3}}{\binom{12}{3}} = \frac{\frac{5!}{0!5!}\frac{7!}{3!4!}}{\frac{12!}{3!9!}} = \frac{1 \times 35}{220} = 0.1591$$

With a probability of zero defective fuses $p(0) = 0.1591$, we conclude that the probability of finding at least one defective fuse must be $1 - 0.1591 = 0.8409$. Hence, there is a reasonably high probability that the inspector will find at least one defective fuse.

The mean and variance of a hypergeometric distribution are as follows:

Expected value for the hypergeometric distribution

$$E(X) = \mu = n\left(\frac{r}{N}\right)$$

(5.17)

Variance for the hypergeometric distribution

$$\text{Var}(X) = \sigma^2 = n\left(\frac{r}{N}\right)\left(1 - \frac{r}{N}\right)\left(\frac{N - n}{N - 1}\right)$$

(5.18)

In the preceding example $n = 3$, $r = 5$ and $N = 12$. Hence, the mean and variance for the number of defective fuses is:

$$\mu = n\left(\frac{r}{N}\right) = 3\left(\frac{5}{12}\right) = 1.25$$

$$\sigma^2 = n\left(\frac{r}{N}\right)\left(1 - \frac{r}{N}\right)\left(\frac{N - n}{N - 1}\right) = 3\left(\frac{5}{12}\right)\left(1 - \frac{5}{12}\right)\left(\frac{12 - 3}{12 - 1}\right) = 0.60$$

The standard deviation is:

$$\sigma = \sqrt{0.60} = 0.77$$

EXERCISES

Methods

37. Suppose $N = 10$ and $r = 3$. Compute the hypergeometric probabilities for the following values of n and x.
   a. $n = 4$, $x = 1$.
   b. $n = 2$, $x = 2$.
   c. $n = 2$, $x = 0$.
   d. $n = 4$, $x = 2$.

38. Suppose $N = 15$ and $r = 4$. What is the probability of $x = 3$ for $n = 10$?

Applications

39. Blackjack, or Twenty-one as it is frequently called, is a popular gambling game played in Monte Carlo casinos. A player is dealt two cards. Face cards (jacks, queens and kings) and tens have a point value of ten. Aces have a point value of 1 or 11. A 52-card deck contains 16 cards with a point value of ten (jacks, queens, kings and tens) and four aces.
   a. What is the probability that both cards dealt are aces or ten-point cards?
   b. What is the probability that both of the cards are aces?
   c. What is the probability that both of the cards have a point value of ten?
   d. A blackjack is a ten-point card and an ace for a value of 21. Use your answers to parts (a), (b) and (c) to determine the probability that a player is dealt a blackjack. (Hint: Part (d) is not a hypergeometric problem. Develop your own logical relationship as to how the hypergeometric probabilities from parts (a), (b) and (c) can be combined to answer this question.)

40. A financial adviser is considering selecting four out of ten new share issues for a client's portfolio but as it turns out three of the ten result in profits and seven in losses.
   a. What would the probability distribution of X, the number of profitable issues in the adviser's selection, have been?
   b. How many profitable issues would the adviser have expected?

41. A machine produces an average of 20 per cent defective items. A batch is accepted if a sample of five items taken from the batch contains no defectives and rejected if the sample contains three or more defectives. Otherwise a second sample is taken.
   What is the probability that a second sample is required?

ONLINE RESOURCES

For the data files, additional questions and answers and the software section for Chapter 5, go to the online platform.

SUMMARY

A random variable provides a numerical description of the outcome of an experiment. The probability distribution for a random variable describes how the probabilities are distributed over the values the random variable can assume. A variety of examples are used to distinguish between discrete and continuous random variables. For any discrete random variable X, the probability distribution is defined by a probability function, denoted by $p(x) = p(X = x)$, which provides the probability associated with each value of the random variable. From the probability function, the expected value, variance and standard deviation for the random variable can be computed and relevant interpretations of these terms are provided.

Particular attention was devoted to the binomial distribution, which can be used to determine the probability of x successes in n trials whenever the experiment has the following properties:

1 The experiment consists of a sequence of n identical trials.
2 Two outcomes are possible on each trial, one called success and the other failure.
3 The probability of a success, π, does not change from trial to trial. Consequently, the probability of failure, $1 - \pi$, does not change from trial to trial.
4 The trials are independent.

Formulae were also presented for the probability function, mean and variance of the binomial distribution.

The Poisson distribution can be used to determine the probability of obtaining x occurrences over an interval of time or space. The necessary assumptions for the Poisson distribution to apply in a given situation are that:

1 The probability of an occurrence of the event is the same for any two intervals of equal length.
2 The occurrence or non-occurrence of the event in any interval is independent of the occurrence or non-occurrence of the event in any other interval.

A third discrete probability distribution, the hypergeometric, was introduced in Section 5.7. Like the binomial, it is used to compute the probability of x successes in n trials. But, in contrast to the binomial, the probability of success changes from trial to trial.

KEY TERMS

Binomial experiment
Binomial probability distribution
Binomial probability function
Bivariate probability distribution
Continuous random variable
Covariance
Discrete random variable
Discrete uniform probability distribution
Empirical discrete distribution
Expected value

Hypergeometric probability distribution
Hypergeometric probability function
Poisson probability distribution
Poisson probability function
Probability distribution
Probability function
Random variable
Standard deviation
Variance

KEY FORMULAE

Discrete uniform probability function

$$p(x) = 1/n \tag{5.3}$$

where:

$n =$ the number of values the random variable may assume

Expected value of a discrete random variable

$$E(X) = \mu = \Sigma x p(x) \tag{5.4}$$

Variance of a discrete random variable

$$Var(X) = \sigma^2 = \Sigma(x - \mu)^2 p(x) \tag{5.5}$$

Covariance of random variables X and Y

$$Cov(X, Y) = \sigma_{XY} = [Var(X + Y) - Var(X) - Var(Y)]/2 \tag{5.6}$$

Correlation between random variables X and Y

$$\rho_{XY} = \frac{\sigma_{XY}}{\sigma_X \sigma_Y} \tag{5.7}$$

Expected value of a linear combination of random variables X and Y

$$E(aX + bY) = aE(X) + bE(Y) \tag{5.8}$$

Variance of a linear combination of two random variables

$$Var(aX + bY) = a^2 Var(X) + b^2 Var(Y) + 2abCov(X, Y) \tag{5.9}$$

Number of experimental outcomes providing exactly x successes in n trials

$$\binom{n}{x} = \frac{n!}{x!(n - x)!} \tag{5.10}$$

Binomial probability function

$$p(x) = \binom{n}{x}\pi^x(1 - \pi)^{(n-x)} \tag{5.12}$$

Expected value for the binomial distribution

$$E(X) = \mu = n\pi \qquad \textbf{(5.13)}$$

Variance for the binomial distribution

$$Var(X) = \sigma^2 = n\pi(1 - \pi) \qquad \textbf{(5.14)}$$

Poisson probability function

$$p(x) = \frac{\mu^x e^{-\mu}}{x!} \qquad \textbf{(5.15)}$$

Hypergeometric probability function

$$p(x) = \frac{\binom{r}{x}\binom{N-r}{n-x}}{\binom{N}{n}} \qquad \textbf{(5.16)}$$

Expected value for the hypergeometric distribution

$$E(X) = \mu = n\left(\frac{r}{N}\right) \qquad \textbf{(5.17)}$$

Variance for the hypergeometric distribution

$$Var(X) = \sigma^2 = n\left(\frac{r}{N}\right)\left(1 - \frac{r}{N}\right)\left(\frac{N-n}{N-1}\right) \qquad \textbf{(5.18)}$$

CASE PROBLEM 1

Adapting a bingo game

Gaming Machines International (GMI) is investigating the adaptation of one of its bingo machine formats to allow for a bonus game facility. With the existing setup, the player has to select seven numbers from the series 1 to 80. Fifteen numbers are then drawn randomly from the 80 available and prizes awarded, according to how many of the 15 coincide with the player's selection, as follows:

Number of 'hits'	Payoff
0	0
1	0
2	0
3	1
4	10
5	100
6	1,000
7	100,000

With the new 'two ball bonus draw' feature, players effectively have the opportunity to improve their prize by buying an extra two balls. Note, however, that the bonus draw is only expected to be available to players who have scored 2, 3, 4 or 5 hits in the main game.

Managerial report

1. Determine the probability characteristics of GMI's original bingo game and calculate the player's expected payoff.

2. Derive corresponding probability details for the proposed bonus game. What is the probability of the player scoring:
   a. 0 hits
   b. 1 hit
   c. 2 hits
   with the extra two balls?

3. Use the results obtained from 2. to revise the probability distribution found for 1. Hence calculate the player's expected payoff in the enhanced game. Comment on how much the player might be charged for the extra gamble.

CASE PROBLEM 2

European airline overbooking

EU Regulation 261/2004 sets the minimum levels of passenger compensation for denied boarding due to overbooking, and extends its coverage to include flight cancellations and long delays.

EA is a small, short-range airline headquartered in Vienna. It has a fleet of small Fokker planes with a capacity of 80 passengers each. They do not have different classes in their planes. In planning for their financial obligations, EA has requested a study of the chances of 'bumping' passengers they have to consider for their overbooking strategy. The airline reports a historical 'no shows' history of 10 to 12 per cent. Compensation has been set at €250 per passenger denied boarding.

Managerial report

Write a report giving the airline some scenarios of their options. Consider scenarios according to their policy of the number of bookings/plane: 80, 85, 89, etc.

1. What percentage of the time should they estimate that their passengers will find a seat when they show up?

2. What percentage of the time may some passengers not find a seat?

3. In each case you consider, find the average amount of loss per plane they have to take into account.

© Pyrosky/iStock

6

Continuous Probability Distributions

LEARNING OBJECTIVES   After reading this chapter and doing the exercises, you should be able to:

1 Understand the difference between how probabilities are computed for discrete and continuous random variables.

2 Compute probability values for a continuous uniform probability distribution and be able to compute the expected value and variance for such a distribution.

3 Compute probabilities using a normal probability distribution. Understand the role of the standard normal distribution in this process.

4 Use the normal distribution to approximate binomial probabilities.

5 Compute probabilities using an exponential probability distribution.

6 Understand the relationship between the Poisson and exponential probability distributions.

In this chapter we turn to the study of continuous random variables. Specifically, we discuss three continuous probability distributions: the uniform, the normal and the exponential. A fundamental difference separates discrete and continuous random variables in terms of how probabilities are computed. For a discrete random variable, the probability function $p(x)$ provides the probability that the random variable assumes a particular value. With continuous random variables the counterpart of the probability function is the probability density function, denoted by $f(x)$. The difference is that the probability density function does not directly provide probabilities. However, the area under the graph of $f(x)$ corresponding to a given interval does provide the probability that the continuous random variable X assumes a value in that interval. So when we compute probabilities for continuous random variables we are computing the probability that the random variable assumes any value in an interval.

One of the implications of the definition of probability for continuous random variables is that the probability of any particular value of the random variable is zero, because the area under the graph of $f(x)$ at any particular point is zero. In Section 6.1 we demonstrate these concepts for a continuous random variable that has a uniform distribution.

Much of the chapter is devoted to describing and showing applications of the normal distribution. The main importance of normal distribution is its extensive use in statistical inference. The chapter closes with a discussion of the exponential distribution.

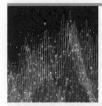

STATISTICS IN PRACTICE
Assessing the effectiveness of new medical procedures

Clinical trials are a vital and commercially very important application of statistics, typically involving the random assignment of patients to two experimental groups. One group receives the treatment of interest, the second a placebo (a dummy treatment that has no effect). To assess the evidence that the probability of success with the treatment will be better than that with the placebo, frequencies a, b, c and d can be collected for a predetermined number of trials according to the following two-way table:

	Treatment	Placebo
Success	a	b
Failure	c	d

and the quantity ('log odds ratio') $X = \log (a/c/b/d)$ calculated. Clearly the larger the value of X obtained the greater the evidence that the treatment is better than the placebo.

In the particular case that the treatment has no effect, the distribution of X can be shown to align very closely to a normal distribution with a mean of zero and standard error (refer to Chapter 7):

$$SE = \sqrt{\frac{1}{a} + \frac{1}{b} + \frac{1}{c} + \frac{1}{d}}$$

Hence, as values of X fall increasingly to the right of the zero this means it should signify stronger and stronger support for belief in the treatment's relative effectiveness.

Interestingly, a similar formulation to this – used by Jacobson and Jakela (2021) on data collected during the recent COVID-19 pandemic – revealed that 15- to 44-year-old males experienced a significant increase in death risk, even though the absolute number of COVID-19 deaths for their particular cohort was small.

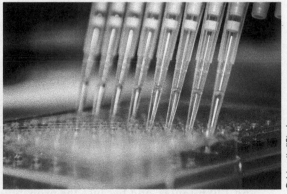

© Jvisentin/iStock

Sources:
Jacobson, S. H. and Jokela, J. A. (2021) 'Beyond COVID-19 deaths during the COVID-19 pandemic in the United States'. *Health Care Management Science*, 24: 661–665.
Szumilas, M. (2010) 'Explaining Odds ratios'. *J. Can. Acad. Child Adolesc. Psychiatry*, 19(3): 227–229.

6.1 UNIFORM PROBABILITY DISTRIBUTION

Consider the random variable X representing the flight time of an aeroplane travelling from Graz to Stansted. Suppose the flight time can be any value in the interval from 120 minutes to 140 minutes. Because the random variable X can assume any value in that interval, X is a continuous rather than a discrete random variable. Let us assume that sufficient actual flight data are available to conclude that the probability of a flight time within any one-minute interval is the same as the probability of a flight time within any other one-minute interval contained in the larger interval from 120 to 140 minutes. With every one-minute interval being equally likely, the random variable X is said to have a **uniform probability distribution**.

If x is any number lying in the range that the random variable X can take then the probability density function, which defines the uniform distribution for the flight-time random variable, is:

$$f(x) = \begin{cases} 1/20 & \text{for } 120 \leq x \leq 140 \\ 0 & \text{elsewhere} \end{cases}$$

Figure 6.1 is a graph of this probability density function. In general, the uniform probability density function for a random variable X is defined by the following formula:

Uniform probability density function

$$f(x) = \begin{cases} \dfrac{1}{b - a} & \text{for } a \leq x \leq b \\ 0 & \text{elsewhere} \end{cases} \tag{6.1}$$

FIGURE 6.1
Uniform probability density function for flight time

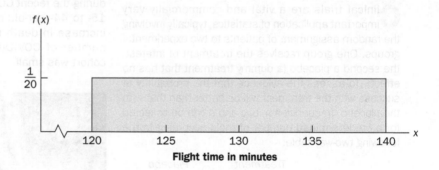

For the flight-time random variable, $a = 120$ and $b = 140$.

As noted in the introduction, for a continuous random variable, we consider probability only in terms of the likelihood that a random variable assumes a value within a specified interval. In the flight time example, we might ask what is the probability that the flight time is between 120 and 130 minutes? That is, what is $P(120 \leq X \leq 130)$? Because the flight time must be between 120 and 140 minutes, and because the probability is described as being uniform over this interval, we feel comfortable saying $P(120 \leq X \leq 130) = 0.50$. In the following subsection we show that this probability can be computed as the area under the graph of $f(x)$ from 120 to 130 (refer to Figure 6.2).

FIGURE 6.2
Area provides probability of flight time between 120 and 130 minutes

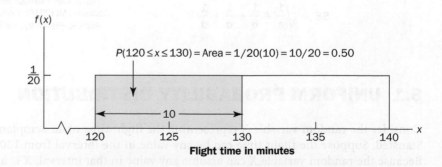

Area as a measure of probability

Let us make an observation about the graph in Figure 6.2. Consider the area under the graph of $f(x)$ in the interval from 120 to 130. The area is rectangular, and the area of a rectangle is simply the width multiplied by the height. With the width of the interval equal to $130 - 120 = 10$ and the height equal to the value of the probability density function $f(x) = 1/20$, we have area = width $\times$ *height* $= 10 \times 1/20 = 10/20 = 0.50$.

What observation can you make about the area under the graph of $f(x)$ and probability? They are identical! Indeed, this observation is valid for all continuous random variables. Once a probability density function $f(x)$ is identified, the probability that X takes a value x between some lower value x_1 and some higher value x_2 can be found by computing the area under the graph of $f(x)$ over the interval from x_1 to x_2.

Given the uniform distribution for flight time and using the interpretation of area as probability, we can answer any number of probability questions about flight times. For example, what is the probability of a flight time between 128 and 136 minutes? The width of the interval is $136 - 128 = 8$. With the uniform height of $f(x) = 1/20$, we see that $P(128 \leq X \leq 136) = 8 \times 1/20 = 0.40$. Note that $P(120 \leq X \leq 140) = 20 \times 1/20 = 1$; that is, the total area under the graph of $f(x)$ is equal to 1. This property holds for all continuous probability distributions and is the analogue of the condition that the sum of the probabilities must equal 1 for a discrete probability function. For a continuous probability density function, we must also require that $f(x) \geq 0$ for all values of x. This requirement is the analogue of the requirement that $p(x) \geq 0$ for discrete probability functions.

Two major differences stand out between the treatment of continuous random variables and the treatment of their discrete counterparts:

1 We no longer talk about the probability of the random variable assuming a particular value. Instead, we talk about the probability of the random variable assuming a value within some given interval.

2 The probability of the random variable assuming a value within some given interval from x_1 to x_2 is defined as the area under the graph of the probability density function between x_1 and x_2. Because a single point corresponds to an interval of zero width, this implies that the probability of a continuous random variable assuming any particular value, is exactly zero. It also means that the probability of a continuous random variable assuming a value in any interval is the same whether or not the endpoints are included.

The calculation of the expected value and variance for a continuous random variable is analogous to that for a discrete random variable. However, because the computational procedure involves integral calculus, we leave the derivation of the appropriate formulae to more advanced texts.

For the uniform continuous probability distribution introduced in this section, the formulae for the expected value and variance are:

$$E(X) = \frac{a + b}{2}$$

$$Var(X) = \frac{(b - a)^2}{12}$$

In these formulae, a is the smallest value and b is the largest value that the random variable may assume.

Applying these formulae to the uniform distribution for flight times from Graz to Stansted, we obtain:

$$E(X) = \frac{(120 + 140)}{2} = 130$$

$$Var(X) = \frac{(140 - 120)^2}{12} = 33.33$$

The standard deviation of flight times can be found by taking the square root of the variance. Hence, $\sigma = 5.77$ minutes.

EXERCISES

Methods

1. The random variable X is known to be uniformly distributed between 1.0 and 1.5.
   a. Show the graph of the probability density function.
   b. Compute $P(X = 1.25)$.
   c. Compute $P(1.0 \leq X \leq 1.25)$.
   d. Compute $P(1.20 < X < 1.5)$.

2. The random variable X is known to be uniformly distributed between 10 and 20.
   a. Show the graph of the probability density function.
   b. Compute $P(X < 15)$.
   c. Compute $P(12 \leq X \leq 18)$.
   d. Compute $E(X)$.
   e. Compute $Var(X)$.

Applications

3. A continuous random variable X has probability density function:

$$f(x) = kx \qquad 0 < x < 2$$
$$0 \text{ } otherwise$$

   a. Determine the value of k.
   b. Find $E(X)$ and $Var(X)$.
   c. What is the probability that X is greater than three standard deviations above the mean?
   d. Find the distribution function $F(X)$ and hence the median of X.

4. Most computer languages include a function that can be used to generate random numbers. In Excel, the RAND function can be used to generate random numbers between 0 and 1. If we let X denote a random number generated using RAND, then X is a continuous random variable with the following probability density function.

$$f(x) = \begin{cases} 1 & \text{for } 0 \leq x \leq 1 \\ 0 & \text{elsewhere} \end{cases}$$

   a. Graph the probability density function.
   b. What is the probability of generating a random number between 0.25 and 0.75?
   c. What is the probability of generating a random number with a value less than or equal to 0.30?
   d. What is the probability of generating a random number with a value greater than 0.60?

5. The electric vehicle manufacturing company Tesla estimates that a driver who commutes 50 miles per day in a Model S will require a nightly charge time of around 1 hour and 45 minutes (105 minutes) to recharge the vehicle's battery (Tesla company website). Assume that the actual recharging time required is uniformly distributed between 90 and 120 minutes.
   a. Give a mathematical expression for the probability density function of battery recharging time for this scenario.
   b. What is the probability that the recharge time will be less than 110 minutes?
   c. What is the probability that the recharge time required is at least 100 minutes?
   d. What is the probability that the recharge time required is between 95 and 110 minutes?

6. The label on a bottle of liquid detergent shows contents to be 12 grams per bottle. The production operation fills the bottle uniformly according to the following probability density function.

$$f(x) = \begin{cases} 8 & \text{for } 11.975 \leq x \leq 12.100 \\ 0 & \text{elsewhere} \end{cases}$$

a. What is the probability that a bottle will be filled with between 12 and 12.05 grams?
b. What is the probability that a bottle will be filled with 12.02 or more grams?
c. Quality control accepts a bottle that is filled to within 0.02 grams of the number of grams shown on the container label. What is the probability that a bottle of this liquid detergent will fail to meet the quality control standard?

7. Suppose we are interested in bidding on a piece of land and we know there is one other bidder. The seller announces that the highest bid in excess of € 10,000 will be accepted. Assume that the competitor's bid X is a random variable that is uniformly distributed between €10,000 and € 15,000.
a. Suppose you bid € 12,000. What is the probability that your bid will be accepted?
b. Suppose you bid € 14,000. What is the probability that your bid will be accepted?
c. What amount should you bid to maximize the probability that you get the property?
d. Suppose you know someone who is willing to pay you € 16,000 for the property. Would you consider bidding less than the amount in part (c)? Why or why not?

6.2 NORMAL PROBABILITY DISTRIBUTION

The most commonly used probability distribution for describing a continuous random variable is the normal probability distribution. The normal distribution has been used in a wide variety of practical applications in which the random variables are heights and weights of people, test scores, scientific measurements, amounts of rainfall and so on. It is also widely used in statistical inference, which is the major topic of the remainder of this book. In such applications, the normal distribution provides a description of the likely results obtained through sampling.

Normal curve

The form, or shape, of the normal distribution is illustrated by the bell-shaped normal curve in Figure 6.3. The probability density function that defines the bell-shaped curve of the normal distribution follows.

Normal probability density function

$$f(x) = \frac{1}{\sigma\sqrt{2\pi}} e^{-(x-\mu)^2/2\sigma^2} \qquad (6.2)$$

where:

μ = mean

σ = standard deviation

π = 3.14159

e = 2.71828

We make several observations about the characteristics of the normal distribution.

1 The entire family of normal distributions is differentiated by its mean μ and its standard deviation σ.
2 The highest point on the normal curve is at the mean, which is also the median and mode of the distribution.

FIGURE 6.3
Bell-shaped curve for the normal distribution

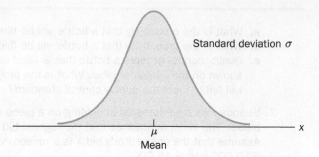

3 The mean of the distribution can be any numerical value: negative, zero or positive. Three normal distributions with the same standard deviation but three different means (−10, 0 and 20) are shown in Figure 6.4.

FIGURE 6.4
Normal distributions with same standard deviation and different means

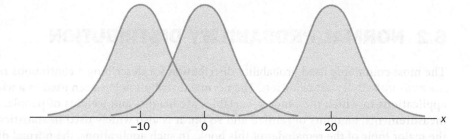

4 The normal distribution is symmetrical, with the shape of the curve to the left of the mean a mirror image of the shape of the curve to the right of the mean. The tails of the curve extend to infinity in both directions and theoretically never touch the horizontal axis. Because it is symmetrical, the normal distribution is not skewed; its skewness measure is zero.

5 The standard deviation determines how flat and wide the curve is. Larger values of the standard deviation result in wider, flatter curves, showing more variability in the data. Two normal distributions with the same mean but with different standard deviations are shown in Figure 6.5.

FIGURE 6.5
Normal distributions with same mean and different standard deviations

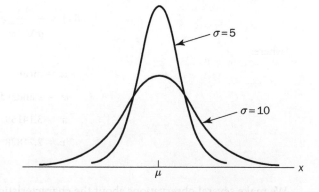

6 Probabilities for the normal random variable are given by areas under the curve. The total area under the curve for the normal distribution is 1. Because the distribution is symmetrical, the area under the curve to the left of the mean is 0.50 and the area under the curve to the right of the mean is 0.50.

7 The percentage of values in some commonly used intervals are:

    a. 68.3 per cent of the values of a normal random variable are within plus or minus one standard deviation of its mean.

    b. 95.4 per cent of the values of a normal random variable are within plus or minus two standard deviations of its mean.

    c. 99.7 per cent of the values of a normal random variable are within plus or minus three standard deviations of its mean.

Figure 6.6 shows properties (a), (b) and (c) graphically.

FIGURE 6.6
Areas under the curve for any normal distribution

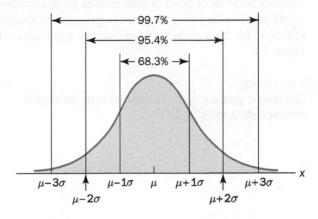

Standard normal probability distribution

A random variable that has a normal distribution with a mean of 0 and a standard deviation of 1 is said to have a **standard normal probability distribution**. The letter Z is commonly used to designate this particular normal random variable. Figure 6.7 is the graph of the standard normal distribution. It has the same general appearance as other normal distributions but with the special properties of $\mu = 0$ and $\sigma = 1$.

FIGURE 6.7
The standard normal distribution

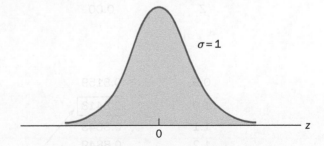

Standard normal density function

$$f(z) = \frac{1}{\sqrt{2\pi}} e^{-z^2}$$

Because $\mu = 0$ and $\sigma = 1$, the formula for the standard normal probability density function is a simpler version of equation (6.2).

As with other continuous random variables, probability calculations with any normal distribution are made by computing areas under the graph of the probability density function. Hence, to find the probability that a normal random variable is within any specific interval, we must compute the area under the normal curve over that interval.

For the standard normal distribution, areas under the normal curve have been computed and are available from Table 1 of Appendix B.

The three types of probabilities we need to compute include (1) the probability that the standard normal random variable Z will be less than or equal to a given value; (2) the probability that Z will take a value between two given values; and (3) the probability that Z will be greater than or equal to a given value. To see how the cumulative probability table for the standard normal distribution can be used to compute these three types of probabilities, let us consider some examples.

We start by showing how to compute the probability that Z is less than or equal to 1.00; that is, $P(Z \leq 1.00)$. This cumulative probability is the area under the normal curve to the left of $z = 1.00$ in Figure 6.8.

FIGURE 6.8
Cumulative probability for standard normal distribution
corresponding to $P(Z \leq 1.00)$

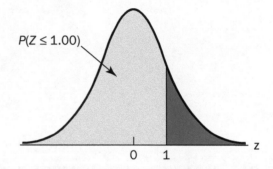

Refer to Table 1 in Appendix B. The cumulative probability corresponding to $z = 1.00$ is the table value located at the intersection of the row labelled 1.0 and the column labelled 0.00. First we find 1.0 in the left column of the table and then find 0.00 in the top row of the table. By looking in the body of the table, we find that the 1.0 row and the 0.00 column intersect at the value of 0.8413; thus, $P(Z \leq 1.00) = 0.8413$. The following excerpt from the probability table shows these steps:

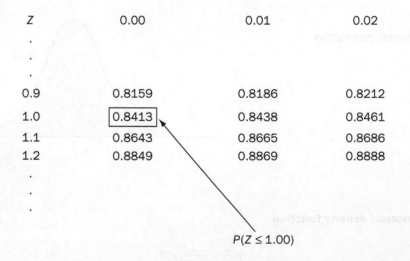

Z	0.00	0.01	0.02
.			
.			
.			
0.9	0.8159	0.8186	0.8212
1.0	0.8413	0.8438	0.8461
1.1	0.8643	0.8665	0.8686
1.2	0.8849	0.8869	0.8888
.			
.			
.			

$P(Z \leq 1.00)$

To illustrate the second type of probability calculation we show how to compute the probability that Z is in the interval between -0.50 and 1.25; that is, $P(-0.50 \leq Z \leq 1.25)$. Figure 6.9 shows this area of probability.

FIGURE 6.9

Cumulative probability for standard normal
distribution corresponding to $P(-0.50 \leq Z \leq 1.25)$

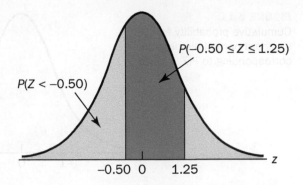

$P(Z < -0.50)$

$P(-0.50 \leq Z \leq 1.25)$

$-0.50 \; 0 \qquad 1.25$

z

Three steps are required to compute this probability. First, we find the area under the normal curve to the left of $z = 1.25$. Second, we find the area under the normal curve to the left of $z = -0.50$. Finally, we subtract the area to the left of $z - 0.50$ from the area to the left of $z = 1.25$ to find $P(-0.5 \leq Z \leq 1.25)$.

To find the area under the normal curve to the left of $z = 1.25$, we first locate the 1.2 row in the standard normal probability table in Appendix B and then move across to the 0.05 column. Because the table value in the 1.2 row and the 0.05 column is 0.8944, $P(Z \leq 1.25) = 0.8944$. Similarly, to find the area under the curve to the left of $z = -0.50$ we use the left-hand page of the table to locate the table value in the -0.5 row and the 0.00 column; with a table value of 0.3085, $P(Z \leq -0.50) = 0.3085$. Hence, $P(-0.50 \leq Z \leq 1.25) = P(Z \leq 1.25) - P(Z \leq -0.50) = 0.8944 - 0.3085 = 0.5859$.

Let us consider another example of computing the probability that Z is in the interval between two given values. Often it is of interest to compute the probability that a normal random variable assumes a value within a certain number of standard deviations of the mean. Suppose we want to compute the probability that the standard normal random variable is within one standard deviation of the mean; that is, $P(-1.00 \leq Z \leq 1.00)$.

To compute this probability we must find the area under the curve between -1.00 and 1.00. Earlier we found that $P(Z \leq 1.00) = 0.8413$. Referring again to Table 1 in Appendix B, we find that the area under the curve to the left of $z = -1.00$ is 0.1587, so $P(Z \leq -1.00) = 0.1587$. Therefore $P(-1.00 \leq Z \leq 1.00) = P(Z \leq 1.00) - P(Z - 1.00) = 0.8413 - 0.1587 = 0.6826$. This probability is shown graphically in Figure 6.10.

FIGURE 6.10

Cumulative probability for standard
normal distribution corresponding
to $P(-1.00 \leq Z \leq 1.00)$

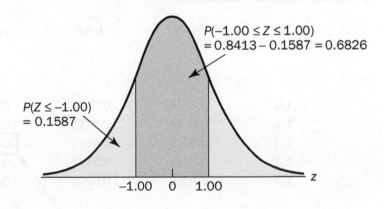

$P(Z \leq -1.00)$
$= 0.1587$

$P(-1.00 \leq Z \leq 1.00)$
$= 0.8413 - 0.1587 = 0.6826$

$-1.00 \quad 0 \quad 1.00$

z

To illustrate how to make the third type of probability computation, suppose we want to compute the probability of obtaining a z value of at least 1.58; that is, $P(Z \geq 1.58)$. The value in the $z = 1.5$ row and the 0.08 column of the cumulative normal table in Appendix B is 0.9429; thus, $P(Z < 1.58) = 0.9429$. However, because the total area under the normal curve is 1, $P(Z \geq 1.58) = 1 - 0.9429 = 0.0571$. This probability is shown in Figure 6.11.

FIGURE 6.11
Cumulative probability for
standard normal distribution
corresponding to $P(Z \geq 1.58)$

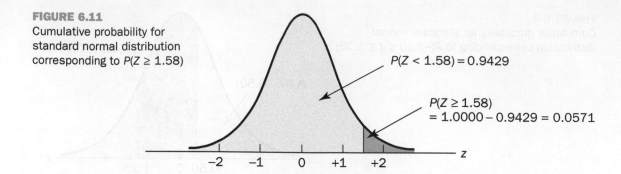

$P(Z < 1.58) = 0.9429$

$P(Z \geq 1.58)$
$= 1.0000 - 0.9429 = 0.0571$

In the preceding illustrations, we showed how to compute probabilities given specified z values. In some situations, we are given a probability and are interested in working backwards to find the corresponding z value. Suppose we want to find a z value such that the probability of obtaining a larger z value is 0.10. Figure 6.12 shows this situation graphically.

FIGURE 6.12
Determining z value such that the
probability of finding a larger value
is 0.10

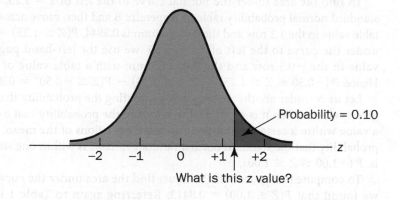

Probability = 0.10

What is this z value?

This problem is the inverse of those in the preceding examples. Previously, we specified the z value of interest and then found the corresponding probability, or area. In this example, we are given the probability, or area, and asked to find the corresponding z value. To do so, we use the standard normal probability table somewhat differently.

z	0.06	0.07	0.08	0.09
.				
.				
.				
1.0	0.8554	0.8577	0.8599	0.8621
1.1	0.8770	0.8790	0.8810	0.8830
1.2	0.8962	0.8980	0.8997	0.9015
1.3	0.9131	0.9147	0.9162	0.9177
1.4	0.9279	0.9292	0.9306	0.9319
.				
.				
.				

Cumulative probability value
closest to 0.9000

Recall that the standard normal probability table gives the area under the curve to the left of a particular z value. We have been given the information that the area in the upper tail of the curve is 0.10. Hence, the area under the curve to the left of the unknown z value must equal 0.9000. Scanning the body of the table, we find 0.8997 is the cumulative probability value closest to 0.9000. The section of the table providing this result is shown above. Reading the z value from the left-most column and the top row of the table, we find that the corresponding z value is 1.28. Hence, an area of approximately 0.9000 (actually 0.8997) will be to the left of $z = 1.28$.* In terms of the question originally asked, the probability is approximately 0.10 that the z value will be larger than 1.28.*

The examples illustrate that the table of areas for the standard normal distribution can be used to find probabilities associated with values of the standard normal random variable Z. Two types of questions can be asked. The first type of question specifies a value, or values, for z and asks us to use the table to determine the corresponding areas, or probabilities.

The second type of question provides an area, or probability, and asks us to use the table to determine the corresponding z value. Hence, we need to be flexible in using the standard normal probability table to answer the desired probability question. In most cases, sketching a graph of the standard normal distribution and shading the appropriate area or probability helps to visualize the situation and aids in determining the correct answer.

Computing probabilities for any normal distribution

The reason for discussing the standard normal distribution so extensively is that probabilities for all normal distributions are computed by using the standard normal distribution. That is, when we have a normal distribution with any mean μ and any standard deviation σ, we answer probability questions about the distribution by first converting to the standard normal distribution. Then we can use the standard normal probability table and the appropriate z values to find the desired probabilities. The formula used to convert any normal random variable X with mean μ and standard deviation σ to the standard normal distribution follows as equation (6.3).

Converting to the standard normal distribution

$$Z = \frac{X - \mu}{\sigma}$$

(6.3)

A value of X equal to the mean μ results in $z = (\mu - \mu)/\sigma = 0$. Hence, we see that a value of X equal to the mean μ of X corresponds to a value of Z at the mean 0 of Z. Now suppose that x is one standard deviation greater than the mean; that is, $x = \mu + \sigma$. Applying equation (6.3), we see that the corresponding z value $[(\mu + \sigma) - \mu]/\sigma = \sigma/\sigma = 1$. Hence, a value of X that is one standard deviation above the mean μ of X corresponds to a z value $= 1$. In other words, we can interpret Z as the number of standard deviations that the normal random variable X is from its mean μ.

To see how this conversion enables us to compute probabilities for any normal distribution, suppose we have a normal distribution with $\mu = 10$ and $\sigma = 2$. What is the probability that the random variable X is between 10 and 14? Using equation (6.3) we see that at $x = 10$, $z = (x - \mu)/\sigma = (10 - 10)/2 = 0$ and that at $x = 14$, $z = (14 - 10)/2 = 4/2 = 2$. Hence, the answer to our question about the probability of X being between 10 and 14 is given by the equivalent probability that Z is between 0 and 2 for the standard normal distribution.

In other words, the probability that we are seeking is the probability that the random variable X is between its mean and two standard deviations greater than the mean. Using $z = 2.00$ and standard normal probability table, we see that $P(Z \le 2) = 0.9772$. Because $P(Z \le 0) = 0.5000$ we can compute $P(0.00 \le Z \le 2.00) = P(Z \le 2) - P(Z \le 0) = 0.9772 - 0.5000 = 0.4772$. Hence the probability that X is between 10 and 14 is 0.4772.

---

* We could use interpolation in the body of the table to get a better approximation of the z value that corresponds to an area of 0.9000. Doing so provides one more decimal place of accuracy and yields a z value of 1.282. However, in most practical situations, sufficient accuracy is obtained by simply using the table value closest to the desired probability.

Greer Tyre Company problem

We turn now to an application of the normal distribution. Suppose the Greer Tyre Company has just developed a new steel-belted radial tyre that will be sold through a national chain of discount stores. Because the tyre is a new product, Greer's managers believe that the kilometres guarantee offered with the tyre will be an important factor in the acceptance of the product. Before finalizing the kilometres guarantee policy, Greer's managers want probability information about the number of kilometres the tyres will last.

From actual road tests with the tyres, Greer's engineering group estimates the mean number of kilometres the tyre will last is $\mu = 36,500$ kilometres and that the standard deviation is $\sigma = 5,000$. In addition, the data collected indicate a normal distribution is a reasonable assumption. What percentage of the tyres can be expected to last more than 40,000 kilometres?

In other words, what is the probability that the number of kilometres the tyre lasts will exceed 40,000? This question can be answered by finding the area of the darkly shaded region in Figure 6.13. At $x = 40,000$, we have:

$$Z = \frac{X - \mu}{\sigma} = \frac{40,000 - 35,500}{5,000} = \frac{3,500}{5,000} = 0.70$$

FIGURE 6.13
Greer Tyre Company kilometres distribution

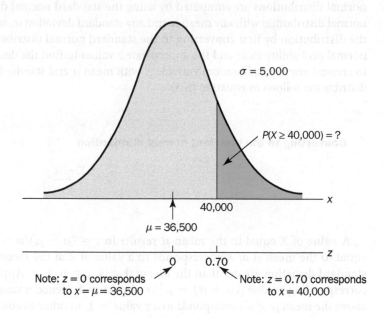

Refer now to the bottom of Figure 6.13. We see that a value of $x = 40,000$ on the Greer Tyre normal distribution corresponds to a value of $z = 0.70$ on the standard normal distribution. Using the standard normal probability table, we see that the area to the left of $z = 0.70$ is 0.7580. Referring again to Figure 6.13, we see that the area to the left of $x = 40,000$ on the Greer Tyre normal distribution is the same. Hence, $1.000 - 0.7580 = 0.2420$ is the probability that X will exceed 40,000. We can conclude that about 24.2 per cent of the tyres will last longer than 40,000 kilometres.

Let us now assume that Greer is considering a guarantee that will provide a discount on replacement tyres if the original tyres do not exceed the number of kilometres stated in the guarantee. What should the guaranteed number of kilometres be if Greer wants no more than 10 per cent of the tyres to be eligible for the discount guarantee? This question is interpreted graphically in Figure 6.14.

According to Figure 6.14, the area under the curve to the left of the unknown guaranteed number of kilometres must be 0.10. So we must find the z value that cuts off an area of 0.10 in the left tail of a standard normal distribution. Using the standard normal probability table, we see that $z = -1.28$ cuts off an area of 0.10 in the lower tail.

FIGURE 6.14
Greer's discount guarantee

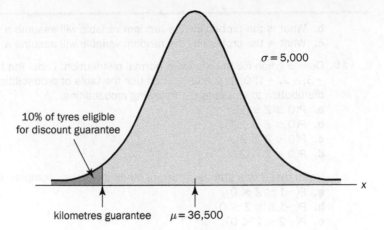

$\sigma = 5{,}000$

10% of tyres eligible
for discount guarantee

kilometres guarantee $\mu = 36{,}500$

Hence $z = -1.28$ is the value of the standard normal variable corresponding to the desired number of kilometres guarantee on the Greer Tyre normal distribution. To find the value of X corresponding to $z = -1.28$, we have:

$$z = \frac{x - \mu}{\sigma} = -1.28$$

$$x - \mu = -1.28\sigma$$

$$x = \mu - 1.28\sigma$$

With $\mu = 36{,}500$ and $\sigma = 5{,}000$,

$$x = 36{,}500 - 1.28 \times 5{,}000 = 30{,}100$$

Hence, a guarantee of 30,100 kilometres will meet the requirement that approximately 10 per cent of the tyres will be eligible for the guarantee. Perhaps, with this information, the firm will set its tyre kilometres guarantee at 30,000 kilometres.

Again, we see the important role that probability distributions play in providing decision-making information. Namely, once a probability distribution is established for a particular application, it can be used quickly and easily to obtain probability information about the problem. Probability does not establish a decision recommendation directly, but it provides information that helps the decision-maker better understand the risks and uncertainties associated with the problem. Ultimately, this information may assist the decision-maker in reaching a good decision.

EXERCISES

Methods

8. Using Figure 6.6 as a guide, sketch a normal curve for a random variable X that has a mean of $\mu = 100$ and a standard deviation of $\sigma = 10$. Label the horizontal axis with values of 70, 80, 90, 100, 110, 120 and 130.

9. A random variable is normally distributed with a mean of $\mu = 50$ and a standard deviation of $\sigma = 5$.
   a. Sketch a normal curve for the probability density function. Label the horizontal axis with values of 35, 40, 45, 50, 55, 60 and 65. Figure 6.6 shows that the normal curve almost touches the horizontal axis at three standard deviations below and at three standard deviations above the mean (in this case at 35 and 65).

   b. What is the probability the random variable will assume a value between 45 and 55?

   c. What is the probability the random variable will assume a value between 40 and 60?

10. Draw a graph for the standard normal distribution. Label the horizontal axis at values of $-3, -2, -1, 0, 1, 2$ and 3. Then use the table of probabilities for the standard normal distribution to compute the following probabilities.

   a. $P(0 \leq Z \leq 1)$.

   b. $P(0 \leq Z \leq 1.5)$.

   c. $P(0 < Z < 2)$.

   d. $P(0 < Z < 2.5)$.

11. Given that Z is a standard normal random variable, compute the following probabilities.

   a. $P(-1 \leq Z \leq 0)$.

   b. $P(-1.5 \leq Z \leq 0)$.

   c. $P(-2 < Z < 0)$.

   d. $P(-2.5 \leq Z \leq 0)$.

   e. $P(-3 \leq Z \leq 0)$.

12. Given that Z is a standard normal random variable, compute the following probabilities.

   a. $P(Z \leq -1.0)$.

   b. $P(Z \geq -1)$.

   c. $P(Z \geq -1.5)$.

   d. $P(-2.5 \leq Z)$.

   e. $P(-3 < Z \leq 0)$.

13. Given that Z is a standard normal random variable, compute the following probabilities.

   a. $P(-1.98 \leq Z \leq 0.49)$.

   b. $P(0.52 \leq Z \leq 1.22)$.

   c. $P(-1.75 \leq Z \leq -1.04)$.

14. Given that Z is a standard normal random variable, find z for each situation.

   a. The area to the left of z is 0.2119.

   b. The area between $-z$ and z is 0.9030.

   c. The area between $-z$ and z is 0.2052.

   d. The area to the left of z is 0.9948.

   e. The area to the right of z is 0.6915.

Applications

15. A machine makes electrical resistors having a mean resistance of 100 ohms with a standard deviation of 5 ohms. It is observed that a firm rejects approximately 11 per cent of the resistors (5.5 per cent over, 5.5 per cent under) because they are not within tolerance limits. Determine what tolerance limits are used by the firm.

16. Joburg Engineering manufactures replacement pistons for various types of engine. Each piston has to meet certain tolerance requirements, otherwise it will not perform satisfactorily when placed in the cylinder. Model A is designed for a particular engine that requires a piston diameter of 6.37 +/− 0.03 cm. Joburg's automatic lathe is currently set to produce pistons with a mean diameter of 6.38 cm and a standard deviation of 0.05 cm.

   a. What percentage of pistons produced on the automatic lathe will meet the specifications?

   b. By how much would the standard deviation of the output have to improve (decrease) in order to double the percentage in (a)?

17. The Barminster Bank is reviewing its service charges and interest paying policies on current accounts. The bank has found that the average daily balance is normally distributed with a mean of £580 and a standard deviation of £140.

a. What percentage of current account customers carry average balances in excess of £800?

b. What percentage of current account customers carry average daily balances between £300 and £700?

c. The bank is considering paying interest to customers carrying average daily balances in excess of a certain amount. If the bank does not want to pay interest to more than 5 per cent of its customers, what is the minimum average daily balance it should be willing to pay interest on?

d. The bank intends to impose service charges on those current account customers whose average daily balance falls below £400. What percentage of customers will pay service charges?

e. If five current account customers are selected at random, what is the probability that at least two of them will pay service charges?

18. The value of customer orders at a mail order company is normally distributed about a mean of £125 with a standard deviation of £21. They have decided that as an attempt to boost sales they will offer to post their products to their customers free provided they place a single order over a stated amount of money. They are unsure at what point they should continue to charge postage but feel that no more than 5 per cent of their customers should be charged.

a. How would you advise them?

b. What percentage of customers currently place orders with a value less than £50?

c. What percentage of their customers currently place orders with a value between £75 and £145?

d. Five per cent of their customers currently place orders below what value?

19. A person must score in the upper 2 per cent of the population on an IQ test to qualify for membership in Mensa, the international high IQ society. If IQ scores are normally distributed with a mean of 100 and a standard deviation of 15, what score must a person have to qualify for Mensa?

6.3 NORMAL APPROXIMATION OF BINOMIAL PROBABILITIES

In Chapter 5, Section 5.5, we presented the discrete binomial distribution. Recall that a binomial experiment consists of a sequence of n identical independent trials with each trial having two possible outcomes: a success or a failure. The probability of a success on a trial is the same for all trials and is denoted by π (Greek pi). The binomial random variable is the number of successes in the n trials, and probability questions pertain to the probability of x successes in the n trials. When the number of trials becomes large, evaluating the binomial probability function by hand or with a calculator is difficult. Hence, when we encounter a binomial distribution problem with a large number of trials, we may want to approximate the binomial distribution. In cases where the number of trials is greater than 20, $n\pi \geq 5$, and $n(1 - \pi) \geq 5$, the normal distribution provides an easy-to-use approximation of binomial probabilities.

When using the normal approximation to the binomial, we set $\mu = n\pi$ and $\sigma = \sqrt{n\pi(1 - \pi)}$ in the definition of the normal curve. Let us illustrate the normal approximation to the binomial by supposing that a particular company has a history of making errors in 10 per cent of its invoices. A sample of 100 invoices has been taken, and we want to compute the probability that 12 invoices contain errors. That is, we want to find the binomial probability of 12 successes in 100 trials.

In applying the normal approximation to the binomial, we set $\mu = n\pi = 100 \times 0.1 = 10$ and $\sigma = \sqrt{n\pi(1 - \pi)} = \sqrt{100 \times 0.1 \times 0.9} = 3$. A normal distribution with $\mu = 10$ and $\sigma = 3$ is shown in Figure 6.15.

FIGURE 6.15
Probability of 12 errors using
the normal approximation to a
binomial probability distribution
with $n = 100$ and $\pi = 0.10$

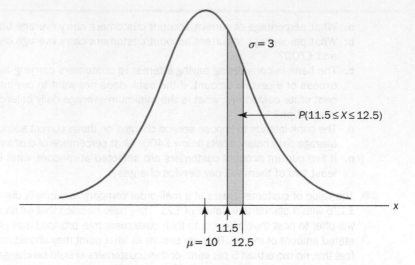

Recall that, with a continuous probability distribution, probabilities are computed as areas under the probability density function. As a result, the probability of any single value for the random variable is zero. Hence to approximate the binomial probability of 12 successes, we must compute the area under the corresponding normal curve between 11.5 and 12.5. The 0.5 that we add and subtract from 12 is called a **continuity correction factor**. It is introduced because a continuous distribution is being used to approximate a discrete distribution. Hence, $P(X = 12)$ for the *discrete* binomial distribution is approximated by $P(11.5 \leq X \leq 12.5)$ for the *continuous* normal distribution.

Converting to the standard normal distribution to compute $P(11.5 \leq X \leq 12.5)$, we have:

$$z = \frac{x - \mu}{\sigma} = \frac{12.5 - 10.0}{3} = 0.83 \text{ at } X = 12.5$$

And:

$$z = \frac{x - \mu}{\sigma} = \frac{11.5 - 10.0}{3} = 0.50 \text{ at } X = 11.5$$

Using the standard normal probability table, we find that the area under the curve (in Figure 6.15) to the left of 12.5 is 0.7967. Similarly, the area under the curve to the left of 11.5 is 0.6915. Therefore, the area between 11.5 and 12.5 is $0.7967 - 0.6915 = 0.1052$. The normal approximation to the probability of 12 successes in 100 trials is 0.1052. Out of interest this approximate value compares with a true binomial probability of 0.0988.

For another illustration, suppose we want to compute the probability of 13 or fewer errors in the sample of 100 invoices. Figure 6.16 shows the area under the normal curve that approximates this probability.

FIGURE 6.16
Probability of 13 or fewer errors
using the normal approximation to
a binomial probability distribution
with $n = 100$ and $\pi = 0.10$

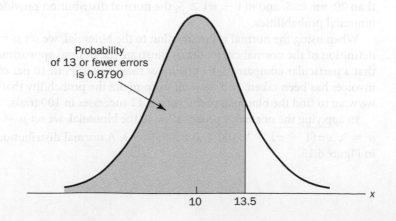

Note that the use of the continuity correction factor results in the value of 13.5 being used to compute the desired probability. The z value corresponding to $x = 13.5$ is:

$$z = \frac{13.5 - 10.0}{3} = 1.17$$

The standard normal probability table shows that the area under the standard normal curve to the left of 1.17 is 0.8790. The area under the normal curve approximating the probability of 13 or fewer errors is given by the shaded portion of the graph in Figure 6.16.

EXERCISES

Methods

20. A binomial probability distribution has $\pi = 0.20$ and $n = 100$.
    a. What is the mean and standard deviation?
    b. Is this a situation in which binomial probabilities can be approximated by the normal probability distribution? Explain.
    c. What is the probability of exactly 24 successes?
    d. What is the probability of 18 to 22 successes?
    e. What is the probability of 15 or fewer successes?

21. Assume a binomial probability distribution has $\pi = 0.60$ and $n = 200$.
    a. What is the mean and standard deviation?
    b. Is this a situation in which binomial probabilities can be approximated by the normal probability distribution? Explain.
    c. What is the probability of 100 to 110 successes?
    d. What is the probability of 130 or more successes?
    e. What is the advantage of using the normal probability distribution to approximate the binomial probabilities? Use part (d) to explain the advantage.

Applications

22. Suppose that of those individuals who play video and computer games, 18 per cent are under 18 years old, 53 per cent are 18–59 years old and 29 per cent are over 59 years old. Use the normal approximation of the binomial distribution to answer the questions below.
    a. For a sample of 800 people who play these games, how many would you expect to be under 18 years of age?
    b. For a sample of 600 people who play these games, what is the probability that fewer than 100 will be under 18 years of age?
    c. For a sample of 800 people who play these games, what is the probability that 200 or more will be over 59 years of age?

23. In 2015, Warren Buffett's Berkshire Hathaway company had been trading for 49 years, 39 years of which he beat the stock market. Some analysts – including Nobel laureates – argue this is simply a matter of luck. What if they are right and Buffett's probability of beating the stock market in any one year was just 0.5?
    a. Assuming this probability of success is true, what is the binomial probability of achieving Buffett's result or better?
    b. Estimate the probability in (a) using the normal approximation. What do you deduce?

6.4 EXPONENTIAL PROBABILITY DISTRIBUTION

The **exponential probability distribution** may be used for random variables such as the time between arrivals at a car wash, the time required to load a truck, the distance between major defects in a highway and so on. The exponential probability density function follows.

Exponential probability density function

$$f(x) = \frac{1}{\mu}e^{-x/\mu} \quad \text{for } x \geq 0, \mu > 0 \tag{6.4}$$

As an example of the exponential distribution, suppose that $X =$ the time it takes to load a truck at the Schips loading dock follows such a distribution. If the mean, or average, time to load a truck is 15 minutes ($\mu = 15$), the appropriate probability density function is:

$$f(x) = \frac{1}{15}e^{-x/15}$$

Figure 6.17 is the graph of this probability density function.

FIGURE 6.17
Exponential distribution for the Schips loading dock example

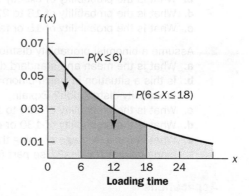

Computing probabilities for the exponential distribution

As with any continuous probability distribution, the area under the curve corresponding to an interval provides the probability that the random variable assumes a value in that interval. In the Schips loading dock example, the probability that loading a truck will take six minutes or less ($X \leq 6$) is defined to be the area under the curve in Figure 6.17 from $x = 0$ to $x = 6$.

Similarly, the probability that loading a truck will take 18 minutes or less ($X \leq 18$) is the area under the curve from $x = 0$ to $x = 18$. Note also that the probability that loading a truck will take between 6 minutes and 18 minutes ($6 \leq X \leq 18$) is given by the area under the curve from $x = 6$ to $x = 18$.

To compute exponential probabilities such as those just described, we use equation (6.5). It provides the cumulative probability of obtaining a value for the exponential random variable of less than or equal to some specific value denoted by x_0.

Exponential distribution: cumulative probabilities

$$P(X \leq x_0) = 1 - e^{-x_0/\mu} \tag{6.5}$$

For the Schips loading dock example, X = loading time and $\mu = 15$, which gives us:

$$P(X \leq x_0) = 1 - e^{-x_0/15}$$

Hence, the probability that loading a truck will take six minutes or less is:

$$P(X \leq 6) = 1 - e^{-6/15} = 0.3297$$

Using equation (6.5), we calculate the probability of loading a truck in 18 minutes or less:

$$P(X \leq 18) = 1 - e^{-18/15} = 0.6988$$

Hence, the probability that loading a truck will take between 6 minutes and 18 minutes is equal to $0.6988 - 0.3297 = 0.3691$. Probabilities for any other interval can be computed similarly.

In the preceding example, the mean time it takes to load a truck is $\mu = 15$ minutes. A property of the exponential distribution is that the mean of the distribution and the standard deviation of the distribution are *equal*. Hence, the standard deviation for the time it takes to load a truck is $\sigma = 15$ minutes.

Relationship between the Poisson and exponential distributions

In Chapter 5, Section 5.6, we introduced the Poisson distribution as a discrete probability distribution that is often useful in examining the number of occurrences of an event over a specified interval of time or space. Recall that the Poisson probability function is:

$$p(x) = \frac{\mu^x e^{-\mu}}{x!}$$

where:

μ = expected value or mean number of occurrences over a specified interval.

The continuous exponential probability distribution is related to the discrete Poisson distribution. If the Poisson distribution provides an appropriate description of the number of occurrences per interval, the exponential distribution provides a description of the length of the interval between occurrences.

To illustrate this relationship, suppose the number of cars that arrive at a car wash during one hour is described by a Poisson probability distribution with a mean of ten cars per hour. The Poisson probability function that gives the probability of X arrivals per hour is:

$$p(x) = \frac{10^x e^{-10}}{x!}$$

Because the average number of arrivals is ten cars per hour, the average time between cars arriving is:

$$\frac{1 \text{ hour}}{10 \text{ cars}} = 0.1 \text{ hour/car}$$

Hence, the corresponding exponential distribution that describes the time between the arrivals has a mean of $\mu = 0.1$ hour per car; as a result, the appropriate exponential probability density function is:

$$f(x) = \frac{1}{0.1} e^{-x/0.1} = 10e^{-10x}$$

EXERCISES

Methods

24. Consider the following exponential probability density function.

$$f(x) = \frac{1}{8}e^{-x/8} \quad for \ x \geq 0$$

   a. Find $P(X \leq 6)$.
   b. Find $P(X \leq 4)$.
   c. Find $P(X \geq 6)$.
   d. Find $P(4 \leq X \leq 6)$.

25. Consider the following exponential probability density function.

$$f(x) = \frac{1}{3}e^{-x/3} \quad for \ x \geq 0$$

   a. Write the formula for $P(X \leq x_0)$.
   b. Find $P(X \leq 2)$.
   c. Find $P(X \geq 3)$.
   d. Find $P(X \leq 5)$.
   e. Find $P(2 \leq X \leq 5)$.

Applications

26. In a parts store in Mumbai, customers arrive randomly. The cashier's service time is random but it is estimated it takes an average of 30 seconds to serve each customer.
   a. What is the probability a customer must wait more than two minutes for service?
   b. Suppose average service time is reduced to 25 seconds. How does this affect the calculation for (a)?

27. The time between arrivals of vehicles at a particular intersection follows an exponential probability distribution with a mean of 12 seconds.
   a. Sketch this exponential probability distribution.
   b. What is the probability that the arrival time between vehicles is 12 seconds or less?
   c. What is the probability that the arrival time between vehicles is six seconds or less?
   d. What is the probability of 30 or more seconds between vehicle arrivals?

28. Battery life between charges for a certain mobile phone is 20 hours when the primary use is talk time, and drops to 7 hours when the phone is primarily used for internet applications over a cellular network. Assume that the battery life in both cases follows an exponential distribution.
   a. Show the probability density function for battery life for this phone when its primary use is talk time.
   b. What is the probability that the battery charge for a randomly selected phone will last no more than 15 hours when its primary use is talk time?
   c. What is the probability that the battery charge for a randomly selected phone will last more than 20 hours when its primary use is talk time?
   d. What is the probability that the battery charge for a randomly selected phone will last no more than 5 hours when its primary use is internet applications?

ONLINE RESOURCES

For the data files, additional questions and answers, and the software section for Chapter 6, go to the online platform.

SUMMARY

This chapter extended the discussion of probability distributions to the case of continuous random variables. The major conceptual difference between discrete and continuous probability distributions involves the method of computing probabilities. With discrete distributions, the probability function $p(x)$ provides the probability that the random variable X assumes various values. With continuous distributions, the probability density function $f(x)$ does not provide probability values directly. Instead, probabilities are given by areas under the curve or graph of $f(x)$. Three continuous probability distributions – the uniform, normal and exponential distributions were the particular focus of the chapter – with detailed examples showing how probabilities could be straightforwardly computed. In addition, relationships between the binomial and normal distributions and also between the Poisson and exponential distributions were established and related probability results exploited.

KEY TERMS

Continuity correction factor
Exponential probability distribution
Normal probability distribution

Probability density function
Standard normal probability distribution
Uniform probability distribution

KEY FORMULAE

Uniform probability density function

$$f(x) = \begin{cases} \dfrac{1}{b - a} & \text{for } a \leq X \leq b \\ 0 & \text{elsewhere} \end{cases} \tag{6.1}$$

Normal probability density function

$$f(x) = \frac{1}{\sigma\sqrt{2\pi}} e^{-(x-\mu)^2/2\sigma^2} \tag{6.2}$$

Converting to the standard normal distribution

$$Z = \frac{X - \mu}{\sigma} \tag{6.3}$$

Exponential probability density function

$$f(x) = \frac{1}{\mu} e^{-x/\mu} \quad \text{for } x \geq 0,\ \mu > 0 \tag{6.4}$$

Exponential distribution: cumulative probabilities

$$p(X \leq x_0) = 1 - e^{-x_0/\mu} \tag{6.5}$$

CASE PROBLEM 1

Prix-Fischer Toys

Prix-Fischer Toys sells a variety of new and innovative children's toys. Management learned that the preholiday season is the best time to introduce a new toy, because many families use this time to look for new ideas for December holiday gifts. When Prix-Fischer discovers a new toy with good market potential, it chooses an October market entry date.

In order to get toys in its stores by October, Prix-Fischer places one-time orders with its manufacturers in June or July of each year. Demand for children's toys can be highly volatile. If a new toy catches on, worry about possible shortages in the market place often increases the demand to high levels and large profits can be realized. However, new toys can also flop, leaving Prix-Fischer stuck with high levels of inventory that must be sold at reduced prices. The most important question the company faces is deciding how many units of a new toy should be purchased to meet anticipated sales demand. If too few are purchased, sales will be lost; if too many are purchased, profits will be reduced because of low prices realized in clearance sales.

For the coming season, Prix-Fischer plans to introduce a new talking bear product called Chattiest Teddy. As usual, Prix-Fischer faces the decision of how many Chattiest Teddy units to order for the coming holiday season. Members of the management team suggested order quantities of 15,000, 18,000, 24,000 or 28,000 units. The wide range of order quantities suggested indicate considerable disagreement concerning the market potential. The product management team asks you for an analysis of the stock-out probabilities for various order quantities, an estimate of the profit potential and to help make an order quantity recommendation.

Prix-Fischer expects to sell Chattiest Teddy for €24 based on a cost of € 16 per unit. If inventory remains after the holiday season, Prix-Fischer will sell all surplus inventory for € 5 per unit. After reviewing the sales history of similar products, Prix-Fischer's senior sales forecaster predicts an expected demand of 20,000 units with a 0.90 probability that demand will be between 10,000 units and 30,000 units.

Managerial report

Prepare a managerial report that addresses the following issues and recommends an order quantity for the Chattiest Teddy product.

1. Use the sales forecaster's prediction to describe a normal probability distribution that can be used to approximate the demand distribution. Sketch the distribution and show its mean and standard deviation.

2. Compute the probability of a stock-out for the order quantities suggested by members of the management team.

3. Compute the projected profit for the order quantities suggested by the management team under three scenarios: worst case in which *sales* = 10,000 units, most likely case in which *sales* = 20,000 units, and best case in which *sales* = 30,000 units.

4. One of Prix-Fischer's managers felt that the profit potential was so great that the order quantity should have a 70 per cent chance of meeting demand and only a 30 per cent chance of any stock-outs. What quantity would be ordered under this policy, and what is the projected profit under the three sales scenarios?

5. Provide your own recommendation for an order quantity and note the associated profit projections. Provide a rationale for your recommendation.

CASE PROBLEM 2

Gebhardt Electronics

Gebhardt Electronics produces a wide variety of transformers that it sells directly to manufacturers of electronics equipment. For one component used in several models of its transformers, Gebhardt uses a 3-foot length of 0.20mm diameter solid wire made of pure Oxygen-Free Electronic (OFE) copper. A flaw in the wire reduces its conductivity and increases the likelihood it will break, and this critical component is difficult to reach and repair after a transformer has been constructed. Therefore, Gebhardt wants to use primarily flawless lengths of wire in making this component. The company is willing to accept no more than a 1 in 20 chance that a 3-foot length taken from a spool will be flawless. Gebhardt also occasionally uses smaller pieces of the same wire in the manufacture of other components, so the 3-foot segments to be used for this component are essentially taken randomly from a long spool of 0.20mm diameter solid OFE copper wire.

Gebhardt is now considering a new supplier for copper wire. This supplier claims that its spools of 0.20mm diameter solid OFE copper wire average 50 inches between flaws. Gebhardt must now determine whether the new supply will be satisfactory if the supplier's claim is valid.

Managerial report

In making this assessment for Gebhardt Electronics, consider the following three questions:

1. If the new supplier does provide spools of 0.20mm solid OFE copper wire that average 50 inches between flaws, how is the length of wire between two consecutive flaws distributed?

2. Using the probability distribution you identified in 1., what is the probability that Gebhardt's criteria will be met (i.e. a 1 in 20 chance that a randomly selected 3-foot segment of wire provided by the new supplier will be flawless)?

3. In inches, what is the minimum mean length between consecutive flaws that would result in satisfaction of Gebhardt's criteria?

4. In inches, what is the minimum mean length between consecutive flaws that would result in a 1 in 100 chance that a randomly selected 3-foot segment of wire provided by the new supplier will be flawless?

© sykono/iStock

7

Sampling and Sampling Distributions

CHAPTER CONTENTS

Statistics in Practice   Sampling 'big data' in Google Analytics

7.1 The EAI sampling problem
7.2 Simple random sampling
7.3 Point estimation
7.4 Introduction to sampling distributions
7.5 Sampling distribution of $\bar{X}$
7.6 Sampling distribution of P
7.7 Big data and standard errors of sampling distributions

LEARNING OBJECTIVES   After reading this chapter and doing the exercises, you should be able to:

1 Explain the terms simple random sample, sampling with replacement and sampling without replacement.

2 Select a simple random sample from a finite population using random number tables.

3 Explain the terms parameter, statistic, point estimator and unbiasedness.

4 Identify relevant point estimators for a population mean, population standard deviation and population proportion.

5 Explain the term sampling distribution.

6 Describe the form and characteristics of the sampling distribution:

6.1 of the sample mean, when the sample size is large or when the population is normal.

6.2 of the sample proportion, when the sample size is large.

In Chapter 1, we defined the terms *element*, *population* and *sample*:

- An *element* is the entity on which data are collected.
- A *population* is the set of all the elements of interest in a study.
- A *sample* is a subset of the population.

STATISTICS IN PRACTICE
Sampling 'big data' in Google Analytics

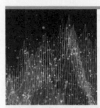

The term *web analytics* refers to the collection, analysis and reporting of data on website usage. Because online commerce is now such a critical activity for many companies, web analytics provide not only a way of improving website functioning but are also a crucial part of the marketing process.

Companies and individuals can use free web analytic software tools, of which Google Analytics (GA) is perhaps the best known. The main menu in GA is divided into four sections: Audience, Acquisition, Behaviour and Conversions. As the names suggest, these sections are set up to provide default reports on, respectively, website users (e.g. new/returning), how they find their way to the website (e.g. search engine, email), their pathways through the website and, where relevant, whether the website visit results in a sale.

The default reports produced by GA are based on 100 per cent of website traffic. However, for ad hoc queries there are circumstances when GA uses a sample of the recorded website traffic, rather than 100 per cent. This is known as 'session sampling'. For Analytics Standard (the free version of GA), the threshold at which session sampling kicks in is 500k sessions in the range of dates being queried. When GA applies session sampling, the user is advised what percentage of sessions has been sampled and has options for adjusting the sampling fraction to increase precision (at the cost of decreasing speed of response) or increase speed of response (at the cost of decreasing precision).

A Google search for 'sampling in Google Analytics' will yield links to GA support pages where session sampling is explained, as well as another type of sampling that can be applied in GA, 'data-collection sampling'. The search will also throw up links where there are criticisms of GA sampling, warnings about the data inaccuracies that might result, and recommendations for techniques and software products that will mitigate or eliminate GA sampling.

Governments and companies routinely make important decisions based on sample data. This chapter examines the basis and practicalities of scientific sampling. After studying the chapter, you should be in a better position to judge whether criticisms of GA sampling are justified.

We collect data by sampling with the aim of making an inference and thereby answering a research question about a population. Numerical characteristics of a population are called parameters (e.g. population mean, population standard deviation). Numerical characteristics of a sample are called sample statistics (e.g. sample mean, sample standard deviation). Primary purposes of statistical inference are to make estimates and test hypotheses about population parameters using sample statistics.

Here are two examples in which samples provide estimates of population parameters:

1　A European tyre manufacturer developed a new car tyre designed to provide an increase in tyre lifetime. To estimate the mean lifetime of the new tyre (in kilometres), the manufacturer selected a sample of 100 new tyres for testing. The test results provided a sample mean of 56,000 km, which provides an estimate of the mean tyre lifetime for the population of new tyres.

2　Officials in an African government were interested in estimating the proportion of registered voters likely to support a proposal for constitutional reform to be put to the electorate in a national referendum. A sample of 5,000 registered voters was selected, of whom 2,810 indicated support for the proposal. An estimate of the proportion of the population of registered voters supporting the proposal was 2,810/5,000 = 0.562.

These two examples illustrate the main reasons for using samples. In Example 1, collecting the data on tyre life involves wearing out each tyre tested. Clearly it is not feasible to test every tyre in the population. A sample is the only realistic way to obtain the tyre lifetime data. In Example 2, it would in principle be possible to contact every registered voter in the population, but the time and cost of doing so are prohibitive. A sample of registered voters is the practical alternative.

It is important to realize that sample results provide only *estimates* of the values of the population characteristics, because the sample contains only a proportion of the population. A sample mean provides an estimate of a population mean, and a sample proportion provides an estimate of a population proportion. Some estimation error can be expected. This chapter provides the basis for determining how large the estimation error might be.

We start by defining two terms. The **sampled population** is the population from which the sample is drawn. A **sampling frame** is a list of the elements in the sampled population. In Example 2 above, the sampled population is all registered voters in the country, and the sampling frame is the list of all registered voters. The number of registered voters is finite, so this is an example of sampling from a finite population. In Section 7.2, we consider how a simple random sample can be selected from a finite population.

The sampled population for Example 1 is more difficult to define. The sample of 100 tyres was from a production process over a specific period. We can think of the sampled population as the conceptual population of all tyres that could be made by the production process under similar conditions to those prevailing at the time of sampling. In this context, the sampled population is considered infinite, making it impossible to construct a sampling frame. In Section 7.2, we consider how to select a random sample in such a situation.

In later sections of this chapter, we show how data from a simple random sample can be used to compute estimates of a population mean, a population standard deviation and a population proportion. We then introduce the important concept of a sampling distribution. Knowledge of the appropriate sampling distribution enables us to make statements about how close the sample estimates might be to the corresponding population parameters.

7.1 THE EAI SAMPLING PROBLEM

EAI

The Personnel Manager for Exports & Imports plc (EAI) has the task of constructing a profile of the company's 2,500 managers. The characteristics of interest include the mean annual salary and the proportion of managers who have completed the company's management training programme. The 2,500 managers are the population for this study. We can find the annual salary and training programme status for each manager by referring to the firm's personnel records. The data file 'EAI' containing this information for all 2,500 managers in the population is on the online platform.

Using this data set and the formulae from Chapter 3, we calculate the population mean and the population standard deviation for the annual salary data.

$$\text{Population mean: } \mu = €51,800$$
$$\text{Population standard deviation: } \sigma = €4,000$$

Let π denote the proportion of the population that completed the training programme. The data set shows that 1,500 of the 2,500 managers did so. Hence $\pi = 1,500/2,500 = 0.60$. The population mean annual salary ($\mu = €51,800$), the population standard deviation for annual salary ($\sigma = €4,000$) and the population proportion that completed the training programme ($\pi = 0.60$) are parameters of the population of EAI managers.

Now, suppose the data on all the EAI managers are *not* readily available in the company's database, and that information must be collected afresh. How can the Personnel Manager obtain estimates of the population parameters by using a sample of managers, rather than all 2,500 managers? Suppose a sample of 30 managers is used. Clearly, the time and the cost of constructing a profile in this situation would be substantially less for 30 managers than for the entire population. If the Personnel Manager could be assured this sample would provide adequate information about the population of 2,500 managers, working with the sample would be preferable to working with the entire population. The cost of collecting information from a sample is generally substantially less than from a population, especially when personal interviews must be conducted to collect the information.

First we consider how to identify a sample of 30 managers.

7.2 SIMPLE RANDOM SAMPLING

Several methods can be used to select a sample from a population. One method that is important in statistical theory, and as a reference point in practical sampling, is simple random sampling. The definition of a simple random sample and the process of selecting such a sample depend on whether the population is *finite* or *infinite*. We first consider sampling from a finite population because this is relevant to the EAI sampling problem.

Sampling from a finite population

> **Simple random sample (finite population)**
>
> A simple random sample of size *n* from a finite population of size *N* is a sample selected such that each possible sample of size *n* has the same probability of being selected.

One procedure for selecting a simple random sample from a finite population is to choose the elements for the sample one at a time in such a way that, at each step, each of the elements remaining in the population has the same probability of being selected.

To select a simple random sample from the population of EAI managers, we first assign each manager a number. We can assign the managers the numbers 1 to 2500 in the order their names appear in the EAI personnel file. Next, we refer to the random numbers shown in Table 7.1. Using the first row of the table, each digit, 6, 3, 2, … , is a random digit with an equal chance of occurring. The random numbers in the table are shown in groups of five for readability. Because the largest number in the population list (2500) has four digits, we shall select random numbers from the table in groups of four digits. We may start the selection of random numbers anywhere in the table and move systematically in a direction of our choice. For simplicity in this example, we shall use the first row of Table 7.1 and move from left to right. The first seven four-digit random numbers are:

6327   1599   8671   7445   1102   1514   1807

These four-digit numbers are equally likely. We use them to give each manager in the population an equal chance of being included in the random sample.

The first number, 6327, is higher than 2500. We discard it because it does not correspond to one of the numbered managers in the population. The second number, 1599, is between 1 and 2500. So, the first manager selected for the random sample is number 1599 on the list of EAI managers. Continuing this process, we ignore the numbers 8671 and 7445 (higher than 2500) before identifying managers numbered 1102, 1514 and 1807 to be included in the random sample. This process continues until we have a simple random sample of 30 EAI managers.

TABLE 7.1 Random numbers

63271	59986	71744	51102	15141	80714	58683	93108	13554	79945
88547	09896	95436	79115	08303	01041	20030	63754	08459	28364
55957	57243	83865	09911	19761	66535	40102	26646	60147	15702
46276	87453	44790	67122	45573	84358	21625	16999	13385	22782
55363	07449	34835	15290	76616	67191	12777	21861	68689	03263
69393	92785	49902	58447	42048	30378	87618	26933	40640	16281
13186	29431	88190	04588	38733	81290	89541	70290	40113	08243
17726	28652	56836	78351	47327	18518	92222	55201	27340	10493
36520	64465	05550	30157	82242	29520	69753	72602	23756	54935
81628	36100	39254	56835	37636	02421	98063	89641	64953	99337
84649	48968	75215	75498	49539	74240	03466	49292	36401	45525
63291	11618	12613	75055	43915	26488	41116	64531	56827	30825
70502	53225	03655	05915	37140	57051	48393	91322	25653	06543
06426	24771	59935	49801	11082	66762	94477	02494	88215	27191
20711	55609	29430	70165	45406	78484	31639	52009	18873	96927
41990	70538	77191	25860	55204	73417	83920	69468	74972	38712
72452	36618	76298	26678	89334	33938	95567	29380	75906	91807
37042	40318	57099	10528	09925	89773	41335	96244	29002	46453
53766	52875	15987	46962	67342	77592	57651	95508	80033	69828
90585	58955	53122	16025	84299	53310	67380	84249	25348	04332
32001	96293	37203	64516	51530	37069	40261	61374	05815	06714
62606	64324	46354	72157	67248	20135	49804	09226	64419	29457
10078	28073	85389	50324	14500	15562	64165	06125	71353	77669
91561	46145	24177	15294	10061	98124	75732	00815	83452	97355
13091	98112	53959	79607	52244	63303	10413	63839	74762	50289

It is possible that a random number we have already used may appear again in the table before the sample has been fully selected. We ignore any such previously used numbers, because we do not want to select the same manager more than once. Selecting a sample in this manner is called **sampling without replacement**. If we selected a sample such that previously used random numbers were acceptable, and specific managers could be included in the sample two or more times, we would be **sampling with replacement**. This is a valid way of constructing a simple random sample, but sampling without replacement is more commonly used. When we refer to simple random sampling, we shall assume that the sampling is without replacement.

Computer-generated random numbers can also be used to implement the random sample selection process. Excel, R, SPSS and Minitab all provide functions for generating random numbers.

The number of different simple random samples of size n that can be selected from a finite population of size N is:

$$\frac{N!}{n!(N-n)!}$$

$N!$, $n!$ and $(N-n)!$ are the factorial computations discussed in Chapter 4. For the EAI problem with $N = 2,500$ and $n = 30$, this expression indicates that approximately 2.75×10^{69} different simple random samples of 30 EAI managers can be selected.

Sampling from an infinite population

In some situations the population is either infinite or so large that for practical purposes it must be treated as infinite. Infinite populations are often associated with a process that operates continuously and makes listing or counting every element in the population impossible or impractical. For example, parts being manufactured on a production line, hits on a company website homepage, telephone calls arriving at a technical support call centre and customers entering stores may all be viewed as coming from an infinite population.

The definition of a simple random sample from an infinite population is:

Simple random sample (infinite population)

A simple random sample from an infinite population is a sample selected such that the following conditions are satisfied:

1. Each element selected comes from the population.
2. Each element is selected independently of other elements.

For example, suppose a coffee shop would like to obtain a profile of its customers by selecting a simple random sample of customers and asking each customer to complete a short questionnaire. The ongoing process of customer visits to the coffee shop can be viewed as representing an infinite population. In selecting a simple random sample of customers, any customer who comes into the coffee shop will satisfy the first requirement above. The second requirement will be satisfied if a sample selection procedure is devised to select the customers independently of each other and ensure that the selection of any one customer does not influence the selection of any other customer. This would not be the case, for instance, if a group of customers were selected because they were sitting at the same table – we might expect this group of customers to exhibit similar characteristics to each other.

EXERCISES

Methods

1. Consider a finite population with five elements A, B, C, D, E. Ten possible simple random samples of size 2 can be selected.
   a. List the ten samples beginning with AB, AC and so on.
   b. What is the probability that each sample of size 2 is selected?
   c. Assume random number 1 corresponds to A, random number 2 corresponds to B and so on. List the simple random sample of size 2 that will be selected by using the random digits 8 0 5 7 5 3 2.

2. Consider a finite population with 350 elements. Using the last three digits of each of the five-digit random numbers below (601, 022, 448, ...), determine the first four elements that will be selected for the simple random sample.

   98001 73022 83448 02147 34229 27553 04147 03203 14203

Applications

3. The EURO STOXX 50 share index is calculated using data for 50 blue-chip companies from 12 Eurozone countries. You wish to select a simple random sample of five companies from the EURO STOXX 50 list. Use the last three digits in column 9 of Table 7.1, beginning with 554. Read down the column and identify the numbers of the five companies that will be selected.

4. A student union is interested in estimating the proportion of students who would like to take a non-credit optional module. A list of names and addresses of the 645 students enrolled during the current semester is available from the registrar's office. Using three-digit random numbers in row 10 of Table 7.1 and moving across the row from left to right, identify the first ten students who would be selected using simple random sampling. The three-digit random numbers begin with 816, 283 and 610.

5. Assume we want to identify a simple random sample of 12 of the 372 doctors practising in a large city. The doctors' names are available from the local health authority. Use the eighth column of five-digit random numbers in Table 7.1 to identify the 12 doctors for the sample. Ignore the first two random digits in each five-digit grouping of the random numbers. This process begins with random number 108 and proceeds down the column of random numbers.

6. Indicate whether the following populations should be considered finite or infinite:
   a. All registered voters in Ireland.
   b. All UHD televisions that could be produced by the Johannesburg factory of TV-M Com.
   c. All orders that could be processed by an online grocery company.
   d. All components that Fibercon plc produced on the second shift on 17 May 2023.

7. Suppose you wish to select a simple random sample of 100 students from a population of students numbered 1 to 10,000 in a registry database.
   a. Using groups of 5 digits from Table 7.1, starting in row 10, column 3 and moving left to right (i.e. starting with 39254, 56835, ...), and the method described in Section 7.2, how many random numbers would you reject whilst selecting the first ten students for the sample?
   b. Can you suggest a selection method that uses the random number table more efficiently, i.e. involves fewer rejected numbers?

7.3 POINT ESTIMATION

We return to the EAI problem. Data on annual salary and management training programme participation for a simple random sample of 30 managers are shown in Table 7.2. The notation $x_1, x_2, \ldots$ is used to denote the annual salary of the first manager in the sample, the second manager in the sample and so on. Completion of the management training programme is indicated by Yes in the relevant column.

To estimate the value of a population parameter, we compute a corresponding sample statistic. For example, to estimate the population mean μ and the population standard deviation σ for the annual salary of EAI managers, we use the Table 7.2 data to calculate the corresponding sample statistics: the sample mean and the sample standard deviation. Using the formulae from Chapter 3, the sample mean is:

$$\bar{x} = \frac{\Sigma x_i}{n} = \frac{1,554,420}{30} = 51,814 \quad (€)$$

The sample standard deviation is:

$$s = \sqrt{\frac{\sum (x_i - \bar{x})^2}{n - 1}} = \sqrt{\frac{325,009,260}{29}} = 3,348 \quad (€)$$

To estimate π, the proportion of managers in the population who completed the management training programme, we use the sample proportion. Let m denote the number of managers in the sample who completed the programme. The data in Table 7.2 show that $m = 19$, so the sample proportion is:

$$p = \frac{m}{n} = \frac{19}{30} = 0.63$$

TABLE 7.2 Annual salary and training programme participation for a simple random sample of 30 EAI managers

Annual salary (€)	Training programme	Annual salary (€)	Training programme
$x_1 = 49,094.30$	Yes	$x_{16} = 51,766.00$	Yes
$x_2 = 53,263.90$	Yes	$x_{17} = 52,541.30$	No
$x_3 = 49,643.50$	Yes	$x_{18} = 44,980.00$	Yes
$x_4 = 49,894.90$	Yes	$x_{19} = 51,932.60$	Yes
$x_5 = 47,621.60$	No	$x_{20} = 52,973.00$	Yes
$x_6 = 55,924.00$	Yes	$x_{21} = 45,120.90$	Yes
$x_7 = 49,092.30$	Yes	$x_{22} = 51,753.00$	Yes
$x_8 = 51,404.40$	Yes	$x_{23} = 54,391.80$	No
$x_9 = 50,957.70$	Yes	$x_{24} = 50,164.20$	No
$x_{10} = 55,109.70$	Yes	$x_{25} = 52,973.60$	No
$x_{11} = 45,922.60$	Yes	$x_{26} = 50,241.30$	No
$x_{12} = 57,268.40$	No	$x_{27} = 52,793.90$	No
$x_{13} = 55,688.80$	Yes	$x_{28} = 50,979.40$	Yes
$x_{14} = 51,564.70$	No	$x_{29} = 55,860.90$	Yes
$x_{15} = 56,188.20$	No	$x_{30} = 57,309.10$	No

These computations are examples of *point estimation*. We refer to the sample mean as the point estimator of the population mean μ, the sample standard deviation as the point estimator of the population standard deviation σ, and the sample proportion as the point estimator of the population proportion π. The numerical value obtained for the sample statistic is called a **point estimate**. For the simple random sample of 30 EAI managers shown in Table 7.2, $\bar{x} = €51,814$ is the point estimate of μ, $s = €3,348$ is the point estimate of σ and $p = 0.63$ is the point estimate of π.

Table 7.3 summarizes the sample results and compares the point estimates to the actual values of the population parameters. The point estimates in Table 7.3 differ somewhat from the corresponding population parameters. This is to be expected because the point estimates are based on a sample, rather than on a census of the entire population. In the next chapter, we shall show how to construct an *interval estimate* to provide information about how close the point estimate may be to the population parameter.

TABLE 7.3 Summary of point estimates obtained from a simple random sample of 30 EAI managers

Population parameter	Parameter value	Point estimator	Point estimate
Population mean annual salary	$\mu = €51,800$	Sample mean annual salary	$\bar{x} = €51,814$
Population standard deviation for annual salary	$\sigma = €4,000$	Sample standard deviation for annual salary	$s = €3,348$
Population proportion who have completed the training programme	$\pi = 0.60$	Sample proportion who have completed the training programme	$p = 0.63$

Practical advice

Much of the rest of this book is about statistical inference. Point estimation is a form of statistical inference where we use a sample statistic to make a single-value estimate of a population parameter. When making sample-based inferences, it is important to have a close correspondence between the sampled population and the target population. The **target population** is the population we want to make inferences about, while the sampled population is the population from which the sample is taken in practice. In this section, we have described the process of drawing a simple random sample from the population of EAI managers and making point estimates of characteristics of that same population. Here the sampled population and the target population are identical, which is the ideal situation. In other cases, it is not as easy to obtain a close correspondence between the sampled and target populations.

Consider the case of a theme park selecting a sample of its customers to learn about characteristics such as age and time spent at the park. Suppose all the sample elements were selected on a day when park attendance was restricted to employees of a large company. The sampled population would consist of employees of that company and members of their families. If the target population we wanted to make inferences about were typical park customers over a typical summer, there might be a substantial difference between the sampled population and the target population. In such a case, we would question the validity of the point estimates being made. The park management would be in the best position to know whether a sample taken on a specific day was likely to be representative of the target population.

Whenever a sample is used to make inferences about a population, we should make sure the study is designed so the sampled population and the target population are in close agreement. Good judgement is a necessary ingredient of sound statistical practice.

EXERCISES

Methods

8. The following data are from a simple random sample:

$$5 \quad 8 \quad 10 \quad 7 \quad 10 \quad 14$$

a. Calculate a point estimate of the population mean.
b. Calculate a point estimate of the population standard deviation.

9. A survey question for a sample of 150 individuals yielded 75 Yes responses, 55 No responses and 20 No Opinion responses.
a. Calculate a point estimate of the proportion in the population who would respond Yes.
b. Calculate a point estimate of the proportion in the population who would respond No.

Applications

10. A private clinic in Dubai is reviewing its fee structure for patient treatments. A sample of ten patients treated recently yields the following fee payments (in Emirati Dirham):

4,380	5,580	2,720	4,920	4,500
4,800	6,450	4,120	4,240	3,810

a. Calculate a point estimate of the mean treatment cost per clinic patient.
b. Calculate a point estimate of the standard deviation of the treatment cost per clinic patient.

11. The data set 'UK Companies' contains data on a sample of 353 companies. These were randomly selected from over 3.5 million companies registered on a publicly available database at Companies House, UK. Use the data set to answer the following questions:
    a. Compute a point estimate of the proportion of registered companies that have not filed accounts.
    b. Compute a point estimate of the proportion of registered companies that are classified as 'Private Limited Companies'.
    c. Compute a point estimate of the proportion of registered companies that have not provided information about SIC (Standard Industrial Classification) code.

12. The American Automobile Association annual automated vehicle survey includes a series of questions about automobile automation technology. The 1,010 interviews completed in the 2021 survey showed that 222 respondents feel manufacturers should focus on developing self-driving vehicles. In addition, 808 respondents want current vehicle safety systems (such as automatic emergency braking and lane-keeping assistance) to work better, and 586 respondents want these systems in the next vehicle they purchase.
    a. Calculate a point estimate of the proportion of respondents who feel manufacturers should focus on developing self-driving vehicles.
    b. Calculate a point estimate of the proportion of respondents who want current vehicle safety systems (such as automatic emergency braking and lane-keeping assistance) to work better.
    c. Calculate a point estimate of the proportion of respondents who want current vehicle safety systems (such as automatic emergency braking and lane-keeping assistance) in the next vehicle they purchase.

7.4 INTRODUCTION TO SAMPLING DISTRIBUTIONS

For the simple random sample of 30 EAI managers in Table 7.2, the point estimate of μ is $\bar{x} = €51,814$ and the point estimate of π is $p = 0.63$. Suppose we select another simple random sample of 30 EAI managers and obtain the following point estimates:

$$\text{Sample mean: } \bar{x} = €52,670$$

$$\text{Sample proportion: } p = 0.70$$

Note that different values of the sample mean and sample proportion were obtained. A second simple random sample of 30 EAI managers cannot be expected to provide precisely the same point estimates as the first sample.

Now, suppose we select simple random samples of 30 EAI managers repeatedly, each time computing the values of the sample mean and sample proportion. Table 7.4 contains a portion of the results obtained for 500 simple random samples, and Table 7.5 shows the frequency and relative frequency distributions for the 500 values. Figure 7.1 shows the relative frequency histogram for the values.

TABLE 7.4 Values $\bar{x}$ and p from 500 simple random samples of 30 EAI managers

Sample number	Sample mean ($\bar{x}$)	Sample proportion (p)
1	51,814	0.63
2	52,670	0.70
3	51,780	0.67
4	51,588	0.53
.	.	.
.	.	.
500	51,752	0.50

TABLE 7.5 Frequency distribution of $\bar{x}$ values from 500 simple random samples of 30 EAI managers

Mean annual salary (€)	Frequency	Relative frequency
49,500.00–49,999.99	2	0.004
50,000.00–50,499.99	16	0.032
50,500.00–50,999.99	52	0.104
51,000.00–51,499.99	101	0.202
51,500.00–51,999.99	133	0.266
52,000.00–52,499.99	110	0.220
52,500.00–52,999.99	54	0.108
53,000.00–53,499.99	26	0.052
53,500.00–53,999.99	6	0.012
Totals	**500**	**1.000**

FIGURE 7.1
Relative frequency histogram
of sample mean values from
500 simple random samples of
size 30 each

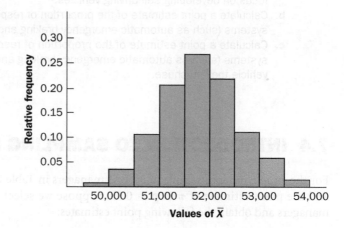

In Chapter 5 we defined a random variable as a numerical description of the outcome of an experiment or trial. If we consider selecting a simple random sample as a trial, the sample mean is a numerical description of its outcome. The sample mean is therefore a random variable. Using the naming conventions described in Chapters 5 and 6 (i.e. capital letters for names of random variables), we denote this random variable $\bar{X}$. Like other random variables, $\bar{X}$ has a mean or expected value, a standard deviation and a probability distribution. Because the various possible values of $\bar{X}$ are the result of different simple random samples, the probability distribution of $\bar{X}$ is called the **sampling distribution** of $\bar{X}$. Knowledge of this sampling distribution will enable us to make probability statements about how close the sample mean is to the population mean μ.

We would need to enumerate every possible sample of 30 managers and compute each sample mean to completely determine the sampling distribution of $\bar{X}$; recall our earlier calculation that there are about 2.75×10^{69} possible samples of size 30 from this population. However, the histogram of 500 $\bar{x}$ values shown in Figure 7.1 gives an approximation of this sampling distribution. We note its bell-shaped appearance. The largest concentration of the $\bar{x}$ values and the mean of the 500 $\bar{x}$ values are near the population mean $\mu = €51,800$. We shall describe the properties of the sampling distribution of $\bar{X}$ more fully in the next section.

The 500 values of the sample proportion are summarized by the relative frequency histogram in Figure 7.2. Like the sample mean, the sample proportion is a random variable, which we denote P. If every possible sample of size 30 were selected from the population and if a value p were computed for each sample, the resulting distribution would be the sampling distribution of P. The relative frequency histogram of the 500 sample values in Figure 7.2 provides a general idea of the appearance of the sampling distribution of P.

In practice, we typically select just one simple random sample from the population for estimating population characteristics. In this section we repeated the sampling process 500 times to illustrate that many samples are possible and that the different samples generate different values $\bar{x}$ and p for the sample statistics $\bar{X}$ and P. The probability distribution of a sample statistic is called the sampling distribution of that statistic. In Section 7.5 we show the characteristics of the sampling distribution of $\overline{X}$. In Section 7.6 we show the characteristics of the sampling distribution of P. The ability to understand the material in subsequent chapters depends heavily on the ability to understand the concept of a sampling distribution and to use the sampling distributions presented in this chapter.

FIGURE 7.2
Relative frequency histogram of sample proportion values from 500 simple random samples of size 30 each

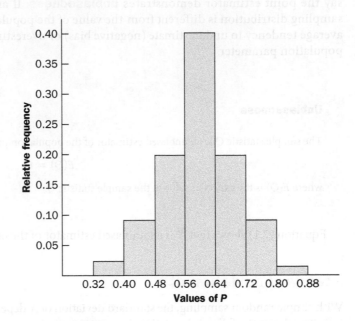

7.5 SAMPLING DISTRIBUTION OF $\overline{X}$

As with other probability distributions we have studied, the sampling distribution of $\overline{X}$ has an expected value or mean, a standard deviation and a characteristic shape or form. We begin by considering the expected value of $\overline{X}$.

> **Sampling distribution of $\overline{X}$**
>
> The sampling distribution of $\overline{X}$ is the probability distribution of all possible values of the sample mean.

Expected value of $\overline{X}$

Consider the $\bar{x}$ values generated by the various possible simple random samples. The mean of all these values is known as the expected value of $\overline{X}$. Let $E(\overline{X})$ represent the expected value of $\overline{X}$, and μ represent the mean of the population from which we are sampling. It can be shown that with simple random sampling, $E(\overline{X})$ and μ are equal.

> **Expected value of $\overline{X}$**
>
> $$E(\overline{X}) = \mu \tag{7.1}$$
>
> where:
>
> $E(\overline{X}) =$ the expected value of $\overline{X}$
>
> $\mu =$ the mean of the population from which the sample is selected

In Section 7.1 we saw that the mean annual salary for the population of EAI managers is $\mu = 51{,}800$. So, equation (7.1) tells us that the mean of all possible sample means for the EAI study is also €51,800.

When the expected value of a point estimator equals the population parameter it estimates, we say the point estimator demonstrates **unbiasedness**. If an estimator is biased, the mean of the sampling distribution is different from the value of the population parameter, so the estimator has an average tendency to underestimate (negative bias) or overestimate (positive bias) the true value of the population parameter.

Unbiasedness

The sample statistic Q is an unbiased estimator of the population parameter θ if

$$E(Q) = \theta$$

where $E(Q)$ is the expected value of the sample statistic Q.

Equation (7.1) shows that $\overline{X}$ is an unbiased estimator of the population mean μ.

Standard deviation of $\overline{X}$

With simple random sampling, the standard deviation of $\overline{X}$ depends on whether the population is finite or infinite. We use the following notation:

$$\sigma_{\overline{X}} = \text{the standard deviation of } \overline{X}$$

$$\sigma = \text{the standard deviation of the population}$$

$$n = \text{the sample size}$$

$$N = \text{the population size}$$

Standard deviation of $\overline{X}$

Finite population	Infinite population

$$\sigma_{\overline{X}} = \left(\frac{\sigma}{\sqrt{n}}\right)\sqrt{\frac{N-n}{N-1}} \qquad\qquad \sigma_{\overline{X}} = \frac{\sigma}{\sqrt{n}} \qquad\qquad \textbf{(7.2)}$$

Comparing the two formulae in (7.2), we see that the factor $\sqrt{(N-n)/(N-1)}$ is required for the finite population case but not for the infinite population case. This factor is commonly referred to as the **finite population correction factor** (fpc). In many practical sampling situations, we find that the population involved, although finite, is large, whereas the sample size is relatively small. In such cases the fpc is close to 1. As a result, the difference between the values of the standard deviation of $\overline{X}$ for the finite and infinite population cases becomes negligible. Then, $\sigma_{\overline{X}} = \sigma/\sqrt{n}$ becomes a good approximation to the standard deviation of $\overline{X}$ even though the population is finite. This observation leads to the following general guideline for computing the standard deviation of $\overline{X}$:

Use the following equation to compute the standard deviation of $\overline{X}$

$$\sigma_{\overline{X}} = \frac{\sigma}{\sqrt{n}} \qquad (7.3)$$

whenever:

1. The population is infinite; or

2. The population is finite *and* the sample size is less than or equal to 5 per cent of the population size; that is, $n/N \leq 0.05$.

In cases where $n/N > 0.05$, the finite population version of formula (7.2) should be used in the computation of $\sigma_{\overline{X}}$. Unless otherwise noted, throughout the text we shall assume that the population size is large, $n/N \leq 0.05$ and equation (7.3) can be used to compute $\sigma_{\overline{X}}$.

To compute $\sigma_{\overline{X}}$, we need to know the standard deviation, σ, of the population. To further emphasize the difference between $\sigma_{\overline{X}}$ and σ, we refer to $\sigma_{\overline{X}}$ as the standard error of the mean. The term standard error is used throughout statistical inference to refer to the standard deviation of a point estimator. Later we shall see that the value of the standard error of the mean is helpful in determining how far the sample mean may be from the population mean.

We return to the EAI example and compute the standard error of the mean associated with simple random samples of 30 EAI managers. In Section 7.1 we saw that the standard deviation of annual salary for the population of 2,500 EAI managers is $\sigma = 4,000$. In this case, the population is finite, with $N = 2,500$. However, with a sample size of 30, we have $n/N = 30/2,500 = 0.012$. Because the sample size is less than 5 per cent of the population size, we shall ignore the fpc and use equation (7.3) to compute the standard error.

$$\sigma_{\overline{X}} = \frac{\sigma}{\sqrt{n}} = \frac{4,000}{\sqrt{30}} = 730.3$$

Form of the sampling distribution of $\overline{X}$

The preceding results concerning the expected value and standard deviation for the sampling distribution of $\overline{X}$ are applicable for any population. The final step in identifying the characteristics of the sampling distribution of $\overline{X}$ is to determine the shape of the sampling distribution. We consider two cases: (1) the population has a normal distribution; and (2) the population does not have a normal distribution.

Population has a normal distribution

In many situations it is reasonable to assume that the population from which we are sampling has a normal, or nearly normal, distribution. When the population has a normal distribution, the sampling distribution of $\overline{X}$ is normally distributed for any sample size.

Population does not have a normal distribution

When the population from which we are selecting a simple random sample does not have a normal distribution, the central limit theorem is helpful in identifying the shape of the sampling distribution of $\overline{X}$.

Central limit theorem

In selecting simple random samples of size n from a population, the sampling distribution of the sample mean $\overline{X}$ can be approximated by a *normal distribution* as the sample size becomes large.

Figure 7.3 shows how the central limit theorem works for three different populations. Each column refers to one of the populations. The top panel shows the populations from which samples have been selected. Clearly, none of the populations is normally distributed. The other panels show empirical estimates of the sampling distribution of $\overline{X}$, constructed in a similar way to the empirical sampling distribution in Figure 7.1 but using a much larger number of samples.

FIGURE 7.3

Illustration of the central limit theorem for three populations

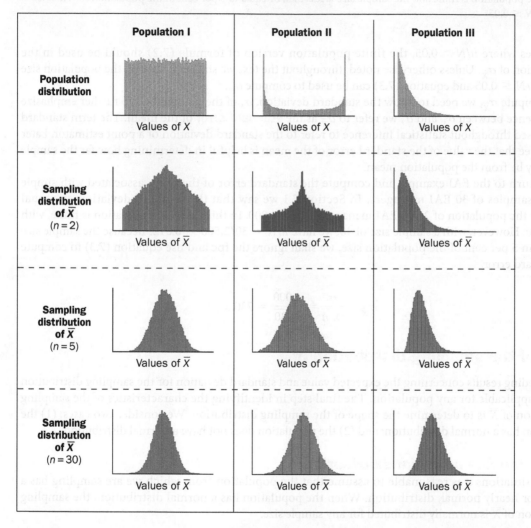

When the samples are of size 2, we see that the sampling distribution begins to take on an appearance different from that of the population distribution. For samples of size 5, we see all three sampling distributions beginning to take on a bell-shaped appearance. Finally, the samples of size 30 show all three sampling distributions to be approximately normally distributed.

For sufficiently large samples, the sampling distribution of $\overline{X}$ can be approximated by a normal distribution. How large must the sample size be before we can assume that the central limit theorem applies? Studies of the sampling distribution of $\overline{X}$ for a variety of populations and a variety of sample sizes have indicated that, for most applications, the sampling distribution of $\overline{X}$ can be approximated by a normal distribution whenever the sample size is 30 or more.

The theoretical proof of the central limit theorem requires independent observations in the sample. This condition is met for infinite populations and for finite populations where sampling is done with replacement. Although the central limit theorem does not directly address sampling without replacement from finite populations, general statistical practice applies the findings of the central limit theorem when the population size is large.

Sampling distribution of $\overline{X}$ for the EAI problem

For the EAI problem, we previously showed that $E(\overline{X}) = €51,800$ and $\sigma_{\overline{X}} = €730.3$. At this point, we do not have any information about the population distribution; it may or may not be normally distributed. If the population has a normal distribution, the sampling distribution of $\overline{X}$ is normally distributed. If the population does not have a normal distribution, the simple random sample of 30 managers and the central limit theorem enable us to conclude that the sampling distribution can be approximated by a normal distribution. In either case, we can proceed with the conclusion that the sampling distribution can be described by the normal distribution shown in Figure 7.4.

FIGURE 7.4

Sampling distribution of $\overline{X}$ for a simple random sample of 30 EAI managers, and the probability of $\overline{X}$ being within 500 of the population mean

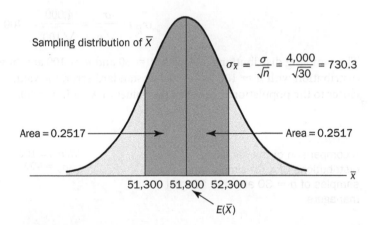

Practical value of the sampling distribution of $\overline{X}$

We are interested in the sampling distribution of $\overline{X}$ because it can be used to provide probability information about the difference between the sample mean and the population mean. Suppose the EAI Personnel Manager believes the sample mean will be an acceptable estimate if it is within €500 of the population mean. It is not possible to guarantee that the sample mean will be within €500 of the population mean. Indeed, Table 7.5 and Figure 7.1 show that some of the 500 sample means differed by more than €2,000 from the population mean. So we must think of the Personnel Manager's requirement in probability terms. What is the probability that the sample mean computed using a simple random sample of 30 EAI managers will be within €500 of the population mean?

We can answer this question using the sampling distribution of $\overline{X}$. Refer back to Figure 7.4. With $\mu = €51,800$, the Personnel Manager wants to know the probability that $\overline{X}$ is between €51,300 and €52,300. The dark shaded area of the sampling distribution shown in Figure 7.4 gives this probability. Because the sampling distribution is normally distributed, with mean 51,800 and standard error of the mean 730.3, we can use the standard normal distribution table to find the probability. At $\overline{X} = 51,300$ we have:

$$z = \frac{51,300 - 51,800}{730.3} = -0.68$$

The standard normal distribution table shows the cumulative probability for $z = -0.68$ as 0.2483. Similar calculations for $\overline{X} = 52,300$ show a cumulative probability for $z = +0.68$ of 0.7517. So the probability that the sample mean is between 51,300 and 52,300 is $0.7517 - 0.2483 = 0.5034$.

These computations show that a simple random sample of 30 EAI managers has a 0.5034 probability of providing a sample mean that is within €500 of the population mean. So there is a $1 - 0.5034 = 0.4966$ probability that the difference between $\overline{X}$ and μ will be more than €500. In other words, a simple random sample of 30 EAI managers has a roughly 50/50 chance of providing a sample mean within the acceptable €500. Perhaps a larger sample size should be considered. We explore this possibility by considering the relationship between the sample size and the sampling distribution of $\overline{X}$.

Relationship between sample size and the sampling distribution of $\overline{X}$

Suppose we select a simple random sample of 100 EAI managers instead of the 30 originally considered. Intuition suggests that a sample mean based on 100 managers should provide a better estimate of the population mean than a sample mean based on 30 managers. To see how much better, let us consider the relationship between the sample size and the sampling distribution of $\overline{X}$.

First note that $E(\overline{X}) = \mu$. Regardless of the sample size n, $\overline{X}$ is an unbiased estimator of μ. However, the standard error of the mean, $\sigma_{\overline{X}}$, is inversely related to the square root of the sample size: its value decreases when the sample size increases. With $n = 30$, the standard error of the mean for the EAI problem is 730.3. With the increase in the sample size to $n = 100$, the standard error of the mean decreases to:

$$\sigma_{\overline{X}} = \frac{\sigma}{\sqrt{n}} = \frac{4,000}{\sqrt{100}} = 400$$

The sampling distributions of $\overline{X}$ with $n = 30$ and $n = 100$ are shown in Figure 7.5. Because the sampling distribution with $n = 100$ has a smaller standard error, the values of $\overline{X}$ have less variation and tend to be closer to the population mean than the values of $\overline{X}$ with $n = 30$.

FIGURE 7.5
A comparison of the sampling distributions of $\overline{X}$ for simple random samples of $n = 30$ and $n = 100$ EAI managers

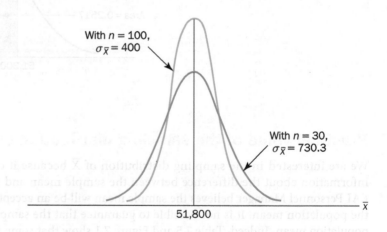

We can use the sampling distribution of $\overline{X}$ for $n = 100$ to compute the probability that a simple random sample of 100 EAI managers will provide a sample mean within €500 of the population mean. As before, we can use the standard normal distribution table to find the probability. At $\overline{X} = 51,300$ (Figure 7.6), we have:

$$z = \frac{51,300 - 51,800}{400} = -1.25$$

We find a cumulative probability for $z = -1.25$ of 0.1056. With a similar calculation for $X = 52,300$, we see that the probability of the sample mean being between 51,300 and 52,300 is $0.8944 - 0.1056 = 0.7888$. By increasing the sample size from 30 to 100 EAI managers, we have increased the probability of obtaining a sample mean within €500 of the population mean from 0.5034 to 0.7888.

The important point in this discussion is that as the sample size increases, the standard error of the mean decreases. As a result, the larger sample size provides a higher probability that the sample mean is within a specified distance of the population mean.

FIGURE 7.6

The probability of a sample mean being within €500 of the population mean when a simple random sample of 100 EAI managers is used

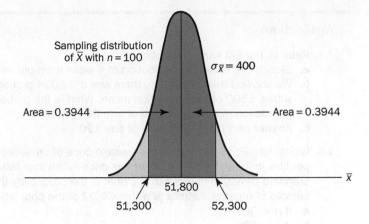

In presenting the sampling distribution of $\bar{X}$ for the EAI problem, we have taken advantage of the fact that the population mean $\mu = 51,800$ and the population standard deviation $\sigma = 4,000$ were known. However, usually these values will be unknown. In Chapter 8 we shall show how the sample mean $\bar{X}$ and the sample standard deviation S are used when μ and σ are unknown.

EXERCISES

Methods

13. A population has a mean of 200 and a standard deviation of 50. A simple random sample of size 100 will be taken and the sample mean will be used to estimate the population mean.
    a. What is the expected value of $\bar{X}$?
    b. What is the standard deviation of $\bar{X}$?
    c. Sketch the sampling distribution of $\bar{X}$.

14. A population has a mean of 200 and a standard deviation of 50. Suppose a simple random sample of size 100 is selected and is used to estimate μ.
    a. What is the probability that the sample mean will be within ±5 of the population mean?
    b. What is the probability that the sample mean will be within ±10 of the population mean?

15. In Exercise 1 earlier in this chapter, you were asked to list the ten possible simple random samples of size 2 from a finite population with five elements A, B, C, D, E. Suppose each of the five elements has a value: €10 for A, €20 for B, €30 for C, €40 for D, €50 for E. Consider the sample mean of these values when a simple random sample of size 2 is selected.
    a. Construct a frequency table showing the sampling distribution of the sample mean.
    b. Construct a chart that shows the sampling distribution graphically.

16. Suppose a simple random sample of size 50 is selected from a population with $\sigma = 25$. Find the value of the standard error of the mean in each of the following cases (use the finite population correction factor if appropriate).
    a. The population size is infinite.
    b. The population size is $N = 50,000$.
    c. The population size is $N = 5,000$.
    d. The population size is $N = 500$.

Applications

17. Refer to the EAI sampling problem.
    a. Sketch the sampling distribution of $\bar{X}$ when a simple random sample of size 60 is used.
    b. We showed that for $n = 30$, there was a 0.5034 probability of obtaining a sample mean within ±500 of the population mean. What is the probability that $\bar{X}$ is within 500 of the population mean if a sample of size 60 is used?
    c. Answer part (b) for a sample of size 120.

18. GlobalPetrolPrices.com gave the average price of unleaded petrol in Germany as €1.778 per litre in August 2022. Assume this price is the population mean, and that the population standard deviation is σ = €0.10. What is the probability that the mean price for a random sample of n petrol stations is within €0.02 of the population mean:
    a. If $n = 30$?
    b. If $n = 50$?
    c. If $n = 100$?
    d. Would you recommend a sample size of 30, 50 or 100 to have at least a 0.95 probability that the sample mean is within €0.02 of the population mean?

19. Three companies carry inventories that differ in size. Firm A's inventory contains 2,000 items, company B's contains 5,000 items and company C's contains 10,000 items. A consultant recommends each firm take a sample of size 50 items from its inventory to estimate the mean cost per item. The population standard deviation for the cost of the items in each company's inventory is σ = €144.
    a. Using the finite population factor, compute the standard error of the sample mean for each of the three companies, given a sample size of 50. Do the different population sizes make any practical difference to the standard error?
    b. What is the probability for each company that the sample mean $\bar{X}$ will be within ±€25 of the population mean μ?

20. A researcher reports survey results by stating that the standard error of the mean is 20. The population standard deviation is 500.
    a. How large was the sample?
    b. What is the probability that the point estimate was within ±25 of the population mean?

21. To estimate the mean age for a population of 4,000 employees in a large company in Kuwait City, a simple random sample of 40 employees is selected.
    a. Would you use the finite population correction factor in calculating the standard error of the mean? Explain.
    b. If the population standard deviation is σ = 8.2, compute the standard error both with and without the finite population correction factor. What is the rationale for ignoring the finite population correction factor whenever $n/N \leq 0.05$?
    c. What is the probability that the sample mean age of the employees will be within ±2 years of the population mean age?

22. Suppose a bank intends to use a simple random sample of 350 accounts to estimate the mean end-of-month debt held by its credit card holders. The bank wishes to have at least a 90 per cent probability that the sample mean end-of-month debt will be within a margin of plus or minus €20 of the population end-of-month debt. What is the largest population standard deviation for end-of-month debt that will allow this requirement to be met?

7.6 SAMPLING DISTRIBUTION OF *P*

The sample proportion P is a point estimator of the population proportion π. The formula for computing the sample proportion is:

$$p = \frac{m}{n}$$

where:

m = the number of elements in the sample that possess the characteristic of interest

n = sample size

The sample proportion P is a random variable and its probability distribution is called the sampling distribution of P.

> **Sampling distribution of P**
>
> The sampling distribution of P is the probability distribution of all possible values of the sample proportion P.

To determine how close the sample proportion is to the population proportion π, we need to understand the properties of the sampling distribution of P: the expected value of P, the standard deviation of P and the shape of the sampling distribution of P.

Expected value of P

The expected value of P (the mean of all possible values of P) is equal to the population proportion π. P is an unbiased estimator of π.

> **Expected value of P**
>
> where: $$E(P) = \pi \tag{7.4}$$
>
> $E(P)$ = the expected value of P
>
> π = the population proportion

In Section 7.1, we noted that the proportion of the population of EAI managers who participated in the company's management training programme was $\pi = 0.60$. The expected value of P is therefore 0.60.

Standard deviation of P

Like the standard deviation of $\overline{X}$, the standard deviation of P depends on whether the population is finite or infinite.

> **Standard deviation of P**
>
> | Finite population | Infinite population | |
> |:---:|:---:|:---:|
> | $\sigma_P = \sqrt{\dfrac{\pi(1-\pi)}{n}}\sqrt{\dfrac{N-n}{N-1}}$ | $\sigma_P = \sqrt{\dfrac{\pi(1-\pi)}{n}}$ | (7.5) |

The only difference between the two formulae in (7.5) is the use of the finite population correction factor $\sqrt{(N-n)/(N-1)}$.

As with the sample mean, the difference between the expressions for the finite and the infinite population becomes negligible if the size of the finite population is large in comparison to the sample size. We follow the same guideline we recommended for the sample mean. That is, if the population is finite with $n/N \leq 0.05$, we shall use $\sigma_P = \sqrt{\pi(1 - \pi)/n}$.

However, if the population is finite with $n/N > 0.05$, the fpc should be used. Again, unless specifically noted, throughout the text we shall assume that the population size is large in relation to the sample size and so the fpc is unnecessary.

In Section 7.5 we used the term standard error of the mean to refer to the standard deviation of $\overline{X}$. We stated that in general the term standard error refers to the standard deviation of a point estimator. Accordingly, for proportions we use *standard error of the proportion* to refer to the standard deviation of P.

We now compute the standard error of the proportion associated with simple random samples of 30 EAI managers. For the EAI study we know that the population proportion of managers who participated in the management training programme is $\pi = 0.60$. With $n/N = 30/2,500 = 0.012$, we shall ignore the fpc when we compute the standard error of the proportion. For the simple random sample of 30 managers, σ_P is:

$$\sigma_P = \sqrt{\frac{\pi(1 - \pi)}{n}} = \sqrt{\frac{0.60(1 - 0.60)}{30}} = 0.0894$$

Form of the sampling distribution of P

The sample proportion is $p = m/n$. For a simple random sample from a large population, the value of m is a binomial random variable indicating the number of elements in the sample with the characteristic of interest. Because n is a constant, the probability of each value of m/n is the same as the binomial probability of m, which means that the sampling distribution of P is also a discrete probability distribution.

In Chapter 6 we showed that a binomial distribution can be approximated by a normal distribution whenever the sample size is large enough to satisfy the following two conditions: $n\pi \geq 5$ and $n(1 - \pi) \geq 5$. Then, the probability of m in the sample proportion, $p = m/n$, can be approximated by a normal distribution. And because n is a constant, the sampling distribution of P can also be approximated by a normal distribution. This approximation is stated as follows:

> The sampling distribution of P can be approximated by a normal distribution whenever $n\pi \geq 5$ and $n(1 - \pi) \geq 5$.

In practical applications, when an estimate of a population proportion is needed, we find that sample sizes are almost always large enough to permit the use of a normal approximation for the sampling distribution of P.

For the EAI sampling problem the population proportion of managers who participated in the training programme is $\pi = 0.60$. With a simple random sample of size 30, we have $n\pi = 30(0.60) = 18$ and $n(1 - \pi) = 30(0.40) = 12$. Consequently, the sampling distribution of P can be approximated by the normal distribution shown in Figure 7.7.

Practical value of the sampling distribution of P

The practical value of the sampling distribution of P is that it can be used to provide probability information about the difference between the sample proportion and the population proportion. For instance, suppose the EAI Personnel Manager wants to know the probability of obtaining a value of P that is within 0.05 of the population proportion of managers who participated in the training programme. That is, what is the probability of obtaining a sample with a sample proportion P between 0.55 and 0.65? The dark shaded area in Figure 7.7 shows this probability. The sampling distribution of P can be approximated by a normal distribution with a mean of 0.60 and a standard error $\sigma_P = 0.0894$. Hence, we find that the standard normal random variable corresponding to $p = 0.55$ has a value of $z = (0.55 - 0.60)/0.0894 = -0.56$. Referring to the standard normal distribution table, we see that the cumulative probability for $z = -0.56$ is 0.2877. Similarly, for $p = 0.56$ we find a cumulative probability of 0.7123. Hence, the probability of selecting a sample that provides a sample proportion P within 0.05 of the population proportion π is $0.7123 - 0.2877 = 0.4246$.

FIGURE 7.7

Sampling distribution of P (proportion of EAI managers who participated in the management training programme)

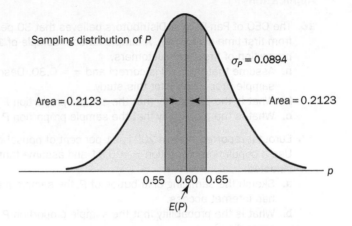

Sampling distribution of P

$\sigma_P = 0.0894$

Area = 0.2123

Area = 0.2123

0.55　0.60　0.65

$E(P)$

p

If we consider increasing the sample size to $n = 100$, the standard error of the sample proportion becomes:

$$\sigma_P = \sqrt{\frac{0.60(1 - 0.60)}{100}} = 0.049$$

The probability of the sample proportion being within 0.05 of the population proportion can now be calculated, again using the standard normal distribution table. At $p = 0.55$, we have $z = (0.55 - 0.60)/0.049 = -1.02$. The cumulative probability for $z = -1.02$ is 0.1539. Similarly, at $p = 0.65$ the cumulative probability is 0.8461. Hence, if the sample size is increased to 100, the probability that the sample proportion P is within 0.05 of the population proportion π will increase to $0.8461 - 0.1539 = 0.6922$.

EXERCISES

Methods

23. A simple random sample of size 100 is selected from a population with $\pi = 0.40$.
    a. What is the expected value of P?
    b. What is the standard error of P?
    c. Sketch the sampling distribution of P.

24. Assume that the population proportion is 0.55. Compute the standard error of the sample proportion, σ_P, for sample sizes of 100, 200, 500 and 1,000. What can you say about the size of the standard error of the proportion as the sample size is increased?

25. The population proportion is 0.30. What is the probability that a sample proportion will be within ± 0.04 of the population proportion for each of the following sample sizes?
   a. $n = 100$
   b. $n = 200$
   c. $n = 500$
   d. $n = 1,000$
   e. What is the advantage of a larger sample size?

Applications

26. The CEO of Pan-Europe Distributors believes that 30 per cent of the firm's orders come from first-time customers. A simple random sample of 100 orders will be used to estimate the proportion of first-time customers.
   a. Assume that the CEO is correct and $\pi = 0.30$. Describe the sampling distribution of the sample proportion P for this study.
   b. What is the probability that the sample proportion P will be between 0.20 and 0.40?
   c. What is the probability that the sample proportion P will be between 0.25 and 0.35?

27. Eurostat reported that, in 2021, 91 per cent of households in Sweden had internet access. Use a population proportion $\pi = 0.91$ and assume that a sample of 300 households will be selected.
   a. Sketch the sampling distribution of P, the sample proportion of Swedish households that had internet access.
   b. What is the probability that the sample proportion P will be within ± 0.03 of the population proportion?
   c. Answer part b. for sample sizes of 600 and 1,000.

28. Advertisers contract with internet service providers and search engines to place ads on websites. They pay a fee based on the number of potential customers who click on their ads. Unfortunately, click fraud, i.e. someone clicking on an ad for the sole purpose of driving up advertising revenue, has become a problem. Forty per cent of advertisers claim to have been a victim of click fraud. Suppose a simple random sample of 380 advertisers is taken to learn about how they are affected by this practice. Assume the population proportion $\pi = 0.40$.
   a. What is the probability the sample proportion will be within ± 0.04 of the population proportion experiencing click fraud?
   b. What is the probability the sample proportion will be greater than 0.45?

29. In 2019, the Ethics Institute published its report *Business Ethics Survey*. Among a sample of 2,253 employees of 19 South African organizations, 73 per cent thought Senior Management in their organization was committed to ethical standards. Assume the population proportion was $\pi = 0.73$, and that P is the sample proportion in a sample of $n = 2,253$.
   a. What is the probability that P will be within plus or minus 0.01 of π?
   b. Answer part b. for a sample size of 1,500.

30. A market research firm conducts telephone surveys with a 40 per cent historical response rate. What is the probability that in a new sample of 400 telephone numbers, at least 150 individuals will cooperate and respond to the questions?

31. @GrrrlPower is an avid Tweep. They have an intuition that about 40 per cent of their tweets are re-tweeted by their followers, but wish to estimate the percentage of re-tweets more accurately by reviewing a random sample of their tweets over the last two months.
   a. How large a sample would you recommend if @GrrrlPower wants at least 90 per cent probability that the sample proportion will be within three percentage points of the 'population' proportion for the last two months?
   b. Suppose @GrrrlPower's intuition that 40 per cent of their tweets are re-tweeted is an overestimate. Would your recommendation in part a. still meet their requirement? Explain your answer briefly.

7.7 BIG DATA AND STANDARD ERRORS OF SAMPLING DISTRIBUTIONS

The purpose of statistical inference is to use sample data to quickly and inexpensively gain insight into some characteristic of a population. Therefore, it is important that we can expect the sample to look like, or be representative of, the population that is being investigated. In practice, individual samples always, to varying degrees, fail to be perfectly representative of the populations from which they have been taken. There are two general reasons a sample may fail to be representative of the population of interest: sampling error and non-sampling error.

Sampling error

One reason a sample may fail to represent the population from which it has been taken is sampling error, or deviation of the sample from the population that results from random sampling. If repeated independent random samples of the same size are collected from the population of interest using a probability sampling technique, on average the samples will be representative of the population. This is the justification for collecting sample data randomly. However, the random collection of sample data does not ensure that any single sample will be perfectly representative of the population of interest. When collecting a sample randomly, the data in the sample cannot be expected to be perfectly representative of the population from which it has been taken. Sampling error is unavoidable when collecting a random sample; this is a risk we must accept when we chose to collect a random sample rather than incur the costs associated with taking a census of the population.

As expressed by equations (7.2) and (7.5), the standard errors of the sampling distributions of the sample mean $\bar{X}$ and the sample proportion P reflect the potential for sampling error when using sample data to estimate the population mean μ and the population proportion π, respectively. As the sample size n increases, the potential impact of extreme values on the statistic decreases, so there is less variation in the potential values of the statistic produced by the sample and the standard errors of these sampling distributions decrease. Because these standard errors reflect the potential for sampling error when using sample data to estimate the population mean μ and the population proportion π, we see that for an extremely large sample there may be little potential for sampling error.

Non-sampling error

Although the standard error of a sampling distribution decreases as the sample size n increases, this does not mean that we can conclude that an extremely large sample will always provide reliable information about the population of interest; this is because sampling error is not the sole reason a sample may fail to represent the target population. Deviations of the sample from the population that occur for reasons other than random sampling are referred to as non-sampling error. Non-sampling error can occur in a sample or a census, and can occur for a variety of reasons.

Consider the online news service PenningtonDailyTimes.com (PDT). Because PDT's primary source of revenue is the sale of advertising, the news service is intent on collecting sample data on the behaviour of visitors to its website in order to support its advertising sales. Prospective advertisers are willing to pay a premium to advertise on websites that have long visit times, so PDT's management is keenly interested in the amount of time customers spend during their visits to PDT's website. Advertisers are also concerned with how frequently visitors to a website click on any of the ads featured on the website, so PDT is also interested in whether visitors to its website clicked on any of the ads featured on the website.

From whom should PDT collect its data? Should it collect data on current visits to PDT's website? Should it attempt to attract new visitors and collect data on these visits? If so, should it measure the time spent at its website by visitors it has attracted from competitors' websites or visitors who do not routinely visit online news sites? The answers to these questions depend on PDT's research objectives. Is the company attempting to evaluate its current market, assess the potential of customers it can attract from competitors, or explore the potential of an entirely new market such as individuals who do not routinely obtain their news from online news services? If the research objective and the population from which the sample is to be drawn are not aligned, the data that PDT collects will not help the company accomplish its research objective. This type of error is referred to as a **coverage error**.

Even when the sample is taken from the appropriate population, non-sampling error can occur when segments of the target population are systematically underrepresented or overrepresented in the sample. This may occur because the study design is flawed or because some segments of the population are either more likely or less likely to respond. Suppose PDT implements a pop-up questionnaire that opens when a visitor leaves the website. Visitors to PDT's website who have installed pop-up blockers will be likely underrepresented, and visitors to the website who have not installed pop-up blockers will likely be overrepresented. If the behaviour of PDT's website visitors who have installed pop-up blockers differs from the behaviours of visitors who have not installed pop-up blockers, attempting to draw conclusions from this sample about how all visitors to the PDT website behave may be misleading. This type of error is referred to as a **non-response error**.

Another potential source of non-sampling error is incorrect measurement of the characteristic of interest. If PDT asks questions that are ambiguous or difficult for respondents to understand, the responses may not accurately reflect how the respondents intended to respond. For example, respondents may be unsure how to respond if PDT asks *Are the news stories on PenningtonDailyTimes.com compelling and accurate?*. How should visitors respond if they feel the news stories on PenningtonDailyTimes.com are compelling but erroneous? What response is appropriate if the respondent feels the news stories on PDT's website are accurate but dull? A similar issue can arise if a question is asked in a biased or leading way. If PDT asks *Many readers find the news stories on PenningtonDailyTimes.com to be compelling and accurate. Do you find the news stories on PenningtonDailyTimes.com to be compelling and accurate?*, the qualifying statement prior to the actual question will likely result in a bias towards positive responses. Incorrect measurement of the characteristic of interest can also occur when respondents provide incorrect answers; this may be due to a respondent's poor recall or unwillingness to respond honestly. This type of error is referred to as a **measurement error**.

Errors that are introduced by interviewers or during the recording and preparation of the data are other types of non-sampling error. These types of error are referred to as interviewer errors and processing errors, respectively.

Non-sampling error can introduce bias into the estimates produced using the sample, and this bias can mislead decision-makers who use the sample data in their decision-making processes. No matter how small or large the sample, we must contend with this limitation of sampling whenever we use sample data to gain insight into a population of interest. Although sampling error decreases as the size of the sample increases, an extremely large sample can still suffer from non-sampling error and fail to be representative of the population of interest. When sampling, care must be taken to ensure that we minimize the introduction of non-sampling error into the data collection process. This can be done by carrying out the following steps:

- Carefully define the target population before collecting sample data, and subsequently design the data collection procedure so that a probability sample is drawn from this target population.
- Carefully design the data collection process and train the data collectors.
- Pretest the data collection procedure to identify and correct for potential sources of non-sampling error prior to final data collection.
- Use stratified random sampling when population-level information about an important qualitative variable is available to ensure that the sample is representative of the population with respect to that qualitative characteristic.
- Use cluster sampling when the population can be divided into heterogeneous subgroups or clusters.
- Use systematic sampling when population-level information about an important quantitative variable is available to ensure that the sample is representative of the population with respect to that quantitative characteristic.

Finally, recognize that every random sample (even an extremely large random sample) will suffer from some degree of sampling error, and eliminating all potential sources of non-sampling error may be impractical. Understanding these limitations of sampling will enable us to be more realistic and pragmatic when interpreting sample data and using sample data to draw conclusions about the target population.

Big data

Recent estimates state that approximately 2.5 quintillion bytes of data are created worldwide each day. This represents a dramatic increase from the estimated 100 gigabytes (GB) of data generated worldwide per day in 1992, the 100 GB of data generated worldwide per hour in 1997, and the 100 GB of data generated worldwide per second in 2002. Every minute, there is an average of 216,000 Instagram posts, 204,000,000 emails sent and 12 hours of footage uploaded to YouTube. Without question, the amount of data that is now generated is overwhelming and this trend is certainly expected to continue.

In each of these cases the data sets that are generated are so large or complex that current data processing capacity and/or analytic methods are not adequate for analyzing the data. Thus, each is an example of big data. There are myriad other sources of big data. Sensors and mobile devices transmit enormous amounts of data. Internet activities, digital processes and social media interactions also produce vast quantities of data.

The amount of data has increased so rapidly that our vocabulary for describing a data set by its size must expand. A few years ago, a petabyte of data seemed almost unimaginably large, but we now routinely describe data in terms of yottabytes. Table 7.6 summarizes terminology for describing the size of data sets.

TABLE 7.6
Terminology for describing the size of data sets

Number of bytes	Metric	Name
$1{,}000^1$	kB	kilobyte
$1{,}000^2$	MB	megabyte
$1{,}000^3$	GB	gigabyte
$1{,}000^4$	TB	terabyte
$1{,}000^5$	PB	petabyte
$1{,}000^6$	EB	exabyte
$1{,}000^7$	ZB	zettabyte
$1{,}000^8$	YB	yottabyte

Understanding what big data is

The processes that generate big data can be described by four attributes or dimensions that are referred to as the four Vs:

- Volume – the amount of data generated
- Variety – the diversity in types and structures of data generated
- Veracity – the reliability of the data generated
- Velocity – the speed at which the data are generated.

A high degree of any of these attributes individually is sufficient to generate big data, and when they occur at high levels simultaneously the resulting amount of data can be overwhelmingly large. Technological advances and improvements in electronic (and often automated) data collection make it easy to collect millions, or even billions, of observations in a relatively short time. Businesses are collecting greater volumes of an increasing variety of data at a higher velocity than ever.

To understand the challenges presented by big data, we consider its structural dimensions. Big data can be tall data; a data set that has so many observations that traditional statistical inference has little meaning. For example, producers of consumer goods collect information on the sentiment expressed in millions of social media posts each day to better understand consumer perceptions of their products. Such data consist of the sentiment expressed (the variable) in millions (or over time, even billions) of social media posts (the observations). Big data can also be wide data; a data set that has so many variables that simultaneous consideration of all variables is infeasible. For example, a high-resolution image can comprise millions or billions of pixels. The data used by facial recognition algorithms consider each pixel in an image when comparing an image to other images in an attempt to find a match. Thus, these algorithms make use of the characteristics of millions or billions of pixels (the variables) for relatively few high-resolution images (the observations). Of course, big data can be both tall and wide, and the resulting data set can again be overwhelmingly large.

Statistics are useful tools for understanding the information embedded in a big data set, but we must be careful when using statistics to analyze big data. It is important that we understand the limitations of statistics when applied to big data and we temper our interpretations accordingly. Because tall data are the most common form of big data used in business, we focus on this structure in the discussions throughout the remainder of this section.

Implications of big data for sampling error

Let's revisit the data collection problem of online news service PDT. Because PDT's primary source of revenue is the sale of advertising, PDT's management is interested in the amount of time customers spend during their visits to PDT's website. From historical data, PDT has estimated that the standard deviation of the time spent by individual customers when they visit the PDT website is $s = 20$ seconds. Table 7.7 shows how the standard error of the sampling distribution of the sample mean time spent by individual customers when they visit the PDT website decreases as the sample size increases.

TABLE 7.7 Standard error of the sample mean $\overline{X}$ when $s = 20$ at various sample sizes n

Sample size n	Standard error $s_{\overline{x}} = s/\sqrt{n}$
10	6.32456
100	2.00000
1,000	0.63246
10,000	0.20000
100,000	0.06325
1,000,000	0.02000
10,000,000	0.00632
100,000,000	0.00200
1,000,000,000	0.00063

PDT also wants to collect information from its sample respondents on whether a visitor to its website clicked on any of the ads featured on the website. From its historical data, PDT knows that 51 per cent of past visitors to its website clicked on an ad featured on the website, so it will use this value as π to estimate the standard error. Table 7.8 shows how the standard error of the sampling distribution of the proportion of the sample that clicked on any of the ads featured on PDT's website decreases as the sample size increases.

A sample of one million or more visitors might seem unrealistic, but keep in mind that Amazon.com had over 2.2 billion visitors in February 2022.

The PDT example illustrates the general relationship between standard errors and the sample size. We see in Table 7.7 that the standard error of the sample mean decreases as the sample size increases. For a sample of $n = 10$, the standard error of the sample mean is 6.32456; when we increase the sample size to $n = 100,000$, the standard error of the sample mean decreases to 0.06325; and at a sample size of $n = 1,000,000,000$, the standard error of the sample mean decreases to only 0.00063. In Table 7.8 we see that the standard error of the sample proportion also decreases as the sample size increases. For a sample of $n = 10$, the standard error of the sample proportion is 0.15808; when we increase the sample size to $n = 100,000$, the standard error of the sample proportion decreases to 0.00158; and at a sample size of $n = 1,000,000,000$, the standard error of the sample mean decreases to only 0.00002. In both Tables 7.7 and 7.8, the standard error when $n = 1,000,000,000$ is *one ten-thousandth of the standard error when $n = 10$.*

TABLE 7.8 Standard error of the sample proportion P when $\pi = 0.51$ at various sample sizes n

Sample size n	Standard error $\sigma_P = \sqrt{p(1 - p)/n}$
10	0.15808
100	0.04999
1,000	0.01581
10,000	0.00500
100,000	0.00158
1,000,000	0.00050
10,000,000	0.00016
100,000,000	0.00005
1,000,000,000	0.00002

Exercises

Methods

32. A population has a mean of 400 and a standard deviation of 100. A sample of size 100,000 will be taken, and the sample mean $\overline{X}$ will be used to estimate the population mean.
    a. What is the expected value of $\overline{X}$?
    b. What is the standard deviation of $\overline{X}$?
    c. Show the sampling distribution of $\overline{X}$.
    d. What does the sampling distribution of $\overline{X}$ show?

33. Assume the population standard deviation is $\sigma = 25$. Compute the standard error of the mean, $\sigma_{\overline{x}}$, for sample sizes of 500,000, 1,000,000, 5,000,000, 10,000,000 and 100,000,000. What can you say about the size of the standard error of the mean as the sample size is increased?

34. A sample of size 100,000 is selected from a population with $\pi = 0.75$.
    a. What is the expected value of P?
    b. What is the standard error of P?
    c. Show the sampling distribution of P.
    d. What does the sampling distribution of P show?

35. Assume that the population proportion is 0.44. Compute the standard error of the proportion, σ_P, for sample sizes of 500,000, 1,000,000, 5,000,000, 10,000,000 and 100,000,000. What can you say about the size of the standard error of the sample proportion as the sample size is increased?

Applications

36. The Bureau of Labour Statistics (BLS) reports that the mean annual number of hours of vacation time earned by blue-collar and service employees who work for small private establishments and have at least 10 years of service is 100. Assume that for this population the standard deviation for the annual number of vacation hours earned is 48. Suppose the BLS would like to select a sample of 15,000 individuals from this population for a follow-up study.
    a. Show the sampling distribution of $\overline{X}$, the sample mean for a sample of 15,000 individuals from this population.
    b. What is the probability that a simple random sample of 15,000 individuals from this population will provide a sample mean that is within one hour of the population mean?
    c. Suppose the mean annual number of hours of vacation time earned for a sample of 15,000 blue-collar and service employees who work for small private establishments and have at least 10 years of service differs from the population mean μ by more than one hour. Considering your results for part (b), how would you interpret this result?

37. *The New York Times* reported that 17.2 million new cars and light trucks were sold in the USA in 2021, and the US Environmental Protection Agency projects the average efficiency for these vehicles to be 25.7 miles per gallon. Assume that the population standard deviation in miles per gallon for these automobiles is $\sigma = 6$.
    a. What is the probability a sample of 70,000 new cars and light trucks sold in the USA in 2021 will provide a sample mean miles per gallon that is within 0.05 miles per gallon of the population mean of 25.2?
    b. What is the probability a sample of 70,000 new cars and light trucks sold in the USA in 2021 will provide a sample mean miles per gallon that is within 0.01 miles per gallon of the population mean of 25.2? Compare this probability to the value computed in part (a).
    c. What is the probability a sample of 90,000 new cars and light trucks sold in the USA in 2017 will provide a sample mean miles per gallon that is within 0.01 of the population mean of 25.2? Comment on the differences between this probability and the value computed in part b.
    d. Suppose the mean miles per gallon for a sample of 70,000 new cars and light trucks sold in the USA in 2021 differs from the population mean μ by more than one gallon. How would you interpret this result?

38. The president of Colossus.com, Inc., believes that 42 per cent of the firm's orders come from customers who have purchased from Colossus.com in the past. A random sample of 108,700 orders from the past six months will be used to estimate the proportion of orders placed by repeat customers.
    a. Assume that Colossus.com's president is correct and the population proportion $\pi = 0.42$. What is the sampling distribution of P for this study?
    b. What is the probability that the sample proportion P will be within ± 0.001 of the population proportion?
    c. What is the probability that the sample proportion P will be within ± 0.0025 of the population proportion? Comment on the difference between this probability and the value computed in part (b).
    d. Suppose the proportion of orders placed by repeat customers for a sample of 108,700 orders from the past six months differs from the population proportion π by more than 0.01. How would you interpret this result?

39. Assume that 49.2 per cent of homes in a large country use landline telephone services.
    a. Suppose a sample of 207,000 homes will be taken to learn about home telephone usage. Specify the distribution form and parameters of the sampling distribution of P, where P is the sample proportion of homes that use landline phone service.
    b. What is the probability that the sample proportion in part a. will be within ± 0.002 of the population proportion?
    c. Suppose a sample of 86,800 homes will be taken to learn about home telephone usage. Show the sampling distribution of P where P is the sample proportion of homes that use landline phone service.
    d. What is the probability that the sample proportion in part (c) will be within ± 0.002 of the population proportion?
    e. Are the probabilities different in parts (b) and (d)? Why or why not?

SUMMARY

In this chapter we presented the concepts of simple random sampling and sampling distributions.

Simple random sampling was defined for sampling without replacement and sampling with replacement. We demonstrated how a simple random sample can be selected and how the sample data can be used to calculate point estimates of population parameters.

Point estimators such as $\overline{X}$ and P are random variables. The probability distribution of such a random variable is called a sampling distribution. We described the sampling distributions of the sample mean $\overline{X}$ and the sample proportion P. We stated that $E(\overline{X}) = \mu$ and $E(P) = \pi$, i.e. they are unbiased estimators of the respective parameters. After giving the standard deviation or standard error formulae for these estimators, we described the conditions necessary for the sampling distributions of $\overline{X}$ and P to follow normal distributions. Finally, we gave examples of how these normal sampling distributions can be used to calculate the probability of $\overline{X}$ or P being within any given distance of μ or π respectively.

KEY TERMS

Central limit theorem
Coverage error
Finite population correction factor
Four Vs
Measurement error
Non-response error
Non-sampling error
Parameters
Point estimate
Sampled population
Sampling distribution

Sampling error
Sampling frame
Sampling with replacement
Sampling without replacement
Simple random sampling
Standard error
Tall data
Target population
Unbiasedness
Wide data

KEY FORMULAE

Expected value of $\overline{X}$

$$E(\overline{X}) = \mu \qquad \textbf{(7.1)}$$

Standard deviation of $\overline{X}$ (standard error)

Finite population	Infinite population	
$\sigma_{\overline{X}} = \left(\dfrac{\sigma}{\sqrt{n}}\right)\sqrt{\dfrac{N-n}{N-1}}$	$\sigma_{\overline{X}} = \dfrac{\sigma}{\sqrt{n}}$	**(7.2)**

Expected value of P

$$E(P) = \pi \qquad \textbf{(7.4)}$$

Standard deviation of P (standard error)

Finite population	Infinite population	
$\sigma_P = \sqrt{\dfrac{\pi(1-\pi)}{n}}\sqrt{\dfrac{N-n}{N-1}}$	$\sigma_P = \sqrt{\dfrac{\pi(1-\pi)}{n}}$	**(7.5)**

CASE PROBLEM

VAPING

Use of e-cigarettes by UK adults

The data file 'VAPING' (online platform) comprises a table constructed from data in an Excel worksheet available on the UK Office for National Statistics (ONS) website. The table contains estimates of the use of e-cigarettes in 2020 by the UK population aged 16 and over, based on responses to the Opinions and Lifestyle Survey (OLS) carried out by the ONS. This survey involved online and face-to-face interviews with over 20,000 respondents (the data collection procedures were modified because of the COVID-19 pandemic). The sample for the OLS was not a simple random sample. It involved a three-stage sampling procedure, involving several stratification variables. Stratification tends to increase precision compared to simple random sampling. (You can read more about stratification in Chapter 22, on the online platform.)

The 'VAPING' data file includes 95% confidence intervals for the various population estimates given (you will learn about confidence intervals in Chapter 8). The first five rows of data have been reproduced below. Row 6 of the worksheet shows, for example, that an estimated 4.6 per cent of the population of men aged 25–34 were daily users of e-cigarettes. This population estimate is based on interviews with a sample of 778 men in this age group. The 95% confidence interval is given as [3.0 per cent, 6.2 per cent]. The width of the interval can be directly related to the sampling distribution of the sample proportion. The interval is based on the central 95 per cent of the sampling distribution, so the width of the interval corresponds to the distance between the sample proportions marking the lowest 2.5 per cent and the highest 2.5 per cent of the sampling distribution.

The aim in this case problem is to examine the degree to which the margins of error implied by the confidence intervals given in the ONS worksheet agree with, or differ from, the margins of error we have discussed in this chapter for simple random samples.

1	Confidence intervals for the adult e-cigarette estimates, Great Britain, 2020					
2						
3	Source: Opinions and Lifestyle Survey, Office for National Statistics					
4	Group	Statistic	Sample size	Point estimate	Lower 95% confidence limit	Upper 95% confidence limit
5	Men aged 16-24	Proportion of population who are daily e-cigarette users	627	4.1	2.2	6.0
6	Men aged 25-34	Proportion of population who are daily e-cigarette users	778	4.6	3.0	6.2
7	Men aged 35-49	Proportion of population who are daily e-cigarette users	1,624	7.0	5.3	8.7
8	Men aged 50-59	Proportion of population who are daily e-cigarette users	1,554	6.6	2.9	10.2
9	Men aged 60 and over	Proportion of population who are daily e-cigarette users	5,824	2.1	1.5	2.6

Workbook last saved: 6m ago

Analyst's report

1. For each of the data rows in the file 'VAPING' (i.e. for each estimated population proportion), calculate a margin of error at the 95% probability level, assuming a simple random sample of individuals.
2. Use a suitable graphical presentation for comparing the margins of error you have calculated in Task 1. with the margins of error implied by the confidence intervals given in the ONS worksheet.
3. Comment on the agreement or disagreement between the two sets of error margins.

Source: www.ons.gov.uk/peoplepopulationandcommunity/ healthandsocialcare/drugusealcoholandsmoking/ datasets/ecigaretteuseingreatbritain (accessed September 2022).

© Yaroslav Litun/iStock

8

Interval Estimation

CHAPTER CONTENTS

Statistics in Practice   How accurate are opinion polls and market research surveys?

LEARNING OBJECTIVES   After reading this chapter and doing the exercises, you should be able to:

1 Explain the purpose of an interval estimate of a population parameter.

2 Explain the terms margin of error, confidence interval, confidence level and confidence coefficient.

3 Construct confidence intervals for a population mean:

   3.1   When the population standard deviation is known, using the normal distribution.

   3.2   When the population standard deviation is unknown, using the t distribution.

4 Construct large-sample confidence intervals for a population proportion.

5 Calculate the sample size required to construct a confidence interval with a given margin of error for a population mean, when the population standard deviation is known.

6 Calculate the sample size required to construct a confidence interval with a given margin of error for a population proportion.

I n Chapter 7, we stated that a point estimator is a sample statistic used to estimate a population parameter. For example, the sample mean is a point estimator of the population mean, and the sample proportion is a point estimator of the population proportion. Because a point estimator cannot be expected to provide the exact value of the population parameter, an interval estimate is often computed. The purpose of an interval estimate is to provide information about how close the point estimate might be to the value of the population parameter. In simple cases, the general form of an interval estimate is:

$$\text{Point estimate} \pm \textbf{Margin of error}$$

In this chapter we show how to compute interval estimates of a population mean μ and a population proportion π. The interval estimates have the same general form:

$$\text{Population mean: } \bar{x} \pm \text{Margin of error}$$
$$\text{Population proportion: } p \pm \text{Margin of error}$$

The sampling distributions of $\bar{X}$ and P play key roles in computing these margins of error.

STATISTICS IN PRACTICE
How accurate are opinion polls and market research surveys?

IPSOS is a worldwide market research and opinion polling company, with global headquarters in Paris. One of the company's research services is the *Global Advisor*. The August 2022 report on the *Global Advisor* survey, covering 28 nations around the world, included results on the IPSOS Global Trustworthiness Index. This index is constructed from respondents' assessments of the trustworthiness of various professions.

Globally, doctors topped the trustworthiness rankings, with 59 per cent of respondents overall rating doctors in their own country as trustworthy. Politicians were at the bottom of the pile, with only 12 per cent of respondents worldwide rating them as trustworthy in their country. However, there were

substantial variations between countries. In Spain, for example, 71 per cent of respondents rated doctors as trustworthy (the highest rating). This compared with 66 per cent in the UK, 60 per cent in Germany, 55 per cent in Sweden, 43 per cent in Japan and 39 per cent in Poland (the lowest rating).

The sample size for the survey was about 3,000 in China, around 1,000 in 10 of the 28 countries (e.g. the UK, Japan), and around 500 in the remaining 17 countries (e.g. Poland, Sweden). How accurate are estimates like these based on sample evidence, and what are the implications of different sample sizes?

Methodological notes in the *Global Advisor* report state that: 'The precision of Ipsos online polls is calculated using a credibility interval with a poll of 1,000 accurate to +/– 3.5 percentage points and of 500 accurate to +/– 5.0 percentage points.' An article on the IPSOS website further explains that the credibility intervals are '95% credibility' intervals, and that the company is 'moving away from the use of "classical" margins of error to demonstrate confidence in the accuracy of our online polling results. We have adopted Bayesian Credibility Intervals as our standard for reporting our confidence in online polling results' (Bayes' theorem, which underlies the Bayesian credibility intervals, was covered in Chapter 4 of this text). This is to take account of the fact that because the *Global Advisor* surveys are conducted online, residents do not all have an equal chance of taking part as they do not have equal opportunities for online access.

In this chapter, you will learn the basis for the 'classical' margins of error, the confidence level associated with them (e.g. 95 per cent) and the calculations that underlie the error margins.

Source: www.ipsos.com/en/ipsos-update-september-2022/.

© Tero Vesalainen/iStock

8.1 POPULATION MEAN: σ KNOWN

To construct an interval estimate of a population mean, either the population standard deviation σ or the sample standard deviation s is used to compute the margin of error. Although σ is rarely known exactly, historical data sometimes give us a close estimate of the population standard deviation prior to sampling. In such cases, the population standard deviation can be considered known for practical purposes. We refer to this situation as the σ known case. In this section we show how a simple random sample can be used to construct an interval estimate of a population mean for the σ known case.

Consider the monthly customer service survey conducted by JJ.COM. The company has a website for taking customer orders and providing follow-up service. Good customer service is critical to the company's success. It prides itself on providing easy online ordering, timely delivery and prompt response to customer enquiries.

JJ.COM's quality assurance team uses a customer survey to measure satisfaction with its website and customer service. Each month, the team emails an online survey to a random sample of customers who placed an order or requested service during the recent month. The online questionnaire asks customers to rate their satisfaction with ease of placing orders, delivery times, accurate order fulfilment and technical advice. The team summarizes each customer's questionnaire by computing an overall satisfaction score x that ranges from 0 (worst possible score) to 100 (best possible). A sample mean customer satisfaction score is then computed.

The sample mean satisfaction score provides a point estimate of the mean satisfaction score μ for the population of all JJ.COM customers in any given month. With this regular measure, JJ.COM can take prompt corrective action in response to a low satisfaction score. The company conducted this survey for many months and consistently obtained an estimate near 12 for the standard deviation of satisfaction scores. Based on these historical data, JJ.COM now assumes a known value of $\sigma = 12$ for the population standard deviation. The historical data also indicate that the population of satisfaction scores follows an approximately normal distribution.

During the most recent month, the quality assurance team surveyed 100 customers ($n = 100$) and obtained a sample mean satisfaction score of $\bar{x} = 72$. This provides a point estimate of the population mean satisfaction score μ. We show how to compute the margin of error for this estimate and construct an interval estimate of the population mean.

JJCOM

Margin of error and the interval estimate

In Chapter 7 we showed how the sampling distribution of the sample mean $\bar{X}$ is used to compute the probability that $\bar{X}$ will be within a given distance of μ. In the JJ.COM example, the historical data show that the population of satisfaction scores is approximately normally distributed with a standard deviation of $\sigma = 12$. So, using what we learned in Chapter 7, we can conclude that the sampling distribution of $\bar{X}$ follows a normal distribution with a standard error of:

$$\sigma_{\bar{X}} = \sigma/\sqrt{n} = 12/\sqrt{100} = 1.2$$

This sampling distribution is shown in Figure 8.1.*

Using the table of cumulative probabilities for the standard normal distribution, we find that 95 per cent of the values of any normally distributed random variable are within ± 1.96 standard deviations of the mean. In the JJ.COM example, we know the sampling distribution of $\bar{X}$ is normal with a standard error of $\sigma_{\bar{X}} = 1.2$. Because $\pm 1.96\sigma_{\bar{X}} = \pm 1.96(1.2) = \pm 2.35$, we conclude that 95 per cent of all possible $\bar{x}$ values, using a sample size of $n = 100$, will be within ± 2.35 units of the population mean μ. Refer to the dark shaded area in Figure 8.1.

---

* The population of satisfaction scores has a normal distribution, so we can conclude that the sampling distribution of $\bar{X}$ is a normal distribution. If the population did not have a normal distribution, we could rely on the central limit theorem, and the sample size of $n = 100$, to conclude that the sampling distribution of $\bar{X}$ is approximately normal. In either case, the sampling distribution would appear as shown in Figure 8.1.

FIGURE 8.1

Sampling distribution of the sample mean satisfaction score from simple random samples of 100 customers, also showing the location of sample means that are within ±2.35 units of μ, and intervals calculated from selected sample means at locations $\bar{x}_1$, $\bar{x}_2$ and $\bar{x}_3$

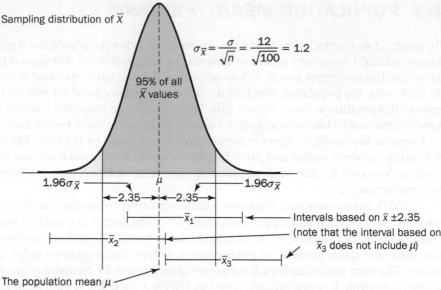

Sampling distribution of $\bar{x}$

$\sigma_{\bar{x}} = \dfrac{\sigma}{\sqrt{n}} = \dfrac{12}{\sqrt{100}} = 1.2$

95% of all $\bar{x}$ values

$1.96\sigma_{\bar{x}}$ μ $1.96\sigma_{\bar{x}}$

←2.35→|←2.35→

$\bar{x}_1$

Intervals based on $\bar{x}$ ±2.35

$\bar{x}_2$

(note that the interval based on $\bar{x}_3$ does not include μ)

$\bar{x}_3$

The population mean μ

We said above that the general form of an interval estimate of the population mean μ is $\bar{x} \pm$ margin of error. For the JJ.COM example, suppose we set the margin of error equal to 2.35 and compute the interval estimate of μ using $\bar{x} \pm 2.35$. To provide an interpretation for this interval estimate, let us consider the values of $\bar{x} \pm 2.35$ that could be obtained if we took three different simple random samples, each consisting of 100 JJ.COM customers.

The first sample mean might turn out to have the value shown as $\bar{x}_1$ in Figure 8.1. In this case, the interval formed by subtracting 2.35 from $\bar{x}_1$ and adding 2.35 to $\bar{x}_1$ includes the population mean μ. Now consider what happens if the second sample mean turns out to have the value shown as $\bar{x}_2$ in Figure 8.1. Although $\bar{x}_2$ differs from $\bar{x}_1$, we see that the interval $\bar{x}_2 \pm 2.35$ also includes the population mean μ. However, consider what happens if the third sample mean turns out to have the value shown as $\bar{x}_3$ in Figure 8.1. In this case, because $\bar{x}_3$ falls in the upper tail of the sampling distribution and is further than 2.35 units from μ, the interval $\bar{x}_3 \pm 2.35$ does not include the population mean μ.

Any sample mean that is within the dark shaded region of Figure 8.1 will provide an interval estimate that includes the population mean μ. Because 95 per cent of all possible sample means are in the dark shaded region, 95 per cent of all intervals formed by subtracting 2.35 from $\bar{x}$ and adding 2.35 to $\bar{x}$ will include the population mean μ.

The general form of an interval estimate of a population mean for the σ known case is:

Interval estimate of a population mean: σ known

$$\bar{x} \pm z_{\alpha/2}\frac{\sigma}{\sqrt{n}} \tag{8.1}$$

where $(1 - \alpha)$ is the confidence coefficient and $z_{\alpha/2}$ is the z value providing an area $\alpha/2$ in the upper tail of the standard normal probability distribution.

Let us use expression (8.1) to construct a 95% confidence interval for the JJ.COM problem. For a 95% confidence interval, the confidence coefficient is $(1 - \alpha) = 0.95$ and so $\alpha = 0.05$, $\alpha/2 = 0.025$ and $z_{0.025} = 1.96$. With the JJ.COM sample mean $\bar{x} = 72$, we obtain:

$$72 \pm 1.96 \frac{12}{\sqrt{100}} = 72 \pm 2.35$$

As we saw above, the margin of error at this level of confidence is ± 2.35. The specific interval estimate of μ based on the data from the most recent month is $72 - 2.35 = 69.65$ to $72 + 2.35 = 74.35$. Because 95 per cent of all the intervals constructed using $\bar{x} \pm 2.35$ will contain the population mean, we say that we are 95 per cent confident that the interval 69.65 to 74.35 includes the population mean μ. We say that this interval has been established at the 95% confidence level. The value 0.95 is referred to as the confidence coefficient, and the interval 69.65 to 74.35 is called the 95% confidence interval.

Although the 95% confidence level is frequently used, other confidence levels such as 90% and 99% may be considered. Values of $z_{\alpha/2}$ for the most commonly used confidence levels are shown in Table 8.1.

TABLE 8.1 Values of $z_{\alpha/2}$ for the most commonly used confidence levels

Confidence level (%)	α	$\alpha/2$	$z_{\alpha/2}$
90	0.10	0.05	1.645
95	0.05	0.025	1.960
99	0.01	0.005	2.576

Using these values and expression (8.1), the 90% confidence interval for the JJ.COM problem is:

$$72 \pm 1.645 \frac{12}{\sqrt{100}} = 72 \pm 1.97$$

At 90% confidence, the margin of error is 1.97 and the confidence interval is $72 - 1.97 = 70.03$, to $72 + 1.97 = 73.97$. Similarly, the 99% confidence interval is:

$$72 \pm 2.576 \frac{12}{\sqrt{100}} = 72 \pm 3.09$$

At 99% confidence, the margin of error is 3.09 and the confidence interval is $72 - 3.09 = 68.93$, to $72 + 3.09 = 75.09$.

Comparing the results for the 90%, 95% and 99% confidence levels, we see that, to have a higher level of confidence, the potential margin of error and consequently the width of the confidence interval must be larger.

Practical advice

If the population follows a normal distribution, the confidence interval provided by expression (8.1) is exact. If expression (8.1) were used repeatedly to generate 95% confidence intervals, 95 per cent of the intervals generated (in the long run) would contain the population mean. If the population does not follow a normal distribution, the confidence interval provided by expression (8.1) will be approximate. In this case, the quality of the approximation depends on both the distribution of the population and the sample size.

In most applications, a sample size of $n \geq 30$ is adequate when using expression (8.1). If the population is not normally distributed but is roughly symmetrical, sample sizes as small as 15 can be expected to provide good approximate confidence intervals. With smaller sample sizes, expression (8.1) should be used only if the analyst believes, or is willing to assume, that the population distribution is at least approximately normal.

EXERCISES

Methods

1. A simple random sample of 40 items results in a sample mean of 25. The population standard deviation is $\sigma = 5$.
   a. What is the value of the standard error of the mean, $\sigma_{\bar{x}}$?
   b. At 95% confidence, what is the margin of error for estimating the population mean?

2. A simple random sample of 50 items from a population with $\sigma = 6$ results in a sample mean of 32.
   a. Construct a 90% confidence interval for the population mean μ.
   b. Construct a 95% confidence interval for μ.
   c. Construct a 99% confidence interval for μ.

3. A simple random sample of 60 items results in a sample mean of 80. The population standard deviation is $\sigma = 15$.
   a. Construct a 95% confidence interval for the population mean μ.
   b. Suppose the same sample mean came from a sample of 120 items. Construct a 95% confidence interval for μ.
   c. What is the effect of a larger sample size on the interval estimate?

4. A 95% confidence interval for a population mean was reported to be 152 to 160. If $\sigma = 15$, what sample size was used in this study?

Applications

5. To estimate the mean amount spent per customer for dinner at a Johannesburg restaurant, data were collected for a sample of 49 customers. Assume a population standard deviation of 40 South African rand (ZAR).
   a. At 95% confidence, what is the margin of error?
   b. If the sample mean is ZAR 186, what is the 95% confidence interval for the population mean?

6. Studies show that massage therapy has a variety of health benefits relative to its cost. A sample of ten typical one-hour massage therapy sessions showed an average charge of €59. The population standard deviation for a one-hour session is $\sigma = €5.50$.
   a. What assumptions about the population should we be willing to make if we calculate a margin of error?
   b. Using 95% confidence, what is the margin of error?
   c. Using 99% confidence, what is the margin of error?

7. A survey of 750 university students found they were paying on average €108 per week in accommodation costs. Assume the population standard deviation for weekly accommodation costs is €22.
   a. Construct a 90% confidence interval estimate of the population mean μ.
   b. Construct a 95% confidence interval estimate of μ.
   c. Construct a 99% confidence interval estimate of μ.
   d. Discuss what happens to the width of the confidence interval as the confidence level is increased. Does this result seem reasonable? Explain.

8.2 POPULATION MEAN: σ UNKNOWN

In most cases in practice, the value of the population standard deviation σ is unknown prior to sampling, so we must use the sample standard deviation s to estimate σ. This is the σ unknown case.

When s is used to estimate σ, the margin of error in estimating the population mean is based on a family of probability distributions known as the t distribution. Each member of the t distribution family is specified by a parameter known as the degrees of freedom (usually abbreviated df). The t distribution with 1 df is a member of the family, as is the t distribution with 2 df, with 3 df and so on. Most members of the t distribution family look quite similar in shape to a standard normal distribution, but have rather fatter tails. In other words, there is rather more probability in the tails of the distribution than for a standard normal distribution. As the number of df increases, the difference between the t distribution and the standard normal distribution becomes smaller and smaller.

Figure 8.2 shows t distributions with 10 and 2 degrees of freedom and their relationship to the standard normal probability distribution. The higher the number of df, the lower is the variability, and the greater the resemblance to the standard normal distribution. Note also that the mean of the t distribution is zero.

FIGURE 8.2

Comparison of the standard normal distribution, t distribution with 10 df and t distribution with 2 df

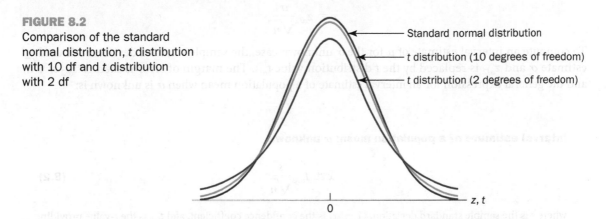

Standard normal distribution

t distribution (10 degrees of freedom)

t distribution (2 degrees of freedom)

0   z, t

Although the mathematical development of the t distribution assumes a normal distribution for the sampled population, research shows that the t distribution can be successfully applied in many situations where the population deviates from normal. Later in this section we provide guidelines for using the t distribution if the population is not normally distributed.

We place a subscript on t to indicate the probability in the upper tail of the t distribution. For example, just as we used $z_{0.025}$ to indicate the z value giving a 0.025 probability in the upper tail of a standard normal distribution, we shall use $t_{0.025}$ to indicate a 0.025 probability in the upper tail of a t distribution. So, the general notation $t_{\alpha/2}$ will represent a t-value with a probability of $\alpha/2$ in the upper tail of the t distribution. Refer to Figure 8.3.

FIGURE 8.3

t distribution with $\alpha/2$ probability in the upper tail

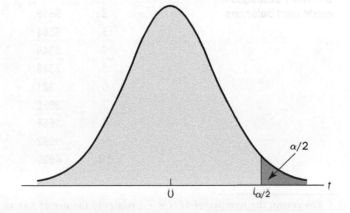

$\alpha/2$

0     $t_{\alpha/2}$ t

Table 2 of Appendix B is a table for the t distribution. Each row in the table corresponds to a separate t distribution with the degrees of freedom shown. For example, for a t distribution with 10 df, $t_{0.025} = 2.228$. Similarly, for a t distribution with 20 df, $t_{0.025} = 2.086$. As the df continue to increase, $t_{0.025}$ approaches $z_{0.025} = 1.96$. The standard normal distribution z values can be found in the infinite degrees of freedom row of the t distribution table (labelled ∞). If the df exceed 100, the infinite df row can be used to approximate the actual t-value. In other words, for more than 100 df, the standard normal z value provides a good approximation to the t-value.

William Sealy Gosset, writing under the pen-name 'Student', was the originator of the t distribution. Gosset, an Oxford graduate in mathematics, worked for the Guinness Brewery in Dublin, Ireland. The distribution is sometimes referred to as 'Student's t distribution'.

Margin of error and the interval estimate

In Section 8.1 we showed that an interval estimate of a population mean for the σ known case is:

$$\bar{x} \pm z_{\alpha/2}\frac{\sigma}{\sqrt{n}}$$

To compute an interval estimate of μ for the σ unknown case, the sample standard deviation s is used to estimate σ and $z_{\alpha/2}$ is replaced by the t distribution value $t_{\alpha/2}$. The margin of error is then $\pm\, t_{\alpha/2}\, s/\sqrt{n}$, and the general expression for an interval estimate of a population mean when σ is unknown is:

Interval estimate of a population mean: σ unknown

$$\bar{x} \pm t_{\alpha/2}\frac{s}{\sqrt{n}} \tag{8.2}$$

where s is the sample standard deviation, $(1 - \alpha)$ is the confidence coefficient, and $t_{\alpha/2}$ is the t-value providing an area of $\alpha/2$ in the upper tail of the t distribution with $n - 1$ df.*

Consider a study designed to estimate the mean credit card debt for a defined population of households. A sample of $n = 85$ households provided the credit card balances in the file 'Balance' on the online platform. The first few rows of this data set are shown in the Excel screenshot in Figure 8.4. For this situation, no previous estimate of the population standard deviation is available. Consequently, the sample data must be used to estimate both the population mean and the population standard deviation.

FIGURE 8.4
First few data rows and
summary statistics for
credit card balances

BALANCE

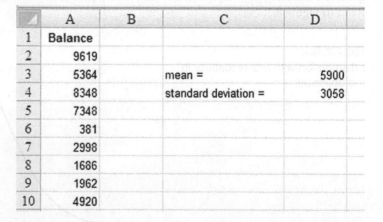

	A	B	C	D
1	Balance			
2	9619			
3	5364		mean =	5900
4	8348		standard deviation =	3058
5	7348			
6	381			
7	2998			
8	1686			
9	1962			
10	4920			

---

* The reason the number of df is $n - 1$ relates to the use of s as an estimate of σ. The equation for calculating s is $s = \sqrt{\Sigma(x_i - \bar{x})^2/(n - 1)}$. Degrees of freedom refers to the number of independent pieces of information in the computation of $\Sigma(x_i - \bar{x})^2$. The n pieces of information involved in computing $\Sigma(x_i - \bar{x})^2$ are as follows: $x_1 - \bar{x}, x_2 - \bar{x}, \cdots, x_n - \bar{x}$. In Section 3.2 we indicated that $\Sigma(x_i - \bar{x}) = 0$. Hence, only $n - 1$ of the $x_i - \bar{x}$ values are independent; that is, if we know $n - 1$ of the values, the remaining value can be determined exactly by using the condition that $\Sigma(x_i - \bar{x}) = 0$. So $n - 1$ is the number of df associated with $\Sigma(x_i - \bar{x})^2$ and hence the number of df for the t distribution in expression (8.2).

Using the data in the 'Balance' file, we compute the sample mean $\bar{x} = $ €5,900 and the sample standard deviation $s = $ €3,058. With 95% confidence and $n - 1 = 84$ df, Table 2 in Appendix B gives $t_{0.025} = 1.989$. We can now use expression (8.2) to compute an interval estimate of the population mean:

$$5,900 \pm 1.989\left(\frac{3,058}{\sqrt{85}}\right) = 5,900 \pm 660$$

The point estimate of the population mean is €5,900, the margin of error is €660 and the 95% confidence interval is $5,900 - 660 = $ €5,240 to $5,900 + 660 = $ €6,560. We are 95 per cent confident that the population mean credit card balance for all households in the defined population is between €5,240 and €6,560.

The procedures used by Excel, R, SPSS and Minitab to construct confidence intervals for a population mean are described in the software guides on the online platform.

Practical advice

If the population follows a normal distribution, the confidence interval provided by expression (8.2) is exact and can be used for any sample size. If the population does not follow a normal distribution, the confidence interval provided by expression (8.2) will be approximate. In this case, the quality of the approximation depends on the distribution of the population and on the sample size.

In most applications, a sample size of $n \geq 30$ is adequate when using expression (8.2). However, if the population distribution is highly skewed or contains outliers, the sample size should be 50 or more. If the population is not normally distributed but is roughly symmetrical, sample sizes as small as 15 can be expected to provide good approximate confidence intervals. With smaller sample sizes, expression (8.2) should only be used if the analyst is confident that the population distribution is at least approximately normal.

Using a small sample

In the following example we construct an interval estimate for a population mean when the sample size is small. An understanding of the distribution of the population becomes a factor in deciding whether the interval estimation procedure provides acceptable results.

Javed Manufacturing is considering a new computer-assisted programme to train maintenance employees to do machine repairs. To fully evaluate the programme, the manufacturing manager requested an estimate of the population mean time required for maintenance employees to complete the training.

A sample of 20 employees is selected, with each employee in the sample completing the training programme. Data on the training time in days for the 20 employees are shown in Table 8.2. Figure 8.5 shows a histogram of the sample data, $n = 20$. What can we say about the distribution of the population based on this histogram? First, the sample data do not support with certainty the conclusion that the distribution of the population is normal, but we do not see any evidence of skewness or outliers. Therefore, using the guidelines in the previous subsection, we conclude that an interval estimate based on the t distribution appears acceptable for the sample of 20 employees.

TABLE 8.2 Training time in days for a sample of 20 Javed Manufacturing employees

52	59	54	52
44	50	42	48
55	54	60	55
54	62	62	57
45	46	43	56

FIGURE 8.5
Histogram of training times for the Javed
Manufacturing sample

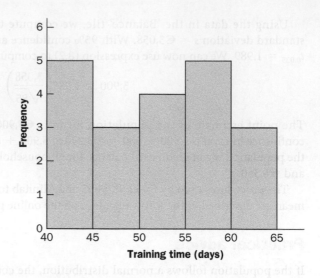

We compute the sample mean and sample standard deviation:

$$\bar{x} = \frac{\Sigma x_i}{n} = \frac{1,050}{20} = 52.5 \text{ days}$$

$$s = \sqrt{\frac{\Sigma(x_i - \bar{x})^2}{n-1}} = \sqrt{\frac{729}{20-1}} = 6.19 \text{ days}$$

For a 95% confidence interval, we use Table 2 from Appendix B and $n-1 = 19$ df to obtain $t_{0.025} = 2.093$. Expression (8.2) provides the interval estimate of the population mean:

$$52.5 \pm 2.093\left(\frac{6.19}{\sqrt{20}}\right) = 52.5 \pm 2.9$$

The point estimate of the population mean is 52.5 days. The margin of error is 2.9 days and the 95% confidence interval is $52.5 - 2.9 = 49.6$ days to $52.5 + 2.9 = 55.4$ days.

Using a histogram of the sample data to learn about the population distribution is rarely conclusive, but in many cases it provides the only information available. The histogram, along with judgement by the analyst, can often be used to decide if expression (8.2) is appropriate for constructing the interval estimate.

Summary of interval estimation procedures

We provided two approaches to computing the margin of error and constructing an interval estimate of a population mean:

- If we can assume a known value for σ, the value of σ and the standard normal distribution are used in expression (8.1).
- If σ is unknown, s and the t distribution are used in expression (8.2).

In most applications, a sample size of $n \geq 30$ is adequate. If the population has a normal or approximately normal distribution, smaller sample sizes may be used. For the σ unknown case a sample size of $n \geq 50$ is recommended if the population distribution may be highly skewed or may have outliers.

EXERCISES

Methods

8. For a *t* distribution with 16 df, find the probability in each region.
   a. To the right of 2.120.
   b. To the left of 1.337.
   c. To the left of −1.746.
   d. To the right of 2.583.
   e. Between −2.120 and 2.120.
   f. Between −1.746 and 1.746.

9. Find the *t*-value(s) for each of the following cases.
   a. Upper-tail probability of 0.025 with 12 df.
   b. Lower-tail probability of 0.05 with 50 df.
   c. Upper-tail probability of 0.01 with 30 df.
   d. Where 90 per cent of the probability falls between these two *t*-values with 25 df.
   e. Where 95 per cent of the probability falls between these two *t*-values with 45 df.

10. The following sample data are from a normal population: 10, 8, 12, 15, 13, 11, 6, 5.
   a. Produce a point estimate of the population mean.
   b. Produce a point estimate of the population standard deviation.
   c. At 95% confidence, calculate the margin of error for estimating the population mean.
   d. Construct a 95% confidence interval for the population mean.

11. A simple random sample with $n = 54$ provided a sample mean of 22.5 and a sample standard deviation of 4.4.
   a. Construct a 90% confidence interval for the population mean μ.
   b. Construct a 95% confidence interval for μ.
   c. Construct a 99% confidence interval for μ.
   d. What happens to the margin of error and the confidence interval as the confidence level is increased?

Applications

12. The mean cost of a meal for two in a mid-range restaurant in Tokyo is $40 (*Numbeo.com* website). How do prices for comparable meals in Hong Kong compare? The file 'HKMEALS' contains the costs for a sample of 42 recent meals for two in Hong Kong mid-range restaurants.
   a. With 95% confidence, what is the margin of error?
   b. What is the 95% confidence interval estimate of the population mean?
   c. How do prices for meals for two in mid-range restaurants in Hong Kong compare with prices for comparable meals in Tokyo restaurants?

HKMEALS

13. Consumption of alcoholic drinks by young women is of concern in the UK and some other European countries. Annual consumption data (in litres) are shown below for a sample of 20 European young women.

ALCOHOL

266	82	199	174	97
170	222	115	130	169
164	102	113	171	0
93	0	93	110	130

Assuming the population is distributed roughly symmetrically, construct a 95% confidence interval for the mean annual consumption of alcoholic beverages by young European women.

14. The International Air Transport Association (IATA) surveys business travellers to develop quality ratings for international airports. The maximum possible rating is 10. Suppose a simple random sample of business travellers is selected and each traveller is asked to provide a rating for Kuwait International Airport. The ratings obtained from the sample of 50 business

IATA

travellers follow. Construct a 95% confidence interval estimate of the population mean rating for Kuwait International.

2	1	8	7	3	1	8	1	7	9	2	9	10	9	7	8	9
1	0	3	0	1	6	2	3	1	6	8	7	7	7	7	7	1
2	5	2	1	2	2	0	2	2	7	0	8	7	0	2	8	

15. A diary study involving a sample of 50 undergraduate business students at a large university finds that the mean time spent per day using social media is 3 hours and 38 minutes. The sample standard deviation of time spent per day is 1 hour 41 minutes.
    a. At 95% confidence, what is the margin of error for estimating μ the population mean time spent per day using social media for undergraduate business students at this university?
    b. What is the 95% confidence interval for μ?

FASTFOOD

16. A sample of 30 fast food restaurants including McDonald's and Burger King were visited. During each visit, the customer went to the drive-through and ordered a basic meal. The time between pulling up to the order kiosk and receiving the filled order was recorded. The times in minutes for the 30 visits are as follows:

| 0.9 | 1.0 | 1.2 | 2.2 | 1.9 | 3.6 | 2.8 | 5.2 | 1.8 | 2.1 | 6.8 | 1.3 | 3.0 | 4.5 | 2.8 |
| 2.3 | 2.7 | 5.7 | 4.8 | 3.5 | 2.6 | 3.3 | 5.0 | 4.0 | 7.2 | 9.1 | 2.8 | 3.6 | 7.3 | 9.0 |

    a. Provide a point estimate of the population mean drive-through time at fast food restaurants.
    b. Calculate a 95% confidence interval estimate of the population mean.
    c. Discuss skewness that may be present in this population. What suggestion would you make for a repeat of this study?

SMART
MEDIA

17. A survey by SmartMedia asked a sample of 200 sales executives to provide data on the number of minutes per day they waste trying to locate computer files and emails that have been poorly named, misfiled or otherwise misplaced. Data consistent with this survey are contained in the data set 'SmartMedia' online.
    a. Use 'SmartMedia' to compute a point estimate of the number of minutes per day sales executives waste trying to locate poorly named, misfiled or otherwise misplaced items.
    b. What is the sample standard deviation?
    c. What is the 95% confidence interval for the population mean number of minutes wasted per day by sales executives?

8.3 DETERMINING THE SAMPLE SIZE

Earlier we commented on the role of sample size in providing good approximate confidence intervals when the population is not normally distributed. In this section, we focus on another aspect of sample size: how to choose a sample size large enough to provide a desired margin of error. To understand this process, we return to the σ known case presented in Section 8.1 where expression (8.1) shows the interval estimate as $\bar{x} \pm z_{\alpha/2}(\sigma/\sqrt{n})$. We see that $z_{\alpha/2}$, the population standard deviation σ and the sample size n combine to determine the margin of error. We can find $z_{\alpha/2}$ once we have selected a confidence coefficient $1 - \alpha$. Then, if we have a value for σ, we can determine the sample size n needed to provide any desired margin of error. Let $E =$ the desired margin of error:

$$E = z_{\alpha/2}\left(\frac{\sigma}{\sqrt{n}}\right)$$

Solving for $\sqrt{n}$, we have:

$$\sqrt{n} = \frac{z_{\alpha/2}\sigma}{E}$$

Squaring both sides of this equation, we obtain equation (8.3) for the sample size that provides the desired margin of error at the chosen confidence level:

Sample size for an interval estimate of a population mean

$$n = \frac{(z_{\alpha/2})^2 \sigma^2}{E^2}$$
(8.3)

In equation (8.3), E is the target margin of error. The value of $z_{\alpha/2}$ follows directly from the chosen confidence level. The most frequently chosen value is 95% confidence ($z_{0.025} = 1.96$), but other levels are sometimes preferred. Equation (8.3) can be used to provide a good sample size recommendation. However, the analyst should use judgement in deciding whether the recommendation given by equation (8.3) needs adjustment.

Use of equation (8.3) requires a value for the population standard deviation σ. Even if σ is unknown, we can use equation (8.3) provided we have a preliminary or *planning value* for σ. In practice, one of the following procedures can be used to provide this planning value:

1 Use an estimate of σ computed from data of previous studies.
2 Use a pilot study to select a preliminary sample, then use the sample standard deviation from the preliminary sample.
3 Use judgement or a 'best guess' for the value of σ. For example, we might begin by estimating the largest and smallest data values in the population. The difference between the largest and smallest values provides an estimate of the range for the data. The range divided by four is often suggested as a rough approximation of the standard deviation and hence an acceptable planning value for σ.

Consider the following example. A travel organization would like to conduct a study to estimate the population mean daily rental cost for a family car in Ireland. The director specifies that the population mean daily rental cost should be estimated with a margin of error of no more than €2, at the 95% confidence level. A study the previous year found a mean cost of approximately €80 per day for renting a family car, with a standard deviation of about €10.

The director specified a desired margin of error of $E \leq 2$. The 95% confidence level indicates $z_{0.025} = 1.96$. We only need a planning value for the population standard deviation σ to compute the required sample size. Using €10 (from the previous study) as the planning value for σ, we obtain:

$$n = \frac{(z_{\alpha/2})^2 \sigma^2}{E^2} = \frac{(1.96)^2 (10)^2}{(2)^2} = 96.04$$

The sample size for the new study needs to be at least 96.04 family car rentals to satisfy the director's €2 margin-of-error requirement. In cases where the computed n is not an integer, we usually round up to the next integer value, in this case 97. Here, the sample size might be rounded for convenience to 100.

EXERCISES

Methods

18. If the population standard deviation is 40, how large a sample should be selected to provide a 95% confidence interval with a margin of error no larger than 10?

19. The range for a set of data is estimated to be 36.
   a. Suggest a planning value for the population standard deviation.
   b. At 95% confidence, how large a sample would provide a margin of error of 3?
   c. At 95% confidence, how large a sample would provide a margin of error of 2?

Applications

20. Refer to the Javed Manufacturing example in Section 8.2. Use 6.2 days as a planning value for the population standard deviation.
   a. Assuming 95% confidence, what sample size would be required to obtain a margin of error of 1.5 days?
   b. If the precision statement were made with 90% confidence, what sample size would be required to obtain a margin of error of 2.0 days?

21. Suppose you are interested in estimating the average cost of staying for one night in a double room in a 3-star hotel in France (outside Paris), with a margin of error of €3. Using €30.00 as the planning value for the population standard deviation, what sample size is recommended for each of the following cases?
   a. A 90% confidence interval estimate of the population mean cost.
   b. A 95% confidence interval estimate of the population mean cost.
   c. A 99% confidence interval estimate of the population mean cost.
   d. When the desired margin of error is fixed, what happens to the sample size as the confidence level is increased? Would you recommend a 99% confidence level be used? Discuss.

22. Suppose the price/earnings (P/E) ratios for stocks listed on a European Stock Exchange have a mean value of 35 and a standard deviation of 18. We want to estimate the population mean P/E ratio for all stocks listed on the exchange. How many stocks should be included in the sample if we want a margin of error of 3? Use 95% confidence.

23. Fuel consumption tests are conducted for a specific car model. If a 95% confidence interval with a margin of error of 0.2 litre per 100 km is desired, how many cars should be used in the test? Assume that preliminary tests indicate the standard deviation is 0.5 litre per 100 km.

24. Annual starting salaries for university graduates with degrees in business administration are generally expected to be between €45,000 and €60,000. Assume that a 95% confidence interval estimate of the population mean annual starting salary is desired. What is the planning value for the population standard deviation? How large a sample should be taken for each of the following margins of error?
   a. €500
   b. €200
   c. €100
   d. Would you recommend trying to obtain the €100 margin of error? Explain.

8.4 POPULATION PROPORTION

We said earlier that the general form of an interval estimate of a population proportion π is $p \pm$ margin of error. The sampling distribution of the sample proportion P plays a key role in computing the margin of error for this interval estimate.

In Chapter 7 we said that the sampling distribution of the sample proportion P can be approximated by a normal distribution whenever $n\pi \geq 5$ and $n(1 - \pi) \geq 5$. Figure 8.6 shows the normal approximation of the sampling distribution of P. The mean of the sampling distribution of P is the population proportion π, and the standard error of P is:

$$\sigma_P = \sqrt{\frac{\pi(1 - \pi)}{n}}$$ (8.4)

If we choose $z_{\alpha/2}\sigma_P$ as the margin of error in an interval estimate of a population proportion, we know that $100(1 - \alpha)$ per cent of the intervals generated will contain the true population proportion. But σ_P cannot be used directly in the computation of the margin of error because π will not be known; π is what we are trying to estimate. So, p is substituted for π and the margin of error for an interval estimate of a population proportion is given by:

$$\text{Margin of error} = z_{\alpha/2}\sqrt{\frac{p(1 - p)}{n}} \tag{8.5}$$

FIGURE 8.6
Normal approximation of the sampling distribution of P

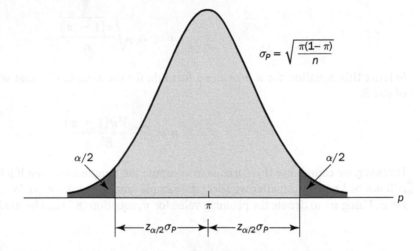

The general expression for an interval estimate of a population proportion is:

Interval estimate of a population proportion

$$p \pm z_{\alpha/2}\sqrt{\frac{p(1 - p)}{n}} \tag{8.6}$$

where $1 - \alpha$ is the confidence coefficient and $z_{\alpha/2}$ is the z value providing an area of $\alpha/2$ in the upper tail of the standard normal distribution.

Consider the following example. A national survey of 900 female gym members was carried out to assess their view on women-only exercise classes. (The data are available in the file 'EXERCISE' on the online platform.) The survey found that 396 of the female gym members were satisfied with the availability of women-only exercise classes. So, the point estimate of the proportion of the population of female gym members who are satisfied is 396/900 = 0.44. Using expression (8.6) and a 95% confidence level,

EXERCISE

$$p \pm z_{\alpha/2}\sqrt{\frac{p(1 - p)}{n}} = 0.44 \pm 1.96\sqrt{\frac{0.44(1 - 0.44)}{900}} = 0.44 \pm 0.0324$$

The margin of error is 0.0324 and the 95% confidence interval estimate of the population proportion is 0.408 to 0.472. Using percentages, the survey results enable us to state with 95% confidence that between 40.8 per cent and 47.2 per cent of all female gym members are satisfied with the availability of women-only classes.

Determining the sample size

The rationale for sample size determination in constructing interval estimates of π is like the rationale used in Section 8.3 to determine the sample size for estimating a population mean.

Previously in this section we said that the margin of error associated with an interval estimate of a population proportion is $z_{\alpha/2}\sqrt{\pi(1-\pi)/n}$, though in practice we must substitute the sample proportion p in place of the unknown population proportion π. The margin of error is based on the values of $z_{\alpha/2}$, the population proportion π or sample proportion p, and the sample size n. Larger sample sizes provide a smaller margin of error and better precision. Let E denote the desired margin of error:

$$E = z_{\alpha/2}\sqrt{\frac{\pi(1-\pi)}{n}}$$

Solving this equation for n provides a formula for the sample size that will provide a margin of error of size E:

$$n = \frac{(z_{\alpha/2})^2\pi(1-\pi)}{E^2}$$

However, we cannot use this formula to compute the sample size, even if p is substituted for π, because p will not be known until after we select the sample (and π is unknown). What we need is a planning value for π. Using π_* to denote the planning value for π, equation (8.7) can be used to compute the sample size:

Sample size for an interval estimate of a population proportion

$$n = \frac{(z_{\alpha/2})^2\pi^*(1-\pi^*)}{E^2} \tag{8.7}$$

In practice, the planning value can be chosen by one of the following procedures:

1 Use the sample proportion from a previous sample of the same or similar units.

2 Use a pilot study to select a preliminary sample. The sample proportion from this sample can be used as the planning value.

3 Use judgement or a 'best guess' for the value of π^*.

4 Use a planning value of $\pi^* = 0.50$.

Let us return to the survey of female gym members and assume there is an interest in conducting a new survey to estimate the current proportion of the population of female gym members who are satisfied with the availability of women-only exercise classes. How large should the sample be if the survey manager wants to estimate the population proportion with a margin of error of 0.025 at 95% confidence? With $E = 0.025$ and $z_{\alpha/2} = 1.96$, we need a planning value π^* to answer the sample size question. Using the previous survey result of $p = 0.44$ as the planning value π^*, equation (8.7) shows that:

$$n = \frac{(z_{\alpha/2})^2\pi^*(1-\pi^*)}{E^2} = \frac{(1.96)^2(0.44)(1-0.44)}{(0.025)^2} = 1{,}514.5$$

Rounding up to the next integer value indicates that a sample of at least 1,515 female gym members is recommended to satisfy the margin of error requirement.

The fourth alternative suggested for selecting a planning value π^* is to use $\pi^* = 0.50$. This π_* value is frequently used when no other information is available. To understand why, note that the numerator of equation (8.7) shows that the sample size is proportional to the quantity $\pi^*(1-\pi^*)$. A larger value for this quantity will result in a larger sample size. Table 8.3 gives some possible values of $\pi^*(1-\pi^*)$. The largest value occurs when $\pi^* = 0.50$. So, when we are uncertain about an appropriate planning value, we know that $\pi^* = 0.50$ will provide the largest sample size recommendation. If the sample proportion turns out to be different from the 0.50 planning value, the margin of error will be smaller than anticipated. In effect, we play it safe by recommending the largest possible sample size.

TABLE 8.3 Some possible values for $\pi^*(1 - \pi^*)$

π^*	$\pi^*(1 - \pi^*)$
0.10	(0.10)(0.90) = 0.09
0.30	(0.30)(0.70) = 0.21
0.40	(0.40)(0.60) = 0.24
0.50	(0.50)(0.50) = 0.25 ⟵ Largest value for $\pi^*(1 - \pi^*)$
0.60	(0.60)(0.40) = 0.24
0.70	(0.70)(0.30) = 0.21
0.90	(0.90)(0.10) = 0.09

In the survey of female gym members example, a planning value of $\pi^* = 0.50$ would have provided the sample size:

$$n = \frac{(z_{\alpha/2})^2 \pi^*(1 - \pi^*)}{E^2} = \frac{(1.96)^2(0.5)(1 - 0.5)}{(0.025)^2} = 1,536.6$$

A slightly larger sample size of 1,537 female gym members would be recommended.

EXERCISES

Methods

25. A simple random sample of 400 individuals provides 100 Yes responses.
    a. Calculate a point estimate of the proportion of the population that would provide Yes responses.
    b. Estimate the standard error of the sample proportion.
    c. Compute a 95% confidence interval for the population proportion.

26. A simple random sample of 800 elements generates a sample proportion $p = 0.70$.
    a. Construct a 90% confidence interval for the population proportion π.
    b. Construct a 95% confidence interval for π.

27. In a survey, the planning value for the population proportion is $\pi^* = 0.35$. How large a sample should be taken to provide a 95% confidence interval with a margin of error of 0.05?

28. At 95% confidence, how large a sample should be taken to obtain a margin of error of 0.03 for the estimation of a population proportion? Assume there are no past data to provide a planning value for π.

Applications

29. A survey of 611 office workers investigated telephone answering practices, including how often each office worker was able to answer incoming calls and how often incoming calls went directly to voicemail. A total of 281 office workers indicated that they never need voicemail and can take every telephone call.
    a. What is the point estimate of the proportion of the population of office workers who can take every telephone call?
    b. At 90% confidence, what is the margin of error?
    c. What is the 90% confidence interval for the proportion of the population of office workers who can take every telephone call?

30. The French market research and polling company CSA carried out surveys to investigate job satisfaction among professionally qualified employees of private companies. A total of 629 professionals were involved in the surveys, of whom 195 said that they were dissatisfied with their employer's recognition of their professional experience.
   a. Produce a point estimate of the proportion of the population of employees who were dissatisfied with their employer's recognition of their professional experience.
   b. At 95% confidence, what is the margin of error?
   c. What is the 95% confidence interval for the proportion of the population of employees who were dissatisfied with their employer's recognition of their professional experience?

31. In 2021, the Pew Research Centre Internet Project conducted a survey of 1,502 internet users that provided a variety of statistics on internet users. For instance, in 2021, 93 per cent of adults in the USA were internet users.
   a. Construct a 95% confidence interval for the proportion of US adults who were internet users in 2021.
   b. The sample survey showed that 77 per cent of its adult US respondents said they have a broadband connection at home. Construct a 95% confidence interval for the proportion of adults in the USA who have a broadband connection at home.
   c. The sample survey showed that 15 per cent of adults in the USA say their smartphones are their primary means of online access at home. Construct a 95% confidence interval for the proportion of adults in the USA who say their smartphones are their primary means of online access at home.
   d. Compare the margin of error for the interval estimates in parts (a), (b) and (c). How is the margin of error related to the sample proportion?

32. A well-known credit card firm wishes to estimate the proportion of credit card holders who carry a non-zero balance at the end of the month and incur an interest charge. Assume that the desired margin of error is 0.03 at 95% confidence.
   a. How large a sample should be selected if it is anticipated that roughly 70 per cent of the firm's cardholders carry a non-zero balance at the end of the month?
   b. How large a sample should be selected if no planning value for the proportion could be specified?

33. Internet users were asked online to rate their satisfaction with the web browser they use most frequently. Of 102,519 respondents, 65,120 indicated they were very satisfied with the web browser they use most frequently.
   a. What is the sample proportion of internet users who are very satisfied with the web browser they use most frequently?
   b. Using 95% confidence, what is the margin of error?
   c. Using the results from parts (a) and (b), construct a 95% confidence interval estimate of the proportion of internet users who are very satisfied with the web browser they use most frequently.

34. The Labour Force Survey published in August 2022 by the UK Office for National Statistics (ONS) indicates that the unemployment rate among UK nationals aged 16+ was 3.5 per cent, compared with 4.9 per cent for non-UK nationals of the same age group. The tables published by ONS show a margin of error at the 95% confidence level of ±0.2 percentage points for the UK nationals estimate, and ±0.9 percentage points for the non-UK nationals estimate.
   a. If you assume the sample taken by ONS was a simple random sample, what do these figures suggest about the sample size for UK nationals? (In fact, the sample design was more complex than a simple random sample.)
   b. Similarly, what do the figures suggest about the sample size for non-UK nationals?

35. As noted at the beginning of this chapter in the Statistics in Practice vignette, the polling organization IPSOS are using 'Bayesian credibility intervals' to characterize the accuracy of their online polling, rather than the 'classical' confidence intervals you have studied in this chapter. An article on the IPSOS website gave the formula below for a 95 per cent Bayesian

credibility interval when estimating a population proportion π. The article commented in respect of this formula: 'Since we want only one measure of precision for all variables in the survey ... we will compute the largest possible credibility interval for any observed sample.'

$$p \pm \frac{1}{\sqrt{n}}$$

How does this compare with the 'classical' confidence intervals you have studied in this chapter? Will the Bayesian credibility interval be narrower, wider or the same width as the 'classical' confidence interval?

8.5 BIG DATA AND CONFIDENCE INTERVALS

We have seen that confidence intervals are powerful tools for making inferences about population parameters. We now consider the ramifications of big data on confidence intervals for means and proportions, and we use the data-collection problem of online news service *PenningtonDailyTimes.com* (PDT). PDT's primary source of revenue is the sale of advertising, so PDT's management is concerned about the time customers spend during their visits to PDT's website and whether visitors click on any of the ads featured on the website.

Big data and the precision of confidence intervals

A review of equations (8.2) and (8.6) shows that confidence intervals for the population mean μ and population proportion π become narrower as the size of the sample increases. Therefore, the potential sampling error also decreases as the sample size increases. To illustrate the rate at which interval estimates narrow for a given confidence level, we consider the online news service *PenningtonDailyTimes.com* (PDT).

Prospective advertisers are willing to pay a premium to advertise on websites that have long visit times, so the time customers spend during their visits to PDT's website has a substantial impact on PDT's advertising revenues. Suppose PDT's management wants to develop a 95% confidence interval estimate of the mean amount of time customers spend during their visits to PDT's website. Table 8.4 shows how the margin of error at the 95% confidence level decreases as the sample size increases when $s = 20$.

TABLE 8.4 Margin of error for interval estimates of the population mean at the 95% confidence level for various sample sizes n

Sample size n	Margin of error $t_{\alpha/2}s_{\bar{x}}$
10	14.30714
100	3.96843
1,000	1.24109
10,000	0.39204
100,000	0.12396
1,000,000	0.03920
10,000,000	0.01240
100,000,000	0.00392
1,000,000,000	0.00124

Suppose that in addition to estimating the population mean amount of time customers spend during their visits to PDT's website, PDT would like to develop a 95% confidence interval estimate of the proportion of its website visitors that click on an ad. Table 8.5 shows how the margin of error for a 95% confidence interval estimate of the population proportion decreases as the sample size increases when the sample proportion is $p = 0.51$.

TABLE 8.5 Margin of error for interval estimates of the population proportion at the 95% confidence level for various sample sizes n

Sample size n	Margin of error $z_{\alpha/2}\sigma_P$
10	0.30984
100	0.09798
1,000	0.03098
10,000	0.00980
100,000	0.00310
1,000,000	0.00098
10,000,000	0.00031
100,000,000	0.00010
1,000,000,000	0.00003

The PDT example illustrates the relationship between the precision of interval estimates and the sample size. We see in Tables 8.4 and 8.5 that at a given confidence level, the margins of error decrease as the sample sizes increase. As a result, if the sample mean time spent by customers when they visit PDT's website is 84.1 seconds, the 95% confidence interval estimate of the population mean time spent by customers when they visit PDT's website decreases from (69.79286, 98.40714) for a sample of $n = 10$ to (83.97604, 84.22396) for a sample of $n = 100,000$ to (84.09876, 84.10124) for a sample of $n = 1,000,000,000$. Similarly, if the sample proportion of its website visitors who clicked on an ad is 0.51, the 95% confidence interval estimate of the population proportion of its website visitors who clicked on an ad decreases from (0.20016, 0.81984) for a sample of $n = 10$ to (0.50690, 0.51310) for a sample of $n = 100,000$ to (0.50997, 0.51003) for a sample of $n = 1,000,000,000$. In both instances, as the sample size becomes extremely large, the margin of error becomes extremely small and the resulting confidence intervals become extremely narrow.

Implications of big data for confidence intervals

Last year the mean time spent by all visitors to the PDT website was 84.0 seconds. Suppose that PDT wants to assess whether the population mean time has changed since last year. PDT now collects a new sample of 1,000,000 visitors to its website and calculates the sample mean time spent by these visitors to the PDT website to be $\bar{x} = 84.1$ seconds. The estimated population standard deviation is $s = 20$ seconds, so the standard error is $s_{\bar{x}} = s/\sqrt{n} = 0.02000$. Furthermore, the sample is sufficiently large to ensure that the sampling distribution of the sample mean will be normally distributed. Thus, the 95% confidence interval estimate of the population mean is

$$\bar{x} \pm t_{\alpha/2}s_{\bar{x}} = 84.1 \pm 0.0392 = (84.06080, 84.13920)$$

What could PDT conclude from these results? There are three possible reasons that PDT's sample mean of 84.1 seconds differs from last year's population mean of 84.0 seconds: (1) sampling error, (2) non-sampling error, or (3) the population mean has changed since last year. The 95% confidence interval estimate of the population mean does not include the value for the mean time spent by all visitors to the PDT website for last year (84.0 seconds), suggesting that the difference between PDT's sample mean for the new sample (84.1 seconds) and the mean from last year (84.0 seconds) is not likely to be exclusively a consequence of sampling error. Non-sampling error is a possible explanation and should be investigated as the results of statistical inference become less reliable as non-sampling error is introduced into the sample data. If PDT determines that it introduced little or no non-sampling error into its sample data, the only remaining plausible explanation for a difference of this magnitude is that the population mean has changed since last year.

If PDT concludes that the sample has provided reliable evidence and the population mean has changed since last year, management must still consider the potential impact of the difference between the sample mean and the mean from last year. If a 0.1 second difference in the time spent by visitors to the PDT website has a consequential effect on what PDT can charge for advertising on its site, this result could have practical business implications for PDT. Otherwise, there may be no practical significance of the 0.1 second difference in the time spent by visitors to the PDT website.

Confidence intervals are extremely useful, but as with any other statistical tool, they are only effective when properly applied. Because interval estimates become increasingly precise as the sample size increases, extremely large samples will yield extremely precise estimates. However, no interval estimate, no matter how precise, will accurately reflect the parameter being estimated unless the sample is relatively free of non-sampling error. Therefore, when using interval estimation, it is always important to carefully consider whether a random sample of the population of interest has been taken.

EXERCISES

Methods

36. Suppose a sample of 10,001 erroneous income tax returns from last year has been taken and is provided in the file 'TaxErrors'. A positive value indicates the taxpayer underpaid and a negative value indicates that the taxpayer overpaid.
    a. What is the sample mean 9 error made on erroneous income tax returns last year?
    b. Using 95% confidence, what is the margin of error?
    c. Using the results from parts (a) and (b), construct the 95% confidence interval estimate of the mean error made on erroneous income tax returns last year.

TAXERRORS

37. According to the US Census Bureau, 2,475,780 people are employed by the federal government in the USA as of 2018. Suppose that a random sample of 3,500 of these federal employees was selected and the number of sick hours each of these employees took last year was collected from an electronic personnel database. The data collected in this survey are provided in the file 'FedSickHours'.
    a. What is the sample mean number of sick hours taken by federal employees last year?
    b. Using 99% confidence, what is the margin of error?
    c. Using the results from parts (a) and (b), calculate the 99% confidence interval estimate of the mean number of sick hours taken by federal employees last year.
    d. If the mean sick hours federal employees took two years ago was 62.2, what would the confidence interval in part (c) lead you to conclude about last year?

FEDSICK HOURS

38. A survey reports that 58 per cent of car drivers admit to speeding. Suppose that a new satellite technology can instantly measure the speed of any vehicle on a particular road and determine whether the vehicle is speeding, and this satellite technology was used to take a sample of 20,000 vehicles at 6.00 p.m. on a recent afternoon. Of these 20,000 vehicles, 9,252 were speeding.
    a. What is the sample proportion of vehicles on this road that were speeding?
    b. Using 99% confidence, what is the margin of error?
    c. Using the results from parts (a) and (b), calculate the 99% confidence interval estimate of the proportion of vehicles on this road that speed.
    d. What does the confidence interval in part (c) lead you to conclude about the survey report?

ONLINE RESOURCES

For the data files, additional questions and answers and the software section for Chapter 8, go to the accompanying online platform.

SUMMARY

In this chapter we introduced the concept of an interval estimate of a population parameter. A point estimator may or may not provide a good estimate of a population parameter. The use of an interval estimate provides a measure of the precision of an estimate. A common form of interval estimate is a confidence interval.

We presented methods for constructing confidence intervals of a population mean and a population proportion. Both are of the form: point estimate ± margin of error. The confidence interval has a confidence coefficient associated with it.

We presented interval estimates for a population mean for two cases. In the σ known case, historical data or other information is used to make an estimate of σ prior to taking a sample. Analysis of the sample data then proceeds based on the assumption that σ is known. In the σ unknown case, the sample data are used to estimate both the population mean and the population standard deviation. In the σ known case, the interval estimation procedure is based on the assumed value σ and the standard normal distribution. In the σ unknown case, the interval estimation procedure uses the sample standard deviation s and the t distribution.

In both cases the quality of the interval estimates obtained depends on the distribution of the population and the sample size. Practical advice about the sample size necessary to obtain good approximations was included in Sections 8.1 and 8.2.

The general form of the interval estimate for a population proportion π is $p \pm$ margin of error. In practice, the sample sizes used for interval estimates of π are generally large. Consequently, the interval estimate is based on the standard normal distribution.

We explained how the margin of error expression can be used to calculate the sample size required to achieve a desired margin of error at a given level of confidence. We did this for two cases: estimating a population mean when the population standard deviation is known, and estimating a population proportion.

KEY TERMS

Confidence coefficient
Confidence interval
Confidence level
Degrees of freedom (df)
Interval estimate

Margin of error
Practical significance
σ known
σ unknown
t distribution

KEY FORMULAE

Interval estimate of a population mean: σ known

$$\bar{x} \pm z_{\alpha/2}\frac{\sigma}{\sqrt{n}} \tag{8.1}$$

Interval estimate of a population mean: σ unknown

$$\bar{x} \pm t_{\alpha/2}\frac{s}{\sqrt{n}} \tag{8.2}$$

Sample size for an interval estimate of a population mean

$$n = \frac{(z_{\alpha/2})^2\sigma^2}{E^2} \tag{8.3}$$

Interval estimate of a population proportion

$$p \pm z_{\alpha/2}\sqrt{\frac{p(1 - p)}{n}}$$ (8.6)

Sample size for an interval estimate of a population proportion

$$n = \frac{(z_{\alpha/2})^2\pi^*(1 - \pi^*)}{E^2}$$ (8.7)

CASE PROBLEM 1

International bank

The manager of a city-centre branch of a well-known international bank commissioned a customer satisfaction survey. The survey investigated three areas of customer satisfaction: experience waiting for service at a till, experience being served at the till and experience of self-service facilities at the branch. Within each of these categories, respondents were asked to give ratings on several aspects of the bank's service. These ratings were then summed to give an overall satisfaction rating in each of the three areas of service. The summed ratings are scaled such that they lie between 0 and 100, with 0 representing extreme dissatisfaction and 100 representing extreme satisfaction. The data file for this case study ('IntnlBank' on the online platform) contains the 0–100 ratings for the three areas of service, together with particulars of respondents' gender (male/female) and whether they would recommend the bank to other people (a simple Yes/No response was required to this question). An SPSS screenshot containing the first few rows of the data file is shown below.

	Waiting	Service	Self_Service	Gender	Recommend
1	55	65	50	Male	No
2	50	80	88	Male	No
3	30	40	44	Male	No
4	65	60	69	Male	Yes
5	55	65	63	Male	No
6	40	60	56	Male	No
7	15	65	38	Male	Yes
8	45	60	56	Male	No
9	55	65	75	Male	No
10	50	50	69	Male	Yes

Managerial report

1. Use descriptive statistics to summarize each of the five variables in the data file (the three service ratings, customer gender and customer recommendation).

2. Calculate 95% and 99% confidence interval estimates of the mean service rating for the population of customers of the branch, for each of the three service areas. Provide a managerial interpretation of each interval estimate.

3. Calculate 90% and 95% confidence interval estimates of the proportion of the branch's customers who would recommend the bank, and of the proportion of the branch's customers who are female. Provide a managerial interpretation of each interval estimate.

4. Suppose the branch manager required an estimate of the percentage of branch customers who would recommend the branch within a margin of error of 3 percentage points. Using 90% confidence, how large should the sample size be?

5. Suppose the branch manager required an estimate of the percentage of branch customers who are female within a margin of error of 5 percentage points. Using 95% confidence, how large should the sample size be?

INTNLBANK

REPAIRS

CASE PROBLEM 2

Consumer Knowhow

Consumer Knowhow is a consumer research organization that conducts surveys designed to evaluate a wide variety of products and services available to consumers. In one study, Consumer Knowhow looked at consumer satisfaction with the performance of cars produced by a major European manufacturer. A questionnaire sent to owners of one of the manufacturer's family cars revealed several complaints about early problems with the satnav system.

© Image Source/iStock

To learn more about the satnav failures, Consumer Knowhow used a sample of satnav repairs/replacements provided by an approved dealership of the car company. The data in the file 'Repairs' are the kilometres driven for 50 cars at the time of satnav failure.

Managerial report

1. Use appropriate descriptive statistics to summarize the satnav failure data.

2. Construct a 95% confidence interval for the mean number of kilometres driven until satnav failure for the population of cars with satnav failure. Provide a managerial interpretation of the interval estimate.

3. Discuss the implication of your statistical findings in relation to the proposition that some owners of the cars experienced early satnav failures.

4. How many repair/replacement records should be sampled if the research company wants the population mean number of kilometres driven until satnav failure to be estimated with a margin of error of 5,000 kilometres? Use 95% confidence.

5. What other information would you like to gather to evaluate the satnav failure problem more fully?

9

Hypothesis Tests

CHAPTER CONTENTS

Statistics in Practice   Hypothesis testing in business and economic research

9.1 Testing a population mean with σ known: one-tailed test
9.2 Testing a population mean with σ known: two-tailed test
9.3 Further discussion of hypothesis-testing fundamentals
9.4 Population mean with σ unknown
9.5 Population proportion
9.6 Type II errors and power
9.7 Big data and hypothesis testing

LEARNING OBJECTIVES   After reading this chapter and doing the exercises, you should be able to:

1 Set up appropriate null and alternative hypotheses for testing research hypotheses and for testing the validity of a claim.

2 Give an account of the logical steps involved in a statistical hypothesis test.

3 Explain the meaning of the terms null hypothesis, alternative hypothesis, Type I error, Type II error, level of significance, p-value and critical value.

4 Construct and interpret hypothesis tests for a population mean:

   4.1   When the population standard deviation is known.

4.2   When the population standard deviation is unknown.

5 Construct and interpret hypothesis tests for a population proportion.

6 Explain the relationship between the construction of hypothesis tests and confidence intervals.

7 Calculate the probability of a Type II error for a hypothesis test of a population mean when the population standard deviation is known.

8 Estimate the sample size required for a hypothesis test of a population mean when the population standard deviation is known.

STATISTICS IN PRACTICE
Hypothesis testing in business and economic research

The accompanying table is taken from a research paper that examined the association between the physical health of employed workers and characteristics of their work and working conditions. It was published in the open-access academic journal PLoS ONE in February 2019.* The statistical analysis in the paper was based on survey data for almost 19,000 respondents, collected in the then 28 countries of the European Union during the Sixth European Working Conditions Survey.

The analysis was a type of regression analysis (Chapters 14 to 16 of this book discuss regression analysis). To convey the results of the analyses, the main inferential tool was the statistical hypothesis test. The accompanying table shows hypothesis test results relating to the effects of changes in the independent variables (the work and working conditions variables, listed down the left column of the table) on the probability of reporting good, fair or bad health (the categorized measure of self-assessed health, SAH).

The part of the table shown here includes 19 of the independent variables, relating to the demands of the job, the psychological environment, job hazards and job recognition. The results of 57 hypothesis tests are reported. Asterisked results (the vast majority) indicate evidence of an association between the dependent variable and the independent variable in question. The original table in the PLoS ONE paper included 35 independent variables and the results of over 100 hypothesis tests (the variables not shown here related to demographics and job characteristics).

In journals in which quantitative research is reported, it is not unusual to find the average number of hypothesis tests reported exceeding 50 per paper. In two previous editions of this book, we used 2011 and 2014 issues of the *British Journal of Management* as examples: papers in the 2011 issue averaged over 50 hypothesis tests per article, while the 2014 issue averaged almost 100 tests per article. Similar results can be found in other academic journals in business and economics, or in finance, psychology and many other fields.

Many of the hypothesis tests used in research papers are those described in Chapters 10 to 18 of this book. In the present chapter, we set the scene by setting out the logic of statistical hypothesis testing and illustrating the logic by describing several simple hypothesis tests.

© yanyong/iStock

* Nappo, N. (2019) 'Is there an association between working conditions and health? An analysis of the Sixth European Working Conditions Survey data.' *PLoS ONE* 14(2): e0211294. Available at doi.org/10.1371/journal.pone.0211294.

The marginal effect (dx/dy) of a change in the regressors on the probability of reporting good, fair and bad health.

Variable	Good SAH			Fair SAH			Bad SAH		
	dx/dy	SE	P>|z|	dx/dy	SE	P>|z|	dx/dy	SE	P>|z|
Job demand									
Howmanyh	0.0017444***	0.0003	0.000	−0.0015735***	0.00027	0.000	−0.0001709 ***	0.00003	0.000
Notimef	−0.0203197***	0.00644	0.002	0.018294***	0.00579	0.002	0.0020257***	0.00066	0.002
Highspeed	−0.0027798**	0.00107	0.010	0.0025074***	0.00097	0.010	0.0002724**	0.00011	0.010
Psychological environment									
Stress1	−0.0341282***	0.01123	0.002	0.0306239***	0.01003	0.002	0.0035043***	0.00122	0.004
Stress2	−0.0026281	0.00907	0.772	0.0023709	0.00819	0.772	0.0002572	0.00089	0.772
Worrying	−0.0378222***	0.00636	0.000	0.0340207***	0.00572	0.000	0.0038015***	0.00068	0.000
Exhausted1	−0.1229589***	0.00966	0.000	0.109152***	0.00848	0.000	0.138069***	0.00142	0.000
Exhausted2	−0.0467442***	0.00836	0.000	0.0420403***	0.00751	0.000	0.0047039***	0.00089	0.000
Satisfied	0.0635434***	0.00899	0.000	−0.0565061***	0.0079	0.000	−0.0070373***	0.00116	0.000
Inforisk	0.0350474***	0.00988	0.000	−0.0313392***	0.00876	0.000	−0.0037082***	0.00114	0.001
Hrisk	−0.085717***	0.00761	0.000	0.0762722***	0.0067	0.000	0.0094447***	0.00105	0.000

Variable	Good SAH			Fair SAH			Bad SAH		
	dx/dy	SE	P>|z|	dx/dy	SE	P>|z|	dx/dy	SE	P>|z|
Job hazard									
Envirconds	0.0013157***	0.00039	0.001	−0.0011868***	0.00035	0.001	−0.0001289***	0.00004	0.001
Physconds	0.0013672***	0.0004	0.001	−0.0012333***	0.00036	0.001	−0.000134***	0.00004	0.001
Job recognition									
Manhelp1	0.0265626**	0.01202	0.027	−0.238959**	0.01078	0.027	−0.0026667**	0.00125	0.033
Manhelp2	0.0190144*	0.0113	0.092	−0.171907*	0.01024	0.093	−0.0018237*	0.00107	0.089
Adcareer1	0.0303639***	0.00718	0.000	−0.274456***	0.00651	0.000	−0.0029182***	0.00069	0.000
Adcareer2	0.0258458***	0.00708	0.000	−0.0234076***	0.00644	0.000	−0.0024382***	0.00066	0.000
Recognition1	0.029917***	0.00851	0.000	−0.269218***	0.00764	0.000	−0.0029952***	0.00089	0.001
Recognition1	0.0091769	0.00838	0.273	−0.0082912	0.00758	0.274	−0.0008858	0.0008	0.267

*** stat. signf. at 1%
** stat. signf. at 5%
* stat. signf. at 10%

Source: doi.org/10.1371/journal.pone.0211294.t002. © 2019 Nunzia Nappo.

In Chapters 7 and 8 we showed how sample data are used to construct point and interval estimates of population parameters. In this chapter we extend the discussion of statistical inference by introducing statistical hypothesis testing, and showing how hypothesis tests can be conducted about a population mean and a population proportion. We begin with two examples of hypothesis testing about a population mean, when the population standard deviation is assumed known (Sections 9.1 and 9.2), using these examples to introduce some of the important concepts involved in hypothesis testing.

9.1 TESTING A POPULATION MEAN WITH σ KNOWN: ONE-TAILED TEST

A statistical hypothesis test has the following logical pattern. We start with a 'working' assumption about the value of a population parameter, such as a population mean. Sample evidence is brought to bear on this working assumption. If we are doing a test on a population mean, the focus of the sample evidence is likely to be the sample mean. An assessment of the sample evidence results in a decision either to reject the working assumption, or not. Because the decision is based on sample evidence, there is a possibility that a valid working assumption will be rejected incorrectly. Similarly, there is a possibility that an invalid working assumption will, incorrectly, not be rejected.

The working assumption is called the null hypothesis, denoted H_0. This is the hypothesis that is tested. Formulating H_0 implies the existence of another hypothesis, the alternative hypothesis H_1, which is the opposite of whatever is stated in H_0. The alternative hypothesis H_1 plays a relatively small part in the test procedure and is not the hypothesis that is tested. It is, though, the hypothesis that is adopted if H_0 is rejected. In research contexts, H_1 is often the hypothesis the researcher is hoping to establish.

We begin by outlining a test about a population mean μ with the following format for H_0 and H_1:

$$H_0: \mu \geq \mu_0$$
$$H_1: \mu < \mu_0$$

For reasons that will become evident later, this format is known as a one-tailed test: more specifically, in this case, a lower-tail test.

Trading Standards Offices (TSOs) periodically conduct statistical studies to test the claims manufacturers make about their products. Suppose a large bottle of Cola is labelled as containing 3 litres. European legislation acknowledges that the bottling process cannot guarantee precisely 3.00 litres of Cola in each bottle, even if the mean volume per bottle, μ, is 3.00 litres for the population of all bottles filled. The legislation interprets the label information as a claim that μ is at least 3.00 litres. We shall show how a TSO can check the claim by doing a statistical hypothesis test.

Formulating H_0 and H_1

The first step is to formulate the null and alternative hypotheses. If the population mean volume μ is at least 3.00 litres, the manufacturer's claim is correct. This establishes the null hypothesis. However, if μ is less than 3.00 litres, the manufacturer's claim is incorrect. This establishes the alternative hypothesis:

$$H_0: \mu \geq 3.00$$

$$H_1: \mu < 3.00$$

The TSO will do the test by taking a random sample of bottles and using the sample evidence to make a decision regarding H_0, the manufacturer's claim. Suppose a sample of 36 bottles is selected and the sample mean $\bar{x}$ is computed as an estimate of the population mean μ. Suppose too that from previous evidence we can assume the population standard deviation is known, with a value of $\sigma = 0.18$ (litres). The previous evidence also shows that the population of bottle contents can be assumed to have a normal distribution.

From the study of sampling distributions in Chapter 7 we know that if the population from which we are sampling is normally distributed, as in this case, the sampling distribution of the sample mean $\bar{X}$ will also be normal in shape. In constructing sampling distributions for hypothesis tests, we assume that H_0 is satisfied as an equality. Figure 9.1 shows the sampling distribution of $\bar{X}$ when $\mu = \mu_0 = 3.00$. Note that $\sigma = 0.18$ and sample size $n = 36$. The standard error of $\bar{X}$ is given by $\sigma_{\bar{x}} = \sigma/\sqrt{n} = 0.18/\sqrt{36} = 0.03$.

FIGURE 9.1

Sampling distribution of $\bar{X}$ for the Cola bottling study when the null hypothesis is true as an equality ($\mu_0 = 3$)

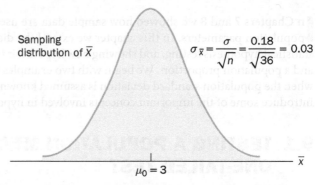

Sampling distribution of $\bar{X}$

$$\sigma_{\bar{x}} = \frac{\sigma}{\sqrt{n}} = \frac{0.18}{\sqrt{36}} = 0.03$$

$\mu_0 = 3$

If the observed value of $\bar{X}$ is less than 3.00 litres, the sample results will tend to cast some doubt on the null hypothesis, because values of $\bar{X}$ below 3.00 are more likely under H_1 than H_0. However, as Figure 9.1 clearly indicates, values of $\bar{X}$ just below 3.00 are not unusual, even if $\mu = 3.00$, because of sampling variability. What we want to know is how much less than 3.00 the sample mean must be before we are willing to conclude that the manufacturer's claim is invalid. In other words, we need to decide how far into the left-hand tail of the sampling distribution the observed $\bar{x}$ needs to be to reject H_0 in favour of H_1.

Deciding the level of significance

Suppose we decide to reject H_0 if $\bar{x}$ falls in the most extreme 5 per cent of the left-hand tail. This decision would imply that, if H_0 is true as an equality, there will be a 5 per cent chance of incorrectly concluding there has been a labelling violation (concluding in favour of H_1). If we make a different decision, say to reject H_0 if $\bar{x}$ falls in the most extreme 1 per cent of the left-hand tail, the chance of incorrectly concluding there has been a labelling violation, if H_0 is true as an equality, will be only 1 per cent. Incorrectly rejecting H_0 when it is true is known in the jargon of hypothesis testing as a Type I error. The value the decision-maker chooses for the probability of making a Type I error is known as the **level of significance** of the test, usually denoted by the Greek letter α (alpha). We discuss Type I errors further in Section 9.3, as well as Type II errors (not rejecting H_0 when H_0 is false).

Level of significance

The level of significance is the probability of making a Type I error when the null hypothesis is true as an equality.

By selecting a particular value for α, the person doing the hypothesis test is controlling the probability of making a Type I error. If the cost of making a Type I error is high, small values of α are preferred. If the cost of making a Type I error is not so high, larger values of α are typically used. Common choices for α are 0.05 and 0.01.

Ideally, the TSO in the Cola bottling study does not want to wrongly accuse the manufacturer of under-filling. Suppose they accept a 1 per cent risk (i.e. a probability of 0.01) of making such an error, if the manufacturer is meeting its weight specifications at $\mu = 3.00$. This means a probability of 0.01 of making a Type I error, so the level of significance for the hypothesis test must be set at $\alpha = 0.01$.

We have now formulated H_0 and H_1, and specified the level of significance α for the test. These are the first two steps required in doing every hypothesis test. The next step is to collect the sample data and compute the value of an appropriate test statistic.

Test statistic

For hypothesis tests about a population mean, with σ known and in conditions where $\overline{X}$ has a normal sampling distribution, $Z = (\overline{X} - \mu_0)/\sigma_{\overline{X}}$ has a standard normal sampling distribution. We can use Z as a **test statistic**, in combination with the standard normal distribution table, to determine whether $\overline{x}$ deviates from the hypothesized value μ_0 sufficiently to justify rejecting H_0. The test statistic used in the σ known case is as follows (note that $\sigma_{\overline{X}} = \sigma/\sqrt{n}$):

> **Test statistic for hypothesis tests about a population mean: σ known**
>
> $$z = \frac{\overline{x} - \mu_0}{\sigma/\sqrt{n}}$$
>
> (9.1)

In the Cola bottling example, an α value of 0.01 has been chosen. Therefore, we are seeking to determine whether the observed value of the test statistic falls in the leftmost 1 per cent of the sampling distribution. There are two approaches that can be used to do this. One approach is to find the value of Z that leaves 1 per cent of the probability in the left-hand tail. This is known as the **critical value**. The observed Z value is compared with the critical value to determine whether or not to reject the null hypothesis. The second, and entirely equivalent, approach is to find the area in the tail of the distribution to the left of the observed Z. This probability is known as the **p-value**. The p-value is compared with α to decide whether or not to reject H_0.

Critical value approach

In the Cola bottling example, the sampling distribution for the test statistic Z is a standard normal distribution. Therefore, the critical value is the z value that corresponds to an area of $\alpha = 0.01$ in the lower tail of this distribution. Using the standard normal distribution table, we find that the critical value is $z = -2.33$ (refer to Figure 9.2). This is the largest value (i.e. least negative value) of the test statistic that will result in the rejection of the null hypothesis. Hence, the rejection rule for $\alpha = 0.01$ is:

COLA

$$\text{Reject } H_0 \text{ if observed } z \leq -2.33$$

Suppose the sample of 36 Cola bottles gives a sample mean $\overline{x} = 2.92$. Is $\overline{x} = 2.92$ small enough to cause us to reject H_0? The test statistic is:

$$z = \frac{\overline{x} - \mu_0}{\sigma/\sqrt{n}} = \frac{2.92 - 3.00}{0.18/\sqrt{36}} = -2.67$$

Because $z = -2.67 < -2.33$, we reject H_0 at $\alpha = 0.01$ and conclude that the Cola manufacturer is under-filling bottles. The sample result is said to be *statistically significant* at the 1 per cent significance level. At the beginning of this section, we said that this test is an example of a **one-tailed test**. This term refers to the fact that the test involves a single critical value in one tail of the sampling distribution (in this case, the lower tail).

FIGURE 9.2
Critical value approach for the Cola
bottling hypothesis test

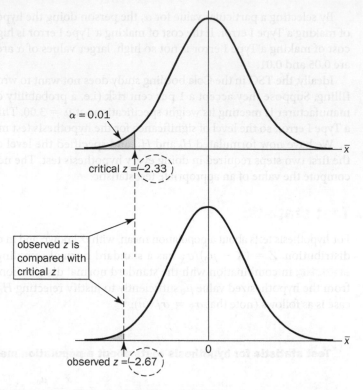

We can generalize the rejection rule for the critical value approach to handle any level of significance. The rejection rule for a lower-tail test is as follows:

> **Rejection rule for a lower-tail test: critical value approach**
>
> $$\text{Reject } H_0 \text{ if } z \leq -z_\alpha$$
>
> where $-z_\alpha$ is the critical value; that is, the Z value that provides an area of α in the lower tail of the standard normal distribution.

p-value approach

In the critical value approach, we compare two values of Z: the observed value calculated from the sample and the value that corresponds to a tail area of α. The p-value approach involves a similar comparison, but instead of comparing two values of Z, we compare the tail areas (probabilities) corresponding to the two Z values. In other words, we compare the tail area for the observed value of Z with the level of significance α. The tail area for the observed value of Z is known as the p-value. The p-value approach has become the preferred method of determining whether H_0 can be rejected, especially when using computer software packages such as SPSS, R, Excel and Minitab. We begin with a formal definition for a p-value.

> **p-value**
>
> The p-value is the probability of getting a value for the test statistic as extreme or more extreme than the value observed in the sample, assuming the null hypothesis to be true as an equality.

Because a *p*-value is a probability, it ranges from 0 to 1. A small *p*-value indicates a sample result that is unusual if H_0 is true. Broadly speaking, small *p*-values lead to rejection of H_0, whereas large *p*-values indicate that there is insufficient evidence to reject H_0.

In the Cola bottling test, we found that $\bar{x} = 2.92$, giving a test statistic $z = -2.67$. The *p*-value is the probability that the test statistic Z is less than or equal to -2.67 (the area under the standard normal curve to the left of $z = -2.67$). Using the standard normal distribution table, we find that the cumulative probability for $z = -2.67$ is 0.00382. This *p*-value indicates a small probability of obtaining a sample mean of $\bar{x} = 2.92$, or smaller, when sampling from a population with $\mu = 3.00$. But is it small enough to cause us to reject H_0? The answer depends upon the level of significance for the test.

As noted previously, the TSO selected a value of 0.01 for the level of significance. The sample of 36 bottles resulted in a *p*-value = 0.0038. Because 0.0038 is less than $\alpha = 0.01$ we reject H_0. Therefore, we find sufficient statistical evidence to reject the null hypothesis at $\alpha = 0.01$ (refer to Figure 9.3).

FIGURE 9.3

p-value approach for the Cola bottling hypothesis test

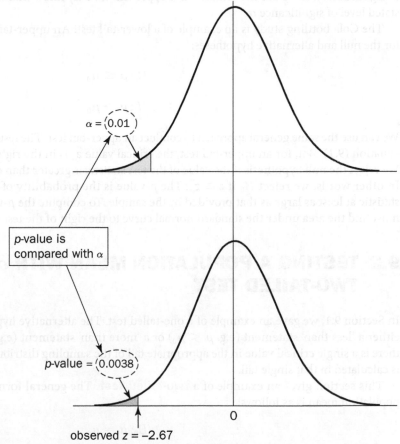

We can now state the general rule for determining whether the null hypothesis can be rejected when using the *p*-value approach. For a level of significance α, the rejection rule using the *p*-value approach is as follows:

Rejection rule using *p*-value

$$\text{Reject } H_0 \text{ if } p\text{-value} \leq \alpha$$

In the Cola bottling test, the *p*-value of 0.0038 resulted in the rejection of the null hypothesis at $\alpha = 0.01$. Moreover, the *p*-value of 0.0038 means we would reject H_0 for any value $\alpha \geq 0.0038$. For this reason, the *p*-value is also called the *observed level of significance* or the *attained level of significance*.

Different decision-makers may express different opinions concerning the cost of making a Type I error and may choose a different level of significance. By providing the p-value as part of the hypothesis testing results, another decision-maker can compare the reported p-value to their own level of significance and possibly make a different decision with respect to rejecting H_0.

The smaller the p-value is, the greater the evidence against H_0, and the more substantial the evidence in favour of H_1. In research articles in academic journals, statistical hypothesis test results are reported using a range of α levels, from $\alpha = 0.10$ (weak evidence against H_0) to 0.001 (very strong evidence against H_0). The most common α level used in academic research reports is 0.05.

The p-value approach and the critical value approach will always lead to the same rejection decision. That is, whenever the p-value is less than or equal to α, the value of the test statistic will be less than or equal to the critical value, as should be evident from comparing Figures 9.2 and 9.3. The advantage of the p-value approach is that the p-value tells us *how* statistically significant the results are (the attained level of significance). If we use the critical value approach, we only know that the results are significant at the stated level of significance α.

The Cola bottling study is an example of a lower-tail test. An upper-tail test has the following format for the null and alternative hypotheses:

$$H_0: \mu \leq \mu_0$$

$$H_1: \mu > \mu_0$$

We can use the same general approach to conduct an upper-tail test. The test statistic is still computed using equation (9.1). But, for an upper-tail test, the critical value z_α is in the right-hand tail of the distribution. We reject the null hypothesis if the value of the test statistic is greater than or equal to the critical value z_α. In other words, we reject H_0 if $z \geq z_\alpha$. The p-value is the probability of obtaining a value for the test statistic at least as large as that provided by the sample. To compute the p-value for the upper-tail test, we must find the area under the standard normal curve to the right of the test statistic.

9.2 TESTING A POPULATION MEAN WITH σ KNOWN: TWO-TAILED TEST

In Section 9.1, we gave an example of a one-tailed test. The alternative hypothesis for a one-tailed test is either a 'less than' statement (e.g. $\mu < \mu_0$) or a 'more than' statement (e.g. $\mu > \mu_0$). As a consequence, there is a single critical value in the appropriate tail of the sampling distribution. Equivalently, the p-value is calculated in that single tail.

This section gives an example of a two-tailed test. The general form for a two-tailed test about a population mean is as follows:

$$H_0: \mu = \mu_0$$

$$H_1: \mu \neq \mu_0$$

We show how to do a two-tailed test about a population mean for the σ known case. As an illustration, we consider the hypothesis-testing situation facing MaxFlight, a high-technology manufacturer of golf balls with an average driving distance of 295 m (metres). Sometimes the process gets out of adjustment and produces golf balls with average distances different from 295 m. When the average distance falls below 295 m, the company worries about losing sales because the golf balls do not provide the advertised distance. However, some of the national golfing associations impose equipment standards for professional competition, and when the average driving distance exceeds 295 m, MaxFlight's balls may be rejected for exceeding the overall distance standard concerning carry and roll.

MaxFlight's quality control programme involves taking periodic samples of 50 golf balls to monitor the manufacturing process. For each sample, a hypothesis test is done to determine whether the process has fallen out of adjustment. Let us formulate the null and alternative hypotheses. We begin by assuming that the process is functioning correctly; that is, the golf balls being produced have a population mean driving distance of 295 m. This assumption establishes the null hypothesis H_0. The alternative hypothesis H_1 is that the mean driving distance is not equal to 295 m.

$$H_0: \mu = 295$$

$$H_1: \mu \neq 295$$

If the sample mean is significantly less than 295 m or significantly greater than 295 m, we will reject H_0. In this case, corrective action will be taken to adjust the manufacturing process. On the other hand, if $\bar{x}$ does not deviate from the hypothesized mean $\mu_0 = 295$ by a significant amount, H_0 will not be rejected and no action will be taken to adjust the manufacturing process.

The quality control team selected $\alpha = 0.05$ as the level of significance for the test. Data from previous tests conducted when the process was known to be in adjustment show that the population standard deviation can be assumed known with a value of $\sigma = 12$. With a sample size of $n = 50$, the standard error of the sample mean is:

$$\sigma_{\bar{x}} = \frac{\sigma}{\sqrt{n}} = \frac{12}{\sqrt{50}} = 1.7$$

Because the sample size is large, the central limit theorem (refer to Chapter 7) allows us to conclude that the sampling distribution of $\bar{X}$ can be approximated by a normal distribution. Figure 9.4 shows the sampling distribution of $\bar{X}$ for the MaxFlight hypothesis test with a hypothesized population mean of $\mu_0 = 295$.

GOLFTEST

FIGURE 9.4
Sampling distribution of $\bar{X}$ for the MaxFlight hypothesis test

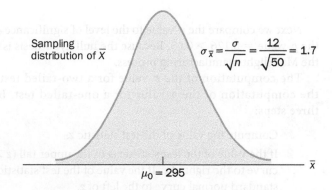

Suppose that a sample of 50 golf balls is selected and that the sample mean is 297.6 m. This sample mean suggests that the population mean may be larger than 295 m. Is this value $\bar{x} = 297.6$ sufficiently larger than 295 to cause us to reject H_0 at the 0.05 level of significance? In the previous section we described two approaches that can be used to answer this question: the p-value approach and the critical value approach.

p-value approach

For a two-tailed test, values of the test statistic in *either* tail of the sampling distribution indicate a lack of support for the null hypothesis. For a two-tailed test, the p-value is the probability of obtaining a value for the test statistic *at least as unlikely* as that provided by the sample, in either tail of the sampling distribution. Let us compute the p-value for the MaxFlight hypothesis test.

For the σ known case, the test statistic Z is a standard normal random variable. Using equation (9.1) with $\bar{x} = 297.6$, the value of the test statistic is:

$$z = \frac{\bar{x} - \mu_0}{\sigma/\sqrt{n}} = \frac{297.6 - 295}{12/\sqrt{50}} = 1.53$$

Now we find the probability of obtaining a value for the test statistic *at least as unlikely* as $z = 1.53$. Clearly values ≥ 1.53 are *at least as unlikely*. But, because this is a two-tailed test, values ≤ -1.53 are also *at least as unlikely* as the value of the test statistic provided by the sample. Referring to Figure 9.5, we see that the two-tailed p-value in this case is given by $P(Z \leq -1.53) + P(Z \geq 1.53)$. Because the normal curve is symmetrical, we can compute this probability by finding the area under the standard normal curve to the left of $z = -1.53$ and doubling it. The table of cumulative probabilities for the standard normal distribution shows that the area to the left of $z = -1.53$ is 0.0630. Doubling this, we find the p-value for the MaxFlight two-tailed hypothesis test is $2(0.0630) = 0.126$.

FIGURE 9.5

p-value for the MaxFlight hypothesis text

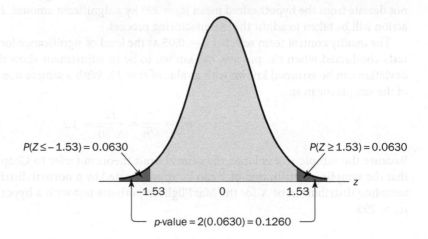

$P(Z \leq -1.53) = 0.0630$ $P(Z \geq 1.53) = 0.0630$

-1.53 0 1.53 z

p-value = $2(0.0630) = 0.1260$

Next we compare the p-value to the level of significance α. With $\alpha = 0.05$, we do not reject H_0 because the p-value = $0.126 > 0.05$. Because the null hypothesis is not rejected, no action will be taken to adjust the MaxFlight manufacturing process.

The computation of the p-value for a two-tailed test may seem a bit confusing as compared to the computation of the p-value for a one-tailed test. But it can be simplified by following these three steps:

1 Compute the value of the test statistic z.

2 If the value of the test statistic is in the upper tail ($z > 0$), find the area under the standard normal curve to the right of z. If the value of the test statistic is in the lower tail, find the area under the standard normal curve to the left of z.

3 Double the tail area, or probability, obtained in step 2 to obtain the p-value.

In practice, the computation of the p-value is done automatically when using computer software such as SPSS, R, Excel and Minitab.

Critical value approach

Now let us see how the test statistic can be compared to critical values to make the hypothesis testing decision for a two-tailed test. Figure 9.6 shows that the critical values for the test will occur in both the lower and upper tails of the standard normal distribution. With a level of significance of $\alpha = 0.05$, the area in each tail beyond the critical values is $\alpha/2 = 0.05/2 = 0.025$. Using the table of probabilities for the standard normal distribution, we find the critical values for the test statistic are $-z_{0.025} = -1.96$ and $z_{0.025} = 1.96$. Using the critical value approach, the two-tailed rejection rule is:

$$\text{Reject } H_0 \text{ if } z \leq -1.96 \text{ or if } z \geq 1.96$$

Because the value of the test statistic for the MaxFlight study is $z = 1.53$, the statistical evidence will not permit us to reject the null hypothesis at the 0.05 level of significance.

FIGURE 9.6
Critical values for the MaxFlight
hypothesis test

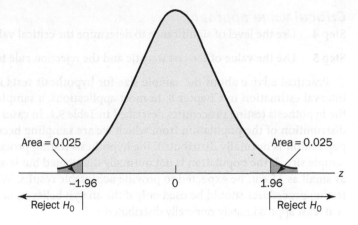

Summary and practical advice

We have presented examples of a lower-tail test and a two-tailed test about a population mean. Based on these examples, we can now summarize the hypothesis testing procedures about a population mean for the σ known case, as shown in Table 9.1. Note that μ_0 is the hypothesized value of the population mean.

The hypothesis testing steps followed in the two examples presented in Sections 9.1 and 9.2 are common to every hypothesis test.

TABLE 9.1 Summary of hypothesis tests about a population mean: σ known case

	Lower-tail test	Upper-tail test	Two-tailed test
Hypotheses	$H_0: \mu \geq \mu_0$ $H_1: \mu < \mu_0$	$H_0: \mu \leq \mu_0$ $H_1: \mu > \mu_0$	$H_0: \mu = \mu_0$ $H_1: \mu \neq \mu_0$
Test statistic	$z = \dfrac{\bar{x} - \mu_0}{\sigma/\sqrt{n}}$	$z = \dfrac{\bar{x} - \mu_0}{\sigma/\sqrt{n}}$	$z = \dfrac{\bar{x} - \mu_0}{\sigma/\sqrt{n}}$
Rejection rule: **p-value approach**	Reject H_0 if p-value $\leq \alpha$	Reject H_0 if p-value $\leq \alpha$	Reject H_0 if p-value $\leq \alpha$
Rejection rule: **critical value approach**	Reject H_0 if $z \leq -z_\alpha$	Reject H_0 if $z \geq z_\alpha$	Reject H_0 if $z \leq -z_{\alpha/2}$ or if $z \geq z_{\alpha/2}$

Steps of hypothesis testing

Step 1   Formulate the null and alternative hypotheses.

Step 2   Specify the level of significance α.

Step 3   Collect the sample data and compute the value of the test statistic.

p-value approach

Step 4   Use the value of the test statistic to compute the p-value.

Step 5   Reject H_0 if the p-value $\leq \alpha$.

Critical value approach

Step 4   Use the level of significance to determine the critical value and the rejection rule.

Step 5   Use the value of the test statistic and the rejection rule to determine whether to reject H_0.

Practical advice about the sample size for hypothesis tests is similar to the advice we provided for interval estimation in Chapter 8. In most applications, a sample size of $n \geq 30$ is adequate when using the hypothesis testing procedures described in Table 9.1. In cases where the sample size is less than 30, the distribution of the population from which we are sampling becomes an important consideration. If the population is normally distributed, the hypothesis testing procedures are exact and can be used for any sample size. If the population is not normally distributed but is at least roughly symmetrical, sample sizes as small as 15 can be expected to provide acceptable results. With smaller sample sizes, the hypothesis testing procedures should be used only if the analyst believes, or is willing to assume, that the population is at least approximately normally distributed.

EXERCISES

Note to students: Some of the exercises ask you to use the *p*-value approach and others ask you to use the critical value approach. Both methods will provide the same hypothesis-testing conclusion. We provide exercises with both methods to give you practice using both. In later sections and in following chapters, we shall generally emphasize the *p*-value approach as the preferred method.

Methods

1. Consider the following hypothesis test:

$$H_0: \mu = 15$$

$$H_1: \mu \neq 15$$

A sample of 50 provided a sample mean of 14.15. The population standard deviation is 3.
   a. Compute the value of the test statistic.
   b. What is the *p*-value?
   c. At $\alpha = 0.05$, what is your conclusion?
   d. What is the rejection rule using the critical value? What is your conclusion?

2. Consider the following hypothesis test:

$$H_0: \mu \geq 20$$

$$H_1: \mu < 20$$

A sample of 50 gave a sample mean of 19.4. The population standard deviation is 2.
   a. Compute the value of the test statistic.
   b. What is the *p*-value?
   c. Using $\alpha = 0.05$, what is your conclusion?
   d. What is the rejection rule using the critical value? What is your conclusion?

3. Consider the following hypothesis test:

$$H_0: \mu \leq 50$$

$$H_1: \mu > 50$$

A sample of 60 is used and the population standard deviation is 8. Use the critical value approach to state your conclusion for each of the following sample results. Use $\alpha = 0.05$.

a. $\bar{x} = 52.5$.

b. $\bar{x} = 51.0$.

c. $\bar{x} = 51.8$.

Applications

4. Houses in the north-east of England tend to be cheaper than the national average. Suppose the national mean sales price for new two-bedroom houses is £181,900. A sample of 40 new two-bedroom house sales in the north-east of England showed a sample mean of £166,400. Consider the null hypothesis H_0: $\mu \geq 181,900$ (implying H_1: $\mu < 181,900$), where μ is the population mean sales price for new two-bedroom houses in the north-east. Assume a known population standard deviation of £33,500.

a. What is the value of the test statistic?

b. What is the p-value?

c. At $\alpha = 0.01$, what is your conclusion?

5. Suppose that the mean length of the working week for a population of workers has been previously reported as 39.2 hours. Suppose a current sample of 112 workers provides a sample mean of 38.5 hours. Consider the null hypothesis H_0: $\mu = 39.2$ (implying H_1: $\mu \neq 39.2$), where μ is the current mean length of the working week. Use a population standard deviation $\sigma = 4.8$ hours.

a. What is the p-value?

b. At $\alpha = 0.05$, can the null hypothesis be rejected? What is your conclusion?

c. Repeat the preceding hypothesis test using the critical value approach.

6. CCN and ActMedia provided a television channel targeted at individuals waiting in supermarket checkout queues. The channel showed news, short features and advertisements. The programme length was based on the assumption that the population mean time a shopper stands in a supermarket checkout queue is 4 minutes. A sample of actual waiting times will be used to test this assumption and determine whether actual mean waiting time differs from this standard. Consider the null hypothesis H_0: $\mu = 4.0$ (implying H_1: $\mu \neq 4.0$), where μ is the population mean waiting time. Assume a population standard deviation $\sigma = 1.6$ minutes.

a. A sample of 30 shoppers showed a sample mean waiting time of 4.5 minutes. What is the p-value?

b. At $\alpha = 0.05$, what is your conclusion?

7. A production line operates with a mean filling weight of 500 grams per container. Over-filling or under-filling presents a serious problem and when detected requires the operator to shut down the production line to readjust the filling mechanism. From past data, a population standard deviation $\sigma = 25$ grams is assumed. A quality control inspector selects a sample of 30 items every hour and at that time makes the decision of whether to shut down the line for readjustment. The level of significance is $\alpha = 0.05$. Consider the null hypothesis H_0: $\mu = 500$ (implying H_1: $\mu \neq 500$), where μ is the population mean filling weight.

a. If a sample mean of 510 grams were found, what is the p-value? What action would you recommend?

b. If a sample mean of 495 grams were found, what is the p-value? What action would you recommend?

c. Use the critical value approach. What is the rejection rule for the preceding hypothesis testing procedure? Repeat parts (b) and (c). Do you reach the same conclusion?

9.3 FURTHER DISCUSSION OF HYPOTHESIS-TESTING FUNDAMENTALS

Formulating null and alternative hypotheses

It is not always self-evident how H_0 and H_1 should be formulated. Care must be taken to structure the hypotheses appropriately so that the hypothesis-testing conclusion provides the information the researcher or decision-maker wants. The context of the situation is very important.

In the chapter introduction, we stated that the null hypothesis H_0 is a tentative assumption about a population parameter such as a population mean or a population proportion. The alternative hypothesis H_1 states the opposite (or counterpart) of the null hypothesis. In some situations, it is easier to identify the alternative hypothesis first and then formulate the null hypothesis. In other situations, it is easier to identify the null hypothesis first and then formulate the alternative hypothesis. We illustrate this in the following examples.

The alternative hypothesis as a research hypothesis

Many applications of hypothesis testing involve an attempt to gather evidence in support of a research hypothesis. Here, it is often best to begin with the alternative hypothesis and make it the conclusion that the researcher hopes to support. As a general guideline, a research hypothesis should be stated as the *alternative hypothesis*.

Consider a particular model of car that currently attains an average fuel consumption of 7.0 litres of fuel per 100 kilometres of driving. A product research group develops a new fuel injection system specifically designed to decrease fuel consumption. To evaluate the new system, several will be manufactured, installed in cars and subjected to controlled driving tests. The product research group is looking for evidence to conclude that the new system *decreases* the mean fuel consumption. In this case, the research hypothesis is that the new fuel injection system will provide a mean fuel consumption below 7.0 litres per 100 kilometres; that is, $\mu < 7.0$. Hence, the appropriate null and alternative hypotheses for the study are:

$$H_0: \mu \geq 7.0$$

$$H_1: \mu < 7.0$$

If the sample results lead to the conclusion to reject H_0, the inference can be made that $H_1: \mu < 7.0$ is true. The researchers have the statistical support to state that the new fuel injection system decreases the mean litres of fuel consumed per 100 kilometres. The production of cars with the new fuel injection system should be considered. However, if the sample results lead to the conclusion that H_0 cannot be rejected, the researchers cannot conclude that the new fuel injection system is better than the current system. Perhaps more research and further testing should be conducted.

The conclusion that the research hypothesis is true is made if the sample data provide sufficient evidence to show that the null hypothesis can be rejected.

Successful companies stay competitive by developing new products, new methods and new systems that are better than those currently available. Before adopting something new, it is desirable to do research to determine if there is statistical support for the conclusion that the new approach is indeed better. In such cases, the research hypothesis is stated as the alternative hypothesis. For example, a new sales force bonus plan is developed in an attempt to increase sales. The alternative hypothesis is that the new bonus plan increases sales. The null hypothesis is that the new bonus plan does not increase sales. A new drug is developed with the goal of lowering blood pressure more effectively than an existing drug. The alternative hypothesis is that the new drug lowers blood pressure more effectively than the existing drug. The null hypothesis is that the new drug does not provide lower blood pressure than the existing drug. In each case, rejection of the null hypothesis H_0 provides statistical support for the research hypothesis. We shall see many examples of hypothesis tests in research situations such as these throughout this chapter and in the remainder of the text.

The null hypothesis as an assumption to be challenged

Not all hypothesis tests involve research hypotheses in the same sense as the examples above. In some applications of hypothesis testing we begin with a belief or an assumption that a statement about the value of a population parameter is true. We then use a hypothesis test to challenge the assumption and determine if there is statistical evidence to conclude that the assumption is incorrect. In these situations, it is helpful to formulate the null hypothesis first. The null hypothesis H_0 expresses the belief or assumption about the value of the population parameter. The alternative hypothesis H_1 is that the belief or assumption is incorrect.

The Cola bottling example in Section 9.1 is such an example. The label on the bottle states that it contains 3 litres. We consider the label correct provided the population mean filling volume for the bottles is *at least* 3.00 litres. In a hypothesis test about the population mean volume per bottle, the TSO gives the manufacturer the benefit of the doubt by beginning with the assumption that the label is correct and stating the null hypothesis as $\mu \geq 3.00$. The challenge to this assumption would imply that the label is incorrect and the bottles are being under-filled. This challenge would be stated as the alternative hypothesis $\mu < 3.00$. The null and alternative hypotheses are:

$$H_0: \mu \geq 3.00$$

$$H_1: \mu < 3.00$$

The TSO selects a sample of bottles, computes the sample mean filling volume and uses the sample results to test the preceding hypotheses. If the sample results lead to the conclusion to reject H_0, the inference that $H_1: \mu < 3.00$ is true can be made. With this statistical support, the TSO is justified in concluding that the label is incorrect and under-filling of the bottles is occurring. Appropriate action to force the manufacturer to comply with labelling standards would be considered. However, if the sample results indicate H_0 cannot be rejected, the assumption that the manufacturer's labelling is correct cannot be rejected. With this conclusion, no action would be taken.

A manufacturer's product or service information is usually assumed to be true and stated as the null hypothesis. The conclusion that the information is incorrect can be made if the null hypothesis is rejected.

Let us now consider a variation of the Cola bottle-filling example by viewing it from the manufacturer's point of view. The bottle-filling operation has been designed to fill soft drink bottles with 3 litres, as stated on the label. The company does not want to under-fill the containers because that could result in an under-filling complaint from customers or a TSO. However, the company does not want to over-fill containers either because putting more soft drink than necessary into the containers would add to costs. The company's goal would be to adjust the bottle-filling operation so that the population mean filling volume per bottle is 3.00 litres.

However, from time to time any production process can get out of adjustment. If this occurs in our example, under-filling or over-filling of the Cola bottles will occur. In either case, the company would like to know about it in order to correct the situation by readjusting the bottle-filling operation to the targeted 3.00 litres. In a hypothesis testing application, we would again begin with the assumption that the production process is operating correctly and state the null hypothesis as $\mu = 3.00$ litres. The alternative hypothesis is that $\mu \neq 3.00$, which indicates either over-filling or under-filling is occurring. The null and alternative hypotheses for the manufacturer's hypothesis test are:

$$H_0: \mu = 3.00$$

$$H_1: \mu \neq 3.00$$

Suppose the Cola manufacturer uses a quality control procedure to periodically select a sample of bottles from the filling operation and compute the sample mean filling volume per bottle. If the sample results lead to the conclusion to reject H_0, the inference is made that $H_1: \mu \neq 3.00$ is true. We conclude that the bottles are not being filled properly and the production process should be adjusted to restore the population mean to 3.00 litres per bottle. However, if the sample results indicate H_0 cannot be rejected, the assumption that the manufacturer's bottle filling operation is functioning properly cannot be rejected. In this case, no further action would be taken and the production operation would continue to run.

The two preceding forms of the soft drink manufacturing hypothesis test show that the null and alternative hypotheses may vary depending upon the point of view of the researcher or decision-maker. To formulate hypotheses correctly it is important to understand the context of the situation and structure the hypotheses to provide the information the researcher or decision-maker wants.

Summary of forms for null and alternative hypotheses

The hypothesis tests in this chapter involve one of two population parameters: the population mean and the population proportion. Depending on the situation, hypothesis tests about a population parameter may take one of three forms. Two include inequalities in the null hypothesis, the third uses only an equality in the null hypothesis. For hypothesis tests involving a population mean, we let μ_0 denote the hypothesized value and choose one of the following three forms for the hypothesis test:

$$H_0: \mu \geq \mu_0 \qquad H_0: \mu \leq \mu_0 \qquad H_0: \mu = \mu_0$$
$$H_1: \mu < \mu_0 \qquad H_1: \mu > \mu_0 \qquad H_1: \mu \neq \mu_0$$

The first two forms are called one-tailed tests. The third form is called a two-tailed test. Note the important point that for all three formats of test, the equality part of the expression (either $\geq$, $\leq$ or $=$) *always* appears in the null hypothesis.

Type I and Type II errors

The null and alternative hypotheses are competing statements about the population. Either H_0 is true or H_1 is true, but not both. Ideally the hypothesis testing procedure should lead to the acceptance of H_0 when H_0 is true and the rejection of H_0 when H_1 is true. However, because hypothesis tests are based on sample information, we must allow for the possibility of errors. Table 9.2 shows the two kinds of errors that can be made in hypothesis testing.

TABLE 9.2 Errors and correct conclusions in hypothesis testing

		Population condition	
		H_0 true	**H_1 true**
Conclusion	**Reject H_0**	Type I error	Correct conclusion
	Do not reject H_0	Correct conclusion	Type II error

The first row of Table 9.2 shows what can happen if the conclusion is to reject H_0. If H_0 is true, we make a **Type I error**; that is, we reject H_0 when it is true. However, if H_1 is true, rejecting H_0 is correct. The second row of Table 9.2 shows what can happen if the conclusion is to not reject H_0. If H_0 is true, this conclusion is correct. However, if H_1 is true, we make a **Type II error**; that is, we fail to reject H_0 when it is false. It is important to emphasize that we never know whether we have made a Type I or Type II error (other than, perhaps, in very exceptional circumstances). The objective is to try to control the probability of error.

Recall the hypothesis-testing illustration discussed above, in which a product research group developed a new fuel injection system designed to decrease the fuel consumption of a particular car model. The hypothesis test was formulated as follows:

$$H_0: \mu \geq 7.0$$
$$H_1: \mu < 7.0$$

The alternative hypothesis indicates that the researchers are looking for sample evidence to support the conclusion that the population mean fuel consumption with the new fuel injection system is less than the current average of 7 litres of fuel per 100 kilometres.

In this application, the Type I error of rejecting H_0 when it is true corresponds to the researchers claiming that the new system reduces fuel consumption ($\mu < 7.0$) when in fact the new system is no better than the current system. In contrast, the Type II error of accepting H_0 when it is false corresponds to the researchers concluding that the new system is no better than the current system ($\mu \geq 7.0$) when in fact the new system reduces fuel consumption.

For the fuel consumption hypothesis test, the null hypothesis is $H_0: \mu \geq 7.0$. The probability of making a Type I error when the null hypothesis is true as an equality, in this case $\mu = 7.0$, is the level of significance of the test, α. In practice, the person conducting the hypothesis test specifies α. If the cost of making a Type I error is high, small values of α are preferred. If the cost of making a Type I error is not so high, larger values of α are typically used. Common choices for α are 0.05 and 0.01. Applications of hypothesis testing that control only for the Type I error are often called *significance tests*. Most applications of hypothesis testing, particularly in academic research, are of this type.

If we decide to accept H_0, and the probability of Type II error has not been controlled, we cannot determine how confident we can be with that decision. Hence statisticians often recommend that in this situation we use the statement 'do not reject H_0' instead of 'accept H_0'. Using the statement 'do not reject H_0' carries the recommendation to withhold both judgement and action. Whenever the probability of making a Type II error has not been determined and controlled, we will not make the statement 'accept H_0'. In such cases, the two conclusions possible are: *do not reject H_0* or *reject H_0*.

In Section 9.6 we shall illustrate procedures for determining and controlling the probability of making a Type II error. If proper controls have been established for this error, action based on the 'accept H_0' conclusion can be appropriate.

Relationship between interval estimation and hypothesis testing

In Chapter 8 we showed how to construct a confidence interval estimate of a population mean. There is a close relationship between the confidence interval and the two-tailed hypothesis test we examined in Section 9.2.

For the σ known case, the confidence interval estimate of a population mean corresponding to a $1 - \alpha$ confidence coefficient is given by:

$$\bar{x} \pm z_{\alpha/2} \frac{\sigma}{\sqrt{n}} \tag{9.2}$$

Doing a hypothesis test requires us first to formulate the hypotheses about the value of a population parameter. In the case of the population mean, the two-tailed test takes the form:

$$H_0: \mu = \mu_0$$

$$H_1: \mu \neq \mu_0$$

where μ_0 is the hypothesized value for the population mean. Using the two-tailed critical value approach, we do not reject H_0 for values of the sample mean that are within $-z_{\alpha/2}$ and $+z_{\alpha/2}$ standard errors of μ_0. Hence, the do-not-reject region for the sample mean in a two-tailed hypothesis test with a level of significance of α is given by:

$$\mu_0 \pm z_{\alpha/2} \frac{\sigma}{\sqrt{n}} \tag{9.3}$$

A close look at expressions (9.2) and (9.3) provides insight into the relationship between the estimation and hypothesis testing approaches to statistical inference. Both procedures require the computation of the values $z_{\alpha/2}$ and $\sigma/\sqrt{n}$. Focusing on α, we see that a confidence coefficient of $(1 - \alpha)$ for interval estimation corresponds to a level of significance of α in hypothesis testing. For example, a 95% confidence interval corresponds to a 0.05 level of significance for hypothesis testing. Furthermore, expressions (9.2) and (9.3) show that, because $z_{\alpha/2}\sigma/\sqrt{n}$ is the plus or minus value for both expressions, if $\bar{x}$ is in the do-not-reject region defined by (9.3), the hypothesized value μ_0 will be in the confidence interval defined by (9.2). Conversely, if the hypothesized value μ_0 is in the confidence interval defined by (9.2), the sample mean will be in the do-not-reject region for the hypothesis H_0: $\mu = \mu_0$ as defined by (9.3). These observations lead to the following procedure for using a confidence interval to conduct a two-tailed hypothesis test.

A confidence interval approach to testing a hypothesis of the form

$$H_0: \mu = \mu_0$$

$$H_1: \mu \neq \mu_0$$

1. Select a simple random sample from the population and use the value of the sample mean to construct the confidence interval for the population mean μ.

$$\bar{x} \pm z_{\alpha/2}\frac{\sigma}{\sqrt{n}}$$

2. If the confidence interval contains the hypothesized value μ_0, do not reject H_0. Otherwise, reject H_0.

An implication of this procedure is that a confidence interval with a confidence coefficient of $1 - \alpha$ includes all those values of μ that will *not* be rejected if they are specified as H_0 in a two-tailed hypothesis test at a significance level of α.

The MaxFlight example involved the following two-tailed test:

$$H_0: \mu = 295$$

$$H_1: \mu \neq 295$$

To test this hypothesis with a level of significance of $\alpha = 0.05$, we sampled 50 golf balls and found a sample mean distance of $\bar{x} = 297.6$ metres. The population standard deviation is $\sigma = 12$. Using these results with $z_{0.025} = 1.96$, we find that the 95% confidence interval estimate of the population mean is:

$$\bar{x} \pm z_{\alpha/2}\frac{\sigma}{\sqrt{n}} = 297.6 \pm 1.96\frac{12}{\sqrt{50}} = 297.6 \pm 3.3$$

This finding enables the quality control manager to conclude with 95% confidence that the mean distance for the population of golf balls is between 294.3 and 300.9 m. Because the hypothesized value for the population mean, $\mu_0 = 295$, is in this interval, the conclusion from the hypothesis test is that the null hypothesis, H_0: $\mu = 295$, cannot be rejected.

Note also that this confidence interval gives a direct indication of how precise the sample is in estimating the population mean. If a disparity of 3.3 m or less between target (295 m) and actual is an issue for MaxFlight, the sample size in the quality control procedure should be increased to give higher precision.

This discussion and example pertain to two-tailed hypothesis tests about a population mean. The same confidence interval and two-tailed hypothesis testing relationship exists for other population parameters. The relationship can also be extended to one-tailed tests about population parameters. Doing so, however, requires the construction of one-sided confidence intervals.

EXERCISES

Methods

8. The managing director of a social media marketing company is considering a new bonus scheme designed to increase sales volume. Currently, the mean sales value per member of the selling team is €13,300 per month. The MD is planning a research study, in which a sample of sales personnel will be allowed to sell under the new bonus scheme for a one-month period, to see whether the scheme increases sales value.
   a. Formulate the null and alternative hypotheses most appropriate for this research situation.
   b. Comment on the conclusion when H_0 cannot be rejected.
   c. Comment on the conclusion when H_0 can be rejected.

9. Refer to Exercise 8. Suppose a sample of $n = 20$ sales personnel yields a sample mean $\bar{x} = €14,100$ sales value per month. Assume the population standard deviation sales per month is $\sigma = €2,700$.
   a. Compute the value of the test statistic.
   b. What is the p-value?
   c. At $\alpha = 0.05$, what is your conclusion?

10. The manager of the Costa Resort Hotel stated that the mean weekend guest bill is €600 or less. A member of the hotel's accounting staff noticed that the total charges for guest bills have been increasing in recent months. The accountant will use a sample of weekend guest bills to test the manager's claim.
   a. Which form of the hypotheses should be used to test the manager's claim? Explain.

$$H_0: \mu \geq 600 \quad H_0: \mu \leq 600 \quad H_0: \mu = 600$$

$$H_1: \mu < 600 \quad H_1: \mu > 600 \quad H_1: \mu \neq 600$$

   b. What conclusion is appropriate when H_0 cannot be rejected?
   c. What conclusion is appropriate when H_0 can be rejected?

11. Refer to Exercise 10. Suppose a sample of $n = 35$ weekend guest bills yields a sample mean $\bar{x} = €629$. Assume the population standard deviation of weekend guest bills is $\sigma = €75$.
   a. Compute the value of the test statistic.
   b. What is the p-value?
   c. At $\alpha = 0.01$, what is your conclusion?

12. The label on a yoghurt pot claims that the yoghurt contains an average of 1 g of fat or less. Answer the following questions for a hypothesis test that could be used to test the claim on the label.
   a. Formulate the appropriate null and alternative hypotheses.
   b. What is the Type I error in this situation? What are the consequences of making this error?
   c. What is the Type II error in this situation? What are the consequences of making this error?

13. A production line is designed to fill cartons with quinoa to a mean weight of 0.75 kg. A sample of cartons is periodically selected and weighed to determine whether under-filling or over-filling is occurring. If the sample data lead to a conclusion of under-filling or over-filling, the production line will be shut down and adjusted.
   a. Formulate the null and alternative hypotheses that will help in deciding whether to shut down and adjust the production line.
   b. Comment on the conclusion and the decision when H_0 cannot be rejected.
   c. Comment on the conclusion and the decision when H_0 can be rejected.

14. Refer to Exercise 13. Suppose a sample of $n = 50$ cartons yields a sample mean $\bar{x} = 0.743$ kg. Assume the population standard deviation of carton weights is $\sigma = 0.009$ kg.
   a. Compute the value of the test statistic.
   b. What is the p-value?
   c. At $\alpha = 0.01$, what is your conclusion?
   d. Construct a 99% confidence interval for the population mean carton weight. Does this support your conclusion from part (c)?

15. Insight Marketing Research bases charges to a client on the assumption that telephone surveys can be completed in a mean time of 15 minutes or less per interview. If a longer mean interview time is necessary, a premium rate is charged. Suppose a sample of 35 interviews shows a sample mean of 17 min. Use $\sigma = 4$ min. Is the premium rate justified?

    a. Formulate the null and alternative hypotheses for this application.

    b. Compute the value of the test statistic.

    c. What is the p-value?

    d. At $\alpha = 0.01$, what is your conclusion?

    e. What is a Type I error in this situation? What are the consequences of making this error?

    f. What is a Type II error in this situation? What are the consequences of making this error?

16. TextRequest reports that 18 to 24-year-old adults send and receive 128 texts every day. Suppose we take a sample of 25 to 34-year-olds to see if their mean number of daily texts differs from the mean for 18 to 24-year-olds reported by TextRequest (taking this figure as a population mean for the 18–24 age group).

    a. State the null and alternative hypotheses we should use to test whether the population mean daily number of texts for 25 to 34-year-olds differs from the population daily mean number of texts for 18 to 24-year-olds.

    b. Suppose a sample of thirty 25 to 34-year-olds showed a sample mean of 118.6 texts per day. Assuming a population standard deviation of 33.17 texts per day, compute the p-value.

    c. With $\alpha = 0.05$ as the level of significance, what is your conclusion?

    d. Repeat the preceding hypothesis test using the critical value approach.

    e. Construct a 95% confidence interval for the population mean number of texts per day by 25 to 34-year-olds. Does this support your hypothesis test conclusion?

9.4 POPULATION MEAN WITH σ UNKNOWN

In this section we describe hypothesis tests about a population mean for the more common situation when σ is unknown. The procedural steps are the same as in the σ known case. However, the test statistic and sampling distribution are different. In Chapter 8, Section 8.2, we showed that interval estimates of a population mean for the σ unknown case are based on a probability distribution known as the t distribution. Hypothesis tests are based on the same distribution. The test statistic, shown below, has a t distribution with $n - 1$ df. The sampling distribution shows somewhat greater variability than in the σ known case because the sample is used to compute estimates of both μ (estimated by $\bar{x}$) and σ (estimated by s).

Test statistic for hypothesis tests about a population mean: σ unknown

$$t = \frac{\bar{x} - \mu_0}{s/\sqrt{n}} \qquad\qquad (9.4)$$

As we said in Chapter 8, using the t distribution is based on an assumption that the population from which we are sampling has a normal distribution. However, research shows that this assumption can be relaxed considerably when the sample size is sufficiently large. We provide practical advice at the end of this section.

One-tailed test

Consider an example of a one-tailed test about a population mean for the σ unknown case. A travel website wants to classify international airports according to the mean rating given by business travellers. A rating scale from 0 to 10 will be used. Airports with a population mean rating greater than 7 will be designated as superior service airports. The website staff surveyed a sample of 60 business travellers at each airport. Suppose the sample for Abu Dhabi International Airport provided a sample mean rating of $\bar{x} = 7.250$ and a sample standard deviation of $s = 1.052$. Do the data indicate that Abu Dhabi should be designated as a superior service airport?

AIRRATING

We want a hypothesis test in which rejection of H_0 will lead to the conclusion that the population mean rating for Abu Dhabi International Airport is *greater* than 7. Therefore, we need an upper-tail test with $H_1: \mu > 7$. The null and alternative hypotheses are:

$$H_0: \mu \le 7$$

$$H_1: \mu > 7$$

We shall use $\alpha = 0.05$ as the level of significance.

Using equation (9.4) with $\bar{x} = 7.250$, $s = 1.052$ and $n = 60$, the value of the test statistic is:

$$t = \frac{\bar{x} - \mu_0}{s/\sqrt{n}} = \frac{7.250 - 7.000}{1.052/\sqrt{60}} = 1.841$$

The sampling distribution has $n - 1 = 60 - 1 = 59$ df. Because the test is an upper-tail test, the p-value is the area under the curve of the t distribution to the right of $t = 1.841$. SPSS, Excel, R and Minitab will determine the exact p-value, and show that it is 0.035. A p-value of 0.035 (<0.05) leads to the rejection of H_0 and to the conclusion that Abu Dhabi should be classified as a superior service airport.

The t distribution table in most textbooks does not contain sufficient detail to determine the exact p-value. For example, Table 2 in Appendix B provides the following information for the t distribution with 59 df.

Area in upper tail	0.20	0.10	0.05	0.025	0.01	0.005
t-value (59 df)	0.848	1.296	1.671	2.001	2.391	2.662

$t = 1.841$

The value $t = 1.841$ is between the tabulated values 1.671 and 2.001. Although the table does not provide the exact p-value, the figures in the row labelled 'Area in upper tail' show that the p-value must be less than 0.05 and greater than 0.025. With $\alpha = 0.05$, this placement is sufficient information to make the decision to reject H_0 and conclude that Abu Dhabi should be classified as a superior service airport.

The critical value approach can also be used. With $\alpha = 0.05$, $t_{0.05} = 1.671$ is the critical value for the test. The rejection rule is therefore:

$$\text{Reject } H_0 \text{ if } t \ge 1.671$$

With the test statistic $t = 1.841$ (>1.671), H_0 is rejected and we conclude that Abu Dhabi can be classified as a superior service airport.

Two-tailed test

To illustrate a two-tailed test about a population mean for the σ unknown case, consider the following example. The company MegaToys manufactures and distributes its products through more than 1,000 retail outlets. In planning production levels for the coming winter season, MegaToys must decide how many units of each product to produce. For this year's most important new gizmo, MegaToys' marketing director is expecting demand to average 140 units per retail outlet. Prior to making the final production decision, MegaToys decided to survey a sample of 25 retailers to gather more information about the likely demand. Each retailer was provided with information about the features of the new product, along with the cost and the suggested selling price, then asked to specify an anticipated order quantity.

The sample data will be used to do the following two-tailed hypothesis test about μ, the population mean order quantity per retail outlet. The level of significance for the test will be $\alpha = 0.05$.

$$H_0: \mu = 140$$

$$H_1: \mu \neq 140$$

If H_0 cannot be rejected, MegaToys will continue its production planning based on the marketing director's estimate that $\mu = 140$ units. However, if H_0 is rejected, MegaToys will re-evaluate its production plan for the product. A two-tailed hypothesis test is used because MegaToys wants to re-evaluate the production plan if μ is less than anticipated or greater than anticipated. Because no historical data are available (it is a new product), the population mean and the population standard deviation must both be estimated, using $\bar{x}$ and s from the sample data.

ORDERS

The 25 retailers provided a sample mean of $\bar{x} = 137.4$ and a sample standard deviation of $s = 11.79$ units. Before going ahead and using the t distribution, the analyst constructed a histogram of the sample data to check on the form of the population distribution. The histogram showed no evidence of skewness or any extreme outliers, so the analyst concluded that using the t distribution with $n - 1 = 24$ df was appropriate. Putting $\bar{x} = 137.4$, $\mu_0 = 140.0$, $s = 1,179$ and $n = 25$ in equation (9.4), the value of the test statistic is:

$$t = \frac{\bar{x} - \mu_0}{s/\sqrt{n}} = \frac{137.4 - 140.0}{11.79/\sqrt{25}} = -1.103$$

Because this is a two-tailed test, the p-value is the area to the left of $t = -1.103$ under the t distribution curve, multiplied by two. Figure 9.7 shows the two areas representing the p-value. Using SPSS, Excel, R or Minitab, we find the exact p-value $= 0.282$. With a level of significance $\alpha = 0.05$, H_0 cannot be rejected. There is insufficient evidence to conclude that MegaToys should change its production plan.

FIGURE 9.7
Area under the curve in both tails provides the p-value

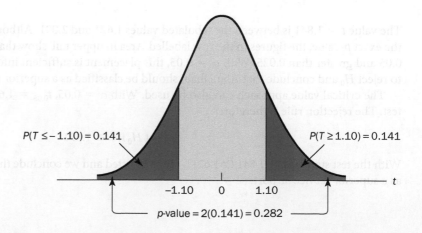

$P(T \leq -1.10) = 0.141$ $P(T \geq 1.10) = 0.141$

−1.10   0   1.10 t

p-value $= 2(0.141) = 0.282$

Table 2 in Appendix B shows the following for the t distribution with 24 df.

Area in upper tail	0.20	0.10	0.05	0.025	0.01	0.005
t-value (24 df)	0.858	1.318	1.711	2.064	2.492	2.797

$$\uparrow$$
$$t = 1.103$$

The t distribution table contains only positive t-values. However, because the distribution is symmetrical, we can estimate the area under the curve to the right of $t = 1.103$ and double it to find the p-value. We see that $t = 1.103$ is between 0.858 and 1.318. From the 'Area in upper tail' row, the area in the tail to the right of $t = 1.103$ is between 0.20 and 0.10. Doubling these amounts, the p-value must be between 0.40 and 0.20. With $\alpha = 0.05$, we know that the p-value is greater than α. Therefore, H_0 cannot be rejected.

The test statistic can also be compared to the critical values to make the hypothesis-testing decision. With $\alpha = 0.05$, $-t_{0.025} = -2.064$ and $t_{0.025} = 2.064$ are the critical values for the two-tailed test. The rejection rule is:

$$\text{Reject } H_0 \text{ if } t \le -2.064 \text{ or if } t \ge 2.064$$

Based on the test statistic $t = -1.103$, H_0 cannot be rejected. This result indicates that MegaToys should continue its production planning based on the expectation that $\mu = 140$, or do further investigation amongst its retailers. A 95% confidence interval for the mean demand is $137.4 \pm (2.064 \times 11.79/5) = [132.5, 142.3]$. Management therefore has a reasonably precise estimate of likely demand at the 95% confidence level (that includes the value 140). A larger sample would narrow the margin of error.

Summary and practical advice

Table 9.3 provides a summary of the hypothesis-testing procedures about a population mean for the σ unknown case. The key difference between the σ unknown case and the σ known case is that s is used, instead of σ, in the computation of the test statistic. Consequently, the sampling distribution of the test statistic is a t distribution rather than standard normal.

The applicability of the procedures based on the t distribution is dependent on the distribution of the population being sampled and the sample size. When the population is normally distributed, the hypothesis tests described in this section provide exact results for any sample size. When the population is not normally distributed, the procedures are approximations. If the population is approximately normal, small sample sizes (e.g. $n < 15$) can provide acceptable results. In situations where the population cannot be approximated by a normal distribution, sample sizes of $n \ge 30$ will provide acceptable results as long as the population is not significantly skewed and does not contain outliers. If the population is significantly skewed or contains outliers, samples sizes approaching 50 are recommended. Sample sizes greater than 50 will provide good results in almost all cases.

TABLE 9.3 Summary of hypothesis tests about a population mean: σ unknown case

	Lower-tail test	Upper-tail test	Two-tailed test
Hypotheses	$H_0: \mu \ge \mu_0$ $H_1: \mu < \mu_0$	$H_0: \mu \le \mu_0$ $H_1: \mu > \mu_0$	$H_0: \mu = \mu_0$ $H_1: \mu \ne \mu_0$
Test statistic	$t = \dfrac{\bar{x} - \mu_0}{s/\sqrt{n}}$	$t = \dfrac{\bar{x} - \mu_0}{s/\sqrt{n}}$	$t = \dfrac{\bar{x} - \mu_0}{s/\sqrt{n}}$
Rejection rule: **p-value approach**	Reject H_0 if $p\text{-value} \le \alpha$	Reject H_0 if $p\text{-value} \le \alpha$	Reject H_0 if $p\text{-value} \le \alpha$
Rejection rule: **critical value approach**	Reject H_0 if $t \le -t_\alpha$	Reject H_0 if $t \ge t_\alpha$	Reject H_0 if $t \le -t_{\alpha/2}$ or if $t \ge t_{\alpha/2}$

EXERCISES

Methods

17. Consider the following hypothesis test:

$$H_0: \mu \leq 12$$

$$H_1: \mu > 12$$

Suppose sample mean $\bar{x} = 14.0$, sample standard deviation $s = 4.32$, sample size $n = 25$.
   a. Compute the value of the test statistic.
   b. What does the t distribution table tell you about the p-value?
   c. At $\alpha = 0.05$, what is your conclusion?
   d. What is the rejection rule using the critical value? What is your conclusion?

18. Consider the following hypothesis test:

$$H_0: \mu = 18$$

$$H_1: \mu \neq 18$$

Suppose sample mean $\bar{x} = 17.0$, sample standard deviation $s = 4.5$, sample size $n = 48$.
   a. Compute the value of the test statistic.
   b. What does the t distribution table tell you about the p-value?
   c. At $\alpha = 0.05$, what is your conclusion?
   d. What is the rejection rule using the critical value? What is your conclusion?

19. Consider the following hypothesis test:

$$H_0: \mu \geq 45$$

$$H_1: \mu < 45$$

Sample size $n = 36$. Identify the p-value and state your conclusion for each of the following sample results, using $\alpha = 0.01$.
   a. $\bar{x} = 44.0$ and $s = 5.2$.
   b. $\bar{x} = 43.0$ and $s = 4.6$.
   c. $\bar{x} = 46.0$ and $s = 5.0$.

Applications

20. Lager can be bought in 330 ml bottles. If a bottle is marked as containing 330 ml, legislation requires that the production batch from which the bottle came must have a mean fill volume of at least 330 ml.
   a. Formulate hypotheses that could be used to determine whether the mean fill volume for a production batch satisfies the legal requirement of being at least 330 ml.
   b. Suppose you take a random sample of 30 bottles from a production line and find that the mean fill for the sample of 30 bottles is 329.5 ml, with a sample standard deviation of 1.9 ml. What is the p-value?
   c. At $\alpha = 0.01$, what is your conclusion?

21. Consider a daily TV programme – like the 10 o'clock news – that over the last calendar year had a mean daily audience of 4.0 million viewers. For a sample of 40 days during the current year, the daily audience was 4.15 million viewers with a sample standard deviation of 0.45 million viewers.
   a. If the TV management company would like to test for a change in mean viewing audience, what statistical hypotheses should be set up?
   b. What is the p-value?
   c. Select your own level of significance. What is your conclusion?
   d. Construct a confidence interval for the mean viewing audience, at a confidence level that complements the significance level you chose in part (c). Comment on your result.

22. A popular pastime among football fans is participation in 'fantasy football' competitions. Participants choose a squad of players, with the objective of increasing the valuation of the squad over the season. Suppose that at the start of the competition, the mean valuation of all available strikers was £14.7 million per player.
    a. Formulate the null and alternative hypotheses that could be used by a football pundit to determine whether midfielders have a higher mean valuation than strikers.
    b. Suppose a random sample of 30 midfielders from the available list had a mean valuation at the start of the competition of £15.80 million, with a sample standard deviation of £2.46 million. On average, by how much did the valuation of midfielders exceed that of strikers?
    c. At $\alpha = 0.05$, what is your conclusion?

23. Most new car models sold in the European Union have to undergo official tests for fuel consumption. Consider a new car model: the official fuel consumption figure for the urban cycle is published as 11.8 litres of fuel per 100 kilometres. A consumer rights organization is interested in examining whether this published figure is truly indicative of urban driving.
    a. State the hypotheses that would enable the consumer rights organization to investigate whether the model's fuel consumption is more than the published 11.8 litres per 100 kilometres.
    b. A sample of 50 fuel consumption tests with the new model showed a sample mean of 12.10 litres per 100 kilometres and a sample standard deviation of 0.92 litre per 100 kilometres. What is the p-value?
    c. What conclusion should be drawn from the sample results? Use $\alpha = 0.01$.
    d. Repeat the preceding hypothesis test using the critical value approach.

24. SuperScapes specializes in custom landscaping for residential areas. Estimated labour costs are based on the number of plantings of trees and shrubs involved. For cost-estimation purposes, managers use two hours of labour time for the planting of a medium-sized tree. Actual times from a sample of ten plantings during the past month follow (times in hours).

    1.7   1.5   2.6   2.2   2.4   2.3   2.6   3.0   1.4   2.3

    With a 0.05 level of significance, test to see whether the mean tree-planting time differs from 2 hours.
    a. State the null and alternative hypotheses.
    b. Compute the sample mean and sample standard deviation.
    c. What is the p-value?
    d. Construct a 95% confidence interval for the mean tree planting time.
    e. What are your conclusions?

9.5 POPULATION PROPORTION

The three possible forms of hypothesis test about a population proportion π are shown below; π_0 denotes the hypothesized value. The first is a lower-tail test, the second an upper-tail test and the third a two-tailed test.

$$H_0: \pi \geq \pi_0 \qquad H_0: \pi \leq \pi_0 \qquad H_0: \pi = \pi_0$$

$$H_1: \pi < \pi_0 \qquad H_1: \pi > \pi_0 \qquad H_1: \pi \neq \pi_0$$

Hypothesis tests about π are based on the difference between the sample proportion p and the hypothesized population proportion π_0. The methods are similar to those for hypothesis tests about a population mean, but the sample proportion and its standard error are used to compute the test statistic.

Consider the situation faced by Aspire gymnasium. Over the past year, 20 per cent of Aspire users were women. In an effort to increase this proportion, Aspire mounted a special promotion aimed at women.

WOMENGYM

One month afterwards, the gym manager requested a sample study to determine whether the proportion of women at Aspire had increased. An upper-tail test with $H_1: \pi > 0.20$ is therefore appropriate. The null and alternative hypotheses for the test are:

$$H_0: \pi \leq 0.20$$

$$H_1: \pi > 0.20$$

If H_0 can be rejected, the results will give statistical support for the conclusion that the proportion of women increased (suggesting the promotion was beneficial). The gym manager specified that a level of significance of $\alpha = 0.05$ be used for the test.

The next step is to select a sample and compute the value of an appropriate test statistic. The sampling distribution of P, the point estimator of the population parameter π, is the basis for constructing the test statistic. When H_0 is true as an equality, the expected value of P equals the hypothesized value π_0; that is, $E(P) = \pi_0$. The standard error of P is given by:

$$\sigma_P = \sqrt{\frac{\pi_0(1 - \pi_0)}{n}}$$

In Chapter 7 we said that if $n\pi \geq 5$ and $n(1 - \pi) \geq 5$, the sampling distribution of P can be approximated by a normal distribution.* Under these conditions, which usually apply in practice, the quantity:

$$Z = \frac{P - \pi_0}{\sigma_P} \tag{9.5}$$

has a standard normal probability distribution. Equation (9.6) gives the test statistic used to do hypothesis tests about a population proportion.

Test statistic for hypothesis tests about a population proportion

$$z = \frac{p - \pi_0}{\sqrt{\dfrac{\pi_0(1 - \pi_0)}{n}}} \tag{9.6}$$

Suppose a random sample of 400 gym users was selected and that 100 of the users were women. The sample proportion of women is $p = 100/400 = 0.25$. Using equation (9.6), the value of the test statistic is:

$$z = \frac{p - \pi_0}{\sqrt{\dfrac{\pi_0(1 - \pi_0)}{n}}} = \frac{0.25 - 0.20}{\sqrt{\dfrac{0.20(1 - 0.20)}{400}}} = \frac{0.05}{0.02} = 2.50$$

Because the Aspire hypothesis test is an upper-tail test, the p-value is the probability that Z is greater than or equal to $z = 2.50$. It is the area under the standard normal curve to the right of $z = 2.50$. Using the standard normal distribution table, we find that the p-value is $(1 - 0.9938) = 0.0062$. Figure 9.8 shows this p-value calculation.

---

* In most applications involving hypothesis tests of a population proportion, sample sizes are large enough to use the normal approximation. The exact sampling distribution of P is discrete, with the probability for each value of P given by the binomial distribution. So hypothesis testing is more complicated for small samples when the normal approximation cannot be used.

FIGURE 9.8

Calculation of the p-value
for the Aspire hypothesis test

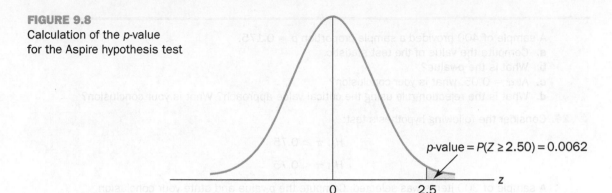

$$p\text{-value} = P(Z \geq 2.50) = 0.0062$$

The gym manager specified a level of significance $\alpha = 0.05$. The p-value 0.0062 (<0.05) gives sufficient statistical evidence to reject H_0 at $\alpha = 0.05$. The test provides statistical support for the conclusion that, following the special promotion, the proportion of women increased at Aspire.

The conclusion can also be reached using the critical value approach. The critical value for $\alpha = 0.05$ is $z_{0.05} = 1.645$. Hence, the rejection rule is to reject H_0 if $z \geq 1.645$. Because $z = 2.50$ (>1.645), H_0 is rejected.

Again, we see that the p-value approach and the critical value approach lead to the same conclusion, but the p-value approach provides more information. With p-value $= 0.0062$, the null hypothesis would be rejected for any level of significance α greater than or equal to 0.0062.

We have illustrated an upper-tail test. Similar procedures can be used for lower-tail and two-tailed tests. Table 9.4 provides a summary of the hypothesis tests about a population proportion.

TABLE 9.4 Summary of hypothesis tests about a population proportion

	Lower-tail test	Upper-tail test	Two-tailed test
Hypotheses	$H_0: \pi \geq \pi_0$ $H_1: \pi < \pi_0$	$H_0: \pi \leq \pi_0$ $H_1: \pi > \pi_0$	$H_0: \pi = \pi_0$ $H_1: \pi \neq \pi_0$
Test statistic	$z = \dfrac{p - \pi_0}{\sqrt{\dfrac{\pi_0(1 - \pi_0)}{n}}}$	$z = \dfrac{p - \pi_0}{\sqrt{\dfrac{\pi_0(1 - \pi_0)}{n}}}$	$z = \dfrac{p - \pi_0}{\sqrt{\dfrac{\pi_0(1 - \pi_0)}{n}}}$
Rejection rule: p-value approach	Reject H_0 if p-value $\leq \alpha$	Reject H_0 if p-value $\leq \alpha$	Reject H_0 if p-value $\leq \alpha$
Rejection rule: critical value approach	Reject H_0 if $z \leq -z_\alpha$	Reject H_0 if $z \geq z_\alpha$	Reject H_0 if $z \leq -z_{\alpha/2}$ or if $z \geq z_{\alpha/2}$

EXERCISES

Methods

25. Consider the following hypothesis test:

$$H_0: \pi = 0.20$$

$$H_1: \pi \neq 0.20$$

A sample of 400 provided a sample proportion $p = 0.175$.
a. Compute the value of the test statistic.
b. What is the p-value?
c. At $\alpha = 0.05$, what is your conclusion?
d. What is the rejection rule using the critical value approach? What is your conclusion?

26. Consider the following hypothesis test:

$$H_0: \pi \geq 0.75$$

$$H_1: \pi < 0.75$$

A sample of 300 items was selected. Compute the p-value and state your conclusion at $\alpha = 0.05$ for each of the following sample results.
a. $p = 0.68$.
b. $p = 0.72$.
c. $p = 0.70$.
d. $p = 0.77$.

Applications

27. An airline promotion to business travellers is based on the assumption that at least two-thirds of business travellers use a laptop computer or tablet on overnight business trips.
a. State the hypotheses that can be used to test the assumption.
b. A survey found that 355 of 546 business travellers used a laptop computer or tablet on overnight business trips. Calculate the p-value for the test.
c. Use $\alpha = 0.05$, What is your conclusion?

WILDLIFE

28. The Wild Life retail chain specializes in outdoor clothing and camping gear. It is considering a promotion that involves mailing discount coupons to all their credit card customers. This promotion will be considered a success if more than 10 per cent of coupon recipients use them. Before going nationwide with the promotion, coupons were sent to a sample of 100 credit card customers.
a. Formulate hypotheses that can be used to test whether the population proportion of those who will use the coupons is sufficient to go national.
b. The file 'WildLife' contains the sample data. Compute a point estimate of the population proportion.
c. Use $\alpha = 0.05$, to conduct your hypothesis test. Should Wild Life go national with the promotion?

29. Members of the millennial generation continue to be dependent on their parents (either living with or otherwise receiving support from parents) into early adulthood. A family research organization has claimed that, in past generations, no more than 30 per cent of individuals aged 18–32 continued to be dependent on their parents. Suppose that a sample of 400 individuals aged 18–32 showed that 136 of them continue to be dependent on their parents.
a. Formulate hypotheses for a test to determine whether the proportion of millennials continuing to be dependent on their parents is higher than for past generations.
b. What is your point estimate of the proportion of millennials that are continuing to be dependent on their parents?
c. What is the p-value provided by the sample data?
d. What is your hypothesis testing conclusion? Use $\alpha = 0.05$ as the level of significance.

30. A study by *Consumer Reports* showed that 64 per cent of supermarket shoppers believe supermarket own brands to be as good as national name brands. To investigate whether this result applies to its own product, the manufacturer of a market-leading tomato ketchup asked a sample of shoppers whether they believed that supermarket ketchup was as good as the market-leading brand ketchup.
a. Formulate the hypotheses that could be used to determine whether the percentage of supermarket shoppers who believe that the supermarket ketchup was as good as the market-leading brand ketchup differed from 64 per cent.
b. If a sample of 100 shoppers showed 52 stating that the supermarket brand was as good as the market-leading brand, what is the p-value?
c. At $\alpha = 0.05$, what is your conclusion?
d. Should the manufacturer of the market leader be pleased with this conclusion? Explain.

31. Microsoft Outlook is a widely used email manager. A Microsoft executive claims that Microsoft Outlook is used by at least 75 per cent of internet users. A sample of internet users will be used to test this claim.
    a. Formulate the hypotheses that can be used to test the claim.
    b. A study reported that Microsoft Outlook is used by 72 per cent of internet users. Assume that the report was based on a sample size of 300 internet users. What is the p-value?
    c. At $\alpha = 0.05$, should the executive's claim of at least 75 per cent be rejected?

32. Last year, 46 per cent of business owners gave a holiday gift to their employees. A survey of business owners conducted this year indicates that 35 per cent plan to provide a holiday gift to their employees. Suppose the survey results are based on a sample of 60 business owners.
    a. How many business owners in the survey plan to provide a holiday gift to their employees this year?
    b. Suppose the business owners in the sample did as they planned. Compute the p-value for a hypothesis test that can be used to determine if the proportion of business owners providing holiday gifts had decreased from last year.
    c. Using a 0.05 level of significance, would you conclude that the proportion of business owners providing gifts had decreased? What is the smallest level of significance for which you could draw such a conclusion?

9.6 TYPE II ERRORS AND POWER

Controlling the probability of Type II error

So far in this chapter we have focused primarily on hypothesis-testing applications that are considered to be significance tests. With a significance test, we control the probability of making a Type I error, by setting α, but not the probability of making a Type II error. If p-value $\leq \alpha$, we conclude 'reject H_0' and declare the results statistically significant. Otherwise, we conclude 'do not reject H_0'. With the latter conclusion, the statistical evidence is considered inconclusive and is usually an indication to postpone a decision or action until further research and testing can be undertaken.

However, if the purpose of a hypothesis test is to make a decision when H_0 is true and a different decision when H_1 is true, the decision-maker may need to take action with both the conclusion *do not reject H_0* and the conclusion *reject H_0*. In such situations, it is desirable to control the probability of making a Type II error. With the probabilities of both types of error controlled, the conclusion from the hypothesis test is either to *accept H_0* or *reject H_0*. In the first case, H_0 is in effect concluded to be true, while in the second case, H_1 is concluded to be true. A decision and appropriate action can be taken when either conclusion is reached.

A good illustration of this situation is lot-acceptance sampling, a topic we shall discuss in more depth in Chapter 20 (on the online platform). For example, a quality control manager must decide whether to accept a shipment of batteries from a supplier or to return the shipment because of poor quality. Assume that design specifications require batteries from the supplier to have a mean useful life of at least 120 hours. To evaluate the quality of an incoming shipment, a sample of batteries will be selected and tested. On the basis of the sample, a decision must be made to accept the shipment of batteries or to return it to the supplier because of poor quality. Let μ denote the mean number of hours of useful life for batteries in the shipment. The null and alternative hypotheses about the population mean is as follows:

$$H_0: \mu \geq 120$$

$$H_1: \mu < 120$$

If H_0 is rejected, H_1 is concluded to be true. This conclusion indicates that the appropriate action is to return the shipment to the supplier. However, if H_0 is not rejected, the decision-maker will accept the shipment as being of satisfactory quality. By not rejecting H_0, a decision with important business consequences has been made.

In such situations, it is desirable that the hypothesis testing procedure should control the probability of making a Type II error, because a decision will be made and action taken when we do not reject H_0. Below we explain how to compute the probability of making a Type II error, and how the sample size can be adjusted to control this probability.

Calculating the probability of Type II error for a test of the population mean

The null and alternative hypotheses in the lot-acceptance example are $H_0: \mu \geq 120$ and $H_1: \mu < 120$. If H_0 is rejected, the shipment will be returned to the supplier because the mean hours of useful life are concluded to be less than the specified 120 hours. If H_0 is not rejected, the shipment will be accepted.

Suppose a level of significance $\alpha = 0.05$ is used. The test statistic in the σ known case is:

$$z = \frac{\bar{x} - \mu_0}{\sigma/\sqrt{n}} = \frac{\bar{x} - 120}{\sigma/\sqrt{n}}$$

Based on the critical value approach and $z_{0.05} = 1.645$, the rejection rule for the lower-tail test is to reject H_0 if $z \leq -1.645$. Suppose a sample of 36 batteries will be selected and, based upon previous testing, the population standard deviation can be assumed known with a value of $\sigma = 12$ hours. The rejection rule indicates that we will reject H_0 if:

$$z = \frac{\bar{x} - 120}{12/\sqrt{36}} \leq -1.645$$

Solving for $\bar{x}$ indicates that we will reject H_0 if:

$$\bar{x} \leq 120 - 1.645\left(\frac{12}{\sqrt{36}}\right) = 116.71$$

Rejecting H_0 when $\bar{x} \leq 116.71$ implies we will accept the shipment whenever $\bar{x} > 116.71$.

We make a Type II error whenever the true shipment mean is less than 120 hours and we decide to accept $H_0: \mu \geq 120$. To compute the probability of making a Type II error, we must therefore select a value of μ less than 120 hours. For example, suppose the shipment is considered to be of very poor quality if the batteries have a mean life of $\mu = 112$ hours. What is the probability of accepting $H_0: \mu \geq 120$, and committing a Type II error, if $\mu = 112$? This is the probability that the sample mean X is greater than 116.71 when $\mu = 112$.

Figure 9.9 shows the sampling distribution of the sample mean when $\mu = 112$. The shaded area in the upper tail gives the probability of $\overline{X} > 116.71$. Using the standard normal distribution, we see that at $\bar{x} = 116.71$:

$$z = \frac{\bar{x} - \mu_0}{\sigma/\sqrt{n}} = \frac{116.71 - 112}{12/\sqrt{36}} = 2.36$$

The standard normal distribution table shows that for $z = 2.36$, the upper-tail area is $1 - 0.0909 = 0.0091$. If the mean of the population is 112 hours, the probability of making a Type II error is 0.0091, less than 1 per cent.

We can repeat these calculations for other values of μ less than 120. Each value of μ results in a different probability of making a Type II error. For example, suppose the shipment of batteries has a mean useful life of $\mu = 115$ hours. Because we will accept H_0 whenever $\bar{x} > 116.71$, the z value for $\mu = 115$ is given by:

$$z = \frac{\bar{x} - \mu_0}{\sigma/\sqrt{n}} = \frac{116.71 - 115}{12/\sqrt{36}} = 0.86$$

FIGURE 9.9
Probability of a Type II error
when $\mu = 112$

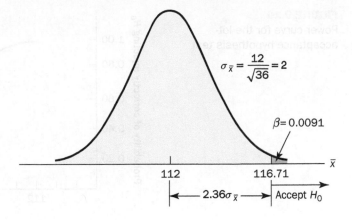

$$\sigma_{\bar{x}} = \frac{12}{\sqrt{36}} = 2$$

$\beta = 0.0091$

112   116.71

$\leftarrow 2.36\sigma_{\bar{x}} \rightarrow$ | Accept H_0

$\bar{x}$

The upper tail area for $z = 0.86$ is $1 - 0.8051 = 0.1949$. The probability of making a Type II error when the true mean is $\mu = 115$ is 0.1949, almost 20 per cent.

The probability of making a Type II error is usually denoted by the Greek letter β. In Table 9.5 we show values of β in the lot-acceptance example for a number of values of μ less than 120. Note that as μ increases towards 120, β increases towards an upper bound of 0.95 (which is $1 - \alpha$). However, as μ decreases to values further below 120, β diminishes. This is what we would expect. When the true population mean μ is close to the null hypothesis value of $\mu = 120$, the probability is high that we will make a Type II error. However, when the true population mean μ is far below the null hypothesis value of 120, the probability is low that we will make a Type II error.

TABLE 9.5 Probability of making a Type II error for the lot-acceptance hypothesis test

Value of μ	$z = \dfrac{116.71 = \mu}{12/\sqrt{36}}$	Probability of a Type II error (β)	Power ($1 - \beta$)
112	2.36	0.0091	0.9909
114	1.36	0.0869	0.9131
115	0.86	0.1949	0.8051
116.71	0.00	0.5000	0.5000
117	−0.15	0.5596	0.4404
118	−0.65	0.7422	0.2578
119.999	−1.645	0.9500	0.0500

The probability of correctly rejecting H_0 when it is false is called the **power** of the test. For any particular value of μ, the power is $1 - \beta$; that is, the probability of correctly rejecting the null hypothesis is 1 minus the probability of making a Type II error. Power values are also listed in Table 9.5. The power associated with each value of μ is shown graphically in Figure 9.10. Such a graph is called a **power curve**. Note that the power curve extends over the values of μ for which H_0 is false.

Another graph, called the *operating characteristic curve* (OCC), is sometimes used to provide information about test characteristics. The OCC plots β against μ. Whereas the power curve shows the probability of correctly rejecting H_0 when H_0 is false ($1 - \beta$), the operating characteristic curve shows the probability of accepting H_0 (and making a Type II error) for the values of μ where H_0 is false.

FIGURE 9.10

Power curve for the lot-acceptance hypothesis test

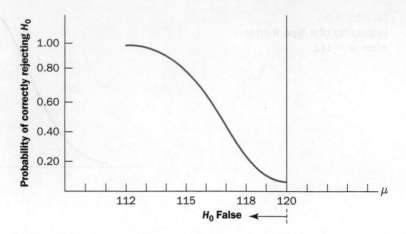

In summary, the following step-by-step procedure can be used to compute the probability of making a Type II error in hypothesis tests about a population mean:

1 Formulate H_0 and H_1.

2 Use the level of significance α and the critical value approach to determine the rejection rule for the test.

3 Use the rejection rule to solve for the value of the sample mean corresponding to the critical value of the test statistic.

4 Use the results from step 3 to state the values of the sample mean that lead to the acceptance of H_0. These values define the acceptance region for the test.

5 Use the sampling distribution of $\overline{X}$ for a value of μ in H_1, and the acceptance region from step 4, to compute the probability that the sample mean will be in the acceptance region. This is the probability of making a Type II error at the chosen value of μ.

Determining the sample size for hypothesis tests about a population mean

The level of significance α specified by the user determines the probability of making a Type I error. By controlling the sample size, the user can also control the probability of making a Type II error. Consider the following lower-tail test about a population mean:

$$H_0: \mu \geq \mu_0$$

$$H_1: \mu < \mu_0$$

The upper panel of Figure 9.11 is the sampling distribution of $\overline{X}$ when H_0 is true with $\mu = \mu_0$. For a lower-tail test, the critical value of the test statistic is $-z_\alpha$. The vertical line labelled c is the corresponding value of $\overline{X}$. We compute c using the following formula:

$$c = \mu_0 - z_\alpha \frac{\sigma}{\sqrt{n}} \tag{9.7}$$

FIGURE 9.11

Determining the sample size for specified levels of the Type I (α) and Type II (β) errors

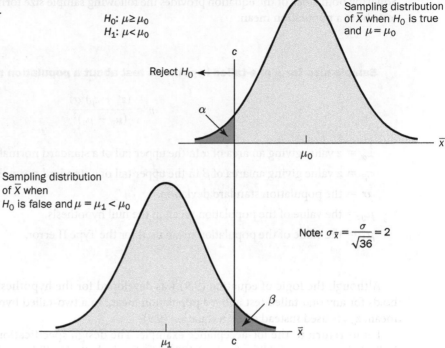

$H_0: \mu \geq \mu_0$
$H_1: \mu < \mu_0$

Sampling distribution of $\overline{X}$ when H_0 is true and $\mu = \mu_0$

Reject H_0

c

α

μ_0

$\overline{X}$

Sampling distribution of $\overline{X}$ when H_0 is false and $\mu = \mu_1 < \mu_0$

Note: $\sigma_{\overline{x}} = \dfrac{\sigma}{\sqrt{36}} = 2$

β

μ_1 c $\overline{X}$

If we reject H_0 when $\overline{x} = c$ the probability of a Type I error will be α.

The lower panel of Figure 9.11 is the sampling distribution of $\overline{X}$ when the alternative hypothesis is true with $\mu = \mu_1 < \mu_0$. The shaded region shows β, the probability of a Type II error if the null hypothesis is accepted when $\overline{x} > c$. We compute c using the following formula:

$$c = \mu_1 + z_\beta \frac{\sigma}{\sqrt{n}} \tag{9.8}$$

where z_β is the z value corresponding to an area of β in the upper tail of the standard normal distribution.

We wish the value c to be such that when we reject H_0 the probability of a Type I error is equal to the chosen α value, and when we accept H_0 the probability of a Type II error is equal to the chosen β value. Therefore, both equations (9.7) and (9.8) must provide the same value for c. Hence:

$$\mu_0 - z_\alpha \frac{\sigma}{\sqrt{n}} = \mu_1 + z_\beta \frac{\sigma}{\sqrt{n}}$$

To determine the required sample size, we first solve for $\sqrt{n}$ as follows:

$$\mu_0 - \mu_1 = z_\alpha \frac{\sigma}{\sqrt{n}} + z_\beta \frac{\sigma}{\sqrt{n}} = \frac{(z_\alpha + z_\beta)\sigma}{\sqrt{n}}$$

and

$$\sqrt{n} = \frac{(z_\alpha + z_\beta)\sigma}{(\mu_0 - \mu_1)}$$

Squaring both sides of the equation provides the following sample size formula for a one-tailed hypothesis test about a population mean:

Sample size for a one-tailed hypothesis test about a population mean

$$n = \frac{(z_\alpha + z_\beta)^2 \sigma^2}{(\mu_0 - \mu_1)^2} \tag{9.9}$$

z_α = z value giving an area of α in the upper tail of a standard normal distribution.
z_β = z value giving an area of β in the upper tail of a standard normal distribution.
σ = the population standard deviation.
μ_0 = the value of the population mean in the null hypothesis.
μ_1 = the value of the population mean used for the Type II error.

Although the logic of equation (9.9) was developed for the hypothesis test shown in Figure 9.11, it holds for any one-tailed test about a population mean. In a two-tailed hypothesis test about a population mean, $z_{\alpha/2}$ is used instead of z_α in equation (9.9).

Let us return to the lot-acceptance example. The design specification for the shipment of batteries indicated a mean useful life of at least 120 hours for the batteries. Shipments were returned if $H_0: \mu \geq 120$ was rejected. Suppose the quality control manager makes the following statements about the allowable probabilities for Type I and Type II errors:

Type I error: If the mean life of the batteries in the shipment is $\mu = 120$, I am willing to risk an $\alpha = 0.05$ probability of rejecting the shipment.

Type II error: If the mean life of the batteries in the shipment is 5 hours under the specification (i.e. $\mu = 115$), I am willing to risk a $\beta = 0.10$ probability of accepting the shipment.

Statements about the allowable error probabilities must be made before the sample size can be determined. These statements are based on the judgement of the manager.

In the example, $\alpha = 0.05$ and $\beta = 0.10$, with corresponding μ values of $\mu_0 = 120$ and $\mu_1 = 115$. Using the standard normal probability distribution, we have $z_{0.05} = 1.645$ and $z_{0.10} = 1.28$. The population standard deviation was assumed to be known at $\sigma = 12$. Using equation (9.9), we find that the recommended sample size is:

$$n = \frac{(1.645 + 1.28)^2 (12)^2}{(120 - 115)^2} = 49.3$$

Rounding up, we recommend a sample size of 50.

Because both the Type I and Type II error probabilities have been controlled at allowable levels with $n = 50$, the quality control manager is now justified in using the *accept H_0* and *reject H_0* statements for the hypothesis test. The accompanying inferences are made with allowable probabilities of making Type I and Type II errors.

We can make three observations about the relationship among α, β and the sample size n:

1 Once two of the three values are known, the other can be computed.
2 For a given level of significance α, increasing the sample size will reduce β.
3 For a given sample size, decreasing α will increase β, whereas increasing α will decrease β.

The third observation should be kept in mind when the probability of a Type II error is not being controlled. Inexperienced users of hypothesis testing often think that smaller values of α are necessarily better. They are better if we are concerned only about making a Type I error. However, for a given sample size, smaller values of α have the disadvantage of increasing the probability of making a Type II error.

EXERCISES

Methods

33. Consider the following hypothesis test:

$$H_0: \mu \geq 10$$
$$H_1: \mu < 10$$

The sample size is 120 and the population standard deviation is known, $\sigma = 5$. Use $\alpha = 0.05$.
  a. If $\mu = 9$, what is the probability of reaching the conclusion *do not reject* H_0?
  b. What type of error would be made if $\mu = 9$ and we conclude that $H_0: \mu \geq 10$ is true?
  c. Suppose the researcher wants to reduce the probability of a Type II error to 0.10 when $\mu = 9$. What sample size is recommended?
  d. What is the probability of making a Type II error if $\mu = 8$?

34. Consider the following hypothesis test:

$$H_0: \mu = 20$$
$$H_1: \mu \neq 20$$

A sample of 200 items will be taken. The population standard deviation is $\sigma = 10$. Use $\alpha = 0.05$. Compute the probability of making a Type II error if the population mean is:
  a. $\mu = 18.0$.
  b. $\mu = 22.5$.
  c. $\mu = 21.0$.
  d. How large a sample should be taken if the researcher is willing to accept a 0.05 probability of making a Type II error when $\mu = 22.0$?

Applications

35. Insight Marketing Research bases charges to a client on the assumption that telephone survey interviews can be completed within 15 minutes or less. If more time is required, a premium rate is charged. With a sample of 35 interviews, a population standard deviation of 4 min, and a level of significance of 0.01, the sample mean will be used to test the null hypothesis $H_0: \mu \leq 15$.
  a. What is your interpretation of the Type II error for this problem? What is its impact on the firm?
  b. What is the probability of making a Type II error when $\mu = 17$ min?
  c. What is the probability of making a Type II error when $\mu = 18$ min?
  d. Sketch the general shape of the power curve for this test.

36. Refer to Exercise 35. Assume the firm selects a sample of 50 interviews and repeat parts (b) and (c). What observation can you make about how increasing the sample size affects the probability of making a Type II error?

37. Millennials magazine states the following hypotheses about the mean age of its subscribers. The population standard deviation is assumed known at $\sigma = 6$ years. The test will be done using $\alpha = 0.05$.

$$H_0: \mu = 28$$

$$H_1: \mu \neq 28$$

   a. What would it mean to make a Type II error in this situation?
   b. If sample size is 100, what is the probability of accepting H_0 for μ equal to 26, 27, 29 and 30?
   c. What is the power of the test at $\mu = 26$? What does this result tell you?
   d. If the manager conducting the test will permit a 0.15 probability of making a Type II error when $\mu = 29$, what sample size should be used?

38. Ethical Investments offered a workplace pension scheme for the employees of a particular company. Ethical estimates that the employees are currently averaging €100 or less per month in pension contributions. A sample of 40 employees will be used to test Ethical's assumption about the current level of pension saving. Assume the employee monthly pension contributions have a standard deviation of €75 and that a 0.05 level of significance will be used in the hypothesis test.
   a. What would it mean to make a Type II error in this situation?
   b. What is the probability of Type II error if the actual mean employee monthly pension contribution is €120?
   c. What is the probability of the Type II error if the actual mean employee monthly pension contribution is €130?
   d. Assume a sample size of 80 employees is used and repeat parts b. and c.

39. A special industrial battery must have a mean life of at least 400 hours. A hypothesis test is to be done with a 0.02 level of significance. If the batteries from a particular production run have an actual mean life of 385 hours, the production manager wants a sampling procedure that only 10 per cent of the time would show erroneously that the batch is acceptable. What sample size is recommended for the hypothesis test? Use 30 hours as an estimate of the population standard deviation.

40. HealthSnacks packs organic roasted peanuts. The hypotheses $H_0: \mu = 120$ and $H_1: \mu \neq 120$ are used to test whether the production process is meeting the target of 120 nuts per pack. Use a 0.05 level of significance for the test and a planning value of 5 for the standard deviation.
   a. If the mean drops to 117 nuts per pack, the firm wants to have a 98 per cent chance of concluding that the target is not being met. How large a sample should be selected?
   b. With your sample size from part (a), what is the probability of concluding that the process is operating satisfactorily, and hence making a Type II error, for each of the following actual mean values: 117, 118, 119, 121, 122 and 123 nuts per pack?

9.7 BIG DATA AND HYPOTHESIS TESTING

We have seen that interval estimates of the population mean μ and the population proportion π narrow as the sample size increases. This occurs because the standard error of the associated sampling distributions decreases as the sample size increases. Now consider the relationship between interval estimation and hypothesis testing that we discussed earlier in this chapter. If we construct a $100(1 - \alpha)\%$ interval estimate for the population mean, we reject $H_0: \mu = \mu_0$ if the $100(1 - \alpha)\%$ interval estimate does not contain μ_0. Thus, for a given level of confidence, as the sample size increases we will reject $H_0: \mu = \mu_0$ for increasingly smaller differences between the sample mean $\overline{X}$ and the hypothesized population mean μ_0. We can see that when the sample size n is very large, almost any difference between the sample mean $\overline{X}$ and the hypothesized population mean μ_0 results in rejection of the null hypothesis.

Big data, hypothesis testing and *p*-values

In this section, we will elaborate how big data affects hypothesis testing and the magnitude of *p*-values. Specifically, we will examine how rapidly the *p*-value associated with a given difference between a point estimate and a hypothesized value of a parameter decreases as the sample size increases.

Let us consider the online news service PenningtonDailyTimes.com (PDT). PDT's primary source of revenue is the sale of advertising, and prospective advertisers are willing to pay a premium to advertise on websites that have long visit times. To promote its news service, PDT's management wants to promise potential advertisers that the mean time spent by customers when they visit the PDT website is greater than last year, that is, more than 84.0 seconds. PDT therefore decides to collect a sample tracking the amount of time spent by individual customers when they visit the PDT website in order to test its null hypothesis $H_0: \mu \leq 84.0$.

For a sample mean of 84.1 seconds and a sample standard deviation of $s = 20$ seconds, Table 9.6 provides the values of the test statistic t and the *p*-values for the test of the null hypothesis $H_0: \mu \leq 84.0$. The *p*-value for this hypothesis test is essentially 0 for all samples in Table 9.6 with at least $n = 1,000,000$.

PDT's management also wants to promise potential advertisers that the proportion of its website visitors who click on an ad this year exceeds the proportion of its website visitors who clicked on an ad last year, which was 0.50. PDT collects information from its sample on whether the visitor to its website clicked on any of the ads featured on the website, and it wants to use these data to test its null hypothesis $H_0: \pi \leq 0.50$.

For a sample proportion of 0.51, Table 9.7 provides the values of the test statistic z and the *p*-values for the test of the null hypothesis $H_0: \pi \leq 0.50$. The *p*-value for this hypothesis test is essentially 0 for all samples in Table 9.7 with at least $n = 100,000$.

We see in Tables 9.6 and 9.7 that the *p*-value associated with a given difference between a point estimate and a hypothesized value of a parameter decreases as the sample size increases. As a result, if the sample mean time spent by customers when they visit PDT's website is 84.1 seconds, PDT's null hypothesis $H_0: \mu \leq 84.0$ is not rejected at $\alpha = 0.01$ for samples with $n \leq 100,000$, and is rejected at $\alpha = 0.01$ for samples with $n \geq 1,000,000$.

Similarly, if the sample proportion of visitors to its website clicked on an ad featured on the website is 0.51, PDT's null hypothesis $H_0: \pi \leq 0.50$ is not rejected at $\alpha = 0.01$ for samples with $n \leq 10,000$, and is rejected at $\alpha = 0.01$ for samples with $n \geq 100,000$. In both instances, as the sample size becomes extremely large the *p*-value associated with the given difference between a point estimate and the hypothesized value of the parameter becomes extremely small.

TABLE 9.6 Values of the test statistic t and the *p*-values for the test of the null hypothesis $H_0: \mu \leq 84.0$ and sample mean $\bar{x} = 84.1$ seconds for various sample sizes n

Sample size n	t	p-value
10	0.01581	0.49386
100	0.05000	0.48011
1,000	0.15811	0.43720
10,000	0.50000	0.30854
100,000	1.58114	0.05692
1,000,000	5.00000	2.87E-07
10,000,000	15.81139	1.30E-56
100,000,000	50.00000	0.00E+00
1,000,000,000	158.11388	0.00E+00

TABLE 9.7 Values of the test statistic z and the p-values for the test of the null hypothesis $H_0: \pi \leq 0.50$ and sample proportion $p = 0.51$ for various sample sizes n

Sample size n	z	p-value
10	0.06325	0.47479
100	0.20000	0.42074
1,000	0.63246	0.26354
10,000	2.00000	0.02275
100,000	6.32456	1.27E-10
1,000,000	20.00000	0.00E+00
10,000,000	63.24555	0.00E+00
100,000,000	200.00000	0.00E+00
1,000,000,000	632.45553	0.00E+00

Implications of big data in hypothesis testing

Suppose PDT collects a sample of 1,000,000 visitors to its website and uses these data to test its null hypotheses $H_0: \mu \leq 84.0$ and $H_0: \pi \leq 0.50$ at the 0.05 level of significance. The sample mean is 84.1 and the sample proportion is 0.51, so the null hypothesis is rejected in both tests, as Tables 9.6 and 9.7 show. As a result, PDT can promise potential advertisers that the mean time spent by individual customers who visit PDT's website exceeds 84.0 seconds and the proportion of individual visitors to its website who click on an ad exceeds 0.50. These results suggest that for each of these hypothesis tests, the difference between the point estimate and the hypothesized value of the parameter being tested is unlikely to be solely a consequence of sampling error. However, the results of any hypothesis test, no matter the sample size, are only reliable if the sample is relatively free of non-sampling error. If non-sampling error is introduced in the data collection process, the likelihood of making a Type I or Type II error may be higher than if the sample data are free of non-sampling error. Therefore, when testing a hypothesis, it is always important to think carefully about whether a random sample of the population of interest has been taken.

If PDT determines that it has introduced little or no non-sampling error into its sample data, the only remaining plausible explanation for these results is that these null hypotheses are false. At this point, PDT and the companies that advertise on the PDT website should also consider whether these statistically significant differences between the point estimates and the hypothesized values of the parameters being tested are of practical significance. Although a 0.1 second increase in the mean time spent by customers when they visit PDT's website is statistically significant, it may not be meaningful to companies that might advertise on the PDT website. Similarly, although an increase of 0.01 in the proportion of visitors to its website that click on an ad is statistically significant, it may not be meaningful to companies that might advertise on the PDT website.

Ultimately, no business decision should be based solely on statistical inference. Practical significance should always be considered in conjunction with statistical significance. This is particularly important when the hypothesis test is based on an extremely large sample because even an extremely small difference between the point estimate and the hypothesized value of the parameter being tested will be statistically significant. When done properly, statistical inference provides evidence that should be considered in combination with information collected from other sources to make the most informed decision possible.

EXERCISES

Methods

41. The government wants to determine if the mean number of business emails sent and received per business day by its employees differs from the mean number of emails sent and received per day by corporate employees, which is 101.5. Suppose the government electronically collects information on the number of business emails sent and received on a randomly selected business day over the past year from each of 10,163 randomly selected government employees. The results are provided in the file 'Email'. Test the government's hypothesis at $\alpha = 0.01$. Discuss the practical significance of the results.

EMAIL

42. CEOs who belong to a popular business-oriented social networking service have an average of 930 connections. Do other members have fewer connections than CEOs? The number of connections for a random sample of 7,515 members who are not CEOs is provided in the file 'SocialNetwork'. Using this sample, test the hypothesis that other members have fewer connections than CEOs at $\alpha = 0.01$. Discuss the practical significance of the results.

SOCIAL NETWORK

43. The American Potato Growers Association (APGA) would like to test the claim that the proportion of fast-food orders this year that include French fries exceeds the proportion of fast-food orders that included French fries last year. Suppose that a random sample of 49,581 electronic receipts for fast-food orders placed this year shows that 31,038 included French fries. Assuming that the proportion of fast-food orders that included French fries last year is 0.62, use this information to test APGA's claim at $\alpha = 0.05$. Discuss the practical significance of the results.

44. According to CNN, 55 per cent of all US smartphone users have used their GPS capability to get directions. Suppose a major provider of wireless telephone services in Canada wants to know how GPS usage by its customers compares with US smartphone users. The company collects usage records for this year for a random sample of 547,192 of its Canadian customers and determines that 302,050 of these customers have used their smartphone's GPS capability this year. Use this data to test whether Canadian smartphone users' GPS usage differs from US smartphone users' GPS usage at $\alpha = 0.01$. Discuss the practical significance of the results.

ONLINE RESOURCES

For the data files, additional questions and answers, and software section for Chapter 9, go to the online platform.

SUMMARY

Hypothesis testing uses sample data to determine whether a statement about the value of a population parameter should or should not be rejected. The hypotheses are two competing statements about a population parameter. One is called the null hypothesis (H_0), and the other is called the alternative hypothesis (H_1).

In all hypothesis tests, a relevant test statistic is calculated using sample data. The test statistic can be used to compute a p-value for the test. A p-value is a probability that gives an indication of the degree to which the sample results disagree with the expectations generated by the null hypothesis. If the p-value is less than or equal to the level of significance, the null hypothesis can be rejected.

(Continued)

Conclusions can also be drawn by comparing the value of the test statistic to a critical value. For lower-tail tests, the null hypothesis is rejected if the value of the test statistic is less than or equal to the critical value. For upper-tail tests, the null hypothesis is rejected if the value of the test statistic is greater than or equal to the critical value. Two-tailed tests involve two critical values: one in the lower tail of the sampling distribution and one in the upper tail. In this case, the null hypothesis is rejected if the value of the test statistic is less than or equal to the critical value in the lower tail or greater than or equal to the critical value in the upper tail.

We illustrated the relationship between hypothesis testing and interval construction in Section 9.3.

When historical data or other information provide a basis for assuming that the population standard deviation is known, the hypothesis testing procedure is based on the standard normal distribution. When σ is unknown, the sample standard deviation s is used to estimate σ and the hypothesis testing procedure is based on the t distribution.

In the case of hypothesis tests about a population proportion, the hypothesis testing procedure uses a test statistic based on the standard normal distribution.

Extensions of hypothesis testing procedures to include an analysis of the Type II error were also presented. In Section 9.6 we showed how to compute the probability of making a Type II error, and how to determine a sample size that will control for both the probability of making a Type I error and a Type II error.

KEY TERMS

Alternative hypothesis	**Power**
Critical value	**Power curve**
Level of significance	**Test statistic**
Null hypothesis	**Two-tailed test**
One-tailed test	**Type I error**
p-value	**Type II error**

KEY FORMULAE

Test statistic for hypothesis tests about a population mean: σ known

$$z = \frac{\bar{x} - \mu_0}{\sigma/\sqrt{n}} \tag{9.1}$$

Test statistic for hypothesis tests about a population mean: σ unknown

$$t = \frac{\bar{x} - \mu_0}{s/\sqrt{n}} \tag{9.4}$$

Test statistic for hypothesis tests about a population proportion

$$z = \frac{p - \pi_0}{\sqrt{\dfrac{\pi_0(1 - \pi_0)}{n}}} \tag{9.6}$$

Sample size for a one-tailed hypothesis test about a population mean

$$n = \frac{(z_\alpha + z_\beta)^2 \sigma^2}{(\mu_0 - \mu_1)^2} \tag{9.9}$$

In a two-tailed test, replace z_α with $z_{\alpha/2}$.

CASE PROBLEM 1

Quality Associates

The consultancy Quality Associates (QA) advises its clients about sampling and statistical procedures that can be used to control their manufacturing processes. One particular client gave QA a sample of 1,000 observations taken during a time when the client's process was operating satisfactorily. The sample standard deviation for these data was 0.21; hence, with so much data, the population standard deviation was assumed to be 0.21. QA then suggested that random samples of size 30 be taken periodically to monitor the process on an ongoing basis. By analyzing the new samples, the client could quickly learn whether the process was operating satisfactorily, and if necessary take corrective action.

The design specification indicated that the mean for the process should be 12.0. The hypothesis test suggested by QA is as follows:

$$H_0: \mu = 12.0$$

$$H_1: \mu \neq 12.0$$

Corrective action will be taken any time H_0 is rejected.

The data set 'Quality' on the online platform contains data from four samples, each of size 30, collected at hourly intervals during the first day of operation of the new statistical control procedure.

Managerial report

1. Do a hypothesis test for each sample, using $\alpha = 0.01$, and determine whether corrective action should be taken. Provide the test statistic and p-value for each test.

2. Compute the standard deviation for each of the four samples. Does the assumption of 0.21 for the population standard deviation appear reasonable?

3. Compute limits for the sample mean $\bar{X}$ around $\mu = 12.0$ such that, as long as a new sample mean is within those limits, the process will be considered to be operating satisfactorily. These limits are referred to as upper and lower control limits for quality control purposes.

4. Discuss the implications of changing the level of significance α to a larger value. What mistake or error could increase if the level of significance is increased?

QUALITY

CASE PROBLEM 2

Ethical behaviour of business students at the World Academy

During the 2008/09 global recession, there were accusations that various bank directors, financial managers and other corporate officers were guilty of unethical behaviour. An article in the *Chronicle of Higher Education* (10 February 2009) suggested that such unethical business behaviour might stem in part from the fact that cheating has become more prevalent among business students. The article reported that 56 per cent of business students admitted to cheating at some time during their academic career, as compared with 47 per cent of non-business students.

Cheating has been a concern of the Dean of the Business Faculty at the World Academy (WA) for several years. Some faculty members believe that cheating is more widespread at the WA than at other universities. To investigate some of these issues, the Dean commissioned a study to assess the current ethical behaviour of business students at the WA. As part of this study, an anonymous exit survey was administered to a sample of 90 business students from this year's graduating class. YES/NO responses to the following questions were used to obtain data regarding three types of cheating.

During your time at the WA, did you ever present work copied off the internet as your own?

During your time at the WA, did you ever copy answers off another student's exam?

During your time at the WA, did you ever collaborate with other students on projects that were supposed to be completed individually.

WORLD
ACADEMY

(Continued)

If a student answered YES to one or more of these questions, they were considered to have been involved in some type of cheating. Below is a table showing a portion of the data.

The complete data set is in the file named 'World Academy' on the accompanying online platform.

Managerial report

Prepare a report for the Dean of the Faculty that summarizes your assessment of the nature of cheating by business students at WA. Be sure to include the following items in your report.

1. Use descriptive statistics to summarize the data and comment on your findings.

2. Construct 95% confidence intervals for the proportion of all students, the proportion of male students and the proportion of female students who were involved in some type of cheating.

3. Do a hypothesis test to determine if the proportion of business students at WA who were involved in some type of cheating is less than that of business students at other institutions, as reported by the *Chronicle of Higher Education*.

WORLD ACADEMY

	Student	Copied_Internet	Copied_Exam	Collaborated	Gender
1	1	No	No	No	Female
2	2	No	No	No	Male
3	3	Yes	No	Yes	Male
4	4	Yes	Yes	No	Male
5	5	No	No	Yes	Male
6	6	Yes	No	No	Female
7	7	Yes	Yes	Yes	Female
8	8	Yes	Yes	Yes	Male
9	9	No	No	No	Male
10	10	Yes	No	No	Female

4. Do a hypothesis test to determine if the proportion of business students at WA who were involved in some form of cheating is less than that of non-business students at other institutions, as reported by the *Chronicle of Higher Education*.

5. What advice would you give to the Dean based on your analysis of the data?

© mediaphotos/iStock

CASE PROBLEM 3

New product development at Baxter Barnes Gelato

Baxter Barnes Gelato (BBG) manufactures and distributes 15 flavours of premium gelato. BBG is known by consumers for its consistent high quality and unique flavours. Its current line of flavours includes ricotta fig, coriander kiwi, clementine mint, gooseberry cobbler and ghost-peppered caramel. BBG adds, at most, one new flavour every two years. It does so only if consumer testing of any of the new flavours it has recently developed in its lab achieves a favourable test response. As the scheduled date on which BBG is to announce its newest flavour approaches, social media typically fills with rumours about the new flavour and consumers of premium ice cream expectantly await BBG's announcement. BBG is due to make an announcement on its newest flavour shortly.

Through its recent introductions of new flavours, BBG management has learned that a new flavour will likely fail to meet the company's sales expectations unless more than three-quarters of premium ice cream consumers indicate, after tasting a sample, that they will consider purchasing the flavour. This year BBG's lab has developed three new flavours: champagne grape, basil cantaloupe and jalapeno

lime. They produced batches of each new flavour and distributed the batches around the nation for use in consumer testing. BBG was ultimately able to secure ratings for each of the three new flavours from a random sample of 21,478 consumers of premium ice cream. Each consumer in the test was provided with a sample of the three new flavours of gelato. For each flavour of gelato, the consumer was asked if they would consider purchasing the flavour. The data collected through this consumer testing are provided in the file 'BBG'.

BBG

Managerial report

1. For the new gelato flavours recently created by BBG's lab, what are the point estimates for the proportion of consumers of premium ice-cream who indicate they would consider purchasing the respective flavours? Explain to BBG management why these results are or are not sufficient for them to make their decision on whether to add any of these new flavours to their existing product line.

2. For each new flavour, conduct a hypothesis test at $\alpha = 0.05$ to determine whether more than three-quarters of premium ice-cream consumers will consider purchasing that new flavour.

3. Interpret the results of your hypothesis tests in Task 2 for BBG management. Which of the new flavours do the results suggest BBG management should add to its current line of flavours? How do you suggest that BBG management proceed?

10

Statistical Inference about Means and Proportions with Two Populations

CHAPTER CONTENTS

Statistics in Practice   Do you prefer ABC 9 or ABC 90?

LEARNING OBJECTIVES   After reading this chapter and doing the exercises, you should be able to:

1 Construct and interpret confidence intervals and hypothesis tests for the difference between two population means, given independent samples from the two populations:

   1.1   When the standard deviations of the two populations are known.

   1.2   When the standard deviations of the two populations are unknown.

2 Construct and interpret confidence intervals and hypothesis tests for the difference between two population means, given matched samples from the two populations.

3 Construct and interpret confidence intervals and hypothesis tests for the difference between two population proportions, given large independent samples from the two populations.

I n Chapters 8 and 9 we showed how to construct interval estimates and do hypothesis tests for a single population mean or a single population proportion. In this chapter we discuss how to construct interval estimates and do hypothesis tests when the focus is on the *difference* between two population means or two population proportions. This is a very common research question. For example, we may want to construct an interval estimate of the difference between the mean starting salary for a population of men and the mean starting salary for a population of women. Or we may want to do a hypothesis test to determine whether there is any difference between the proportion of defective parts in a population of parts produced by supplier A and the proportion of defective parts in a population of parts produced by supplier B.

We begin by showing how to construct interval estimates and do hypothesis tests for the difference between two population means when the population standard deviations are assumed known.

STATISTICS IN PRACTICE
Do you prefer ABC 9 or ABC 90?

HP 600 and BMW 535i are examples of *alphanumeric* brand names: they comprise a sequence of letters and numbers. In a paper in the journal *Social Behaviour and Personality*, three researchers from universities in Wuhan, China, reported on studies into the effect on consumer preferences of number size in alphanumeric brand names (Feng, Wang and Rui, 2019). They reviewed previous research indicating that larger numbers receive more favourable consumer evaluations than smaller numbers. However, they hypothesized that this effect may be reversed for luxury brands.

Their paper reported on two experimental studies designed to test their hypothesis. In the first study, participants were divided randomly into two groups of about 30. They were given information about a 'target', fictitious luxury perfume brand with the name PIW-00, followed by a number, and about a 'reference' brand PIW-0052. They were then asked to give a rating of the target brand on a scale from 1 to 7, with higher ratings indicating stronger preference. For the high-number group, the target brand was named PIW-0093; for the low-number group the target brand was named PIW-009. The mean preference

rating for the high-number group was 4.56, compared with a mean of 5.03 for the low-number group. This difference was declared statistically significant beyond the 5 per cent level, using a statistical hypothesis test known as the independent-samples *t* test. This result offered support to the researchers' hypothesis.

In this chapter, you will learn how to construct interval estimates and do hypothesis tests about means and proportions with two populations. The independent-samples *t* test used in the consumer preferences research is an example of such a test.

Source:
Feng, W., Wang, T. & Rui, G. (2019) Influence of number magnitude in luxury brand names on consumer preference. *Social Behaviour and Personality*, 47(5). Available at doi.org.10.2224/sbp.7486, accessed September 2022.

10.1 INFERENCES ABOUT THE DIFFERENCE BETWEEN TWO POPULATION MEANS: σ_1 AND σ_2 KNOWN

Let μ_1 denote the mean of population 1 and μ_2 denote the mean of population 2. We focus on inferences about the difference between the means: $\mu_1 - \mu_2$. We select a simple random sample of n_1 units from population 1 and a second simple random sample of n_2 units from population 2. The two samples, taken separately and randomly from each population, are called independent simple random samples. In this section, we assume the two population standard deviations, σ_1 and σ_2, are known prior to collecting the samples. We refer to this as the σ_1 and σ_2 known case. In the following example we show how to compute a margin of error and construct an interval estimate of the difference between the two population means in the σ_1 and σ_2 known case. In practice, this is a relatively uncommon situation, but it will lead us to the more frequently encountered case when σ_1 and σ_2 are unknown.

Interval estimation of $\mu_1 - \mu_2$

Suppose a retailer such as Currys.com (selling TVs, tablets, computers and other consumer electronics) operates two stores in Dublin, Ireland. One store is in the inner city and the other is in an out-of-town retail park. The regional manager noticed that products selling well in one store do not always sell well in the other. The manager believes this may be due to differences in customer demographics at the two locations. Customers may differ in age, education, income and so on. Suppose the manager asks us to investigate the difference between the mean ages of the customers who shop at the two stores.

Let us define population 1 as all customers who shop at the inner-city store and population 2 as all customers who shop at the out-of-town store.

$$\mu_1 = \text{mean age of population 1}$$
$$\mu_2 = \text{mean age of population 2}$$

The difference between the two population means is $\mu_1 - \mu_2$. To estimate $\mu_1 - \mu_2$, we shall select simple random samples of n_1 and n_2 customers from populations 1 and 2 respectively. We then compute the two sample means.

$\bar{x}_1 = $ sample mean age for the simple random sample of n_1 inner-city customers

$\bar{x}_2 = $ sample mean age for the simple random sample of n_2 out-of-town customers

The point estimator of the difference between the two populations is the difference between the sample means.

Point estimator of the difference between two population means

$$\bar{X}_1 - \bar{X}_2 \tag{10.1}$$

Figure 10.1 provides a schematic overview of the process used to estimate the difference between two population means based on two independent simple random samples.

FIGURE 10.1

Estimating the difference between two population means

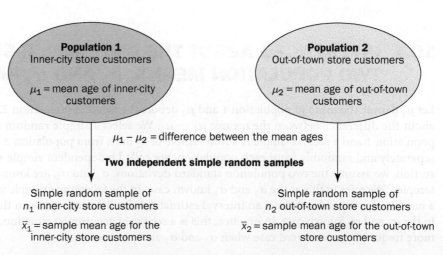

Population 1
Inner-city store customers

$\mu_1 = $ mean age of inner-city customers

Population 2
Out-of-town store customers

$\mu_2 = $ mean age of out-of-town customers

$\mu_1 - \mu_2 = $ difference between the mean ages

Two independent simple random samples

Simple random sample of n_1 inner-city store customers

Simple random sample of n_2 out-of-town store customers

$\bar{x}_1 = $ sample mean age for the inner-city store customers

$\bar{x}_2 = $ sample mean age for the out-of-town store customers

$\bar{x}_1 - \bar{x}_2 = $ point estimate of $\mu_1 - \mu_2$

Like other point estimators, the point estimator $\overline{X}_1 - \overline{X}_2$ has a standard error that summarizes the variation in the sampling distribution of the estimator. With two independent simple random samples, the standard error of $\overline{X}_1 - \overline{X}_2$ is as follows:

Standard error of $\overline{X}_1 - \overline{X}_2$

$$\sigma_{\overline{X}_1 - \overline{X}_2} = \sqrt{\frac{\sigma_1^2}{n_1} + \frac{\sigma_2^2}{n_2}} \qquad (10.2)$$

If both populations have a normal distribution, or if the sample sizes are large enough to use a normal approximation for the sampling distributions of $\overline{X}_1$ and $\overline{X}_2$, the sampling distribution of $\overline{X}_1 - \overline{X}_2$ will be normal (or approximately normal), with a mean of $\mu_1 - \mu_2$.

As we showed in Chapter 8, an interval estimate is given by a point estimate $\pm$ a margin of error. In the case of estimating the difference between two population means, an interval estimate will take the form $(\overline{x}_1 - \overline{x}_2) \pm$ margin of error. When the sampling distribution of $\overline{X}_1 - \overline{X}_2$ is a normal distribution, we can write the margin of error as follows:

$$\text{Margin of error} = z_{\alpha/2}\sigma_{\overline{X}_1 - \overline{X}_2} = z_{\alpha/2}\sqrt{\frac{\sigma_1^2}{n_1} + \frac{\sigma_2^2}{n_2}} \qquad (10.3)$$

Therefore, the interval estimate of the difference between two population means is as follows:

Interval estimate of the difference between two population means: σ_1 and σ_2 known

$$(\overline{x}_1 - \overline{x}_2) \pm z_{\alpha/2}\sqrt{\frac{\sigma_1^2}{n_1} + \frac{\sigma_2^2}{n_2}} \qquad (10.4)$$

where $1 - \alpha$ is the confidence coefficient.

We return to the example of the Dublin retailer. Based on data from previous customer demographic studies, the two population standard deviations are assumed known, $\sigma_1 = 9$ years and $\sigma_2 = 10$ years. The data collected from the two independent simple random samples of the retailer's customers provided the following results:

	Inner-city store	Out-of-town store
Sample size	$n_1 = 36$	$n_2 = 49$
Sample mean	$\overline{x}_1 = 36$ years	$\overline{x}_2 = 35$ years

Using expression (10.1), we find that the point estimate of the difference between the mean ages of the two populations is $\overline{x}_1 - \overline{x}_2 = 40 - 35 = 5$ years. We estimate that the mean age of customers at the inner-city store is five years greater than the mean age of the out-of-town customers. We can now use expression (10.4) to compute the margin of error and provide the interval estimate of $\mu_1 - \mu_2$. Using 95% confidence and $z_{\alpha/2} = z_{0.025} = 1.96$, we have:

$$(\overline{x}_1 - \overline{x}_2) \pm z_{\alpha/2}\sqrt{\frac{\sigma_1^2}{n_1} + \frac{\sigma_2^2}{n_2}} = (40 - 35) \pm 1.96\sqrt{\frac{(9)^2}{36} + \frac{(10)^2}{49}} = 5 \pm 4.1$$

The margin of error is 4.1 years and the 95% confidence interval estimate of the difference between the two population means is $5 - 4.1 = 0.9$ years to $5 + 4.1 = 9.1$ years.

Hypothesis tests about $\mu_1 - \mu_2$

Let us consider hypothesis tests about the difference between two population means. Using D_0 to denote the hypothesized difference between μ_1 and μ_2, the three forms for a hypothesis test are as follows:

$$
\begin{array}{lll}
H_0: \mu_1 - \mu_2 \geq D_0 & H_0: \mu_1 - \mu_2 \leq D_0 & H_0: \mu_1 - \mu_2 = D_0 \\
H_1: \mu_1 - \mu_2 < D_0 & H_1: \mu_1 - \mu_2 > D_0 & H_1: \mu_1 - \mu_2 \neq D_0
\end{array}
$$

In most applications, $D_0 = 0$. Using the two-tailed test as an example, when $D_0 = 0$ the null hypothesis is $H_0: \mu_1 - \mu_2 = 0$, that is, the null hypothesis is that μ_1 and μ_2 are equal. Rejection of H_0 leads to the conclusion that $H_1: \mu_1 - \mu_2 \neq 0$ is true, i.e. μ_1 and μ_2 are not equal.

The steps presented in Chapter 9 for doing hypothesis tests are applicable here. We choose a level of significance α, compute the value of the test statistic and find the p-value to determine whether the null hypothesis should be rejected. With two independent simple random samples, we showed that the point estimator $\overline{X}_1 - \overline{X}_2$ has a standard error $\sigma_{\overline{X}_1 - \overline{X}_2}$ given by equation (10.2), and the distribution of $\overline{X}_1 - \overline{X}_2$ can be described by a normal distribution for sufficiently large samples or for normally distributed populations. In this case, the test statistic for the difference between two population means when σ_1 and σ_2 are known is as follows:

Test statistic for hypothesis tests about $\mu_1 - \mu_2$: σ_1 and σ_2 known

$$
z = \frac{(\overline{x}_1 - \overline{x}_2) - D_0}{\sqrt{\dfrac{\sigma_1^2}{n_1} + \dfrac{\sigma_2^2}{n_2}}}
$$

(10.5)

Here is an example. As part of a study to evaluate differences in education quality between two training centres, a standardized examination is given to individuals trained there. The difference between the mean examination scores is used to assess quality differences between the centres. The population means for the two centres are as follows:

μ_1 = the mean examination score for the population of individuals trained at centre A

μ_2 = the mean examination score for the population of individuals trained at centre B

We begin with the tentative assumption that no difference exists between the average training quality provided at the two centres. In terms of the mean examination scores, the null hypothesis is that $\mu_1 - \mu_2 = 0$. If sample evidence leads to the rejection of this hypothesis, we shall conclude that the mean examination scores differ for the two populations. This conclusion indicates a quality differential between the two centres and suggests that a follow-up study investigating the reason for the differential may be warranted. The null and alternative hypotheses for this two-tailed test are written as follows:

$$
H_0: \mu_1 - \mu_2 = 0
$$
$$
H_1: \mu_1 - \mu_2 \neq 0
$$

EXAMSCORES

The standardized examination given previously in a variety of settings always resulted in an examination score standard deviation near 10 points. We shall use this information to assume that the population standard deviations are known with $\sigma_1 = \sigma_2 = 10$. An $\alpha = 0.05$ level of significance is specified for the test.

Independent simple random samples of $n_1 = 30$ individuals from training centre A and $n_2 = 40$ individuals from training centre B are taken. The respective sample means are $\bar{x}_1 = 82$ and $\bar{x}_2 = 78$. Do these data suggest a difference between the population means at the two training centres? To help answer this question, we compute the test statistic using equation (10.5):

$$z = \frac{(\bar{x}_1 - \bar{x}_2) - D_0}{\sqrt{\dfrac{\sigma_1^2}{n_1} + \dfrac{\sigma_2^2}{n_2}}} = \frac{(82 - 78) - 0}{\sqrt{\dfrac{(10)^2}{30} + \dfrac{(10)^2}{40}}} = 1.66$$

Next, we compute the p-value for this two-tailed test. As the test statistic z is in the upper tail, we first find the probability to the right of $z = 1.66$. Using the standard normal distribution table, the cumulative probability for $z = 1.66$ is 0.9515, so the area in the upper tail is $1 - 0.9515 = 0.0485$. Because this test is a two-tailed test, we must double the tail area: p-value $= 2(0.0485) = 0.0970$. Following the usual rule to reject H_0 if p-value $\leq \alpha$, we see that the p-value of 0.0970 does not allow us to reject H_0 at the 0.05 level of significance. The sample results do not provide sufficient evidence to conclude that the training centres differ in quality.

In this chapter we shall use the p-value approach to hypothesis testing as described in Chapter 9. However, if you prefer, the test statistic and the critical value rejection rule can be used. With $\alpha = 0.05$ and $z_{\alpha/2} = z_{0.025} = 1.96$, the rejection rule using the critical value approach would be to reject H_0 if $z \leq -1.96$ or if $z \geq 1.96$. With $z = 1.66$, we reach the same 'do not reject H_0' conclusion.

In the preceding example, we demonstrated a two-tailed hypothesis test about the difference between two population means. Lower-tail and upper-tail tests can also be considered. These tests use the same test statistic as given in equation (10.5). The procedure for computing the p-value, and the rejection rules for these one-tailed tests, follow the same principles as those presented in Chapter 9.

Practical advice

In most applications of the interval estimation and hypothesis testing procedures presented in this section, random samples with $n_1 \geq 30$ and $n_2 \geq 30$ are adequate. In cases where either or both sample sizes are less than 30, the distributions of the populations become important considerations. In general, with smaller sample sizes, it is more important for the analyst to be satisfied that the distributions of the two populations are at least approximately normal.

EXERCISES

Methods

1. Consider the following results for two independent random samples taken from two populations.

Sample 1	Sample 2
$n_1 = 50$	$n_2 = 35$
$\bar{x}_1 = 13.6$	$\bar{x}_2 = 11.6$
$\sigma_1 = 2.2$	$\sigma_2 = 3.0$

   a. What is the point estimate of the difference between the two population means?
   b. Construct a 90% confidence interval for the difference between the two population means.
   c. Construct a 95% confidence interval for the difference between the two population means.

2. Consider the following hypothesis test.

$$H_0: \mu_1 - \mu_2 \leq 0$$

$$H_1: \mu_1 - \mu_2 > 0$$

The following results are for two independent samples taken from the two populations.

Sample 1	Sample 2
$n_1 = 40$	$n_2 = 50$
$\bar{x}_1 = 25.2$	$\bar{x}_2 = 22.8$
$\sigma_1 = 5.2$	$\sigma_2 = 6.0$

a. What is the value of the test statistic?
b. What is the p-value?
c. With $\alpha = 0.05$, what is your hypothesis testing conclusion?

3. Consider the following hypothesis test.

$$H_0: \mu_1 - \mu_2 = 0$$

$$H_1: \mu_1 - \mu_2 \neq 0$$

The following results are for two independent samples taken from the two populations.

Sample 1	Sample 2
$n_1 = 80$	$n_2 = 70$
$\bar{x}_1 = 104$	$\bar{x}_2 = 106$
$\sigma_1 = 8.4$	$\sigma_2 = 7.6$

a. What is the value of the test statistic?
b. What is the p-value?
c. With $\alpha = 0.05$, what is your hypothesis testing conclusion?

Applications

4. A study of wage differentials between men and women reported that one of the reasons wages for men are higher than wages for women is that men tend to have more years of work experience than women. Assume that the following sample summaries show the years of experience for each group.

Men	Women
$n_1 = 100$	$n_2 = 85$
$\bar{x}_1 = 14.9$ years	$\bar{x}_2 = 10.3$ years
$\sigma_1 = 5.2$ years	$\sigma_2 = 3.8$ years

a. What is the point estimate of the difference between the two population means?
b. At 95% confidence, what is the margin of error?
c. What is the 95% confidence interval estimate of the difference between the two population means?

5. The Dublin retailer age study (used as an example above) provided the following data on the ages of customers from independent random samples taken at the two store locations.

Inner-city store	Out-of-town store
$n_1 = 36$	$n_2 = 49$
$\bar{x}_1 = 40$ years	$\bar{x}_2 = 35$ years
$\sigma_1 = 9$ years	$\sigma_2 = 10$ years

a. State the hypotheses that could be used to investigate a difference between the population mean ages at the two stores.
   b. What is the value of the test statistic?
   c. What is the p-value?
   d. At $\alpha = 0.05$, what is your conclusion?

6. Consider the following results from a survey looking at how much people spend on gifts on Valentine's Day (14 February). The average expenditure of 40 men was €135.67, and the average expenditure of 30 women was €68.64. Based on past surveys, the standard deviation for men is assumed to be €35, and the standard deviation for women is assumed to be €20. Do men and women differ in the average amounts they spend?
   a. What is the point estimate of the difference between the population mean expenditure for men and the population mean expenditure for women?
   b. Calculate the margin of error at 99% confidence and construct a 99% confidence interval for the difference between the two population means.

10.2 INFERENCES ABOUT THE DIFFERENCE BETWEEN TWO POPULATION MEANS: σ_1 AND σ_2 UNKNOWN

In this section we consider inferences about $\mu_1 - \mu_2$ when the two population standard deviations are unknown. This is the situation more commonly encountered in practice. In this case, we use the sample standard deviations, s_1 and s_2, to estimate the unknown σ_1 and σ_2. The interval estimation and hypothesis testing procedures are based on the t distribution rather than the standard normal distribution.

Interval estimation of $\mu_1 - \mu_2$

Platinum Bank is conducting a study designed to identify differences between account practices by customers holding one of its two current account offerings: (1) its Current Interest account (interest paid on positive balances, small monthly account fee) and (2) its Free Banking account (no interest, no monthly fee). A simple random sample of 28 Current Interest account holders is selected and an independent simple random sample of 22 Free Banking current account holders is selected. The account balance is recorded for each of the accounts. A summary of the account balances follows:

CURR ACCT

	Current Interest	Free Banking
Sample size	$n_1 = 28$	$n_2 = 22$
Sample mean	$\bar{x}_1 = $ €1,025	$\bar{x}_2 = $ €910
Sample standard deviation	$s_1 = $ €150	$s_2 = $ €125

Platinum Bank would like to estimate the difference between the mean account balances maintained by the population of Current Interest customers and the population of Free Banking customers. In Section 10.1, we provided the following interval estimate for the case when the population standard deviations, σ_1 and σ_2, are known:

$$(\bar{x}_1 - \bar{x}_2) \pm z_{\alpha/2}\sqrt{\frac{\sigma_1^2}{n_1} + \frac{\sigma_2^2}{n_2}}$$

With σ_1 and σ_2 unknown, we shall use the sample standard deviations s_1 and s_2 to estimate σ_1 and σ_2 and replace $z_{\alpha/2}$ with $t_{\alpha/2}$. As a result, the interval estimate of the difference between two population means is given by the following expression:

Interval estimate of the difference between two population means: σ_1 and σ_2 unknown

$$(\bar{x}_1 - \bar{x}_2) \pm t_{\alpha/2}\sqrt{\frac{s_1^2}{n_1} + \frac{s_2^2}{n_2}} \tag{10.6}$$

where $1 - \alpha$ is the confidence coefficient.

In this expression, the use of the t distribution is an approximation, but it provides good results and is relatively easy to use. The only difficulty in using expression (10.6) is determining the appropriate degrees of freedom for $t_{\alpha/2}$. The formula used is as follows:

Degrees of freedom for the t distribution using two independent random samples

$$df = \frac{\left(\dfrac{s_1^2}{n_1} + \dfrac{s_2^2}{n_2}\right)^2}{\left(\dfrac{1}{n_1 - 1}\right)\left(\dfrac{s_1^2}{n_1}\right)^2 + \left(\dfrac{1}{n_2 - 1}\right)\left(\dfrac{s_2^2}{n_2}\right)^2} \tag{10.7}$$

We return to the Platinum Bank example. The sample data show $n_1 = 28$, $\bar{x}_1 = €1,025$, $s_1 = €150$ for the Current Interest sample, and $n_2 = 22$, $\bar{x}_2 = €910$, $s_2 = €125$ for the Free Banking sample. The calculation for degrees of freedom for $t_{\alpha/2}$ is as follows:

$$df = \frac{\left(\dfrac{s_1^2}{n_1} + \dfrac{s_2^2}{n_2}\right)^2}{\left(\dfrac{1}{n_1 - 1}\right)\left(\dfrac{s_1^2}{n_1}\right)^2 + \left(\dfrac{1}{n_2 - 1}\right)\left(\dfrac{s_2^2}{n_2}\right)^2} = \frac{\left(\dfrac{(150)^2}{28} + \dfrac{(125)^2}{22}\right)^2}{\left(\dfrac{1}{28 - 1}\right)\left(\dfrac{(150)^2}{28}\right)^2 + \left(\dfrac{1}{22 - 1}\right)\left(\dfrac{(125)^2}{22}\right)^2} = 47.8$$

We round the non-integer degrees of freedom *down* to 47 to provide a larger t-value and a slightly more conservative interval estimate. Using the t distribution table with 47 df, we find $t_{0.025} = 2.012$. Using expression (10.6), we construct the 95% confidence interval estimate of the difference between the two population means as follows:

$$(\bar{x}_1 - \bar{x}_2) \pm t_{\alpha/2}\sqrt{\frac{s_1^2}{n_1} + \frac{s_2^2}{n_2}} = (1,025 - 910) \pm 2.012\sqrt{\frac{(150)^2}{28} + \frac{(125)^2}{22}} = 115 \pm 78$$

The point estimate of the difference between the population mean account balances of the two types of account is € 115. The margin of error is € 78, and the 95% confidence interval estimate of the difference between the two population means is $115 - 78 = €37$ to $115 + 78 = €193$.

The computation of df (equation (10.7)) is cumbersome if you are doing the calculation by hand. Note that the terms s_1^2/n_1 and s_2^2/n_2 appear in both expression (10.6) and in equation (10.7), and therefore need to be computed only once to evaluate both (10.6) and (10.7). Statistical software packages compute the appropriate df automatically.

Hypothesis tests about $\mu_1 - \mu_2$

Let us now consider hypothesis tests for $\mu_1 - \mu_2$ when the population standard deviations σ_1 and σ_2 are unknown. Letting D_0 denote the hypothesized value for $\mu_1 - \mu_2$, Section 10.1 showed that the test statistic used for the case where σ_1 and σ_2 are known is as follows. The test statistic, Z, follows the standard normal distribution:

$$z = \frac{(\bar{x}_1 - \bar{x}_2) - D_0}{\sqrt{\dfrac{\sigma_1^2}{n_1} + \dfrac{\sigma_2^2}{n_2}}}$$

When σ_1 and σ_2 are unknown, we use s_1 as an estimate of σ_1 and s_2 as an estimate of σ_2. Substituting these sample standard deviations for σ_1 and σ_2 gives the following test statistic when σ_1 and σ_2 are unknown:

Test statistic for hypothesis tests about $\mu_1 - \mu_2$: σ_1 and σ_2 unknown

$$t = \frac{(\bar{x}_1 - \bar{x}_2) - D_0}{\sqrt{\dfrac{s_1^2}{n_1} + \dfrac{s_2^2}{n_2}}} \qquad \text{(10.8)}$$

The degrees of freedom for t are given by equation (10.7).

Consider an example involving a new piece of computer software intended to help app developers reduce the time required to design, construct and implement smartphone apps. To evaluate the benefits of the new software, two samples of app developers are randomly selected, each sample comprising 12 developers. The developers are given specifications for a hypothetical smartphone app. Then the developers in one sample are instructed to produce the app using current technology. The developers in the other sample are trained in the use of the new software package and instructed to use it to produce the smartphone app.

This study involves two populations: a conceptual population of app developers using the current technology and a conceptual population of app developers using the new software package. In terms of the time required to complete the smartphone app design project, the population means are as follows:

μ_1 = the mean project completion time for app developers using the current technology

μ_2 = the mean project completion time for app developers using the new software package

The researcher in charge of the new software evaluation project hopes to show that the new software package will provide a shorter mean project completion time, i.e. the researcher is looking for evidence to conclude that μ_2 is less than μ_1. In this case, $\mu_1 - \mu_2$ will be greater than zero. The research hypothesis $\mu_1 - \mu_2 > 0$ is stated as the alternative hypothesis. The hypothesis test becomes:

$$H_0: \mu_1 - \mu_2 \leq 0$$

$$H_1: \mu_1 - \mu_2 > 0$$

We shall use $\alpha = 0.05$ as the level of significance. Suppose that the 24 app developers in the study achieve the results shown in Table 10.1. Using the test statistic in equation (10.8), we have:

$$t = \frac{(\bar{x}_1 - \bar{x}_2) - D_0}{\sqrt{\dfrac{s_1^2}{n_1} + \dfrac{s_2^2}{n_2}}} = \frac{(325 - 286) - 0}{\sqrt{\dfrac{(40)^2}{12} + \dfrac{(44)^2}{12}}} = 2.272$$

SOFTWARE
TEST

TABLE 10.1 Completion time data and summary statistics for the software testing study

	Current technology	New software
	300	274
	280	220
	344	308
	385	336
	372	198
	360	300
	288	315
	321	258
	376	318
	290	310
	301	332
	283	263
Summary statistics		
Sample size	$n_1 = 12$	$n_2 = 12$
Sample mean	$\bar{x}_1 = 325$ hours	$\bar{x}_2 = 286$ hours
Sample standard deviation	$s_1 = 40$ hours	$s_2 = 44$ hours

Computing the degrees of freedom using equation (10.7), we have:

$$df = \frac{\left(\dfrac{s_1^2}{n_1} + \dfrac{s_2^2}{n_2}\right)^2}{\left(\dfrac{1}{n_1 - 1}\right)\left(\dfrac{s_1^2}{n_1}\right)^2 + \left(\dfrac{1}{n_2 - 1}\right)\left(\dfrac{s_2^2}{n_2}\right)^2} = \frac{\left(\dfrac{(40)^2}{12} + \dfrac{(44)^2}{12}\right)^2}{\left(\dfrac{1}{12 - 1}\right)\left(\dfrac{(40)^2}{12}\right)^2 + \left(\dfrac{1}{12 - 1}\right)\left(\dfrac{(44)^2}{12}\right)^2} = 21.8$$

Rounding down, we shall use a t distribution with 21 df. This row of the t distribution table is as follows:

Area in upper tail	0.20	0.10	0.05	0.025	0.01	0.005
t-value (21 df)	0.859	1.323	1.721	2.080	2.518	2.831

$$t = 2.272$$

With an upper-tail test, the p-value is the probability in the upper tail to the right of $t = 2.272$. From the above results, we see that the p-value is between 0.025 and 0.01. Hence, the p-value is less than $\alpha = 0.05$ and H_0 is rejected. The sample results enable the researcher to conclude that $\mu_1 - \mu_2 > 0$ or $\mu_1 > \mu_2$. The research study supports the conclusion that the new software package provides a lower population mean completion time.

Practical advice

The interval estimation and hypothesis testing procedures presented in this section are robust and can be used with relatively small sample sizes. In most applications, equal or nearly equal sample sizes such that the total sample size $n_1 + n_2$ is at least 20 can be expected to provide very good results even if the populations are not normal. Larger sample sizes are recommended if the distributions of the populations are highly skewed or contain outliers. Smaller sample sizes should be used only if the analyst is satisfied that the distributions of the populations are at least approximately normal.

Another approach sometimes used to make inferences about the difference between two population means when σ_1 and σ_2 are unknown assumes that the two population standard deviations are equal. You will find this approach as an option in Excel, R, SPSS and Minitab. Under the assumption of equal population variances, the two sample standard deviations are combined to provide the following 'pooled' sample variance s^2:

$$s^2 = \frac{(n_1 - 1)s_1^2 + (n_2 - 1)s_2^2}{n_1 + n_2 - 2}$$

The t test statistic becomes:

$$t = \frac{(\bar{x}_1 - \bar{x}_2) - D_0}{s\sqrt{\dfrac{1}{n_1} + \dfrac{1}{n_2}}}$$

and has $n_1 + n_2 - 2$ df. At this point, the computation of the p-value and the interpretation of the sample results are identical to the procedures discussed earlier in this section. A difficulty with this procedure is that the assumption of equal population standard deviations is usually difficult to verify. Unequal population standard deviations are frequently encountered. Using the pooled procedure may not provide satisfactory results especially if the sample sizes n_1 and n_2 are quite different. The t procedure that we presented in this section does not require the assumption of equal population standard deviations and can be applied whether the population standard deviations are equal or not. It is a more general procedure and is recommended for most applications.

EXERCISES

Methods

7. Consider the following results for independent random samples taken from two populations.

Sample 1	Sample 2
$n_1 = 20$	$n_2 = 30$
$\bar{x}_1 = 22.5$	$\bar{x}_2 = 20.1$
$s_1 = 2.5$	$s_2 = 4.8$

  a. What is the point estimate of the difference between the two population means?
  b. What are the df for the t distribution?
  c. At 95% confidence, what is the margin of error?
  d. What is the 95% confidence interval for the difference between the two population means?

8. Consider the following hypothesis test.

$$H_0: \mu_1 - \mu_2 = 0$$
$$H_1: \mu_1 - \mu_2 \neq 0$$

The following results are from independent samples taken from two populations.

Sample 1	Sample 2
$n_1 = 35$	$n_2 = 40$
$\bar{x}_1 = 13.6$	$\bar{x}_2 = 10.1$
$s_1 = 5.2$	$s_2 = 8.5$

   a. What is the value of the test statistic?
   b. What are the df for the t distribution?
   c. What is the p-value?
   d. At the 0.05 significance level, what is your conclusion?

9. Consider the following data for two independent random samples taken from two normal populations.

Sample 1	10	7	13	7	9	8
Sample 2	8	7	8	4	6	9

   a. Compute the two sample means.
   b. Compute the two sample standard deviations.
   c. What is the point estimate of the difference between the two population means?
   d. What is the 90% confidence interval estimate of the difference between the two population means?

Applications

AIRPORTS

10. The International Air Transport Association surveyed business travellers to determine ratings of various international airports. The maximum possible score was 10. Suppose 50 business travellers were asked to rate airport L and 50 other business travellers were asked to rate airport M. The rating scores follow.

Airport L

10 9 6 7 8 7 9 8 10 7 6 5 7 3 5 6 8 7 10 8 4 7 8 6 9
 9 5 3 1 8 9 6 8 5 4 6 10 9 8 3 2 7 9 5 3 10 3 5 10 8

Airport M

 6 4 6 8 7 7 6 3 3 8 10 4 8 7 8 7 5 9 5 8 4 3 8 5 5
 4 4 4 8 4 5 6 2 5 9 9 8 4 8 9 9 5 9 7 8 3 10 8 9 6

Construct a 95% confidence interval estimate of the difference between the mean ratings of the airports L and M.

MONTHLY INCOMES

11. The data in the file 'MONTHLY INCOMES' are based on a survey carried out in Cairo, Egypt, and described in Case Problem 3, Chapter 12. The figures are monthly incomes (in Egyptian pounds) of a sample of 25 men and 20 women. All were in work and had education to a similar level.
   a. What is the point estimate of the difference between mean monthly incomes for the two populations?
   b. Construct a 95% confidence interval estimate of the difference between the two population means.
   c. Does there appear to be any difference in mean monthly incomes for men and women? Explain.
   d. If this comparison were aimed at shedding light on gender equality/inequality, with respect to incomes, what other factors might need to be taken into consideration?

12. In a study of family economic status in Kerman city, Iran (published in the *African Journal of Business Management*), a comparison was made between the household incomes of families headed by men and families headed by women. The summary statistics for total household monthly income (converted to US dollars) are shown in the table below.

	Head of household	
	Men	*Women*
Sample size	360	30
Sample mean	1,202	850
Sample standard deviation	1,320	842

Consider the research hypothesis that households headed by men have higher mean income than households headed by women.

a. Formulate the hypotheses that can be used to determine whether the sample data support the hypothesis that households headed by men have higher mean income than households headed by women.

b. What is the point estimate of the difference between the means for the two populations?

c. Compute the p-value for the hypothesis test.

d. At $\alpha = 0.05$, what is your conclusion?

13. Periodically, Merrill Lynch customers are asked to evaluate Merrill Lynch financial consultants and services. Higher ratings on the client satisfaction survey indicate better service, with 7 the maximum service rating. Independent samples of service ratings for two financial consultants in the Dubai office are summarized here. Consultant A has ten years of experience while consultant B has one year of experience. Using $\alpha = 0.05$, test whether the consultant with more experience has the higher population mean service rating.

Consultant A	Consultant B
$n_1 = 16$	$n_2 = 10$
$\bar{x}_1 = 6.82$	$\bar{x}_2 = 6.25$
$s_1 = 0.64$	$s_2 = 0.75$

a. State the null and alternative hypotheses.

b. Compute the value of the test statistic.

c. What is the p-value?

d. What is your conclusion?

14. Safegate Foods is redesigning the checkouts in its supermarkets throughout the country and is considering two designs. Tests on customer checkout times conducted at two stores where the two new systems have been installed result in the following summary of the data.

System A	System B
$n_1 = 120$	$n_2 = 100$
$\bar{x}_1 = 4.1$ minutes	$\bar{x}_2 = 3.4$ minutes
$s_1 = 2.2$ minutes	$s_2 = 1.5$ minutes

Test at the 0.05 level of significance to determine whether the population mean checkout times of the two systems differ. Which system is preferred?

15. Over time, there are changes in hotel room pricing (*Lodging Magazine*), but is there a difference between the changes for hotel prices in Europe and changes for US hotel prices? The file 'IntHotels' contains changes in hotel prices for 47 major European cities and 53 major US cities.

a. On the basis of the sample results, can we conclude that the mean change in hotel rates in Europe and the USA are different? Set out appropriate null and alternative hypotheses.

b. Use $\alpha = 0.01$. What is your conclusion?

INTHOTELS

16. In a large study of the financial structure of small- and medium-sized companies in Italy (published in the *African Journal of Business Management*), one of the company characteristics calculated was the debt to total assets ratio. Sample means and sample standard deviations for this ratio are shown in the table below, for two sectors of the Italian economy, manufacturing and construction.

	Debt/Total assets ratio	
	Manufacturing	Construction
Sample mean	0.64	0.78
Sample standard deviation	0.17	0.16

a. Formulate the hypotheses that can be used to test whether the population mean debt to total assets ratio is the same in both sectors of the economy.

b. Suppose a sample of 80 companies in the manufacturing sector was used to produce the sample statistics in the table, and a sample of 50 companies in the construction sector. Using $\alpha = 0.05$, do the hypothesis test you have set up in (a). What is your conclusion?

c. What is the point estimate of the difference in population mean debt to total assets ratio between these two sectors of the economy? Provide a 95% confidence interval estimate of the difference.

10.3 INFERENCES ABOUT THE DIFFERENCE BETWEEN TWO POPULATION MEANS: MATCHED SAMPLES

Suppose employees at a manufacturing company can use two different methods to complete a production task. To maximize output, the company wants to identify the method with the smaller population mean completion time. Let μ_1 and μ_2 denote the population mean completion times for production methods 1 and 2 respectively. With no preliminary indication of the preferred production method, we begin by tentatively assuming that the two production methods have the same population mean completion time. The null hypothesis is $H_0: \mu_1 - \mu_2 = 0$. If this hypothesis is rejected, we can conclude that the population mean completion times differ. In this case, the method providing the smaller mean completion time would be recommended. The null and alternative hypotheses are written as follows:

$$H_0: \mu_1 - \mu_2 = 0$$

$$H_1: \mu_1 - \mu_2 \neq 0$$

In choosing the sampling procedure to collect production time data and test the hypotheses, we consider two alternative designs. One is based on **independent samples** and the other is based on **matched samples**.

1 *Independent samples design:* A simple random sample of workers is selected and each worker in the sample uses method 1. A second independent simple random sample of workers is selected and each worker in this sample uses method 2. The test of the difference between population means is based on the procedures in Section 10.2.

2 *Matched samples design:* One simple random sample of workers is selected. Each worker first uses one method and then uses the other method. The order of the two methods is assigned randomly to the workers, with some workers performing method 1 first and others performing method 2 first. Each worker provides a pair of data values, one value for method 1 and another value for method 2.

In the matched samples design the two production methods are tested using the same workers. Because this eliminates a major source of sampling variation, the matched samples design often leads to a smaller sampling error than the independent samples design.

We demonstrate the analysis of a matched samples design by taking it as the method used to examine the difference between population means for the two production methods. A random sample of six workers is used. The data on completion times for the six workers are given in Table 10.2. Each worker provides a pair of data values, one for each production method. Note that the last column contains the difference in completion times d_i for each worker in the sample.

MATCHED

TABLE 10.2 Task completion times for a matched samples design

Worker	Completion time, method 1 (min)	Completion time, method 2 (min)	Difference in completion times (d_i)
1	6.0	5.4	0.6
2	5.0	5.2	−0.2
3	7.0	6.5	0.5
4	6.2	5.9	0.3
5	6.0	6.0	0.0
6	6.4	5.8	0.6

In the analysis of the matched samples design, the key step is to consider only the column of differences. Therefore, we have six data values (0.6, −0.2, 0.5, 0.3, 0.0, 0.6). These will be used to analyze the difference between the population means of the two production methods.

Let μ_d = the mean of the *difference* values for the population of workers. The d notation is a reminder that the matched samples provides *difference* data for the units of analysis (in this example, individual workers). The null and alternative hypotheses are rewritten below using this notation. If H_0 is rejected, we can conclude that the population mean completion times differ:

$$H_0: \mu_d = 0$$

$$H_1: \mu_d \neq 0$$

Other than the use of the d notation, the formulae for the sample mean and sample standard deviation are the same ones used previously in the text:

$$\bar{d} = \frac{\sum d_i}{n} = \frac{1.8}{8} = 0.30$$

$$s_d = \sqrt{\frac{\sum (d_i - \bar{d})^2}{n - 1}} = \sqrt{\frac{0.56}{5}} = 0.335$$

With the small sample of $n = 6$ workers, we need to assume that the population of differences has a normal distribution. This assumption is necessary so that we may use the t distribution for hypothesis testing and interval estimation procedures. Sample size guidelines for using the t distribution were presented in Chapters 8 and 9. Based on this assumption, the following test statistic has a t distribution with $n - 1$ degrees of freedom:

Test statistic for hypothesis test involving matched samples

$$t = \frac{\bar{d} - \mu_d}{s_d / \sqrt{n}} \tag{10.9}$$

We use equation (10.9) for the test $H_0: \mu_d = 0$ and $H_1: \mu_d \neq 0$, using $\alpha = 0.05$. Substituting the sample results $\bar{d} = 0.30$, $s_d = 0.335$ and $n = 6$ into equation (10.9), we compute the value of the test statistic.

$$t = \frac{\bar{d} - \mu_d}{s_d / \sqrt{n}} = \frac{0.30 - 0}{0.335 / \sqrt{6}} = 2.194$$

Now we compute the p-value for this two-tailed test. The test statistic $t = 2.194$ (> 0) is in the upper tail of the t distribution. The probability in the upper tail is found by using the t distribution table with $n - 1 = 6 - 1 = 5$ df. Information from the 5 df row of the table is as follows:

Area in upper tail	0.20	0.10	0.05	0.025	0.01	0.005
t-value (5 df)	0.920	1.476	2.015	2.571	3.365	4.032

$t = 2.194$

The probability in the upper tail is between 0.05 and 0.025. As this test is a two-tailed test, we double these values to conclude that the p-value is between 0.10 and 0.05. This p-value is greater than $\alpha = 0.05$, so the null hypothesis $H_0: \mu_d = 0$ is not rejected. SPSS, R, Excel and Minitab show the p-value as 0.080.

In addition, we can construct an interval estimate of the difference between the two population means by using the single population methodology of Chapter 8. At 95% confidence, the calculation is as follows:

$$\bar{d} \pm t_{0.025}\frac{s_d}{\sqrt{n}} = 0.30 \pm 2.571\left(\frac{0.335}{\sqrt{6}}\right) = 0.30 \pm 0.35$$

The margin of error is 0.35 and the 95% confidence interval for the difference between the population means of the two production methods is -0.05 minute to $+0.65$ minute. Note that, in agreement with the non-significant hypothesis test result, this interval includes the value 0.

Workers completed the production task using first one method and then the other method. This example illustrates a matched samples design in which each sampled element (worker) provides a pair of data values. It is also possible to use different but 'similar' elements to provide the pair of data values. For example, a worker at one location could be matched with a similar worker at another location (similarity based on age, education, sex, experience, etc.). The pairs of workers would provide the difference data that could be used in the matched samples analysis.

A matched samples procedure for inferences about two population means generally provides better precision than the independent samples approach, and in that respect is preferable. However, in some applications the design is not feasible, for example because there will be 'carry-over' effects if the same sample elements are used to provide two data values. Or perhaps the time and cost associated with matching are excessive.

EXERCISES

Methods

17. Consider the following hypothesis test.

$$H_0: \mu_d \leq 0$$
$$H_1: \mu_d > 0$$

The following data are from matched samples taken from two populations.

	Population	
Element	1	2
1	21	20
2	28	26
3	18	18
4	20	20
5	26	24

   a. Compute the difference value for each element.
   b. Compute $\bar{d}$.
   c. Compute the standard deviation s_d.
   d. Conduct a hypothesis test using $\alpha = 0.05$. What is your conclusion?

18. The following data are from matched samples taken from two populations.

	Population	
Element	1	2
1	11	8
2	7	8
3	9	6
4	12	7
5	13	10
6	15	15
7	15	14

   a. Compute the difference value for each element.
   b. Compute $\bar{d}$.
   c. Compute the standard deviation s_d.
   d. What is the point estimate of the difference between the two population means?
   e. Provide a 95% confidence interval for the difference between the two population means.

Applications

19. In recent years, a growing array of entertainment options has been competing for consumer time. Researchers used a sample of 15 individuals and collected data on the hours per week spent watching catch-up TV and hours per week spent watching YouTube videos.

Individual	Catch-up TV	YouTube	Individual	Catch-up TV	YouTube
1	22	25	9	21	21
2	8	10	10	23	23
3	25	29	11	14	15
4	22	19	12	14	18
5	12	13	13	14	17
6	26	28	14	16	15
7	22	23	15	24	23
8	19	21			

   a. What is the sample mean number of hours per week spent watching catch-up TV? What is the sample mean number of hours per week spent watching YouTube videos? Which medium has the greater usage?
   b. Use a 0.05 level of significance and test for a difference between the population mean usage for catch-up TV and YouTube.

20. A market research firm used a sample of individuals to rate the purchase potential of a new product before and after the individuals saw a new Twitter ad about the product. The purchase potential ratings were based on a 0 to 10 scale, with higher values indicating a higher purchase potential. The null hypothesis stated that the mean rating 'after' would be less than or equal to the mean rating 'before'. Rejection of this hypothesis would show that the ad improved the mean purchase potential rating. Use $\alpha = 0.05$ and the following data to test the hypothesis and comment on the value of the ad.

| | Purchase rating | | | Purchase rating | |
Individual	After	Before	Individual	After	Before
1	6	5	5	3	5
2	6	4	6	9	8
3	7	7	7	7	5
4	4	3	8	6	6

21. Scores in the first and fourth (final) rounds for a sample of 20 golfers who competed in PGA tournaments are shown in the following table. Suppose you would like to determine if the mean score for the first round of a PGA Tour event is significantly different from the mean score for the fourth and final round. Does the pressure of playing in the final round cause scores to go up? Or does the increased player concentration cause scores to come down?

Player	First round	Final round	Player	First round	Final round
Michael Leitzig	70	72	Aron Price	72	72
Scott Verplank	71	72	Charles Howell	72	70
D A Points	70	75	Jason Dufner	70	73
Jerry Kelly	72	71	Mike Weir	70	77
Soren Hansen	70	69	Carl Pettersson	68	70
D J Trahan	67	67	Bo Van Pelt	68	65
Bubba Watson	71	67	Ernie Els	71	70
Retief Goosen	68	75	Cameron Beckman	70	68
Jeff Klauk	67	73	Nick Watney	69	68
Kenny Perry	70	69	Tommy Armour III	67	71

a. Use $\alpha = 0.10$ to test for a statistically significantly difference between the population means for first-round and fourth-round scores. What is the *p*-value? What is your conclusion?

b. What is the point estimate of the difference between the two population means? For which round is the population mean score lower?

c. What is the margin of error for a 90% confidence interval estimate for the difference between the population means? Could this confidence interval have been used to test the hypothesis in (a)? Explain.

22. A survey was made of Amazon 'First Read' subscribers to ascertain whether they spend more time reading on e-readers such as Kindle than they do reading print books. Assume a sample of 15 respondents provided the following data on weekly hours of reading on e-readers and weekly hours of reading print books. Using a 0.05 level of significance, can you conclude that Amazon 'First Read' subscribers spend more hours per week reading on e-readers than they do reading print books?

READING

Respondent	e-Reader	Print	Respondent	e-Reader	Print
1	10	6	9	4	7
2	14	16	10	8	8
3	16	8	11	16	5
4	18	10	12	5	10
5	15	10	13	8	3
6	14	8	14	19	10
7	10	14	15	11	6
8	12	14			

10.4 INFERENCES ABOUT THE DIFFERENCE BETWEEN TWO POPULATION PROPORTIONS

Let π_1 and π_2 denote the population proportions for populations 1 and 2 respectively. We next consider inferences about the difference $\pi_1 - \pi_2$ between the two population proportions. We shall select two independent random samples consisting of n_1 units from population 1 and n_2 units from population 2.

Interval estimation of $\pi_1 - \pi_2$

An accountancy firm specializing in the preparation of income tax returns is interested in comparing the quality of work at two of its regional offices. The firm will be able to estimate the proportion of erroneous returns by randomly selecting samples of tax returns prepared at each office and verifying their accuracy.

$\pi_1 =$ proportion of erroneous returns for population 1 (office 1)

$\pi_2 =$ proportion of erroneous returns for population 2 (office 2)

$P_1 =$ sample proportion for a simple random sample from population 1

$P_2 =$ sample proportion for a simple random sample from population 2

The difference between the two population proportions is $\pi_1 - \pi_2$. The point estimator of $\pi_1 - \pi_2$ is the difference between the sample proportions of two independent simple random samples:

Point estimator of the difference between two population proportions

$$P_1 - P_2 \tag{10.10}$$

As with other point estimators, the point estimator $P_1 - P_2$ has a sampling distribution that reflects the possible values of $P_1 - P_2$ if we repeatedly took two independent random samples. The mean of this sampling distribution is $\pi_1 - \pi_2$ and the standard error of $P_1 - P_2$ is:

Standard error of $P_1 - P_2$

$$\sigma_{P_1 - P_2} = \sqrt{\frac{\pi_1(1 - \pi_1)}{n_1} + \frac{\pi_2(1 - \pi_2)}{n_2}} \tag{10.11}$$

If the sample sizes are large enough that $n_1\pi_1$, $n_1(1 - \pi_1)$, $n_2\pi_2$ and $n_2(1 - \pi_2)$ are all greater than or equal to five, the sampling distribution of $P_1 - P_2$ can be approximated by a normal distribution.

In the estimation of the difference between two population proportions, an interval estimate will take the form $(p_1 - p_2) \pm$ margin of error. We would like to use $z_{\alpha/2}\sigma_{P_1 - P_2}$ as the margin of error, but $\sigma_{P_1 - P_2}$ given by equation (10.11) cannot be used directly because the two population proportions, π_1 and π_2, are unknown. Using the sample proportions p_1 and p_2 to estimate the corresponding population parameters, the estimated margin of error is:

$$\text{Margin of error} = z_{\alpha/2}\sqrt{\frac{p_1(1 - p_1)}{n_1} + \frac{p_2(1 - p_2)}{n_2}} \tag{10.12}$$

The general form of an interval estimate of the difference between two population proportions is:

Interval estimate of the difference between two population proportions

$$(p_1 - p_2) \pm z_{\alpha/2} \sqrt{\frac{p_1(1 - p_1)}{n_1} + \frac{p_2(1 - p_2)}{n_2}}$$ (10.13)

where $1 - \alpha$ is the confidence coefficient.

Suppose that in the tax returns example, we find that independent simple random samples from the two offices provide the following information:

Office 1	Office 2
$n_1 = 250$	$n_1 = 300$
Number of returns with errors = 35	Number of returns with errors = 27

The sample proportions for the two offices are:

$$p_1 = \frac{35}{250} = 0.14 \qquad p_2 = \frac{27}{300} = 0.09$$

TAXPREP

The point estimate of the difference between the proportions of erroneous tax returns for the two populations is $p_1 - p_2 = 0.14 - 0.09 = 0.05$. We estimate that Office 1 has a 0.05, or 5 percentage points, greater error rate than Office 2.

Expression (10.13) can now be used to provide a margin of error and interval estimate of the difference between the two population proportions. Using a 90% confidence interval with $z_{\alpha/2} = z_{0.05} = 1.645$, we have:

$$(p_1 - p_2) \pm z_{\alpha/2} \sqrt{\frac{p_1(1 - p_1)}{n_1} + \frac{p_2(1 - p_2)}{n_2}}$$

$$= (0.14 - 0.09) \pm 1.645 \sqrt{\frac{0.14(1 - 0.14)}{250} + \frac{0.09(1 - 0.09)}{300}} = 0.05 \pm 0.045$$

The margin of error is 0.045, and the 90% confidence interval is 0.005 to 0.095.

Hypothesis tests about $\pi_1 - \pi_2$

Let us now consider hypothesis tests about the difference between the proportions of two populations. The three forms for a hypothesis test are as follows:

$$H_0: \pi_1 - \pi_2 \geq D \qquad H_0: \pi_1 - \pi_2 \leq D \qquad H_0: \pi_1 - \pi_2 = D_0$$
$$H_1: \pi_1 - \pi_2 < D \qquad H_1: \pi_1 - \pi_2 > D \qquad H_1: \pi_1 - \pi_2 \neq D_0$$

As with the hypothesis tests in Sections 10.1 to 10.3, D_0 will usually be zero. In such cases, if we assume H_0 is true as an equality, we have $\pi_1 - \pi_2 = 0$, or $\pi_1 = \pi_2$. The test statistic is based on the sampling distribution of the point estimator $P_1 - P_2$.

In equation (10.11), we showed that the standard error of $P_1 - P_2$ is given by:

$$\sigma_{P_1 - P_2} = \sqrt{\frac{\pi_1(1 - \pi_1)}{n_1} + \frac{\pi_2(1 - \pi_2)}{n_2}}$$

Under the assumption that H_0 is true as an equality, the population proportions are equal and $\pi_1 = \pi_2 = \pi$. In this case, $\sigma_{P_1 - P_2}$ becomes:

Standard error of $P_1 - P_2$ when $\pi_1 = \pi_2 = \pi$

$$\sigma_{P_1 - P_2} = \sqrt{\frac{\pi(1 - \pi)}{n_1} + \frac{\pi(1 - \pi)}{n_2}} = \sqrt{\pi(1 - \pi)\left(\frac{1}{n_1} + \frac{1}{n_2}\right)} \tag{10.14}$$

With π unknown, we pool, or combine, the point estimates from the two samples (p_1 and p_2) to obtain a single point estimate of π as follows:

Pooled estimate of π when $\pi_1 = \pi_2 = \pi$

$$p = \frac{n_1 p_1 + n_2 p_2}{n_1 + n_2} \tag{10.15}$$

This **pooled estimate of π** is a weighted average of p_1 and p_2.

Substituting p for π in equation (10.14), we obtain an estimate of $\sigma_{P_1 - P_2}$, which is used in the test statistic.

Test statistic for hypothesis tests about $\pi_1 = \pi_2$

$$z = \frac{(p_1 - p_2) - D_0}{\sqrt{p(1 - p)\left(\frac{1}{n_1} + \frac{1}{n_2}\right)}} \tag{10.16}$$

This test statistic applies to large sample situations where $n_1 \pi_1$, $n_1(1 - \pi_1)$, $n_2 \pi_2$ and $n_2(1 - \pi_2)$ are all greater than or equal to five.

Let us assume in the tax returns example that the firm wants to use a hypothesis test to determine whether the error proportions differ between the two offices. A two-tailed test is required. The null and alternative hypotheses are as follows:

$$H_0: \pi_1 - \pi_2 = 0$$
$$H_1: \pi_1 - \pi_2 \neq 0$$

If H_0 is rejected, the firm can conclude that the error proportions at the two offices differ. For illustration purposes, we shall use $\alpha = 0.10$ as the level of significance.

The sample data previously collected showed $p_1 = 0.14$ for the $n_1 = 250$ returns sampled at Office 1 and $p_2 = 0.09$ for the $n_2 = 300$ returns sampled at Office 2. The pooled estimate of π is:

$$p = \frac{n_1 p_1 + n_2 p_2}{n_1 + n_2} = \frac{250(0.14) + 300(0.09)}{250 + 300} = 0.1127$$

Using this pooled estimate and the difference between the sample proportions, the value of the test statistic is as follows:

$$z = \frac{(p_1 - p_2) - D_0}{\sqrt{p(1 - p)\left(\frac{1}{n_1} + \frac{1}{n_2}\right)}} = \frac{(0.14 - 0.09) - 0}{\sqrt{0.1127(1 - 0.1127)\left(\frac{1}{250} + \frac{1}{300}\right)}} = 1.85$$

To compute the p-value for this two-tailed test, we first note that $z = 1.85$ is in the upper tail of the standard normal distribution. Using the standard normal distribution table, we find the probability in the upper tail for $z = 1.85$ is $1 - 0.9678 = 0.0322$. Doubling this area for a two-tailed test, we find the p-value $2(0.0322) = 0.0644$. With the p-value less than $\alpha = 0.10$, H_0 is rejected at the 0.10 level of significance. The firm can conclude that the error rates differ between the two offices. This hypothesis test conclusion is consistent with the earlier interval estimation results that showed the interval estimate of the difference between the population error rates at the two offices to be 0.005 to 0.095, with Office 1 having the higher error rate.

EXERCISES

Methods

23. Consider the following results for independent samples taken from two populations.

Sample 1	Sample 2
$n_1 = 400$	$n_2 = 300$
$p_1 = 0.48$	$p_2 = 0.36$

a. What is the point estimate of the difference between the two population proportions?
b. Construct a 90% confidence interval for the difference between the two population proportions.
c. Construct a 95% confidence interval for the difference between the two population proportions.

24. Consider the hypothesis test:

$$H_0 : \pi_1 - \pi_2 \leq 0$$
$$H_1 : \pi_1 - \pi_2 > 0$$

The following results are for independent samples taken from the two populations.

Sample 1	Sample 2
$n_1 = 200$	$n_2 = 300$
$p_1 = 0.22$	$p_2 = 0.10$

a. What is the p-value?
b. With $\alpha = 0.05$, what is your hypothesis testing conclusion?

Applications

25. A *Businessweek/Harris* poll asked senior executives at large corporations their opinions about the economic outlook for the future. One question was, 'Do you think that there will be an increase in the number of full-time employees at your company over the next 12 months?' In the current survey, 220 of 400 executives answered Yes, while in a previous year's survey, 192 of 400 executives had answered Yes. Provide a 95% confidence interval estimate for the difference between the proportions at the two points in time. What is your interpretation of the interval estimate?

26. In a test of the quality of two TV ads, each was shown in a separate test area six times over a one-week period. The following week a telephone survey was conducted to identify individuals who had seen the ads. Those individuals were asked to state the primary message in the ads. The following results were recorded.

	Ad A	Ad B
Number who saw ad	150	200
Number who recalled message	63	60

    a. Use $\alpha = 0.05$ and test the hypothesis that there is no difference in the recall proportions for the two ads.
    b. Compute a 95% confidence interval for the difference between the recall proportions for the two populations.

27. *Forbes* reports that females trust recommendations from Pinterest more than recommendations from any other social network platform. But does trust in Pinterest differ by sex? The following sample data show the number of females and males who stated in a recent sample that they trust recommendations made on Pinterest.

	Females	Males
Sample size	150	170
Trust recommendations made on Pinterest	117	102

    a. What is the point estimate of the proportion of females who trust recommendations made on Pinterest?
    b. What is the point estimate of the proportion of males who trust recommendations made on Pinterest?
    c. Construct a 95% confidence interval estimate of the difference between the proportion of women and men who trust recommendations made on Pinterest.

28. A large car insurance company selected samples of single and married male policyholders and recorded the number who made an insurance claim over the preceding three-year period.

Single policyholders	Married policyholders
$n_1 = 400$	$n_2 = 900$
Number making claims = 76	Number making claims = 90

    a. Using $\alpha = 0.05$, test whether the claim rates differ between single and married male policyholders.
    b. Provide a 95% confidence interval for the difference between the proportions for the two populations.

29. The Adecco Workplace Insights Survey sampled male and female workers and asked if they expected to get a rise or promotion this year. Suppose the survey sampled 200 male workers and 200 female workers. If 104 of the male workers replied Yes and 74 of the female workers replied Yes, are the results statistically significant in that you can conclude a greater proportion of male workers are expecting to get a rise or a promotion this year?
    a. State the hypothesis test in terms of the population proportion of male workers and the population proportion of female workers.
    b. What is the sample proportion for male workers? For female workers?
    c. Use a 0.01 level of significance. What is the *p*-value and what is your conclusion?

ONLINE RESOURCES

For the data files, additional questions and answers, and software section for Chapter 10, go to the online platform.

SUMMARY

In this chapter we discussed procedures for constructing interval estimates and doing hypothesis tests involving two populations. First, we showed how to make inferences about the difference between two population means when independent simple random samples are selected. We considered the case where the population standard deviations, σ_1 and σ_2, could be assumed known. The standard normal distribution Z was used to construct the interval estimate and calculate the p-value for hypothesis tests. We then considered the case where the population standard deviations were unknown and estimated by the sample standard deviations s_1 and s_2. In this case, the t distribution was used in constructing the interval estimate and doing hypothesis tests.

Inferences about the difference between two population means were then discussed for the matched samples design. In the matched samples design each element provides a pair of data values, one from each population. The differences between the paired data values are then used in the statistical analysis. The matched samples design is generally preferred to the independent samples design, when it is feasible, because the matched samples procedure usually improves the precision of the estimate.

Finally, interval estimation and hypothesis testing about the difference between two population proportions were discussed. For large independent samples, these procedures are based on use of the normal distribution.

KEY TERMS

Independent samples   Pooled estimate of π
Matched samples

KEY FORMULAE

Point estimator of the difference between two population means

$$\bar{X}_1 - \bar{X}_2 \tag{10.1}$$

Standard error of $\bar{X}_1 - \bar{X}_2$

$$\sigma_{\bar{X}_1 - \bar{X}_2} = \sqrt{\frac{\sigma_1^2}{n_1} + \frac{\sigma_2^2}{n_2}} \tag{10.2}$$

Interval estimate of the difference between two population means: σ_1 and σ_2 known

$$(\bar{x}_1 - \bar{x}_2) \pm z_{a/2}\sqrt{\frac{\sigma_1^2}{n_1} + \frac{\sigma_2^2}{n_2}} \tag{10.4}$$

Test statistic for hypothesis tests about $\mu_1 - \mu_2$: σ_1 and σ_2 known

$$z = \frac{(\bar{x}_1 - \bar{x}_2) - D_0}{\sqrt{\dfrac{\sigma_1^2}{n_1} + \dfrac{\sigma_2^2}{n_2}}} \qquad (10.5)$$

Interval estimate of the difference between two population means: σ_1 and σ_2 unknown

$$(\bar{x}_1 - \bar{x}_2) \pm t_{\alpha/2}\sqrt{\frac{s_1^2}{n_1} + \frac{s_2^2}{n_2}} \qquad (10.6)$$

Degrees of freedom for the t distribution using two independent random samples

$$df = \frac{\left(\dfrac{s_1^2}{n_1} + \dfrac{s_2^2}{n_2}\right)^2}{\left(\dfrac{1}{n_1 - 1}\right)\left(\dfrac{s_1^2}{n_1}\right)^2 + \left(\dfrac{1}{n_2 - 1}\right)\left(\dfrac{s_2^2}{n_2}\right)^2} \qquad (10.7)$$

Test statistic for hypothesis tests about $\mu_1 - \mu_2$: σ_1 and σ_2 unknown

$$t = \frac{(\bar{x}_1 - \bar{x}_2) - D_0}{\sqrt{\dfrac{s_1^2}{n_1} + \dfrac{s_2^2}{n_2}}} \qquad (10.8)$$

Test statistic for hypothesis test involving matched samples

$$t = \frac{\bar{d} - \mu_d}{s_d/\sqrt{n}} \qquad (10.9)$$

Point estimator of the difference between two population proportions

$$P_1 - P_2 \qquad (10.10)$$

Standard error of $P_1 - P_2$

$$\sigma_{P_1 - P_2} = \sqrt{\frac{\pi_1(1 - \pi_1)}{n_1} + \frac{\pi_2(1 - \pi_2)}{n_2}} \qquad (10.11)$$

Interval estimate of the difference between two population proportions

$$(p_1 - p_2) \pm z_{\alpha/2}\sqrt{\frac{p_1(1 - p_1)}{n_1} + \frac{p_2(1 - p_2)}{n_2}} \qquad (10.13)$$

Standard error of $P_1 - P_2$ when $\pi_1 = \pi_2 = \pi$

$$\sigma_{P_1 - P_2} = \sqrt{\pi(1 - \pi)\left(\frac{1}{n_1} + \frac{1}{n_2}\right)} \qquad (10.14)$$

Pooled estimate of π when $\pi_1 = \pi_2 = \pi$

$$p = \frac{n_1 p_1 + n_2 p_2}{n_1 + n_2} \qquad (10.15)$$

Test statistic for hypothesis tests about $\pi_1 - \pi_2$

$$z = \frac{(p_1 - p_2) - D_0}{\sqrt{p(1 - p)\left(\dfrac{1}{n_1} + \dfrac{1}{n_2}\right)}} \qquad (10.16)$$

CASE PROBLEM

Brand origin recognition accuracy (BORA)

The academic literature on consumer behaviour and marketing includes research looking into the effects of country of origin on consumers' perceptions of, and willingness to buy, products and brands. This research has examined concepts such as consumer cosmopolitanism, consumer ethnocentricity and consumer patriotism, and their effects on consumer behaviour. Some of the more recent research has attempted to take into account consumers' recognition of brand origin, which is typically far from perfect. Marketing researchers refer to this as brand origin recognition accuracy (BORA).

The file 'BORA' on the online platform contains results from a survey of residents in the north-west of England, who were asked to identify the country of origin of 16 well-known brands. These included Tesco, Aldi, Dyson and Samsung. Eight of the brands were of UK origin, and eight of non-UK origin. The data in the file are for 100 respondents to the survey. They comprise demographic characteristics of the survey respondents, as well as figures for the number of UK brands correctly identified as being of UK origin, the number of non-UK brands correctly identified as being of non-UK origin, and the number of non-UK brands for which the specific country of origin was correctly identified. In the data file these latter three variables are labelled respectively as BORA_UK, BORA_non_UK and BORA_Country. The first few lines of data have been reproduced below.

Analyst's report

Prepare a marketing background report that addresses the following.

1. Summarize the demographic characteristics of the survey respondents.

2. Summarize the distribution of the 'BORA_UK', 'BORA_non_UK' and 'BORA_Country' figures, and produce estimates of average brand origin recognition accuracy for the population from which these respondents were selected.

3. Is there evidence that, in the population from which the respondents were selected, the average level of recognition of the UK brands was different from that for the non-UK brands?

4. Is there evidence that, in the population from which the respondents were selected, there were differences in average brand origin recognition accuracy between men and women, or between older people and younger people?

© alexsl /iStock

Gender	Age	Education	BORA_UK	BORA_non_UK	BORA_Country
Female	35–44...	Bachelor's degree	7	6	3
Male	35–44...	A Level	5	8	1
Male	18–24...	A Level	1	8	3
Female	55–64...	A Level	5	5	2
Female	25–34...	A Level	8	8	6
Male	35–44...	Master's degree...	7	8	5
Male	18–24...	GCSE	6	7	2
Male	55–64...	Master's degree...	5	8	7
Male	18–24...	Bachelor's degree	8	7	5
Female	35–44...	Bachelor's degree	5	8	3

BORA

11

Inferences about Population Variances

CHAPTER CONTENTS

Statistics in Practice Effects of global events on oil prices

11.1 Inferences about a population variance
11.2 Inferences comparing two population variances

LEARNING OBJECTIVES   After reading this chapter and doing the exercises, you should be able to:

1 Construct confidence intervals for a population standard deviation or population variance, using the chi-squared distribution.

2 Carry out and interpret the results of hypothesis tests for a population standard

deviation or population variance, using the chi-squared distribution.

3 Carry out and interpret the results of hypothesis tests to compare two population standard deviations or population variances, using the *F* distribution.

In Chapters 7 to 10, we examined inferential methods for population means and population proportions. In this chapter we look at population variances and population standard deviations.

In many manufacturing processes, it is extremely important in maintaining quality to control the process variance. Consider a production process filling packs with long-grain rice, for example. The filling mechanism is adjusted so the mean filling weight is 500 g per pack. In addition, the variance of the filling weights is critical. Even with the filling mechanism properly adjusted for the mean of 500 g, we cannot expect every pack to contain precisely 500 g. By selecting a sample of packs, we can compute a sample variance for the weight per pack, which will provide an estimate of the variance for the population of packs being filled by the production process. If the sample variance is modest, the production process will be continued. However, if the sample variance is excessive, over-filling and under-filling may be occurring, even though the mean is correct at 500 g. In this case, the filling mechanism will be readjusted in an attempt to reduce the filling variance for the packs.

In the first section we consider inferences about the variance of a single population. Subsequently, we shall discuss procedures for comparing the variances of two populations.

STATISTICS IN PRACTICE
Effects of global events on oil prices

Though pressure is mounting in many countries for a shift towards a 'greener' future in respect of energy production and consumption, oil remains a vital commodity around the globe. In 2021, global production of oil averaged 77 million barrels per day.

The price of oil is consequently an important economic factor, and one that is easily influenced by world events, as clearly evidenced in recent years. The effects of such events are seen not only in the average price of oil but also in the hour-to-hour or day-to-day volatility in price. In a 2020 paper in *Energy Research Letters* (vol. 1(2)), Devpura and Narayan compared hourly oil prices in the six months before the first global COVID-19 case (December 2019) with those in the subsequent six months. They noted that the mean price of oil fell by over 30 per cent between the two periods, while the volatility, as measured by the standard deviation of prices, increased sixfold.

The chart below shows daily prices of Brent crude oil during a 10-month period from October 2021 to July 2022, straddling the start of the Russian invasion of Ukraine on 24 February 2022. Visually, it is clear that the average oil price was higher in the five months after the start of the conflict than in the five months before. Additionally, it is evident from the chart that the day-to-day variability in price was higher after the start of the conflict than before.

Chapter 10 looked at methods for examining differences between two mean values. The present chapter turns to estimation and testing of standard deviations and variances.

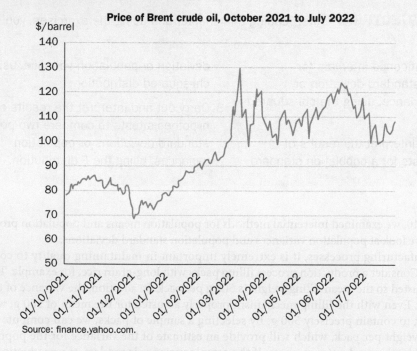

Price of Brent crude oil, October 2021 to July 2022

Source: finance.yahoo.com.

11.1 INFERENCES ABOUT A POPULATION VARIANCE

Sample variance is calculated as follows (refer to Chapter 3 if you need a reminder):

$$s^2 = \frac{\Sigma(x_i - \bar{x})^2}{n - 1} \tag{11.1}$$

The sample variance (S^2) is a point estimator of the population variance σ^2. To make inferences about σ^2, the sampling distribution of the quantity $(n - 1)S^2/\sigma^2$ can be used, under appropriate circumstances.

Sampling distribution of $(n - 1)S^2/\sigma^2$

When a simple random sample of size n is selected from a normally distributed population, the sampling distribution of

$$\frac{(n - 1)S^2}{\sigma^2} \tag{11.2}$$

has a chi-squared (χ^2) distribution with $n - 1$ degrees of freedom (df).

Chi-squared is often written as χ^2 (χ is the Greek letter chi). Figure 11.1 shows examples of the χ^2 distribution. When sampling from a normally distributed population, we can use this distribution to construct interval estimates and do hypothesis tests about the population variance.

Table 3 of Appendix B is a table of probabilities for the χ^2 distribution.

FIGURE 11.1
Examples of the sampling
distribution of $(n - 1)S^2/\sigma^2$
(chi-squared distribution)

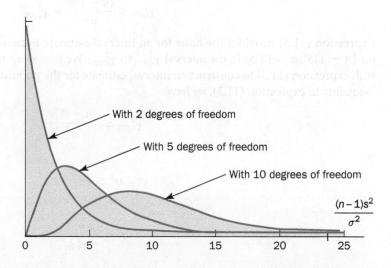

With 2 degrees of freedom

With 5 degrees of freedom

With 10 degrees of freedom

$\dfrac{(n-1)s^2}{\sigma^2}$

0   5   10   15   20   25

Interval estimation

Suppose we are interested in estimating the population variance for the production filling process described above. A sample of 20 packs is taken and the sample variance for the filling weights is found to be $s^2 = 2.50$ (this is measured in grams^2). However, we cannot expect the variance of a sample of 20 packs to provide the exact value of the variance for the population of packs filled by the production process. Our interest is in constructing an interval estimate for the population variance.

We shall use the notation χ^2_α to denote the value for the chi-squared random variable that gives a probability of α to the *right* of the χ^2_α value. For example, in Figure 11.2 the χ^2 distribution with 19 df is shown, with $\chi^2_{0.025} = 32.852$ indicating that 2.5 per cent of the χ^2 values are to the right of 32.852, and

$\chi^2_{0.975} = 8.907$ indicating that 97.5 per cent of the χ^2 values are to the right of 8.907. Refer to Table 3 of Appendix B and verify that these χ^2 values with 19 df are correct (19th row of the table).

From Figure 11.2 we see that 0.95, or 95 per cent, of the χ^2 values are between $\chi^2_{0.975}$ and $\chi^2_{0.025}$. That is, there is a 0.95 probability of obtaining a χ^2 value such that:

$$\chi^2_{0.975} \leq \chi^2 \leq \chi^2_{0.025}$$

FIGURE 11.2
A chi-squared distribution
with 19 df

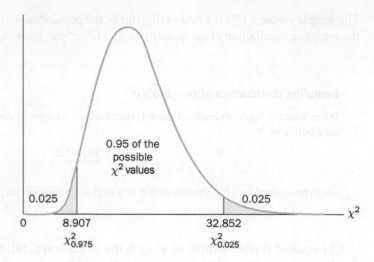

We stated in expression (11.2) that the random variable $(n - 1)S^2/\sigma^2$ follows a χ^2 distribution, therefore we can substitute $(n - 1)s^2/\sigma^2$ for χ^2 and write:

$$\chi^2_{0.975} \leq \frac{(n - 1)s^2}{\sigma^2} \leq \chi^2_{0.025} \tag{11.3}$$

Expression (11.3) provides the basis for an interval estimate because 95 per cent of all possible values for $(n - 1)S^2/\sigma^2$ will be in the interval $\chi^2_{0.975}$ to $\chi^2_{0.025}$. We now need to do some algebraic manipulations with expression (11.3) to construct an interval estimate for the population variance σ^2. Using the leftmost inequality in expression (11.3), we have:

$$\chi^2_{0.975} \leq \frac{(n - 1)s^2}{\sigma^2}$$

So,

$$\chi^2_{0.975}\sigma^2 \leq (n - 1)s^2$$

or

$$\sigma^2 \leq \frac{(n - 1)s^2}{\chi^2_{0.975}} \tag{11.4}$$

Doing similar algebraic manipulations with the rightmost inequality in expression (11.3) gives:

$$\frac{(n - 1)s^2}{\chi^2_{0.025}} \leq \sigma^2 \tag{11.5}$$

Expressions (11.4) and (11.5) can be combined to provide:

$$\frac{(n - 1)s^2}{\chi^2_{0.025}} \leq \sigma^2 \leq \frac{(n - 1)s^2}{\chi^2_{0.975}} \tag{11.6}$$

Because expression (11.3) is true for 95 per cent of the $(n - 1)s^2/\sigma^2$ values, expression (11.6) provides a 95% confidence interval estimate for the population variance σ^2.

We return to the problem of providing an interval estimate for the population variance of filling weights. The sample of 20 packs provided a sample variance $s^2 = 2.50$. With a sample size of 20, we have 19 df. As shown in Figure 11.2, we have already determined that $\chi^2_{0.975} = 8.907$ and $\chi^2_{0.025} = 32.852$. Using these values in expression (11.6) provides the following interval estimate for the population variance:

$$\frac{(19)(2.50)}{32.852} \leq \sigma^2 \leq \frac{(19)(2.50)}{8.907}$$

or

$$1.45 \leq \sigma^2 \leq 5.33$$

Taking the square root of these values provides the following 95% confidence interval for the population standard deviation:

$$1.20 \leq \sigma \leq 2.31$$

As $\chi^2_{0.975} = 8.907$ and $\chi^2_{0.025} = 32.852$ were used, the interval estimate has a 0.95 confidence coefficient. Extending expression (11.6) to the general case of any confidence coefficient, we have the following interval estimate of a population variance:

Interval estimate of a population variance

$$\frac{(n-1)s^2}{\chi^2_{\alpha/2}} \leq \sigma^2 \leq \frac{(n-1)s^2}{\chi^2_{1-\alpha/2}} \tag{11.7}$$

where the χ^2 values are based on a χ^2 distribution with $n - 1$ df and where $1 - \alpha$ is the confidence coefficient.

Hypothesis testing

The three possible forms for a hypothesis test about a population variance follow. Here, σ_0^2 denotes the hypothesized value for the population variance:

$$H_0: \sigma^2 \geq \sigma_0^2 \qquad H_0: \sigma^2 \leq \sigma_0^2 \qquad H_0: \sigma^2 = \sigma_0^2$$
$$H_1: \sigma^2 < \sigma_0^2 \qquad H_1: \sigma^2 > \sigma_0^2 \qquad H_1: \sigma^2 \neq \sigma_0^2$$

These are like the three forms we used for one-tailed and two-tailed hypothesis tests about population means and proportions in Chapters 9 and 10.

Hypothesis tests about a population variance use σ_0^2 and the sample variance s^2 to compute the value of a χ^2 test statistic. Assuming the population has a normal distribution, the test statistic is:

Test statistic for hypothesis tests about a population variance

$$\chi^2 = \frac{(n-1)S^2}{\sigma_0^2} \tag{11.8}$$

The test statistic has a χ^2 distribution with $n - 1$ df.

After computing the value of the χ^2 test statistic, either the p-value approach or the critical value approach may be used to determine whether the null hypothesis can be rejected.

Here is an example. The EuroBus Company wants to promote an image of reliability by maintaining consistent schedules. The company would like arrival times at bus stops to have low variability. The company guideline specifies an arrival time variance of 4.0 or less when arrival times are measured in minutes.

The following hypothesis test is formulated to help the company determine whether the arrival time population variance is excessive.

$$H_0: \sigma^2 \leq 4.0$$

$$H_1: \sigma^2 > 4.0$$

In provisionally assuming H_0 is true, we are assuming the population variance of arrival times is within the company guideline. We reject H_0 if the sample evidence indicates that the population variance exceeds the guideline. In this case, follow-up steps should be taken to reduce the population variance. We conduct the hypothesis test using a level of significance of $\alpha = 0.05$.

Suppose a random sample of 24 bus arrivals taken at a city-centre bus stop provides a sample variance $s^2 = 4.9$. We shall assume the population distribution of arrival times is approximately normal. The value of the test statistic is as follows:

$$\chi^2 = \frac{(n-1)s^2}{\sigma_0^2} = \frac{(24-1)4.9}{4.0} = 28.175$$

The χ^2 distribution with $n - 1 = 24 - 1 = 23$ df is shown in Figure 11.3. As this is an upper-tail test, the area under the curve to the right of the test statistic $\chi^2 = 28.175$ is the p-value for the test.

FIGURE 11.3
Chi-squared distribution for the EuroBus example

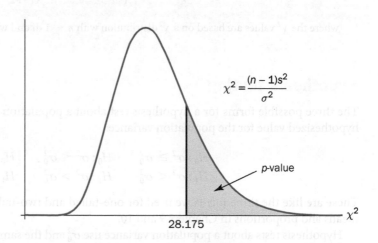

$$\chi^2 = \frac{(n-1)s^2}{\sigma^2}$$

p-value

28.175

Like the t distribution table, the χ^2 distribution table does not contain sufficient detail to enable us to determine the p-value exactly. However, we can use the table to obtain a range for the p-value. Using Table 3 of Appendix B, we find the following information for a χ^2 distribution with 23 df.

Area in upper tail	0.10	0.05	0.025	0.01
χ^2 value (23 df)	32.007	35.172	38.076	41.638

$\chi^2 = 28.175$

Because $\chi^2 = 28.175$ is less than 32.007, the probability in the upper tail (the p-value) is greater than 0.10. With the p-value $> \alpha = 0.05$, we cannot reject the null hypothesis. The sample does not provide convincing evidence that the population variance of the arrival times is excessive.

Because of the difficulty of determining the exact p-value from the χ^2 distribution table, a computer software package such as SPSS, R, Excel or Minitab is helpful. The guides on the online platform describe the procedures, showing that with 23 df, $\chi^2 = 28.175$ gives a p-value $= 0.209$.

As with other hypothesis testing procedures, the critical value approach can also be used to draw the conclusion. With $\alpha = 0.05$, $\chi^2_{0.05}$ provides the critical value for the upper-tail hypothesis test. Using Table 3 of Appendix B and 23 df, $\chi^2_{0.05} = 35.172$. Consequently, the rejection rule for the bus arrival time example is as follows:

$$\text{Reject } H_0 \text{ if } \chi^2 \geq 35.172$$

As the value of the test statistic is $\chi^2 = 28.175$, we cannot reject the null hypothesis.

In practice, upper-tail tests like this one are the most frequently used tests about a population variance. In situations involving arrival times, production times, filling weights, part dimensions and so on, low variances are desirable, whereas large variances are unacceptable. The most common test format is to test the null hypothesis that the population variance is less than or equal to the maximum allowable value against the alternative hypothesis that the population variance is greater than the maximum allowable value. With this test structure, corrective action will be taken whenever rejection of the null hypothesis indicates that the population variance is excessive.

As we saw with population means and proportions, other forms of hypothesis test can be done. We shall demonstrate a two-tailed test about a population variance by considering a situation faced by a car driver licensing authority. Historically, the variance in test scores for individuals applying for driving licences has been $\sigma^2 = 100$. A new examination with a new style of online test question has been developed. The licensing authority would like the variance in the test scores for the new examination to remain at the historical level. To evaluate the variance in the new examination test scores, the following two-tailed hypothesis test has been proposed:

$$H_0: \sigma^2 = 100$$

$$H_1: \sigma^2 \neq 100$$

Rejection of H_0 will indicate that a change in the variance has occurred and suggest that some questions in the new examination may need revision to make the variance of the new test scores close to the variance of the old test scores.

A random sample of 30 applicants for driving licences is given the new version of the examination. The sample provides a sample variance $s^2 = 162$. We shall use a level of significance $\alpha = 0.05$ to do the hypothesis test. The value of the χ^2 test statistic is as follows:

$$\chi^2 = \frac{(n-1)s^2}{\sigma_0^2} = \frac{(30-1)162}{100} = 46.980$$

Now, we compute the p-value. Using Table 3 of Appendix B and $n - 1 = 30 - 1 = 29$ df, we find the following:

Area in upper tail	0.10	0.05	0.025	0.01
χ^2 value (29 df)	39.087	42.557	45.722	49.588

$$\chi^2 = 46.980$$

The test statistic value $\chi^2 = 46.980$ gives a probability between 0.025 and 0.01 in the upper tail of the χ^2 distribution. Doubling these values shows that the two-tailed p-value is between 0.05 and 0.02. SPSS, Excel, R or Minitab can be used to show the exact p-value $= 0.0374$. With p-value $< \alpha = 0.05$, we reject H_0 and conclude that the new examination test scores have a population variance different from (in this case, greater than) the historical variance of $\sigma^2 = 100$.

A summary of the hypothesis testing procedures for a population variance is shown in Table 11.1.

TABLE 11.1 Summary of hypothesis tests about a population variance

	Lower-tail test	Upper-tail test	Two-tailed test
Hypotheses	$H_0: \sigma^2 \geq \sigma_0^2$	$H_0: \sigma^2 \leq \sigma_0^2$	$H_0: \sigma^2 = \sigma_0^2$
	$H_1: \sigma^2 < \sigma_0^2$	$H_1: \sigma^2 > \sigma_0^2$	$H_1: \sigma^2 \neq \sigma_0^2$
Test statistic	$\chi^2 = \dfrac{(n-1)S^2}{\sigma_0^2}$	$\chi^2 = \dfrac{(n-1)S^2}{\sigma_0^2}$	$\chi^2 = \dfrac{(n-1)S^2}{\sigma_0^2}$
Rejection rule: p-value approach	Reject H_0 if p-value $\leq \alpha$	Reject H_0 if p-value $\leq \alpha$	Reject H_0 if p-value $\leq \alpha$
Rejection rule: critical value approach	Reject H_0 if $\chi^2 \leq \chi_{1-\alpha}^2$	Reject H_0 if $\chi^2 \geq \chi_{\alpha}^2$	Reject H_0 if $\chi^2 \leq \chi_{1-\alpha/2}^2$ or if $\chi^2 \geq \chi_{\alpha/2}^2$

EXERCISES

Methods

1. Find the following χ^2 distribution values from Table 3 of Appendix B.
   a. $\chi_{0.05}^2$ with df = 5.
   b. $\chi_{0.025}^2$ with df = 15.
   c. $\chi_{0.975}^2$ with df = 20.
   d. $\chi_{0.01}^2$ with df = 10.
   e. $\chi_{0.95}^2$ with df = 18.

2. A sample of 20 items provides a sample standard deviation of 5.0.
   a. Construct a 90% confidence interval estimate of the population variance σ^2.
   b. Construct a 95% confidence interval estimate of σ^2.
   c. Construct a 95% confidence interval estimate of the population standard deviation σ.

3. A sample of 16 items provides a sample standard deviation of 9.5. Complete the following hypothesis test using $\alpha = 0.05$. What is your conclusion? Use both the p-value approach and the critical value approach.

$$H_0: \sigma^2 \leq 50$$

$$H_1: \sigma^2 > 50$$

Applications

4. The variance in drug weights is critical for the pharmaceutical industry. For a specific drug, with weights measured in grams, a sample of 18 units provided a sample variance of $s^2 = 0.36$.
   a. Construct a 90% confidence interval estimate of the population variance for the weight of this drug.
   b. Construct a 90% confidence interval estimate of the population standard deviation.

5. Amazon.com is testing the use of drones to deliver packages for same-day delivery. In order to quote narrow time windows, the variability in delivery times must be sufficiently small. Consider a sample of 24 drone deliveries with a sample variance of $s^2 = 0.81$.
   a. Construct a 90% confidence interval estimate of the population variance for the drone delivery time.
   b. Construct a 90% confidence interval estimate of the population standard deviation.

6. Because of staffing decisions, managers of the Worldview Hotel are interested in the variability in the number of rooms occupied per day during the busy season of the year. A sample of 20 days of operation shows a sample mean of 290 rooms occupied per day and a sample standard deviation of 30 rooms.
   a. What is the point estimate of the population variance?
   b. Provide a 90% confidence interval estimate of the population variance.
   c. Provide a 90% confidence interval estimate of the population standard deviation.

7. The Dow Jones Industrial Average (DJIA) is the most widely known US stock market index. Consider a day when the DJIA rose by nearly 150 points. The following table shows the stock price changes for a sample of 12 DJIA companies on that day.
   a. Compute the sample variance for the daily price change.
   b. Compute the sample standard deviation for the price change.
   c. Construct 95% confidence interval estimates of the population variance and the population standard deviation.

Company	Price change ($)	Company	Price change ($)
Aflac	0.81	Johnson & Johnson	1.46
Altice USA	0.41	Loews Corporation	0.92
Bank of America	−0.05	Nokia Corporation	0.21
Diageo plc	1.32	Sempra Energy	0.97
Fluor Corporation	2.37	Sunoco LP	0.52
Goodrich Petroleum	0.30	Tyson Foods, Inc	0.12

8. In the file 'Travel' on the online platform, there are estimated daily living costs (in euros) for a businessman travelling to 20 major cities. The estimates include a single room at a four-star hotel, beverages, breakfast, taxi fares and incidental costs.
   a. Compute the sample mean.
   b. Compute the sample standard deviation.
   c. Compute a 95% confidence interval for the population standard deviation.

TRAVEL

9. *Consumer Reports* uses a 100-point customer satisfaction score to rate major chain stores. Assume that from past experience with the satisfaction rating score, a population standard deviation of $\sigma = 12$ is expected. In one *Consumer Reports* survey, Costco was the only chain store to earn an outstanding rating for overall quality. A sample of 15 Costco customer satisfaction scores is as follows:

   95 90 83 75 95 98 80 83 82 93 86 80 94 64 62

   a. What is the sample mean customer satisfaction score for Costco?
   b. What is the sample variance?
   c. What is the sample standard deviation?
   d. Construct a hypothesis test to determine whether the population standard deviation of $\sigma = 12$ should be rejected for Costco. At the 0.05 level of significance, what is your conclusion?

10. Part variability is critical in the manufacturing of ball bearings. Large variances in the size of the ball bearings cause bearing failure and rapid wear. Production standards call for a maximum variance of 0.0025 when the bearing sizes are measured in millimetres. A sample of 15 bearings shows a sample standard deviation of 0.066 mm.
    a. Use $\alpha = 0.10$ to determine whether the sample indicates that the maximum acceptable variance is being exceeded.
    b. Compute a 90% confidence interval estimate for the variance of the ball bearings in the population.

11. Suppose that any investment with an annualized standard deviation of percentage returns greater than 20 per cent is classified as 'high-risk'. The annualized standard deviation of percentage returns for the MSCI Emerging Markets index, based on a sample of size 36, is 25.2 per cent. Construct a hypothesis test that can be used to determine whether an investment based on the movements in the MSCI index would be classified as 'high-risk'. With a 0.05 level of significance, what is your conclusion?

12. The sample standard deviation for the number of passengers taking a particular airline flight is 8. A 95% confidence interval estimate of the population standard deviation is 5.86 passengers to 12.62 passengers.
    a. Was a sample size of 10 or 15 used in the statistical analysis?
    b. Suppose the sample standard deviation of s = 8 had been based on a sample of 25 flights. What change would you expect in the confidence interval for the population standard deviation? Compute a 95% confidence interval estimate of σ with a sample size of 25.

11.2 INFERENCES COMPARING TWO POPULATION VARIANCES

We may want to compare the variances in product quality resulting from two different production processes, the variances in assembly times for two assembly methods or the variances in temperatures for two heating devices. In making such comparisons, we shall be using data collected from two independent random samples, one from population 1 and the other from population 2. The two sample variances S_1^2 and S_2^2 will be the basis for making inferences comparing the two population variances σ_1^2 and σ_2^2. Whenever the variances of two normally distributed populations are equal $(\sigma_1^2 = \sigma_2^2)$, the sampling distribution of the ratio of the two sample variances is as follows:

Sampling distribution of S_1^2/S_2^2 when $\sigma_1^2 = \sigma_2^2$

When independent simple random samples of sizes n_1 and n_2 are selected from two normally distributed populations with equal variances, the sampling distribution of:

$$\frac{S_1^2}{S_2^2}$$

(11.9)

has an F distribution with $n_1 - 1$ df for the numerator and $n_2 - 1$ df for the denominator. S_1^2 is the sample variance for the random sample of n_1 items from population 1, and S_2^2 is the sample variance for the random sample of n_2 items from population 2.

Figure 11.4 is a graph of the F distribution with 20 df for both the numerator and denominator. The graph indicates that F values are never negative, and the F distribution is not symmetrical. The precise shape of any specific F distribution depends on its numerator and denominator degrees of freedom.

We shall use F_α to denote the value of F that gives a probability of α in the upper tail of the distribution. For example, as noted in Figure 11.4, $F_{0.05}$ identifies the upper tail area of 0.05 for an F distribution with

20 df for both numerator and denominator. The specific value of $F_{0.05}$ can be found by referring to the F distribution table, Table 4 of Appendix B. Using 20 df for the numerator, 20 df for the denominator and the row corresponding to a probability of 0.05 in the upper tail, we find $F_{0.05} = 2.12$. Note that the table can be used to find F values for upper tail areas of 0.10, 0.05, 0.025 and 0.01.

FIGURE 11.4

F distribution with 20 df for the numerator and 20 df for the denominator

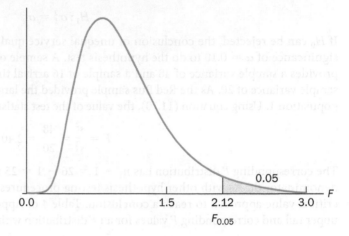

We now show how the F distribution can be used to do a hypothesis test about the equality of two population variances. The hypotheses are stated as follows:

$$H_0: \sigma_1^2 = \sigma_2^2$$

$$H_1: \sigma_1^2 \neq \sigma_2^2$$

We make the tentative assumption that the population variances are equal. If H_0 is rejected, we shall draw the conclusion that the population variances are not equal.

The hypothesis test requires two independent random samples, one from each population. The two sample variances are computed. We refer to the population providing the *larger* sample variance as population 1. This notational device is adopted to simplify the use of the probability table for the F distribution (more below). Sample size n_1 and sample variance s_1^2 correspond to population 1. Sample size n_2 and sample variance s_2^2 correspond to population 2. Based on the assumption that both populations have a normal distribution, the ratio of sample variances provides the following test statistic F.

Test statistic for hypothesis tests about population variances with $\sigma_1^2 = \sigma_2^2$

$$F = \frac{S_1^2}{S_2^2} \qquad\qquad (11.10)$$

Denoting the population with the larger sample variance as population 1, the test statistic has an F distribution with $n_1 - 1$ df for the numerator and $n_2 - 1$ df for the denominator.

Because the test statistic F is constructed with the larger sample variance in the numerator, the value of the test statistic will be in the upper tail of the F distribution. Therefore, the F distribution table (Table 4 of Appendix B) need only provide upper-tail probabilities.

We now consider an example. New Century Schools is renewing its school bus service contract for the coming year and must select one of two bus companies, the Red Bus Company or the Route One Company. We shall assume that the two companies have similar performance for average punctuality (i.e. mean arrival time) and use the variance of the arrival times as a primary measure of the bus service quality. Low variance values indicate the more consistent and higher quality service. If the variances of arrival times associated with the two services are equal, New Century Schools' managers will select the

SCHOOL BUS

company offering the better financial terms. However, if the sample data on bus arrival times provide evidence of a difference between the two variances, the managers may want to give special consideration to the company with the better or lower variance service. The appropriate hypotheses follow.

$$H_0: \sigma_1^2 = \sigma_2^2$$

$$H_1: \sigma_1^2 \neq \sigma_2^2$$

If H_0 can be rejected, the conclusion of unequal service quality is appropriate. We shall use a level of significance of $\alpha = 0.10$ to do the hypothesis test. A sample of 26 arrival times for the Red Bus service provides a sample variance of 48 and a sample of 16 arrival times for the Route One service provides a sample variance of 20. As the Red Bus sample provided the larger sample variance, we denote Red Bus as population 1. Using equation (11.10), the value of the test statistic is:

$$F = \frac{s_1^2}{s_2^2} = \frac{48}{20} = 2.40$$

The corresponding F distribution has $n_1 - 1 = 26 - 1 = 25$ numerator df and $n_2 - 1 = 16 - 1 = 15$ denominator df. As with other hypothesis testing procedures, we can use the p-value approach or the critical value approach to reach a conclusion. Table 4 of Appendix B shows the following areas in the upper tail and corresponding F values for an F distribution with 25 numerator df and 15 denominator df.

Area in upper tail	0.10	0.05	0.025	0.01
F value (df$_1$ = 25, df$_2$ = 15)	1.89	2.28	2.69	3.28

$$\uparrow$$
$$F = 2.40$$

Because $F = 2.40$ is between 2.28 and 2.69, the probability in the upper tail of the distribution is between 0.05 and 0.025. Since this is a two-tailed test, we double the upper-tail probability, which results in a p-value between 0.10 and 0.05. For this test, we selected $\alpha = 0.10$ as the level of significance, which gives us a p-value $< \alpha = 0.10$. Hence, the null hypothesis is rejected. This finding leads to the conclusion that the two bus services differ in terms of arrival time variances. The recommendation is that the New Century Schools' managers give special consideration to the better or lower variance service offered by the Route One Company.

We can use R, Excel, SPSS or Minitab to show that the test statistic $F = 2.40$ provides a two-tailed p-value = 0.0811. With $0.0811 < \alpha = 0.10$, the null hypothesis of equal population variances is rejected.

Using the critical value approach to do the two-tailed hypothesis test at $\alpha = 0.10$, there are critical values for a probability of $\alpha/2 = 0.10/2 = 0.05$ in each tail of the distribution. Because the value of the test statistic computed using equation (11.10) will always be in the upper tail, we need to determine only the upper-tail critical value. From Table 4 of Appendix B, we see that $F_{0.05} = 2.28$. So, even though we use a two-tailed test, the rejection rule is stated as follows:

$$\text{Reject } H_0 \text{ if } F \geq 2.28$$

As the test statistic $F = 2.40$ is greater than 2.28, we reject H_0 and conclude that the two bus services differ in terms of arrival time variances.

One-tailed tests involving two population variances are also possible. In this case, we use the F distribution to assess whether one population variance is significantly greater than the other. If we are using tables of the F distribution to determine the p-value or the critical value, a one-tailed hypothesis test about two population variances will always be formulated as an *upper*-tail test:

$$H_0: \sigma_1^2 \leq \sigma_2^2$$

$$H_1: \sigma_1^2 > \sigma_2^2$$

This form of the hypothesis test always places the p-value and the critical value in the upper tail of the F distribution. As a result, only upper-tail F values will be needed, simplifying both the computations and the table for the F distribution.

As an example of a one-tailed test, consider a public opinion survey. Samples of 31 men and 41 women were used to study attitudes about current political issues. The researcher wants to test whether women show a greater variation in attitude on political issues than men. In the form of the one-tailed hypothesis test given previously, women will be denoted as population 1 and men will be denoted as population 2. The hypothesis test will be stated as follows:

$$H_0: \sigma^2_{women} \leq \sigma^2_{men}$$

$$H_1: \sigma^2_{women} > \sigma^2_{men}$$

Rejection of H_0 will give the researcher the statistical support necessary to conclude that women show a greater variation in attitude on political issues.

With the sample variance for women in the numerator and the sample variance for men in the denominator, the F distribution will have $n_1 - 1 = 41 - 1 = 40$ numerator df and $n_2 - 1 = 31 - 1 = 30$ denominator df. We shall use a level of significance $\alpha = 0.05$ for the hypothesis test. The survey results provide a sample variance $s_1^2 = 120$ for women and a sample variance $s_2^2 = 80$ for men. The test statistic is as follows:

$$F = \frac{s_1^2}{s_2^2} = \frac{120}{80} = 1.50$$

Referring to Table 4 in Appendix B, we find that an F distribution with 40 numerator df and 30 denominator df has $F_{0.10} = 1.57$. Because the test statistic $F = 1.50$ is less than 1.57, the probability in the upper tail must be greater than 0.10. Hence, we can conclude that the p-value is greater than 0.10. Using R, SPSS, Excel or Minitab provides a p-value $= 0.1256$. Because the p-value $> \alpha = 0.05$, H_0 cannot be rejected. Hence, the sample results do not provide convincing evidence that women show greater variation in attitude on political issues than men.

Table 11.2 provides a summary of hypothesis tests about two population variances. Research confirms that the F distribution is sensitive to the assumption of normally distributed populations. The F distribution should not be used unless it is reasonable to assume that both populations are at least approximately normally distributed.

TABLE 11.2 Summary of hypothesis tests comparing two population variances

	Upper-tail test	Two-tailed test
Hypotheses	$H_0: \sigma_1^2 \leq \sigma_2^2$	$H_0: \sigma_1^2 = \sigma_2^2$
	$H_1: \sigma_1^2 > \sigma_2^2$	$H_1: \sigma_1^2 \neq \sigma_2^2$
	Note: Population 1 has the larger sample variance	
Test statistic	$F = \dfrac{S_1^2}{S_2^2}$	$F = \dfrac{S_1^2}{S_2^2}$
Rejection rule: p-value approach	Reject H_0 if p-value $\leq \alpha$	Reject H_0 if p-value $\leq \alpha$
Rejection rule: critical value approach	Reject H_0 if $F \geq F_\alpha$	Reject H_0 if $F \geq F_{\alpha/2}$

EXERCISES

Methods

13. Find the following F distribution values from Table 4 of Appendix B.
    a. $F_{0.05}$ with df 5 and 10.
    b. $F_{0.025}$ with df 20 and 15.
    c. $F_{0.01}$ with df 8 and 12.
    d. $F_{0.10}$ with df 10 and 20.

14. A sample of 16 items from population 1 has a sample variance $s_1^2 = 5.8$ and a sample of 21 items from population 2 has a sample variance $s_2^2 = 2.4$. Complete the following hypothesis test at the 0.05 level of significance.

    $$H_0 : \sigma_1^2 \leq \sigma_2^2$$
    $$H_1 : \sigma_1^2 > \sigma_2^2$$

    a. What is your conclusion using the p-value approach?
    b. Repeat the test using the critical value approach.

15. Consider the following hypothesis test.

    $$H_0 : \sigma_1^2 = \sigma_2^2$$
    $$H_1 : \sigma_1^2 \neq \sigma_2^2$$

    a. What is your conclusion if $n_1 = 21$, $s_1^2 = 8.2$, $n_2 = 26$, $s_2^2 = 4.0$? Use $\alpha = 0.05$ and the p-value approach.
    b. Repeat the test using the critical value approach.

Applications

16. The average annual repair cost for a car tends to increase as the car gets older. A researcher is interested in finding out whether the variance of annual repair costs also increases with car age. A sample ($n = 26$) of 8-year-old cars showed a sample standard deviation for annual repair costs of £170, compared with a sample standard deviation of £100 for a sample ($n = 25$) of 4-year-old cars.
    a. Suppose the research hypothesis is that the variance in annual repair costs is larger for the older cars. State the null and alternative hypotheses for an appropriate hypothesis test.
    b. At a 0.01 level of significance, what is your conclusion? What is the p-value? Discuss the reasonableness of your findings.

17. Based on data provided by a salary survey, the variance in annual salaries for seniors in accounting firms is approximately 2.1 and the variance in annual salaries for managers in accounting firms is approximately 11.1. The salary data are in thousands of euros. The salary data were based on samples of 25 seniors and 26 managers. Test the hypothesis that the population variances in the salaries are equal. At a 0.05 level of significance, what is your conclusion?

18. Battery life is an important issue for many smartphone owners. Public health studies have examined 'low-battery anxiety' and acute anxiety called *nomophobia* that results when a smartphone user's phone battery charge runs low and then dies. Battery life between charges for the Samsung Galaxy S9 averages 31 hours when the primary use is talk time and 10 hours when the primary use is internet applications. Because the mean hours for talk time usage are greater than the mean hours for internet usage, the question was raised as to whether the variance in hours of usage is also greater when the primary use is talk time. Sample data showing battery life between charges for the two applications follows:

Primary use: Talking
35.8 22.2 24.0 32.6 18.5 42.5 28.0 23.8 30.0 22.8 20.3 35.5

Primary use: Internet
14.0 12.5 16.4 11.9 9.9 3.1 5.4 11.0 15.2 4.0 4.7

a. Formulate hypotheses about the two population variances that can be used to determine if the population variance in battery life is greater for the talk time application.
b. What are the standard deviations of battery life for the two samples?
c. Conduct the hypothesis test and compute the *p*-value. Using a 0.05 level of significance, what is your conclusion?

19. Two new assembly methods are tested and the variances in assembly times are reported. Using $\alpha = 0.10$, test for equality of the two population variances.

	Method A	Method B
Sample size	$n_1 = 31$	$n_2 = 25$
Sample variation	$s_1^2 = 25$	$s_2^2 = 12$

20. A research hypothesis is that the variance of stopping distances of cars on wet roads is greater than the variance of stopping distances of cars on dry roads. In the research study, 16 cars travelling at the same speeds are tested for stopping distances on wet roads and 16 cars are tested for stopping distances on dry roads. On wet roads, the standard deviation of stopping distances is 10.0 metres. On dry roads, the standard deviation is 5.0 metres.
a. At a 0.05 level of significance, do the sample data justify the conclusion that the variance in stopping distances on wet roads is greater than the variance in stopping distances on dry roads? What is the *p*-value?
b. What are the implications of your statistical conclusions in terms of driving safety recommendations?

21. The grade point averages of 352 students who completed a college course in financial accounting have a standard deviation of 0.940. The grade point averages of 73 students who dropped out of the same course have a standard deviation of 0.797. Do the data indicate a difference between the variances of grade point averages for students who completed a financial accounting course and students who dropped out? Use a 0.05 level of significance.

Note: $F_{0.025}$ with 351 and 72 df is 1.466.

22. A large variance in a production process often signals an improvement opportunity by finding ways to reduce the process variance. The file 'Bags' on the online platform contains data for two machines that fill bags with powder. The file has 25 bag weights for Machine 1 and 22 bag weights for Machine 2. Do a statistical test to determine whether there is a significant difference between the variances in the bag weights for the two machines. Use a 0.05 level of significance. What is your conclusion? Which machine, if either, provides the greater opportunity for quality improvements?

BAGS

ONLINE RESOURCES

For the data files, additional questions and answers, and the software section for Chapter 11, visit the online platform.

SUMMARY

In this chapter we presented statistical procedures for making inferences about population variances. We introduced two new probability distributions: the chi-squared (χ^2) distribution and the F distribution. The χ^2 distribution can be used as the basis for interval estimation and hypothesis tests about the variance of a normally distributed population. We illustrated the use of the F distribution in hypothesis tests comparing the variances of two normally distributed populations. With independent simple random samples of sizes n_1 and n_2 selected from two normally distributed populations with equal variances, the sampling distribution of the ratio of the two sample variances has an F distribution with $n_1 - 1$ df for the numerator and $n_2 - 1$ df for the denominator.

KEY FORMULAE

Interval estimate of a population variance

$$\frac{(n-1)s^2}{\chi^2_{\alpha/2}} \leq \sigma^2 \leq \frac{(n-1)s^2}{\chi^2_{1-\alpha/2}}$$ (11.7)

Test statistic for hypothesis tests about a population variance

$$\chi^2 = \frac{(n-1)S^2}{\sigma_0^2}$$ (11.8)

Test statistic for hypothesis tests about population variances with $\sigma_1^2 = \sigma_2^2$

$$F = \frac{S_1^2}{S_2^2}$$ (11.10)

TRAINING

CASE PROBLEM 1

Air Force training programme

An Air Force introductory course in electronics uses a personalized system of instruction in which each student views a videotaped lecture and is then given a programmed instruction text. The students work independently with the text until they have completed the training and passed a test. A concern is the varying pace at which the students complete this portion of their training programme. Some students can cover the programmed instruction text relatively quickly, whereas other students work much longer with the text and require additional time to complete the course. The fast students wait until the slow students complete the introductory

course before the entire group proceeds with other aspects of their training.

A proposed alternative system involves use of computer-assisted instruction. In this method, all students view the same videotaped lecture and then each is assigned to a computer terminal for further instruction. The computer guides the student, working independently, through the self-training portion of the course.

To compare the proposed and current methods of instruction, a new class of 122 students was assigned randomly to one of the two methods. One group of 61 students used the current programmed-text method and the other group of 61 students used the proposed computer-assisted method. The course completion time in hours was recorded for each student in the study. The following data are provided in the data set 'Training'.

Course completion times (hours) for current training method

76 76 77 74 76 74 74 77 72 78 73 78 75 80 79 72 69 79 72 70 70 81
76 78 72 82 72 73 71 70 77 78 73 79 82 65 77 79 73 76 81 69 75 75
77 79 76 78 76 76 73 77 84 74 74 69 79 66 70 74 72

Course completion times (hours) for proposed computer-assisted method

74 75 77 78 74 80 73 73 78 76 76 74 77 69 76 75 72 75 72 76 72 77
73 77 69 77 75 76 74 77 75 78 72 77 78 78 76 75 76 76 75 76 80 77
76 75 73 77 77 77 79 75 75 72 82 76 76 74 72 78 71

Managerial report

1. Use appropriate descriptive statistics to summarize the training time data for each method. What similarities or differences do you observe from the sample data?

2. Conduct a hypothesis test on the difference between the population means for the two methods. Discuss your findings.

3. Compute the standard deviation and variance for each training method. Conduct a hypothesis test about the equality of population variances for the two training methods. Discuss your findings.

4. What conclusion can you reach about any differences between the two methods? What is your recommendation? Explain.

5. Can you suggest other data or testing that might be desirable before making a final decision on the training programme to be used in the future?

© bfk92/iStock

CASE PROBLEM 2

Meticulous Drill & Reamer

Meticulous Drill & Reamer (MD&R) specializes in drilling and boring precise holes in hard metals (e.g. steel alloys, tungsten carbide and titanium). The company recently contracted to drill holes with three-centimetre diameters in large carbon-steel alloy disks, and it will need to purchase a special drill to complete this job. MD&R has eliminated from consideration all but two of the drills it has been assessing: Davis Drills' T2005 and Worth IndustrialTools' AZ100. These manufacturers have each agreed to allow MD&R to use a T2005 and an AZ100 for one week to determine which drill it will purchase. During the one-week trial, MD&R uses each of these drills to drill 31 holes with a target diameter of three centimetres in one large carbon-steel alloy disk, then measures the diameter of each hole and records the results. MD&R's results are provided in the table that follows and are available in the 'MeticulousDrills' data file.

MD&R wants to consider the accuracy (closeness of the diameter to three centimetres) as well as the precision (the variance of the diameter) of the holes drilled by the T2005 and the AZ100 when deciding which model to purchase.

In making this assessment for MD&R, consider the following four questions:

1. Are the holes drilled by the T2005 or the AZ100 more accurate? That is, which model of drill produces holes with a mean diameter closer to three centimetres? Is a hypothesis test comparing these two means necessary? Why or why not?

METICULOUS
DRILLS

(Continued)

2. Are the holes drilled by the T2005 or the AZ100 more precise? That is, which model of drill produces holes with a smaller variance?

3. Conduct a test of the hypothesis that the T2005 and the AZ100 are equally precise (that is, have equal variances) at $\alpha = 0.05$. Discuss your findings.

4. Which drill do you recommend to MD&R? Why?

Hole diameter (cm)					
T2005	AZ100	T2005	AZ100	T2005	AZ100
3.06	2.91	3.05	2.97	3.04	3.06
3.04	3.31	3.01	3.05	3.01	3.25
3.13	2.82	2.73	2.95	2.95	2.82
3.01	3.01	3.12	2.92	3.14	3.22
2.95	2.94	3.04	2.71	3.31	2.93
3.02	3.17	3.10	2.77	3.01	3.24
3.02	3.25	3.02	2.73	2.93	2.77
3.12	3.39	2.92	3.18	3.00	2.94
3.00	3.22	3.01	2.95	3.04	3.31
3.04	2.97	3.15	2.86		
3.03	2.93	2.69	3.16		

© Christian Camus/iStock

12

Tests of Goodness of Fit and Independence

CHAPTER CONTENTS

Statistics in Practice   Pan-European and National lotteries

LEARNING OBJECTIVES   After studying this chapter and doing the exercises, you should be able to:

1 Construct and interpret the results of goodness of fit tests, using the chi-squared distribution, for several situations:

   1.1   A multinomial population.

   1.2   A Poisson distribution.

1.3   A normal distribution.

1.4   A test of independence in a two-way contingency table.

In Chapter 11 we showed how the chi-squared distribution can be used in estimation and in hypothesis tests about a population variance. In the present chapter, we introduce hypothesis-testing procedures based on the χ^2 distribution for data in one-dimensional and two-dimensional frequency tables. Like other hypothesis-testing procedures, these tests compare sample results with those expected when the null hypothesis is true.

In the following section we introduce a goodness of fit test for a multinomial population. Later we show goodness of fit tests for the Poisson and normal distributions, and then discuss the test for independence in two-way contingency tables.

12.1 GOODNESS OF FIT TEST: A MULTINOMIAL POPULATION

Suppose each element of a population is assigned to one, and only one, of several classes or categories. Such a population is a **multinomial population**. The multinomial distribution can be thought of as an extension of the binomial distribution to three or more categories of outcomes. On each trial of a multinomial experiment, just one of the outcomes occurs. Each trial of the experiment is assumed to be independent of all others, and the probabilities of the outcomes remain the same at each trial.

STATISTICS IN PRACTICE
Pan-European and National lotteries

Every week, hundreds of millions of people across Europe pay to take a small gamble in the hope of becoming an instant millionaire. They do this by buying one or more tickets in a national lottery or a pan-European lottery. In 2019, the market-leading EuroMillions lottery sold a total of almost 11 billion tickets over the year. European Lotteries, the umbrella organization for national lotteries in Europe, reported the 2020 'gross gaming revenue' (sales minus prizes) of its 69 members as over €20 billion.

The precise details of the game, or gamble, differ from lottery to lottery, but the general principle is that each ticket buyer chooses several numbers from a prescribed set – for example the numbers 1 to 59 in the UK National Lottery. The jackpot winner (or winners) is the ticket holder whose chosen numbers exactly match those picked out from the full set by a 'randomizing device' on the day the lottery is decided. The randomizing device is usually a moderately sophisticated and TV-friendly piece of machinery that thoroughly mixes a set of numbered balls and picks out balls one by one. The objective is to give each ball an equal probability of being picked, so that every possible combination of numbers has equal probability.

Checks are periodically made to provide assurance on this principle of fairness. An independent body usually makes the checks. For example, the Centre for the Study of Gambling at the University of Salford, UK reported to the National Lotteries Commission in January 2010 on the randomness of the EuroMillions draws. In the report, comparisons were made between the actual frequencies with which individual balls have been drawn and the frequencies expected assuming fairness or randomness. In statistical parlance, these are known as goodness of fit tests, more specifically as chi-squared tests.

In this chapter you will learn how chi-squared tests like those in the EuroMillions report are done.

© vatrushka67 / iStock

Consider a market share study. Over the past year market shares stabilized at 30 per cent for company A, 50 per cent for company B and 20 per cent for company C. Recently, company C developed a 'new and improved' product to replace its current market offering. Company C retained market research analysts Schott MR to assess whether the new product would alter market shares.

In this case, the population of interest is a multinomial population. Each customer is classified as buying from company A, company B or company C, so we have a multinomial population with three possible outcomes. We use the notation π_A, π_B and π_C to represent the market shares for companies A, B and C respectively.

Schott MR will conduct a sample survey and find the sample proportion preferring each company's product. A hypothesis test will then be done to assess whether the new product will lead to a change in market shares. The null and alternative hypotheses are:

$$H_0: \pi_A = 0.30, \pi_B = 0.50, \pi_C = 0.20$$
$$H_1: \text{The population proportions are not } \pi_A = 0.30, \pi_B = 0.50, \pi_C = 0.20$$

If the sample results lead to the rejection of H_0, Schott MR will have evidence suggesting that the introduction of the new product will affect market shares.

Schott MR has used a consumer panel of 200 customers for the study, in which each customer is asked to specify a purchase preference for one of three alternatives: company A's product, company B's product or company C's new product. This is equivalent to a multinomial experiment with 200 trials. The 200 responses are summarized here:

	Observed frequency	
Company A's product	Company B's product	Company C's new product
48	98	54

We now do a **goodness of fit test** to assess whether the sample of 200 customer purchase preferences is consistent with the null hypothesis. The goodness of fit test compares the sample of *observed* results with the *expected* results, the latter calculated under the assumption that the null hypothesis is true. The next step is therefore to compute expected purchase preferences for the 200 customers under the assumption that $\pi_A = 0.30$, $\pi_B = 0.50$ and $\pi_C = 0.20$. The expected frequency for each category is found by multiplying the sample size of 200 by the hypothesized proportion for the category.

	Expected frequency	
Company A's product	Company B's product	Company C's new product
$200(0.30) = 60$	$200(0.50) = 100$	$200(0.20) = 40$

The goodness of fit test now focuses on the differences between the observed frequencies and the expected frequencies. Large differences between observed and expected frequencies cast doubt on the assumption that the hypothesized proportions (market shares) are correct. The following test statistic is used to assess the implications of the differences between the observed and expected frequencies.

Test statistic for goodness of fit

$$\chi^2 = \sum_{i=1}^{k} \frac{(f_i - e_i)^2}{e_i} \tag{12.1}$$

where:

f_i = observed frequency for category i

e_i = expected frequency for category i

k = the number of categories

Note: The test statistic has a χ^2 distribution with $k - 1$ degrees of freedom (df), provided the expected frequencies are 5 or more for all categories.

There is an alternative way of expressing the formula for the χ^2 statistic, which under some circumstances is more convenient for calculation purposes. This is:

$$\chi^2 = \sum_{i=1}^{k} \frac{f_i^2}{e_i} - n \tag{12.2}$$

where $n = \sum_{i=1}^{k} f_i = \sum_{i=1}^{k} e_i$, the total sample size.

In the Schott MR example, we use the sample data to test the hypothesis that the multinomial population has the proportions $\pi_A = 0.30$, $\pi_B = 0.50$ and $\pi_C = 0.20$. We shall use level of significance $\alpha = 0.05$. The computation of the χ^2 test statistic is shown in Table 12.1, giving $\chi^2 = 7.34$. The calculations are also shown in the Excel worksheet 'Market Share', using both equation (12.1) and equation (12.2). Using (12.2), $\sum(f_i^2/e_i) = 207.34$. Subtracting $n = 200$ gives $\chi^2 = 7.34$.

MARKET
SHARE

TABLE 12.1 Computation of the χ^2 test statistic for the market share study

	Hypothesized proportion	Obs freq (f_i)	Exp freq (e_i)	Difference $(f_i - e_i)$	$(f_i - e_i)^2$	$(f_i - e_i)^2/e_i$
Company A	0.30	48	60	−12	144	2.400
Company B	0.50	98	100	−2	4	0.040
Company C	0.20	54	40	14	196	4.900
Total		200	200			$\chi^2 = 7.340$

We shall reject the null hypothesis if the differences between the observed and expected frequencies are large, which in turn will result in a large value for the test statistic. Hence the goodness of fit test will always be an upper-tail test. With $k - 1 = 3 - 1 = 2$ df, the χ^2 table (Table 3 of Appendix B) provides the following:

Area in upper tail	0.10	0.05	0.025	0.01
χ^2 value (2 df)	4.605	5.991	7.378	9.210

$$\chi^2 = 7.340$$

The test statistic $\chi^2 = 7.340$ is between 5.991 and 7.378 (very close to 7.378), so the corresponding upper-tail area or p-value must be between 0.05 and 0.025 (very close to 0.025). With p-value $< \alpha = 0.05$, we reject H_0 and conclude that the introduction of the new product by company C may alter the current market share structure. SPSS, R, Excel or Minitab can be used to show that $\chi^2 = 7.340$ gives a p-value $= 0.0255$ (refer to the software guides on the online platform).

Instead of using the p-value, we could use the critical value approach to draw the same conclusion. With $\alpha = 0.05$ and 2 df, the critical value for the test statistic is $\chi^2 = 5.991$. The upper tail rejection rule is: Reject H_0 if $\chi^2 \geq 5.991$. With $\chi^2 = 7.340 > 5.991$, we reject H_0. The p-value approach and critical value approach provide the same conclusion.

Although the test itself does not directly tell us about *how* market shares may change, we can compare the observed and expected frequencies descriptively to get an idea of the change in market structure. We see that the observed frequency of 54 for company C is larger than the expected frequency of 40. As the latter was based on current market shares, the larger observed frequency suggests that the new product will have a positive effect on company C's market share. Similar comparisons for the other two companies suggest that company C's gain in market share will hurt company A more than company B.

Here are the steps for doing a goodness of fit test for a hypothesized multinomial population distribution:

Multinomial distribution goodness of fit test: a summary

1. State the null and alternative hypotheses.

H_0: The population follows a multinomial distribution with specified probabilities for each of the k categories

H_1: The population does not follow a multinomial distribution with the specified probabilities for each of the k categories

2. Select a random sample and record the observed frequencies f_i for each category.
3. Assume the null hypothesis is true and determine the expected frequency e_i in each category by multiplying the category probability by the sample size.
4. Compute the value of the test statistic.

5. Rejection rule:

p-value approach:	Reject H_0 if p-value $\leq \alpha$
Critical value approach:	Reject H_0 if $\chi^2 \geq \chi^2_\alpha$

where α is the level of significance for the test and there are $k - 1$ df.

EXERCISES

Methods

1. Do the following χ^2 goodness of fit test.

$$H_0: \pi_A = 0.40, \pi_B = 0.40, \pi_C = 0.20$$

$$H_1: \text{The population proportions are not } \pi_A = 0.40, \pi_B = 0.40, \pi_C = 0.20$$

A sample of size 200 yielded 60 in category A, 120 in category B and 20 in category C. Use $\alpha = 0.01$ and test to see whether the proportions are as stated in H_0.
   a. Use the p-value approach.
   b. Repeat the test using the critical value approach.

2. Suppose we have a multinomial population with four categories: A, B, C and D. The null hypothesis is that the proportion of items is the same in every category, i.e.

$$H_0: \pi_A = \pi_B = \pi_C = \pi_D = 0.25$$

A sample of size 300 yielded the following results.

   A: 85   B: 95   C: 50   D: 70

Use $\alpha = 0.05$ to determine whether H_0 should be rejected. What is the p-value?

Applications

3. The Mars company manufactures M&Ms, one of the most popular sweet treats in the world. The milk chocolate sweets come in a variety of colours: blue, brown, green, orange, red and yellow. The overall proportions for the colours are 0.24 blue, 0.13 brown, 0.20 green, 0.16 orange, 0.13 red and 0.14 yellow. In a sampling study, several bags of M&M milk chocolates were opened and the following colour counts were obtained:

Blue	Brown	Green	Orange	Red	Yellow
105	72	89	84	70	80

Use a 0.05 level of significance and the sample data to test the hypothesis that the overall proportions for the colours are as stated above. What is your conclusion?

4. How well do airline companies serve their customers? A study by *Business Week* showed the following customer ratings: 3 per cent excellent, 28 per cent good, 45 per cent fair and 24 per cent poor. In a follow-up study of service by telephone companies, assume that a sample of 400 adults found the following customer ratings: 24 excellent, 124 good, 172 fair and 80 poor. Taking the figures from the *Business Week* study as 'population' values, is the distribution of the customer ratings for telephone companies different from the distribution of customer ratings for airline companies? Test with $\alpha = 0.01$. What is your conclusion?

5. In setting sales quotas, the marketing manager of a multinational company assumes that order potentials are the same for each of four sales territories in the Middle East. A sample of 200 sales follows. Should the manager's assumption be rejected? Use $\alpha = 0.05$.

Sales territories			
1	2	3	4
60	45	59	36

6. A community park will open soon in a large European city. A sample of 210 individuals are asked to state their preference for when they would most like to visit the park. The sample results follow.

Monday	Tuesday	Wednesday	Thursday	Friday	Saturday	Sunday
20	30	30	25	35	20	50

In developing a staffing plan, should the park manager plan on the same number of individuals visiting the park each day? Support your conclusion with a statistical test. Use $\alpha = 0.05$.

7. The results of *ComputerWorld's* Annual Job Satisfaction exercise showed that 28 per cent of information systems managers are very satisfied with their job, 46 per cent are somewhat satisfied, 12 per cent are neither satisfied nor dissatisfied, 10 per cent are somewhat dissatisfied and 4 per cent are very dissatisfied. Suppose that a sample of 500 computer programmers yielded the following results.

Category	Number of respondents
Very satisfied	105
Somewhat satisfied	235
Neither	55
Somewhat dissatisfied	90
Very dissatisfied	15

Taking the *ComputerWorld* figures as 'population' values, use $\alpha = 0.05$ and test to determine whether the job satisfaction for computer programmers is different from the job satisfaction for information systems managers.

12.2 GOODNESS OF FIT TEST: POISSON AND NORMAL DISTRIBUTIONS

In the previous section, we introduced the χ^2 goodness of fit test for a multinomial population. In general, the χ^2 goodness of fit test can be used with any hypothesized probability distribution. In this section we illustrate for cases in which the population is hypothesized to have a Poisson or a normal distribution. The test follows the same general procedure as in Section 12.1.

Poisson distribution

Consider the arrival of customers at the Mediterranean Food Market. Because of recent staffing problems, the Mediterranean's managers asked a local consultancy to assist with the scheduling of checkout assistants. After reviewing the checkout operation, the consultancy will make a recommendation for a scheduling procedure. The procedure, based on a mathematical analysis of waiting times, is applicable only if the number of customer arrivals follows the Poisson distribution. Therefore, before the scheduling process is implemented, data on customer arrivals must be collected and a statistical test done to see whether an assumption of a Poisson distribution for arrivals is reasonable.

We define the arrivals at the store in terms of the *number of customers* entering the store during 5-minute intervals. The following null and alternative hypotheses are appropriate:

H_0: The number of customers entering the store during 5-min intervals has a Poisson probability distribution

H_1: The number of customers entering the store during 5-min intervals does not have a Poisson distribution

If a sample of customer arrivals provides insufficient evidence to reject H_0, the Mediterranean will proceed with the implementation of the consultancy's scheduling procedure. However, if the sample leads to the rejection of H_0, the assumption of the Poisson distribution for the arrivals cannot be made and other scheduling procedures will be considered.

To test the assumption of a Poisson distribution for the number of arrivals during weekday morning hours, a store assistant randomly selects a sample, $n = 128$, of 5-min intervals during weekday mornings over a 3-week period. For each 5-min interval in the sample, the store assistant records the number of customer arrivals. The store assistant then summarizes the data by counting the number of 5-min intervals with no arrivals, the number of 5-min intervals with one arrival and so on. These data are summarized in Table 12.2.

TABLE 12.2 Observed frequency of the Mediterranean's customer arrivals for a sample ($n = 128$) of 5-min intervals

Number of customers arriving	Observed frequency	Number of customers arriving	Observed frequency
0	2	6	22
1	8	7	16
2	10	8	12
3	12	9	6
4	18		
5	22	**Total**	**128**

To do the goodness of fit test, we need to consider the expected frequency for each of the ten categories, under the assumption that the Poisson distribution of arrivals is true. The Poisson probability function, introduced in Chapter 5, is:

$$p(X = x) = \frac{\mu^x e^{-\mu}}{x!}$$

In this function, μ represents the mean or expected number of customers arriving per 5-min interval, X is a random variable indicating the number of customers arriving during a 5-min interval and $p(X = x)$ is the probability that exactly x customers will arrive in a 5-min interval.

To use equation (12.2), we must obtain an estimate of μ, the mean number of customer arrivals during a 5-min time interval. The sample mean for the data in Table 12.2 provides this estimate. With no customers arriving in two 5-min time intervals, one customer arriving in eight 5-min time intervals and so on, the total number of customers who arrived during the sample of 128 5-min time intervals is given by $0(2) + 1(8) + 2(10) + \ldots + 9(6) = 640$. The 640 customer arrivals over the sample of 128 intervals provide an estimated mean arrival rate of $640/128 = 5$ customers per 5-min interval. With this value for the mean of the distribution, an estimate of the Poisson probability function for the Mediterranean Food Market is:

$$p(X = x) = \frac{5^x e^{-5}}{x!}$$

This probability function can be evaluated for different x values to determine the probability associated with each category of arrivals (refer to Chapter 5 if you need a reminder). These probabilities are given in Table 12.3. For example, the probability of zero customers arriving during a 5-min interval is $p(0) = 0.0067$, the probability of one customer arriving during a 5-min interval is $p(1) = 0.0337$ and so on. As we saw in Section 12.1, the expected frequencies for the categories are found by multiplying the probabilities by the sample size. For example, the expected number of periods with zero arrivals is given by $(0.0067)(128) = 0.86$, the expected number of periods with one arrival is given by $(0.0337)(128) = 4.31$ and so on.

TABLE 12.3   Expected frequency of Mediterranean's customer arrivals, assuming a Poisson distribution with $\mu = 5$

Number of customers arriving (x)	Poisson probability p(x)	Expected number of 5-min time intervals with x arrivals, 128p(x)
0	0.0067	0.86
1	0.0337	4.31
2	0.0842	10.78
3	0.1404	17.97
4	0.1755	22.46
5	0.1755	22.46
6	0.1462	18.71
7	0.1044	13.36
8	0.0653	8.36
9	0.0363	4.65
10 or more	0.0318	4.07
Total	**1.0000**	**127.99**

In Table 12.3, four of the categories have an expected frequency less than five. This condition violates the requirements for use of the χ^2 distribution. However, adjacent categories can be combined to satisfy the 'at least five' expected frequency requirement. We shall combine 0 and 1 into a single category, and then combine 9 with '10 or more' into another single category. Table 12.4 shows the observed and expected frequencies after combining categories.

TABLE 12.4   Observed and expected frequencies for the Mediterranean's customer arrivals after combining categories, and computation of the χ^2 test statistic

Number of customers arriving (x)	Obs freq (f_i)	Exp freq (e_i)	Difference (f_i − e_i)	(f_i − e_i)²	(f_i − e_i)²/e_i
0 or 1	10	5.17	4.83	23.33	4.51
2	10	10.78	−0.78	0.61	0.06
3	12	17.97	−5.97	35.64	1.98
4	18	22.46	−4.46	19.89	0.89
5	22	22.46	−0.46	0.21	0.01
6	22	18.71	3.29	10.82	0.58
7	16	13.36	2.64	6.97	0.52
8	12	8.36	3.64	13.25	1.58
9 or more	6	8.72	−2.72	7.34	0.84
Total	**128**	**127.99**			$\chi^2 = 10.98$

As in Section 12.1, the goodness of fit test focuses on the differences between observed and expected frequencies, $f_i - e_i$. The calculations are shown in Table 12.4. The value of the test statistic is $\chi^2 = 10.98$. The calculations using both equation (12.1) and equation (12.2) are in the Excel worksheet 'Mediterranean'. Using (12.2), $\Sigma(f_i^2/e_i) = 138.98$. Subtracting $n = 128$ gives $\chi^2 = 10.98$. In general, the χ^2 distribution for a goodness of fit test has $k - p - 1$ df, where k is the number of categories and p is the number of population parameters estimated from the sample data. Table 12.4 shows $k = 9$ categories. As the sample data were used to estimate the mean of the Poisson distribution, $p = 1$. Hence, there are $k - p - 1 = 9 - 1 - 1 = 7$ df.

MEDITER-RANEAN

Suppose we test the null hypothesis with a 0.05 level of significance. We need to determine the p-value for the test statistic $\chi^2 = 10.96$ by finding the area in the upper tail of a χ^2 distribution with 7 df. Using Table 3 of Appendix B, we find that $\chi^2 = 10.98$ provides an area in the upper tail greater than 0.10. So we know that the p-value is greater than 0.10. SPSS, R, Excel or Minitab shows p-value $= 0.1403$. With p-value $> \alpha = 0.10$, we cannot reject H_0. The assumption of a Poisson probability distribution for weekday morning customer arrivals cannot be rejected. As a result, the Mediterranean's management may proceed with the consulting firm's scheduling procedure for weekday mornings.

Poisson distribution goodness of fit test: a summary

1. State the null and alternative hypotheses.

H_0: The population has a Poisson distribution

H_1: The population does not have a Poisson distribution

2. Select a random sample and
   a. Record the observed frequency f_i for each value of the Poisson random variable.
   b. Compute the mean number of occurrences.
3. Compute the expected frequency of occurrences e_i for each value of the Poisson random variable. Multiply the sample size by the Poisson probability of occurrence for each value of the Poisson random variable. If there are fewer than five expected occurrences for some values, combine adjacent values and reduce the number of categories as necessary.
4. Compute the value of the test statistic.

$$\chi^2 = \sum_{i=1}^{k} \frac{(f_i - e_i)^2}{e_i}$$

5. Rejection rule:

   p-value approach: Reject H_0 if p-value $\leq \alpha$

   Critical value approach: Reject H_0 if $\chi^2 \geq \chi_\alpha^2$

   where α is the level of significance for the test, and there are $k - 2$ df. (This assumes the Poisson mean has been estimated from the sample data. If the Poisson mean has been determined on theoretical grounds, the number of df is $k - 1$.)

Normal distribution

A χ^2 goodness of fit test can also be used for a hypothesized normal distribution. It is like the procedure for the Poisson distribution. However, because the normal distribution is continuous, we must modify the way the categories are defined and how the expected frequencies are computed.

Consider the job applicant test data for the digital marketing company DigiProm, listed in Table 12.5. DigiProm hires approximately 400 new employees annually for its various offices located in Europe and the Middle East. The personnel manager asks whether a normal distribution applies for the population of test scores. If such a distribution can be used, the distribution would be helpful in evaluating specific test scores; that is, scores in the upper 20 per cent, lower 40 per cent and so on, could be identified quickly. Hence, we want to test the null hypothesis that the population of test scores has a normal distribution.

TABLE 12.5 DigiProm employee aptitude test scores for 50 randomly chosen job applicants

71	65	54	93	60	86	70	70	73	73
55	63	56	62	76	54	82	79	76	68
53	58	85	80	56	61	64	65	62	90
69	76	79	77	54	64	74	65	65	61
56	63	80	56	71	79	84	66	61	61

We first use the data in Table 12.5 to calculate estimates of the mean and standard deviation of the normal distribution that will be considered in the null hypothesis. We use the sample mean and the sample standard deviation as point estimators of the mean and standard deviation of the normal distribution. The calculations follow.

$$\bar{x} = \frac{\Sigma x_i}{n} = \frac{3{,}421}{50} = 68.42$$

$$s = \sqrt{\frac{\Sigma (x_i - \bar{x})^2}{n - 1}} = \sqrt{\frac{5{,}310.04}{49}} = 10.41$$

DIGIPROM

Using these values, we state the following hypotheses about the distribution of the job applicant test scores.

H_0: The population of test scores has a normal distribution with mean 68.42 and standard deviation 10.41

H_1: The population of test scores does not have a normal distribution with mean 68.42 and standard deviation 10.41

Now we look at how to define the categories for a goodness of fit test involving a normal distribution. For the discrete probability distribution in the Poisson distribution test, the categories were readily defined in terms of the number of customers arriving, such as 0, 1, 2 and so on. However, with the continuous normal probability distribution, we must use a different procedure for defining the categories. We need to define the categories in terms of *intervals* of test scores.

Recall the rule of thumb for an expected frequency of at least five in each interval or category. We define the categories of test scores such that the expected frequencies will be at least five for each category. With a sample size of 50, one way of establishing categories is to divide the normal distribution into ten equal-probability intervals (refer to Figure 12.1). With a sample size of 50, we would expect five outcomes in each interval or category and the rule of thumb for expected frequencies would be satisfied.

When the normal probability distribution is assumed, the standard normal distribution tables can be used to determine the category boundaries. First consider the test score cutting off the lowest 10 per cent of the test scores. From Table 1 of Appendix B we find that the z value for this test score is -1.28. Therefore, the test score $x = 68.42 - 1.28(10.41) = 55.10$ provides this cut-off value for the lowest 10 per cent of the scores. For the lowest 20 per cent, we find $z = -0.84$ and so $x = 68.42 - 0.84(10.41) = 59.68$. Working through the normal distribution in that way provides the following test score values:

Lower 10%: $68.42 - 1.28(10.41) = 55.10$
Lower 20%: $68.42 - 0.84(10.41) = 59.68$
Lower 30%: $68.42 - 0.52(10.41) = 63.01$
Lower 40%: $68.42 - 0.25(10.41) = 65.82$
Mid-score: $68.42 - 0(10.41) = 68.42$
Upper 40%: $68.42 + 0.25(10.41) = 71.02$
Upper 30%: $68.42 + 0.52(10.41) = 73.83$
Upper 20%: $68.42 + 0.84(10.41) = 77.16$
Upper 10%: $68.42 + 1.28(10.41) = 81.74$

These cut-off or interval boundary points are identified in Figure 12.1.

FIGURE 12.1

Normal distribution for the DigiProm example with ten equal-probability intervals

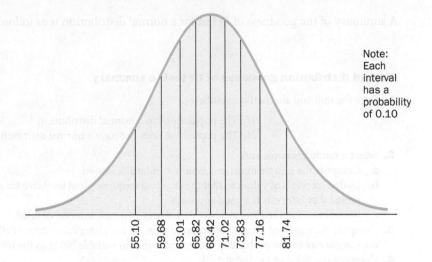

Note: Each interval has a probability of 0.10

55.10 59.68 63.01 65.82 68.42 71.02 73.83 77.16 81.74

We can now return to the sample data of Table 12.5 and determine the observed frequencies for the categories. The results are in Table 12.6. The goodness of fit calculations now proceed exactly as before. We compare the observed and expected results by computing a χ^2 value. The computations are also shown in Table 12.6. We see that the value of the test statistic is $\chi^2 = 7.2$.

TABLE 12.6 Observed and expected frequencies for DigiProm job applicant test scores, and computation of the χ^2 test statistic

Test score interval	Obs freq (f_i)	Exp freq (e_i)	Difference $(f_i - e_i)$	$(f_i - e_i)^2$	$(f_i - e_i)^2/e_i$
Less than 55.10	5	5	0	0	0.0
55.10 to 59.67	5	5	0	0	0.0
59.68 to 63.00	9	5	4	16	3.2
63.01 to 65.81	6	5	1	1	0.2
65.82 to 68.41	2	5	−3	9	1.8
68.42 to 71.01	5	5	0	0	0.0
71.02 to 73.82	2	5	−3	9	1.8
73.83 to 77.15	5	5	0	0	0.0
77.16 to 81.73	5	5	0	0	0.0
81.74 and over	6	5	1	1	0.2
Totals	**50**	**50**			$\chi^2 = 7.2$

To determine whether the computed χ^2 value of 7.2 is large enough to reject H_0, we need to refer to the appropriate χ^2 distribution tables. Using the rule for computing the number of degrees of freedom for the goodness of fit test, we have $k - p - 1 = 10 - 2 - 1 = 7$ df based on $k = 10$ categories and $p = 2$ parameters (mean and standard deviation) estimated from the sample data.

Suppose we do the test with $\alpha = 0.10$. To test this hypothesis, we need to determine the p-value for the test statistic $\chi^2 = 7.2$ by finding the area in the upper tail of a χ^2 distribution with 7 df. Using Table 3 of Appendix B, we find that $\chi^2 = 7.2$ provides an area in the upper tail greater than 0.10. So we know that the p-value is greater than 0.10. Excel, R, SPSS or Minitab shows p-value $= 0.4084$.

With p-value $> \alpha = 0.10$, the hypothesis that the probability distribution for the DigiProm job applicant test scores is a normal distribution cannot be rejected. The normal distribution may be applied to assist in the interpretation of test scores.

A summary of the goodness of fit test for a normal distribution is as follows:

Normal distribution goodness of fit test: a summary

1. State the null and alternative hypotheses.

H_0: The population has a normal distribution
H_1: The population does not have a normal distribution

2. Select a random sample and
   a. Compute the sample mean and sample standard deviation.
   b. Define intervals of values so that the expected frequency is at least five for each interval. Using equal probability intervals is a good approach.
   c. Record the observed frequency of data values f_i in each interval defined.
3. Compute the expected number of occurrences e_i for each interval of values defined in step 2(b). Multiply the sample size by the probability of a normal random variable being in the interval.
4. Compute the value of the test statistic.

$$\chi^2 = \sum_{i=1}^{k} \frac{(f_i - e_i)^2}{e_i}$$

5. Rejection rule:

p-value approach: Reject H_0 if p-value $\leq \alpha$

Critical value approach: Reject H_0 if $\chi^2 \geq \chi_{\alpha}^2$

where α is the level of significance for the test, and there are $k - 3$ df. (This assumes the mean and standard deviation for the normal distribution have been estimated from the sample data. If the normal parameters have been determined on theoretical grounds, the number of df is $k - 1$.)

EXERCISES

Methods

8. Data on the number of occurrences per time interval and observed frequencies follow. Use a goodness of fit test with $\alpha = 0.05$ to see whether the data fit a Poisson distribution.

Number of occurrences	Observed frequency
0	39
1	30
2	30
3	18
4	3

9. The following data are believed to have come from a normal distribution. Use a goodness of fit test with $\alpha = 0.05$ to test this claim.

17	23	22	24	19	23	18	22	20	13	11	21	18	20	21
21	18	15	24	23	23	43	29	27	26	30	28	33	23	29

Applications

10. Use a goodness of fit test and the following data to test whether the weekly demand for a specific product in a white-goods store is normally distributed. Use $\alpha = 0.10$. The sample mean is 24.5 and the sample standard deviation is 3.0.

18	20	22	27	22	25	22	27	25	24
26	23	20	24	26	27	25	19	21	25
26	25	31	29	25	25	28	26	28	24

11. In the UK National Lottery, the jackpot prize is divided between ticket holders who match all six numbers drawn in the lottery. The table below shows the frequency distribution for the number of jackpot winners in each draw, in the year (104 draws) preceding the increase in the number of balls from 49 to 59 (refer to Case Problem 2 at the end of the chapter).
    a. Test whether the prize distribution follows a Poisson distribution (use $\alpha = 0.05$).
    b. How would you expect the frequency distribution for the number of jackpot winners per draw to have changed following the increase in the number of balls?

Number of jackpot winners	Observed frequency
0	46
1	35
2	12
3	9
4	2
Total	104

12. One of the products offered to savers by the UK Government is the Premium Bond. No interest is paid on Premium Bonds. Instead, each £1 bond is entered in a monthly lottery, with a small chance of winning a tax-free cash prize. The table below shows the prize history over a five-year period (60 monthly lotteries) of an individual who has the maximum permitted holding of £50,000 Premium Bonds.
    a. What is the mean number of prizes won per month?
    b. Calculate an estimate of the probability that a single £1 bond wins a prize in any month.
    c. Test whether the prize distribution follows a Poisson distribution, using $\alpha = 0.10$.

Number of prizes won	Observed frequency (months)
0	13
1	24
2	14
3	5
4	2
6	1
10	1

13. A random sample of final examination grades for a college course in Middle-East studies is as follows:

55	85	72	99	48	71	88	70	59	98
80	74	93	85	74	82	90	71	83	60
95	77	84	73	63	72	95	79	51	85
76	81	78	65	75	87	86	70	80	64

Using $\alpha = 0.05$, assess whether the distribution of grades in the population is normal.

12.3 TEST OF INDEPENDENCE

Another important application of the χ^2 goodness of fit test involves testing for the independence of two qualitative (categorical) variables. Consider a study conducted by the Millennium Brewery, which manufactures and distributes three types of beer: pilsner, export and dark beer. In an analysis of the market segments for the three beers, the firm's market researchers raised the question of whether preferences for the three beers differ between male and female beer drinkers. If beer preference is independent of gender, a single advertising campaign will be devised for all the Millennium beers. However, if beer preference depends on the gender of the beer drinker, the firm will tailor its promotions to different target markets.

A test of independence addresses the question of whether the beer preference (pilsner, export or dark) is independent of the gender of the beer drinker (male, female). The hypotheses for this test are:

H_0: Beer preference is independent of the gender of the beer drinker
H_1: Beer preference is not independent of the gender of the beer drinker

Table 12.7 can be used to describe the situation. The population under study is all male and female beer drinkers. A sample can be selected from this population and each individual asked to state their preference among the three Millennium beers. Each individual in the sample will be classified in one of the six cells in the table. For example, an individual may be a male preferring export (cell (1,2)), a female preferring pilsner (cell (2,1)), a female preferring dark beer (cell (2,3)) and so on.

TABLE 12.7   Contingency table for beer preference and gender of beer drinker

Gender	Beer preference		
	Pilsner	Export	Dark
Male	cell(1,1)	cell(1,2)	cell(1,3)
Female	cell(2,1)	cell(2,2)	cell(2,3)

As we have listed all possible combinations of beer preference and gender – in other words, all possible contingencies – Table 12.7 is called a **contingency table**. The test of independence is sometimes referred to as a *contingency table test*.

Suppose a simple random sample of 150 beer drinkers is selected. After tasting each beer, the individuals in the sample are asked to state their first-choice preference. The cross tabulation in Table 12.8 summarizes the responses. The data for the test of independence are collected in terms of counts or frequencies for each cell or category. Of the 150 individuals in the sample, 20 were men favouring pilsner, 40 were men favouring export, 20 were men favouring dark beer and so on. The data in Table 12.8 are the observed frequencies for the six classes or categories.

TABLE 12.8   Sample results for beer preferences of male and female beer drinkers (observed frequencies)

Gender	Beer preference			
	Pilsner	Export	Dark	Total
Male	20	40	20	**80**
Female	30	30	10	**70**
Total	**50**	**70**	**30**	**150**

If we can determine the expected frequencies under the assumption of independence between beer preference and gender of the beer drinker, we can use the χ^2 distribution to determine whether there is a statistically significant difference between observed and expected frequencies.

Expected frequencies for the cells of the contingency table are based on the following rationale. We assume the null hypothesis of independence between beer preference and gender of the beer drinker is true. We note that in the entire sample of 150 beer drinkers, a total of 50 prefer pilsner, 70 prefer export and 30 prefer dark beer. In terms of fractions, 50/150 of the beer drinkers prefer pilsner, 70/150 prefer export and 30/150 prefer dark beer. If the *independence* assumption is valid, these fractions must be applicable to both male and female beer drinkers. So we would expect the sample of 80 male beer drinkers to contain $80(50/150) = 26.67$ who prefer pilsner, $80(70/150) = 37.33$ who prefer export, and $80(30/150) = 16$ who prefer dark beer. Application of the same fractions to the 70 female beer drinkers provides the expected frequencies shown in Table 12.9.

TABLE 12.9 Expected frequencies if beer preference is independent of the gender of the beer drinker

Gender	Beer preference			
	Pilsner	Export	Dark	Total
Male	26.67	37.33	16.00	80
Female	23.33	32.67	14.00	70
Total	50	70	30	150

Let e_{ij} denote the expected frequency for the category in row i and column j of the contingency table. Consider the expected frequency calculation e_{12} for males (row $i = 1$) who prefer export (column $j = 2$). The argument above showed that $e_{12} = 80(70/150) = 37.33$. The 80 is the total number of males (row 1 total), 70 is the total number of individuals who prefer export (column 2 total) and 150 is the total sample size. Hence, we see that:

$$e_{12} = \frac{(\text{Row 1 Total})(\text{Column 2 Total})}{\text{Sample Size}}$$

Generalization of this equation shows that the following formula provides the expected frequencies for a contingency table in the test of independence:

Expected frequencies for contingency tables under the assumption of independence

$$e_{ij} = \frac{(\text{Row } i \text{ Total})(\text{Column } j \text{ Total})}{\text{Sample Size}} \tag{12.3}$$

Using this formula for male beer drinkers who prefer dark beer, we find an expected frequency of $e_{13} = (80)(30)/(150) = 16.00$, as shown in Table 12.9. Use equation (12.3) to verify the other expected frequencies shown in Table 12.9.

The χ^2 value based on the observed and expected frequencies is computed as follows. The double summation in equation (12.4) is used to indicate that the calculation must be made for all the cells (all combinations of rows and columns) in the contingency table.

Test statistic for independence

$$\chi^2 = \sum_i \sum_j \frac{(f_{ij} - e_{ij})^2}{e_{ij}} \tag{12.4}$$

where:

f_{ij} = observed frequency for contingency table category in row i and column j

e_{ij} = expected frequency for contingency table category in row i and column j on the assumption of independence

Note: With n rows and m columns in the contingency table, the test statistic has a χ^2 distribution with $(n - 1)(m - 1)$ df provided that the expected frequencies are five or more for all categories.

The expected frequencies are five or more for each category. We therefore proceed with the computation of the χ^2 test statistic, as shown in Table 12.10. We see that the value of the test statistic is $\chi^2 = 6.12$.

TABLE 12.10 Computation of the chi-squared test statistic for determining whether beer preference is independent of the gender of the beer drinker

Gender	Beer preference	Obs freq (f_{ij})	Exp freq (e_{ij})	Difference ($f_{ij} - e_{ij}$)	$(f_{ij} - e_{ij})^2$	$(f_{ij} - e_{ij})^2/e_{ij}$
Male	Pilsner	20	26.67	−6.67	44.44	1.67
Male	Export	40	37.33	2.67	7.11	0.19
Male	Dark	20	16.00	4.00	16.00	1.00
Female	Pilsner	30	23.33	6.67	44.44	1.90
Female	Export	30	32.67	−2.67	7.11	0.22
Female	Dark	10	14.00	−4.00	16.00	1.14
	Total	**150**				$\chi^2 = 6.12$

The number of degrees of freedom for the appropriate χ^2 distribution is computed by multiplying the number of rows minus one by the number of columns minus one. With two rows and three columns, we have $(2 - 1)(3 - 1) = 2$ df. Like other goodness of fit tests, the test for independence rejects H_0 if the differences between observed and expected frequencies provide a large value for the test statistic, and so is an upper-tail test. Using the χ^2 table (Table 3 of Appendix B), we find that the upper-tail area or p-value at $\chi^2 = 6.12$ is between 0.025 and 0.05. At the 0.05 level of significance, p-value $< \alpha = 0.05$. We reject the null hypothesis of independence and conclude that beer preference is not independent of the gender of the beer drinker.

Computer software packages such as SPSS, R, Excel and Minitab can simplify the computations for a test of independence and provide the p-value for the test (refer to the software guides on the online platform). In the Millennium Brewery example, Excel, R, SPSS or Minitab show p-value $= 0.0468$.

The test itself does not tell us directly about the nature of the dependence between beer preference and gender, but we can compare the observed and expected frequencies descriptively to get an idea. Refer to Tables 12.8 and 12.9. Male beer drinkers have higher observed than expected frequencies for both export and dark beer, whereas female beer drinkers have a higher observed than expected frequency only for pilsner. These observations give us insight into the beer preference differences between male and female beer drinkers.

Here are the steps in a contingency table test of independence:

Test of independence: a summary

1. State the null and alternative hypotheses.

H_0 : the column variable is independent of the row variable

H_1 : the column variable is not independent of the row variable

2. Select a random sample and record the observed frequencies for each cell of the contingency table.
3. Use equation (12.3) to compute the expected frequency for each cell.
4. Use equation (12.4) to compute the value of the test statistic.
5. Rejection rule:

p-value approach: Reject H_0 if p-value $\leq \alpha$

Critical value approach: Reject H_0 if $\chi^2 \geq \chi_\alpha^2$

where α is the level of significance for the test, with n rows and m columns providing $(n - 1)(m - 1)$ df.

Note: the test statistic for the χ^2 tests in this chapter requires an expected frequency of five or more for each category. When a category has fewer than five, it is often appropriate to combine two adjacent rows or columns to obtain an expected frequency of five or more in each category.

EXERCISES

Methods

14. The following 2×3 contingency table contains observed frequencies for a sample of 200. Test for independence of the row and column variables using the χ^2 test with $\alpha = 0.05$.

	Column variable		
Row variable	A	B	C
P	20	44	50
Q	30	26	30

15. The following 3×3 contingency table contains observed frequencies for a sample of 240. Test for independence of the row and column variables using the χ^2 test with $\alpha = 0.05$.

	Column variable		
Row variable	A	B	C
P	20	30	20
Q	30	60	25
R	10	15	30

Applications

16. The Barna Group conducted a survey about church attendance. The survey respondents were asked about their church attendance and asked to indicate their age. Use the sample data to determine whether church attendance is independent of age. Using a 0.05 level of significance, what is the p-value and what is your conclusion? What conclusion can you draw about church attendance as individuals grow older?

	Age group			
Church attender	20–29	30–39	40–49	50–59
Yes	31	63	94	72
No	69	87	106	78

17. A Pew Research Centre survey asked respondents if they would rather live in a place with a slower pace of life or a place with a faster pace of life. The survey also asked the respondent's sex. Consider the following sample data.

	Sex	
Preferred pace of life	Male	Female
Slower	230	218
No preference	20	24
Faster	90	48

a. Is the preferred pace of life independent of sex? Using a 0.05 level of significance, what is the p-value and what is your conclusion?

b. Discuss any differences between the preferences of men and women.

18. A *Financial Times/Harris Poll* surveyed people in six countries to assess attitudes towards a variety of forms of energy. The data in the following table are a portion of the poll's findings concerning whether people favour or oppose the building of new nuclear power plants.

	Country					
	UK	France	Italy	Spain	Germany	USA
Strongly favour	141	161	298	133	128	204
Favour more than oppose	348	366	309	222	272	326
Oppose more than favour	381	334	219	311	322	316
Strongly oppose	217	215	219	443	389	174

   a. How large was the sample in this poll?

   b. Carry out a hypothesis test to determine whether people's attitude towards building new nuclear power plants is independent of country. What is your conclusion?

   c. Using the percentage of respondents who 'strongly favour' and 'favour more than oppose', which country has the most favourable attitude towards building new nuclear power plants? Which country has the least favourable attitude?

19. The table below shows figures from a survey done in the Netherlands to examine the relationship between the length of time Dutch offenders had spent in prison and the likelihood of re-offending on release (*Crime and Delinquency*, 2018, 64(8)). The study focused on Netherlands-born men aged 18–65 and recorded re-offences up to six months after release.

	Length of time in prison				
Re-offence?*	1–6 wks	6–12 wks	12–16 wks	16–24 wks	>24 wks
Yes	105	143	79	67	83
No	179	253	187	157	214

* Within 6 months of release.

   a. Calculate either row or column percentages and comment on any apparent relationship between length of time in prison and likelihood of re-offending on release.

   b. Using $\alpha = 0.05$, test for independence between length of time in prison and likelihood of re-offending on release.

20. In a study carried out in Jordan to examine adolescents' perceptions of fast food (*Nutrition and Health*, 2017, 23(1)), participants (15–18 year old students) were asked whether they considered each of a range of food items to be fast food, or not fast food. The table below shows the responses of males and females to the food item 'Pizza'.

	Fast food	Not fast food
Males	239	161
Females	259	136

Test the hypothesis that perceptions of pizza as fast food, or not, are independent of respondent gender, using $\alpha = 0.05$. What is your conclusion?

21. Visa studied how frequently consumers of various age groups use plastic cards (debit and credit cards) when making purchases. Sample data for 300 customers show the use of plastic cards by four age groups.

Payment	Age group			
	18–24	25–34	35–44	45 and over
Plastic	21	27	27	36
Cash or cheque	21	36	42	90

a. Test for the independence between method of payment and age group. What is the *p*-value? Using $\alpha = 0.05$, what is your conclusion?

b. If method of payment and age group are not independent, what observation can you make about how different age groups use plastic to make purchases?

c. What implications does this study have for companies such as Visa and MasterCard?

22. In some African countries, vitamin A deficiency among children is a major health issue. One important strategy aimed at mitigating the problem is to promote and encourage the consumption of orange-flesh squash pulp (OFSP). Below is a table based on results from an interview survey done in several rural areas of Zambia, and reported in *Food and Nutrition Bulletin* (2018, 39(1)). The table shows a frequency distribution relating to the reported consumption of OFSP, for households with children under five years of age and for households with no children under five.

	Never eat OFSP	For those who eat OFSP, number of days in last 7 when eaten		
		0	1–3	4-plus
Children under 5 in household	16	12	100	55
No children under 5 in household	23	9	46	34

Does there appear to be a relationship between the presence of children under five years old in the household and the frequency that OFSP was eaten? Support your conclusion with a statistical test using $\alpha = 0.05$.

23. The five most visited art museums in the world, as listed by *The Art Newspaper* Visitor Figures Survey, are the Louvre Museum, the National Museum in China, the Metropolitan Museum of Art, the Vatican Museums and the British Museum. Which of these five museums would visitors most frequently rate as spectacular? Samples of recent visitors of each of these museums were taken, and the results of these samples follow.

	Louvre Museum	National Museum in China	Metropolitan Museum of Art	Vatican Museums	British Museum
Spectacular	113	88	94	98	96
Not spectacular	37	44	46	72	64

a. Use the sample data to calculate the point estimate of the population proportion of visitors who rated each of these museums as spectacular.

b. Carry out a hypothesis test to determine if the population proportion of visitors who rated the museum as spectacular is equal for these five museums. Using a 0.05 level of significance, what is the *p*-value and what is your conclusion?

ONLINE RESOURCES

For the data files, additional questions and answers, and the software section for Chapter 12, go to the online platform.

SUMMARY

The purpose of a goodness of fit test is to determine whether a hypothesized probability distribution can be used as a model for a population of interest. The computations for the goodness of fit test involve comparing observed frequencies from a sample with expected frequencies when the hypothesized probability distribution is assumed true. A chi-squared distribution is used to determine whether the differences between observed and expected frequencies are large enough to reject the hypothesized probability distribution.

In this chapter we illustrated the goodness of fit test for a multinomial distribution, for a Poisson and for a normal distribution.

A test of independence between two categorical variables defining a contingency table is an extension of the same methodology.

KEY TERMS

Contingency table   Multinomial population
Goodness of fit test

KEY FORMULAE

Test statistic for goodness of fit

$$\chi^2 = \sum_{i=1}^{k} \frac{(f_i - e_i)^2}{e_i} \tag{12.1}$$

Expected frequencies for contingency tables under the assumption of independence

$$e_{ij} = \frac{(\text{Row } i \text{ Total})(\text{Column } j \text{ Total})}{\text{Sample Size}} \tag{12.3}$$

Test statistic for test of independence in a contingency table

$$\chi^2 = \sum_i \sum_j \frac{(f_{ij} - e_{ij})^2}{e_{ij}} \tag{12.4}$$

CASE PROBLEM 1

Evaluation of Management School website pages

A group of MSc students at an international university conducted a survey to assess the students' views regarding the web pages of the university's Management School. Among the questions in the survey were items that asked respondents to express agreement or disagreement with the following statements:

1. The Management School web pages are attractive for prospective students.

2. I find it easy to navigate the Management School web pages.

3. There is up-to-date information about courses on the Management School web pages.

4. If I were to recommend the university to someone else, I would suggest that they go to the Management School web pages.

Responses were originally given on a five-point scale, but in the data file on the online platform ('Web Pages'), the responses have been recoded as binary variables. For each questionnaire item, those who agreed or agreed strongly with the statement have been grouped into one category (Agree). Those who disagreed, disagreed strongly, were indifferent or opted for a 'Don't know' response, have been grouped into a second category (Don't Agree). The data file also contains particulars of respondent gender (male/female) and level of study (undergraduate or postgraduate). A screenshot of the first few rows of the data file is shown below.

Managerial report

1. Use descriptive statistics to summarize the data from this study. What are your preliminary conclusions about the independence of the response (Agree or Don't Agree) and gender for each of the four items? What are your preliminary conclusions about the independence of the response (Agree or Don't Agree) and level of study for each of the four items?

2. For each of the four items, test for the independence of the response (Agree or Don't Agree) and gender. Use $\alpha = 0.05$.

3. For each of the four items, test for the independence of the response (Agree or Don't Agree) and level of study. Use $\alpha = 0.05$.

4. Does it appear that views regarding the web pages are consistent for students of both genders and both levels of study? Explain.

WEB PAGES

Gender	Study level	Attractiveness	Navigation	Up-to-date	Referrals
Female	Undergraduate	Don't Agree	Agree	Agree	Agree
Female	Undergraduate	Agree	Agree	Agree	Agree
Male	Undergraduate	Don't Agree	Don't Agree	Don't Agree	Don't Agree
Male	Undergraduate	Agree	Agree	Agree	Agree
Male	Undergraduate	Agree	Agree	Agree	Agree
Female	Undergraduate	Don't Agree	Don't Agree	Agree	Agree
Male	Undergraduate	Don't Agree	Agree	Agree	Agree
Male	Undergraduate	Agree	Agree	Agree	Agree
Male	Undergraduate	Don't Agree	Agree	Agree	Agree

CASE PROBLEM 2

© Talaj/iStock

LOTTO
2019–2022

Checking for randomness in Lotto draws

In the UK National Lottery (drawn on Wednesdays and Saturdays each week), six main balls, followed by a 'bonus' ball, are selected randomly from a set of balls numbered 1, 2, 3, ... , 59. The file 'Lotto 2019–2022' on the online platform contains details of the draws between January 2019 and December 2022.

The first few rows of the data have been reproduced below. In addition to showing the numbers drawn in the game each time, and the order in which they were drawn, the file also gives details of the day on which the draw took place, the machine that was used to do the draw and the set of balls that was used. Several similar machines are used for the draws: Arthur, Guinevere, etc., and eight sets of balls are used.

Analyst's report

1. Use an appropriate hypothesis test to assess whether there is any evidence of non-randomness in the first ball drawn. Similarly, test for non-randomness in the second ball drawn, third ball drawn, ..., sixth ball drawn.

2. Use an appropriate hypothesis test to assess whether there is any evidence of non-randomness overall in the drawing of the 59 numbers (regardless of the order of selection).

3. Use an appropriate hypothesis test to assess whether there is evidence of any dependence between the numbers drawn and the day on which the draw is made.

4. Use an appropriate hypothesis test to assess whether there is evidence of any dependence between the numbers drawn and the machine on which the draw is made.

5. Use an appropriate hypothesis test to assess whether there is evidence of any dependence between the numbers drawn and the set of balls that is used.

Day	Day	Month	Year	Ball 1	Ball 2	Ball 3	Ball 4	Ball 5	Ball 6	Bonus ball	Machine	Ball Set
Wed	2	Jan	2019	8	40	53	35	10	9	21	Lancelot	1
Sat	5	Jan	2019	28	32	18	25	47	27	6	Guinevere	4
Wed	9	Jan	2019	57	36	51	17	3	42	18	Lancelot	5
Sat	12	Jan	2019	13	49	55	43	32	29	26	Guinevere	2
Wed	16	Jan	2019	54	44	48	12	55	1	8	Guinevere	7
Sat	19	Jan	2019	11	52	22	17	31	15	45	Guinevere	8
Wed	23	Jan	2019	58	41	5	45	30	27	38	Guinevere	5
Sat	26	Jan	2019	42	47	29	56	40	51	45	Lancelot	7
Wed	30	Jan	2019	58	47	7	43	22	31	2	Guinevere	8

CASE PROBLEM 3

Development of human capital in Egypt

Globalization has been a dominant feature of business and economic development over recent decades. This case problem uses data collected by Dr Mohga Bassim (University of Buckingham, UK) as part of a study of globalization and its effects on, among other things, the development of human capital and women's economic empowerment. Part of the study drew on the evidence collected in a survey of over 2,000 people of working age in Cairo, Egypt.

The data set for the case study is a subset of the Cairo survey data, comprising survey respondents who were in work and had at least some university education (though not necessarily to the level of a Bachelor's degree). The variables in the data file are: gender (male/female), age group, level of education, length of experience in the current company, length of experience before joining the current company, salary level (in Egyptian pounds per month). All variables are categorized. The first few rows of data (from R) are shown below.

The focus of the case problem is the examination of salary level and its correlates.

Analyst's report

1. Produce graphical displays showing the distribution of each of the variables in the data file. Comment briefly on your results.

2. Using chi-squared tests and appropriate descriptive statistics, examine the association between salary level, on the one hand, and each of the other variables in the file, on the other. Report your results fully.

3. Examine the association between gender (male/female) and the other four non-salary variables. Report your results fully.

4. In light of your results from (2) and (3), what can you conclude about gender (male/female) equality with respect to salary level?

CAIRO SURVEY

	Gender	Age	Education	Exp_with_Co	Exp_Outside	Income
1	Female	26 – 30	University Education	0 – 5	0 – 5	less than 1 thousand
2	Female	21 – 25	University Education	0 – 5	0 – 5	less than 1 thousand
3	Female	26 – 30	University Education	0 – 5	0 – 5	less than 1 thousand
4	Female	31 – 35	University Education	6 – 10	0 – 5	less than 1 thousand
5	Female	26 – 30	University Education	6 – 10	0 – 5	less than 1 thousand
6	Female	26 – 30	University Education	6 – 10	0 – 5	less than 1 thousand
7	Female	31 – 35	University Degree and Diploma	11 – 15	6 – 10	less than 1 thousand
8	Male	36 – 40	University Education	6 – 10	0 – 5	less than 1 thousand
9	Male	31 – 35	University Education	0 – 5	0 – 5	1 – 2 thousand
10	Male	31 – 35	University Education	6 – 10	0 – 5	less than 1 thousand

The authors are grateful to Dr Mohga Bassim, University of Buckingham, UK, for permission to use the data she collected for her study.

13

Experimental Design and Analysis of Variance

CHAPTER CONTENTS

Statistics in Practice   Product customization and manufacturing trade-offs

LEARNING OBJECTIVES   After reading this chapter and doing the exercises, you should be able to:

1 Understand the basics of experimental design and how the analysis of variance procedure can be used to determine if the means of two or more populations are equal.

2 Know the assumptions necessary to use the analysis of variance procedure.

3 Understand the use of the F distribution in performing the analysis of variance procedure.

4 Know how to set up an ANOVA table and interpret the entries in the table.

5 Use output from computer software packages to solve analysis of variance problems.

6 Know how to use Fisher's least significant difference (LSD) procedure and Fisher's LSD with the Bonferroni adjustment to conduct statistical comparisons between pairs of population means.

7 Understand the difference between a completely randomized design, a randomized block design and factorial experiments.

8 Know the definition of the following terms: comparisonwise Type I error rate; experimentwise Type I error rate; factor; level; treatment; partitioning; blocking; main effect; interaction; replication.

I n Chapter 1 we stated that statistical studies can be classified as either experimental or observational. In an experimental statistical study, an experiment is conducted to generate the data. An experiment begins with identifying a variable of interest. Then one or more other variables, thought to be related, are identified and controlled, and data are collected about how those variables influence the variable of interest.

STATISTICS IN PRACTICE
Product customization and
manufacturing trade-offs

The findings suggest that customization is not cost-free and that the advent of mass customization – or 'niche' customization as Franke and Hader (2014) prefer to term it – is unlikely to see the end of trade-offs with other key priorities.

The analysis of variance technique was used by Squire *et al.* (2005) in a study to investigate trade-offs between product customization and other manufacturing priorities. A total of 102 UK manufacturers from eight industrial sectors were involved in the research. Three levels of customization were considered: full customization where customer input was incorporated at the product design or fabrication stages; partial customization with customer input incorporated into product assembly or delivery stages; and standard products which did not incorporate any customer input at all.

The impact of customization was considered against four competitive imperatives: cost, quality, delivery and volume flexibility.

It was found that the degree of customization had a significant effect on delivery (both in terms of speed and lead times); also on manufacturers' costs, although not design, component, delivery and servicing costs.

Sources:
Franke, N. and Hader, C. (2014) 'Mass or only "niche customization"? Why we should interpret configuration toolkits as learning instruments'. *Journal of Product Innovation Management* 31(6): 1214–1234.
Squire, B., Brown, S., Readman, J. and Bessant, J. (2005) 'The impact of mass customization on manufacturing trade-offs'. *Production and Operations Management Journal* 15(1): 10–21.

In an observational study, data are usually obtained through sample surveys and not a controlled experiment. Good design principles are still employed, but the rigorous controls associated with an experimental statistical study are often not possible.

For instance, in a study of the relationship between smoking and lung cancer the researcher cannot assign a smoking habit to subjects. The researcher is restricted to simply observing the effects of smoking on people who already smoke and the effects of not smoking on people who do not already smoke.

In this chapter we introduce three types of experimental designs: a completely randomized design, a randomized block design and a factorial experiment. For each design we show how a statistical procedure known as analysis of variance (ANOVA) can be used to analyze the data available. ANOVA can also be used to analyze the data obtained through an observation study. For instance, we will see that the ANOVA procedure used for a completely randomized experimental design also works for testing the equality of two or more population means when data are obtained through an observational study. In later chapters we will see that ANOVA plays a key role in analyzing the results of regression studies involving both experimental and observational data.

In Section 13.1, we introduce the basic principles of an experimental study and show how they are employed in a completely randomized design. In Section 13.2, we then show how ANOVA can be used to analyze the data from a completely randomized experimental design. In later sections we discuss multiple comparison procedures and two other widely used experimental designs: the randomized block design and the factorial experiment.

13.1 AN INTRODUCTION TO EXPERIMENTAL DESIGN AND ANALYSIS OF VARIANCE

As an example of an experimental statistical study, let us consider the problem facing the Chemitech company. Chemitech has developed a new filtration system for municipal water supplies.

The components for the new filtration system will be purchased from several suppliers, and Chemitech will assemble the components at its plant in North Saxony. The industrial engineering group is responsible for determining the best assembly method for the new filtration system. After considering a variety of possible approaches, the group narrows the alternatives to three: method A, method B and method C. These methods differ in the sequence of steps used to assemble the system. Managers at Chemitech wish to determine which assembly method can produce the greatest number of filtration systems per week.

In the Chemitech experiment, assembly method is the independent variable or **factor**. As three assembly methods correspond to this factor, we say that three treatments are associated with this experiment; each **treatment** corresponds to one of the three assembly methods. The Chemitech problem is an example of a **single-factor experiment**; it involves one qualitative factor (method of assembly). More complex experiments may consist of multiple factors; some factors may be qualitative and others may be quantitative.

The three assembly methods or treatments define the three populations of interest for the Chemitech experiment. One population is all Chemitech employees who use assembly method A, another is those who use method B and the third is those who use method C. Note that for each population the dependent or **response variable** is the number of filtration systems assembled per week, and the primary statistical objective of the experiment is to determine whether the mean number of units produced per week is the same for all three populations (methods).

Suppose a random sample of three employees is selected from all assembly workers at the Chemitech production facility. In experimental design terminology, the three randomly selected workers are the **experimental units**. The experimental design that we will use for the Chemitech problem is called a **completely randomized design**. This type of design requires that each of the three assembly methods or treatments be assigned randomly to one of the experimental units or workers. For example, method A might be randomly assigned to the second worker, method B to the first worker and method C to the third worker. The concept of *randomization,* as illustrated in this example, is an important principle of all experimental designs.

Note that this experiment would result in only one measurement or number of units assembled for each treatment. To obtain additional data for each assembly method, we must repeat or replicate the basic experimental process. Suppose, for example, that instead of selecting just three workers at random we selected 15 workers and then randomly assigned each of the three treatments to five of the workers. Because each method of assembly is assigned to five workers, we say that five replicates have been obtained. The process of *replication* is another important principle of experimental design. Figure 13.1 shows the completely randomized design for the Chemitech experiment.

FIGURE 13.1
Completely randomized design for evaluating the Chemitech assembly method experiment

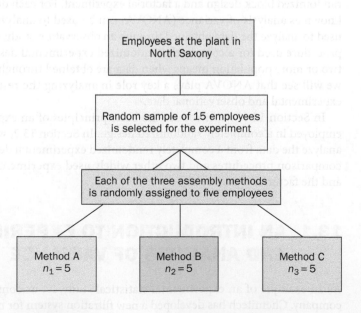

Data collection

Once we are satisfied with the experimental design, we proceed by collecting and analyzing the data. In the Chemitech case, the employees would be instructed in how to perform the assembly method assigned to them and would then begin assembling the new filtration systems using that method. After this assignment and training, the number of units assembled by each employee during one week is as shown in Table 13.1. The sample means, sample variances and sample standard deviations for each assembly method are also provided. Therefore, the sample mean number of units produced using method A is 62; the sample mean using method B is 66; and the sample mean using method C is 52. From these data, method B appears to result in higher production rates than either of the other methods.

TABLE 13.1   Number of units produced by 15 workers

	Method		
	A	**B**	**C**
	58	58	48
	64	69	57
	55	71	59
	66	64	47
	67	68	49
Sample mean	62	66	52
Sample variance	27.5	26.5	31.0
Sample standard deviation	5.244	5.148	5.568

The real issue is whether the three sample means observed are different enough for us to conclude that the means of the populations corresponding to the three methods of assembly are different. To write this question in statistical terms, we introduce the following notation:

μ_1 = mean number of units produced per week using method A

μ_2 = mean number of units produced per week using method B

μ_3 = mean number of units produced per week using method C

Although we will never know the actual values of μ_1, μ_2 and μ_3, we want to use the sample means to test the following hypotheses:

$$H_0: \mu_1 = \mu_2 = \mu_3$$
$$H_1: \text{Not all population means are equal}$$

CHEMITECH

As we will demonstrate shortly, analysis of variance (ANOVA) is the statistical procedure used to determine whether the observed differences in the three sample means are large enough to reject H_0.

Assumptions for analysis of variance

Three assumptions are required to use analysis of variance:

1. **For each population, the response variable is normally distributed.** Implication: In the Chemitech experiment the number of units produced per week (response variable) must be normally distributed for each assembly method.
2. **The variance of the response variable, denoted σ^2, is the same for all of the populations.** Implication: In the Chemitech experiment, the variance of the number of units produced per week must be the same for each assembly method.
3. **The observations must be independent.** Implication: In the Chemitech experiment, the number of units produced per week for each employee must be independent of the number of units produced per week for any other employee.

Analysis of variance: a conceptual overview

If the means for the three populations are equal, we would expect the three sample means to be close together. In fact, the closer the three sample means are to one another, the more evidence we have for the conclusion that the population means are equal. Alternatively, the more the sample means differ, the more evidence we have for the conclusion that the population means are not equal. In other words, if the variability among the sample means is 'small' it supports H_0; if the variability among the sample means is 'large' it supports H_1.

If the null hypothesis, $H_0: \mu_1 = \mu_2 = \mu_3$, is true, we can use the variability among the sample means to develop an estimate of σ^2. First, note that if the assumptions for analysis of variance are satisfied, each sample will have come from the same normal distribution with mean μ and variance σ^2. Recall from Chapter 7 that the sampling distribution of the sample mean $\bar{x}$ for a simple random sample of size n from a normal population will be normally distributed with mean μ and variance σ^2/n. Figure 13.2 illustrates such a sampling distribution.

FIGURE 13.2
Sampling distribution of $\bar{X}$ given H_0 is true

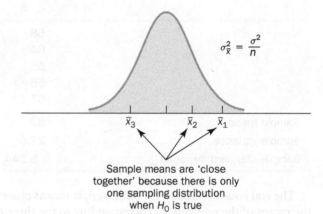

Therefore, if the null hypothesis is true, we can think of each of the three sample means, $\bar{x}_1 = 62$, $\bar{x}_2 = 66$ and $\bar{x}_3 = 52$ from Table 13.1, as values drawn at random from the sampling distribution shown in Figure 13.2. In this case, the mean and variance of the three $\bar{x}$ values can be used to estimate the mean and variance of the sampling distribution. When the sample sizes are equal, as in the Chemitech experiment, the best estimate of the mean of the sampling distribution of $\bar{x}$ is the mean or average of the sample means. Thus, in the Chemitech experiment, an estimate of the mean of the sampling distribution of $\bar{x}$ is $(62 + 66 + 52)/3 = 60$. We refer to this estimate as the *overall sample mean*. An estimate of the variance of the sampling distribution of $\bar{x}$, $\sigma_{\bar{x}}^2$ is provided by the variance of the three sample means:

$$s_{\bar{x}}^2 = \frac{(62 - 60)^2 + (66 - 60)^2 + (52 - 60)^2}{3 - 1} = \frac{104}{2} = 52$$

Because $\sigma_{\bar{x}}^2 = \sigma^2/n$, solving for σ^2 gives:

$$\sigma^2 = n\sigma_{\bar{x}}^2$$

Hence:

$$\text{Estimate of } \sigma^2 = n \left(\text{Estimate of } \sigma_{\bar{x}}^2 \right) = ns_{\bar{x}}^2 = 5(52) = 260$$

The result, $ns_{\bar{x}}^2 = 260$, is referred to as the *between-treatments* estimate of σ^2.

The between-treatments estimate of σ^2 is based on the assumption that the null hypothesis is true. In this case, each sample comes from the same population, and there is only one sampling distribution of $\bar{X}$. To illustrate what happens when H_0 is false, suppose the population means all differ. Note that because the three samples are from normal populations with different means, they will result in three different sampling distributions. Figure 13.3 shows that, in this case, the sample means are not as close together as they were when H_0 was true. Therefore, $s_{\bar{x}}^2$ will be larger, causing the between-treatments estimate of σ^2 to be larger.

FIGURE 13.3
Sampling distribution of $\bar{X}$ given H_0 is false

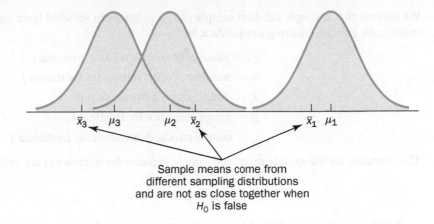

$\bar{x}_3$ μ_3 μ_2 $\bar{x}_2$ $\bar{x}_1$ μ_1

Sample means come from
different sampling distributions
and are not as close together when
H_0 is false

In general, when the population means are not equal, the between-treatments estimate will overestimate the population variance σ^2.

The variation within each of the samples also has an effect on the conclusion we reach in analysis of variance. When a simple random sample is selected from each population, each of the sample variances provides an unbiased estimate of σ^2. Hence, we can combine or pool the individual estimates of σ^2 into one overall estimate. The estimate of σ^2 obtained in this way is called the *pooled* or *within-treatments* estimate of σ^2. Because each sample variance provides an estimate of σ^2 based only on the variation within each sample, the within-treatments estimate of σ^2 is not affected by whether the population means are equal.

When the sample sizes are equal, the within-treatments estimate of σ^2 can be obtained by computing the average of the individual sample variances. For the Chemitech experiment we obtain:

$$\text{Within-treatments estimate of } \sigma^2 = \frac{27.5 + 26.5 + 31.0}{3} = \frac{85}{3} = 28.33$$

In the Chemitech experiment, the between-treatments estimate of σ^2 (260) is much larger than the within-treatments estimate of σ^2 (28.33). In fact, the ratio of these two estimates is $260/28.33 = 9.18$. Recall, however, that the between-treatments approach provides a good estimate of σ^2 only if the null hypothesis is true; if the null hypothesis is false, the between-treatments approach overestimates σ^2. The within-treatments approach provides a good estimate of σ^2 in either case. Therefore, if the null hypothesis is true, the two estimates will be similar and their ratio will be close to 1. If the null hypothesis is false, the between-treatments estimate will be larger than the within-treatments estimate, and their ratio will be large. In the next section we will show how large this ratio must be to reject H_0.

In summary, the logic behind ANOVA is based on the development of two independent estimates of the common population variance σ^2. One estimate of σ^2 is based on the variability among the sample means themselves, and the other estimate of σ^2 is based on the variability of the data within each sample. By comparing these two estimates of σ^2, we will be able to determine whether the population means are equal.

13.2 ANALYSIS OF VARIANCE AND THE COMPLETELY RANDOMIZED DESIGN

In this section we show how analysis of variance can be used to test for the equality of k population means for a completely randomized design. The general form of the hypotheses tested is:

$$H_0: \mu_1 = \mu_2 \cdots = \mu_k$$

$$H_1: \text{Not all population means are equal}$$

where:

$$\mu_j = \text{mean of the } j\text{th population}$$

We assume that a simple random sample of size n_j has been selected from each of the k populations or treatments. For the resulting sample data, let:

$$x_{ij} = \text{value of observation } i \text{ for treatment } j$$
$$n_j = \text{number of observations for treatment } j$$
$$\bar{x}_j = \text{sample mean for treatment } j$$
$$s_j^2 = \text{sample variance for treatment } j$$
$$s_j = \text{sample standard deviation for treatment } j$$

The formulae for the sample mean and sample variance for treatment j are as follows:

Testing for the equality of k population means sample mean for treatment j

$$\bar{x}_j = \frac{\sum_{i=1}^{n_j} x_{ij}}{n_j} \tag{13.1}$$

Sample variance for treatment j

$$s_j^2 = \frac{\sum_{i=1}^{n_j} (x_{ij} - \bar{x}_j)^2}{n_j - 1} \tag{13.2}$$

The overall sample mean, denoted $\bar{\bar{x}}$, is the sum of all the observations divided by the total number of observations. That is:

Overall sample mean

$$\bar{\bar{x}} = \frac{\sum_{j=1}^{k} \sum_{i=1}^{n_j} x_{ij}}{n_T} \tag{13.3}$$

where:

$$n_T = n_1 + n_2 + \ldots + n_k \tag{13.4}$$

If the size of each sample is n, $n_T = kn$; in this case equation (13.3) reduces to:

$$\bar{\bar{x}} = \frac{\sum_{j=1}^{k} \sum_{i=1}^{n_j} x_{ij}}{kn} = \frac{\sum_{j=1}^{k} \sum_{i=1}^{n_j} x_{ij}/n}{k} = \frac{\sum_{j=1}^{k} \bar{x}_j}{k} \tag{13.5}$$

In other words, whenever the sample sizes are the same, the overall sample mean is just the average of the k sample means.

As each sample in the Chemitech experiment consists of $n = 5$ observations, the overall sample mean can be computed by using equation (13.5). For the data in Table 13.1 we obtained the following result:

$$\bar{\bar{x}} = \frac{63 + 66 + 52}{3} = 60$$

If the null hypothesis is true ($\mu_1 = \mu_2 = \mu_3 = \mu$), the overall sample mean of 60 is the best estimate of the population mean μ.

Between-treatments estimate of population variance

In the preceding section, we introduced the concept of a between-treatments estimate of σ^2 and showed how to compute it when the sample sizes were equal. This estimate of σ^2 is called the *mean square due to treatments* and is denoted MSTR. The general formula for computing MSTR is:

$$MSTR = \frac{\sum_{j=1}^{k} n_j(\bar{x}_j - \bar{\bar{x}})^2}{k - 1} \tag{13.6}$$

The numerator in equation (13.6) is called the *sum of squares due to treatments* and is denoted SSTR. The denominator, $k - 1$, represents the degrees of freedom associated with SSTR. Hence, the mean square due to treatments can be computed using the following formula:

Mean square due to treatments

$$MSTR = \frac{SSTR}{k - 1} \tag{13.7}$$

where:

$$SSTR = \sum_{j=1}^{k} n_j(\bar{x}_j - \bar{\bar{x}})^2 \tag{13.8}$$

If H_0 is true, MSTR provides an unbiased estimate of σ^2. However, if the means of the k populations are not equal, MSTR is not an unbiased estimate of σ^2; in fact, in that case, MSTR should overestimate σ^2. For the Chemitech data in Table 13.1, we obtain the following results:

$$SSTR = \sum_{j=1}^{k} n_j(\bar{x}_j - \bar{\bar{x}})^2 = 5(62 - 60)^2 + 5(66 - 60)^2 + 5(52 - 60)^2 = 520$$

$$MSTR = \frac{SSTR}{k - 1} = \frac{520}{2} = 260$$

Within-treatments estimate of population variance

Earlier we introduced the concept of a within-treatments estimate of σ^2 and showed how to compute it when the sample sizes were equal. This estimate of σ^2 is called the *mean square due to error* and is denoted MSE. The general formula for computing MSE is:

$$MSE = \frac{\sum_{j=1}^{k} (n_j - 1)s_j^2}{n_T - k} \tag{13.9}$$

The numerator in equation (13.9) is called the *sum of squares due to error* and is denoted SSE. The denominator of MSE is referred to as the degrees of freedom associated with SSE. Hence, the formula for MSE can also be stated as follows:

Mean square due to error

$$MSE = \frac{SSE}{n_T - k} \tag{13.10}$$

where:

$$SSE = \sum_{j=1}^{k} (n_j - 1)s_j^2 \tag{13.11}$$

Note that MSE is based on the variation within each of the treatments; it is not influenced by whether the null hypothesis is true. Therefore, MSE always provides an unbiased estimate of σ^2.

For the Chemitech data in Table 13.1 we obtain the following results:

$$SSE = \sum_{j=1}^{k} (n_j - 1)s_j^2 = (5 - 1)27.5 + (5 - 1)26.5 + (5 - 1)31 = 340$$

$$MSE = \frac{SSE}{n_T - k} = \frac{340}{15 - 3} = \frac{340}{12} = 28.33$$

Comparing the variance estimates: the *F* test

If the null hypothesis is true, MSTR and MSE provide two independent, unbiased estimates of σ^2. Based on the material covered in Chapter 11 we know that, for normal populations, the sampling distribution of the ratio of two independent estimates of σ^2 follows an *F* distribution. Hence, if the null hypothesis is true and the ANOVA assumptions are valid, the sampling distribution of MSTR/MSE is an *F* distribution with numerator degrees of freedom equal to $k - 1$ and denominator degrees of freedom equal to $n_T - k$. In other words, if the null hypothesis is true, the value of MSTR/MSE should appear to have been selected from this *F* distribution.

However, if the null hypothesis is false, the value of MSTR/MSE will be inflated because MSTR overestimates σ^2. Hence, we will reject H_0 if the resulting value of MSTR/MSE appears to be too large to have been selected from an *F* distribution with $k - 1$ numerator degrees of freedom and $n_T - k$ denominator degrees of freedom. As the decision to reject H_0 is based on the value of MSTR/MSE, the test statistic used to test for the equality of k population means is as follows:

Test statistic for the equality of *k* population means

$$F = \frac{MSTR}{MSE} \tag{13.12}$$

The test statistic follows an *F* distribution with $k - 1$ degrees of freedom in the numerator and $n_T - k$ degrees of freedom in the denominator.

Let us return to the Chemitech experiment and use a level of significance $\alpha = 0.05$ to conduct the hypothesis test. The value of the test statistic is:

$$F = \frac{MSTR}{MSE} = \frac{260}{28.33} = 9.18$$

The numerator degrees of freedom is $k - 1 = 3 - 1 = 2$ and the denominator degrees of freedom is $n_T - k = 15 - 3 = 12$. As we will only reject the null hypothesis for large values of the test statistic, the

p-value is the upper tail area of the F distribution to the right of the test statistic $F = 9.18$. Figure 13.4 shows the sampling distribution of $F = \text{MSTR/MSE}$, the value of the test statistic and the upper tail area that is the p-value for the hypothesis test.

FIGURE 13.4
Computation of p-value using the sampling distribution of MSTR/MSE

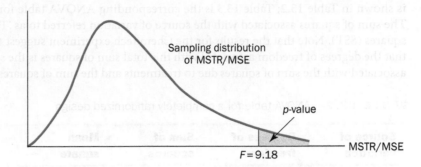

From Table 4 of Appendix B we find the following areas in the upper tail of an F distribution with two numerator degrees of freedom and 12 denominator degrees of freedom.

Area in upper tail	0.10	0.05	0.025	0.01
F value (df1 = 2, df2 = 12)	2.81	3.89	5.10	6.93

$$F = 9.18$$

Because $F = 9.18$ is greater than 6.93, the area in the upper tail at $F = 9.18$ is less than 0.01. Therefore, the p-value is less than 0.01. R, Excel, SPSS or Minitab can be used to show that the exact p-value is 0.004. With p-value $\leq \alpha = 0.05$, H_0 is rejected. The test provides sufficient evidence to conclude that the means of the three populations are not equal. In other words, analysis of variance supports the conclusion that the population mean number of units produced per week for the three assembly methods is not equal.

As with other hypothesis testing procedures, the critical value approach may also be used. With $\alpha = 0.05$, the critical F value occurs with an area of 0.05 in the upper tail of an F distribution with 2 and 12 degrees of freedom. From the F distribution table, we find $F_{0.05} = 3.89$. Hence, the appropriate upper tail rejection rule for the Chemitech experiment is:

$$\text{Reject } H_0 \text{ if } \quad F \geq 3.89$$

With $F = 9.18$, we reject H_0 and conclude that the means of the three populations are not equal. A summary of the overall procedure for testing for the equality of k population means is as follows:

Test for the equality of k population means

$$H_0: \mu_1 = \mu_2 = \ldots = \mu_k$$
$$H_1: \text{Not all population means are equal}$$

Test statistic

$$F = \frac{\text{MSTR}}{\text{MSE}}$$

Rejection rule

$$p\text{-value approach: } \textit{Reject } H_0 \textit{ if } p\text{-value} \leq \alpha$$
$$\text{Critical value approach: Reject } H_0 \textit{ if } F \geq F_\alpha$$

where the value of F_α is based on an F distribution with $k - 1$ numerator degrees of freedom and $n_T - k$ denominator degrees of freedom.

ANOVA table

The results of the preceding calculations can be displayed conveniently in a table referred to as the analysis of variance or **ANOVA table**. The general form of the ANOVA table for a completely randomized design is shown in Table 13.2; Table 13.3 is the corresponding ANOVA table for the Chemitech experiment. The sum of squares associated with the source of variation referred to as 'Total' is called the total sum of squares (SST). Note that the results for the Chemitech experiment suggest that SST = SSTR + SSE, and that the degrees of freedom associated with this total sum of squares is the sum of the degrees of freedom associated with the sum of squares due to treatments and the sum of squares due to error.

TABLE 13.2　ANOVA table for a completely randomized design

Source of variation	Degrees of freedom	Sum of squares	Mean square	F	p-value
Treatments	$k - 1$	SSTR	$MSTR = \dfrac{SSTR}{k - 1}$	$\dfrac{MSTR}{MSE}$	
Error	$n_T - k$	SSE	$MSE = \dfrac{SSE}{n_T - k}$		
Total	$n_T - 1$	SST			

TABLE 13.3　Analysis of variance table for the Chemitech experiment

Source of variation	Degrees of freedom	Sum of squares	Mean square	F	p-value
Treatments	2	520	260.00	9.18	0.004
Error	12	340	28.33		
Total	14	860			

We point out that SST divided by its degrees of freedom $n_T - 1$ is nothing more than the overall sample variance that would be obtained if we treated the entire set of 15 observations as one data set. With the entire data set as one sample, the formula for computing the total sum of squares, SST, is:

Total sum of squares

$$SST = \sum_{j=1}^{k} \sum_{i=1}^{n_j} (x_{ij} - \bar{\bar{x}})^2 \tag{13.13}$$

It can be shown that the results we observed for the ANOVA table for the Chemitech experiment also apply to other problems. That is:

Partitioning of sum of squares

$$SST = SSTR + SSE \tag{13.14}$$

In other words, SST can be partitioned into two sums of squares: the sum of squares due to treatments and the sum of squares due to error.

Note also that the degrees of freedom corresponding to SST, $n_T - 1$, can be partitioned into the degrees of freedom corresponding to SSTR, $k - 1$, and the degrees of freedom corresponding to SSE, $n_T - k$. The analysis of variance can be viewed as the process of partitioning the total sum of squares and the degrees of freedom into their corresponding sources: treatments and error. Dividing the sum of squares by the appropriate degrees of freedom provides the variance estimates and dividing the variance estimates, the F value, from which the p-value can be derived to test the hypothesis of equal population means.

Computer results for analysis of variance

Using statistical computer packages, analysis of variance computations with large sample sizes or a large number of populations can be performed easily. In Figure 13.5 we show computer output for the Chemitech experiment. The first part of the computer output contains the familiar ANOVA table format.

FIGURE 13.5
Output for the Chemitech experiment analysis of variance

Source	DF	Adj SS	Adj MS	F value	p-value
Factor	2	520.0	260.00	9.18	0.004
Error	12	340.0	28.33		
Total	14	860.0			

Model Summary

s	R-sq	R-sq(adj)
5.32291	60.47%	53.88%

Means

Factor	N	Mean	StDev	95% CI
Method A	5	62.00	5.24	(56.81, 67.19)
Method B	5	66.00	5.15	(60.81, 71.19)
Method C	5	52.00	5.57	(46.81, 57.19)

Pooled StDev = 5.32291

Note that following the ANOVA table the computer output contains the respective sample sizes, the sample means and the standard deviations. In addition, the calculation provides a figure that shows individual 95% confidence interval estimates of each population mean. In developing these confidence interval estimates, the computer uses MSE as the estimate of σ^2. Therefore, the square root of MSE provides the best estimate of the population standard deviation σ. This estimate of σ on the computer output is the Pooled StDev; it is equal to 5.323. To provide an illustration of how these interval estimates are developed, we compute a 95% confidence interval estimate of the population mean for method A.

From our study of interval estimation in Chapter 8, we know that the general form of an interval estimate of a population mean is:

$$\bar{x} \pm t_{\alpha/2}\frac{s}{\sqrt{n}} \qquad \textbf{(13.15)}$$

where s is the estimate of the population standard deviation σ. As the best estimate of σ is provided by the Pooled StDev, we use a value of 5.323 for s in expression (13.15). The degrees of freedom for the t-value is 12, the degrees of freedom associated with the error sum of squares. Hence, with $t_{0.025} = 2.179$ we obtain:

$$62 \pm 2.179\frac{5.323}{\sqrt{5}} = 62 \pm 5.19$$

Therefore, the individual 95% confidence interval for method A goes from $62 - 5.19 = 56.81$ to $62 + 5.19 = 67.19$. Because the sample sizes are equal for the Chemitech experiment, the individual confidence intervals for methods B and C are also constructed by adding and subtracting 5.19 from each sample mean.

Therefore, in the figure we see that the widths of the confidence intervals are the same.

Testing for the equality of k population means: an observational study

National Computer Products (NCP) manufactures printers and fax machines at plants located in Ayr, Dusseldorf and Stockholm. To measure how much employees at these plants know about quality management, a random sample of six employees was selected from each plant and the employees selected were given a quality awareness examination. The examination scores for these 18 employees are shown in Table 13.4. The sample means, sample variances and sample standard deviations for each group are also provided. Managers want to use these data to test the hypothesis that the mean examination score is the same for all three plants.

TABLE 13.4   Examination scores for 18 employees

	Plant 1 Ayr	Plant 2 Dusseldorf	Plant 3 Stockholm
	85	71	59
	75	75	64
	82	73	62
	76	74	69
	71	69	75
	85	82	67
Sample mean	79	74	66
Sample variance	34	20	32
Sample standard deviation	5.83	4.47	5.66

NCP

We define population 1 as all employees at the Ayr plant, population 2 as all employees at the Dusseldorf plant and population 3 as all employees at the Stockholm plant. Let:

$$\mu_1 = \text{mean examination score for population 1}$$
$$\mu_2 = \text{mean examination score for population 2}$$
$$\mu_3 = \text{mean examination score for population 3}$$

Although we will never know the actual values of μ_1, μ_2 and μ_3, we want to use the sample results to test the following hypotheses:

$$H_0: \mu_1 = \mu_2 = \mu_3$$
$$H_1: \text{Not all population means are equal}$$

Note that the hypothesis test for the NCP observational study is exactly the same as the hypothesis test for the Chemitech experiment. Indeed, the same analysis of variance methodology we used to analyze the Chemitech experiment can also be used to analyze the data from the NCP observational study.

Even though the same ANOVA methodology is used for the analysis, it is worth noting how the NCP observational statistical study differs from the Chemitech experimental statistical study. The individuals who conducted the NCP study had no control over how the plants were assigned to individual employees. That is, the plants were already in operation and a particular employee worked at one of the three plants. All that NCP could do was to select a random sample of six employees from each plant and administer the quality awareness examination. To be classified as an experimental study, NCP would have had to be able to randomly select 18 employees and then assign the plants to each employee in a random fashion.

EXERCISES

Methods

1. The following data are from a completely randomized design:

	Treatment		
	A	B	C
	162	142	126
	142	156	122
	165	124	138
	145	142	140
	148	136	150
	174	152	128
Sample mean	156	142	134
Sample variance	164.4	131.2	110.4

   a. Compute the sum of squares between treatments.
   b. Compute the mean square between treatments.
   c. Compute the sum of squares due to error.
   d. Compute the mean square due to error.
   e. Set up the ANOVA table for this problem.
   f. At the $\alpha = 0.05$ level of significance, test whether the means for the three treatments are equal.

2. In a completely randomized design, seven experimental units were used for each of the five levels of the factor. Complete the following ANOVA table.

Source of variation	Degrees of freedom	Sum of squares	Mean square	F	p-value
Treatments		300			
Error					
Total		460			

3. Refer to Exercise 2.
   a. What hypotheses are implied in this problem?
   b. At the $\alpha = 0.05$ level of significance, can we reject the null hypothesis in (a)? Explain.

4. In an experiment designed to test the output levels of three different treatments, the following results were obtained: SST = 400, SSTR = 150, $n_T = 19$. Set up the ANOVA table and test for any significant difference between the mean output levels of the three treatments. Use $\alpha = 0.05$.

5. In a completely randomized design, 12 experimental units were used for the first treatment, 15 for the second treatment and 20 for the third treatment. Complete the following analysis of variance. At a 0.05 level of significance, is there a significant difference between the treatments?

Source of variation	Degrees of freedom	Sum of squares	Mean square	F	p-value
Treatments		1,200			
Error					
Total		1,800			

EXER6

6. Develop the analysis of variance computations for the following completely randomized design. At $\alpha = 0.05$, is there a significant difference between the treatment means?

	Treatment		
	A	B	C
	136	107	92
	120	114	82
	113	125	85
	107	104	101
	131	107	89
	114	109	117
	129	97	110
	102	114	120
		104	98
		89	106
$\bar{x}_j$	119	107	100
s_j^2	146.86	96.44	173.78

Applications

7. Three different methods for assembling a product were proposed by an industrial engineer. To investigate the number of units assembled correctly with each method, 30 employees were randomly selected and randomly assigned to the three proposed methods in such a way that each method was used by 10 workers.

   The number of units assembled correctly was recorded, and the analysis of variance procedure was applied to the resulting data set. The following results were obtained:

   $$SST = 10{,}800; \quad SSTR = 4{,}560.$$

   a. Set up the ANOVA table for this problem.
   b. Use $\alpha = 0.05$ to test for any significant difference in the means for the three assembly methods.

8. A fast food house is to introduce a new takeaway product. To help determine the effect of price on product sales, it is first test-marketed in different geographical areas at three contrasting price levels. Over a four-week period the average number of products sold per day was recorded as follows:

	Price level		
	1	2	3
	925	910	860
	850	845	935
	930	905	820
	955	860	845

Is there significant evidence that mean sales differ as a result of the price levels tested?

9. A study investigated the perception of corporate ethical values among individuals specializing in marketing. Use $\alpha = 0.05$ and the following data (higher scores indicate higher ethical values) to test for significant differences in perception between the three groups.

Marketing managers	Marketing research	Advertising
6	5	6
5	5	7
4	4	6
5	4	5
6	5	6
4	4	6

10. A study reported in the *Journal of Small Business Management* concluded that self-employed individuals experience higher job stress than individuals who are not self-employed. In this study job stress was assessed with a 15-item scale designed to measure various aspects of ambiguity and role conflict. Ratings for each of the 15 items were made using a scale with 1–5 response options ranging from strong agreement to strong disagreement. The sum of the ratings for the 15 items for each individual surveyed is between 15 and 75, with higher values indicating a higher degree of job stress. Suppose that a similar approach, using a 20-item scale with 1–5 response options, was used to measure the job stress of individuals for 15 randomly selected property agents, 15 architects and 15 stockbrokers. The results obtained are as follows:

Property agent	Architect	Stockbroker
81	43	65
48	63	48
68	60	57
69	52	91
54	54	70
62	77	67
76	68	83
56	57	75
61	61	53
65	80	71
64	50	54
69	37	72
83	73	65
85	84	58
75	58	58

Use $\alpha = 0.05$ to test for any significant difference in job stress between the three professions.

11. Four different paints are advertised as having the same drying time. To check the manufacturer's claims, five samples were tested for each of the paints. The time in minutes until the paint was dry enough for a second coat to be applied was recorded. The following data were obtained:

PAINT

Paint 1	Paint 2	Paint 3	Paint 4
128	144	133	150
137	133	143	142
135	142	137	135
124	146	136	140
141	130	131	153

At the $\alpha = 0.05$ level of significance, test to see whether the mean drying time is the same for each type of paint.

ALLOYS

12. The following data relate to the tensile strength of samples of drive shafts, made from different alloys, tested to destruction:

	Alloy		
A	B	C	D
5.8	8.3	4.9	8.7
6.3	6.2	5.2	6.5
5.6	7.9	4.8	9.2
7.2	8.5	5.0	7.3
5.9	7.5	5.3	6.8

Carry out an appropriate analysis of these data, stating all relevant assumptions. Do the alloys differ significantly in their tensile strength properties and if so how?

13.3 MULTIPLE COMPARISON PROCEDURES

When we use analysis of variance to test whether the means of k populations are equal, rejection of the null hypothesis allows us to conclude only that the population means are *not all equal*. In some cases we will want to go a step further and determine where the differences between means occur. The purpose of this section is to show how **multiple comparison procedures** can be used to conduct statistical comparisons between pairs of population means.

Fisher's LSD

Suppose that analysis of variance provides statistical evidence to reject the null hypothesis of equal population means. In this case, Fisher's least significant difference (LSD) procedure can be used to determine where the differences occur. To illustrate the use of Fisher's LSD procedure in making pairwise comparisons of population means, recall the Chemitech experiment introduced in Section 13.1. Using analysis of variance, we concluded that the mean number of units produced per week is not the same for the three assembly methods. In this case, the follow-up question is: We believe the assembly methods differ, but where do the differences occur? That is, do the means of populations 1 and 2 differ? Or those of populations 1 and 3? Or those of populations 2 and 3?

In Chapter 10 we presented a statistical procedure for testing the hypothesis that the means of two populations are equal. With a slight modification in how we estimate the population variance, Fisher's LSD procedure is based on the t test statistic presented for the two-population case. The following details summarize Fisher's LSD procedure:

Fisher's LSD procedure

$$H_0: \mu_i = \mu_j$$
$$H_1: \mu_i \neq \mu_j$$

Test statistic

$$t = \frac{\bar{x}_i - x_j}{\sqrt{MSE\left(\dfrac{1}{n_i} + \dfrac{1}{n_j}\right)}}$$ (13.16)

Rejection rule

$$p\text{-value approach: Reject } H_0 \text{ if } p\text{-value} \leq \alpha$$
$$\text{Critical value: Reject } H_0 \text{ if } t \leq -t_{\alpha/2} \text{ or } t \geq t_{\alpha/2}$$

where the value of $t_{\alpha/2}$ is based on a t distribution with $n_T - k$ degrees of freedom.

Let us now apply this procedure to determine whether there is a significant difference between the means of population 1 (method A) and population 2 (method B) at the $\alpha = 0.05$ level of significance. Table 13.1 showed that the sample mean is 62 for method A and 66 for method B. Table 13.3 showed that the value of MSE is 28.33, the estimate of σ^2 and is based on 12 degrees of freedom. For the Chemitech data the value of the test statistic is:

$$t = \frac{62 - 66}{\sqrt{28.33\left(\frac{1}{5} + \frac{1}{5}\right)}} = -1.19$$

As we have a two-tailed test, the p-value is two times the area under the curve for the t distribution to the left of $t = -1.19$. Using Table 2 in Appendix B, the t distribution table for 12 degrees of freedom provides the following information:

Area in upper tail	0.20	0.10	0.05	0.025	0.01	0.005
t-value (12 df)	0.873	1.356	1.782	2.179	2.681	3.055

$$\uparrow$$
$$t = 1.19$$

The t distribution table only contains positive t-values. As the t distribution is symmetrical, however, we can find the area under the curve to the right of $t = 1.19$ and double it to find the p-value corresponding to $t = -1.19$. We see that $t = 1.19$ is between 0.20 and 0.10. Doubling these amounts, we see that the p-value must be between 0.40 and 0.20. Excel or Minitab can be used to show that the exact p-value is 0.2571. As the p-value is greater than $\alpha = 0.05$, we cannot reject the null hypothesis. Hence, we cannot conclude that the population mean number of units produced per week for method A is different from the population mean for method B.

Many practitioners find it easier to determine how large the difference between the sample means must be to reject H_0. In this case the test statistic is $\bar{x}_i - \bar{x}_j$ and the test is conducted by the following procedure:

Fisher's LSD procedure based on the test statistic $\bar{x}_i - \bar{x}_j$

$$H_0 : \mu_i = \mu_j$$
$$H_1 : \mu_i \neq \mu_j$$

Test statistic

$$\bar{x}_i - \bar{x}_j$$

Rejection rule at a level of significance

$$\text{Reject } H_0 \text{ if } |\bar{x}_i - \bar{x}_j| \geq \text{LSD}$$

where:

$$\text{LSD} = t_{\alpha/2}\sqrt{\text{MSE}\left(\frac{1}{n_i} + \frac{1}{n_j}\right)} \qquad \textbf{(13.17)}$$

For the Chemitech experiment the value of LSD is:

$$LSD = t_{\alpha/2}\sqrt{MSE\left(\frac{1}{n_i} + \frac{1}{n_j}\right)}$$

Note that when the sample sizes are equal, only one value for LSD is computed. In such cases we can simply compare the magnitude of the difference between any two sample means with the value of LSD. For example, the difference between the sample means for population 1 (method A) and population 3 (method C) is $62 - 52 = 10$. This difference is greater than LSD = 7.34, which means we can reject the null hypothesis that the population mean number of units produced per week for method A is equal to the population mean for method C. Similarly, with the difference between the sample means for populations 2 and 3 of $66 - 52 = 14 > 7.34$, we can also reject the hypothesis that the population mean for method B is equal to the population mean for method C. In effect, our conclusion is that methods A and B both differ from method C.

Fisher's LSD can also be used to develop a confidence interval estimate of the difference between the means of two populations. The details are as follows:

Confidence interval estimate of the difference between two population means using Fisher's LSD procedure

$$\bar{x}_i - \bar{x}_j \pm LSD \qquad (13.18)$$

where:

$$LSD = t_{\alpha/2}\sqrt{MSE\left(\frac{1}{n_i} + \frac{1}{n_j}\right)} \qquad (13.19)$$

and $t_{\alpha/2}$ is based on a t distribution with $n_T - k$ degrees of freedom.

If the confidence interval in expression (13.18) includes the value zero, we cannot reject the hypothesis that the two population means are equal. However, if the confidence interval does not include the value zero, we conclude that there is a difference between the population means. For the Chemitech experiment, recall that LSD = 7.34 (corresponding to $t_{.025} = 2.179$). Therefore, a 95% confidence interval estimate of the difference between the means of populations 1 and 2 is $62 - 66 \pm 7.34 = -4 \pm 7.34 = -11.34$ to 3.34; and because this interval includes zero, we cannot reject the hypothesis that the two population means are equal.

Type I error rates

We began the discussion of Fisher's LSD procedure with the premise that analysis of variance gave us statistical evidence to reject the null hypothesis of equal population means. We showed how Fisher's LSD procedure can be used in such cases to determine where the differences occur. Technically, it is referred to as a *protected* or *restricted* LSD test because it is employed only if we first find a significant F value by using analysis of variance. To see why this distinction is important in multiple comparison tests, we need to explain the difference between a *comparisonwise* Type I error rate and an *experimentwise* Type I error rate.

In the Chemitech experiment we used Fisher's LSD procedure to make three pairwise comparisons.

Test 1	Test 2	Test 3
$H_0: \mu_1 = \mu_2$	$H_0: \mu_1 = \mu_3$	$H_0: \mu_2 = \mu_3$
$H_1: \mu_1 \neq \mu_2$	$H_1: \mu_1 \neq \mu_3$	$H_1: \mu_2 \neq \mu_3$

In each case, we used a level of significance of $\alpha = 0.05$. Therefore, for each test, if the null hypothesis is true, the probability that we will make a Type I error is $\alpha = 0.05$, hence the probability that we will not make a Type I error on each test is $1 - 0.05 = 0.95$. In discussing multiple comparison procedures we refer to this probability of a Type I error ($\alpha = 0\,05$) as the comparisonwise Type I error rate. Comparisonwise Type I error rates indicate the level of significance associated with a single pairwise comparison.

Let us now consider a slightly different question. What is the probability that in making three pairwise comparisons, we will commit a Type I error on at least one of the three tests? To answer this question, note that the probability that we will not make a Type I error on any of the three tests is $(0.95)(0.95)(0.95) = 0.8574$.[*] Therefore, the probability of making at least one Type I error is $1 - 0.8574 = 0.1426$. When we use Fisher's LSD procedure to make all three pairwise comparisons, the Type I error rate associated with this approach is not 0.05, but actually 0.1426; we refer to this error rate as the *overall* or experimentwise Type I error rate. To avoid confusion, we denote the experimentwise Type I error rate as EW.

The experimentwise Type I error rate gets larger for problems with more populations. For example, a problem with five populations has ten possible pairwise comparisons. If we tested all possible pairwise comparisons by using Fisher's LSD with a comparisonwise error rate of $\alpha = 0.05$, the experimentwise Type I error rate would be $1 - (1 - 0.05)^{10} = 0.40$. In such cases, practitioners look to alternatives that provide better control over the experimentwise error rate.

One alternative for controlling the overall experimentwise error rate, referred to as the Bonferroni adjustment, involves using a smaller comparisonwise error rate for each test. For example, if we want to test C pairwise comparisons and want the maximum probability of making a Type I error for the overall experiment to be α_{EW}, we simply use a comparisonwise error rate equal to α_{EW}/C. In the Chemitech experiment, if we want to use Fisher's LSD procedure to test all three pairwise comparisons with a maximum experimentwise error rate of $\alpha_{EW} = 0.05$, we set the comparisonwise error rate to be $\alpha = 0.05/3 = 0.017$. For a problem with five populations and ten possible pairwise comparisons, the Bonferroni adjustment would suggest a comparisonwise error rate of $0.05/10 = 0.005$. Recall from our discussion of hypothesis testing in Chapter 9 that for a fixed sample size, any decrease in the probability of making a Type I error will result in an increase in the probability of making a Type II error, which corresponds to accepting the hypothesis that the two population means are equal when in fact they are not equal. As a result, many practitioners are reluctant to perform individual tests with a low comparisonwise Type I error rate because of the increased risk of making a Type II error.

Several other procedures, such as Tukey's procedure and Duncan's multiple range test, have been developed to help in such situations. However, there is considerable controversy in the statistical community as to which procedure is 'best'. The truth is that no one procedure is best for all types of problem.

EXERCISES

Methods

13. The following data are from a completely randomized design.

	Treatment A	Treatment B	Treatment C
	32	44	33
	30	43	36
	30	44	35
	26	46	36
	32	48	40
Sample mean	30	45	36
Sample variance	6.00	4.00	6.50

---

[*] The assumption is that the three tests are independent, and hence the joint probability of the three events can be obtained by simply multiplying the individual probabilities. In fact, the three tests are not independent because MSE is used in each test; therefore, the error involved is even greater than that shown.

a. At the $\alpha = 0.05$ level of significance, can we reject the null hypothesis that the means of the three treatments are equal?

b. Use Fisher's LSD procedure to test whether there is a significant difference between the means for treatments A and B, treatments A and C and treatments B and C. Use $\alpha = 0.05$.

c. Use Fisher's LSD procedure to develop a 95% confidence interval estimate of the difference between the means of treatments A and B.

14. The following data are from a completely randomized design. In the following calculations, use $\alpha = 0.05$.

	Treatment 1	Treatment 2	Treatment 3
	63	82	69
	47	72	54
	54	88	61
	40	66	61
$\bar{x}_j$	51	77	58
s_j^2	96.67	97.34	81.99

a. Use analysis of variance to test for a significant difference between the means of the three treatments.

b. Use Fisher's LSD procedure to determine which means are different.

Applications

15. To test whether the mean time needed to mix a batch of material is the same for machines produced by three manufacturers, the Jacobs Chemical Company obtained the following data on the time (in minutes) needed to mix the material.

Manufacturer		
1	2	3
20	28	20
26	26	19
24	31	23
22	27	22

a. Use these data to test whether the population mean times for mixing a batch of material differ for the three manufacturers. Use $\alpha = 0.05$.

b. At the $\alpha = 0.05$ level of significance, use Fisher's LSD procedure to test for the equality of the means for manufacturers 1 and 3. What conclusion can you draw after carrying out this test?

16. Refer to Exercise 15. Use Fisher's LSD procedure to develop a 95% confidence interval estimate of the difference between the means for manufacturer 1 and manufacturer 2.

17. To test for any significant difference in the number of hours between breakdowns for four machines, the following data were obtained.

Machine 1	Machine 2	Machine 3	Machine 4
6.4	8.7	11.1	9.9
7.8	7.4	10.3	12.8
5.3	9.4	9.7	12.1
7.4	10.1	10.3	10.8
8.4	9.2	9.2	11.3
7.3	9.8	8.8	11.5

a. At the $\alpha = 0.05$ level of significance, what is the difference, if any, in the population mean times between the four machines?

b. Use Fisher's LSD procedure to test for the equality of the means for machines 2 and 4. Use a 0.05 level of significance.

c. Use the Bonferroni adjustment to test for a significant difference between all pairs of means. Assume that a maximum overall experimentwise error rate of 0.05 is desired.

13.4 RANDOMIZED BLOCK DESIGN

Thus far we have considered the completely randomized experimental design. Recall that to test for a difference between treatment means, we computed an F value by using the ratio:

F test statistic

$$F = \frac{\text{MSTR}}{\text{MSE}}$$

(13.20)

A problem can arise whenever differences due to extraneous factors (ones not considered in the experiment) cause the MSE term in this ratio to become large. In such cases, the F value in equation (13.20) can become small, signalling no difference between treatment means when in fact such a difference exists.

In this section we present an experimental design known as a randomized block design. Its purpose is to control some of the extraneous sources of variation by removing such variation from the MSE term. This design tends to provide a better estimate of the true error variance and leads to a more powerful hypothesis test in terms of the ability to detect differences between treatment means. To illustrate, let us consider a stress study for air traffic controllers.

Air traffic controller stress test

A study measuring the fatigue and stress of air traffic controllers resulted in proposals for modification and redesign of the controller's work station. After consideration of several designs for the work station, three specific alternatives are selected as having the best potential for reducing controller stress. The key question is: to what extent do the three alternatives differ in terms of their effect on controller stress? To answer this question, we need to design an experiment that will provide measurements of air traffic controller stress under each alternative.

In a completely randomized design, a random sample of controllers would be assigned to each work station alternative. However, controllers are believed to differ substantially in their ability to handle stressful situations. What is high stress to one controller might be only moderate or even low stress to another. Hence, when considering the within-group source of variation (MSE), we must realize that this variation includes both random error and error due to individual controller differences. In fact, managers expected controller variability to be a major contributor to the MSE term.

One way to separate the effect of the individual differences is to use a randomized block design. Such a design will identify the variability stemming from individual controller differences and remove it from the MSE term. The randomized block design calls for a single sample of controllers. Each controller in the sample is tested with each of the three work station alternatives. In experimental design terminology, the work station is the *factor of interest* and the controllers are the *blocks*. The three treatments or populations associated with the work station factor correspond to the three work station alternatives. For simplicity, we refer to the work station alternatives as system A, system B and system C.

The *randomized* aspect of the randomized block design is the random order in which the treatments (systems) are assigned to the controllers. If every controller were to test the three systems in the same order, any observed difference in systems might be due to the order of the test rather than to true differences in the systems.

To provide the necessary data, the three work station alternatives were installed at the Berlin Control Centre. Six controllers were selected at random and assigned to operate each of the systems. A follow-up interview and a medical examination of each controller participating in the study provided a measure of the stress for each controller on each system. The data are reported in Table 13.5.

TABLE 13.5   A randomized block design for the air traffic controller stress test

		Treatments		
		System A	**System B**	**System C**
	Controller 1	15	15	18
	Controller 2	14	14	14
Blocks	Controller 3	10	11	15
	Controller 4	13	12	17
	Controller 5	16	13	16
	Controller 6	13	13	13

AIRTRAFFIC

Table 13.6 is a summary of the stress data collected. In this table we include column totals (treatments) and row totals (blocks) as well as some sample means that will be helpful in making the sum of squares computations for the ANOVA procedure. As lower stress values are viewed as better, the sample data seem to favour system B with its mean stress rating of 13. However, the usual question remains: do the sample results justify the conclusion that the population mean stress levels for the three systems differ? That is, are the differences statistically significant? An analysis of variance computation similar to the one performed for the completely randomized design can be used to answer this statistical question.

TABLE 13.6   Summary of stress data for the air traffic controller stress test

		Treatments			Row or block totals	Block means
		System A	**System B**	**System C**		
	Controller 1	15	15	18	48	$\bar{x}_{1.} = 48/3 = 16.0$
	Controller 2	14	14	14	42	$\bar{x}_{2.} = 42/3 = 14.0$
Blocks	Controller 3	10	11	15	36	$\bar{x}_{3.} = 36/3 = 12.0$
	Controller 4	13	12	17	42	$\bar{x}_{4.} = 42/3 = 15.0$
	Controller 5	16	13	16	45	$\bar{x}_{5.} = 45/3 = 15.0$
	Controller 6	13	13	13	39	$\bar{x}_{6.} = 39/3 = 13.0$
Column or treatment totals		81	78	93	252	$\bar{\bar{x}} = \dfrac{252}{18} = 14.0$
Treatment means		$\bar{x}_{.1} = \dfrac{81}{6}$	$\bar{x}_{.2} = \dfrac{78}{6}$	$\bar{x}_{.3} = \dfrac{93}{6}$		
		$= 13.5$	$= 13.0$	$= 15.5$		

ANOVA procedure

The ANOVA procedure for the randomized block design requires us to partition the sum of squares total (SST) into three groups: sum of squares due to treatments (SSTR), sum of squares due to blocks (SSBL) and sum of squares due to error (SSE). The formula for this partitioning follows.

$$SST = SSTR + SSBL + SSE \qquad (13.21)$$

This sum of squares partition is summarized in the ANOVA table for the randomized block design as shown in Table 13.7. The notation used in the table is:

$$k = \text{the number of treatments}$$
$$b = \text{the number of blocks}$$
$$n_T = \text{the total sample size } n_T = kb$$

TABLE 13.7 ANOVA table for the randomized block design with k treatments and b blocks

Source of variation	Degrees of freedom	Sum of squares	Mean square	F
Treatments	$k - 1$	SSTR	$MSTR = \dfrac{SSTR}{k - 1}$	$\dfrac{MSTR}{MSE}$
Blocks	$b - 1$	SSBL	$MSBL = \dfrac{SSBL}{b - 1}$	
Error	$(k - 1)(b - 1)$	SSE	$MSE = \dfrac{SSE}{(k - 1)(b - 1)}$	
Total	$n_T - 1$	SST		

Note that the ANOVA table also shows how the $n_T - 1$ total degrees of freedom are partitioned such that $k - 1$ degrees of freedom go to treatments, $b - 1$ go to blocks and $(k - 1)(b - 1)$ go to the error term. The mean square column shows the sum of squares divided by the degrees of freedom, and $F = MSTR/MSE$ is the F ratio used to test for a significant difference between the treatment means. The primary contribution of the randomized block design is that, by including blocks, we remove the individual controller differences from the MSE term and obtain a more powerful test for the stress differences in the three work station alternatives.

Computations and conclusions

To compute the F statistic needed to test for a difference between treatment means with a randomized block design, we need to compute MSTR and MSE. To calculate these two mean squares, we must first compute SSTR and SSE; in doing so, we will also compute SSBL and SST. To simplify the presentation, we perform the calculations in four steps. In addition to k, b and n_T as previously defined, the following notation is used:

$$x_{ij} = \text{value of the observation corresponding to treatment } j \text{ in block } i$$
$$\bar{x}_{.j} = \text{sample mean of the } j \text{ th treatment}$$
$$\bar{x}_{i.} = \text{sample mean for the } i \text{ th block}$$
$$\bar{\bar{x}} = \text{overall sample mean}$$

Step 1. Compute the total sum of squares (SST).

$$SST = \sum_{i=1}^{b}\sum_{j=1}^{k}(x_{ij} - \bar{\bar{x}})^2 \tag{13.22}$$

Step 2. Compute the sum of squares due to treatments (SSTR).

$$SSTR = b\sum_{j=1}^{k}(\bar{x}_j - \bar{\bar{x}})^2 \tag{13.23}$$

Step 3. Compute the sum of squares due to blocks (SSBL).

$$SSBL = k\sum_{i=1}^{b}(\bar{x}_{i.} - \bar{\bar{x}})^2 \tag{13.24}$$

Step 4. Compute the sum of squares due to error (SSE).

$$SSE = SST - SSTR - SSBL \tag{13.25}$$

For the air traffic controller data in Table 13.6, these steps lead to the following sums of squares:

Step 1. $SST = (15-14)^2 + (15-14)^2 + (18-14)^2 + \ldots + (13-14)^2 = 70$

Step 2. $SSTR = [(13.5 - 14)^2 + (13.0 - 14)^2 + (15.5 - 14)^2] = 21$

Step 3. $SSBL = 3[(16 - 14)^2 + (14 - 14)^2 + (12 - 14)^2 + (14 - 14)^2 + (15 - 14)^2 + (13 - 14)^2] = 30$

Step 4. $SSE = 70 - 21 - 30 = 19$

These sums of squares divided by their degrees of freedom provide the corresponding mean square values shown in Table 13.8.

TABLE 13.8   ANOVA table for the air traffic controller stress test

Source of variation	Degrees of freedom	Sum of squares	Mean square	F	p-value
Treatments	2	21	10.5	10.5/1.9 = 5.53	0.024
Blocks	5	30	6.0		
Error	10	19	1.9		
Total	17	70			

Let us use a level of significance $\alpha = 0.05$ to conduct the hypothesis test. The value of the test statistic is:

$$F = \frac{MSTR}{MSE} = \frac{10.5}{1.9} = 5.53$$

The numerator degrees of freedom is $k - 1 = 3 - 1 = 2$ and the denominator degrees of freedom is $(k - 1)(b - 1) = (3 - 1)(6 - 1) = 10$. From Table 4 of Appendix B we find that with the degrees of freedom 2 and 10, $F = 5.53$ is between $F_{0.025} = 5.46$ and $F_{0.01} = 7.56$. As a result, the area in the upper tail, or the p-value, is between 0.01 and 0.025. Alternatively, we can use R, Excel, SPSS or Minitab to show that the exact p-value for $F = 5.53$ is 0.024. With p-value $\leq \alpha = 0.05$, we reject the null hypothesis $H_0 : \mu_1 = \mu_2 = \mu_3$ and conclude that the population mean stress levels differ for the three work station alternatives.

Some general comments can be made about the randomized block design. The experimental design described in this section is a *complete* block design; the word 'complete' indicates that each block is subjected to all *k* treatments. That is, all controllers (blocks) were tested with all three systems (treatments). Experimental designs in which some but not all treatments are applied to each block are referred to as *incomplete* block designs. A discussion of incomplete block designs is beyond the scope of this text.

As each controller in the air traffic controller stress test was required to use all three systems, this approach guarantees a complete block design. In some cases, however, blocking is carried out with 'similar' experimental units in each block. For example, assume that in a pretest of air traffic controllers, the population of controllers was divided into two groups ranging from extremely high-stress individuals to extremely low-stress individuals.

The blocking could still be accomplished by having three controllers from each of the stress classifications participate in the study. Each block would then consist of three controllers in the same stress group. The randomized aspect of the block design would be the random assignment of the three controllers in each block to the three systems.

Finally, note that the ANOVA table shown in Table 13.7 provides an *F* value to test for treatment effects but *not* for blocks. The reason is that the experiment was designed to test a single factor – work station design. The blocking based on individual stress differences was conducted to remove such variation from the MSE term. However, the study was not designed to test specifically for individual differences in stress.

Some analysts compute $F = MSB/MSE$ and use that statistic to test for significance of the blocks. Then they use the result as a guide to whether the same type of blocking would be desired in future experiments. However, if individual stress difference is to be a factor in the study, a different experimental design should be used. A test of significance on blocks should not be performed as a basis for a conclusion about a second factor.

EXERCISES

Methods

18. Consider the experimental results for the following randomized block design. Make the calculations necessary to set up the analysis of variance table.

	Treatments		
	A	B	C
	10	9	8
	12	6	5
Blocks	18	15	14
	20	18	18
	8	7	8

Use $\alpha = 0.05$ to test for any significant differences.

19. The following data were obtained for a randomized block design involving five treatments and three blocks: SST = 430, SSTR = 310, SSBL = 85. Set up the ANOVA table and test for any significant differences. Use $\alpha = 0.05$.

20. An experiment has been conducted for four treatments with eight blocks. Complete the following analysis of variance table.

Source of variation	Degrees of freedom	Sum of squares	Mean square	F
Treatments		900		
Blocks		400		
Error				
Total		1,800		

Use $\alpha = 0.05$ to test for any significant differences.

Applications

PACKAGING

21. A Cape Town company is considering changing the packaging of one of its iconic brands. Three alternative packages were tested along with the present packaging in three different outlet channels for a period of a month. Sales in thousands of rands are as follows:

	Package			
Outlet channel	I (present)	II	III	IV
1	79	68	85	79
2	74	68	103	87
3	67	69	89	91

Analyze the data as originating from a randomized complete block design. What are your conclusions?

22. A car dealer conducted a test to determine if the time in minutes needed to complete a minor engine tune-up depends on whether a computerized engine analyzer or an electronic analyzer is used. Because tune-up time varies between compact, intermediate and full-sized cars, the three types of cars were used as blocks in the experiment. The data obtained follow.

		Analyzer	
		Computerized	Electronic
	Compact	50	42
Car	Intermediate	55	44
	Full-sized	63	46

Use $\alpha = 0.05$ to test for any significant differences.

23. A textile mill produces a silicone proofed fabric for making into rainwear. The chemist in charge thinks that a silicone solution of about 12 per cent strength should yield a fabric with maximum waterproofing index. They also suspected there may be some batch to batch variation because of slight differences in the cloth. To allow for this possibility five different strengths of solution were used on each of the three different batches of fabric. The following values of waterproofing index were obtained:

		Strength of silicone solution (%)				
		6	9	12	15	18
	1	20.8	20.6	22.0	22.6	20.9
Fabric	2	19.4	21.2	21.8	23.9	22.4
	3	19.9	21.1	22.7	22.7	22.1

Using $\alpha = 0.05$, carry out an appropriate test of these data and comment on the chemist's initial beliefs.

WHEAT

24. An agricultural experiment to compare five different strains of wheat, A, B, C, D, E was carried out using five randomized blocks each linked to a particular cultivation regime. The resulting yields (kg) are as follows:

			Strain		
Block	A	B	C	D	E
1	36.5	47.1	53.5	37.1	46.5
2	48.4	43	58	40.9	41.3
3	50.9	52.4	66	47.9	48.8
4	60.9	65.5	67.1	56.1	55.7
5	46.3	50	58	44.5	45.3

Analyze these data appropriately. What do you conclude?

13.5 FACTORIAL EXPERIMENT

The experimental designs we have considered thus far enable us to draw statistical conclusions about one factor. However, in some experiments we want to draw conclusions about more than one variable or factor. A *factorial experiment* is an experimental design that allows simultaneous conclusions about two or more factors. The term *factorial* is used because the experimental conditions include all possible combinations of the factors. For example, for *a* levels of factor A and *b* levels of factor B, the experiment will involve collecting data on *ab* treatment combinations. In this section we will show the analysis for a two-factor factorial experiment. The basic approach can be extended to experiments involving more than two factors.

As an illustration of a two-factor factorial experiment, we will consider a study involving the Graduate Management Admissions Test (GMAT), a standardized test used by graduate schools of business to evaluate an applicant's ability to pursue a graduate programme in that field. Scores on the GMAT range from 200 to 800, with higher scores implying higher aptitude.

In an attempt to improve students' performance on the GMAT, a major Spanish university is considering offering the following three GMAT preparation programmes:

1 A three-hour review session covering the types of questions generally asked on the GMAT.

2 A one-day programme covering relevant exam material, along with the taking and grading of a sample exam.

3 An intensive ten-week course involving the identification of each student's weaknesses and the setting up of individualized programmes for improvement.

Hence, one factor in this study is the GMAT preparation programme, which has three treatments: three-hour review, one-day programme and ten-week course. Before selecting the preparation programme to adopt, further study will be conducted to determine how the proposed programmes affect GMAT scores.

The GMAT is usually taken by students from three colleges: the College of Business, the College of Engineering and the College of Arts and Sciences. Therefore, a second factor of interest in the experiment is whether a student's undergraduate college affects the GMAT score. This second factor, undergraduate college, also has three treatments: business, engineering, and arts and sciences. The factorial design for this experiment with three treatments corresponding to factor A, the preparation programme, and three treatments corresponding to factor B, the undergraduate college, will have a total of 3 × 3 = 9 treatment combinations. These treatment combinations or experimental conditions are summarized in Table 13.9.

TABLE 13.9   Nine treatment combinations for the two-factor GMAT experiment

		Factor B: College		
		Business	**Engineering**	**Arts and Sciences**
Factor A:	Three-hour review	1	2	3
Preparation	One-day programme	4	5	6
Programme	Ten-week course	7	8	9

Assume that a sample of two students will be selected corresponding to each of the nine treatment combinations shown in Table 13.9: two business students will take the three-hour review, two will take the one-day programme and two will take the ten-week course. In addition, two engineering students and two arts and sciences students will take each of the three preparation programmes. In experimental design terminology, the sample size of two for each treatment combination indicates that we have two replications. Additional replications and a larger sample size could easily be used, but we elect to minimize the computational aspects for this illustration.

This experimental design requires that six students who plan to attend graduate school be randomly selected from *each* of the three undergraduate colleges. Then two students from each college should be assigned randomly to each preparation programme, resulting in a total of 18 students being used in the study.

Assume that the randomly selected students participated in the preparation programmes and then took the GMAT. The scores obtained are reported in Table 13.10.

TABLE 13.10   GMAT scores for the two-factor experiment

		Factor B: College		
		Business	**Engineering**	**Arts and Sciences**
	Three-hour review	500	540	480
Factor A:		580	460	400
Preparation	One-day programme	460	560	420
Programme		540	620	480
	Ten-week course	560	600	480
		600	580	410

The analysis of variance computations with the data in Table 13.10 will provide answers to the following questions:

- **Main effect (factor A):** Do the preparation programmes differ in terms of effect on GMAT scores?
- **Main effect (factor B):** Do the undergraduate colleges differ in terms of effect on GMAT scores?
- **Interaction effect (factors A and B):** Do students in some colleges do better on one type of preparation programme whereas others do better on a different type of preparation programme?

The term interaction refers to a new effect that we can now study because we used a factorial experiment. If the interaction effect has a significant impact on the GMAT scores, we can conclude that the effect of the type of preparation programme depends on the undergraduate college.

ANOVA procedure

The ANOVA procedure for the two-factor factorial experiment requires us to partition the sum of squares total (SST) into four groups: sum of squares for factor A (SSA), sum of squares for factor B (SSB), sum of squares for interaction (SSAB) and sum of squares due to error (SSE). The formula for this partitioning is as follows:

$$SST = SSA + SSB = SSAB + SSE \tag{13.26}$$

The partitioning of the sum of squares and degrees of freedom is summarized in Table 13.11. The following notation is used:

a = number of levels of factor A

b = number of levels of factor B

r = number of replications

n_T = total number of observations taken in the experiment $n_T = abr$

TABLE 13.11   ANOVA table for the two-factor factorial experiment with r replications

Source of variation	Degrees of freedom	Sum of squares	Mean square	F
Factor A	$a - 1$	SSA	$MSA = \dfrac{SSA}{a-1}$	$\dfrac{MSA}{MSE}$
Factor B	$b - 1$	SSB	$MSB = \dfrac{SSB}{b-1}$	$\dfrac{MSB}{MSE}$
Interaction	$(a-1)(b-1)$	SSAB	$MSAB = \dfrac{SSAB}{(a-1)(b-1)}$	$\dfrac{MSAB}{MSE}$
Error	$ab(r-1)$	SSE	$MSE = \dfrac{SSE}{ab(r-1)}$	
Total	$n_T - 1$	SST		

Computations and conclusions

To compute the F statistics needed to test for the significance of factor A, factor B and interaction, we need to compute MSA, MSB, MSAB and MSE. To calculate these four mean squares, we must first compute SSA, SSB, SSAB and SSE; in doing so we will also compute SST. To simplify the presentation, we perform the calculations in five steps. In addition to a, b, r and n_T as previously defined, the following notation is used:

x_{ijk} = observation corresponding to the kth replicate taken from treatment i of factor A and treatment j of factor B

$\bar{x}_{i.}$ = sample mean for the observations in treatment i factor A

$\bar{x}_{.j}$ = sample mean for the observations in treatment j factor B

$\bar{x}_{ij}$ = sample mean for the observations corresponding to the combination of treatment i factor A and treatment j factor B

$\bar{\bar{x}}$ = overall sample mean of all n_T observations

Step 1. Compute the total sum of squares.

$$SST = \sum_{i=1}^{a} \sum_{j=1}^{b} \sum_{k=1}^{r} (x_{ijk} - \bar{\bar{x}})^2 \tag{13.27}$$

Step 2. Compute the sum of squares for factor A.

$$SSA = br \sum_{i=1}^{a} (\bar{x}_{i.} - \bar{\bar{x}})^2 \tag{13.28}$$

Step 3. Compute the sum of squares for factor B.

$$SSB = ar \sum_{j=1}^{b} (\bar{x}_{j} - \bar{\bar{x}})^2 \tag{13.29}$$

Step 4. Compute the sum of squares for interaction.

$$SSAB = r \sum_{i=1}^{a} \sum_{j=1}^{b} (\bar{x}_{ij} - \bar{x}_{i.} - \bar{x}_{j} - \bar{\bar{x}})^2 \tag{13.30}$$

Step 5. Compute the sum of squares due to error.

$$SSE = SST - SSA - SSB - SSAB \tag{13.31}$$

Table 13.12 reports the data collected in the experiment and the various sums that will help us with the sum of squares computations. Using equations (13.27) to (13.31), we calculate the following sums of squares for the GMAT two-factor factorial experiment:

Step 1. $SST = (500 - 515)^2 + (580 - 515)^2 + (540 - 515)^2 + \ldots + (410 - 515)^2 = 82{,}450$

Step 2. $SSA = (3)(2)[(493.33 - 515)^2 + (513.33 - 515)^2 + (538.33 - 515)^2] = 6{,}100$

Step 3. $SSB = (3)(2)[(540 - 515)^2 + (560 - 515)^2 + (445 - 515)^2] = 45{,}300$

Step 4. $SSAB = 2[(540 - 493.33 - 540 + 515)^2 + (500 - 493.33 - 560 + 515)^2 + \ldots + (445 - 538.33 - 445 + 515)^2] = 11{,}200$

Step 5. $SSE = 82{,}450 - 6{,}100 - 45{,}300 - 11{,}200 = 19{,}850$

These sums of squares divided by their corresponding degrees of freedom provide the appropriate mean square values for testing the two main effects (preparation programme and undergraduate college) and the interaction effect.

Due to the computational effort involved in any modest- to large-size factorial experiment, the computer usually plays an important role in performing the analysis of variance computations shown above and in the calculation of the p-values used to make the hypothesis testing decisions. Table 13.13 shows the computer output for the analysis of variance for the GMAT two-factor factorial experiment. For this output we set the level of significance $\alpha = 0.05$ as the benchmark for subsequent hypothesis tests.

As the p-value $= 0.299$ for factor A is greater than $\alpha = 0.05$, we deduce that there is no significant difference in the mean GMAT test scores for the three preparation programmes. However, for the undergraduate college effect, the p-value of 0.005 is less than $\alpha = 0.05$, so we infer that there is a significant difference in the mean GMAT test scores between the three undergraduate colleges. Finally, because the p-value of 0.350 for the interaction effect is greater than $\alpha = 0.05$, there is no significant interaction effect. Therefore, the study provides no reason to believe that the three preparation programmes differ in their ability to prepare students from the different colleges for the GMAT.

TABLE 13.12 GMAT summary data for the two-factor experiment

Treatment combination totals	Factor B: College			Row totals	Factor A means
	Business	Engineering	Arts and Sciences		
Factor A: Preparation programme — Three-hour review	500 580 1,080 $\bar{x}_{11}=\dfrac{1,080}{2}=540$ 460	540 460 1,000 $\bar{x}_{12}=\dfrac{1,000}{2}=500$ 560	480 400 880 $\bar{x}_{13}=\dfrac{880}{2}=440$ 420	2,960	$\bar{x}_{1.}=\dfrac{2,960}{6}=493.33$
One-day programme	540 1,000 $\bar{x}_{21}=\dfrac{1,000}{2}=500$ 560 600 1,160	620 1,180 $\bar{x}_{22}=\dfrac{1,180}{2}=590$ 600 580 1,180	480 900 $\bar{x}_{23}=\dfrac{900}{2}=450$ 480 410 890	3,080	$\bar{x}_{2.}=\dfrac{3,080}{6}=513.33$
10-week course	$\bar{x}_{31}=\dfrac{1,160}{2}=580$	$\bar{x}_{32}=\dfrac{1,180}{2}=590$	$\bar{x}_{33}=\dfrac{890}{2}=445$	3,230	$\bar{x}_{3.}=\dfrac{3,230}{6}=538.33$
Column totals	3,240	3,360	2,670	9,270 → Overall total	
Factor B means	$\bar{x}_{1}=\dfrac{3,240}{6}=540$	$\bar{x}_{2}=\dfrac{3,360}{6}=560$	$\bar{x}_{3}=\dfrac{2,670}{6}=445$	$\bar{x}=\dfrac{9,270}{18}=515$	

TABLE 13.13   Output for the GMAT two-factor design

Source	DF	SS	MS	F	P
Factor A	2	6,100	3,050	1.38	0.299
Factor B	2	45,300	22,650	10.27	0.005
Interaction	4	11,200	2,800	1.27	0.350
Error	9	19,850	2,206		
Total	17	82,450			

Undergraduate college was found to be a significant factor. Checking the calculations in Table 13.12, we see that the sample means are: Business students $\bar{x}_1 = 540 = 540$, Engineering students $\bar{x}_2 = 560 = 560$, and Arts and Sciences students $\bar{x}_3 = 445 = 445$. Tests on individual treatment means can be conducted; yet after reviewing the three sample means, we would anticipate no difference in preparation for business and engineering graduates. However, the arts and sciences students appear to be significantly less prepared for the GMAT than students in the other colleges. Perhaps this observation will lead the university to consider other options for assisting these students in preparing for the GMAT.

EXERCISES

Methods

25. A factorial experiment involving two levels of factor A and three levels of factor B resulted in the following data.

		Factor B		
		Level 1	Level 2	Level 3
		135	90	75
	Level 1	165	66	93
Factor A		125	127	120
	Level 2	95	105	136

Test for any significant main effects and any interaction. Use $\alpha = 0.05$.

26. The calculations for a factorial experiment involving four levels of factor A, three levels of factor B, and three replications resulted in the following data: SST = 280, SSA = 26, SSB = 23, SSAB = 175. Set up the ANOVA table and test for any significant main effects and any interaction effect. Use $\alpha = 0.05$.

Applications

27. A factorial experiment was designed to test for differences in the time needed to translate English into foreign languages using two computerized translator systems. Results for translation into French, Spanish and German were as follows:

	Language		
	Spanish	French	German
System 1	8	10	12
	12	14	16
System 2	6	14	16
	10	16	22

Perform an appropriate analysis of these data. State your assumptions. What are your conclusions?

28. The following data (in days) were collected to investigate whether the length of stay in hospital for a particular medical condition is affected by either the gender (male/female) of the patient or the hospital:

	Hospital					
	1		2		3	
Female patients	28	36	14	7	22	25
	35	32	9	7	20	30
	28	38	10	16	24	32
Male patients	35	36	3	5	18	7
	31	33	7	9	15	11
	27	34	4	6	9	10

Analyze the data as arising from a 2×3 factorial design. What are your conclusions?

29. Based on a 2018 study, the average elapsed time between when a user navigates to a website on a mobile device until its main content is available was 14.6 seconds. This is more than a 20 per cent increase from 2017 (*searchenginejournal.com*). Responsiveness is certainly an important feature of any website and is perhaps even more important on a mobile device. What other web design factors need to be considered for a mobile device to make it more user-friendly? Among other things, navigation menu placement and amount of text entry required are important on a mobile device. The following data provide the time (in seconds) it took randomly selected students (two for each factor combination) to perform a pre-specified task with the different combinations of navigation menu placement and amount of text entry required.

		Amount of text entry required	
		Low	High
	Right	8	12
		12	8
Navigation menu position	Middle	22	36
		14	20
	Left	10	18
		18	14

Use the ANOVA procedure for factorial designs to test for any significant effects resulting from navigation menu position and amount of text entry required. Use $\alpha = 0.05$.

30. Aircraft primer paints are applied to aluminium surfaces by two methods: dipping and spraying. The purpose of the primer is to improve paint adhesion and some parts can be primed using either application method. An engineering group is interested in learning if three different primers differ in their adhesive properties. A factorial experiment was performed to investigate as much. For each combination of primer type and application method, three specimens were painted, then a finish paint was applied and the adhesion force measured. Data were as follows (high values correspond with high adhesiveness and vice versa):

| | Application method | |
Primer	Dipping	Spraying
	4.0	5.4
1	4.5	4.9
	4.3	5.6
	5.6	5.8
2	4.9	6.1
	5.4	6.3
	3.8	5.5
3	3.7	5.0
	4.0	5.0

Analyze these data appropriately.

ONLINE RESOURCES

For the data files, additional questions and answers, and the software section for Chapter 13, go to the online platform.

SUMMARY

In this chapter we showed how analysis of variance can be used to test for differences between means of several populations or treatments. We introduced the completely randomized design, the randomized block design and the two-factor factorial experiment. The completely randomized design and the randomized block design are used to draw conclusions about differences in the means of a single factor. The primary purpose of blocking in the randomized block design is to remove extraneous sources of variation from the error term. Such blocking provides a better estimate of the true error variance and a better test to determine whether the population or treatment means of the factor differ significantly.

We showed that the basis for the statistical tests used in analysis of variance and experimental design is the development of two independent estimates of the population variance σ^2. In the single-factor case, one estimator is based on the variation between the treatments; this estimator provides an unbiased estimate of σ^2 only if the means $\mu_1, \mu_2, \ldots, \mu_k$ are all equal. A second estimator of σ^2 is based on the variation of the observations within each sample; this estimator will always provide an unbiased estimate of σ^2. By computing the ratio of these two estimators (the F statistic) we developed a rejection rule for determining whether to reject the null hypothesis that the population or treatment means are equal. In all the experimental designs considered, the partitioning of the sum of squares and degrees of freedom into their various sources enabled us to compute the appropriate values for the analysis of variance calculations and tests. We also showed how Fisher's LSD procedure and the Bonferroni adjustment can be used to perform pairwise comparisons to determine which means are different.

KEY TERMS

ANOVA table
Blocking
Comparisonwise Type I error rate
Completely randomized design
Experimental units
Experimentwise Type I error rate
Factor
Factorial experiments

Interaction
Multiple comparison procedures
Partitioning
Randomized block design
Replications
Response variable
Single-factor experiment
Treatment

KEY FORMULAE

Completely randomized design sample mean for treatment j

$$\bar{x}_j = \frac{\sum_{i=1}^{n_j} x_{ij}}{n_j} \tag{13.1}$$

Sample variance for treatment j

$$s_j^2 = \frac{\sum_{i=1}^{n_j} (x_{ij} - \bar{x}_j)^2}{n_j - 1} \tag{13.2}$$

Overall sample mean

$$\bar{\bar{x}} = \frac{\sum_{j=1}^{k} \sum_{i=1}^{n_j} x_{ij}}{n_T} \tag{13.3}$$

$$n_T = n_1 + n_2 + \ldots + n_k \tag{13.4}$$

Mean square due to treatments

$$MSTR = \frac{SSTR}{k - 1} \tag{13.7}$$

Sum of squares due to treatments

$$SSTR = \sum_{j=1}^{k} n_j(\bar{x}_j - \bar{\bar{x}})^2 \tag{13.8}$$

Mean square due to error

$$MSE = \frac{SSE}{n_T - k} \tag{13.10}$$

Sum of squares due to error

$$SSE = \sum_{j=1}^{k} (n_j - 1)s_j^2 \tag{13.11}$$

Test statistic for the equality of k population means

$$F = \frac{\text{MSTR}}{\text{MSE}} \tag{13.12}$$

Total sum of squares

$$\text{SST} = \sum_{j=1}^{k} \sum_{i=1}^{n_j} (x_{ij} - \bar{\bar{x}})^2 \tag{13.13}$$

Partitioning of sum of squares

$$\text{SST} = \text{SSTR} + \text{SSE} \tag{13.14}$$

Multiple comparison procedures test statistic for Fisher's LSD procedure

$$t = \frac{\bar{x}_i - \bar{x}_j}{\sqrt{\text{MSE}\left(\dfrac{1}{n_i} + \dfrac{1}{n_j}\right)}} \tag{13.16}$$

Fisher's LSD

$$\text{LSD} = t_{a/2}\sqrt{\text{MSE}\left(\frac{1}{n_i} + \frac{1}{n_j}\right)} \tag{13.17}$$

Randomized block design total sum of squares

$$\text{SST} = \sum_{i=1}^{b} \sum_{j=1}^{k} (x_{ij} - \bar{\bar{x}})^2 \tag{13.22}$$

Sum of squares due to treatments

$$\text{SSTR} = b\sum_{j=1}^{k} (\bar{x}_{j} - \bar{\bar{x}})^2 \tag{13.23}$$

Sum of squares due to blocks

$$\text{SSBL} = k\sum_{i=1}^{b} (\bar{x}_{i.} - \bar{\bar{x}})^2 \tag{13.24}$$

Sum of squares due to error

$$\text{SSE} = \text{SST} - \text{SSTR} - \text{SSBL} \tag{13.25}$$

Factorial experiments total sum of squares

$$\text{SST} = \sum_{i=1}^{a} \sum_{j=1}^{b} \sum_{k=1}^{r} (x_{ijk} - \bar{\bar{x}})^2 \tag{13.27}$$

Sum of squares for factor A

$$SSA = br \sum_{i=1}^{a} (\bar{x}_{i.} - \bar{\bar{x}})^2 \qquad (13.28)$$

Sum of squares for factor B

$$SSB = ar \sum_{j=1}^{b} (\bar{x}_{.j} - \bar{\bar{x}})^2 \qquad (13.29)$$

Sum of squares for interaction

$$SSAB = r \sum_{i=1}^{a} \sum_{j=1}^{b} (\bar{x}_{ij} - \bar{x}_{i.} - \bar{x}_{.j} - \bar{\bar{x}})^2 \qquad (13.30)$$

Sum of squares due to error

$$SSE = SST - SSA - SSB - SSAB \qquad (13.31)$$

CASE PROBLEM 1

Product design testing

An engineering manager has been designated the task of evaluating a commercial device subject to marked variations in temperature. Three different types of component are being considered for the device. When the device is manufactured and is shipped to the field, the manager has no control over the temperature extremes that the device will encounter, but knows from experience that temperature is an important factor in relation to the component's life. Notwithstanding this, temperature can be controlled in the laboratory for the purposes of the test.

The engineering manager arranges for all three components to be tested at the temperature levels: −10°C, 20°C and 50°C, as these temperature levels are consistent with the product end-use environment. Four components are tested for each combination of type and temperature, and all 36 tests are run in random

order. The resulting observed component life data are presented in Table 1.

Managerial report

1. What are the effects of the chosen factors on the life of the component?

2. Do any components have a consistently long life regardless of temperature?

3. What recommendation would you make to the engineering manager?

© CasarsaGuru/iStock

A product component is tested for its capability of enduring extreme heat

(Continued)

DEVICE

TABLE 1   Component lifetimes (000 of hours)

Type	Temperature (°C)					
	−10		20		50	
1	3.12	3.70	0.82	0.96	0.48	1.68
	1.80	4.32	1.92	1.80	1.97	1.39
2	3.60	4.51	3.02	2.93	0.60	1.68
	3.82	3.02	2.54	2.76	1.39	1.08
3	3.31	2.64	4.18	2.88	2.30	2.50
	4.03	3.84	3.60	3.34	1.97	1.44

CASE PROBLEM 2

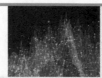

TourisTopia Travel

TourisTopia Travel (Triple T) is an online travel agency that specializes in trips to exotic locations around the world for groups of ten or more travellers. The agency's marketing manager has been working on a major revision of the homepage of Triple T's website. The content for the homepage has been selected and the only remaining decisions involve the selection of the background colour (white, green or pink) and the type of font (Arial, Calibri or Tahoma).

Triple T's IT group has designed prototype homepages featuring every combination of these background colours and fonts, and it has implemented computer code that will randomly direct each website visitor to one of these prototype homepages. For three weeks, the prototype homepage to which each visitor was directed and the amount of time in seconds spent on the website during each visit were recorded. Ten visitors to each of the prototype homepages were then selected randomly; the complete data set for these visitors is available in the data file 'TourisTopia'.

TOURISTOPIA

Managerial report

1. Use descriptive statistics to summarize the data from Triple T's study. Based on descriptive statistics, what are your preliminary conclusions about whether the time spent by visitors to the Triple T website differs by background colour or font? What are your preliminary conclusions about whether time spent by visitors to the Triple T website differs by different combinations of background colour and font?

2. Has Triple T used an observational study or a controlled experiment for its data collection? Explain.

3. Use the data from Triple T's study to test the hypothesis that the time spent by visitors to the Triple T website is equal for the three background colours. Include both factors and their interaction in the ANOVA model, and use $\alpha = 0.05$.

4. Use the data from Triple T's study to test the hypothesis that the time spent by visitors to the Triple T website is equal for the three fonts. Include both factors and their interaction in the ANOVA model, and use $\alpha = 0.05$.

5. Use the data from Triple T's study to test the hypothesis that time spent by visitors to the Triple T website is equal for the nine

combinations of background colour and font. Include both factors and their interaction in the ANOVA model, and use $\alpha = 0.05$.

6. Do your analyses of the data – following steps 3–5 above – provide evidence that the time spent by visitors to the Triple T website differs by background colour, font or combination of background colour and font? In addition, how would you advise the agency?

14
Simple Linear Regression

CHAPTER CONTENTS

Statistics in Practice   Foreign direct investment (FDI) in China

LEARNING OBJECTIVES   After reading this chapter and doing the exercises, you should be able to:

1 Understand how regression analysis can be used to develop an equation that estimates mathematically how two variables are related.

2 Understand the differences between the regression model, the regression equation and the estimated regression equation.

3 Know how to fit an estimated regression equation to a set of sample data based upon the least squares method.

4 Determine how good a fit is provided by the estimated regression equation and compute the sample correlation coefficient from the regression analysis output.

5 Understand the assumptions necessary for statistical inference and test for a significant relationship.

6 Know how to develop confidence interval estimates of the mean value of Y and an individual value of Y for a given value of X.

7 Learn how to use a residual plot to make a judgement as to the validity of the regression assumptions, recognize outliers and identify influential observations.

8 Use the Durbin–Watson test to test for autocorrelation.

9 Know the definition of the following terms: independent and dependent variable; simple linear regression; regression model; regression equation and estimated regression equation; scatter diagram; coefficient of determination; standard error of the estimate; confidence interval; prediction interval; residual plot; standardized residual plot; outlier; influential observation; leverage.

Managerial decisions are often based on the relationship between two or more variables. For example, after considering the relationship between advertising expenditures and sales, a marketing manager might attempt to predict sales for a given level of advertising expenditure. In another case, a public utility might use the relationship between the daily high temperature and the demand for electricity to predict electricity usage on the basis of next month's anticipated daily high temperatures. Sometimes a manager will rely on intuition to judge how two variables are related. However, if data can be obtained, the procedure, *regression analysis,* can be used to develop an equation showing how the variables are related mathematically.

In regression terminology, the variable being predicted is called the dependent variable. The variable being used to predict the value of the dependent variable is called the independent variable. For example, in analyzing the effect of advertising expenditures on sales, a marketing manager's desire to predict sales would suggest making sales the dependent variable. Advertising expenditure would be the independent variable used to help predict sales. In statistical notation, Y denotes the dependent variable and X denotes the independent variable.

STATISTICS IN PRACTICE
Foreign direct investment (FDI) in China

Foreign direct investment (FDI) in China increased by 3.3 per cent year-on-year to CNY 701 billion ($101 billion) in the first 10 months of 2018. Historically, FDI has often been used to forecast gross domestic product (GDP) using simple regression analysis. For example, in 2002, Foster estimated regression equations across 31 Chinese provinces as follows:

$$\hat{y} = 1.1m + 21.7x \qquad 1990–1993$$

$$\hat{y} = 2.1m + 8.9x \qquad 1995–1998$$

$$\hat{y} = 3.3m + 14.6x \qquad 2000–2003$$

where: $\hat{y}$ = estimated GDP

 x = FDI

These results broadly accord with those of Ma (2009) showing that GDP growth in China accelerated between 1992 and 1996 and after 2001 – in line with corresponding FDI inflows.

© Richard Ellis/Alamy

Sources:
www.tradingeconomics.com/china/foreign-direct-investment (accessed 21 November 2018).
Foster, M. J. (2002) 'On evaluation of FDI's: Principles, actualities and possibilities'. *International Journal of Management and Decision-Making* 3(1): 67–82.
Ma, X. (2009) 'An empirical analysis of the impact of FDI on China's economic growth'. *International Journal of Business and Management* 4(6): 76–80.

In this chapter we consider a type of regression analysis involving one independent variable and one dependent variable in which the relationship between the variables is approximated by a straight line. This is known as simple linear regression. Regression analysis involving two or more independent variables is called *multiple regression analysis*; multiple regression including cases involving curvilinear relationships are covered in Chapters 15 and 16.

14.1 SIMPLE LINEAR REGRESSION MODEL

Armand's Pizza Parlours is a chain of Italian food restaurants located in northern Italy. Armand's most successful locations are near college campuses. The managers believe that quarterly sales for these restaurants (denoted by Y) are related positively to the size of the student population (denoted by X); that is, restaurants near campuses with a large student population tend to generate more sales than those located near campuses with a small student population. Using regression analysis, we can develop an equation showing how the dependent variable Y is related to the independent variable X.

Regression model and regression equation

In the Armand's Pizza Parlours example, the population consists of all the Armand's restaurants.

For every restaurant in the population, there is a value x of X (student population) and a corresponding value y of Y (quarterly sales). The equation that describes how Y is related to x and an error term is called the regression model. The regression model used in simple linear regression is as follows:

Simple linear regression model

$$Y = \beta_0 + \beta_1 x + \varepsilon \qquad (14.1)$$

β_0 and β_1 are referred to as the parameters of the model, and ε (the Greek letter epsilon) is a random variable referred to as the *error term*. The error term ε accounts for the variability in Y that cannot be explained by the linear relationship between X and Y.

The population of all Armand's restaurants can also be viewed as a collection of subpopulations, one for each distinct value of X. For example, one subpopulation consists of all Armand's restaurants located near college campuses with 8,000 students; another subpopulation consists of all Armand's restaurants located near college campuses with 9,000 students and so on. Each subpopulation has a corresponding distribution of Y values. Thus, a distribution of Y values is associated with restaurants located near campuses with 8,000 students, a distribution of Y values is associated with restaurants located near campuses with 9,000 students and so on. Each distribution of Y values has its own mean or expected value. The equation that describes how the expected value of Y – denoted by $E(Y)$ or equivalently $E(Y \mid X = x)$ – is related to x is called the regression equation. The regression equation for simple linear regression is as follows:

Simple linear regression equation

$$E(Y) = \beta_0 + \beta_1 x \qquad (14.2)$$

The graph of the simple linear regression equation is a straight line; β_0 is the y-intercept of the regression line; β_1 is the slope and $E(Y)$ is the mean or expected value of Y for a given value of X.

Examples of possible regression lines are shown in Figure 14.1. The regression line in Panel A shows that the mean value of Y is related positively to X, with larger values of $E(Y)$ associated with larger values of X. The regression line in Panel B shows the mean value of Y is related negatively to X, with smaller values of $E(Y)$ associated with larger values of X. The regression line in Panel C shows the case in which the mean value of Y is not related to X; that is, the mean value of Y is the same for every value of X.

FIGURE 14.1
Possible regression lines in simple linear regression

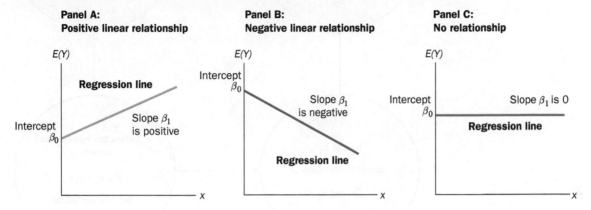

Estimated regression equation

If the values of the population parameters β_0 and β_1 were known, we could use equation (14.2) to compute the mean value of Y for a given value of X. In practice, the parameter values are not known and must be estimated using sample data. Sample statistics (denoted b_0 and b_1) are computed as estimates of the population parameters β_0 and β_1. Substituting the values of the sample statistics b_0 and b_1 for β_0 and β_1 in the regression equation, we obtain the **estimated regression equation**. The estimated regression equation for simple linear regression is as follows:

> **Estimated simple linear regression equation**
>
> $$\hat{y} = b_0 + b_1x \qquad (14.3)$$

The graph of the estimated simple linear regression equation is called the *estimated regression line*; b_0 is the y intercept and b_1 is the slope. In the next section, we show how the least squares method can be used to compute the values of b_0 and b_1 in the estimated regression equation.

In general, $\hat{y}$ is the point estimator of $E(Y)$, the mean value of Y for a given value of X. Thus, to estimate the mean or expected value of quarterly sales for all restaurants located near campuses with 10,000 students, Armand's would substitute the value of 10,000 for x in equation (14.3).

In some cases, however, Armand's may be more interested in predicting sales for one particular restaurant. For example, suppose Armand's would like to predict quarterly sales for the restaurant located near Cabot College, a school with 10,000 students.

As it turns out, the best estimate of Y for a given value of X is also provided by $\hat{y}$. Thus, to predict quarterly sales for the restaurant located near Cabot College, Armand's would also substitute the value of 10,000 for x in equation (14.3). As the value of $\hat{y}$ provides both a point estimate of $E(Y)$ and an individual value of Y for a given value of X, we will refer to $\hat{y}$ simply as the *estimated value of Y*.

Figure 14.2 provides a summary of the estimation process for simple linear regression.

FIGURE 14.2
The estimation process
in simple linear
regression

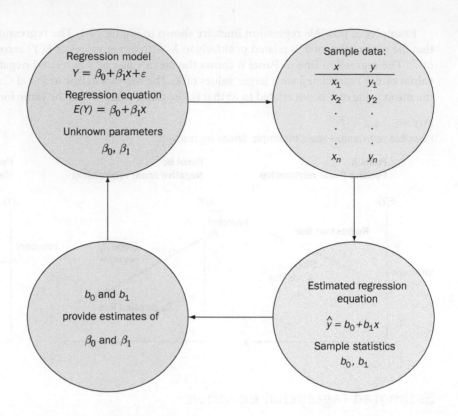

14.2 LEAST SQUARES METHOD

The **least squares method** is a procedure for using sample data to find the estimated regression equation. To illustrate the least squares method, suppose data were collected from a sample of ten Armand's Pizza Parlour restaurants located near college campuses. For the ith observation or restaurant in the sample, x_i is the size of the student population (in thousands) and y_i is the quarterly sales (in thousands of euros). The values of x_i and y_i for the ten restaurants in the sample are summarized in Table 14.1. We see that restaurant 1, with $x_1 = 2$ and $y_1 = 58$, is near a campus with 2,000 students and has quarterly sales of €58,000. Restaurant 2, with $x_2 = 6$ and $y_2 = 105$, is near a campus with 6,000 students and has quarterly sales of €105,000. The largest sales value is for restaurant 10, which is near a campus with 26,000 students and has quarterly sales of €202,000.

ARMANDS

TABLE 14.1   Student population and quarterly sales data for ten Armand's Pizza Parlours

Restaurant i	Student population (000) x_i	Quarterly sales (€000) y_i
1	2	58
2	6	105
3	8	88
4	8	118
5	12	117
6	16	137
7	20	157
8	20	169
9	22	149
10	26	202

Figure 14.3 is a scatter diagram ('scattergram') of the data in Table 14.1. Student population is shown on the horizontal axis and quarterly sales are shown on the vertical axis. Scatter diagrams for regression analysis are constructed with the values of independent variable X on the horizontal axis and the dependent variable Y on the vertical axis. The scatter diagram enables us to observe the data graphically and to draw preliminary conclusions about the possible relationship between the variables.

FIGURE 14.3

Scatter diagram of student population and quarterly sales for Armand's Pizza Parlours

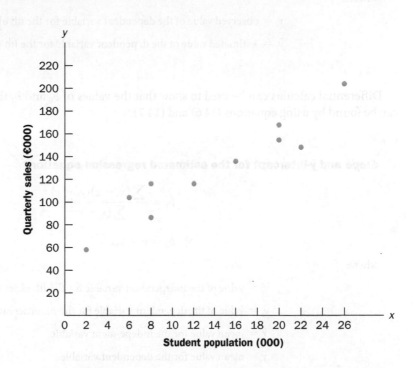

What preliminary conclusions can be drawn from Figure 14.3? Quarterly sales appear to be higher at campuses with larger student populations. In addition, for these data the relationship between the size of the student population and quarterly sales appears to be approximated by a straight line; indeed, a positive linear relationship is indicated between X and Y.

We therefore choose the simple linear regression model to represent the relationship between quarterly sales and student population. Given that choice, our next task is to use the sample data in Table 14.1 to determine the values of b_0 and b_1 in the estimated simple linear regression equation. For the ith restaurant, the estimated regression equation provides:

$$\hat{y}_i = b_0 + b_1 x_i \tag{14.4}$$

where:

$\hat{y}_i$ = the estimated value *of* quarterly sales €000s for the ith restaurant

b_0 = the intercept of the estimated regression line

b_1 = the slope of the estimated regression line

x_i = size of the student population in 000s for the ith restaurant

Every restaurant in the sample will have an observed value of sales y_i and an estimated value of sales $\hat{y}_i$. For the estimated regression line to provide a good fit to the data, we want the differences between the observed sales values and the estimated sales values to be small.

The least squares method uses the sample data to provide the values of b_0 and b_1 that minimize the *sum of the squares of the deviations* between the observed values of the dependent variable y_i and the estimated values of the dependent variable. The criterion for the least squares method is given by expression (14.5).

Least squares criterion

$$\text{Min} \sum (y_i - \hat{y}_i)^2 \tag{14.5}$$

where:

y_i = observed value of the dependent variable for the ith observation

$\hat{y}_i$ = estimated value of the dependent variable for the ith observation

Differential calculus can be used to show that the values of b_0 and b_1 that minimize expression (14.5) can be found by using equations (14.6) and (14.7).

Slope and y-intercept for the estimated regression equation*

$$b_1 = \frac{\sum (x_i - \bar{x})(y_i - \bar{y})}{\sum (x - \bar{x})^2} \tag{14.6}$$

$$b_0 = \bar{y} - b_1\bar{x} \tag{14.7}$$

where:

x_i = value of the independent variable for the ith observation

y_i = value of the dependent variable for the ith observation

$\bar{x}$ = mean value for the independent variable

$\bar{y}$ = mean value for the dependent variable

n = total number of observations

Some of the calculations necessary to develop the least squares estimated regression equation for Armand's Pizza Parlours are shown in Table 14.2. With the sample of ten restaurants, we have $n = 10$ observations. As equations (14.6) and (14.7) require $\bar{x}$ and $\bar{y}$ we begin the calculations by computing $\bar{x}$ and $\bar{y}$:

$$\bar{x} = \frac{\sum x_i}{n} = \frac{140}{10} = 14$$

$$\bar{y} = \frac{\sum y_i}{n} = \frac{1,300}{10} = 130$$

Using equations (14.6) and (14.7) and the information in Table 14.2, we can compute the slope and intercept of the estimated regression equation for Armand's Pizza Parlours. The calculation of the slope (b_1) proceeds as follows:

$$b_1 = \frac{\sum (x_i - \bar{x})(y_i - \bar{y})}{\sum (x_i - \bar{x})^2}$$

$$= \frac{2,840}{568} = 5$$

---

* An alternative formula for b_1 is:

$$b_1 = \frac{\sum x_i y_i - \left(\sum x_i \sum y_i\right)/n}{\sum x^2 - \left(\sum x_i\right)^2/n}$$

This form of equation (14.6) is often recommended when using a calculator to compute b_1.

TABLE 14.2 Calculations for the least squares estimated regression equation for Armand's Pizza Parlours

Restaurant i	x_i	y_i	$x_i - \bar{x}$	$y_i - \bar{y}$	$(x_i - \bar{x})(y_i - \bar{y})$	$(x_i - \bar{x})^2$
1	2	58	-12	-72	864	144
2	6	105	-8	-25	200	64
3	8	88	-6	-42	252	36
4	8	118	-6	-12	72	36
5	12	117	-2	-13	26	4
6	16	137	2	7	14	4
7	20	157	6	27	162	36
8	20	169	6	39	234	36
9	22	149	8	19	152	64
10	26	202	12	72	864	144
Totals	**140**	**1,300**			**2,840**	**568**
	$\sum x_i$	$\sum y_i$			$\sum (x_i - \bar{x})(y_i - \bar{y})$	$\sum (x_i - \bar{x})^2$

The calculation of the y intercept (b_0) is as follows:

$$b_0 = \bar{y} - b_1\bar{x}$$
$$= 130 - 5(14)$$
$$= 60$$

Thus, the estimated regression equation is:

$$\hat{y} = 60 + 5x$$

Figure 14.4 shows the graph of this equation on the scatter diagram.

FIGURE 14.4
Graph of the estimated
regression equation for
Armand's Pizza Parlours
$\hat{y} = 60 + 5x$

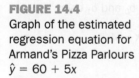

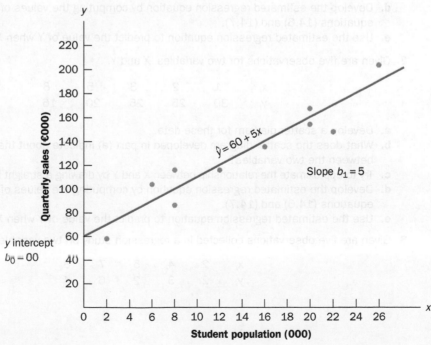

The slope of the estimated regression equation ($b_1 = 5$) is positive, implying that as student population increases, sales increase. In fact, we can conclude (based on sales measured in €000 and student population in thousands) that an increase in the student population of 1,000 is associated with an increase of €5,000 in expected sales; that is, quarterly sales are expected to increase by €5 per student.

If we believe the least squares estimated regression equation adequately describes the relationship between X and Y, it would seem reasonable to use the estimated regression equation to predict the value of Y for a given value of X.

For example, if we wanted to predict quarterly sales for a restaurant to be located near a campus with 16,000 students, we would compute:

$$\hat{y} = 60 + 5(16) = 140$$

Therefore, we would predict quarterly sales of €140,000 for this restaurant. In the following sections we will discuss methods for assessing the appropriateness of using the estimated regression equation for estimation and prediction.

EXERCISES

Methods

1. Given are five observations for two variables, X and Y.

x_i	1	2	3	4	5
y_i	3	7	5	11	14

   a. Develop a scatter diagram for these data.
   b. What does the scatter diagram developed in part (a) indicate about the relationship between the two variables?
   c. Try to approximate the relationship between X and Y by drawing a straight line through the data.
   d. Develop the estimated regression equation by computing the values of b_0 and b_1 using equations (14.6) and (14.7).
   e. Use the estimated regression equation to predict the value of Y when $X = 4$.

2. Given are five observations for two variables, X and Y.

x_i	1	2	3	5	8
y_i	30	25	25	20	16

   a. Develop a scatter diagram for these data.
   b. What does the scatter diagram developed in part (a) indicate about the relationship between the two variables?
   c. Try to approximate the relationship between X and Y by drawing a straight line through the data.
   d. Develop the estimated regression equation by computing the values of b_0 and b_1 using equations (14.6) and (14.7).
   e. Use the estimated regression equation to predict the value of Y when $X = 6$.

3. Given are five observations collected in a regression study on two variables.

x_i	2	4	5	7	8
y_i	2	3	2	6	4

a. Develop a scatter diagram for these data.
b. Develop the estimated regression equation for these data.
c. Use the estimated regression equation to predict the value of Y when $X = 4$.

Applications

4. In a factory the number of shifts worked per month is related to output as shown below:

Months	Shifts worked	Units made
1	50	352
2	70	555
3	25	207
4	55	508
5	20	48
6	60	498
7	40	310
8	30	160

a. Produce a scattergram for the data with shifts worked as the independent variable.
b. What does the scattergram developed for (a) indicate about the relationship between the two variables?
c. Try to approximate the relationship between shifts worked and units made by drawing a straight line through the plot in (a). Calculate the equivalent least squares regression equation. If 45 shifts are planned next month, what is your forecast of corresponding output?

5. The Dow Jones Industrial Average (DJIA) and the Standard & Poor's 500 (S&P) indexes are both used as measures of overall movement in the stock market. The DJIA is based on the price movements of 30 large companies; the S&P 500 is an index composed of 500 stocks. Some say the S&P 500 is a better measure of stock market performance because it is broader based. The closing prices for the DJIA and the S&P 500 for ten months, beginning with 18 January 2018, is as follows:

DOW S&P

Date	DJIA	S&P
18-Jan-18	26,149.39	2,798.03
18-Feb-18	25,029.20	2,732.22
18-Mar-18	24,103.11	2,712.92
18-Apr-18	24,163.15	2,708.64
18-May-18	24,413.84	2,712.97
18-Jun-18	24,271.41	2,773.75
18-Jul-18	25,415.19	2,815.62
18-Aug-18	25,964.82	2,850.13
18-Sep-18	26,458.31	2,904.31
18-Oct-18	25,115.76	2,768.78

a. Develop a scatter diagram for these data with DJIA as the independent variable.
b. Develop the least squares estimated regression equation.
c. Suppose the closing price for the DJIA is 27,000. Estimate the closing price for the S&P 500.

6. A large city hospital conducted a study to investigate the relationship between the number of unauthorized days that employees were absent per year and the distance (kilometres) between home and work for the employees. A sample of ten employees was selected and the following data were collected.

ABSENT

Distance to work (kilometres)	Number of days absent
1	8
3	5
4	8
6	7
8	6
10	3
12	5
14	2
14	4
18	2

a. Develop a scatter diagram for these data. Does a linear relationship appear reasonable? Explain.
b. Develop the least squares estimated regression equation that relates the distance to work to the number of days absent.
c. Predict the number of days absent for an employee who lives 5 kilometres from the hospital.

14.3 COEFFICIENT OF DETERMINATION

For the Armand's Pizza Parlours example, we developed the estimated regression equation $\hat{y} = 60 + 5x$ to approximate the linear relationship between the size of student population X and quarterly sales Y. A question now is: how well does estimated regression equation fit the data? In this section, we show that **coefficient of determination** provides a measure of the goodness of fit for the estimated regression equation.

For the ith observation, the difference between the observed value of the dependent variable, y_i, and the estimated value of the dependent variable, $\hat{y}_i$, is called the **ith residual**. The ith residual represents the error in using $\hat{y}_i$ to estimate y_i. Thus, for the ith observation, the residual is $y_i - \hat{y}_i$. The sum of squares of these residuals or errors is the quantity that is minimized by the least squares method. This quantity, also known as the *sum of squares due to error*, is denoted by SSE.

Sum of squares due to error

$$\text{SSE} = \sum (y_i - \hat{y}_i)^2 \tag{14.8}$$

The value of SSE is a measure of the error in using the least squares regression equation to estimate the values of the dependent variable in the sample.

In Table 14.3 we show the calculations required to compute the sum of squares due to error for the Armand's Pizza Parlours example. For instance, for restaurant 1 the values of the independent and dependent variables are $x_1 = 2$ and $y_1 = 58$. Using the estimated regression equation, we find that the estimated value of quarterly sales for restaurant 1 is $\hat{y}_1 = 60 + 5(2) = 70$.

Thus, the error in using $y_1 - \hat{y}_1 = 58 - 70 = -12$. The squared error $(-12)^2 = 144$, is shown in the last column of Table 14.3. After computing and squaring the residuals for each restaurant in the sample, we sum them to obtain SSE = 1,530. Thus, SSE = 1,530 measures the error in using the estimated regression equation $\hat{y}_1 = 60 + 5x$ to predict sales.

TABLE 14.3 Calculation of SSE for Armand's Pizza Parlours

Restaurant i	x_i = Student population (000)	y_i = Quarterly sales (€000)	Predicted sales $\hat{y}_i = 60 + 5x_i$	Error $y_i - \hat{y}_i$	Squared error $(y_i - \hat{y}_i)^2$
1	2	58	70	−12	144
2	6	105	90	15	225
3	8	88	100	−12	144
4	8	118	100	18	324
5	12	117	120	−3	9
6	16	137	140	−3	9
7	20	157	160	−3	9
8	20	169	160	9	81
9	22	149	170	−21	441
10	26	202	190	12	144
					SSE = 1,530

Now suppose we are asked to develop an estimate of quarterly sales without knowledge of the size of the student population. Without knowledge of any related variables, we would use the sample mean as an estimate of quarterly sales at any given restaurant. Table 14.2 shows that for the sales data, $\sum y_i = 1,300$. Hence, the mean value of quarterly sales for the sample of ten Armand's restaurants is $\bar{y} = \sum y/n = 1,300/10 = 130$.

In Table 14.4 we show the sum of squared deviations obtained by using the sample mean $\bar{y} = 130$ to estimate the value of quarterly sales for each restaurant in the sample. For the ith restaurant in the sample, the difference $y_i - \bar{y}$ provides a measure of the error involved in using $\bar{y}$ to estimate sales. The corresponding sum of squares, called the *total sum of squares*, is denoted SST.

Total sum of squares

$$\text{SST} = \sum (y_1 - \bar{y})^2 \tag{14.9}$$

TABLE 14.4 Computation of the total sum of squares for Armand's Pizza Parlours

Restaurant i	x_i = Student population (000)	y_i = Quarterly sales (€000)	Deviation $y_i - \bar{y}$	Squared deviation $(y_i - \bar{y})^2$
1	2	58	−72	5,184
2	6	105	−25	625
3	8	88	−42	1,764
4	8	118	−12	144
5	12	117	−13	169
6	16	137	7	49
7	20	157	27	729
8	20	169	39	1,521
9	22	149	19	361
10	26	202	72	5,184
				SST = 15,730

The sum at the bottom of the last column in Table 14.4 is the total sum of squares for Armand's Pizza Parlours; it is SST = 15,730.

In Figure 14.5 we show the estimated regression line $\hat{y}_i = 60 + 5x$ and the line corresponding to $\bar{y} = 130$. Note that the points cluster more closely around the estimated regression line than they do about the line $\bar{y} = 130$.

For example, for the tenth restaurant in the sample we see that the error is much larger when $\bar{y} = 130$ is used as an estimate of y_{10} than when $\hat{y}_i = 60 + 5(26) = 190$ is used. We can think of SST as a measure of how well the observations cluster about the $\bar{y}$ line and SSE as a measure of how well the observations cluster about the $\hat{y}$ line.

FIGURE 14.5

Deviations about the estimated regression line and the line $y = \bar{y}$ for Armand's Pizza Parlours

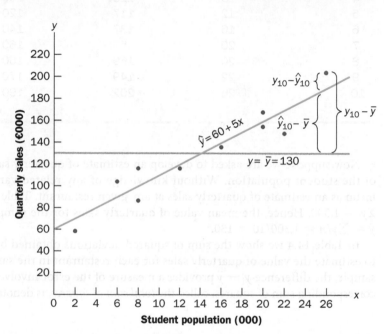

To measure how much the $\hat{y}$ values on the estimated regression line deviate from $\bar{y}$, another sum of squares is computed. This sum of squares, called the *sum of squares due to regression*, is denoted SSR.

Sum of squares due to regression

$$SSR = \sum (\hat{y}_i - \bar{y})^2 \tag{14.10}$$

From the preceding discussion, we should expect that SST, SSR and SSE are related. Indeed, the relationship between these three sums of squares provides one of the most important results in statistics.

Relationship between SST, SSR and SSE

$$SST = SSR + SSE \tag{14.11}$$

where:

SST = total sum of squares
SSR = sum of squares due to regression
SSE = sum of squares due to error

Equation (14.11) shows that the total sum of squares can be partitioned into two components: the regression sum of squares and the sum of squares due to error. Hence, if the values of any two of these sum of squares are known, the third sum of squares can be computed easily. For instance, in the Armand's Pizza Parlours example, we already know that SSE = 1,530 and SST = 15,730; therefore, solving for SSR in equation (14.11), we find that the sum of squares due to regression is:

$$SSR = SST - SSE = 15,730 - 1,530 = 14,200$$

Now let us see how the three sums of squares, SST, SSR and SSE, can be used to provide a measure of the goodness of fit for the estimated regression equation. The estimated regression equation would provide a perfect fit if every value of the dependent variable y_i happened to lie on the estimated regression line. In this case, $y_i - \hat{y}_i$ would be zero for each observation, resulting in SSE = 0. As SST = SSR + SSE, we see that for a perfect fit SSR must equal SST and the ratio (SSR/SST) must equal one. Poorer fits will result in larger values for SSE. Solving for SSE in equation (14.11), we see that SSE = SST − SSR. Hence, the largest value for SSE (which corresponds to the poorest fit) occurs when SSR = 0 and SSE = SST in other words when the ratio SSR/SST = 0. The ratio SSR/SST is used to evaluate the goodness of fit for the estimated regression equation. This ratio is called the *coefficient of determination* and is denoted by r^2, where $0 \leq r^2 \leq 1$.

Coefficient of determination

$$r^2 = \frac{SSR}{SST} \qquad \textbf{(14.12)}$$

For the Armand's Pizza Parlours' example, the value of the coefficient of determination is:

$$r^2 = \frac{SSR}{SST} = \frac{14,200}{15,730} = 0.9027$$

When we express the coefficient of determination as a percentage, r^2 can be interpreted as the percentage of the total sum of squares that can be explained by using the estimated regression equation. For Armand's Pizza Parlours, we can conclude that 90.27 per cent of the total sum of squares can be explained by using the estimated regression equation $\hat{y} = 60 + 5x$ to predict quarterly sales. In other words, 90.27 per cent of the variability in sales can be explained by the linear relationship between the size of the student population and sales. We should be pleased to find such a good fit for the estimated regression equation.

Correlation coefficient

In Chapter 3 we introduced the correlation coefficient as a descriptive measure of the strength of linear association between two variables, X and Y. Values of the correlation coefficient are always between −1 and +1. A value of +1 indicates that the two variables X and Y are perfectly related in a positive linear sense. That is, all data points are on a straight line that has a positive slope. A value of −1 indicates that X and Y are perfectly related in a negative linear sense, with all data points on a straight line that has a negative slope. Values of the correlation coefficient close to zero indicate that X and Y are not linearly related.

In Section 3.5 we presented the equation for computing the sample correlation coefficient. If a regression analysis has already been performed and the coefficient of determination r^2 computed, the sample correlation coefficient can be computed as follows:

Sample correlation coefficient

$$r_{XY} = (\text{sign of } b_1)\sqrt{\text{Coefficient of determination}}$$
$$= (\text{sign of } b_1)\sqrt{r^2} \qquad\qquad\qquad\qquad \textbf{(14.13)}$$

where:

$$b_1 = \text{the slope of the estimated regression equation } \hat{y} = b_0 + b_1 x$$

The sign for the sample correlation coefficient is positive if the estimated regression equation has a positive slope $(b_1 > 0)$ and negative if the estimated regression equation has a negative slope $(b_1 < 0)$.

For the Armand's Pizza Parlour example, the value of the coefficient of determination corresponding to the estimated regression equation $\hat{y} = 60 + 5x$ is 0.9027. As the slope of the estimated regression equation is positive, equation (14.13) shows that the sample correlation coefficient is equal to $\sqrt{0.9027} = 0.9501$.

With a sample correlation coefficient of $r_{XY} = 0.9501$, we would conclude that a strong positive linear association exists between X and Y.

In the case of a linear relationship between two variables, both the coefficient of determination and the sample correlation coefficient provide measures of the strength of the relationship. The coefficient of determination provides a measure between zero and one whereas the sample correlation coefficient provides a measure between -1 and $+1$. Although the sample correlation coefficient is restricted to a linear relationship between two variables, the coefficient of determination can be used for nonlinear relationships and for relationships that have two or more independent variables. Thus, the coefficient of determination provides a wider range of applicability.

EXERCISES

Methods

7. The data from Exercise 1 follow:

x_i	1	2	3	4	5
y_i	3	7	5	11	14

The estimated regression equation for these data is $\hat{y} = 0.20 + 2.60x$.
 a. Compute SSE, SST and SSR using equations (14.8), (14.9) and (14.10).
 b. Compute the coefficient of determination r^2. Comment on the goodness of fit.
 c. Compute the sample correlation coefficient.

8. For the data in Exercise 2 it can be shown that the estimated regression equation is $\hat{y} = 30.33 - 1.88x$.
 a. Compute SSE, SST and SSR.
 b. Compute the coefficient of determination r^2. Comment on the goodness of fit.
 c. Compute the sample correlation coefficient.

9. For the data in Exercise 3 it can be shown the estimated regression equation for these data is
   $\hat{y} = 0.75 + 0.51x$.
   a. What percentage of the total sum of squares can be accounted for by the estimated regression equation?
   b. What is the value of the sample correlation coefficient?

Applications

10. The estimated regression equation for the data in Exercise 5 can be shown to be
    $\hat{y} = 1130.3 + 0.0656x$. What percentage of the total sum of squares can be accounted for by the estimated regression equation?
    Comment on the goodness of fit. What is the sample correlation coefficient?

DOW S&P

11. An investment manager studying haulage companies calculates for a random sample of six such firms, the percentage capital investment in vehicles and the profit before tax as a percentage of turnover, with the following results:

% Capital investment, vehicles	37	47	10	22	41	25
% Profit	14	21	-5	16	19	8

    a. Calculate the coefficient of determination.
    b. Carry out a linear regression analysis for the data.
    c. Hence estimate the percentage profit when the percentage capital investment in vehicles is:
       (i) 30%.
       (ii) 90%.

12. *PC World* provided details for ten of the most economical laser printers (*PC World*, February 2019). The following data show the maximum printing speed in approximate pages per minute (ppm) and the price (in euros including 20 per cent value added tax) for each printer.

Name	Speed (ppm)	Price (€)
HP ENVY 5032	10	67.79
Brother DCP1502	20	79.09
HP Laser M15W	18	90.39
Epson XP355	10	90.39
HP Laserjet Pro M28W	18	112.99
CANON PIXMA MG5750	7	112.99
HP Laserjet Pro MFP M24dw	21	146.90
Samsung Express C430W	18	146.90
Brother HL 1212WVP	22	175.50
HP Laserjet Pro M254dw	21	214.69

    a. Develop the estimated regression equation with speed as the independent variable.
    b. Compute r^2. What percentage of the variation in cost can be explained by the printing speed?
    c. What is the sample correlation coefficient between speed and price? Does it reflect a strong or weak relationship between printing speed and cost?

14.4 MODEL ASSUMPTIONS

We saw in the previous section that the value of the coefficient of determination (r^2) is a measure of the goodness of fit of the estimated regression equation. However, even with a large value of r^2, the estimated regression equation should not be used until further analysis of the suitability of the assumed model has been conducted. An important step in determining whether the assumed model is appropriate involves testing for the significance of the relationship. The tests of significance in regression analysis are based on the following assumptions about the error term ε.

Assumptions about the error term ε in the regression model

$$Y = \beta_0 + \beta_1 x + \varepsilon$$

1. The error term ε is a random variable with a mean or expected value of zero; that is, $E(\varepsilon) = 0$.

Implication: β_0 and β_1 are constants, therefore $E(\beta_0) = \beta_0$ and $E(\beta_1) = \beta_1$; thus, for a given value x of X, the expected value of Y is:

$$E(Y) = \beta_0 + \beta_1 x \qquad \qquad (14.14)$$

As we indicated previously, equation (14.14) is referred to as the regression equation.

2. The variance of ε, denoted by σ^2, is the same for all values of X.

Implication: The variance of Y about the regression line equals σ^2 and is the same for all values of X.

3. The values of ε are independent.

Implication: The value of ε for a particular value of X is not related to the value of ε for any other value of X; thus, the value of Y for a particular value of X is not related to the value of any other value of X.

4. The error term ε is a normally distributed random variable.

Implication: Because Y is a linear function of ε, Y is also a normally distributed random variable.

Figure 14.6 illustrates the model assumptions and their implications; note that in this graphical interpretation, the value of $E(Y)$ changes according to the specific value of X considered. However, regardless of the X value, the probability distribution of ε and hence the probability distributions of Y are normally distributed, each with the same variance. The specific value of the error ε at any particular point depends on whether the actual value of Y is greater than or less than $E(Y)$.

FIGURE 14.6

Assumptions for the regression model

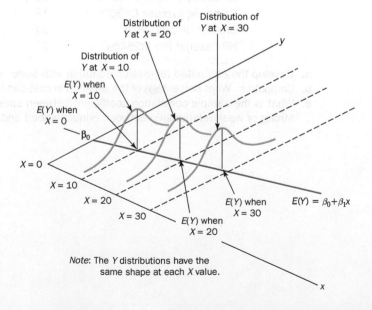

Note: The Y distributions have the same shape at each X value.

At this point, we must keep in mind that we are also making an assumption or hypothesis about the form of the relationship between X and Y. That is, we assume that a straight line represented by $\beta_0 + \beta_1 x$ is the basis for the relationship between the variables. We must not lose sight of the fact that some other model, for instance $Y = \beta_0 + \beta_1 x^2 + \varepsilon$ may turn out to be a better model for the underlying relationship.

14.5 TESTING FOR SIGNIFICANCE

In a simple linear regression equation, the mean or expected value of $E(Y) = \beta_0 + \beta_1 x$. If the value of $\beta_1 = 0$, $E(Y) = \beta_0 + (0)x = \beta_0$. In this case, the mean value of Y does not depend on the value of X and hence we would conclude that X and Y are not linearly related. Alternatively, if the value of β_1 is not equal to zero, we would conclude that the two variables are related. Thus, to test for a significant regression relationship, we must conduct a hypothesis test to determine whether the value of β_1 is zero. Two tests are commonly used. Both require an estimate of σ^2, the variance of ε in the regression model.

Estimate of σ^2

From the regression model and its assumptions we can conclude that σ^2 also represents the variance of the Y values about the regression line. Recall that the deviations of the Y values about the estimated regression line are called residuals. Thus, SSE, the sum of squared residuals, is a measure of the variability of the actual observations about the estimated regression line.

With $\hat{y}_i = b_0 + b_1 x_i$, SSE can be written as:

$$SSE = \Sigma(y_i - \hat{y})^2 = \Sigma(y_i - b_0 - b_1 x_i)^2$$

Every sum of squares is associated with a number called its degrees of freedom. Statisticians have shown that SSE has $n - 2$ degrees of freedom because two parameters (β_0 and β_1) must be estimated to compute SSE. Thus, the mean square is computed by dividing SSE by $n - 2$. The **mean square error (MSE)** provides an unbiased estimator of σ^2 and is denoted s^2.

Mean square error (estimate of σ^2)

$$s^2 = MSE = \frac{SSE}{n - 2} \tag{14.15}$$

In Section 14.3 we showed that for the Armand's Pizza Parlours' example, SSE $= 1,530$; hence:

$$s^2 = MSE = \frac{1,530}{8} = 191.25$$

provides an unbiased estimate of σ^2.

To estimate σ we take the square root of s^2. The resulting value, s, is referred to as the **standard error of the estimate**.

Standard error of the estimate

$$s = \sqrt{MSE} = \sqrt{\frac{SSE}{n - 2}} \tag{14.16}$$

For the Armand's Pizza Parlours' example, $s = \sqrt{MSE} = \sqrt{191.25} = 13.829$. In the following discussion, we use the standard error of the estimate in the tests for a significant relationship between X and Y.

t test

The simple linear regression model is $Y = \beta_0 + \beta_1 x + \varepsilon$. If X and Y are linearly related, we must have $\beta_1 \neq 0$. The purpose of the t test is to see whether we can conclude that $\beta_1 \neq 0$.

Consider the rival hypotheses about the parameter β_1:

$$H_0: \beta_1 = 0$$

$$H_1: \beta_1 \neq 0$$

If H_0 is rejected, we conclude that $\beta_1 \neq 0$ and that a linear relationship exists between the two variables. However, if H_0 cannot be rejected, we have insufficient evidence to conclude such a relationship exists. The properties of the sampling distribution of b_1, the least squares estimator of β_1, provide the basis for the hypothesis test.

First, let us consider what would happen if we used a different random sample for the same regression study. For example, suppose that Armand's Pizza Parlours used the sales records of a different sample of ten restaurants. A regression analysis of this new sample might result in an estimated regression equation similar to our previous estimated regression equation $\hat{y} = 60 + 5x$. However, it is doubtful that we would obtain exactly the same equation (with an intercept of exactly 60 and a slope of exactly 5). Indeed, b_0 and b_1, the least squares estimators, are sample statistics with their own sampling distributions. The properties of the sampling distribution of b_1 are as follows:

Sampling distribution of b_1

Expected value

$$E(b_1) = \beta_1$$

Standard deviation

$$\sigma_{b_1} = \frac{\sigma}{\sqrt{\Sigma(x_i - \bar{x})^2}} \tag{14.17}$$

Distribution form
Normal

Note that the expected value of b_1 is equal to β_1, so b_1 is an unbiased estimator of β_1. As we do not know the value of σ, we estimate σ_{b_1} by s_{b_1} where s_{b_1} is derived by substituting s for σ in equation (14.17).

Estimated standard deviation of b_1

$$s_{b_1} = \frac{s}{\sqrt{\Sigma(x_i - \bar{x})^2}} \tag{14.18}$$

For Armand's Pizza Parlours, $s = 13.829$. Hence, using $\Sigma(x_i - \bar{x})^2 = 568$ as shown in Table 14.2, we have:

$$s_{b_1} = \frac{13.829}{\sqrt{568}} = 0.5803$$

as the estimated standard deviation of b_1. The t test for a significant relationship is based on the fact that the test statistic:

$$\frac{b_1 - \beta_1}{s_{b_1}}$$

follows a t distribution with $n - 2$ degrees of freedom. If the null hypothesis is true, then $\beta_1 = 0$ and $t = b_1/s_{b_1}$.

Suppose we conduct this test of significance for Armand's Pizza Parlours at the $\alpha = 0.01$ level of significance. The test statistic is:

$$t = \frac{b_1}{s_{b_1}} = \frac{5}{0.5803} = 8.62$$

The t distribution Table 2 in Appendix B shows that with $n - 2 = 10 - 2 = 8$ degrees of freedom, $t = 3.355$ provides an area of 0.005 in the upper tail. Thus, the area in the upper tail of the t distribution corresponding to the test statistic $t = 8.62$ must be less than 0.005. As this test is a two-tailed test, we double this value to conclude that the p-value associated with $t = 8.62$ must be less than $2(0.005) = 0.01$. As the p-value is less than $\alpha = 0.01$, we reject H_0 and conclude that β_1 is not equal to zero. This evidence is sufficient to conclude that a significant relationship exists between the student population and quarterly sales. A summary of the t test for significance in simple linear regression is as follows:

t Test for significance in simple linear regression

$$H_0: \beta_1 = 0$$
$$H_1: \beta_1 \neq 0$$

Test statistic

$$t = \frac{b_1}{s_{b_1}} \tag{14.19}$$

Rejection rule

p-value approach: Reject H_0 if p-value $\leq \alpha$

Critical value approach: Reject H_0 if $t \leq -t_{\alpha/2}$ or if $t \geq t_{\alpha/2}$

where $t_{\alpha/2}$ is based on a t distribution with $n - 2$ degrees of freedom.

Confidence interval for β_1

The form of a confidence interval for β_1 is as follows:

$$b_1 \pm t_{\alpha/2}s_{b_1}$$

The point estimator is b_1 and the margin of error is $t_{\alpha/2}s_{b_1}$. The confidence coefficient associated with this interval is $1 - \alpha$, and $t_{\alpha/2}$ is the t-value providing an area of $\alpha/2$ in the upper tail of a t distribution with $n - 2$ degrees of freedom. For example, suppose that we wanted to develop a 99% confidence interval estimate of β_1 for Armand's Pizza Parlours. From Table 2 of Appendix B we find that the t-value corresponding to $\alpha = 0.01$ and $n - 2 = 10 - 2 = 8$ degrees of freedom is $t_{0.005} = 3.355$. Thus, the 99% confidence interval estimate of β_1 is:

$$b_1 \pm t_{\alpha/2}s_{b_1} = 5 \pm 3.355(0.5803) = 5 \pm 1.95$$

or 3.05 to 6.95.

Note that as 0, the hypothesized value of β_1 under H_0, is not included in the confidence interval (3.05 to 6.95), we can reject H_0 and conclude that a significant statistical relationship exists between the size of the student population and quarterly sales. In general, a confidence interval can be used to test any two-sided hypothesis about β_1. If the hypothesized value of β_1 is contained in the confidence interval, do not reject H_0. Otherwise, reject H_0.

F test

An F test, based on the F probability distribution, can also be used to test for significance in regression. With only one independent variable, the F test will provide the same conclusion as the t test; that is, if the t test indicates $\beta_1 \neq 0$ and hence a significant relationship, the F test will also indicate a significant relationship.* But with more than one independent variable, only the F test can be used to test for an overall significant relationship.

The logic behind the use of the F test for determining whether the regression relationship is statistically significant is based on the development of two independent estimates of σ^2. We explained how MSE provides an estimate of σ^2. If the null hypothesis $H_0: \beta_1 = 0$ is true, the sum of squares due to regression, SSR, divided by its degrees of freedom provides another independent estimate of σ^2. This estimate is called the *mean square due to regression*, or simply the *mean square regression*, and is denoted MSR. In general:

$$\text{MSR} = \frac{\text{SSR}}{\text{Regression degrees of freedom}}$$

For the models we consider in this text, the regression degrees of freedom is always equal to the number of independent variables in the model.

Mean square regression

$$\text{MSR} = \frac{\text{SSR}}{\text{Number of independent variables}} \tag{14.20}$$

As we consider only regression models with one independent variable in this chapter, we have MSR = SSR/1 = SSR. Hence, for Armand's Pizza Parlours, MSR = SSR = 14,200.

If the null hypothesis $(H_0: \beta_1 = 0)$ is true, MSR and MSE are two independent estimates of σ^2 and the sampling distribution of MSR/MSE follows an F distribution with numerator degrees of freedom equal to one and denominator degrees of freedom equal to $n - 2$. Therefore, when $\beta_1 = 0$, the value of MSR/MSE should be close to one. However, if the null hypothesis is false $(\beta_1 \neq 0)$, MSR will overestimate σ^2 and the value of MSR/MSE will be inflated; thus, large values of MSR/MSE lead to the rejection of H_0 and the conclusion that the relationship between X and Y is statistically significant.

For the Armand's Pizza Parlours example, the F statistic takes the value:

$$F = \frac{\text{MSR}}{\text{MSE}} = \frac{14,200}{191.25} = 74.25$$

The F distribution table (Table 4 of Appendix B) shows that with one degree of freedom in the numerator and $n - 2 = 10 - 2 = 8$ degrees of freedom in the denominator, $F = 11.26$ provides an area of 0.01 in the upper tail. Thus, the area in the upper tail of the F distribution corresponding to the test statistic $F = 74.25$ must be less than 0.01. In other words the p-value must be less than 0.01. As the p-value is less than $\alpha = 0.01$, we reject H_0 and conclude that a significant relationship

---

* In fact $F = t^2$ for a simple regression model.

exists between the size of the student population and quarterly sales. A summary of the F test for significance in simple linear regression is as follows:

F test for significance in simple linear regression

$$H_0: \beta_1 = 0$$
$$H_1: \beta_1 \neq 0$$

Test statistic

$$F = \frac{MSR}{MSE} \qquad (14.21)$$

Rejection rule

p-value approach: Reject H_0 if p-value $\leq \alpha$
Critical value approach: Reject H_0 if $F \geq F_\alpha$

where F_α is based on a F distribution with one degree of freedom in the numerator and $n - 2$ degrees of freedom in the denominator.

In Chapter 13 we covered analysis of variance (ANOVA) and showed how an ANOVA table could be used to provide a convenient summary of the computational aspects of analysis of variance. A similar ANOVA table can be used to summarize the results of the F test for significance in regression. Table 14.5 is the general form of the ANOVA table for simple linear regression.

Table 14.6 is the ANOVA table with the F test computations performed for Armand's Pizza Parlours. Regression, Error and Total are the labels for the three sources of variation, with SSR, SSE and SST appearing as the corresponding sum of squares in column 3. The degrees of freedom, 1 for SSR, $n - 2$ for SSE and $n - 1$ for SST, are shown in column 2. Column 4 contains the values of MSR and MSE and column 5 contains the value of $F = $ MSR/MSE. Almost all computer printouts of regression analysis include an ANOVA table summary and the F test for significance.

TABLE 14.5 General form of the ANOVA table for simple linear regression

Source of variation	Degrees of freedom	Sum of squares	Mean square	F
Regression	1	SSR	$MSR = \dfrac{SSR}{1}$	$\dfrac{MSR}{MSE}$
Error	$n - 2$	SSE	$MSE = \dfrac{SSE}{n - 2}$	
Total	$n - 2$	SST		

TABLE 14.6 ANOVA table for the Armand's Pizza Parlours' problem

Source of variation	Degrees of freedom	Sum of squares	Mean square	F
Regression	1	14,200	$\dfrac{14,200}{1} = 14,200$	$\dfrac{14,200}{191.25} = 74.2$
Error	8	1,530	$\dfrac{1,530}{8} = 191.25$	
Total	9	15,730		

Some cautions about the interpretation of significance tests

Rejecting the null hypothesis $H_0: \beta_1 = 0$ and concluding that the relationship between X and Y is significant does not enable us to conclude that a cause-and-effect relationship is present between X and Y. Concluding a cause-and-effect relationship is warranted only if the analyst can provide some type of theoretical justification that the relationship is in fact causal. In the Armand's Pizza Parlours' example, we can conclude that there is a significant relationship between the size of the student population X and quarterly sales Y; moreover, the estimated regression equation $\hat{y} = 60 + 5x$ provides the least squares estimate of the relationship. We cannot, however, conclude that changes in student population X cause changes in quarterly sales Y just because we identified a statistically significant relationship. The appropriateness of such a cause-and-effect conclusion is left to supporting theoretical justification and to good judgement on the part of the analyst. Armand's managers felt that increases in the student population were a likely cause of increased quarterly sales.

Thus, the result of the significance test enabled them to conclude that a cause-and-effect relationship was present.

In addition, just because we are able to reject $H_0: \beta_1 = 0$ and demonstrate statistical significance it does not enable us to conclude that the relationship between X and Y is linear. We can state only that X and Y are related and that a linear relationship explains a significant proportion of the variability in Y over the range of values for X observed in the sample. Figure 14.7 illustrates this situation. The test for significance calls for the rejection of the null hypothesis $H_0 \beta_1 = 0$ and leads to the conclusion that X and Y are significantly related, but the figure shows that the actual relationship between X and Y is not linear. Although the linear approximation provided by $\hat{y} = b_0 + b_1 x$ is good over the range of X values observed in the sample, it becomes poor for X values outside that range.

Given a significant relationship, we should feel confident in using the estimated regression equation for predictions corresponding to X values within the range of the X values observed in the sample. For Armand's Pizza Parlours, this range corresponds to values of X between 2 and 26. Unless other reasons indicate that the model is valid beyond this range, predictions outside the range of the independent variable should be made with caution. For Armand's Pizza Parlours, because the regression relationship has been found significant at the 0.01 level, we should feel confident using it to predict sales for restaurants where the associated student population is between 2,000 and 26,000.

FIGURE 14.7
Example of a linear approximation of a nonlinear relationship

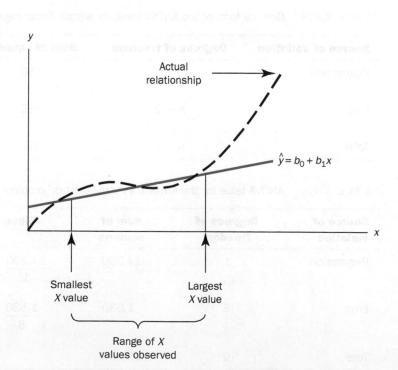

EXERCISES

Methods

13. For the data from Exercise 1:
    a. Compute the mean square error using equation (14.15).
    b. Compute the standard error of the estimate using equation (14.16).
    c. Compute the estimated standard deviation of b_1 using equation (14.18).
    d. Use the t test to test the following hypotheses $(\alpha = 0.05)$:

    $$H_0: \beta_1 = 0$$
    $$H_1: \beta_1 \neq 0$$

    e. Use the F test to test the hypotheses in part (d) at a 0.05 level of significance. Present the results in the analysis of variance table format.

14. For the data from Exercise 2:
    a. Compute the mean square error using equation (14.15).
    b. Compute the standard error of the estimate using equation (14.16).
    c. Compute the estimated standard deviation of b_1 using equation (14.18).
    d. Use the t test to test the following hypotheses $(\alpha = 0.05)$:

    $$H_0: \beta_1 = 0$$
    $$H_1: \beta_1 \neq 0$$

    e. Use the F test to test the hypotheses in part (d) at a 0.05 level of significance. Present the results in the analysis of variance table format.

15. For the data from Exercise 3:
    a. What is the value of the standard error of the estimate?
    b. Test for a significant relationship by using the t test. Use $\alpha = 0.05$.
    c. Perform the F test to test for a significant relationship. Use $\alpha = 0.05$. What is your conclusion?

Applications

16. In an effort to determine the relationship between annual wages for employees and the number of days absent from work due to sickness, a large corporation's HR department studied the records for a random sample of employees. The resultant data are as follows:

Employee	Annual wages (€000)	Days absent
A	15.7	4
B	17.2	3
C	13.8	6
D	24.2	5
E	15.0	3
F	12.7	12
G	13.8	5
H	18.7	1
I	10.8	12
J	11.8	11
K	25.4	2
L	17.2	4

Carry out an appropriate analysis of this information. What inferences do you draw? (Use $\alpha = 0.05$.)

ONLINEEDU

17. One of the biggest changes in higher education in recent years has been the growth of online universities. The Online Education Database is an independent organization whose mission is to build a comprehensive list of the top accredited online colleges. The following table shows the retention rate (%) and the graduation rate (%) for 29 online colleges.

Retention rate (%)	Graduation rate (%)
7	25
51	25
4	28
29	32
33	33
47	33
63	34
45	36
60	36
62	36
67	36
65	37
78	37
75	38
54	39
45	41
38	44
51	45
69	46
60	47
37	48
63	50
73	51
78	52
48	53
95	55
68	56
100	57
100	61

a. Develop a scatter diagram with retention rate as the independent variable. What does the scatter diagram indicate about the relationship between the two variables?

b. Develop the estimated regression equation.

c. Use the *t* test to determine whether there is a significant relationship. Use $\alpha = 0.05$.

d. Did the estimated regression equation provide a good fit?

14.6 USING THE ESTIMATED REGRESSION EQUATION FOR ESTIMATION AND PREDICTION

When using the simple linear regression model we are making an assumption about the relationship between X and Y. We then use the least squares method to obtain the estimated simple linear regression equation. If a significant relationship exists between X and Y, and the coefficient of determination shows that the fit is good, the estimated regression equation should be useful for estimation and prediction.

Point estimation

In the Armand's Pizza Parlours' example, the estimated regression equation $\hat{y} = 60 + 5x$ provides an estimate of the relationship between the size of the student population X and quarterly sales Y. We can use the estimated regression equation to develop a point estimate of either the mean value of Y or an individual value of Y corresponding to a given value of X. For instance, suppose Armand's managers want a point estimate of the mean quarterly sales for all restaurants located near college campuses with 10,000 students. Using the estimated regression equation $\hat{y} = 60 + 5x$, we see that for $X = 10$ (or 10,000 students), $\hat{y} = 60 + 5(10) = 110$. Thus, a point estimate of the mean quarterly sales for all restaurants located near campuses with 10,000 students is €110,000.

Now suppose Armand's managers want to predict sales for an individual restaurant located near Cabot College, a school with 10,000 students. Then, as the point estimate for an individual value of Y is the same as the point estimate for the mean value of Y we would predict quarterly sales of $\hat{y} = 60 + 5(10) = 110$ or €110,000 for this one restaurant.

Interval estimation

Point estimates do not provide any information about the precision associated with an estimate. For that we must develop interval estimates much like those in Chapters 10 and 11. The first type of interval estimate, a confidence interval, is an interval estimate of the *mean value of Y* for a given value of X. The second type of interval estimate, a prediction interval is used whenever we want an interval estimate of an *individual value* of Y for a given value of X. The point estimate of the mean value of Y is the same as the point estimate of an individual value of Y. But the interval estimates we obtain in the two cases are different. The margin of error is larger for a prediction interval.

Confidence interval for the mean value of Y

The estimated regression equation provides a point estimate of the mean value of Y for a given value of X. In developing the confidence interval, we will use the following notation:

$x_p =$ the particular or given value of the independent variable X
$Y_p =$ the dependent variable Y corresponding to the given x_p
$E(Y_p) =$ the mean or expected value of the dependent variable Y_p corresponding to the given x_p
$\hat{y}_p = b_0 + b_1 x_p =$ the point estimate of $E(Y_p)$ when $X = x_p$

Using this notation to estimate the mean sales for all Armand's restaurants located near a campus with 10,000 students, we have $x_p = 10$, and $E(Y_p)$ denotes the unknown mean value of sales for all restaurants where $x_p = 10$. The point estimate of $E(Y_p)$ is given by $\hat{y}_p = 60 + 5(10) = 110$.

In general, we cannot expect $\hat{y}_p$ to equal $E(Y_p)$ exactly. If we want to make an inference about how close $\hat{y}_p$ is to the true mean value $E(Y_p)$, we will have to estimate the variance of $\hat{y}_p$. The formula for estimating the variance of $\hat{y}_p$ given x_p, is given by:

$$s_{\hat{y}_p}^2 = s^2 \left[\frac{1}{n} + \frac{(x_p - \bar{x})^2}{\Sigma(x_i - \bar{x})^2} \right]$$

Following on, it can be shown:

Confidence interval for $E(Y_p)$

$$\hat{y}_p \pm t_{\alpha/2}s\sqrt{\frac{1}{n} + \frac{(x_p - \bar{x})^2}{\Sigma(x_i - \bar{x})^2}} \tag{14.22}$$

where the confidence coefficient is $1 - \alpha$ and $t_{\alpha/2}$ is based on a t distribution with $n - 2$ degrees of freedom.

Using expression (14.22) to develop a 95% confidence interval of the mean quarterly sales for all Armand's restaurants located near campuses with 10,000 students, we need the value of t for $\alpha/2 = 0.025$ and $n - 2 = 10 - 2 = 8$ degrees of freedom. Using Table 2 of Appendix B, we have $t_{0.025} = 2.306$. Thus, with $\hat{y}_p = 110$, the 95% confidence interval estimate is:

$$\hat{y}_p \pm t_{\alpha/2}s\sqrt{\frac{1}{n} + \frac{(x_p - \bar{x})^2}{\Sigma(x_i - \bar{x})^2}}$$

$$110 \pm 2.306 \times 13.829\sqrt{\frac{1}{10} + \frac{(10 - 14)^2}{568}}$$

$$= 110 \pm 11.415$$

In euros, the 95% confidence interval for the mean quarterly sales of all restaurants near campuses with 10,000 students is €110,000 ± €11,415. Therefore, the 95% confidence interval for the mean quarterly sales when the student population is 10,000 is €98,585 to €121,415.

Note that the estimated standard deviation of $\hat{y}_p$ is smallest when $x_p = \bar{x}$ so that the quantity $x_p - \bar{x} = 0$. In this case, the estimated standard deviation of $\hat{y}_p$ becomes:

$$s\sqrt{\frac{1}{n} + \frac{(x_p - \bar{x})^2}{\Sigma(x_i - \bar{x})^2}} = s\sqrt{\frac{1}{n}}$$

This result implies that the best or most precise estimate of the mean value of Y occurs when $x_p = \bar{x}$. But, the further x_p is from $x_p = \bar{x}$ the larger $x_p = \bar{x}$ becomes and thus the wider confidence intervals will be for the mean value of Y. This pattern is shown graphically in Figure 14.8.

Prediction interval for an individual value of Y

Suppose that instead of estimating the mean value of sales for all Armand's restaurants located near campuses with 10,000 students, we wish to estimate the sales for an individual restaurant located near Cabot College, a school with 10,000 students.

As was mentioned previously, the point estimate of y_p, the value of Y corresponding to the given x_p, is provided by the estimated regression equation $\hat{y}_p = b_0 + b_1 x_p$. For the restaurant at Cabot College, we have $x_p = 10$ and a corresponding predicted quarterly sales of $\hat{y}_p = 60 + 5(10) = 110$ or €110,000.

This value is the same as the point estimate of the mean sales for all restaurants located near campuses with 10,000 students.

FIGURE 14.8
Confidence intervals for the mean sales Y at given values of student population x

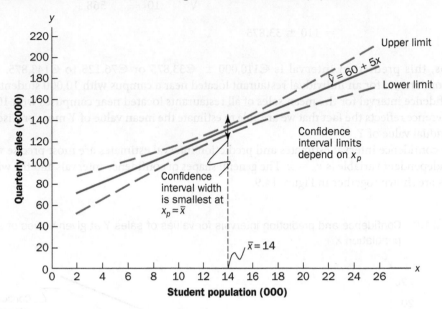

To develop a prediction interval, we must next determine the variance associated with using $\hat{y}_p$ as an estimate of an individual value of Y when $X = x_p$. This variance is made up of the sum of the following two components:

1. The variance of individual Y values about the mean $E(Y_p)$, an estimate of which is given by s^2.
2. The variance associated with using $\hat{y}_p$ to estimate $E(Y_p)$, an estimate of which is given by:

$$s^2_{\hat{y}_p} = s^2 \left[\frac{1}{n} + \frac{(x_p - \bar{x})^2}{\Sigma(x_i - \bar{x})^2} \right]$$

Thus the formula for estimating the variance of an individual value of Y_p, is:

$$s^2 + s^2_{\hat{y}_p} = s^2 + s^2 \left[\frac{1}{n} + \frac{(x_p - \bar{x})^2}{\Sigma(x_i - \bar{x})^2} \right] = s^2 \left[1 + \frac{1}{n} + \frac{(x_p - \bar{x})^2}{\Sigma(x_i - \bar{x})^2} \right]$$

The general expression for a prediction interval is as follows:

Prediction interval for y_p

$$\hat{y}_p \pm t_{\alpha/2} s \sqrt{1 + \frac{1}{n} + \frac{(x_p - \bar{x})^2}{\Sigma(x_i - \bar{x})^2}} \tag{14.23}$$

where the confidence coefficient is $1 - \alpha$ and $t_{\alpha/2}$ is based on a t distribution with $n - 2$ degrees of freedom.

Thus the 95% prediction interval of sales for one specific restaurant located near a campus with 10,000 students is:

$$\hat{y}_p \pm t_{\alpha/2} s \sqrt{1 + \frac{1}{n} + \frac{(x_p - \bar{x})^2}{\Sigma(x_i - \bar{x})^2}}$$

$$= 110 \pm 2.306 \times 13.829 \sqrt{1 + \frac{1}{10} + \frac{(10 - 14)^2}{568}}$$

$$= 110 \pm 33.875$$

In euros, this prediction interval is €110,000 ± €33,875 or €76,125 to €143,875. Note that the prediction interval for an individual restaurant located near a campus with 10,000 students is wider than the confidence interval for the mean sales of all restaurants located near campuses with 10,000 students. The difference reflects the fact that we are able to estimate the mean value of Y more precisely than we can an individual value of Y.

Both confidence interval estimates and prediction interval estimates are most precise when the value of the independent variable is $x_p = \bar{x}$. The general shapes of confidence intervals and the wider prediction intervals are shown together in Figure 14.9.

FIGURE 14.9 Confidence and prediction intervals for values of sales Y at given values of student population X

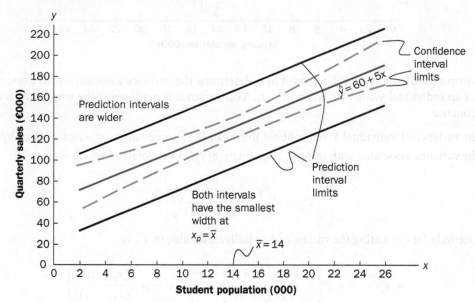

EXERCISES

Methods

18. For the data from Exercise 1:
   a. Use expression (14.22) to develop a 95% confidence interval for the expected value of Y when $X = 4$.
   b. Use expression (14.23) to develop a 95% prediction interval for Y when $X = 4$.

19. For the data from Exercise 2:
   a. Estimate the standard deviation of $\hat{y}_p$ when $X = 3$.
   b. Develop a 95% confidence interval for the expected value of Y when $X = 3$.
   c. Estimate the standard deviation of an individual value of Y when $X = 3$.
   d. Develop a 95% prediction interval for Y when $X = 3$.

20. For the data from Exercise 3:
Develop the 95% confidence and prediction intervals when $X = 3$. Explain why these two intervals are different.

Applications

21. A company is introducing a job evaluation scheme in which jobs are graded by points for skill, danger, responsibility and so on. Monthly pay scales are then drawn up according to the number of points allocated and other factors such as experience and local conditions. To date the company has applied this scheme to jobs with details as follows:

Job	Points	Pay (€)
A	250	10,000
B	1,250	30,500
C	350	12,500
D	950	25,000
E	500	15,000
F	600	16,000
G	750	20,000
H	1,400	32,500
I	800	21,000

a. Draw a scattergram for the data and comment on any patterns that this might reveal.
b. Calculate the correlation coefficient between points and pay. How would you interpret this?
c. Estimate a linear regression model for the data.
d. By superimposing the fitted line on your earlier scattergram, comment on the success of your modelling.
e. Provide 99% confidence and prediction intervals for the monthly pay of a job graded at 1,500 points.

14.7 COMPUTER SOLUTION

Performing the regression analysis computations without the help of a computer can be quite time-consuming. In this section we discuss how the computational burden can be minimized by using a computer software package such as Excel.

Although computer software layouts may differ, the simple regression output shown in Figure 4.10 is fairly typical. Interpreting each of the highlighted portions of this output in turn:

1 The ANOVA table is printed below the heading Analysis of Variance. The label Error is used for the error source of variation. Note that DF is an abbreviation for degrees of freedom and that MSR is given in the Regression row under the column Adj MS as 14,200 and MSE is given in the Error row under Adj MS as 191.2. The ratio of these two values provides the F value of 74.25 and the corresponding p-value of 0.000. Because the p-value is zero (to three decimal places), the relationship between Sales and Population is judged statistically significant.

2 Under the heading Model Summary, the standard error of the estimate, $s = 13.8293$, is given as well as information about the goodness of fit. Note that 'R-sq = 90.27%' is the coefficient of determination expressed as a percentage. The value 'R-Sq(adj) = 89.06%' is discussed in Chapter 15.

3 A table is printed that shows the values of the coefficients b_0 and b_1, the standard deviation of each coefficient, the t-value obtained by dividing each coefficient value by its standard deviation, and the p-value associated with the t test. This appears under the heading Coefficients. Because the p-value is zero (to three decimal places), the sample results indicate that the null hypothesis $(H_0: \beta_1 = 0)$ should be rejected.

Alternatively, we could compare 8.62 (located in the t-value column) to the appropriate critical value. This procedure for the t test was described in Section 14.5.

4 Under the heading Regression Equation, the estimated regression equation is given:

$$\text{Sales} = 60.00 + 5.000 \text{ Population}$$

5 The 95% confidence interval estimate of the expected sales and the 95% prediction interval estimate of sales for an individual restaurant located near a campus with 10,000 students are printed below the ANOVA table. The confidence interval is (98.5830, 121.4417) and the prediction interval is (76.1275, 143.873) as we showed in Section 14.6.

FIGURE 14.10

Output for the Armand's Pizza Parlours' problem

```
Analysis of Variance

Source        DF    Adj SS    Adj MS   F-value  p-value  ⎫
Regression     1  14,200.0  14,200.0    74.25    0.000   ⎪
Error          8   1,530.0     191.2                     ⎬ ◄── ANOVA
Total          9  15,730.0                              ⎭      table


Model Summary

      s    R-sq  R-sq(adj)
 13.8293  90.27%    89.06%


Coefficients

Term         Coef  SE Coef  t-value  p-value
Constant    60.00     9.23     6.50    0.000
Population  5.000    0.580     8.62    0.000

Regression Equation

Sales - 60.00 + 5.000 Population ◄── Estimated Regression
                                     Equation
Prediction for Sales

Variable     Setting

Population      10

Fit   SE Fit      95% CI            95% PI
110  4.95099  (98.5830, 121.417) (76.1275, 143.873) ◄── Interval
                                                          Estimates
```

EXERCISES

Applications

22. The commercial division of the Supreme real estate firm in Cyprus is conducting a regression analysis of the relationship between X, annual gross rents (in thousands of euros), and Y, selling price (in thousands of euros) for apartment buildings. Data were collected on several properties recently sold and the following selective computer output was obtained.

The regression equation is
$Y = 20.00 + 7.0210 X$

Coefficients

Term	Coef	SE Coef	t-value
Constant	20.000	3.221	6.21
X	7.210	1.363	5.29

Analysis of Variance

Source	DF	SS
Regression	1	41,587.3
Error	7	
Total	8	51,984.1

    a. How many apartment buildings were in the sample?
    b. Write the estimated regression equation.
    c. What is the value of S_{b_0}?
    d. Use the F statistic to test the significance of the relationship at a 0.05 level of significance.
    e. Estimate the selling price of an apartment building with gross annual rents of €50,000.

23. What follows is a portion of the computer output for a regression analysis relating Y = maintenance expense (euros per month) to X = usage (hours per week) of a particular brand of computer terminal.

```
Analysis of Variance

SOURCE          DF          Adj SS          Adj MS
Regression       1        1,575.76        1,575.76
Error            8          349.14           43.64
Total            9        1,924.90


Predictor        Coef          SE Coef
Constant       6.1092          0.9361
X              0.8951          0.1490


Regression Equation
Y = 6.1092 + 0.8951 X
```

    a. What is the estimated regression equation?
    b. Use a t test to determine whether monthly maintenance expense is related to usage at the 0.05 level of significance.
    c. Use the estimated regression equation to predict mean monthly maintenance expense for any terminal that is used 25 hours per week.

24. Cheri is a production manager for a small manufacturing shop and is interested in developing a predictive model to estimate the time to produce an order of a given size, that is, the total time to produce a certain quantity of the product. Cheri has collected data on the total time to produce 30 different orders of various quantities in the file 'Setup'.
    a. Develop a scatter diagram with quantity as the independent variable.
    b. What does the scatter diagram developed in (a) indicate about the relationship between the two variables?
    c. Develop the estimated regression equation. Interpret the intercept and slope.
    d. Test for a significant relationship. Use $\alpha = 0.05$.
    e. Did the estimated regression equation provide a good fit?

SETUP

14.8 RESIDUAL ANALYSIS: VALIDATING MODEL ASSUMPTIONS

As we noted previously, the *residual* for observation i is the difference between the observed value of the dependent variable (y_i) and the estimated value of the dependent variable ($\hat{y}_i$).

Residual for observation i

$$y_i - \hat{y}_i \qquad \text{(14.24)}$$

where:

y_i is the observed value of the dependent variable

$\hat{y}_i$ is the estimated value of the dependent variable

In other words, the ith residual is the error resulting from using the estimated regression equation to predict the value of the dependent variable. The residuals for the Armand's Pizza Parlours' example are shown in Table 14.7. The observed values of the dependent variable are in the second column and the estimated values of the dependent variable, obtained using the estimated regression equation $\hat{y} = 60 + 5x$, are in the third column. An analysis of the corresponding residuals in the fourth column will help determine whether the assumptions made about the regression model are appropriate.

TABLE 14.7 Residuals for Armand's Pizza Parlours

Student population x_i	Sales y_i	Estimated sales $\hat{y}_i + 60 - 5x_i$	Residuals $y_i - \hat{y}_i$
2	58	70	−12
6	105	90	15
8	88	100	−12
8	118	100	18
12	117	120	−3
16	137	140	−3
20	157	160	−3
20	169	160	9
22	149	170	−21
26	202	190	12

Recall that for the Armand's Pizza Parlours' example it was assumed the simple linear regression model took the form:

$$Y = \beta_0 + \beta_1 x + \varepsilon \qquad \text{(14.25)}$$

In other words we assumed quarterly sales (Y) to be a linear function of the size of the student population (X) plus an error term ε. In Section 14.4 we made the following assumptions about the error term ε:

1 $E(\varepsilon) = 0$.

2 The variance of ε, denoted by σ^2, is the same for all values of X.

3 The values of ε are independent.

4 The error term ε has a normal distribution.

These assumptions provide the theoretical basis for the t test and the F test used to determine whether the relationship between X and Y is significant, and for the confidence and prediction interval estimates presented in Section 14.6. If the assumptions about the error term ε appear questionable, the hypothesis tests about the significance of the regression relationship and the interval estimation results may not be valid.

The residuals provide the best information about ε; hence an analysis of the residuals is an important step in determining whether the assumptions for ε are appropriate. Much of residual analysis is based on an examination of graphical plots. In this section, we discuss the following residual plots:

1 A plot of the residuals against values of the independent variable X.
2 A plot of residuals against the predicted values $\hat{y}$ of the dependent variable.
3 A standardized residual plot.
4 A normal probability plot.

Residual plot against X

A residual plot against the independent variable X is a graph in which the values of the independent variable are represented by the horizontal axis and the corresponding residual values are represented by the vertical axis. A point is plotted for each residual.

The first coordinate for each point is given by the value of x_i and the second coordinate is given by the corresponding value of the residual $y_i - \hat{y}_i$. For a residual plot against X with the Armand's Pizza Parlours' data from Table 14.7, the coordinates of the first point are $(2, -12)$, corresponding to $x_1 = 2$ and $y_1 - \hat{y}_1 = -12$ the coordinates of the second point are $(6, 15)$, corresponding to $x_2 = 6$ and $y_2 - \hat{y}_2 = 15$ and so on. Figure 14.11 shows the resulting residual plot.

Before interpreting the results for this residual plot, let us consider some general patterns that might be observed in any residual plot. Three examples appear in Figure 14.12.

FIGURE 14.11

Plot of the residuals against the independent variable for Armand's Pizza Parlours

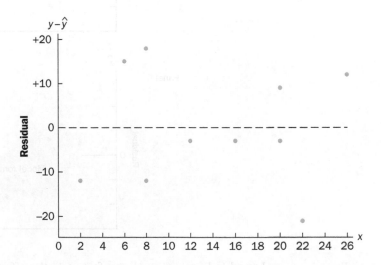

If the assumption that the variance of ε is the same for all values of X, and the assumed regression model is an adequate representation of the relationship between the variables, the residual plot should give an overall impression of a horizontal band of points such as the one in Panel A of Figure 14.12. However, if the variance of ε is not the same for all values of X – for example, if variability about the regression line is greater for larger values of X – a pattern such as the one in Panel B of Figure 14.12 could be observed. In this case, the assumption of a constant variance of ε is violated. Another possible residual plot is shown in Panel C. In this case, we would conclude that the assumed regression model is not an adequate representation of the relationship between the variables. A curvilinear regression model or multiple regression model should be considered.

Now let us return to the residual plot for Armand's Pizza Parlours shown in Figure 14.11. The residuals appear to approximate the horizontal pattern in Panel A of Figure 14.12. Hence, we conclude that the residual plot does not provide evidence that the assumptions made for Armand's regression model should be challenged. At this point, we are confident in the conclusion that Armand's simple linear regression model is valid.

FIGURE 14.12
Residual plots from
three regression studies

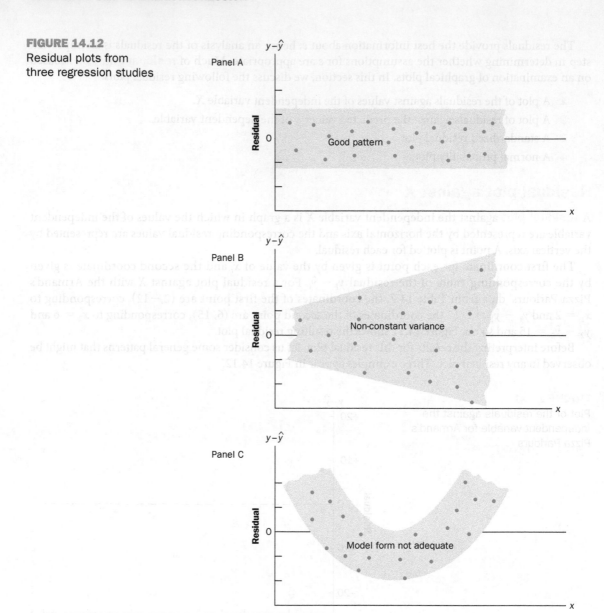

Experience and good judgement are always factors in the effective interpretation of residual plots. Seldom does a residual plot conform precisely to one of the patterns in Figure 14.12. Yet analysts who frequently conduct regression studies and frequently review residual plots become adept at understanding the differences between patterns that are reasonable and patterns that indicate the assumptions of the model should be questioned. A residual plot against x provides one technique to assess the validity of the assumptions for a regression model.

Residual plot against $\hat{y}$

Another residual plot represents the predicted value of the dependent variable $\hat{y}$ on the horizontal axis and the residual values on the vertical axis. A point is plotted for each residual. The first coordinate for each point is given by $\hat{y}_i$ and the second coordinate is given by the corresponding value of the ith residual $y_i - \hat{y}_i$. With the Armand's data from Table 14.7, the coordinates of the first point are (70, 12), corresponding to $\hat{y}_1 = 70$ and $y_1 - \hat{y}_1 = -12$; the coordinates of the second point are (90, 15) and so on. Figure 14.13 provides the residual plot. Note that the pattern of this residual plot is the same as the pattern of the residual plot against the independent variable X.

FIGURE 14.13

Plot of the residuals against the predicted values $\hat{y}$ for Armand's Pizza Parlours

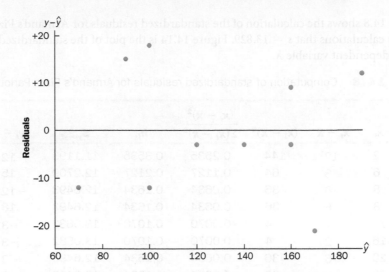

It is not a pattern that would lead us to question model assumptions. For simple linear regression, both the residual plot against X and the residual plot against $\hat{y}$ provide the same pattern. For multiple regression analysis, the residual plot against $\hat{y}$ is more widely used because of the presence of more than one independent variable.

Standardized residuals

Many of the residual plots provided by computer software packages use a standardized version of the residuals. As demonstrated in preceding chapters, a random variable is standardized by subtracting its mean and dividing the result by its standard deviation. The mean of least squares residuals is zero. Thus, simply dividing each residual by its standard deviation provides the standardized residual.

It can be shown that the standard deviation of residual i depends on the standard error of the estimate s and the corresponding value of the independent variable x_i.

Note that equation (14.26) shows that the standard deviation of the ith residual depends on x_i because of the presence of h_i in the formula.* Once the standard deviation of each residual is calculated, we can compute the standardized residual by dividing each residual by its corresponding standard deviation.

Standard deviation of the ith residual**

$$s_{y_i - \hat{y}_i} = s\sqrt{1 - h_i}$$ (14.26)

where:

$s_{y_i - \hat{y}_i}$ = the standard deviation of residual i

s = the standard error of the estimate

$$h_i = \frac{1}{n} + \frac{(x_i - \bar{x})^2}{\Sigma(x_i - \bar{x})^2}$$ (14.27)

Standardized residual for observation i

$$\frac{y_i - \hat{y}_i}{s_{y_i - \hat{y}_i}}$$ (14.28)

---

* h_i is referred to as the *leverage of observation i*. Leverage will be discussed further when we consider influential observations in Section 14.10.

** This equation actually provides an estimate of the standard deviation of the ith residual, because s is used instead of σ.

Table 14.8 shows the calculation of the standardized residuals for Armand's Pizza Parlours. Recall from previous calculations that $s = 13.829$. Figure 14.14 is the plot of the standardized residuals against values of the independent variable X.

TABLE 14.8   Computation of standardized residuals for Armand's Pizza Parlours

i	x_i	$x_i - \bar{x}$	$(x_i - \bar{x})^2$	$\dfrac{(x_i - \bar{x})^2}{\Sigma(x_i - \bar{x})^2}$	h_i	$s_{yi - \hat{y}i}$	$y_i - \hat{y}_i$	Standardized residual
1	2	−12	144	0.2535	0.3535	11.1193	−12	−1.0792
2	6	−8	64	0.1127	0.2127	12.2709	15	1.2224
3	8	−6	36	0.0634	0.1634	12.6493	−12	−0.9487
4	8	−6	36	0.0634	0.1634	12.6493	18	1.4230
5	12	−2	4	0.0070	0.1070	13.0682	−3	−0.2296
6	16	2	4	0.0070	0.1070	13.0682	−3	−0.2296
7	20	6	36	0.0634	0.1634	12.6493	−3	−0.2372
8	20	6	36	0.0634	0.1634	12.6493	9	0.7115
9	22	8	64	0.1127	0.2127	12.2709	−21	−1.7114
10	26	12	144	0.2535	0.3535	11.1193	12	1.0792
		Total	**568**					

Note: the values of the residuals were computed in Table 14.7.

The standardized residual plot can provide insight into the assumption that the error term ε has a normal distribution. If this assumption is satisfied, the distribution of the standardized residuals should appear to come from a standard normal probability distribution.*

Thus, when looking at a standardized residual plot, we should expect to see approximately 95% of the standardized residuals between −2 and +2. We see in Figure 14.14 that for the Armand's example all standardized residuals are between −2 and +2. Therefore, on the basis of the standardized residuals, this plot gives us no reason to question the assumption that ε has a normal distribution.

Because of the effort required to compute the estimated values $\hat{y}$, the residuals and the standardized residuals, most statistical packages provide these values as optional regression output. But residual plots can be easily obtained. For large problems computer packages are the only practical means of acquiring the residual plots discussed in this section.

FIGURE 14.14
Plot of the standardized residuals against the independent variable X for Armand's Pizza Parlours

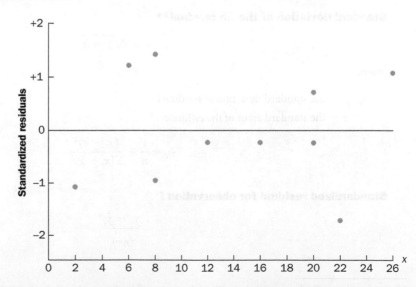

* Because s is used instead of σ in equation (14.26), the probability distribution of the standardized residuals is not technically normal. However, in most regression studies, the sample size is large enough that a normal approximation is very good.

Normal probability plot

An approach for determining the validity of the assumption that the error term has a normal distribution is the normal probability plot. To show how a normal probability plot is developed, we introduce the concept of *normal scores.*

Suppose ten values are selected randomly from a normal probability distribution with a mean of zero and a standard deviation of one, and that the sampling process is repeated over and over with the values in each sample of ten ordered from smallest to largest. For now, let us consider only the smallest value in each sample. The random variable representing the smallest value obtained in repeated sampling is called the first-order statistic.

Statisticians show that for samples of size ten from a standard normal probability distribution, the expected value of the first-order statistic is -1.55. This expected value is called a normal score. For the case with a sample of size $n = 10$, there are ten order statistics and ten normal scores. In general, a data set consisting of n observations will have n order statistics and hence n normal scores.

Let us now show how the ten normal scores can be used to determine whether the standardized residuals for Armand's Pizza Parlours appear to come from a standard normal probability distribution. We begin by ordering the ten standardized residuals from Table 14.8. The ten normal scores and the ordered standardized residuals are shown together in Table 14.9. If the normality assumption is satisfied, the smallest standardized residual should be close to the smallest normal score, the next smallest standardized residual should be close to the next smallest normal score and so on. If we were to develop a plot with the normal scores on the horizontal axis and the corresponding standardized residuals on the vertical axis, the plotted points should cluster closely around a 45-degree line passing through the origin if the standardized residuals are approximately normally distributed. Such a plot is referred to as a *normal probability plot.*

TABLE 14.9   Normal scores for $n = 10$ and ordered standardized residuals for Armand's Pizza Parlours

Ordered statistic	Normal scores	Standardized residuals
1	−1.55	−1.7114
2	−1.00	−1.0792
3	−0.65	−0.9487
4	−0.37	−0.2372
5	−0.12	−0.2296
6	0.12	0.2296
7	0.37	0.7115
8	0.65	1.0792
9	1.00	1.2224
10	1.55	1.4230

Figure 14.15 is the normal probability plot for the Armand's Pizza Parlours example. Judgement is used to determine whether the pattern observed deviates from the line sufficiently to conclude that the standardized residuals are not from a standard normal probability distribution. In Figure 14.15, we see that the points are grouped closely about the line. We therefore conclude that the assumption of the error term having a normal probability distribution is reasonable. In general, the more closely the points are clustered about the 45-degree line, the stronger the evidence supporting the normality assumption. Any substantial curvature in the normal probability plot is evidence that the residuals have not come from a normal distribution. Normal scores and the associated normal probability plot can be obtained easily from statistical packages.

FIGURE 14.15
Normal probability plot for Armand's
Pizza Parlours

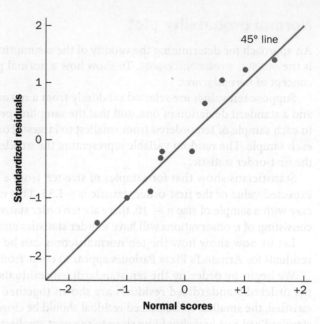

14.9 RESIDUAL ANALYSIS: AUTOCORRELATION

In the last section we showed how residual plots can be used to detect violations of assumptions about the error term ε in the regression model. In many regression studies, particularly involving data collected over time, a special type of correlation between the error terms can cause problems; it is called **serial correlation** or **autocorrelation**. In this section we show how the **Durbin–Watson test** can be used to detect significant autocorrelation.

Autocorrelation and the Durbin–Watson test

Often, the data used for regression studies in business and economics are collected over time. It is not uncommon for the value of Y at time t, denoted by y_t, to be related to the value of Y at previous time periods. In such cases, we say autocorrelation (also called serial correlation) is present in the data. If the value of Y in time period t is related to its value in time period $t - 1$, first-order autocorrelation is present. If the value of Y in time period t is related to the value of Y in time period $t - 2$, second-order autocorrelation is present and so on.

When autocorrelation is present, one of the assumptions of the regression model is violated: the error terms are not independent. In the case of first-order autocorrelation, the error at time t, denoted ε_t, will be related to the error at time period $t - 1$, denoted ε_{t-1}. Two cases of first-order autocorrelation are illustrated in Figure 14.16.

FIGURE 14.16
Two data sets with
first-order
autocorrelation

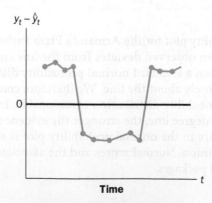

Panel A. Positive autocorrelation

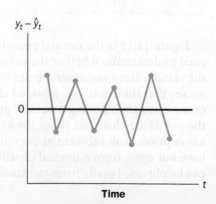

Panel B. Negative autocorrelation

Panel A is the case of positive autocorrelation; panel B is the case of negative autocorrelation. With positive autocorrelation we expect a positive residual in one period to be followed by a positive residual in the next period, a negative residual in one period to be followed by a negative residual in the next period and so on. With negative autocorrelation, we expect a positive residual in one period to be followed by a negative residual in the next period, then a positive residual and so on. When autocorrelation is present, serious errors can be made in performing tests of statistical significance based upon the assumed regression model. It is therefore important to be able to detect autocorrelation and take corrective action. We will show how the Durbin–Watson statistic can be used to detect first-order autocorrelation.

Suppose the values of ε are not independent but are related as shown in equation 14.29.

First-order autocorrelation

$$\varepsilon_t = \rho\varepsilon_{t-1} + z_t \tag{14.29}$$

In equation 14.29, ρ is a parameter with an absolute value less than one and z_t is a normally and independently distributed random variable with a mean of zero and a variance of σ^2. We see that if $\rho = 0$, the error terms are not related, and each has a mean of zero and a variance of σ^2. In this case, there is no autocorrelation and the regression assumptions are satisfied. If $\rho > 0$, we have positive autocorrelation; if $\rho < 0$, we have negative autocorrelation. In either of these cases, the regression assumptions about the error term are violated.

The Durbin–Watson test for autocorrelation uses the residuals to determine whether $\rho = 0$. To simplify the notation for the Durbin–Watson statistic, we denote the tth residual by $e_t = y_t - \hat{y}_t$. The Durbin–Watson test statistic is computed as follows.

Durbin–Watson test statistic

$$d = \frac{\sum\limits_{t=2}^{n} (e_t - e_{t-1})^2}{\sum\limits_{t=1}^{n} e_t^2} \tag{14.30}$$

If successive values of the residuals are close together (positive autocorrelation), the value of the Durbin–Watson test statistic will be small. If successive values of the residuals are far apart (negative autocorrelation), the value of the Durbin–Watson statistic will be large.

The Durbin–Watson test statistic ranges in value from zero to four, with a value of two indicating no autocorrelation is present. Durbin and Watson developed tables that can be used to determine when their test statistic indicates the presence of autocorrelation. Table 5 in Appendix B shows lower and upper bounds (d_L and d_U) for hypothesis tests using $\alpha = 0.05$, $\alpha = 0.025$ and $\alpha = 0.01$; n denotes the number of observations. The null hypothesis to be tested is always that there is no autocorrelation:

$$H_0: \rho = 0$$

The alternative hypothesis to test for positive autocorrelation is:

$$H_1: \rho > 0$$

The alternative hypothesis to test for negative autocorrelation is:

$$H_1: \rho < 0$$

A two-sided test is also possible. In this case the alternative hypothesis is:

$$H_1: \rho \neq 0$$

Figure 14.17 shows how the values of d_L and d_U in Table 5 in Appendix B are used to test for autocorrelation.

FIGURE 14.17
Hypothesis test for
autocorrelation using the
Durbin–Watson test

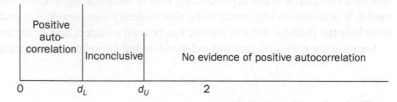

Panel A. Test for positive autocorrelation

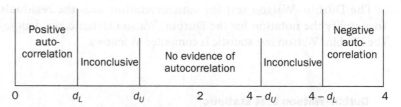

Panel B. Test for negative autocorrelation

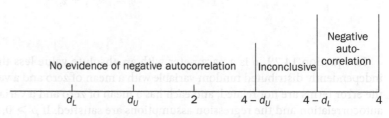

Panel C. Two-sided test for autocorrelation

Panel A illustrates the test for positive autocorrelation. If $d < d_L$, we conclude that positive autocorrelation is present. If $d_L \leq d \leq d_U$, we say the test is inconclusive. If $d > d_U$, we conclude that there is no evidence of positive autocorrelation.

Panel B illustrates the test for negative autocorrelation. If $d > 4 - d_L$, we conclude that negative autocorrelation is present. If $4 - d_U \leq d \leq 4 - d_L$, we say the test is inconclusive. If $d < 4 - d_U$, we conclude that there is no evidence of negative autocorrelation.

Note: entries in Table 5 in Appendix B are the critical values for a one-tailed Durbin–Watson test for autocorrelation. For a two-tailed test, the level of significance is doubled.

Panel C illustrates the two-sided test. If $d < d_L$ or $d > 4 - d_L$, we reject H_0 and conclude that autocorrelation is present. If $d_L \leq d \leq d_U$ or $4 - d_U \leq d \leq 4 - d_L$, we say the test is inconclusive. If $d_U \leq d \leq 4 - d_U$, we conclude that there is no evidence of autocorrelation.

If significant autocorrelation is identified, we should investigate whether we omitted one or more key independent variables that have time-ordered effects on the dependent variable. If no such variables can be identified, including an independent variable that measures the time of the observation (for instance, the value of this variable could be one for the first observation, two for the second observation and so on) this will sometimes eliminate or reduce the autocorrelation. When these attempts to reduce or remove autocorrelation do not work, transformations on the dependent or independent variables can prove helpful; a discussion of such transformations can be found in more advanced texts on regression analysis.

Note that the Durbin–Watson tables list the smallest sample size as 15. The reason is that the test is generally inconclusive for smaller sample sizes; in fact, many statisticians believe the sample size should be at least 50 for the test to produce worthwhile results.

EXERCISES

Methods

25. Given are data for two variables, X and Y.

x_i	6	11	15	18	20
y_i	6	8	12	20	30

a. Develop an estimated regression equation for these data.
b. Compute the residuals.
c. Develop a plot of the residuals against the independent variable X. Do the assumptions about the error terms seem to be satisfied?
d. Compute the standardized residuals.
e. Develop a plot of the standardized residuals against $\hat{y}$. What conclusions can you draw from this plot?

26. The following data were used in a regression study:

Observation	x_i	y_i	Observation	x_i	y_i
1	2	4	6	7	6
2	3	5	7	7	9
3	4	4	8	8	5
4	5	6	9	9	11
5	7	4			

a. Develop an estimated regression equation for these data.
b. Construct a plot of the residuals. Do the assumptions about the error term seem to be appropriate?

Applications

27. A doctor has access to historical data as follows:

	Vehicles per 100 population	Road deaths per 100,000 population
Great Britain	31	14
Belgium	32	29
Denmark	30	22
France	47	32
Germany	30	25
Irish Republic	19	20
Italy	36	21
Netherlands	40	22
Canada	47	30
USA	58	35

a. First, identifying the X and Y variables appropriately, use the method of least squares to develop a straight-line approximation of the relationship between the two variables.
b. Test whether vehicles and road deaths are related at a 0.05 level of significance.
c. Prepare a residual plot of $y - \hat{y}$ versus $\hat{y}$. Use the result from part (a) to obtain the values of $\hat{y}$.
d. What conclusions can you draw from the residual analysis? Should this model be used, or should we look for a better one?

CAMRY

28. Toyota Camry is one of the world's best-selling cars. The cost of a previously owned Camry depends upon many factors, including the model year, mileage and condition. To investigate the relationship between the car's mileage and the sales price for a used Camry, the following data show the mileage and sale price for 19 sales.

Miles (000)	Price ($100)
22	16.2
29	16.0
36	13.8
47	11.5
63	12.5
77	12.9
73	11.2
87	13.0
92	11.8
101	10.8
110	8.3
28	12.5
59	11.1
68	15.0
68	12.2
91	13.0
42	15.6
65	12.7
110	8.3

a. Develop a scatter diagram with the car mileage on the horizontal axis and the price on the vertical axis.
b. What does the scatter diagram developed in (a) indicate about the relationship between the two variables?
c. Develop the estimated regression equation that could be used to predict the price ($000) given the miles (000).
d. Use the t test to assess whether there is a significant relationship at the 0.05 level of significance.
e. Did the estimated regression equation provide a good fit? Explain.
f. Provide an interpretation for the slope of the estimated regression equation.
g. Suppose that you are considering purchasing a previously owned Camry that has been driven 60,000 miles. Using the estimated regression equation developed in (c), predict the price for this car. Is this the price you would offer the seller?

14.10 RESIDUAL ANALYSIS: OUTLIERS AND INFLUENTIAL OBSERVATIONS

In Section 14.8 we showed how residual analysis could be used to determine when violations of assumptions about the regression model occur. In this section, we discuss how residual analysis can be used to identify observations that can be classified as outliers or as being especially influential in determining the estimated regression equation. Some steps that should be taken when such observations occur are discussed.

Detecting outliers

Figure 14.18 is a scatter diagram for a data set that contains an outlier, a data point (observation) that does not fit the trend shown by the remaining data. Outliers represent observations that are suspect and warrant careful examination. They may represent erroneous data; if so, the data should be corrected. They may signal a violation of model assumptions; if so, another model should be considered. Finally, they may simply be unusual values that occurred by chance. In this case, they should be retained.

FIGURE 14.18

A data set with an outlier

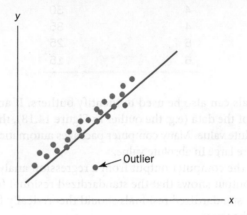

To illustrate the process of detecting outliers, consider the data set in Table 14.10; Figure 14.19 is a scatter diagram. Except for observation 4 ($x_4 = 3$, $y_4 = 75$), a pattern suggesting a negative linear relationship is apparent. Indeed, given the pattern of the rest of the data, we would expect y_4 to be much smaller and hence would identify the corresponding observation as an outlier. For the case of simple linear regression, one can often detect outliers by simply examining the scatter diagram.

FIGURE 14.19

Scatter diagram for outlier data set

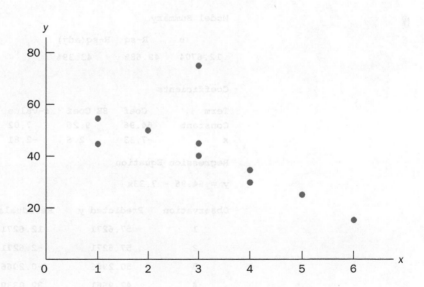

TABLE 14.10
Data set illustrating the effect of
an outlier

x_i	y_i
1	45
1	55
2	50
3	75
3	40
3	45
4	30
4	35
5	25
6	15

The standardized residuals can also be used to identify outliers. If an observation deviates greatly from the pattern of the rest of the data (e.g. the outlier in Figure 14.18), the corresponding standardized residual will be large in absolute value. Many computer packages automatically identify observations with standardized residuals that are large in absolute value.

In Figure 14.20 we show the computer output from a regression analysis of the data in Table 14.10. The next to last line of the output shows that the standardized residual for observation 4 is 2.67. With normally distributed errors, standardized residuals should theoretically fall outside the range −2 to +2 approximately 5 per cent of the time.

FIGURE 14.20
Output for regression analysis
of the outlier data set

Analysis of Variance

Source	DF	Adj SS	Adj MS	F-value	p-value
Regression	1	1,268.2	1,268.2	7.90	0.023
Error	8	1,284.3	160.5		
Total	9	2,552.5			

Model Summary

s	R-sq	R-sq(adj)
12.6704	49.68%	43.39%

Coefficients

Term	Coef	SE Coef	t-value	p-value
Constant	64.96	9.26	7.02	0.000
x	−7.33	2.6	−2.81	0.023

Regression Equation

y = 64.96 − 7.33x

Observation	Predicted y	Residuals	Standard Residuals
1	57.6271	−12.6271	−1.0570
2	57.6271	−2.6271	−0.2199
3	50.2966	−0.2966	−0.0248
4	42.9661	32.0339	2.6816
5	42.9661	−2.9661	−0.2483
6	42.9661	2.0339	0.1703
7	35.6356	−5.6356	−0.4718
8	35.6356	−0.6356	−0.0532
9	28.3051	−3.3051	−0.2767
10	20.9746	−5.9746	−0.5001

In deciding how to handle an outlier, we should first check to see whether it is a valid observation. Perhaps an error was made in initially recording the data or in entering the data into the computer file. For example, suppose that in checking the data for the outlier in Table 14.10, we find an error; the correct value for observation 4 is $x_4 = 3$, $y_4 = 30$. Figure 14.21 shows computer output obtained after correction of the value of y_4. We see that using the incorrect data value substantially affected the goodness of fit. With the correct data, the value of R-sq increased from 49.7 per cent to 83.8 per cent and the value of b_0 decreased from 64.958 to 59.237. The slope of the line changed from -7.331 to -6.949. The identification of the outlier enabled us to correct the data error and improve the regression results.

FIGURE 14.21

Output for the revised outlier data set

Analysis of Variance

Source	DF	Adj SS	Adj MS	F-value	p-value
Regression	1	1,139.66	1,139.66	41.38	0.000
Error	8	220.34	27.54		
Total	9	1,360.00			

Model Summary

s	R-sq	R-sq(adj)
5.24808	83.80%	81.77%

Coefficients

Term	Coef	SE Coef	t-value	P-value
Constant	59.24	3.83	15.45	0.000
x	-6.95	1.08	-6.43	0.000

Regression Equation

y = 59.24 - 6.95x

Detecting influential observations

Sometimes one or more observations exert a strong influence on the results obtained. Figure 14.22 shows an example of an **influential observation** in simple linear regression.

FIGURE 14.22

A data set with an influential observation

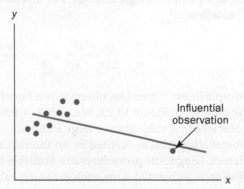

Influential observations can be identified from a scatter diagram when only one independent variable is present. An influential observation may be an outlier (an observation with a Y value that deviates substantially from the trend), it may correspond to an X value far away from its mean (e.g. refer to Figure 14.22), or it may be caused by a combination of the two (a somewhat off-trend Y value and a somewhat extreme X value).

Clearly, this one observation is much more influential in determining the estimated regression line than any of the others; dropping one of the other observations from the data set would have little effect on the estimated regression equation. Influential observations can be identified from a scatter diagram when only one independent variable is present.

Because influential observations may have such a dramatic effect on the estimated regression equation, they must be examined carefully. We should first check to make sure that no error was made in collecting or recording the data. If an error occurred, it can be corrected and a new estimated regression equation can be developed. If the observation is valid, we might consider ourselves fortunate to have it. Such a point, if valid, can contribute to a better understanding of the appropriate model and can lead to a better estimated regression equation. The presence of the influential observation in Figure 14.22, if valid, would suggest trying to obtain data on intermediate values of X to understand better the relationship between X and Y.

Observations with extreme values for the independent variables are called **high leverage points**. The influential observation in Figure 14.22 is a point with high leverage. The leverage of an observation is determined by how far the values of the independent variables are from their mean values. For the single-independent-variable case, the leverage of the ith observation, denoted h_i, can be computed by using equation (14.31).

Leverage of observation i

$$h_i = \frac{1}{n} + \frac{(x_i - \bar{x})^2}{\Sigma(x_i - \bar{x})^2}$$

(14.31)

From the formula, it is clear that the further x_i is from its mean x, the higher the leverage of observation i.

Many statistical packages automatically identify observations with high leverage as part of the standard regression output. As an illustration of how statistical packages identify points with high leverage, consider the data plotted in Figure 14.23.

Clearly, observation 7 ($X = 70$, $Y = 100$) is one with an extreme value of X. Hence, we would expect it to be identified as a point with high leverage. For this observation, the leverage is computed by using equation (14.31) as follows:

$$h_7 = \frac{1}{n} + \frac{(x_7 - \bar{x})^2}{\Sigma(x_i - \bar{x})^2} = \frac{1}{7} + \frac{(70 - 24.286)^2}{2,621.43} = 0.94$$

For the case of simple linear regression, observations have high leverage if $h_i > 6/n$ or 0.99, whichever is smaller for the data set in Figure 14.23, $6/n = 6/7 = 0.86$. Because $h_7 = 0.94 > 0.86$, observation 7 is identified as one whose X value gives it large influence.

Influential observations that are caused by an interaction of large residuals and high leverage can be difficult to detect. Diagnostic procedures are available that take both into account in determining when an observation is influential. One such measure, called Cook's D statistic, will be discussed in Chapter 15.

FIGURE 14.23
Scatter diagram for
the data set with a
high leverage
observation

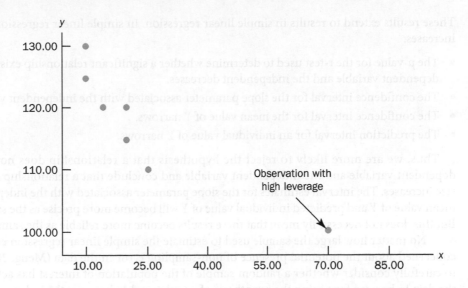

EXERCISES

Methods

29. Consider the following data for two variables, X and Y.

x_i	135	110	130	145	175	160	120
y_i	145	100	120	120	130	130	110

  a. Compute the standardized residuals for these data. Do there appear to be any outliers in the data? Explain.
  b. Plot the standardized residuals against $\hat{y}$. Does this plot reveal any outliers?
  c. Develop a scatter diagram for these data. Does the scatter diagram indicate any outliers in the data? In general, what implications does a finding of outliers have for simple linear regression?

30. Consider the following data for two variables, X and Y.

x_i	4	5	7	8	10	12	12	22
y_i	12	14	16	15	18	20	24	19

  a. Compute the standardized residuals for these data. Do there appear to be any outliers in the data? Explain.
  b. Compute the leverage values for these data. Do there appear to be any influential observations in these data? Explain.
  c. Develop a scatter diagram for these data. Does the scatter diagram indicate any influential observations? Explain.

14.11 PRACTICAL ADVICE: BIG DATA AND HYPOTHESIS TESTING IN SIMPLE LINEAR REGRESSION

In Chapter 7, we observed that the standard errors of the sampling distributions of the sample mean $\overline{X}$ (shown in equation (7.2)) and the sample proportion P (shown in equation (7.5)) decrease as the sample size increases. In Chapters 8 and 9, we observed that this narrower confidence interval estimates μ and π and smaller p-values for the tests of the hypotheses $H_0: \mu \leq \mu_0$ and $H_0: \pi \leq \pi_0$ as the sample size increases.

These results extend to results in simple linear regression. In simple linear regression, as the sample size increases:

- The *p*-value for the *t*-test used to determine whether a significant relationship exists between the dependent variable and the independent decreases.
- The confidence interval for the slope parameter associated with the independent variable narrows.
- The confidence interval for the mean value of *Y* narrows.
- The prediction interval for an individual value of *Y* narrows.

Thus, we are more likely to reject the hypothesis that a relationship does not exist between the dependent variable and the independent variable and conclude that a relationship exists as the sample size increases. The interval estimates for the slope parameter associated with the independent variable, the mean value of *Y* and predicted individual value of *Y* will become more precise as the sample size increases. But this does not necessarily mean that these results become more reliable as the sample size increases.

No matter how large the sample used to estimate the simple linear regression equation, we must be concerned about the potential presence of non-sampling error in the data (Meng, 2018). It is important to carefully consider whether a random sample of the population of interest has actually been taken. If the data to be used for testing the hypothesis of no relationship between the independent and dependent variable are corrupted by non-sampling error, the likelihood of making a Type I or Type II error may be higher than if the sample data are free of non-sampling error. If the relationship between the independent and dependent variables is statistically significant, it is also important to consider whether the relationship in the simple linear regression equation is of *practical* significance.

Although simple linear regression is an extremely powerful statistical tool, it provides evidence that should be considered only in combination with information collected from other sources to make the most informed decision possible. No business decision should be based exclusively on inference in simple linear regression. Non-sampling error may lead to misleading results, and practical significance should always be considered in conjunction with statistical significance. This is particularly important when a hypothesis test is based on an extremely large sample because *p*-values in such cases can be extremely small. However, when executed properly, inference based on simple linear regression can be an important component in the business decision-making process.

ONLINE RESOURCES

For data files, additional questions and answers, and the software section for Chapter 14, please go to the online platform.

SUMMARY

In this chapter we showed how regression analysis can be used to determine how a dependent variable *Y* is related to an independent variable *X*. In simple linear regression, the regression model is $Y = \beta_0 + \beta_1 x + \varepsilon$. The simple linear regression equation $E(Y) = \beta_0 + \beta_1 x$ describes how the mean or expected value of *Y* is related to a particular value of *X*. We used sample data and the least squares method to develop the estimated regression equation $\hat{y} = b_0 + b_1 x$ for a given value *x* of *X*. In effect, b_0 and b_1 are the sample statistics used to estimate the unknown model parameters β_0 and β_1.

The coefficient of determination was presented as a measure of the goodness of fit for the estimated regression equation; it can be interpreted as the proportion of the variation in the dependent variable *Y* that can be explained by the estimated regression equation. We reviewed correlation as a descriptive measure of the strength of a linear relationship between two variables.

The assumptions about the regression model and its associated error term ϵ were discussed, and *t* and *F* tests, based on those assumptions, were presented as a means for determining whether the relationship between two variables is statistically significant. We showed how to use the estimated regression equation to develop confidence interval estimates of the mean value of Y and prediction interval estimates of individual values of Y.

The chapter concluded with sections on the computer solution of regression problems, the use of residual analysis to validate the model assumptions – in particular, autocorrelation, the identification of outliers and influential observations – and practical advice concerning the limitations of regression modelling in the presence of big data.

KEY TERMS

Autocorrelation
Coefficient of determination
Dependent variable
Durbin–Watson test
Estimated regression equation
High leverage points
Independent variable
Influential observation
ith residual
Least squares method
Mean square error (MSE)

Normal probability plot
Prediction interval
Regression equation
Regression model
Residual analysis
Residual plot
Serial correlation
Simple linear regression
Standard error of the estimate
Standardized residual

KEY FORMULAE

Simple linear regression model

$$Y = \beta_0 + \beta_1 x + \varepsilon \tag{14.1}$$

Simple linear regression equation

$$E(Y) = \beta_0 + \beta_1 x \tag{14.2}$$

Estimated simple linear regression equation

$$\hat{y} = b_0 + b_1 x \tag{14.3}$$

Least squares criterion

$$\text{Min} \sum (y_i - \hat{y}_i)^2 \tag{14.5}$$

Slope and y-intercept for the estimated regression equation

$$b_1 = \frac{\sum (x_i - x)(y_i - \bar{y})}{\sum (x_i - \bar{x})^2} \tag{14.6}$$

$$b_0 = \bar{y} - b_1 \bar{x} \tag{14.7}$$

Sum of squares due to error

$$SSE = \sum (y_i - \hat{y}_i)^2 \tag{14.8}$$

Total sum of squares

$$SST = \sum (y_1 - \bar{y})^2 \tag{14.9}$$

Sum of squares due to regression

$$SSR = \sum (\hat{y}_i - \bar{y})^2 \tag{14.10}$$

Relationship among SST, SSR and SSE

$$SST = SSR + SSE \tag{14.11}$$

Coefficient of determination

$$r^2 = \frac{SSR}{SST} \tag{14.12}$$

Sample correlation coefficient

$$r_{XY} = (\text{sign of } b_1)\sqrt{\text{Coefficient of determination}} \tag{14.13}$$
$$= (\text{sign of } b_1)\sqrt{r^2}$$

Mean square error (estimate of s²)

$$s^2 = MSE = \frac{SSE}{n - 2} \tag{14.15}$$

Standard error of the estimate

$$s = \sqrt{MSE} = \sqrt{\frac{SSE}{n - 2}} \tag{14.16}$$

Standard deviation of b_1

$$\sigma_{b1} = \frac{\sigma}{\sqrt{\sum (x_i - \bar{x})^2}} \tag{14.17}$$

Estimated standard deviation of b_1

$$s_{b_1} = \frac{s}{\sqrt{\sum (x_i - \bar{x})^2}} \tag{14.18}$$

t test statistic

$$t = \frac{b_1}{s_{b_1}} \tag{14.19}$$

Mean square regression

$$MSR = \frac{SSR}{\text{Number of independent variables}}$$ (14.20)

F test statistic

$$F = \frac{MSR}{MSE}$$ (14.21)

Confidence interval for $E(Y_p)$

$$\hat{y}_p \pm t_{\alpha/2}s\sqrt{\frac{1}{n} + \frac{(x_p - \bar{x})^2}{\sum(x_i - \bar{x})^2}}$$ (14.22)

Prediction interval for y_p

$$\hat{y}_p \pm t_{\alpha/2}s\sqrt{1 + \frac{1}{n} + \frac{(x_p - \bar{x})^2}{\sum(x_i - \bar{x})^2}}$$ (14.23)

Residual for observation i

$$y_i - \hat{y}_i$$ (14.24)

Standard deviation of the ith residual

$$s_{y_i - \hat{y}_i} = s\sqrt{1 - h_i}$$ (14.26)

Standardized residual for observation i

$$\frac{y_i - \hat{y}_i}{s_{y_i - \hat{y}_i}}$$ (14.28)

First-order autocorrelation

$$\varepsilon_t = \rho\varepsilon_{t-1} + z_t$$ (14.29)

Durbin–Watson test statistic

$$d = \frac{\sum_{t=2}^{n}(e_t - e_{t-1})^2}{\sum_{t=1}^{n}e_t^2}$$ (14.30)

Leverage of observation i

$$h_i = \frac{1}{n} + \frac{(x_i - \bar{x})^2}{\sum(x_i - \bar{x})^2}$$ (14.31)

CASE PROBLEM 1

Investigating the relationship between weight loss and triglyceride level reduction*

Epidemiological studies have shown that there is a relationship between raised blood levels of triglyceride and coronary heart disease, but it is not certain how important a risk factor triglycerides are. It is believed that exercise and lower consumption of fatty acids can help to reduce triglyceride levels.**

In 1998 Knoll Pharmaceuticals received authorization to market sibutramine for the treatment of obesity in the USA. One of their suite of studies involved 35 obese patients who followed a treatment regime comprising a combination of diet, exercise and drug treatment.

TRIGLYCERIDE

Each patient's weight and triglyceride level were recorded at the start (known as *baseline*) and at week 8. The information recorded for each patient was:

- Patient ID
- Weight at baseline (kg)
- Weight at week 8 (kg)
- Triglyceride level at baseline (mg/dl)
- Triglyceride level at week 8 (mg/dl).

Triglyceride

The results are shown below.

Managerial report

1. Are weight loss and triglyceride level reduction (linearly) correlated?

2. Is there a linear relationship between weight loss and triglyceride level reduction?

3. How can a more detailed regression analysis be undertaken?

Patient ID	Weight at baseline	Weight at week 8	Triglyceride level at baseline	Triglyceride level at week 8
201	84.0	82.4	90	131
202	88.8	87.0	137	82
203	87.0	81.8	182	152
204	84.5	80.4	72	72
205	69.4	69.0	143	126
206	104.7	102.0	96	157
207	90.0	87.6	115	88
208	89.4	86.8	124	123
209	95.2	92.8	188	255
210	108.1	100.9	167	87
211	93.9	90.2	143	213
212	83.4	75.0	143	102
213	104.4	102.9	276	313
214	103.7	95.7	84	84
215	99.2	99.2	142	135
216	95.6	88.5	64	114
217	126.0	123.2	226	152
218	103.7	95.5	199	120
219	133.1	130.8	212	156
220	85.0	80.0	268	250
221	83.8	77.9	111	107

* Source: STARS (Creation of Statistics Resources project), University of Coventry, UK (Project Director: Colin James).
** Triglycerides are lipids (fats) which are formed from glycerol and fatty acids. They can be absorbed into the body from food intake, particularly from fatty food, or produced in the body itself when the uptake of energy (food) exceeds the expenditure (exercise). Triglycerides provide the principal energy store for the body. Compared with carbohydrates or proteins, triglycerides produce a substantially higher number of calories per gram.

222	104.5	98.3	132	117
223	76.8	73.2	165	96
224	90.5	88.9	57	63
225	106.9	103.7	163	131
226	81.5	78.9	111	54
227	96.5	94.9	300	241
228	103.0	97.2	192	124
229	127.5	124.7	176	215
230	103.2	102.0	146	138
231	113.5	115.0	446	795
232	107.0	99.2	232	63
233	106.0	103.5	255	204
234	114.9	105.3	187	144
235	103.4	96.0	154	96

CASE PROBLEM 2

Measuring stock market risk

One measure of the risk or volatility of an individual stock is the standard deviation of the total return (capital appreciation plus dividends) over several periods of time. Although the standard deviation is easy to compute, it does not take into account the extent to which the price of a given stock varies as a function of a standard market index, such as the S&P 500. As a result, many financial analysts prefer to use another measure of risk referred to as *beta*.

Betas for individual stocks are determined by simple linear regression. The dependent variable is the total return for the stock and the independent variable is the total return for the stock market. For this Case Problem we will use the S&P 500 index as the measure of the total return for the stock market, and an estimated regression equation will be developed using monthly data. The beta for the stock is the slope of the estimated regression equation (b_1). The data contained in the file named 'Beta' provide the total return (capital appreciation plus dividends) over 36 months for eight widely traded common stocks and the S&P 500.

Beta

The value of beta for the stock market will always be 1; thus, stocks that tend to rise and fall with the stock market will also have a beta close to 1. Betas greater than 1 indicate that the stock is more volatile than the market, and betas less than 1 indicate that the stock is less volatile than the market. For instance, if a stock has a beta of 1.4, it is 40 per cent *more* volatile than the market, and if a stock has a beta of 0.4, it is 60 per cent *less* volatile than the market.

Managerial report

You have been assigned to analyze the risk characteristics of these stocks. Prepare a report that includes but is not limited to the following items.

(Continued)

BETA

1. Compute descriptive statistics for each stock and the S&P 500. Comment on your results. Which stocks are the most volatile?

2. Compute the value of beta for each stock. Which of these stocks would you expect to

perform best in an up market? Which would you expect to hold their value best in a down market?

3. Comment on how much of the return for the individual stocks is explained by the market.

© shapecharge/iStock

CASE PROBLEM 3

DYSLEXIA

Can we detect dyslexia?

Data were collected on 34 pre-school children and then in follow-up tests (on the same children) three years later when they were seven years old.

Scores were obtained from a variety of tests on all the children at age four when they were at nursery school. The tests were:

- Knowledge of vocabulary, measured by the British Picture Vocabulary Test (BPVT) in three versions – as raw scores, standardized scores and percentile norms.
- Another vocabulary test – non-word repetition.
- Motor skills, where the children were scored on the time in seconds to complete five different peg board tests.
- Knowledge of prepositions, scored as the number correct out of ten.
- Three tests on the use of rhyming, scored as the number correct out of ten.

Three years later the same children were given a reading test, from which a reading deficiency was calculated as Reading Age – Chronological Age (in months), this being known as Reading Age Deficiency (RAD). The children were then classified into 'poor' or 'normal' readers, depending on their RAD scores. Poor reading ability is taken as an indication of potential dyslexia.

One purpose of this study is to identify which of the tests at age four might be used as predictors of poor reading ability, which in turn is a possible indication of dyslexia.

Data

The data set 'Dyslexia' contains 18 variables:

- Child Code   an identification number for each child (1–34)
- Sex   m for male, f for female

The BPVT scores:

BPVT raw   the raw score
BPVT std   the standardized score
BPVT % norm   cumulative percentage scores
Non-wd repn   score for non-word repetition

Scores in motor skills:

- Pegboard set1 to Pegboard set5 — the time taken to complete each test
- Mean — child's average over the pegboard tests
- Preps score — knowledge of prepositions (6–10)

Scores in rhyming tests (2–10):

- Rhyme set1
- Rhyme set2
- Rhyme set3
- RAD
- Poor/Normal   RAD scores, categorized as 1 = normal, 2 = poor

Details for ten records from the data set are shown below.

Child code	Sex	BPVT raw	BPVT std	BPVT % norm	Non-wd repn	Pegboard set1	Pegboard set2	Pegboard set3	Pegboard set4	Pegboard set5
1	m	29	88	22	15	20.21	28.78	28.04	20.00	24.37
2	m	21	77	6	11	26.34	26.20	20.35	28.25	20.87
3	m	50	107	68	17	21.13	19.88	17.63	16.25	19.76
4	m	23	80	9	5	16.46	16.47	16.63	14.16	17.25
5	f	35	91	28	13	17.88	15.13	17.81	18.41	15.99
6	m	36	97	42	16	20.41	18.64	17.03	16.69	14.47
7	f	47	109	72	25	21.31	18.06	28.00	21.88	18.03
8	m	32	92	30	12	14.57	14.22	13.47	12.29	18.38
9	f	38	101	52	14	22.07	22.69	21.19	22.72	20.62
10	f	44	105	63	15	16.40	14.48	13.83	17.59	34.68

Child code	Mean	Preps score	Rhyme set1	Rhyme set2	Rhyme set3	RAD	Poor/ Normal
1	24.3	6	5	5	5	−6.50	P
2	24.4	9	3	3	4	−7.33	P
3	18.9	10	9	8	*	49.33	N
4	16.2	7	4	6	4	−11.00	P
5	17.0	10	10	6	6	−2.67	N
6	17.5	10	6	5	5	−8.33	P
7	21.5	8	9	10	10	26.33	N
8	14.6	10	8	6	3	9.00	N
9	21.9	9	10	10	7	2.67	N
10	19.4	10	7	8	4	9.67	N

Managerial report

1. Is there a (linear) relationship between scores in tests at ages four and seven?
2. Can we predict RAD from scores at age four?

© Rat0C37/iStock

CASE PROBLEM 4

Selecting a point-and-shoot digital camera

Consumer Reports tested 166 different point-and-shoot digital cameras. Based upon factors such as the number of megapixels, weight (g), image quality and ease of use, they developed an overall score for each camera tested. The overall score ranges from 0 to 100, with higher scores indicating better overall test results.

Selecting a camera with many options can be a difficult process, and price is certainly a key issue for most consumers. By spending more, will a consumer really get a superior camera? And, do cameras that have more megapixels, a factor often considered to be a good measure of picture quality, cost more than cameras with fewer megapixels? Table 14.11 shows the brand, average retail price ($), number of megapixels, weight (g) and the overall score for 13 Canon and 15 Nikon subcompact cameras tested by *Consumer Reports* (*Consumer Reports* website).

CAMERAS

TABLE 14.11

Observation	Brand	Price ($)	Megapixels	Weight (g)	Score
1	Canon	300	10	198	66
2	Canon	200	12	142	66
3	Canon	300	12	198	65
4	Canon	200	10	170	62
5	Canon	180	12	142	62
6	Canon	200	12	198	61
7	Canon	200	14	142	60
8	Canon	130	10	198	60
9	Canon	130	12	142	59
10	Canon	110	16	142	55
11	Canon	90	14	142	52
12	Canon	100	10	170	51
13	Canon	90	12	198	46
14	Nikon	270	16	142	65
15	Nikon	300	16	198	63
16	Nikon	200	14	170	61
17	Nikon	400	14	198	59
18	Nikon	120	14	142	57
19	Nikon	170	16	170	56
20	Nikon	150	12	142	56
21	Nikon	230	14	170	55
22	Nikon	180	12	170	53
23	Nikon	130	12	170	53
24	Nikon	80	12	198	52
25	Nikon	80	14	198	50
26	Nikon	100	12	113	46
27	Nikon	110	12	142	45
28	Nikon	130	14	113	42

Managerial report

1. Develop numerical summaries of the data.

2. Using overall score as the dependent variable, develop three scatter diagrams, one using price as the independent variable, one using the number of megapixels as the independent variable, and one using weight as the independent variable. Which of the three independent variables appears to be the best predictor of overall score?

3. Using simple linear regression, develop an estimated regression equation that could be used to predict the overall score given the price of the camera. For this estimated regression equation, perform an analysis of the residuals and discuss your findings and conclusions.

4. Analyze the data using only the observations for the Canon cameras. Discuss the appropriateness of using simple linear regression and make any recommendations regarding the prediction of overall score using just the price of the camera.

© Denniro/iStock

15

Multiple Regression

CHAPTER CONTENTS

Statistics in Practice   Jura

LEARNING OBJECTIVES   After reading this chapter and doing the exercises, you should be able to:

1 Understand how multiple regression analysis can be used to develop relationships involving one dependent variable and several independent variables.

2 Interpret the coefficients in a multiple regression analysis.

3 Appreciate the background assumptions necessary to conduct statistical tests involving the hypothesized regression model.

4 Understand the role of computer packages in performing multiple regression analysis.

5 Interpret and use computer output to develop the estimated regression equation.

6 Determine how good a fit is provided by the estimated regression equation.

7 Test the significance of the regression equation.

8 Understand how multicollinearity affects multiple regression analysis.

9 Understand how residual analysis can be used to make a judgement as to the appropriateness of the model, identify outliers and determine which observations are influential.

10 Understand how logistic regression is used for regression analyses involving a binary dependent variable.

n Chapter 14 we presented simple linear regression and demonstrated its use in developing an estimated regression equation that describes the relationship between two variables. Recall that the variable being predicted or explained is called the dependent variable and the variable being used to predict or explain the dependent variable is called the independent variable. In this chapter we continue our study of regression analysis by considering situations involving two or more independent variables. This subject area, called multiple regression analysis, enables us to consider more than one potential predictor and thus obtain better estimates than are possible with simple linear regression.

STATISTICS IN PRACTICE
Jura

P_{1it}/P_{2it} is the relative price between route 1 and route 2 to i in year t

F_{1it}/F_{2it} is the relative frequency between route 1 and route 2 to i in year t

J_{1it}/J_{2it} is the relative journey time between route 1 and route 2 to i in year t

Jura is a large island (380 km²) off the south west coast of Scotland, famous for its malt whisky and the large deer population that wander the quartz mountains ('the Paps') that dominate the landscape. With a population of a mere 461 it has one of the lowest population densities of any place in the UK. Until recently, Jura was only accessible via the adjoining island, Islay, which has three ferry services a day – crossings taking about two hours. However, because Jura is only four miles from the mainland it has been suggested that a direct car ferry taking less than half an hour would be preferable and more economical than existing provisions.

In exploring the case for an alternative service, Riddington (1996) arrives at a number of alternative mathematical formulations that essentially reduce to multiple regression analysis. In particular, using historical data that also encompasses other inner Hebridean islands of Arran, Bute, Mull and Skye, he obtains the estimated binary logistic regression model for the ferry to Jura:

Route 1 is the proposed short route to Jura and route 2 is the existing route from Kennacraig to Port Ellen. Following the Scottish Ferries Review in 2012 it was proposed to divert the Kennacraig ferry on Saturdays to Port Askaig for the benefit of Jura residents. Though a direct service has since been established, clearly current ferry provisions to the island are far from ideal. Unfortunately, planned vessel replacements appear to pose even greater problems for travel to Jura in the future.

© Vedad Ceric/iStock

$$Log_e \frac{Q_{1it}}{Q_{2it}} = 6.48 - 0.89 \frac{P_{1it}}{P_{2it}}$$

$$+ 0.129 \frac{F_{1it}}{F_{2it}} - 6.18 \frac{J_{1it}}{J_{2it}}$$

where:

Q_{1it}/Q_{2it} is the number of cars travelling by route 1 relative to the number travelling by route 2 to island i in year t

Sources:

Argyllshire Advertiser (2022) 'Crisis point for Jura's failing ferry service', 20 May.

Riddington, G. (1996) 'How many for the ferry boat?' *OR Insight* 9(2): 26–32.

Scottish Daily Express (2022) 'Ferry fiasco deepens as Ferguson Marine vessels cannot berth at more than half of Scottish ports', 5 November.

The Oban Times (2021) 'Continual unreliability of Jura ferry makes a mockery of council's lifeline service', 30 November.

Transport Scotland (19 December 2012) Scottish Ferry Services: Ferries Plan (2013–2022).

15.1 MULTIPLE REGRESSION MODEL

Multiple regression analysis is the study of how a dependent variable Y is related to two or more independent variables. In the general case, we will use p to denote the number of independent variables.

Regression model and regression equation

The concepts of a regression model and a regression equation introduced in the preceding chapter are applicable in the multiple regression case. The equation that describes how the dependent variable Y is related to the independent variables $X_1, X_2, \ldots X_p$ and an error term is called the **multiple regression model**. We begin with the assumption that the multiple regression model takes the following form:

Multiple regression model

$$Y = \beta_0 + \beta_1 x_1 + \beta_2 x_2 + \cdots + \beta_p x_p + \varepsilon \qquad \textbf{(15.1)}$$

where: $X_1 = x_1, X_2 = x_2, \ldots, X_p = x_p$

In the multiple regression model, $\beta_0, \beta_1, \ldots, \beta_p$, are the parameters and ε (the Greek letter epsilon) is a random variable. A close examination of this model reveals that Y is a linear function of $x_1, x_2, \ldots, x_p$ (the $\beta_0 + \beta_1 x_1 + \beta_2 x_2 + \ldots + \beta_p x_p$ part) plus an error term ε. The error term accounts for the variability in Y that cannot be explained by the linear effect of the p independent variables.

In Section 15.4 we will discuss the assumptions for the multiple regression model and ε. One of the assumptions is that the mean or expected value of ε is zero. A consequence of this assumption is that the mean or expected value of Y, denoted $E(Y)$, is equal to $\beta_0 + \beta_1 x_1 + \beta_2 x_2 + \ldots + \beta_p x_p$ The equation that describes how the mean value of Y is related to $x_1, x_2, \ldots x_p$ is called the **multiple regression equation**.

Multiple regression equation

$$E(Y) = \beta_0 + \beta_1 x_1 + \beta_2 x_2 + \cdots + \beta_p x_p \qquad \textbf{(15.2)}$$

Estimated multiple regression equation

If the values of $\beta_0, \beta_1, \ldots, \beta_p$ were known, equation (15.2) could be used to compute the mean value of Y at given values of $x_1, x_2, \ldots x_p$. Unfortunately, these parameter values will not, in general, be known and must be estimated from sample data. A simple random sample is used to compute sample statistics $b_0, b_1, \ldots, b_p$ that are used as the point estimators of the parameters $\beta_0, \beta_1, \ldots, \beta_p$. These sample statistics provide the following **estimated multiple regression equation**.

Estimated multiple regression equation

$$\hat{y} = b_0 + b_1 x_1 + b_2 x_2 + \cdots + b_p x_p \qquad \textbf{(15.3)}$$

where:

$$b_0, b_1, \ldots, b_p \text{ are the estimates of } \beta_0, \beta_1, \ldots, \beta_p$$

$$\hat{y} = \text{estimated value of the dependent variable}$$

The estimation process for multiple regression is shown in Figure 15.1.

FIGURE 15.1
The estimation
process for
multiple regression

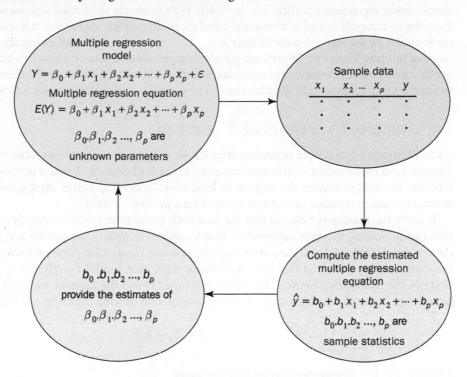

15.2 LEAST SQUARES METHOD

In Chapter 14 we used the least squares method to develop the estimated regression equation that best approximated the straight-line relationship between the dependent and independent variables. This same approach is used to develop the estimated multiple regression equation. The least squares criterion is restated as follows:

Least squares criterion

$$\min \sum (y_i - \hat{y}_i)^2 \tag{15.4}$$

where:

y_i = observed value of the dependent variable for the ith observation

$\hat{y}_i$ = estimated value of the dependent variable for the ith observation

The estimated values of the dependent variable are computed by using the estimated multiple regression equation:

$$\hat{y} = b_0 + b_1 x_1 + b_2 x_2 + \cdots + b_p x_p$$

As expression (15.4) shows, the least squares method uses sample data to provide the values of $b_0, b_1, \ldots, b_p$ that make the sum of squared residuals {the deviations between the observed values of the dependent variable (y_i) and the estimated values of the dependent variable $\hat{y}_i$} a minimum.

In Chapter 14 we presented formulae for computing the least squares estimators b_0 and b_1 for the estimated simple linear regression equation $\hat{y} = b_0 + b_1x$. With relatively small data sets, we were able to use those formulae to compute b_0 and b_1 by manual calculations. In multiple regression, however, the presentation of the formulae for the regression coefficients b_0, b_1, $\ldots$, b_p involves the use of matrix algebra and is beyond the scope of this text. Therefore, in presenting multiple regression, we focus on how computer software packages can be used to obtain the estimated regression equation and other information. The emphasis will be on how to interpret the computer output rather than on how to make the multiple regression computations.

An example: Eurodistributor Company

As an illustration of multiple regression analysis, we will consider a problem faced by the Eurodistributor Company, an independent distribution company in the Netherlands. A major portion of Eurodistributor's business involves deliveries throughout its local area. To develop better work schedules, the company's managers want to estimate the total daily travel time for their drivers.

Initially the managers believed that the total daily travel time would be closely related to the distance travelled in making the daily deliveries. A simple random sample of ten driving assignments provided the data shown in Table 15.1 and the scatter diagram shown in Figure 15.2. After reviewing this scatter diagram, the managers hypothesized that the simple linear regression model $Y = \beta_0 + \beta_1 x_1 + \varepsilon$ could be used to describe the relationship between the total travel time (Y) and the distance travelled (X_1). To estimate the parameters β_0 and β_1, the least squares method was used to develop the estimated regression equation:

$$\hat{y} = b_0 + b_1x_1 \tag{15.5}$$

TABLE 15.1   Preliminary data for Eurodistributor

Driving assignment	X_1 = Distance travelled (kilometres)	Y = Travel time (hours)
1	100	9.3
2	50	4.8
3	100	8.9
4	100	6.5
5	50	4.2
6	80	6.2
7	75	7.4
8	65	6.0
9	90	7.6
10	90	6.1

FIGURE 15.2
Scatter diagram of preliminary data for Eurodistributor

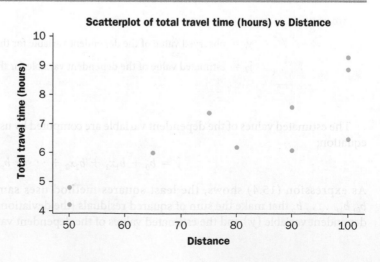

In Figure 15.3, we show the computer output from applying simple linear regression to the data in Table 15.1. The estimated regression equation is:

$$\hat{y} = 1.27 + 0.0678x_1$$

FIGURE 15.3
Output for Eurodistributor with one independent variable

```
Analysis of Variance

Source        DF    Adj SS    Adj MS    F-value    p-value
Regression     1    15.871    15.8713   15.81      0.004
Error          8     8.029     1.0036
Total          9    23.900

Model Summary

      S      R-sq    R-sq(adj)
1.00179    66.41%     62.21%

Coefficients

Term         Coef    SE Coef    t-value    p-value
Constant     1.27       1.40       0.91      0.390
Kilometres   0.0678     0.0171     3.98      0.004

Regression Equation

Time = 1.27 + 0.0678 Kilometres
```

At the 0.05 level of significance, the F value of 15.81 and its corresponding p-value of 0.004 indicate that the relationship is significant; that is, we can reject H_0: $\beta_1 = 0$ because the p-value is less than $\alpha = 0.05$. Thus, we can conclude that the relationship between the total travel time and the distance travelled is significant; longer travel times are associated with more distance. With a coefficient of determination (expressed as a percentage) of R-sq = 66.4 per cent, we see that 66.4 per cent of the variability in travel time can be explained by the linear effect of the distance travelled. This finding is fairly good, but the managers might want to consider adding a second independent variable to explain some of the remaining variability in the dependent variable.

In attempting to identify another independent variable, the managers felt that the number of deliveries could also contribute to the total travel time. The Eurodistributor data, with the number of deliveries added, are shown in Table 15.2.

TABLE 15.2 Data for Eurodistributor with distance (X_1) and number of deliveries (X_2) as the independent variables

Driving assignment	X_1 = Distance travelled (kilometres)	X_2 = Number of deliveries	Y = Travel time (hours)
1	100	4	9.3
2	50	3	4.8
3	100	4	8.9
4	100	2	6.5
5	50	2	4.2
6	80	2	6.2
7	75	3	7.4
8	65	4	6.0
9	90	3	7.6
10	90	2	6.1

The computer solution with both distance (X_1) and number of deliveries (X_2) as independent variables is shown in Figure 15.4. The estimated regression equation is:

$$\hat{y} = -0.869 + 0.0611x_1 + 0.923x_2 \tag{15.6}$$

FIGURE 15.4
Output for Eurodistributor
with two independent
variables

```
Analysis of Variance

Source        DF      Adj SS    Adj MS    F-value   p-value
Regression     2     21.6006   10.8003     32.88     0.000
Error          7      2.2994    0.3285
Total          9     23.900


Model Summary

         S     R-sq    R-sq(adj)
  0.573142   90.38%      87.63%


Coefficients

Term          Coef     SE Coef   t-value   p-value
Constant    -0.869      0.952     -0.91     0.392
Kilometres   0.06113    0.00989    6.18     0.000
Deliveries   0.923      0.221      4.18     0.004


Regression Equation

Time = -0.869 + 0.06113 Kilometres + 0.923 Deliveries
```

In the next section we will discuss the use of the coefficient of multiple determination in measuring how good a fit is provided by this estimated regression equation. Before doing so, let us examine more carefully the values of $b_1 = 0.0611$ and $b_2 = 0.923$ in equation (15.6).

Note on interpretation of coefficients

One observation can be made at this point about the relationship between the estimated regression equation with only the distance as an independent variable, and the equation that includes the number of deliveries as a second independent variable. The value of b_1 is not the same in both cases. In simple linear regression, we interpret b_1 as an estimate of the change in Y for a one-unit change in the independent variable. In multiple regression analysis, this interpretation must be modified somewhat. That is, in multiple regression analysis, we interpret each regression coefficient as follows: b_i represents an estimate of the change in Y corresponding to a one-unit change in X_i when all other independent variables are held constant.

In the Eurodistributor example involving two independent variables, $b_1 = 0.0611$. Thus, 0.0611 hours is an estimate of the expected increase in travel time corresponding to an increase of one kilometre in the distance travelled when the number of deliveries is held constant. Similarly, because $b_2 = 0.923$, an estimate of the expected increase in travel time corresponding to an increase of one delivery when the distance travelled is held constant is 0.923 hours.

EXERCISES

Note to students: The exercises involving data in this and subsequent sections were designed to be solved using a computer software package.

Methods

1. The estimated regression equation for a model involving two independent variables and ten observations follows.

$$\hat{y} = 29.1270 + 0.5906x_1 + 0.4980x_2$$

   a. Interpret b_1 and b_2 in this estimated regression equation.
   b. Estimate Y when $X_1 = 180$ and $X_2 = 310$.

2. Consider the following data for a dependent variable Y and two independent variables, X_1 and X_2.

EXER2

X_1	X_2	Y
30	12	94
47	10	108
25	17	112
51	16	178
40	5	94
51	19	175
74	7	170
36	12	117
59	13	142
76	16	211

   a. Develop an estimated regression equation relating Y to X_1. Estimate Y if $X_1 = 45$.
   b. Develop an estimated regression equation relating Y to X_2. Estimate Y if $X_2 = 15$.
   c. Develop an estimated regression equation relating Y to X_1 and X_2. Estimate Y if $X_1 = 45$ and $X_2 = 15$.

3. In a regression analysis involving 30 observations, the following estimated regression equation was obtained.

$$\hat{y} = 17.6 + 03.8x_1 - 2.3x_2 + 7.6x_3 + 2.7x_4$$

   a. Interpret b_1, b_2 b_3 and b_4 in this estimated regression equation.
   b. Estimate Y when $X_1 = 10$, $X_2 = 5$, $X_3 = 1$ and $X_4 = 2$.

Applications

4. The stack loss plant data of Brownlee (1965) contains 21 days of measurements from a plant's oxidation of ammonia to nitric acid. The nitric oxide pollutants are captured in an absorption tower. Details of variables are as follows:

   - Y = LOSS = ten times the percentage of ammonia going into the plant that escapes from the absorption column.
   - X_1 = AIRFLOW = Rate of operation of the plant.
   - X_2 = TEMP = Cooling water temperature in the absorption tower.
   - X_3 = ACID = Acid concentration of circulating acid minus 50 times.

The following estimated regression equation relating LOSS to AIRFLOW and TEMP was given.

$$\hat{y} = -50.359 + 0.671x_1 + 1.295x_2$$

a. Estimate sales resulting from an AIRFLOW of 60 and a TEMP of 20.
b. Interpret b_1 and b_2 in this estimated regression equation.

5. A shoe shop developed the following estimated regression equation relating sales to inventory investment and advertising expenditures:

$$\hat{y} = 25 + 10x_1 + 8x_2$$

where:

X_1 = inventory investment (€000)
X_2 = advertising expenditures (€000)
Y = sales (€000)

a. Predict the sales resulting from a €15,000 investment in inventory and an advertising budget of €10,000.
b. Interpret b_1 and b_2 in this estimated regression equation.

6. *PC Magazine* provided ratings for several characteristics of computer monitors, including an overall rating (PC Magazine website). The following data show the rating for contrast ratio, resolution and the overall rating for ten monitors tested using a 0–100 point scale. The highest rated monitor was the BenQ BL3201PH, with an overall rating of 87.

Model	Contrast ratio	Resolution	Overall rating
BenQ BL3201PH	78	89	87
AOC U2868PQU	98	87	86
NEC MultiSync PA322UHD	84	82	85
Acer XB280HK	78	77	82
Asus ROG Swift PG278Q	65	82	82
AOC E1759Fwu	57	78	82
Dell UltraSharp UZ2715H	56	83	81
NEC MultiSync EA244UHD	77	75	79
HP DreamColor Z27x	47	81	77
Dell UltraSharp UZ2315H	55	70	76

Source: PC Magazine website, April, 2015 (www.pcmag.com/reviews/monitors).

MONITOR
RATINGS

a. Develop the estimated regression equation that can be used to predict the overall rating using the contrast ratio rating.
b. Develop the estimated regression equation that can be used to predict the overall rating using both the contrast ratio rating and the resolution rating.
c. Predict the overall rating for a computer monitor computer that has a contrast ratio rating of 85 and a resolution rating of 74.

15.3 MULTIPLE COEFFICIENT OF DETERMINATION

In simple linear regression we showed that the total sum of squares can be partitioned into two components: the sum of squares due to regression and the sum of squares due to error.

The same procedure applies to the sum of squares in multiple regression.

Relationship between SST, SSR and SSE

$$SST = SSR + SSE \tag{15.7}$$

where:

$$SST = \text{total sum of squares} = \Sigma (y_i - \bar{y})^2$$
$$SSR = \text{sum of squares due to regression} = \Sigma (\hat{y}_i - \bar{y})^2$$
$$SSE = \text{sum of squares due to error} = \Sigma (y_i - \hat{y}_i)^2$$

Due to the computational difficulty in computing the three sums of squares, we rely on computer packages to determine those values. The analysis of variance part of the computer output in Figure 15.4 shows the three values for the Eurodistributor problem with two independent variables: SST = 23.900, SSR = 21.601 and SSE = 2.299. With only one independent variable (distance travelled), the computer output in Figure 15.3 shows that SST = 23.900, SSR = 15.871 and SSE = 8.029. The value of SST is the same in both cases because it does not depend on $\hat{y}$ but SSR increases and SSE decreases when a second independent variable (number of deliveries) is added. The implication is that the estimated multiple regression equation provides a better fit for the observed data.

In Chapter 14, we used the coefficient of determination, $R^2 = SSR/SST$, to measure the goodness of fit for the estimated regression equation. The same concept applies to multiple regression. The term multiple coefficient of determination indicates that we are measuring the goodness of fit for the estimated multiple regression equation. The multiple coefficient of determination, denoted R^2, is computed as follows:

Multiple coefficient of determination

$$R^2 = \frac{SSR}{SST} \tag{15.8}$$

The multiple coefficient of determination can be interpreted as the proportion of the variability in the dependent variable that can be explained by the estimated multiple regression equation. Hence, when multiplied by 100, it can be interpreted as the percentage of the variability in Y that can be explained by the estimated regression equation.

In the two-independent-variable Eurodistributor example, with SSR = 21.601 and SST = 23.900, we have:

$$R^2 = \frac{21.601}{23.900} = 0.904$$

Therefore, 90.4 per cent of the variability in travel time Y is explained by the estimated multiple regression equation with distance and number of deliveries as the independent variables. In Figure 15.4, we see that the multiple coefficient of determination is also provided by the computer output; it is denoted by R-sq = 90.4 per cent.

Figure 15.3 shows that the R-sq value for the estimated regression equation with only one independent variable, distance travelled (X_1), is 66.4 per cent. Thus, the percentage of the variability in travel times that is explained by the estimated regression equation increases from 66.4 per cent to 90.4 per cent when number of deliveries is added as a second independent variable. In general, R^2 increases as independent variables are added to the model.

Many analysts prefer adjusting R^2 for the number of independent variables to avoid overestimating the impact of adding an independent variable on the amount of variability explained by the estimated regression equation. With n denoting the number of observations and p denoting the number of independent variables, the **adjusted multiple coefficient of determination** is computed as follows:

Adjusted multiple coefficient of determination

$$\text{adj } R^2 = 1 - (1 - R^2)\frac{n - 1}{n - p - 1} \qquad (15.9)$$

For the Eurodistributor example with $n = 10$ and $p = 2$, we have:

$$\text{adj } R^2 = 1 - (1 - 0.904)\frac{10 - 1}{10 - 2 - 1} = 0.88$$

Therefore, after adjusting for the two independent variables, we have an adjusted multiple coefficient of determination of 0.88. This value, allowing for rounding, corresponds with the value in the computer output in Figure 15.4 of R-sq(adj) = 87.6 per cent.

EXERCISES

Methods

7. In Exercise 1, the following estimated regression equation based on ten observations was presented.

$$\hat{y} = 29.1270 + 0.5906x_1 + 0.4980x_2$$

The values of SST and SSR are 6,724.125 and 6,216.375, respectively.
   a. Find SSE.
   b. Compute R^2.
   c. Compute adj R^2.
   d. Comment on the goodness of fit.

8. In Exercise 2, ten observations were provided for a dependent variable Y and two independent variables X_1 and X_2; for these data $SST = 15,182.9$ and $SSR = 14,052.2$.
   a. Compute R^2.
   b. Compute adj R^2.
   c. Does the estimated regression equation explain a large amount of the variability in the data? Explain.

9. In Exercise 3, the following estimated regression equation based on 30 observations was presented.

$$\hat{y} = 17.6 + 3.8x_1 - 2.3x_2 + 7.6x_3 + 2.7x_4$$

   The values of SST and SSR are 1,805 and 1,760, respectively.
   a. Compute R^2.
   b. Compute adj R^2.
   c. Comment on the goodness of fit.

Applications

10. In Exercise 4, the following estimated regression equation relating LOSS (Y) to AIRFLOW (X_1) and TEMP (X_2) was given.

$$\hat{y} = -50.359 + 0.671x_1 + 1.295x_2$$

   For these data $SST = 2,069.238$ and $SSR = 1,880.443$.
   a. For the estimated regression equation given, compute R^2.
   b. Compute adj R^2.
   c. Does the model appear to explain a large amount of variability in the data? Explain.

11. In Exercise 5, the following estimated regression equation relating sales to inventory investment and advertising expenditures was given.

$$\hat{y} = 25 + 10x_1 + 8x_2$$

   The data used to develop the model came from a survey of ten shops; for those data, SST = 16,000 and SSR = 12,000.
   a. For the estimated regression equation given, compute R^2.
   b. Compute adj R^2.
   c. Does the model appear to explain a large amount of variability in the data? Explain.

15.4 MODEL ASSUMPTIONS

In Section 15.1 we introduced the following multiple regression model:

Multiple regression model

$$Y = \beta_0 + \beta_1 x_1 + \beta_2 x_2 + \cdots + \beta_p x_p + \varepsilon \qquad \text{(15.10)}$$

The assumptions about the error term ε in the multiple regression model parallel those for the simple linear regression model.

Assumptions about the error term in the multiple regression model

$$Y = \beta_0 + \beta_1 x_1 + \beta x_2 + \cdots + \beta_p x_p + \varepsilon$$

1. The error ε is a random variable with mean or expected value of zero; that is, $E(\varepsilon) = 0$.

Implication: For given values of $X_1, X_2, \ldots X_p$, the expected, or average, value of Y is given by:

$$E(Y) = \beta_0 + \beta_1 x_1 + \beta_2 x_2 + \cdots + \beta_p x_p \qquad \textbf{(15.11)}$$

Equation (15.11) is the multiple regression equation we introduced in Section 15.1. In this equation, $E(Y)$ represents the average of all possible values of Y that might occur for the given values of $X_1, X_2, \ldots, X_p$.

2. The variance of ε is denoted by σ^2 and is the same for all values of the independent variables $X_1, X_2, \ldots, X_p$.

Implication: The variance of Y about the regression line equals σ^2 and is the same for all values of $X_1, X_2, \ldots, X_p$.

3. The values of ε are independent.

Implication: The size of the error for a particular set of values for the independent variables is not related to the size of the error for any other set of values.

4. The error ε is a normally distributed random variable reflecting the deviation between the Y value and the expected value of Y given by $\beta_0 + \beta_1 x_1 + \beta_2 x_2 + \cdots + \beta_p x_p$.

Implication: Because $\beta_0, \beta_1, \ldots, \beta_p$ are constants for the given values of $x_1, x_2, \ldots x_p$, the dependent variable Y is also a normally distributed random variable.

To obtain more insight into the form of the relationship given by equation (15.11), consider the following two-independent-variable multiple regression equation:

$$E(Y) = \beta_0 + \beta_1 x_1 + \beta_2 x_2$$

The graph of this equation is a plane in three-dimensional space. Figure 15.5 provides an example of such a graph. Note that the value of ε shown is the difference between the actual Y value and the expected value of y, $E(Y)$, when $X_1 = x_1^*$ and $X_2 = x_2^*$.

FIGURE 15.5

Graph of the regression equation for multiple regression analysis with two independent variables

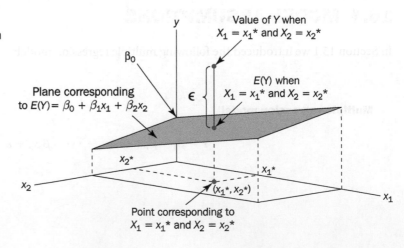

In regression analysis, the term *response variable* is often used in place of the term *dependent variable*. Furthermore, since the multiple regression equation generates a plane or surface, its graph is called a *response surface*.

15.5 TESTING FOR SIGNIFICANCE

In this section we show how to conduct significance tests for a multiple regression relationship.

The significance tests we used in simple linear regression were a t test and an F test. In simple linear regression, both tests provide the same conclusion: that is, if the null hypothesis is rejected, we conclude that the slope parameter $\beta_1 \neq 0$. In multiple regression, the t test and the F test have different purposes.

1 The F test is used to determine whether a significant relationship exists between the dependent variable and the set of all the independent variables; we will refer to the F test as the test for *overall significance*.

2 If the F test shows an overall significance, the t test is used to determine whether each of the individual independent variables is significant. A separate t test is conducted for each of the independent variables in the model; we refer to each of these t tests as a test for *individual significance*.

In the material that follows, we will explain the F test and the t test and apply each to the Eurodistributor Company example.

F test

Given the multiple regression model defined in equation (15.1)

$$Y = \beta_0 + \beta_1 x_1 + \beta_2 x_2 + \cdots + \beta_p x_p + \varepsilon$$

the hypotheses for the F test can be written as follows:

$$H_0: \beta_1 = \beta_2 = \ldots \ldots = \beta_p = 0$$

H_1: One or more of the parameters is not equal to zero

If H_0 is rejected, the test gives us sufficient statistical evidence to conclude that one or more of the parameters is not equal to zero and that the overall relationship between Y and the set of independent variables $X_1, X_2, \ldots X_p$ is significant. However, if H_0 cannot be rejected, we deduce there is not sufficient evidence to conclude that a significant relationship is present.

Before confirming the steps involved in performing the F test, it might be helpful if we first review the concept of *mean square*. A mean square is a sum of squares divided by its corresponding degrees of freedom. In the multiple regression case, the total sum of squares has $n - 1$ degrees of freedom, the sum of squares due to regression (SSR) has p degrees of freedom, and the sum of squares due to error has $n - p - 1$ degrees of freedom. Hence, the mean square due to regression (MSR) is:

Mean square regression

$$\text{MSR} = \frac{\text{SSR}}{p} \qquad \qquad \textbf{(15.12)}$$

and:

Mean square error

$$\text{MSE} = s^2 = \frac{\text{SSE}}{n - p - 1} \tag{15.13}$$

As has already been acknowledged in Chapter 14, MSE provides an unbiased estimate of σ^2, the variance of the error term ε. If $H_0: \beta_1 = \beta_2 = \ldots \ldots = \beta_p = 0$ is true, MSR also provides an unbiased estimate of σ^2, and the value of MSR/MSE should be close to 1. However, if H_0 is false, MSR overestimates σ^2 and the value of MSR/MSE becomes larger. To determine how large the value of MSR/MSE must be to reject H_0, we make use of the fact that if H_0 is true and the assumptions about the multiple regression model are valid, the sampling distribution of MSR/MSE is an F distribution with p degrees of freedom in the numerator and $n - p - 1$ in the denominator. A summary of the F test for significance in multiple regression follows.

F test for overall significance

$$H_0: \beta_1 = \beta_2 = \ldots = \beta_p = 0$$
$$H_1: \text{One or more of the parameters is not equal to zero}$$

Test statistic

$$F = \frac{\text{MSR}}{\text{MSE}} \tag{15.14}$$

Rejection rule

p-value approach:	Reject H_0 if p-value $\leq \alpha$
Critical value approach:	Reject H_0 if $F \geq F_\alpha$

where F_α is based on an F distribution with p degrees of freedom in the numerator and $n - p - 1$ degrees of freedom in the denominator.

Applying the F test to the Eurodistributor Company multiple regression problem with two independent variables, the hypotheses can be written as follows:

$$H_0: \beta_1 = \beta_2 = 0$$
$$H_1: \beta_1 \text{ and/or } \beta_2 \text{ is not equal to zero}$$

Figure 15.6 reproduces the earlier computer output for the multiple regression model with distance (X_1) and number of deliveries (X_2) as the two independent variables. In the analysis of variance part of the output, we see that MSR = 10.8 and MSE = 0.328. Using equation (15.14), we obtain the test statistic:

$$F = \frac{10.8}{0.328} = 32.9$$

Note that the F value on the computer output is $F = 32.88$; the value we calculated differs because we used rounded values for MSR and MSE in the calculation. Using $\alpha = 0.01$, the p-value $= 0.000$ in the last column of the analysis of variance table (Figure 15.6) indicates that we can reject $H_0: \beta_1 = \beta_2 = 0$ because the p-value is less than $\alpha = 0.01$. Alternatively, Table 4 of Appendix B shows that with two degrees of freedom in the numerator and seven degrees of freedom in the denominator, $F_{0.01} = 9.55$. With $32.9 > 9.55$, we reject $H_0: \beta_1 = \beta_2 = 0$ and conclude that a significant relationship is present between travel time Y and the two independent variables, distance and number of deliveries.

FIGURE 15.6
Output for Eurodistributor with two independent variables, distance (X_1) and number of deliveries (X_2)

```
Analysis of Variance

Source        DF    Adj SS    Adj MS    F-value   p-value
Regression     2   21.6006   10.8003     32.88     0.000
Error          7    2.2994    0.3285
Total          9   23.900

Model Summary

      S     R-sq   R-sq(adj)
0.573142   90.38%    87.63%

Coefficients

Term           Coef   SE Coef   t-value   p-value
Constant     -0.869     0.952     -0.91     0.392
Kilometres  0.06113   0.00989      6.18     0.000
Deliveries    0.923     0.221      4.18     0.004

Regression Equation

Time = -0.869 + 0.06113 Kilometres + 0.923 Deliveries
```

As noted previously, the mean square error provides an unbiased estimate of σ^2, the variance of the error term ε. Referring to Figure 15.6, we see that the estimate of σ^2 is MSE = 0.328. The square root of MSE is the estimate of the standard deviation of the error term. As defined in Section 14.5, this standard deviation is called the standard error of the estimate and is denoted s. Hence, we have $s = \sqrt{\text{MSE}} = \sqrt{0.328} = 0.573$. Note that the value of the standard error of the estimate appears in the computer output in Figure 15.6.

Table 15.3 is the general analysis of variance (ANOVA) table that provides the F test results for a multiple regression model.

TABLE 15.3   ANOVA table for a multiple regression model with p independent variables

Source	Degrees of freedom	Sum of squares	Mean square	F
Regression	p	SSR	$MSR = \dfrac{SSR}{p}$	$F = \dfrac{MSR}{MSE}$
Error	$n - p - 1$	SSE	$MSE = \dfrac{SSE}{n - p - 1}$	
Total	$n - 1$	SST		

The value of the F test statistic appears in the last column and can be compared to F_α with p degrees of freedom in the numerator and $n - p - 1$ degrees of freedom in the denominator to make the hypothesis test conclusion.

By reviewing the computer output for Eurodistributor Company in Figure 15.6, we see that the computer analysis of variance table contains this information. In addition, the computer output provides the p-value corresponding to the F test statistic.

t test

If the F test shows that the multiple regression relationship is significant, a t test can be conducted to determine the significance of each of the individual parameters. The t test for individual significance is as follows:

---

t test for individual significance

For any parameter β_i

$$H_0: \beta_i = 0$$
$$H_1: \beta_i \neq 0$$

Test statistic

$$t = \frac{b_i}{s_{b_i}} \tag{15.15}$$

Rejection rule

p-value approach: Reject H_0 if p-value $\leq \alpha$
Critical value approach: Reject H_0 if $t \leq t_{\alpha/2}$ or if $t \geq t_{\alpha/2}$
where $t_{\alpha/2}$ is based on a t distribution with $n - p - 1$ degrees of freedom.

In the test statistic, s_{b_i} is the estimate of the standard deviation of b_i. The value of s_{b_i} will be provided by the computer software package.

Let us conduct the t test for the Eurodistributor regression problem. Refer to the section of Figure 15.6 that shows the computer output for the t-ratio calculations. Values of b_1, b_2, s_{b_2} and s_{b_2} are as follows.

$$b_1 = 0.06113 \qquad s_{b_1} = 0.00989$$
$$b_2 = 0.923 \qquad s_{b_2} = 0.221$$

Using equation (15.15), we obtain the test statistic for the hypotheses involving parameters β_1 and β_2.

$$t = 0.06113/0.00989 = 6.18$$
$$t = 0.923/0.221 = 4.18$$

Note that both of these t-ratio values and the corresponding p-values are provided by the computer output in Figure 15.6. Using $\alpha = 0.01$, the p-values of 0.000 and 0.004 from the computer output indicate that we can reject $H_0: \beta_1 = 0$ and $H_0: \beta_2 = 0$. Hence, both parameters are statistically significant. Alternatively, Table 2 of Appendix B shows that with $n - p - 1 = 10 - 2 - 1 = 7$ degrees of freedom, $t_{0.005} = 3.499$. With $6.18 > 3.499$, we reject $H_0: \beta_1 = 0$. Similarly, with $4.18 > 3.499$, we reject $H_0: \beta_2 = 0$.

Multicollinearity

In multiple regression analysis, **multicollinearity** refers to the correlation between the independent variables. We used the term independent variable in regression analysis to refer to any variable being used to predict or explain the value of the dependent variable. The term does not mean, however, that the independent variables themselves are independent in any statistical sense. On the contrary, most independent variables in a multiple regression problem are correlated to some degree with one another. For example, in the Eurodistributor example involving the two independent variables X_1 (distance) and X_2 (number of deliveries), we could treat the distance as the dependent variable and the number of deliveries as the independent variable to determine whether those two variables are themselves related. We could then compute the sample correlation coefficient to determine the extent to which the variables are related. Doing so yields:

$$\text{Pearson correlation of distance and deliveries} = 0.162$$

which suggests only a small degree of linear association exists between the two variables. The implication from this would be that multicollinearity is not a problem for the data. If however the association had been more pronounced, the resultant multicollinearity might seriously have jeopardized the estimation of the model.

To provide a better perspective of the potential problems of multicollinearity, let us consider a modification of the Eurodistributor example. Instead of X_2 being the number of deliveries, let X_2 denote the number of litres of petrol consumed. Clearly, X_1 (the distance) and X_2 are related, that is, we know that the number of litres of petrol used depends on the distance travelled. Hence, we would conclude logically that X_1 and X_2 are highly correlated independent variables.

Assume that we obtain the equation $\hat{y} = b_0 + b_1 x_1 + b_2 x_2$ and find that the F test shows the relationship to be significant. Then suppose we conduct a t test on β_1 to determine whether $\beta_1 = 0$, and we cannot reject $H_0: \beta_1 = 0$. Does this result mean that travel time is not related to distance? Not necessarily. What it probably means is that with X_2 already in the model, X_1 does not make a significant contribution to determining the value of Y. This interpretation makes sense in our example; if we know the amount of petrol consumed, we do not gain much additional information useful in predicting Y by knowing the distance. Similarly, a t test might lead us to conclude $\beta_2 = 0$ on the grounds that, with X_1 in the model, knowledge of the amount of petrol consumed does not add much.

One useful way of detecting multicollinearity is to calculate the variance inflation factor (VIF) for each independent variable (X_j) in the model. The VIF is defined as:

Variance inflation factor

$$\text{VIF}(X_j) \frac{1}{1 - R_j^2} \tag{15.16}$$

where R_j^2 is the coefficient of determination obtained when X_j $(j = 1, 2, \ldots, p)$ is regressed on all remaining independent variables in the model.

If X_j is not correlated with other predictors, $R_j^2 = 0$ and $\text{VIF} \approx 1$. Correspondingly, if R_j^2 is close to 1, the VIF will be very large. Typically VIF values of ten or more are regarded as problematic.

For the Eurodistributor data, the VIF for X_1 (and also X_2 by symmetry) would be:

$$\text{VIF}(X_j) = \frac{1}{1 - 0.162^2} = 1.027$$

signifying, as before, there is no problem with multicollinearity.

To summarize, for t tests associated with testing for the significance of individual parameters, the difficulty caused by multicollinearity is that it is possible to conclude that none of the individual parameters is significantly different from zero when an F test on the overall multiple regression equation indicates there is a significant relationship. This problem is avoided, however, when little correlation between the independent variables exists.

If possible, every attempt should be made to avoid including independent variables that are highly correlated. In practice, however, strict adherence to this policy is not always possible. When decision-makers have reason to believe substantial multicollinearity is present, they must realize that separating the effects of the individual independent variables on the dependent variable is difficult.

EXERCISES

Methods

12. In Exercise 1, the following estimated regression equation based on ten observations was presented.

$$\hat{y} = 29.1270 + 0.5906x_1 + 0.4980x_2$$

Here SST = 6,724.125, SSR = 6,216.375, s_{b_1} = 0.0813 and s_{b_2} = 0.0567

a. Compute MSR and MSE.
b. Compute F and perform the appropriate F test. Use $\alpha = 0.05$.
c. Perform a t test for the significance of β_1. Use $\alpha = 0.05$.
d. Perform a t test for the significance of β_2. Use $\alpha = 0.05$.

EXER2

13. Refer to the data presented in Exercise 2. The estimated regression equation for these data is

$$\hat{y} = -18.4 + 2.01x_1 + 4.74x_2$$

Here SST = 15,182.9, SSR = 14,052.2, s_{b_1} = 0.2471 and s_{b_2} = 0.9484

a. Test for a significant relationship between X_1, X_2 and Y. Use $\alpha = 0.05$.
b. Is β_1 significant? Use $\alpha = 0.05$.
c. Is β_2 significant? Use $\alpha = 0.05$.

14. In Exercise 4, the following estimated regression equation relating LOSS (Y) to AIRFLOW (X_1) and TEMP (X_2) was given.

$$\hat{y} = -50.359 + 0.671x_1 + 1.295x_2$$

For these data SST = 2,069.238 and SSR = 1,880.443.

Compute SSE, MSE and MSR.
     Use an F test and a 0.05 level of significance to determine whether there is a relationship between the variables.

Applications

15. In Exercise 5, the following estimated regression equation relating sales to inventory investment and advertising expenditures was given.

$$\hat{y} = 25 + 10x_1 + 8x_2$$

The data used to develop the model came from a survey of 10 shops; for these data SST = 16,000 and SSR = 12,000.

a. Compute SSE, MSE and MSR.
b. Use an F test and a 0.05 level of significance to determine whether there is a relationship among the variables.

AUTORESALE

16. The Honda Accord was named the best midsized car for resale value for 2018 by the Kelley Blue Book (Kelley Blue Book website). The file 'AutoResale' contains mileage, age and selling price for a sample of 33 Honda Accords.

a. Develop an estimated regression equation that predicts the selling price of a used Honda Accord given the mileage and age of the car.
b. Is multicollinearity an issue for this model? Find the correlation between the independent variables to answer this question.
c. Use the F test to determine the overall significance of the relationship. What is your conclusion at the 0.05 level of significance?
d. Use the t test to determine the significance of each independent variable. What is your conclusion at the 0.05 level of significance?

15.6 USING THE ESTIMATED REGRESSION EQUATION FOR ESTIMATION AND PREDICTION

The procedures for estimating the mean value of Y and predicting an individual value of Y in multiple regression are similar to those in regression analysis involving one independent variable. First, recall that in Chapter 14 we showed the point estimate of the expected value of Y for a given value of X was the same as the point estimate of an individual value of Y. In both cases, we used $\hat{y} = b_0 + b_1x$ as the point estimate.

In multiple regression we use the same procedure. That is, we substitute the given values of $X_1, X_2, \ldots X_p$ into the estimated regression equation and use the corresponding value of $\hat{y}$ as the point estimate. Suppose that for the Eurodistributor example we want to use the estimated regression equation involving X_1 (distance) and X_2 (number of deliveries) to develop two interval estimates:

1 A *confidence interval* of the mean travel time for all trucks that travel 100 kilometres and make two deliveries.

2 A *prediction interval* of the travel time for *one specific* truck that travels 100 kilometres and makes two deliveries.

Using the estimated regression equation $\hat{y} = -0.869 + 0.0611x_1 + 0.923x_2$ with $X_1 = 100$ and $X_2 = 2$, we obtain the following value of $\hat{y}$:

$$\hat{y} = -0.869 + 0.0611(100) + 0.923(2) = 7.09$$

Hence, the point estimate of travel time in both cases is approximately seven hours.

To develop interval estimates for the mean value of Y and for an individual value of Y, we use a procedure similar to that for regression analysis involving one independent variable.

The formulae required are beyond the scope of the text, but computer packages for multiple regression analysis will often provide confidence intervals once the values of $X_1, X_2, \ldots X_p$ are specified by the user. In Table 15.4 we show the 95% confidence and prediction intervals for the Eurodistributor example for selected values of X_1 and X_2; these values were obtained using a statistics software package. Note that the interval estimate for an individual value of Y is wider than the interval estimate for the expected value of Y. This difference simply reflects the fact that for given values of X_1 and X_2 we can estimate the mean travel time for all trucks with more precision than we can predict the travel time for one specific truck.

TABLE 15.4 The 95% confidence and prediction intervals for Eurodistributor

Value of X_1	Value of X_2	95% Confidence interval		95% Prediction interval	
		Lower limit	Upper limit	Lower limit	Upper limit
100	4	8.135	9.742	7.363	10.514
50	3	4.127	5.789	3.369	6.548
100	4	8.135	9.742	7.363	10.514
100	2	6.258	7.925	5.500	8.683
50	2	3.146	4.924	2.414	5.656
80	2	5.232	6.505	4.372	7.366
75	3	6.037	6.936	5.059	7.915
65	4	5.960	7.637	5.205	8.392
90	3	6.917	7.891	5.964	8.844
90	2	5.776	7.184	4.953	8.007
75	4	6.669	8.152	5.865	8.955

EXERCISES

Methods

EXER2

17. In Exercise 1, the following estimated regression equation based on ten observations was presented.

$$\hat{y} = 29.1270 + 0.5906x_1 + 0.4980x_2$$

a. Develop a point estimate of the mean value of Y when $X_1 = 180$ and $X_2 = 310$.
b. Develop a point estimate for an individual value of Y when $X_1 = 180$ and $X_2 = 310$.

18. Refer to the data in Exercise 2. The estimated regression equation for those data is

$$\hat{y} = -18.4 + 2.01x_1 + 4.74x_2$$

a. Develop a 95% confidence interval for the mean value of Y when $X_1 = 45$ and $X_2 = 15$.
b. Develop a 95% prediction interval for Y when $X_1 = 45$ and $X_2 = 15$.

Applications

19. Refer to Exercise 16. From the estimated regression equation from (a):
a. Estimate the selling price of a four-year-old Honda Accord with mileage of 40,000 miles.
b. Develop a 95% confidence interval for the selling price of a car with the data in (a).
c. Develop a 95% prediction interval for the selling price of a particular car having the data in part a.

15.7 CATEGORICAL INDEPENDENT VARIABLES

Thus far, the examples we considered involved quantitative independent variables such as distance travelled and number of deliveries. In many situations, however, we must work with **categorical independent variables** such as sex (male, female), method of payment (cash, credit card, cheque) and so on. The purpose of this section is to show how categorical variables are handled in regression analysis. To illustrate the use and interpretation of a categorical independent variable, we will consider a problem facing the managers of Johansson Filtration.

An example: Johansson Filtration

Johansson Filtration provides maintenance services for water-filtration systems throughout southern Denmark. Customers contact Johansson with requests for maintenance service on their water-filtration systems. To estimate the service time and the service cost, Johansson's managers wish to predict the repair time necessary for each maintenance request. Hence, repair time in hours is the dependent variable. Repair time is believed to be related to two factors: the number of months since the last maintenance service and the type of repair problem (mechanical or electrical). Data for a sample of ten service calls are reported in Table 15.5.

TABLE 15.5   Data for the Johansson Filtration example

Service call	Months since last service	Type of repair	Repair time in hours
1	2	electrical	2.9
2	6	mechanical	3.0
3	8	electrical	4.8
4	3	mechanical	1.8
5	2	electrical	2.9
6	7	electrical	4.9
7	9	mechanical	4.2
8	8	mechanical	4.8
9	4	electrical	4.4
10	6	electrical	4.5

Let Y denote the repair time in hours and X_1 denote the number of months since the last maintenance service. The regression model that uses only X_1 to predict Y is:

$$Y = \beta_0 + \beta_1 x_1 + \varepsilon$$

Using a statistics software package to develop the estimated regression equation, we obtained the output shown in Figure 15.7. The estimated regression equation is:

$$\hat{y} = 2.15 + 0.304 x_1 \tag{15.17}$$

JOHANSSON

FIGURE 15.7

Output for Johansson Filtration with months since last service (X_1) as the independent variable

```
Analysis of Variance

Source       DF    Adj SS    Adj MS    F-value    p-value
Regression    1     5.596    5.5960       9.17      0.016
Error         8     4.880    0.6100
Total         9    10.476

Model Summary

       S    R-sq    R-sq(adj)
0.781022  53.42%      47.59%

Coefficients

Term                              Coef   SE Coef   t-value   p-value
Constant                         2.147     0.605      3.55     0.008
Months Since Last Service        0.304     0.100      3.03     0.016

Regression Equation

Repair Time (hours) = 2.147 + 0.304 Months Since Last Service
```

At the 0.05 level of significance, the p-value of 0.016 for the t (or F) test indicates that the number of months since the last service is significantly related to repair time. R-sq = 53.4 per cent indicates that X_1 alone explains 53.4 per cent of the variability in repair time.

To incorporate the type of failure into the regression model, we define the following variable:

$$X_2 = 0 \text{ if the type of repair is mechanical}$$
$$X_2 = 1 \text{ if the type of repair is electrical}$$

In regression analysis X_2 is called a **dummy variable** or *indicator variable*. Using this dummy variable, we can write the multiple regression model as:

$$Y = \beta_0 + \beta_1 x_1 + \beta_2 x_2 + \varepsilon$$

Table 15.6 is the revised data set that includes the values of the dummy variable. Using statistics software and the data in Table 15.6, we can develop estimates of the model parameters. The computer output in Figure 15.8 shows that the estimated multiple regression equation is:

$$\hat{y} = 0.93 + 0.388x_1 + 1.26x_2 \tag{15.18}$$

TABLE 15.6 Data for the Johansson Filtration example with type of repair indicated by a dummy variable ($X_2 = 0$ for mechanical; $X_2 = 1$ for electrical)

Customer	Months since last service (X_1)	Type of repair (X_2)	Repair time in hours (Y)
1	2	1	2.9
2	6	0	3.0
3	8	1	4.8
4	3	0	1.8
5	2	1	2.9
6	7	1	4.9
7	9	0	4.2
8	8	0	4.8
9	4	1	4.4
10	6	1	4.5

FIGURE 15.8
Output for Johansson Filtration with months since last service (X_1) and type of repair (X_2) as the independent variables

Analysis of Variance

Source	DF	Adj SS	Adj MS	F-value	p-value
Regression	2	9.0009	4.50046	21.36	0.001
Error	7	1.4751	0.21073		
Total	9	10.4760			

Model Summary

S	R-sq	R-sq(adj)
0.459048	85.92%	81.90%

Coefficients

Term	Coef	SE Coef	t-value	p-value
Constant	0.930	0.467	1.99	0.087
Months Since Last Service	0.3876	0.0626	6.20	0.000
Type of Repair	1.263	0.314	4.02	0.005

Regression Equation

Repair Time (hours) = -0.930 + 0.3876
Month Since Last Service + 1.263 Type of Repair

At the 0.05 level of significance, the p-value of 0.001 associated with the F test ($F = 21.36$) indicates that the regression relationship is significant. The t test part of the printout in Figure 15.8 shows that both months since last service (p-value = 0.000) and type of repair (p-value = 0.005) are statistically significant. In addition, R-sq $= 85.9$ per cent and R-sq(adj) $= 81.9$ per cent indicate that the estimated regression equation does a good job of explaining the variability in repair times. Thus, equation (15.18) should prove helpful in estimating the repair time necessary for the various service calls.

Interpreting the parameters

The multiple regression equation for the Johansson Filtration example is:

$$E(Y) = \beta_0 + \beta_1 x_1 + \beta_2 x_2 \qquad \textbf{(15.19)}$$

To understand how to interpret the parameters β_0, β_1 and β_2 when a categorical variable is present, consider the case when $X_2 = 0$ (mechanical repair). Using $E(Y \mid \text{mechanical})$ to denote the mean or expected value of repair time *given* a mechanical repair, we have:

$$E(Y \mid \text{mechanical}) = \beta_0 + \beta_1 x_1 + \beta_2(0) = \beta_0 + \beta_1 x_1 \qquad \textbf{(15.20)}$$

Similarly, for an electrical repair $(X_2 = 1)$, we have:

$$\begin{aligned} E(Y \mid \text{electrical}) &= \beta_0 + \beta_1 x_1 + \beta_2(1) = \beta_0 + \beta_1 x_1 + \beta_2 \\ &= (\beta_0 + \beta_2) + \beta_1 x_1 \end{aligned} \qquad \textbf{(15.21)}$$

Comparing equations (15.20) and (15.21), we see that the mean repair time is a linear function of X_1 for both mechanical and electrical repairs. The slope of both equations is β_1, but the y-intercept differs. The y-intercept is β_0 in equation (15.20) for mechanical repairs and $(\beta_0 + \beta_2)$ in equation (15.21) for electrical repairs. The interpretation of β_2 is that it indicates the difference between the mean repair time for an electrical repair and the mean repair time for a mechanical repair.

If β_2 is positive, the mean repair time for an electrical repair will be greater than that for a mechanical repair; if β_2 is negative, the mean repair time for an electrical repair will be less than that for a mechanical repair. Finally, if $\beta_2 = 0$, there is no difference in the mean repair time between electrical and mechanical repairs, and the type of repair is not related to the repair time.

Using the estimated multiple regression equation $\hat{y} = 0.93 + 0.388 x_1 + 1.26 x_2$, we see that 0.93 is the estimate of β_0 and 1.26 is the estimate of β_2. Thus, when $X_2 = 0$ (mechanical repair):

$$\hat{y} = 0.93 + 0.388 x_1 \qquad \textbf{(15.22)}$$

and when $X_2 = 1$ (electrical repair):

$$\begin{aligned} \hat{y} &= 0.93 + 0.388 x_1 + 1.26(1) \\ &= 2.19 + 0.388 x_1 \end{aligned} \qquad \textbf{(15.23)}$$

In effect, the use of a dummy variable for type of repair provides two equations that can be used to predict the repair time, one corresponding to mechanical repairs and one corresponding to electrical repairs. In addition, with $b_2 = 1.26$, we learn that, on average, electrical repairs require 1.26 hours longer than mechanical repairs.

Figure 15.9 is the plot of the Johansson data from Table 15.6. Repair time in hours (Y) is represented by the vertical axis and months since last service (X_1) is represented by the horizontal axis. A data point for a mechanical repair is indicated by an M and a data point for an electrical repair is indicated by an E. Equations (15.22) and (15.23) are plotted on the graph to show graphically the two equations that can be used to predict the repair time, one corresponding to mechanical repairs and one corresponding to electrical repairs.

FIGURE 15.9
Scatter diagram for the Johansson Filtration repair data from Table 15.6

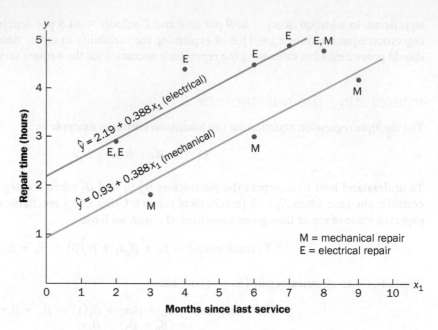

More complex categorical variables

As the categorical variable for the Johansson Filtration example had two levels (mechanical and electrical), defining a dummy variable with zero indicating a mechanical repair and one indicating an electrical repair was easy. However, when a categorical variable has more than two levels, care must be taken in both defining and interpreting the dummy variables. As we will show, if a categorical variable has k levels, $k - 1$ dummy variables are required, with each dummy variable being coded as 0 or 1.

For example, suppose a manufacturer of copy machines organized the sales territories for a particular area into three regions: A, B and C. The managers want to use regression analysis to help predict the number of copiers sold per week. With the number of units sold as the dependent variable, they are considering several independent variables (the number of sales personnel, advertising expenditures and so on). Suppose the managers believe sales region is also an important factor in predicting the number of copiers sold. As the sales region is a categorical variable with three levels, A, B and C, we will need $3 - 1 = 2$ dummy variables to represent the sales region. Each variable can be coded 0 or 1 as follows:

$$X_1 = \begin{cases} 1 & \text{if sales region B} \\ 0 & \text{otherwise} \end{cases}$$

$$X_2 = \begin{cases} 1 & \text{if sales region C} \\ 0 & \text{otherwise} \end{cases}$$

With this definition, we have the following values of X_1 and X_2:

Region	X_1	X_2
A	0	0
B	1	0
C	0	1

Observations corresponding to region A would be coded $X_1 = 0$, $X_2 = 0$; observations corresponding to region B would be coded $X_1 = 1$, $X_2 = 0$; and observations corresponding to region C would be coded $X_1 = 0$, $X_2 = 1$.

The regression equation relating the expected value of the number of units sold, $E(Y)$, to the dummy variables would be written as:

$$E(Y) = \beta_0 + \beta_1 x_1 + \beta_2 x_2$$

To help us interpret the parameters β_0, β_1 and β_2, consider the following three variations of the regression equation.

$$E(Y \mid \text{region A}) = \beta_0 + \beta_1(0) + \beta_2(0) = \beta_0$$
$$E(Y \mid \text{region B}) = \beta_0 + \beta_1(1) + \beta_2(0) = \beta_0 + \beta_1$$
$$E(Y \mid \text{region C}) = \beta_0 + \beta_1(0) + \beta_2(1) = \beta_0 + \beta_2$$

Therefore, β_0 is the mean or expected value of sales for region A; β_1 is the difference between the mean number of units sold in region B and the mean number of units sold in region A; and β_2 is the difference between the mean number of units sold in region C and the mean number of units sold in region A.

Two dummy variables were required because sales region is a categorical variable with three levels. But the assignment of $X_1 = 0$, $X_2 = 0$ to indicate region A, $X_1 = 1$, $X_2 = 0$ to indicate region B, and $X_1 = 0$, $X_2 = 1$ to indicate region C was arbitrary. For example, we could have chosen $X_1 = 1$, $X_2 = 0$ to indicate region A, $X_1 = 0$, $X_2 = 0$ to indicate region B, and $X_1 = 0$, $X_2 = 1$ to indicate region C. In that case, β_1 would have been interpreted as the mean difference between regions A and B and β_2 as the mean difference between regions A and C.

EXERCISES

Methods

20. Consider a regression study involving a dependent variable Y, a quantitative independent variable X_1 and a categorical variable with two levels (level 1 and level 2).
    a. Write a multiple regression equation relating X_1 and the categorical variable to Y.
    b. What is the expected value of Y corresponding to level 1 of the categorical variable?
    c. What is the expected value of Y corresponding to level 2 of the categorical variable?
    d. Interpret the parameters in your regression equation.

21. Consider a regression study involving a dependent variable Y, a quantitative independent variable X_1, and a categorical independent variable with three possible levels (level 1, level 2 and level 3).
    a. How many dummy variables are required to represent the categorical variable?
    b. Write a multiple regression equation relating X_1 and the categorical variable to Y.
    c. Interpret the parameters in your regression equation.

Applications

22. Management proposed the following regression model to predict the effect of physical exercise on pulse in an experiment involving 92 participants:

$$Y = \beta_0 + \beta_1 x_1 + \beta_2 x_2 + \beta_3 x_3 + \varepsilon$$

where:

$Y = $ Pulse 2 $ = $ second pulse reading taken at end of experiment
$x_1 = $ Pulse 1 $ = $ initial resting pulse reading
$x_2 = $ Ran $ = 1$ if individual ran on the spot for one minute, 2 if they did not (this was decided randomly)
$x_3 = $ Sex $ = 1$ if male, 2 if female

The following estimated regression equation was developed using a statistics software package:

$$\hat{y} = 42.62 + 0.812 x_1 - 20.1 x_2 + 7.8 x_3$$

a. What is the amount of the expected value of Pulse 2 attributable to x_3?
b. Predict Pulse 2 for a female participant who ran on the spot for one minute and had an initial pulse reading of 70 bpm (beats per minute).
c. Predict Pulse 2 for a male participant who did not run on the spot for one minute and had an initial pulse reading of 60 bpm (beats per minute).

REPAIR

23. Refer to the Johansson Filtration problem introduced in this section. Suppose that in addition to information on the number of months since the machine was serviced and whether a mechanical or an electrical failure had occurred, the managers obtained a list showing which engineer performed the service. The revised data follow.

Repair time in hours	Months since last service	Type of repair	Engineer
2.9	2	Electrical	Heinz Kolb
3.0	6	Mechanical	Heinz Kolb
4.8	8	Electrical	Wolfgang Linz
1.8	3	Mechanical	Heinz Kolb
2.9	2	Electrical	Heinz Kolb
4.9	7	Electrical	Wolfgang Linz
4.2	9	Mechanical	Wolfgang Linz
4.8	8	Mechanical	Wolfgang Linz
4.4	4	Electrical	Wolfgang Linz
4.5	6	Electrical	Heinz Kolb

a. Ignore for now the months since the last maintenance service (X_1) and the engineer who performed the service. Develop the estimated simple linear regression equation to predict the repair time (Y) given the type of repair (X_2). Recall that $X_2 = 0$ if the type of repair is mechanical and 1 if the type of repair is electrical.
b. Does the equation that you developed in part (a) provide a good fit for the observed data? Explain.
c. Ignore for now the months since the last maintenance service and the type of repair associated with the machine. Develop the estimated simple linear regression equation to predict the repair time given the engineer who performed the service. Let $X_3 = 0$ if Heinz Kolb performed the service and $X_3 = 1$ if Wolfgang Linz performed the service.
d. Does the equation that you developed in part (c) provide a good fit for the observed data? Explain.

24. In a multiple regression analysis by McIntyre (1994), tar, nicotine and weight are considered as possible predictors of carbon monoxide (CO) content for 25 different brands of cigarette. Details of variables and data follow.

Brand	The cigarette brand
Tar	The tar content (in mg)
Nicotine	The nicotine content (in mg)
Weight	The weight (in g)
CO	The carbon monoxide (CO) content (in mg)

Brand	Tar	Nicotine	Weight	CO
Alpine	14.1	0.86	0.9853	13.6
Benson & Hedges	16.0	1.06	1.0938	16.6
Bull Durham	29.8	2.03	1.1650	23.5
Camel Lights	8.0	0.67	0.9280	10.2
Carlton	4.1	0.40	0.9462	5.4
Chesterfield	15.0	1.04	0.8885	15.0
Golden Lights	8.8	0.76	1.0267	9.0

Brand	Tar	Nicotine	Weight	CO
Kent	12.4	0.95	0.9225	12.3
Kool	16.6	1.12	0.9372	16.3
L&M	14.9	1.02	0.8858	15.4
Lark Lights	13.7	1.01	0.9643	13.0
Marlboro	15.1	0.90	0.9316	14.4
Merit	7.8	0.57	0.9705	10.0
Multi Filter	11.4	0.78	1.1240	10.2
Newport Lights	9.0	0.74	0.8517	9.5
Now	1.0	0.13	0.7851	1.5
Old Gold	17.0	1.26	0.9186	18.5
Pall Mall Light	12.8	1.08	1.0395	12.6
Raleigh	15.8	0.96	0.9573	17.5
Salem Ultra	4.5	0.42	0.9106	4.9
Tareyton	14.5	1.01	1.0070	15.9
True	7.3	0.61	0.9806	8.5
Viceroy Rich Light	8.6	0.69	0.9693	10.6
Virginia Slims	15.2	1.02	0.9496	13.9
Winston Lights	12.0	0.82	1.1184	14.9

a. Examine correlations between variables in the study and hence assess the possibility of problems of multicollinearity affecting any subsequent regression model involving independent variables tar and nicotine.

b. Thus develop an estimated multiple regression equation using an appropriate number of the independent variables featured in the study.

c. Are your predictors statistically significant? Use $\alpha = 0.05$. What explanation can you give for the results observed?

CIGARETTES

25. Management proposed the following regression model to predict sales at a fast food outlet:

$$Y = \beta_0 + \beta_1 x_1 + \beta_2 x_2 + \beta_3 x_3 + \varepsilon$$

where

$X_1 =$ number of competitors within one kilometre

$X_2 =$ population within one kilometre (1,000s)

$X_3 = \begin{cases} 1 \text{ if drive-up window present} \\ 0 \text{ otherwise} \end{cases}$

$Y =$ sales ($1,000s)

The following estimated regression equation was developed after 20 outlets were surveyed:

$$\hat{y} = 10.1 - 4.2x_1 + 6.8x_2 + 15.3x_3$$

a. What is the expected amount of sales attributable to the drive-up window?

b. Predict sales for a store with two competitors, a population of 8,000 within one kilometre and no drive-up window.

c. Predict sales for a store with one competitor, a population of 3,000 within one kilometre and a drive-up window.

15.8 RESIDUAL ANALYSIS

In Chapter 14 we pointed out that standardized residuals were frequently used in residuals plots and in the identification of outliers. The general formula for the standardized residual for observation i is as follows:

Standardized residual for observation i

$$\frac{y_i - \hat{y}_i}{s_{y_i - \hat{y}_i}}$$

(15.24)

where:

$s_{y_i - \hat{y}_i}$ = the standard deviation of residual i

The general formula for the standard deviation of residual i is defined as follows:

Standard deviation of residual i

$$s_{y_i - \hat{y}_i} = s\sqrt{1 - h_1}$$

(15.25)

where:

s = standard error of the estimate
h_i = leverage of observation i

As we saw in Section 14.10, the **leverage** of an observation provides a measure of how far the value of an independent variable is from its mean. The computation of h_i, $s_{y_i - \hat{y}_i}$ and hence the standardized residual for observation i in multiple regression analysis is too complex to be done by hand. However, the standardized residuals can be easily obtained as part of the output from statistical software packages. Table 15.7 lists the predicted values, the residuals and the standardized residuals for the Eurodistributor example presented previously in this chapter; we obtained these values by using a statistical software package. The predicted values in the table are based on the estimated regression equation:

$$\hat{y} = -0.869 + 0.0611x_1 + 0.923x_2$$

TABLE 15.7 Residuals and standardized residuals for the Eurodistributor regression analysis

Distance travelled (x_1)	Deliveries (x_2)	Travel time (Y)	Predicted value ($\hat{y}$)	Residual ($y - \hat{y}$)	Standardized residual
100	4	9.3	8.93846	0.361540	0.78344
50	3	4.8	4.95830	−0.158305	−0.34962
100	4	8.9	8.93846	−0.038460	−0.08334
100	2	6.5	7.09161	−0.591609	−1.30929
50	2	4.2	4.03488	0.165121	0.38167
80	2	6.2	5.86892	0.331083	0.65431
75	3	7.4	6.48667	0.913330	1.68917
65	4	6.0	6.79875	−0.798749	−1.77372
90	3	7.6	7.40369	0.196311	0.36703
90	2	6.1	6.48026	−0.380263	−0.77639

The standardized residuals and the predicted values of *Y* from Table 15.7 are used in the standardized residual plot in Figure 15.10.

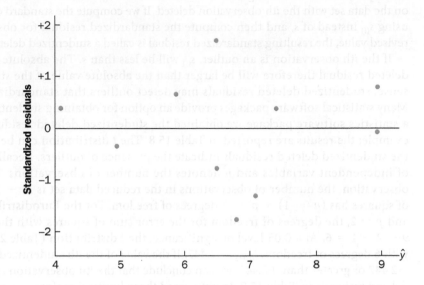

FIGURE 15.10
Standardized residual plot for Eurodistributor

This standardized residual plot does not indicate any unusual abnormalities. Also, all of the standardized residuals are between −2 and +2; hence, we have no reason to question the assumption that the error term ε is normally distributed. We conclude that the model assumptions are reasonable.

A normal probability plot can also be used to determine whether the distribution of ε appears to be normal. The procedure and interpretation for a normal probability plot were discussed in Section 14.8. The same procedure is appropriate for multiple regression. Again, we would use a statistical software package to perform the computations and provide the normal probability plot.

Detecting outliers

An outlier is an observation that is unusual in comparison with the other data; in other words, an outlier does not fit the pattern of the other data. In Chapter 14 we showed an example of an outlier and discussed how standardized residuals can be used to detect outliers.

Statistics software classifies an observation as an outlier if the value of its standardized residual is less than −2 or greater than +2. Applying this rule to the standardized residuals for the Eurodistributor example (refer to Table 15.7), we do not detect any outliers in the data set.

In general, the presence of one or more outliers in a data set tends to increase *s*, the standard error of the estimate, and hence increase $s_{y_i - \hat{y}_i}$, the standard deviation of residual *i*. As $s_{y_i - \hat{y}_i}$ appears in the denominator of the expression for the standardized residual (15.24), the size of the standardized residual will decrease as *s* increases.

As a result, even though a residual may be unusually large, the large denominator in expression (15.24) may cause the standardized residual rule to fail to identify the observation as being an outlier. We can circumvent this difficulty by using a form of standardized residuals called **studentized deleted residuals**.

Studentized deleted residuals and outliers

Suppose the ith observation is deleted from the data set and a new estimated regression equation is developed with the remaining $n - 1$ observations. Let $s_{(i)}$ denote the standard error of the estimate based on the data set with the ith observation deleted. If we compute the standard deviation of residual i (15.25) using $s_{(i)}$ instead of s, and then compute the standardized residual for observation i (15.24) using the revised value, the resulting standardized residual is called a studentized deleted residual.

If the ith observation is an outlier, $s_{(i)}$ will be less than s. The absolute value of the ith studentized deleted residual therefore will be larger than the absolute value of the standardized residual. In this sense, studentized deleted residuals may detect outliers that standardized residuals do not detect. Many statistical software packages provide an option for obtaining studentized deleted residuals. Using a statistics software package we obtained the studentized deleted residuals for the Eurodistributor example; the results are reported in Table 15.8. The t distribution can be used to determine whether the studentized deleted residuals indicate the presence of outliers. Recall that p denotes the number of independent variables and n denotes the number of observations. Hence, if we delete the ith observation, the number of observations in the reduced data set is $n - 1$; in this case the error sum of squares has $(n - 1) - p - 1$ degrees of freedom. For the Eurodistributor example with $n = 10$ and $p = 2$, the degrees of freedom for the error sum of squares with the ith observation deleted is $9 - 2 - 1 = 6$. At a 0.05 level of significance, the t distribution (Table 2 of Appendix B) shows that with 6 degrees of freedom, $t_{0.025} = 2.447$. If the value of the ith studentized deleted residual is less than -2.447 or greater than $+2.447$, we can conclude that the ith observation is an outlier. The studentized deleted residuals in Table 15.8 do not exceed those limits; therefore, we conclude that outliers are not present in the data set.

TABLE 15.8   Studentized deleted residuals for Eurodistributor

Distance travelled (X_1)	Deliveries (X_2)	Travel time (Y)	Standardized residual	Studentized deleted residual
100	4	9.3	0.78344	0.75938
50	3	4.8	-0.34962	-0.32654
100	4	8.9	-0.08334	-0.07720
100	2	6.5	-1.30929	-1.39494
50	2	4.2	0.38167	0.35709
80	2	6.2	0.65431	0.62519
75	3	7.4	1.68917	2.03187
65	4	6.0	-1.77372	-2.21314
90	3	7.6	0.36703	0.34312
90	2	6.1	-0.77639	-0.75190

Influential observations

In Chapter 14, Section 14.10, we discussed how the leverage of an observation can be used to identify observations for which the value of the independent variable may have a strong influence on the regression results. Statistics software computes the leverage values and uses the rule of thumb:

$$h_i > 3(p + 1)/n$$

to identify influential observations. For the Eurodistributor example with $p = 2$ independent variables and $n = 10$ observations, the critical value for leverage is $3(2 + 1)/10 = 0.9$. The leverage values for the Eurodistributor example obtained by using a statistics software package are reported in Table 15.9. As h_i does not exceed 0.9 anywhere here, no influential observations in the data set are detected.

TABLE 15.9 Leverage and Cook's distance measures for Eurodistributor

Distance travelled (X_1)	Deliveries (X_2)	Travel time (Y)	Leverage (h_i)	Cook's D (D_i)
100	4	9.3	0.351704	0.110994
50	3	4.8	0.375863	0.024536
100	4	8.9	0.351704	0.001256
100	2	6.5	0.378451	0.347923
50	2	4.2	0.430220	0.036663
80	2	6.2	0.220557	0.040381
75	3	7.4	0.110009	0.117561
65	4	6.0	0.382657	0.650029
90	3	7.6	0.129098	0.006656
90	2	6.1	0.269737	0.074217

Using Cook's distance measure to identify influential observations

A problem that can arise in using leverage to identify influential observations is that an observation can be identified as having high leverage and not necessarily be influential in terms of the resulting estimated regression equation.

Table 15.10 shows a data set consisting of eight observations and their corresponding leverage values. As the leverage for the eighth observation is $0.91 > 0.75$ (the critical leverage value), this observation is identified as influential. Before reaching any final conclusions, however, let us consider the situation from a different perspective.

TABLE 15.10
Data set illustrating potential problem using the leverage criterion

x_i	y_i	Leverage h_i
1	18	0.204170
1	21	0.204170
2	22	0.164205
3	21	0.138141
4	23	0.125977
4	24	0.125977
5	26	0.127715
15	39	0.909644

Figure 15.11 shows the scatter diagram and the estimated regression equation corresponding to the data set in Table 15.10. We used a statistics software package to obtain the following estimated regression equation for these data:

$$\hat{y} = 18.2 + 1.39x$$

The straight line in Figure 15.11 is the graph of this equation. Now, let us delete the highly leveraged observation $X = 15$, $Y = 39$ from the data set and fit a new estimated regression equation to the remaining seven observations; the new estimated regression equation is:

$$\hat{y} = 18.1 + 1.42x$$

We note that the y-intercept and slope of the new estimated regression equation are not fundamentally different from the values obtained by using all the data. Although the leverage criterion identified the eighth observation as influential, this observation clearly had little influence on the results obtained. Thus, in some situations using only leverage to identify influential observations can lead to wrong conclusions.

Cook's distance measure uses both the leverage of observation i, h_i, and the residual for observation $i,(y_i - \hat{y}_i)$, to determine whether the observation is influential.

FIGURE 15.11
Scatter diagram for the data set in Table 15.10

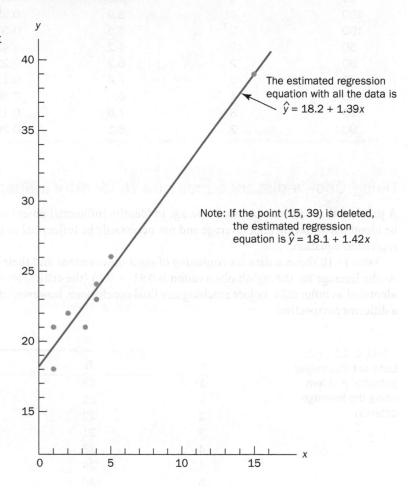

The estimated regression equation with all the data is $\hat{y} = 18.2 + 1.39x$

Note: If the point (15, 39) is deleted, the estimated regression equation is $\hat{y} = 18.1 + 1.42x$

Cook's distance measure

$$D_i = \frac{(y_i - \hat{y}_i)^2 h_i}{(p - 1)s^2(1 - h_i)^2} \qquad (15.26)$$

where:

$$D_i = \text{Cook's distance measure for observation } i$$
$$y_i - \hat{y}_i = \text{the residual for observation } i$$
$$h_i = \text{the leverage for observation } i$$
$$p = \text{the number of independent variables}$$
$$s = \text{the standard error of the estimate}$$

The value of Cook's distance measure will be large and indicate an influential observation if the residual or the leverage is large. As a rule of thumb, values of $D_i > 1$ indicate that the ith observation is influential and should be studied further. The last column of Table 15.9 provides Cook's distance measure for the Eurodistributor problem. Observation 8 with $D_i = 0.650029$ has the most influence. However, applying the rule $D_i > 1$, we should not be concerned about the presence of influential observations in the Eurodistributor data set. Note that in the case of the the eighth observation in the Table 15.10 dataset, the Cook's distance is 0.056, so again influence does not appear to be an issue here either.

EXERCISES

Methods

26. Data for two variables, X and Y, follow.

x_i	1	2	3	4	5
y_i	3	7	5	11	14

   a. Develop the estimated regression equation for these data.
   b. Plot the standardized residuals versus $\hat{y}$. Do there appear to be any outliers in these data? Explain.
   c. Compute the studentized deleted residuals for these data. At the 0.05 level of significance, can any of these observations be classified as an outlier? Explain.

27. Data for two variables, X and Y, follow.

x_i	22	24	26	28	40
y_i	12	21	31	35	70

   a. Develop the estimated regression equation for these data.
   b. Compute the studentized deleted residuals for these data. At the 0.05 level of significance, can any of these observations be classified as an outlier? Explain.
   c. Compute the leverage values for these data. Do there appear to be any influential observations in these data? Explain.
   d. Compute Cook's distance measure for these data. Are any observations influential? Explain.

Applications

28. A regression analysis of ice cream sales (Kadiyala, 1970) is based on the following variables:

Period	The week of the study
Consumption	The ice cream consumption (in pints per capita)
Price	The price of ice cream (in dollars)
Income	The weekly family income (in dollars)
Temp	The mean temperature (in degrees F)

with corresponding statistic software results as follows:

Regression

Model summary[b]

Model	R	R square	Adjusted R square	Std. error of the estimate	Durbin–Watson
1	0.848[a]	0.719	0.687	0.03683	1.021

[a] Predictors: (Constant), Temperature, Price, Income
[b] Dependent variable: Consumption

ANOVA[b]

Model		Sum of squares	df	Mean square	F	Sig.
1	Regression	0.090	3	0.030	22.175	0.000[a]
	Residual	0.035	26	0.001		
	Total	0.126	29			

[a] Predictors: (Constant), Temperature, Price, Income
[b] Dependent variable: Consumption

Coefficients[a]

Model		Unstandardized coefficients		Standardized coefficients	t	Sig.
		B	Std. error	Beta		
1	(Constant)	0.197	0.270		0.730	0.472
	Price	−1.044	0.834	−0.132	−1.252	0.222
	Income	0.003	0.001	0.314	2.824	0.009
	Temperature	0.003	0.000	0.863	7.762	0.000

[a] Dependent variable: Consumption

Charts

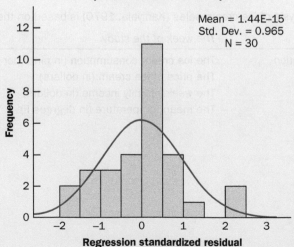

Histogram
Dependent variable: Consumption

Mean = 1.44E–15
Std. Dev. = 0.965
N = 30

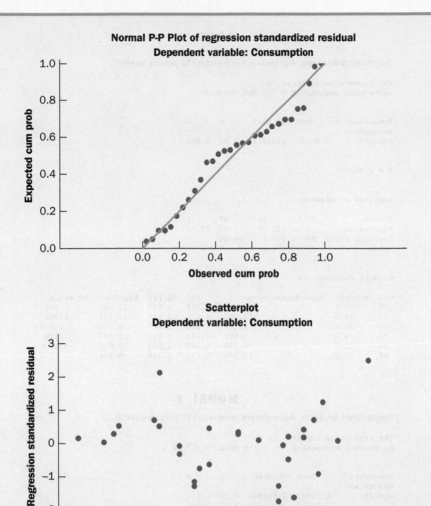

Normal P-P Plot of regression standardized residual
Dependent variable: Consumption

Scatterplot
Dependent variable: Consumption

Explain this computer output, carrying out any additional tests you think necessary or appropriate. State your assumptions.

29. In a study by Tomkinson and Freeman (2011), 178 dissertations were assessed by two markers. In 57 cases where the difference in marks was greater than or equal to 8 marks, a third marker was brought in to resolve the conflict and determine an agreed mark (*Agreedmark*). The average mark for the first and second markers (*meanmark12*) was calculated, also the absolute difference (*absdiff*) between the marks for Marker 1 and Marker 2.

In MODEL 1 the dependent variable *Agreedmark-meanmark12* (= Agreed mark – meanmark12) is regressed on the *absdiff* variable.

Subsequently it was found that one third marker had a much lower mean mark than other third markers. To allow for this 'rogue' third marker, a revised model (MODEL 2) was fitted to the data set with the results given below. Note that D here is a dummy variable which takes the value 1 for the rogue marker in question (corresponding to cases 10–15 in the data set) and 0 for all other third markers.

a. Explain this computer output, carrying out any additional tests you think necessary or appropriate.

b. Which model (if either) do you prefer and why? State your assumptions.

MODEL 1

Regression Analysis: Agreedmark-meanmark12 versus absdiff

```
The regression equation is
Agreedmark-meanmark12 = - 0.211 absdiff


Predictor      Coef  SE Coef      T      P
Noconstant
absdiff     -0.21143  0.04021  -5.26  0.000


S = 4.96618


Analysis of Variance

Source       DF       SS      MS      F      P
Regression    1   681.87  681.87  27.65  0.000
Residual Error 56 1,381.13  24.66
Total         57 2,063.00


Unusual Observations

Obs  absdiff  Agreedmark-meanmark12     Fit  SE Fit  Residual  St Resid
  7    17.0                  8.500  -3.594   0.684    12.094     2.46R
 10    24.0                -18.500  -5.074   0.965   -13.426    -2.76R
 20    35.5                 -0.750  -7.506   1.427     6.756     1.42 X
 26    16.0                  7.000  -3.383   0.643    10.383     2.11R
 33    31.5                 -6.750  -6.660   1.267    -0.090    -0.02 X
 47    29.0                -11.500  -6.132   1.166    -5.368    -1.11 X
```

MODEL 2

Regression Analysis: Agreedmark-meanmark12 versus absdiff, D

```
The regression equation is
Agreedmark-meanmark12 = - 0.158 absdiff - 7.52 D


Predictor      Coef  SE Coef      T      P
Noconstant
absdiff     -0.15795  0.03854  -4.10  0.000
D             -7.519    1.943  -3.87  0.000


S = 4.44270


Analysis of Variance

Source       DF       SS      MS      F      P
Regression    2   977.43  488.72  24.76  0.000
Residual Error 55 1,085.57  19.74
Total         57 2,063.00


Source   DF  Seq SS
absdiff   1  681.87
D         1  295.56


Unusual Observations

Obs  absdiff  Agreedmark-meanmark12     Fit  SE Fit  Residual  St Resid
  7    17.0                  8.500  -2.685   0.655    11.185     2.55R
  9    27.0                -13.500 -11.783   1.846    -1.717    -0.42 X
 10    24.0                -18.500 -11.310   1.828    -7.190    -1.78 X
 11    12.0                 -6.500  -9.414   1.829     2.914     0.72 X
 12    10.0                 -2.000  -9.098   1.840     7.098     1.76 X
 13    26.5                -14.750 -11.704   1.842    -3.046    -0.75 X
 14     9.0                 -7.000  -8.940   1.847     1.940     0.48 X
 26    16.0                  7.000  -2.527   0.617     9.527     2.17R
 34    20.0                -13.000  -3.159   0.771    -9.841    -2.25R


R denotes an observation with a large standardized residual.
X denotes an observation whose X value gives it large leverage.
```

15.9 LOGISTIC REGRESSION

In many regression applications the dependent variable may only assume two discrete values. For instance, a bank might like to develop an estimated regression equation for predicting whether a person will be approved for a credit card. The dependent variable can be coded as $Y = 1$ if the bank approves the request for a credit card and $Y = 0$ if the bank rejects the request for a credit card. Using logistic regression we can estimate the probability that the bank will approve the request for a credit card given a particular set of values for the chosen independent variables.

Consider an application of logistic regression involving a direct mail promotion being used by Stamm Stores. Stamm owns and operates a national chain of fashion stores. Five thousand copies of an expensive four-colour sales catalogue have been printed, and each catalogue includes a coupon that provides a €50 discount on purchases of €200 or more.

The catalogues are expensive and Stamm would like to send them to only those customers who have the highest probability of making a €200 purchase using the discount coupons.

Management thinks that annual spending at Stamm Stores and whether a customer has a Stamm credit card are two variables that might be helpful in predicting whether a customer who receives the catalogue will use the coupon to make a €200 purchase. Stamm conducted a pilot study using a random sample of 50 Stamm credit card customers and 50 other customers who do not have a Stamm credit card. Stamm sent the catalogue to each of the 100 customers selected. At the end of a test period, Stamm noted whether the customer made a purchase (coded 1 if the customer made a purchase and 0 if not). The sample data for the first ten catalogue recipients are shown in Table 15.11. The amount each customer spent last year at Stamm is shown in thousands of euros and the credit card information has been coded as 1 if the customer has a Stamm credit card and 0 if not. In the Purchase column a 1 is recorded if the sampled customer used the €50 discount coupon to make a purchase of €200 or more.

TABLE 15.11 Sample data for Stamm Stores

Customer	Annual spending (€000)	Stamm card	Purchase
1	2.291	1	0
2	3.215	1	0
3	2.135	1	0
4	3.924	0	0
5	2.528	1	0
6	2.473	0	1
7	2.384	0	0
8	7.076	0	0
9	1.182	1	1
10	3.345	0	0

STAMM

We might think of building a multiple regression model using the data in Table 15.11 to help Stamm predict whether a catalogue recipient will make a purchase. We would use Annual spending and Stamm card as independent variables and Purchase as the dependent variable.

Because the dependent variable may only assume the values of 0 or 1, however, the ordinary multiple regression model is not applicable. This example shows the type of situation for which logistic regression was developed. Let us see how logistic regression can be used to help Stamm predict which type of customer is most likely to take advantage of their promotion.

Logistic regression equation

In many ways logistic regression is like ordinary regression. It requires a dependent variable, Y, and one or more independent variables. In multiple regression analysis, the mean or expected value of Y, is referred to as the multiple regression equation.

$$E(Y) = \beta_0 + \beta_1 x_1 + \beta_2 x_2 + \cdots + \beta_p x_p \tag{15.27}$$

In logistic regression, statistical theory as well as practice has shown that the relationship between $E(Y)$ and $X_1, X_2, \ldots X_p$ is better described by the following nonlinear equation:

Logistic regression equation

$$E(Y) = \frac{e^{\beta_0 + \beta_1 x_1 + \beta_2 x_2 + \ldots + \beta_p x_p}}{1 + e^{\beta_0 + \beta_1 x_1 + \beta_2 x_2 + \ldots + \beta_p x_p}} \tag{15.28}$$

If the two values of the dependent variable Y are coded as 0 or 1, the value of $E(Y)$ in equation (15.28) provides the *probability* that $Y = 1$ given a particular set of values for the independent variables $X_1, X_2, \ldots X_p$. Because of the interpretation of $E(Y)$ as a probability, the logistic regression equation is often written as follows:

Interpretation of $E(Y)$ as a probability in logistic regression

$$E(Y) = P(Y = 1 \mid x_1, x_2, \ldots x_p) \tag{15.29}$$

To provide a better understanding of the characteristics of the logistic regression equation, suppose the model involves only one independent variable X and the values of the model parameters are $\beta_0 = -7$ and $\beta_1 = 3$. The logistic regression equation corresponding to these parameter values is:

$$E(Y) = P(Y = 1 \mid x) = \frac{e^{\beta_0 + \beta_1 x}}{1 + e^{\beta_0 + \beta_1 x}} = \frac{e^{-7 + 3x}}{1 + e^{-7 + 3x}} \tag{15.30}$$

Figure 15.12 shows a graph of equation (15.30). Note that the graph is S-shaped. The value of $E(Y)$ ranges from 0 to 1, with the value of $E(Y)$ gradually approaching 1 as the value of X becomes larger and the value of $E(Y)$ approaching 0 as the value of X becomes smaller. Note also that the values of $E(Y)$, representing probability, increase fairly rapidly as X increases from 2 to 3. The fact that the values of $E(Y)$ range from 0 to 1 and that the curve is S-shaped makes equation (15.30) ideally suited to model the probability the dependent variable is equal to 1.

FIGURE 15.12
Logistic regression equation
for $\beta_0 = -7$ and $\beta_1 = 3$

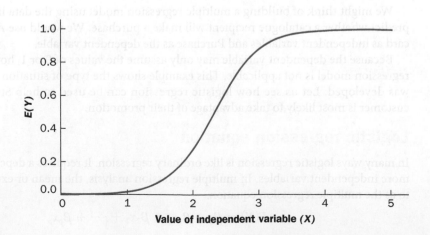

Estimating the logistic regression equation

In simple linear and multiple regression the least squares method is used to compute $b_0, b_1, \ldots, b_p$ as estimates of the model parameters $(\beta_0, \beta_1, \ldots, \beta_p)$. The nonlinear form of the logistic regression equation makes the method of computing estimates more complex and beyond the scope of this text. We will use computer software to provide the estimates. The estimated logistic regression equation is:

Estimated logistic regression equation

$$\hat{y} = \text{estimate of } P(Y = 1 \mid x_1, x_2, \ldots x_p) = \frac{e^{b_0 + b_1 x_1 + b_2 x_2 + \ldots + b_p x_p}}{1 + e^{b_0 + b_1 x_1 + b_2 x_2 + \ldots + b_p x_p}} \qquad \textbf{(15.31)}$$

Here $\hat{y}$ provides an estimate of the probability that $Y = 1$, given a particular set of values for the independent variables.

Let us now return to the Stamm Stores example. The variables in the study are defined as follows:

$$Y = \begin{cases} 0 & \text{if the customer made no purchase during the test period} \\ 1 & \text{if the customer made a purchase during the test period} \end{cases}$$

X_1 = annual spending at Stamm Stores (€000)

$$X_2 = \begin{cases} 0 & \text{if the customer does not have a Stamm credit card} \\ 1 & \text{if the customer has a Stamm credit card} \end{cases}$$

Therefore, we choose a logistic regression equation with two independent variables.

$$E(Y) = \frac{e^{\beta_0 + \beta_1 x_1 + \beta_2 x_2}}{1 + e^{\beta_0 + \beta_1 x_1 + \beta_2 x_2}} \qquad \textbf{(15.32)}$$

Using the sample data (refer to Table 15.11), a computer binary logistic regression procedure was used to compute estimates of the model parameters β_0, β_1 and β_2. A portion of the output obtained is shown in Figure 15.13. We see that $b_0 = -2.1464$, $b_1 = 0.3416$ and $b_2 = 1.0987$. Thus, the estimated logistic regression equation is:

$$\hat{y} = \frac{e^{b_0 + b_1 x_1 + \cdots + b_p b_p}}{1 + e^{b_0 + b_1 x_1 + \cdots + b_p x_p}} = \frac{e^{-2.1464 + 0.3416 x_1 + 1.0987 x_2}}{1 + e^{-2.1464 + 0.3416 x_1 + 1.0987 x_2}} \qquad \textbf{(15.33)}$$

We can now use equation (15.33) to estimate the probability of making a purchase for a particular type of customer. For example, to estimate the probability of making a purchase for customers that spend €2,000 annually and do not have a Stamm credit card, we substitute $X_1 = 2$ and $X_2 = 0$ into equation (15.33).

$$\hat{y} = \frac{e^{-2.1464 + 0.3416(2) + 1.0987(0)}}{1 + e^{-2.1464 + 0.3416(2) + 1.0987(0)}} = \frac{e^{-1.4632}}{1 + e^{-1.4632}} = \frac{0.2315}{1.2315} = 0.1880$$

Thus, an estimate of the probability of making a purchase for this particular group of customers is approximately 0.19. Similarly, to estimate the probability of making a purchase for customers that spent €2,000 last year and have a Stamm credit card, we substitute $X_1 = 2$ and $X_2 = 1$ into equation (15.33).

$$\hat{y} = \frac{e^{-2.1464 + 0.3416(2) + 1.0987(1)}}{1 + e^{-2.1464 + 0.3416(2) + 1.0987(1)}} = \frac{e^{-0.3645}}{1 + e^{-0.3645}} = \frac{0.6945}{1.6945} = 0.4099$$

Thus, for this group of customers, the probability of making a purchase is approximately 0.41. It appears that the probability of making a purchase is much higher for customers with a Stamm credit card. Before reaching any conclusions, however, we need to assess the statistical significance of our model.

FIGURE 15.13
Logistic regression output for the Stamm Stores example

Significance Tests

Term	Degrees of freedom	χ^2	p-value
Whole Model	2	13.63	0.0011
Spending	1	7.56	0.0060
Card	1	6.41	0.0013

Parameter Estimates

Term	Estimate	Standard Error
Intercept	-2.146	0.577
Spending	0.342	0.129
Cars	1.099	0.440

Odds Ratios

Term	Odds Ratio	Lower 95%	Upper 95%
Spending	1.4073	1.0936	1.8109
Cars	3.0000	1.2550	7.1730

Testing for significance

Testing for significance in logistic regression is similar to testing for significance in multiple regression. First we conduct a test for overall significance. For the Stamm Stores example, the hypotheses for the test of overall significance is as follows:

$$H_0: \beta_1 = \beta_2 = 0$$
$$H_1: \beta_1 \text{ and/or } \beta_2 \text{ is not equal to zero}$$

The test for overall significance is based upon the value of a G test statistic shown in Figure 15.13 as the 'Adj Dev' for Regression in the Deviance table. This is commonly referred to as the 'Deviance Statistic'. If the null hypothesis is true, the sampling distribution of G follows a chi-square distribution with degrees of freedom equal to the number of independent variables in the model. Although the computation of G is beyond the scope of this book, the value of G and its corresponding p-value are provided as part of the computer's binary logistic regression output. Referring to Figure 15.13, we see that the value of G is 13.63, its degrees of freedom are 2 and its p-value is 0.001. At any level of significance $\alpha \geq 0.001$, we would therefore reject the null hypothesis and conclude that the overall model is significant.

If the G test shows an overall significance, we use the Adj Dev details provided for individual independent variables to see if they make a significant contribution to the overall model. For the independent variables X_i, the hypotheses are:

$$H_0: \beta_i = 0$$
$$H_1: \beta_i \neq 0$$

Suppose we use $\alpha = 0.05$ to test for the significance of the independent variables in the Stamm model. For the independent variable X_1 the Adj Dev value is 7.556 and the corresponding p-value is 0.006. Thus, at the 0.05 level of significance we can reject $H_0: \beta_1 = 0$. In a similar fashion we can also reject $H_0: \beta_2 = 0$ because the p-value corresponding to Adj Dev = 6.410 is 0.011. Hence, at the 0.05 level of significance, both independent variables are statistically significant.

Managerial use

We now use the estimated logistic regression equation to make a decision recommendation concerning the Stamm Stores catalogue promotion. For Stamm Stores, we already computed:

$$P(Y = 1 \mid X_1 = 2, X_2 = 1) = 0.4099 \text{ and } P(Y = 1 \mid X_1 = 2, X_2 = 0) = 0.1880$$

These probabilities indicate that for customers with annual spending of €2,000, the presence of a Stamm credit card increases the probability of making a purchase using the discount coupon. In Table 15.12 we show estimated probabilities for values of annual spending ranging from €1,000 to €7,000 for both customers who have a Stamm credit card and customers who do not have a Stamm credit card. How can Stamm use this information to better target customers for the new promotion?

TABLE 15.12   Estimated probabilities for Stamm Stores

		Annual spending						
		€1,000	**€2,000**	**€3,000**	**€4,000**	**€5,000**	**€6,000**	**€7,000**
Credit card	Yes	0.3307	0.4102	0.4948	0.5796	0.6599	0.7320	0.7936
	No	0.1414	0.1881	0.2460	0.3148	0.3927	0.4765	0.5617

Suppose Stamm wants to send the promotional catalogue only to customers who have a 0.40 or higher probability of making a purchase. Using the estimated probabilities in Table 15.12, Stamm's promotion strategy would be:

Customers who have a Stamm credit card: Send the catalogue to every customer that spent €2,000 or more last year.
Customers who do not have a Stamm credit card: Send the catalogue to every customer that spent €6,000 or more last year.

Looking at the estimated probabilities further, we see that the probability of making a purchase for customers who do not have a Stamm credit card, but spend €5,000 annually, is 0.3921. Thus, Stamm may want to consider revising this strategy by including those customers who do not have a credit card as long as they spent €5,000 or more last year.

Interpreting the logistic regression equation

Interpreting a regression equation involves relating the independent variables to the business question that the equation was developed to answer. With logistic regression, it is difficult to interpret the relation between the independent variables and the probability that $Y = 1$ directly because the logistic regression equation is nonlinear. However, statisticians have shown that the relationship can be interpreted indirectly using a concept called the odds ratio.

The **odds in favour of an event occurring** is defined as the probability the event will occur divided by the probability the event will not occur. In logistic regression the event of interest is always $Y = 1$. Given a particular set of values for the independent variables, the odds in favour of $Y = 1$ can be calculated as follows:

$$\text{Odds} = \frac{P(Y = 1 \mid X_1, X_2, \ldots X_p)}{P(Y = 0 \mid X_1, X_2, \ldots X_p)} = \frac{P(Y = 1 \mid X_1, X_2, \ldots X_p)}{1 - P(Y = 1 \mid X_1, X_2, \ldots X_p)} \tag{15.34}$$

The **odds ratio** measures the impact on the odds of a one-unit increase in only one of the independent variables. The odds ratio is the odds that $Y = 1$ given that one of the independent variables has been increased by one unit (odds$_1$) divided by the odds that $Y = 1$ given no change in the values for the independent variables (odds$_0$).

> **Odds ratio**
>
> $$\text{Odds ratio} = \frac{\text{Odds}_1}{\text{Odds}_0}$$ (15.35)

For example, suppose we want to compare the odds of making a purchase for customers who spend €2,000 annually and have a Stamm credit card ($X_1 = 2$ and $X_2 = 1$) to the odds of making a purchase for customers who spend €2,000 annually and do not have a Stamm credit card ($X_1 = 2$ and $X_2 = 0$). We are interested in interpreting the effect of a one-unit increase in the independent variable X_2. In this case:

$$\text{Odds}_1 = \frac{P(Y = 1 | X_1 = 2, X_2 = 1)}{1 - P(Y = 1 | X_1 = 2, X_2 = 1)}$$

and:

$$\text{Odds}_0 = \frac{P(Y = 1 | X_1 = 2, X_2 = 0)}{1 - P(Y = 1 | X_1 = 2, X_2 = 0)}$$

Previously we showed that an estimate of the probability that $Y = 1$ given $X_1 = 2$ and $X_2 = 1$ is 0.4099, and an estimate of the probability that $Y = 1$ given $X_1 = 2$ and $X_2 = 0$ is 0.1880. Thus,

$$\text{Estimate of odds}_1 = \frac{0.4099}{1 - 0.4099} = 0.6946$$

and:

$$\text{Estimate of odds}_0 = \frac{0.1880}{1 - 0.1880} = 0.2315$$

yielding the:

$$\text{Estimated odds ratio} = \frac{0.6946}{0.2315} = 3.00$$

Thus, we can conclude that the estimated odds in favour of making a purchase for customers who spent €2,000 last year and have a Stamm credit card are three times greater than the estimated odds in favour of making a purchase for customers who spent €2,000 last year and do not have a Stamm credit card.

The odds ratio for each independent variable is computed while holding all the other independent variables constant. But it does not matter what constant values are used for the other independent variables. For instance, if we computed the odds ratio for the Stamm credit card variable (X_2) using €3,000 instead of €2,000 as the value for the annual spending variable (X_1), we would still obtain the same value for the estimated odds ratio (3.00). Thus, we can conclude that the estimated odds of making a purchase for customers who have a Stamm credit card are three times greater than the estimated odds of making a purchase for customers who do not have a Stamm credit card.

The odds ratio is standard output for logistic regression software packages. Refer to the computer output in Figure 15.13. The column headed by Odds Ratio contains the estimated odds ratios for each of the independent variables. The estimated odds ratio for X_1 is 1.41 and the estimated odds ratio for X_2 is 3.00. We already showed how to interpret the estimated odds ratio for the binary independent variable X_2. Let us now consider the interpretation of the estimated odds ratio for the continuous independent variable X_1.

The value of 1.41 in the Odds Ratio column of the computer output tells us that the estimated odds in favour of making a purchase for customers who spent €3,000 last year is 1.41 times greater than the estimated odds in favour of making a purchase for customers who spent €2,000 last year. Moreover, this interpretation is true for any one-unit change in X_1.

For instance, the estimated odds in favour of making a purchase for someone who spent €5,000 last year is 1.41 times greater than the odds in favour of making a purchase for a customer who spent €4,000 last year. But suppose we are interested in the change in the odds for an increase of more than one unit for an independent variable. Note that X_1 can range from 1 to 7. The odds ratio as printed by the computer output does not answer this question.

To answer this question we must explore the relationship between the odds ratio and the regression coefficients.

A unique relationship exists between the odds ratio for a variable and its corresponding regression coefficient. For each independent variable in a logistic regression equation it can be shown that:

$$\text{Odds ratio} = e^{b_i}$$

To illustrate this relationship, consider the independent variable X_1 in the Stamm example. The estimated odds ratio for X_1 is:

$$\text{Estimated odds ratio} = e^{b_1} = e^{0.3416} = 1.41$$

Similarly, the estimated odds ratio for X_2 is:

$$\text{Estimated odds ratio} = e^{b_2} = e^{1.0987} = 3.00$$

This relationship between the odds ratio and the coefficients of the independent variables makes it easy to compute estimates of the odds ratios once we develop estimates of the model parameters. Moreover, it also provides us with the ability to investigate changes in the odds ratio of more than or less than one unit for a continuous independent variable.

The odds ratio for an independent variable represents the change in the odds for a one unit change in the independent variable holding all the other independent variables constant. Suppose that we want to consider the effect of a change of more than one unit, say c units. For instance, suppose in the Stamm example that we want to compare the odds of making a purchase for customers who spend €5,000 annually ($X_1 = 5$) to the odds of making a purchase for customers who spend €2,000 annually ($X_1 = 2$). In this case $c = 5 - 2 = 3$ and the corresponding estimated odds ratio is:

$$e^{cb} = e^{3(0.3416)} = e^{1.0248} = 2.79$$

This result indicates that the estimated odds of making a purchase for customers who spend €5,000 annually is 2.79 times greater than the estimated odds of making a purchase for customers who spend €2,000 annually. In other words, the estimated odds ratio for an increase of €3,000 in annual spending is 2.79.

In general, the odds ratio enables us to compare the odds for two different events. If the value of the odds ratio is 1, the odds for both events are the same. Thus, if the independent variable we are considering (such as Stamm credit card status) has a positive impact on the probability of the event occurring, the corresponding odds ratio will be greater than 1. Most logistic regression software packages provide a confidence interval for the odds ratio. The computer output in Figure 15.13 provides a 95% confidence interval for each of the odds ratios. For example, the point estimate of the odds ratio for X_1 is 1.41 and the

95% confidence interval is 1.09 to 1.81. As the confidence interval does not contain the value of 1, we can conclude that X_1 has a significant effect on the odds ratio. Similarly, the 95% confidence interval for the odds ratio for X_2 is 1.25 to 7.17. Because this interval does not contain the value of 1, we can also conclude that X_2 has a significant effect on the odds ratio.

Logit transformation

An interesting relationship can be observed between the odds in favour of $Y = 1$ and the exponent for e in the logistic regression equation. It can be shown that:

$$\ln(\text{odds}) = \beta_0 + \beta_1 x_1 + \beta_2 x_2 + \cdots + \beta_p x_p$$

This equation shows that the natural logarithm of the odds in favour of $Y = 1$ is a linear function of the independent variables. This linear function is called the **logit**. We will use the notation $g(x_1, x_2, \ldots x_p)$ to denote the logit.

> **Logit**
>
> $$g(x_1, x_2, \ldots, x_p) = \beta_0 + \beta_1 x_1 + \beta_2 x_2 + \cdots + \beta_p x_p \qquad \textbf{(15.36)}$$

Substituting $g(x_1, x_2, \ldots x_p)$ for $\beta_0 + \beta_1 x_1 + \beta_2 x_2 + \cdots + \beta_p x_p$ in equation (15.28), we can write the logistic regression equation as:

$$E(Y) = \frac{e^{g(x_1, x_2, \ldots, x_p)}}{1 + e^{g(x_1, x_2, \ldots, x_p)}} \qquad \textbf{(15.37)}$$

Once we estimate the parameters in the logistic regression equation, we can compute the **estimated logit**:

> **Estimated logit**
>
> $$\hat{g}(x_1, x_2, \ldots, x_p) = b_0 + b_1 x_1 + b_2 x_2 + \cdots + b_p x_p \qquad \textbf{(15.38)}$$

Therefore, in terms of the estimated logit, the estimated regression equation is:

$$\hat{y} = \frac{e^{b_0 + b_1 x_1 + b_2 x_2 + \cdots + b_p x_p}}{1 + e^{b_0 + b_1 x_1 + b_2 x_2 + \cdots + b_p x_p}} = \frac{e^{\hat{g}(x_1, x_2, \ldots, x_p)}}{1 + e^{\hat{g}(x_1, x_2, \ldots, x_p)}}$$

For the Stamm Stores example, the estimated logit is:

$$\hat{g} = (x_1, x_2) = -2.1464 + 0.3416 x_1 + 1.0987 x_2$$

and the estimated regression equation is:

$$\hat{y} = \frac{e^{\hat{g}(x_1, x_2)}}{1 + e^{\hat{g}(x_1, x_2)}} = \frac{e^{-2.1464 + 0.3416 x_1 + 1.0987 x_2}}{1 + e^{-2.1464 + 0.3416 x_1 + 1.0987 x_2}}$$

Therefore, because of the unique relationship between the estimated logit and the estimated logistic regression equation, we can compute the estimated probabilities for Stamm Stores by dividing $e^{\hat{g}(x_1, x_2)}$ by $1 + e^{\hat{g}(x_1, x_2)}$.

EXERCISES

Applications

30. Refer to the Stamm Stores example introduced in this section. The dependent variable is coded as $Y = 1$ if the customer makes a purchase and 0 if not.

Suppose that the only information available to help predict whether the customer will make a purchase is the customer's credit card status, coded as $X = 1$ if the customer has a Stamm credit card and $X = 0$ if not.

a. Write the logistic regression equation relating X to Y.
b. What is the interpretation of $E(Y)$ when $X = 0$?
c. For the Stamm data in Table 15.11, use a computer software package to compute the estimated logit.
d. Use the estimated logit computed in part (c) to compute an estimate of the probability of making a purchase for customers who do not have a Stamm credit card and an estimate of the probability of making a purchase for customers who have a Stamm credit card.
e. What is the estimate of the odds ratio? What is its interpretation?

31. In Table 15.12 we provided estimates of the probability of a purchase in the Stamm Stores catalogue promotion. A different value is obtained for each combination of values for the independent variables.

a. Compute the odds in favour of a purchase for a customer with annual spending of €4,000 who does not have a Stamm credit card ($X_1 = 4, X_2 = 0$).
b. Use the information in Table 15.12 and part (a) to compute the odds ratio for the Stamm credit card variable X_2 holding annual spending constant at $X_1 = 4$.
c. In the text, the odds ratio for the credit card variable was computed using the information in the €2,000 column of Table 15.12. Did you get the same value for the odds ratio in part (b)?

32. Community Bank would like to increase the number of customers who use payroll direct deposit. Management is considering a new sales campaign that will require each branch manager to call each customer who does not currently use payroll direct deposit. As an incentive to sign up for payroll direct deposit, each customer contacted will be offered free banking for two years. Because of the time and cost associated with the new campaign, management would like to focus their efforts on customers who have the highest probability of signing up for payroll direct deposit. Management believes that the average monthly balance in a customer's current account may be a useful predictor of whether the customer will sign up for direct payroll deposit. To investigate the relationship between these two variables, Community Bank tried the new campaign using a sample of 50 current account customers that do not currently use payroll direct deposit. The sample data show the average monthly current account balance (in hundreds of euros) and whether the customer contacted signed up for payroll direct deposit (coded 1 if the customer signed up for payroll direct deposit and 0 if not). The data are contained in the data set named 'Bank' on the companion online platform; a portion of the data follows.

Customer	X Monthly balance	Y Direct deposit
1	1.22	0
2	1.56	0
3	2.10	0
4	2.25	0
5	2.89	0
6	3.55	0
7	3.56	0
8	3.65	1

(Continued)

Customer	X Monthly balance	Y Direct deposit
.	.	.
.	.	.
.	.	.
48	18.45	1
49	24.98	0
50	26.05	1

BANK

a. Write the logistic regression equation relating X to Y.
b. For the Community Bank data, use a computer software package to compute the estimated logistic regression equation.
c. Conduct a test of significance using the G test statistic. Use $\alpha = 0.05$.
d. Estimate the probability that customers with an average monthly balance of €1,000 will sign up for direct payroll deposit.
e. Suppose Community Bank only wants to contact customers who have a 0.50 or higher probability of signing up for direct payroll deposit. What is the average monthly balance required to achieve this level of probability?
f. What is the estimate of the odds ratio? What is its interpretation?

33. Prior to the *Challenger* tragedy on 28 January 1986, after each launch of the space shuttle the solid rocket boosters were recovered from the ocean and inspected. Of the previous 24 shuttle launches, 7 had incidents of damage to the joints, 16 had no incidents of damage and 1 was unknown because the boosters were not recovered after launch.

In trying to explain the damage to the joints it was thought that temperature at the time of launch could be a contributing factor.

For the data that follow, a 1 represents damage to joints, and a 0 represents no damage.

SHUTTLE

Temp	Damage	Temp	Damage	Temp	Damage
66	0	57	1	70	0
70	1	63	1	81	0
69	0	70	1	76	0
68	0	78	0	79	0
67	0	67	0	75	1
72	0	53	1	76	0
73	0	67	0	58	1
70	0	75	0		

a. Fit a logistic regression model to these data and obtain a plot of the data and fitted curve.
b. Conduct a test of significance using the G test statistic. Use $\alpha = 0.05$.
c. Estimate the probability of damage for a temperature of 50.
d. What is the estimate of the odds ratio? How would you interpret it?

34. The Tyre Rack maintains an independent consumer survey to help drivers help each other by sharing their long-term tyre experiences. The data contained in the file named TyreRatings show survey results for 68 all-season tyres (Tyre Rack website, 21 March 2012). Performance traits are rated using the following 10-point scale.

Superior		Excellent		Good		Fair		Unacceptable	
10	9	8	7	6	5	4	3	2	1

The values for the variable labelled Wet are the average of the ratings for each tyre's wet traction performance and the values for the variable labeled Noise are the average of the ratings for the noise level generated by each tyre.

Respondents were also asked whether they would buy the tyre again using the following 10-point scale:

Definitely	Probably	Possibly	Probably Not	Definitely Not
10 9	8 7	6 5	4 3	2 1

The values for the variable labelled Buy Again are the average of the buy-again responses. For the purposes of the analysis below, a binary dependent variable was created as follows:

$$Purchase = \begin{cases} 1 & \text{if the value of the Buy-Again variable is 7 or greater} \\ 0 & \text{if the value of the Buy-Again variable is less than 7} \end{cases}$$

Thus, if Purchase = 1, the respondent would be considered to probably or definitely buy the tyre again.

By undertaking an appropriate computer analysis:

a. Determine the logistic regression equation relating $X_1 =$ Wet performance rating and $X_2 =$ Noise performance rating to $Y =$ Purchase.

b. Explain the test results and the meaning of the odds ratio and confidence interval (CI) results provided.

c. Use the estimated logit to compute an estimate of the probability that a customer will probably or definitely purchase a particular tyre again with a Wet performance rating of 8 and a Noise performance rating of 8.

d. Suppose that the Wet and Noise performance ratings were 7. How does that affect the probability that a customer will probably or definitely purchase a particular tyre again with these performance ratings?

e. If you were the CEO of a tyre company, what do the results for parts (c) and (d) tell you?

TYRE
RATINGS

15.10 PRACTICAL ADVICE: BIG DATA AND HYPOTHESIS TESTING IN MULTIPLE REGRESSION

In Chapter 14, we observed that in simple linear regression, the p-value for the test of the hypothesis $H_0: \beta_1 = 0$ decreases as the sample size increases. Likewise, as the sample size increases:

- The p-value for the F test used to determine whether a significant relationship exists between the dependent variable and the set of all independent variables in the regression model decreases.
- The p-value for each t test used to determine whether a significant relationship exists between the dependent variable and an individual independent variable in the regression model decreases.
- The confidence interval for the slope parameter associated with each individual independent variable narrows.
- The confidence interval for the mean value of Y narrows.
- The prediction interval for an individual value of Y narrows.

Thus, the interval estimates for the slope parameter associated with each individual independent variable, the mean value of Y and predicted individual value of Y will become more precise as the sample size increases. In consequence, we are more likely to reject the hypothesis that a relationship does not exist between the dependent variable and the set of all individual independent variables in the model as the sample size increases. Similarly, for each individual independent variable, we are more likely to reject the hypothesis that a relationship does not exist between the dependent variable and the individual independent variable as the sample size increases. Even when severe multicollinearity is present, if the sample is sufficiently large, independent variables may each have a significant relationship with the dependent variable. However, this does not necessarily signify that corresponding tests have greater reliability.

No matter how large the sample used to estimate the multiple regression model, we must be concerned about the potential presence of non-sampling error in the data. It is important to carefully consider whether a random sample of the population of interest has actually been taken. If non-sampling error is present in the data collection process, the likelihood of encountering problematic Type I or Type II errors in hypothesis tests may be higher than when the sample data are free of non-sampling error. Furthermore, multicollinearity may cause the estimated slope coefficients to be misleading; this problem persists as the size of the sample used to estimate the multiple regression model increases. Finally, it is important to consider whether the statistically significant relationship(s) in the multiple regression model are of practical significance.

Although multiple regression is an extremely powerful statistical tool, no business decision should be based exclusively on hypothesis testing in multiple regression. Non-sampling error may lead to misleading results. If severe multicollinearity is present, we must be cautious about interpreting the estimated slope coefficients. And practical significance should always be considered in conjunction with statistical significance; this is particularly important when a hypothesis test is based on an extremely large sample because p-values in such cases can be extremely small. When executed properly, hypothesis tests in multiple regression provide evidence that should be considered in conjunction with information collected from other sources to make the most informed decision possible.

ONLINE RESOURCES

For data files, additional questions and answers, and the software section for Chapter 15, visit the online platform.

SUMMARY

In this chapter, we introduced multiple regression analysis as an extension of the simple linear regression analysis technique presented in Chapter 14. Multiple regression analysis enables us to understand how a dependent variable is related to two or more independent variables. The regression equation $E(Y) = \beta_0 + \beta_1 x_1 + \beta_2 x_2 + \ldots + \beta_p x_p$ shows that the expected value or mean value of the dependent variable Y is related to the values of independent variables $X_1, X_2, \ldots, X_p$. Sample data and the least squares method are used to develop the estimated regression equation $\hat{y} = b_0 + b_1 x_1 + b_2 x_2 + \ldots + b_p x_p$. In effect $b_0, b_1, b_2, \ldots, b_p$ are sample statistics used to estimate the unknown model parameters $\beta_0, \beta_1, \beta_2, \ldots, \beta_p$. Computer printouts were used throughout the chapter to emphasize the fact that statistical software packages are the only realistic means of performing the computations required in multiple regression analysis.

The multiple coefficient of determination was presented as a measure of the goodness of fit of the estimated regression equation. It determines the proportion of the variation of Y that can be explained by the estimated regression equation. The adjusted multiple coefficient of determination is a similar measure of goodness of fit that adjusts for the number of independent variables and thus avoids overestimating the impact of adding more independent variables. Model assumptions for multiple regression are shown to parallel those for simple regression analysis.

An F test and a t test were presented as ways of determining statistically whether the relationship between the variables is significant. The F test is used to determine whether there is a significant overall relationship between the dependent variable and the set of all independent variables. The t test is used to determine whether there is a significant relationship between the dependent variable and an individual independent variable given the other independent variables in the regression model. Correlation between the independent variables, known as multicollinearity, was also discussed.

The section on categorical independent variables showed how dummy variables can be used to incorporate categorical data into multiple regression analysis. Following on, the role of residual analysis for validating model assumptions, detecting outliers and identifying influential observations was shown. Standardized residuals, leverage, studentized deleted residuals and Cook's distance measure were discussed. Next, details on how logistic regression can be used to model situations in which the dependent variable may only assume two values. The chapter concluded with practical advice on the limitations of multiple regression modelling for big data applications.

KEY TERMS

Adjusted multiple coefficient of determination
Categorical independent variables
Cook's distance measure
Dummy variable
Estimated logistic regression equation
Estimated logit
Estimated multiple regression equation
Influential observation
Least squares method
Leverage
Logistic regression equation

Logit
Multicollinearity
Multiple coefficient of determination
Multiple regression analysis
Multiple regression equation
Multiple regression model
Odds in favour of an event occurring
Odds ratio
Outlier
Studentized deleted residuals
Variance inflation factor

KEY FORMULAE

Multiple regression model

$$Y = \beta_0 + \beta_1 x_1 + \beta_2 x_2 + \ldots + \beta_p x_p + \varepsilon \tag{15.1}$$

Multiple regression equation

$$E(Y) = \beta_0 + \beta_1 x_1 + \beta_2 x_2 + \ldots + \beta_p x_p \tag{15.2}$$

Estimated multiple regression equation

$$\hat{y} = b_0 + b_1 x_1 + b_2 x_2 + \ldots + b_p x_p \tag{15.3}$$

Least squares criterion

$$\min \sum (y_i - \hat{y}_i)^2 \tag{15.4}$$

Relationship between SST, SSR and SSE

$$SST = SSR + SSE \tag{15.7}$$

Multiple coefficient of determination

$$R^2 = \frac{SSR}{SST} \tag{15.8}$$

Adjusted multiple coefficient of determination

$$\text{adj} R^2 = 1 - (1 - R^2)\frac{n-1}{n-p-1} \tag{15.9}$$

Mean square regression

$$\text{MSR} = \frac{\text{SSR}}{p} \tag{15.12}$$

Mean square error

$$\text{MSE} = s^2 = \frac{\text{SSE}}{n - p - 1} \tag{15.13}$$

F test statistic

$$F = \frac{\text{MSR}}{\text{MSE}} \tag{15.14}$$

t Test statistic

$$t = \frac{b_i}{s_{b_i}} \tag{15.15}$$

Variance inflation factor

$$\text{VIF}(X_j) = \frac{1}{1 - R_j^2} \tag{15.16}$$

Standardized residual for observation i

$$\frac{y_i - \hat{y}_i}{s_{y_i - \hat{y}_i}} \tag{15.24}$$

Standard deviation of residual i

$$s_{y_i - \hat{y}_i} = s\sqrt{1 - h_1} \tag{15.25}$$

Cook's distance measure

$$D_i = \frac{(y_i - \hat{y}_i)^2 h_i}{(p - 1)s^2(1 - h_i)^2} \tag{15.26}$$

Logistic regression equation

$$E(Y) = \frac{e^{\beta_0 + \beta_1 x_1 + \beta_2 x_2 + \ldots + \beta_p x_p}}{1 + e^{\beta_0 + \beta_1 x_1 + \beta_2 x_2 + \ldots + \beta_p x_p}} \tag{15.28}$$

Interpretation of E(Y) as a probability in logistic regression

$$E(Y) = P(Y = 1 \mid x_1, x_2, \ldots x_p) \tag{15.29}$$

Estimated logistic regression equation

$$\hat{y} = \text{estimate of } P(Y = 1 | x_1, x_2, \ldots x_p) = \frac{e^{b_0 + b_1 x_1 + b_2 x_2 + \ldots + b_p x_p}}{1 + e^{b_0 + b_1 x_1 + b_2 x_2 + \ldots + b_p x_p}} \tag{15.31}$$

Odds ratio

$$\text{Odds ratio} = \frac{\text{Odds}_1}{\text{Odds}_0} \tag{15.35}$$

Logit

$$g(x_1, x_2, \ldots, x_p) = \beta_0 + \beta_1 x_1 + \beta_2 x_2 + \ldots + \beta_p x_p \tag{15.36}$$

Estimated logit

$$\hat{g}(x_1, x_2, \ldots, x_p) = b_0 + b_1 x_1 + b_2 x_2 + \cdots + b_p x_p \tag{15.38}$$

CASE PROBLEM 1

P/E ratios

Valuation is one of the most important aspects of business. Frequently, although an absolute valuation (e.g. $100 million) would be desirable, relative valuation (e.g. company A is better than Company B) is enough for investment decision making. When deciding to perform a relative valuation, it is necessary to decide on what attributes to compare; of the many possibilities, price-to-earnings ratios (P/Es) are perhaps the most frequently used. This ratio typically is calculated using data on a per share basis:

$$\text{P/E ratio} = \frac{\text{Market price per share}}{\text{Earnings per share}}$$

Other things being equal, the higher the price-to-earnings ratio, the higher the expected future income relative to the reported income.

© shih-wei/iStock

Managerial report

A portfolio manager in a leading brokerage firm has asked you to develop a model that can help them to allocate funds between the various international markets. Theoretically, the job is easy – invest in undervalued markets and sell any assets in overvalued markets. P/E ratios can be used to identify over-/undervalued markets.

Three variables thought to influence the P/E ratio are:

FUNDS

(Continued)

1. price-to-book value (PBV)[1]
2. return on equity (ROE)[2]
3. the effective tax rate (Tax)[3]

Formulate and estimate a multiple regression model using the data provided. In your report,

you should help the manager understand each of the estimated regression coefficients, the standard error of estimate, and the co-efficient of determination.

Data are available in a file called 'Funds' on the online platform. Below is a part of the table.

Criteria for inclusion: Publicly traded firms with $ market cap > $50 million

Country	Number of firms	P/E	PBV	Return on equity (%)	Effective tax rate (%)
Argentina	43	14.10	1.67	−11.48	10.30
Australia	419	28.93	4.78	11.32	22.37
Austria	68	41.81	2.00	7.54	22.41

[1] A ratio used to compare a stock's market value to its book value. It is calculated by dividing the current closing price of the stock by the latest quarter's book value (book value is simply total assets minus intangible assets and liabilities). A lower PBV ratio could mean that the stock is undervalued. However, it could also mean that something is fundamentally wrong with the company.
[2] Essentially, ROE reveals how much profit a company generates with the money shareholders have invested in it. The ROE is useful for comparing the profitability of a company to that of other firms in the same industry. Investors usually look for companies with ROEs that are high and growing.
[3] Actual income tax paid divided by net taxable income before taxes.

CASE PROBLEM 2

Indicators of poverty in Mexico

As part of a government contract, researchers in Mexico have been studying associations between poverty and cultural indicators across 32 distinct geographical provinces. In particular, three different types of poverty were considered: **feeding poverty** – the inability of obtaining a 'basic-basket', even though the full income might be spent on it; **capability poverty** – the inability of income to cover, in addition to the 'basic-basket', expenses related to health care and education, even if the full income is spent on them; and **asset poverty** – the insufficiency of income for meeting, besides the 'basic-basket', expenses associated with health, clothing, education, housing and transportation expenses (Azevedo and Robles, 2013). In addition, data were collected on the following variables:

Access to the Oportunidades program (a government anti-poverty initiative)
Number of archaeological sites
Number of public libraries
Number of deaths
Number of murders
Unemployment rate
Average hours per week watching TV

MEXICAN POVERTY 2000

MEXICAN POVERTY 2010

Separate data sets were constructed for 2000 and 2010 – refer to the accompanying data details.

Managerial report

1. Investigate associations between the three poverty indices and the other variables for each data set. Comment on inconsistencies in the dataset if and where appropriate.

2. Explore possible regression models for predicting each of the three poverty indices.

What light do your analyses in 1 and 2 shed on possible changes in poverty levels between 2000 and 2010?

© KIKILOMBO/iStock

Source:
Azevedo, V. and M. Robles (2013) 'Multidimensional targeting: Identifying beneficiaries of conditional cash transfer programs'. *Social Indicators Research* 112(2): 447–475.

16
Regression Analysis: Model Building

CHAPTER CONTENTS

Statistics in Practice BASF, Ludwigshafen, Germany

16.1 General linear model
16.2 Determining when to add or delete variables
16.3 Variable selection procedures

LEARNING OBJECTIVES After reading this chapter and doing the exercises, you should be able to:

1 Appreciate how the general linear model can be used to model problems involving curvilinear relationships.

2 Understand the concept of interaction and how it can be accounted for in the general linear model.

3 Understand how an *F* test can be used to determine when to add or delete one or more variables.

4 Appreciate the complexities involved in solving larger regression analysis problems.

5 Understand how variable selection procedures can be used to choose a set of independent variables for an estimated regression equation.

Model building in regression analysis is the process of developing an estimated regression equation that describes the relationship between a dependent variable and one or more independent variables. The major issues in model building are finding an effective functional form of the relationship and selecting the independent variables to be included in the model. In Section 16.1 we establish the framework for model building by introducing the concept of a general linear model. Section 16.2, which provides the foundation for the more sophisticated computer-based procedures, introduces a general approach for determining when to add or delete independent variables. In Section 16.3 a larger regression problem involving 8 independent variables and 25 observations; is used to illustrate variable selection procedures including stepwise regression, the forward selection procedure, the backward elimination procedure and best-subsets regression.

STATISTICS IN PRACTICE
BASF
Ludwigshafen, Germany

Researchers used regression analysis to develop an optimal feed composition for poultry growers

BASF SE is a German multinational chemical company comprising subsidiaries and joint ventures in more than 80 countries with customers in over 190 countries. The animal nutrition division of BASF is a global supplier of innovative feed additives for livestock, aquaculture and companion animals.

BASF's animal nutrition division manufactures and markets a methionine supplement used in poultry, swine, and cattle feed products. Because poultry growers work with high volumes and low profit margins, cost-effective poultry feed products with the best possible nutrition value are needed. Optimal feed composition will result in rapid growth and high final body weight for a given level of feed intake. The chemical industry works closely with poultry growers to optimize poultry feed products. Ultimately, success depends on keeping the cost of poultry low in comparison with the cost of beef and other meat products.

Researchers used regression analysis to model the relationship between body weight Y and the amount of methionine X added to the poultry feed. Initially, the following simple linear estimated regression equation was developed.

$$\hat{y} = 0.21 + 0.42x$$

This estimated regression equation proved statistically significant; however, the analysis of the residuals indicated that a curvilinear relationship would be a better model of the relationship between body weight and methionine.

Further research showed that although small amounts of methionine tended to increase body weight, at some point body weight levelled off and additional amounts of the methionine were of little or no benefit. In fact, when the amount of methionine increased beyond nutritional requirements, body weight tended to decline. The following estimated multiple regression equation was used to model the curvilinear relationship between body weight and methionine.

$$\hat{y} = -1.89 + 1.32x - 0.506x^2$$

The regression model determines the optimal level of methionine to be used in poultry feed products.

In this chapter, we will extend the discussion of regression analysis by showing how curvilinear models can be developed. In addition, we will describe a variety of tools that help determine which independent variables lead to the best estimated regression equation.

16.1 GENERAL LINEAR MODEL

Suppose we collected data for one dependent variable Y and k independent variables $X_1, X_2, \ldots X_k$. Our objective is to use these data to develop an estimated regression equation that provides the best relationship between the dependent and independent variables. As a general framework for developing more complex relationships between the independent variables we introduce the concept of the general linear model involving p independent variables.

General linear model

$$Y = \beta_0 + \beta_1 z_1 + \beta_2 z_2 + \ldots + \beta_p z_p + \varepsilon \tag{16.1}$$

In equation (16.1), each of the independent variables Z_j (where $j = 1, 2, \ldots, p$) is a function of $X_1, X_2, \ldots, X_k$ (the variables for which data are collected). In some cases, each Z_j may be a function of only one X variable. The simplest case is when we collect data for just one variable X_1 and want to estimate Y by using a straight-line relationship. In this case $Z_1 = X_1$ and equation (16.1) becomes:

$$Y = \beta_0 + \beta_1 x_1 + \varepsilon \tag{16.2}$$

Equation (16.2) is the simple linear regression model introduced in Chapter 14 with the exception that the independent variable is labelled X_1 instead of X. In the statistical modelling literature, this model is called a *simple first-order model with one predictor variable*.

Modelling curvilinear relationships

More complex types of relationships can be modelled with equation (16.1). To illustrate, let us consider the problem facing Reynard Ltd, a manufacturer of industrial scales and laboratory equipment. Managers at Reynard want to investigate the relationship between length of employment of their salespeople and the number of electronic laboratory scales sold. Table 16.1 gives the number of scales sold by 15 randomly selected salespeople for the most recent sales period and the number of months each salesperson has been employed by the firm. Figure 16.1 is the scatter diagram for these data. The scatter diagram indicates a possible curvilinear relationship between the length of time employed and the number of units sold. Before considering how to develop a curvilinear relationship for Reynard, let us consider the computer output in Figure 16.2 corresponding to a simple first-order model; the estimated regression is:

$$\text{Sales} = 111 + 2.38 \text{ Months}$$

where:

$$\text{Sales} = \text{number of electronic laboratory scales sold}$$
$$\text{Months} = \text{the number of months the salesperson has been employed}$$

Figure 16.3 is the corresponding standardized residual plot. Although the computer output shows that the relationship is significant (p-value = 0.000) and that a linear relationship explains a high percentage of the variability in sales (R-sq = 78.1 per cent), the standardized residual plot suggests that a curvilinear relationship is needed.

REYNARD

TABLE 16.1
Data for the Reynard example

Months employed	Scales sold
41	275
106	296
76	317
104	376
22	162
12	150
85	367
111	308
40	189
51	235
9	83
12	112
6	67
56	325
19	189

FIGURE 16.1
Scatter diagram for the
Reynard example

To account for the curvilinear relationship, we set $Z_1 = X_1$ and $Z_2 = X_1^2$ in equation (16.1) to obtain the model:

$$Y = \beta_0 + \beta_1 x_1 + \beta_2 x_1^2 + \varepsilon \qquad \textbf{(16.3)}$$

This model is called a *second-order model with one predictor variable*. To develop an estimated regression equation corresponding to this second-order model, the statistical software package we are using needs the original data in Table 16.1, as well as the data corresponding to adding a second independent variable that is the square of the number of months the employee has been with the firm. In Figure 16.4 we show the computer output corresponding to the second-order model; the estimated regression equation is:

Sales $= 45.3 + 6.34$ Months $- 0.0345$ MonthsSq

where:

MonthsSq = the square of the number of months the salesperson has been employed

FIGURE 16.2
Output for the Reynard
example: first-order model

Analysis of Variance

Source	DF	Adj SS	Adj MS	F-value	p-value
Regression	1	113,783	113,783	46.41	0.000
Error	13	31,874	2,452		
Total	14	115,657			

Model Summary

S	R-sq	R-sq(adj)
49.5158	78.12%	76.43%

Coefficients

Term	Coef	SE Coef	t-value	p-value
Constant	111.2	21.6	5.14	0.000
Months	2.377	0.349	6.81	0.000

Regression Equation

Sales = 111.2 + 2.377 Months

FIGURE 16.3
Standardized residual plot
for the Reynard example:
first-order model

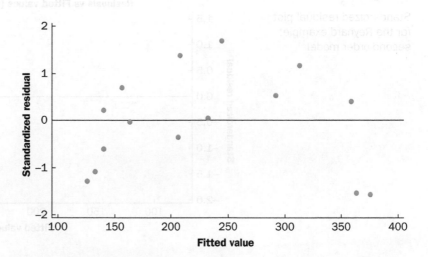

Figure 16.5 is the corresponding standardized residual plot. It shows that the previous curvilinear pattern has been removed. At the 0.05 level of significance, the computer output shows that the overall model is significant (p-value for the F test is 0.000); note also that the p-value corresponding to the t-ratio for MonthsSq (p-value = 0.002) is less than 0.05, and hence we can conclude that adding MonthsSq to the model involving Months is significant. With an R-sq(adj) value of 88.6 per cent, we should be pleased with the fit provided by this estimated regression equation. More important, however, is seeing how easy it is to handle curvilinear relationships in regression analysis.

Clearly, many types of relationships can be modelled by using equation (16.1). The regression techniques with which we have been working are definitely not limited to linear, or straight-line, relationships. In multiple regression analysis the word *linear* in the term 'general linear model' refers only to the fact that $\beta_0, \beta_1, \ldots, \beta_p$ all have exponents of 1; it does not imply that the relationship between Y and the X_is is linear. Indeed, in this section we have seen one example of how equation (16.1) can be used to model a curvilinear relationship.

FIGURE 16.4
Output for the Reynard
example: second-order model

Analysis of Variance

Source	DF	Adj SS	Adj MS	F-value	p-value
Regression	2	131,413	65,706.5	55.36	0.000
Error	12	14,244	1,187.0		
Total	14	145,657			

Model Summary

S	R-sq	R-sq(adj)
34.4528	90.22%	88.59%

Coefficients

Term	Coef	SE Coef	t-value	p-value
Constant	45.3	22.8	1.99	0.070
Months	6.34	1.06	6.00	0.000
MonthsSq	-0.03449	0.00895	-3.85	0.002

Regression Equation

Sales = 45.3 + 6.34 Months - 0.03449 MonthsSq

FIGURE 16.5
Standardized residual plot
for the Reynard example:
second-order model

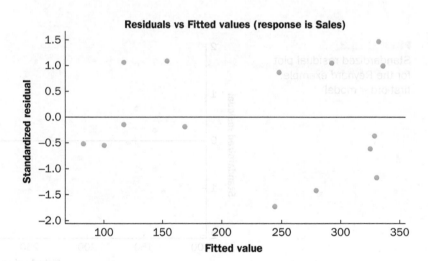

Interaction

To provide an illustration of interaction and what it means, let us review the regression study conducted by Veneto Care for one of its new shampoo products. Two factors believed to have the most influence on sales are unit selling price and advertising expenditure.

To investigate the effects of these two variables on sales, prices of €2.00, €2.50 and €3.00 were paired with advertising expenditures of €50,000 and €100,000 in 24 test markets. Refer to Figure 16.6. The unit sales (in thousands) that were observed are reported in Table 16.2.

Table 16.3 is a summary of these data. Note that the mean sales corresponding to a price of €2.00 and an advertising expenditure of €50,000 is 461,000, and the mean sales corresponding to a price of €2.00 and an advertising expenditure of €100,000 is 808,000. Hence, with price held constant at €2.00, the difference in mean sales between advertising expenditures of €50,000 and €100,000 is 808,000 − 461,000 = 347,000 units.

TABLE 16.2   Data for the Veneto Care example

Price	Advertising expenditure (€000)	Sales (000)	Price	Advertising expenditure (€000)	Sales (000)
€2.00	50	478	€2.00	100	810
€2.50	50	373	€2.50	100	653
€3.00	50	335	€3.00	100	345
€2.00	50	473	€2.00	100	832
€2.50	50	358	€2.50	100	641
€3.00	50	329	€3.00	100	372
€2.00	50	456	€2.00	100	800
€2.50	50	360	€2.50	100	620
€3.00	50	322	€3.00	100	390
€2.00	50	437	€2.00	100	790
€2.50	50	365	€2.50	100	670
€3.00	50	342	€3.00	100	393

TABLE 16.3   Mean unit sales (000) for the Veneto Care example

| | | Price | | |
		€2.00	€2.50	€3.00
Advertising	€50,000	461	364	332
Expenditure	€100,000	808	646	375

Mean sales of 808,000 units when price = €2.00 and advertising expenditure = €100,000

When the price of the product is €2.50, the difference in mean sales is 646,000 − 364,000 = 282,000 units. Finally, when the price is €3.00, the difference in mean sales is 375,000 − 332,000 = 43,000 units. Clearly, the difference in mean sales between advertising expenditures of €50,000 and €100,000 depends on the price of the product. In other words, at higher selling prices, the effect of increased advertising expenditure diminishes. These observations provide evidence of interaction between the price and advertising expenditure variables (Figure 16.6).

When interaction between two variables is present, we cannot study the effect of one variable on the response Y independently of the other variable. In other words, meaningful conclusions can be developed only if we consider the joint effect that both variables have on the response.

To account for the effect of interaction, we will use the following regression model:

$$Y = \beta_0 + \beta_1 x_1 + \beta_2 x_2 + \beta_3 x_1 x_2 + \varepsilon \tag{16.4}$$

where:

$$Y = \text{unit sales (000s)}$$
$$X_1 = \text{price (€)}$$
$$X_2 = \text{advertising expenditure (€000s)}$$

Note that equation (16.4) reflects Veneto's belief that the number of units sold depends linearly on selling price and advertising expenditure (accounted for by the $\beta_1 x_1$ and $\beta_2 x_2$ terms), and that there is interaction between the two variables (accounted for by the $\beta_3 x_1 x_2$ term).

To develop an estimated regression equation, a general linear model involving three independent variables (Z_1, Z_2 and Z_3) was used.

$$Y = \beta_0 + \beta_1 z_1 + \beta_2 z_2 + \beta_3 z_3 + \varepsilon \tag{16.5}$$

FIGURE 16.6
Sample mean unit sales (000)
as a function of selling price
and advertising expenditure

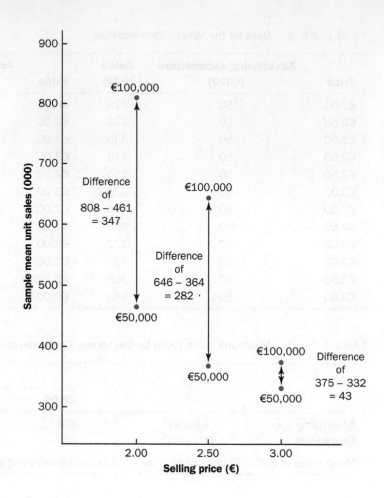

where:

$$z_1 = x_1$$
$$z_2 = x_2$$
$$z_3 = x_1 x_2$$

Figure 16.7 is the computer output corresponding to the interaction model for the Veneto Care example. The resulting estimated regression equation is:

$$\text{Sales} = -276 + 175\,\text{Price} + 19.7\,\text{AdvExp} - 6.08\,\text{PriceAdv}$$

where:

$$\text{Sales} = \text{unit sales (000)}$$
$$\text{Price} = \text{price of the product (€)}$$
$$\text{AdvExp} = \text{advertising expenditure (€000)}$$
$$\text{PriceAdv} = \text{interaction term (Price} \times \text{AdvExp)}$$

Because the model is significant (p-value for the F test is 0.000) and the p-value corresponding to the t test for PriceAdv is 0.000, we conclude that interaction is significant given the linear effect of the price of the product and the advertising expenditure. Thus, the regression results show that the effect of advertising expenditure on sales depends on the price.

FIGURE 16.7
Output for the Veneto Care example

```
Analysis of Variance

Source        DF    Adj SS    Adj MS    F-value   p-value
Regression     3   709,316   236,439    297.87    0.000
Error         20    15,875       794
Total         23   725,191

Model Summary

      S      R-sq   R-sq(adj)
28.1739    97.81%     97.48%

Coefficients

Term         Coef   SE Coef   t-value   p-value
Constant     -276       113     -2.44     0.024
Prices      175.0      44.5      3.93     0.001
Advert      19.68      1.43     13.79     0.000
PriceAdvert -6.080     0.563   -10.79     0.000

Regression Equation

Sales = -276 + 175.0 Price + 19.68 Advert - 6.080 PriceAdvert
```

Transformations involving the dependent variable

In showing how the general linear model can be used to model a variety of possible relationships between the independent variables and the dependent variable, we have focused attention on transformations involving one or more of the independent variables. Often it is worthwhile to consider transformations involving the dependent variable Y. As an illustration of when we might want to transform the dependent variable, consider the data in Table 16.4, which shows the kilometres-per-litre (KPL) ratings and weights (kg) for 12 cars.

TABLE 16.4
Kilometres-per-litre ratings and weights for 12 cars

Weight	Kilometres per litre
1,038	10.2
958	10.3
989	12.1
1,110	9.9
919	11.8
1,226	9.3
1,205	8.5
955	10.8
1,463	6.4
1,457	6.9
1,636	5.1
1,310	7.4

KPL

The scatter diagram in Figure 16.8 indicates a negative linear relationship between these two variables. Therefore, we use a simple first-order model to relate the two variables. The computer output is shown in Figure 16.9; the resulting estimated regression equation is:

$$KPL = 19.8 - 0.00907\ Weight$$

where:

KPL = kilometres-per-litre rating

$Weight$ = weight of the car in kilograms

FIGURE 16.8

Scatter diagram for the kilometres-per-litre problem

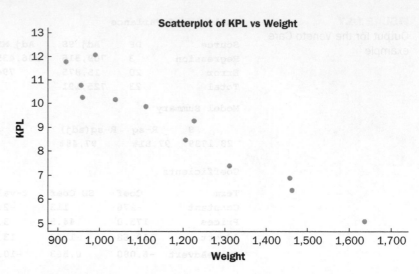

Scatterplot of KPL vs Weight

FIGURE 16.9

Output for the kilometres-per-litre problem

```
Analysis of Variance

Source        DF     Adj SS    Adj MS    F-value   p-value
Regression    -1     50.403    50.403    145.43    0.000
Error         10      3.466    0.3466
Total         11     53.869

Model Summary

    S        R-sq    R-sq(adj)
0.588710    93.57%    92.92%

Coefficients

Term        Coef      SE Coef    t-value   p-value
Constant    19.838     0.910      21.80    0.000
Weight     -0.009068   0.000752  -12.06    0.000

Regression Equation

KPL = 19.838 - 0.009068 Weight
```

The model is significant (*p*-value for the *F* test is 0.000) and the fit is very good (*R*-sq = 93.6 per cent). However, we note in Figure 16.9 that observation 3 is identified as having a large standardized residual.

Figure 16.10 is the standardized residual plot corresponding to the first-order model. The pattern we observe does not look like the horizontal band we should expect to find if the assumptions about the error term are valid. Instead, the variability in the residuals appears to increase as the value increases. In other words, we see the wedge-shaped pattern referred to in Chapter 14 as being indicative of a non-constant variance. We are not justified in reaching any conclusions about the statistical significance of the resulting estimated regression equation when the underlying assumptions for the tests of significance do not appear to be satisfied.

Often the problem of non-constant variance can be corrected by transforming the dependent variable to a different scale.

For instance, if we work with the logarithm of the dependent variable instead of the original dependent variable, the effect will be to compress the values of the dependent variable and thus diminish the effects of non-constant variance.

Most statistical packages provide the ability to apply logarithmic transformations using either the base 10 (common logarithm) or the base $e = 2.71828 \ldots$ (natural logarithm). We applied a natural logarithmic transformation to the kilometres-per-litre data and developed the estimated regression equation relating weight to the natural logarithm of KPL. The regression results obtained by using the natural logarithm of kilometres-per-litre as the dependent variable, labelled Log$_e$ KPL in the output, are shown in Figure 16.11; Figure 16.12 is the corresponding standardized residual plot.

FIGURE 16.10
Standardized residual plot for the kilometres-per-litre problem

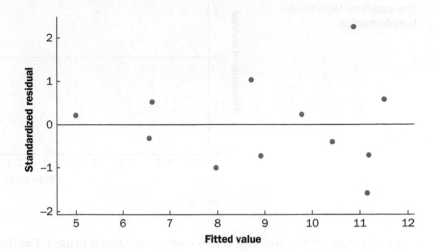

FIGURE 16.11
Output for the kilometres-per-litre problem: logarithmic transformation

Analysis of Variance

Source	DF	Adj SS	Adj MS	F-value	p-value
Regression	1	0.74221	0.742211	185.37	0.000
Error	10	0.04004	0.004004		
Total	11	0.78225			

Model Summary

S	R-sq	R-sq(adj)
0.0632759	94.88%	92.52%

Coefficients

Term	Coef	SE Coef	t-value	p-value
Constant	3.4811	0.0978	35.60	0.000
Weight	-0.001100	0.000081	-13.62	0.000

Regression Equation

LogeKPL = 3.4811 - 0.001100 Weight

Looking at the residual plot in Figure 16.12, we see that the wedge-shaped pattern has now disappeared. Moreover, none of the observations is identified as having a large standardized residual. The model with the logarithm of kilometres per litre as the dependent variable is statistically significant and provides an excellent fit to the observed data. Hence, we would recommend using the estimated regression equation:

$$\text{Log}_e \text{ KPL} = 3.49 - 0.00110 \text{ Weight}$$

To estimate the kilometres-per-litre rating for a car that weighs 1,500 kilograms, we first develop an estimate of the logarithm of the kilometres-per-litre rating:

$$\text{Log}_e \text{ KPL} = 3.49 - 0.00110 \, (1,500) = 1.84$$

The kilometres-per-litre estimate is obtained by finding the number whose natural logarithm is 1.84. Using a calculator with an exponential function, or raising e to the power 1.84, we obtain 6.2 kilometres per litre.

FIGURE 16.12
Standardized residual plot for the kilometres-per-litre problem: logarithmic transformation

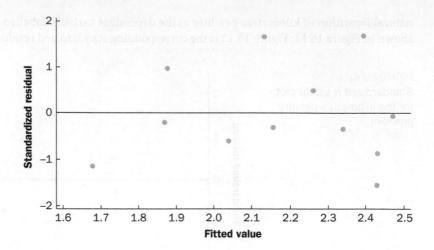

Another approach to problems of non-constant variance is to use $1/Y$ as the dependent variable instead of Y. This type of transformation is called a *reciprocal transformation*. For instance, if the dependent variable is measured in kilometres per litre, the reciprocal transformation would result in a new dependent variable whose units would be 1/(kilometres per litre) or litres per kilometre. In general, there is no way to determine whether a logarithmic transformation or a reciprocal transformation will perform better without actually trying each of them.

When considering a model with a transformed dependent variable, care must be taken when comparing it to a model that uses the untransformed (original) dependent variable or to a model that uses a different transformation of the dependent variable. Reconsider the kilometres-per-litre example in this section. The $R^2 = 93.57$ per cent of the model

$$\text{KPL} = 19.838 - 0.009068 \text{ Weight}$$

in Figure 16.9 should not be compared to the $R^2 = 94.88$ per cent of the model

$$\text{Log}_e \text{ KPL} = 3.4811 - 0.0011 \text{ Weight}$$

in Figure 16.11. This comparison is not valid because the units used to compute the R^2 in Figure 16.9 are measured in kilometres per litre, but the units used to compute the R^2 in Figure 16.11 are measured in (natural) logged kilometres per litre. Due to this difference in scale, the R^2 measures are not comparable.

To compare the fit of a model with a transformed dependent variable with one based on the original dependent variable, the estimated transformed model can be transformed back into the units of the original dependent variable. The fit can then be compared to a model based on the original dependent variable. (In the case of the kilometres-per-litre example, transforming back would entail raising each side of the fitted regression equation given in Figure 16.11 to the power of e to obtain:

$$\text{KPL} = e^{3.3811 - 0.0011 \text{ Weight}} = 32.5 e^{-0.0011 \text{ Weight}}$$

Forecasts from the latter model can now be compared with those based on the model in Figure 16.9.)

Nonlinear models that are intrinsically linear

Models in which the parameters $(\beta_0, \beta_1, \ldots, \beta_p)$ have exponents other than 1 are called nonlinear models. The exponential model involves the following regression equation:

$$E(Y) = \beta_0 \beta_1^x \qquad \textbf{(16.6)}$$

This model is appropriate when the dependent variable Y increases or decreases by a constant percentage, instead of by a fixed amount, as X increases.

As an example, suppose sales for a product Y are related to advertising expenditure X (in thousands of euros) according to the following exponential model:

$$E(Y) = 500 (1.2)^x$$

Thus: for $X = 1, E(Y) = 500(1.2)^1 = 600$; for $X = 2, E(Y) = 500(1.2)^2 = 720$; and for $X = 3$, $E(Y) = 500 (1.2)^3 = 864$ Note that $E(Y)$ is not increasing by a constant amount in this case, but by a constant percentage; the percentage increase is 20 per cent.

We can transform this nonlinear model to a linear model by taking the logarithm of both sides of equation (16.6).

$$\log E(Y) = \log \beta_0 + x \log \beta_1 \tag{16.7}$$

Now if we let $y' = \log E(Y)$, $\beta_0' = \log \beta_0$, and $\beta_1' = \log \beta_1$ we can rewrite equation (16.7) as:

$$y' = \beta_0' + \beta_1' x \tag{16.8}$$

It is clear that the formulae for simple linear regression can now be used to develop estimates of β_0' and β_1'. Denoting the estimates as β_0' and β_1' leads to the following estimated regression equation:

$$\hat{y}' = b_0' + b_1' x \tag{16.9}$$

To obtain predictions of the original dependent variable Y given a value of X, we would first substitute the value of X into equation (16.8) and compute $\hat{y}$. The antilog of $\hat{y}'$ would be the prediction of the expected value of Y.

Many nonlinear models cannot be transformed into an equivalent linear model. However, such models have had limited use in business and economic applications. Furthermore, the mathematical background needed for study of such models is beyond the scope of this text.

EXERCISES

Methods

1. Consider the following data for two variables, X and Y.

x	22	24	26	30	35	40
y	12	21	33	35	40	36

a. Develop an estimated regression equation for the data of the form $\hat{y} = b_0 + b_1 x$.
b. Use the results from part (a) to test for a significant relationship between X and Y. Use $\alpha = 0.05$.
c. Develop a scatter diagram for the data. Does the scatter diagram suggest an estimated regression equation of the form $\hat{y} = b_0 + b_1 x + b_2 x^2$? Explain.
d. Develop an estimated regression equation for the data of the form $\hat{y} = b_0 + b_1 x + b_2 x^2$.
e. Refer to part (d). Is the relationship between X, X^2 and Y significant? Use $\alpha = 0.05$.
f. Predict the value of Y when $X = 25$.

2. Consider the following data for two variables, X and Y.

x	9	32	18	15	26
y	10	20	21	16	22

a. Develop an estimated regression equation for the data of the form $\hat{y} = b_0 + b_1x$. Comment on the adequacy of this equation for predicting Y.

b. Develop an estimated regression equation for the data of the form $\hat{y} = b_0 + b_1x + b_2x^2$. Comment on the adequacy of this equation for predicting Y.

c. Predict the value of Y when $X = 20$.

3. Consider the following data for two variables, X and Y.

x	2	3	4	5	7	7	7	8	9
y	4	5	4	6	4	6	9	5	11

a. Does there appear to be a linear relationship between X and Y? Explain.

b. Develop the estimated regression equation relating X and Y.

c. Plot the standardized residuals versus the estimated regression equation developed in part (b). Do the model assumptions appear to be satisfied? Explain.

d. Perform a logarithmic transformation on the dependent variable Y. Develop an estimated regression equation using the transformed dependent variable. Do the model assumptions appear to be satisfied by using the transformed dependent variable? Does a reciprocal transformation work better in this case? Explain.

Applications

4. The table below lists the number of people (millions) living with HIV globally (www.avert.org/global-hiv-and-aids-statistics) from 2013 to 2017.

	Number of people
Year	living with HIV (m)
2013	35.2
2014	35.9
2015	36.7
2016	36.7
2017	36.9

a. Plot the data, letting $X = 0$ correspond to the year 2013. Find a linear $\hat{y} = b_0 + b_1x$ that models the data.

b. Plot the function on the graph with the data and determine how well the graph fits the data.

5. In working further with the problem of Exercise 4, statisticians suggested the use of the following curvilinear estimated regression equation.

$$\hat{y} = b_0 + b_1x + b_2x^2$$

a. Use the data of Exercise 4 to determine the estimated regression equation.

b. Use $\alpha = 0.01$ to test for a significant relationship.

6. An international study of life expectancy by Rossman (1994) covers the following variables:

LifeExp	Life expectancy in years
People.per.TV	Average number of people per TV
People.per.Dr	Average number of people per physician
LifeExp.Male	Male life expectancy in years
LifeExp.Female	Female life expectancy in years

With data details as follows:

LIFE
EXPECTANCY

	LifeExp	People.per.TV	People.per.Dr	LifeExp.Male	LifeExp.Female
Argentina	70.5	4	370	74	67
Bangladesh	53.5	315	6,166	53	54
Brazil	65	4	684	68	62
Canada	76.5	1.7	449	80	73
China	70	8	643	72	68
Colombia	71	5.6	1,551	74	68
Egypt	60.5	15	616	61	60
Ethiopia	51.5	503	36,660	53	50
France	78	2.6	403	82	74
Germany	76	2.6	346	79	73
India	57.5	44	2,471	58	57
Indonesia	61	24	7,427	63	59
Iran	64.5	23	2,992	65	64
Italy	78.5	3.8	233	82	75
Japan	79	1.8	609	82	76
Kenya	61	96	7,615	63	59
Korea.North	70	90	370	73	67
Korea.South	70	4.9	1,066	73	67
Mexico	72	6.6	600	76	68
Morocco	64.5	21	4,873	66	63
Myanmar	54.5	592	3,485	56	53
Pakistan	56.5	73	2,364	57	56
Peru	64.5	14	1,106	67	62
Philippines	64.5	8.8	1,062	67	62
Poland	73	3.9	480	77	69
Romania	72	6	559	75	69
Russia	69	3.2	259	74	64
South.Africa	64	11	1,340	67	61
Spain	78.5	2.6	275	82	75
Sudan	53	23	12,550	54	52
Taiwan	75	3.2	965	78	72
Tanzania	52.5	NA	25,229	55	50
Thailand	68.5	11	4,883	71	66
Turkey	70	5	1,189	72	68
Ukraine	70.5	3	226	75	66
UK	76	3	611	79	73
USA	75.5	1.3	404	79	72
Venezuela	74.5	5.6	576	78	71
Vietnam	65	29	3,096	67	63
Zaire	54	NA	23,193	56	52

(Note that the average number of people per TV is not given for Tanzania and Zaire.)

   a. Develop scatter diagrams for these data, treating LifeExp as the dependent variable.
   b. Does a simple linear model appear to be appropriate? Explain.
   c. Estimate simple regression equations for the data accordingly. Which do you prefer and why?

MEDIA

7. To assess the reliability of computer media, *Choice* magazine (www.choice.com.au) has obtained data by:

Price (AU$)	Paid in April 2005
Pack	Number of disks in the pack
Media	One of CD (CD), DVD (DVD-R) or DVDRW (DVD+/− RW)

with details as follows:

Price	Pack	Media	Price	Pack	Media
0.48	50	CD	1.85	10	DVD
0.60	25	CD	0.72	25	DVD
0.64	25	CD	2.28	10	DVD
0.50	50	CD	2.34	5	DVD
0.89	10	CD	2.40	10	DVD
0.89	10	CD	1.49	5	DVD
1.20	10	CD	3.60	5	DVDRW
1.30	10	CD	5.00	10	DVDRW
1.29	10	CD	2.79	5	DVDRW
0.50	10	CD	2.79	10	DVDRW
0.57	50	DVD	4.37	5	DVDRW
2.60	10	DVD	1.50	10	DVDRW
1.59	10	DVD	2.50	5	DVDRW
1.85	10	DVD	3.90	10	DVDRW

a. Develop scatter diagrams for these data with pack and media as potential independent variables.

b. Does a simple or multiple linear regression model appear to be appropriate?

c. Develop an estimated regression equation for the data you believe will best explain the relationship between these variables.

8. In Europe the number of internet users varies widely from country to country. In 1999, 44.3 per cent of all Swedes used the internet, while in France the audience was less than 10 per cent. The disparities are expected to persist even though internet usage is expected to grow dramatically over the next several years. The following table shows the number of internet users in 2011 and in 2018 for selected European countries. (www.internetworldstats.com/)

INTERNET
2018

	% Internet users	
	2011	2018
Austria	74.8	87.9
Belgium	81.4	94.4
Denmark	89.0	96.9
Finland	88.6	94.3
France	77.2	92.6
Germany	82.7	96.2
Ireland	66.8	92.7
Netherlands	89.5	95.9
Norway	97.2	99.2
Spain	65.6	92.6
Sweden	92.9	96.7
Switzerland	84.2	91.0
UK	84.5	94.7

a. Develop a scatter diagram of the data using the 2011 internet user percentage as the independent variable. Does a simple linear regression model appear to be appropriate? Discuss.

b. Develop an estimated multiple regression equation with X = the number of 2011 internet users and X^2 as the two independent variables.

c. Consider the nonlinear relationship shown by equation (16.6). Use logarithms to develop an estimated regression equation for this model.

d. Do you prefer the estimated regression equation developed in (b) or (c)? Explain.

16.2 DETERMINING WHEN TO ADD OR DELETE VARIABLES

In this section we will show how an F test can be used to determine whether it is advantageous to add one or more independent variables to a multiple regression model. This test is based on a determination of the amount of reduction in the error sum of squares resulting from adding one or more independent variables to the model. We will first illustrate how the test can be used in the context of the Eurodistributor Company example covered in Chapter 15. Eurodistributor was looking to improve how it creates delivery schedules for its drivers. To this effect, the managers there sought to develop an estimated regression equation to predict total daily travel time for trucks using two independent variables: kilometres travelled and number of deliveries. With kilometres travelled X_1 as the only independent variable, the least squares procedure provided the following estimated regression equation:

$$\hat{y} = 1.27 + 0.0678x_1$$

The error sum of squares for this model was SSE = 8.029. When X_2, the number of deliveries, was added as a second independent variable, we obtained the following estimated regression equation:

$$\hat{y} = -0.869 + 0.06113x_1 + 0.923x_2$$

The error sum of squares for this model was SSE = 2.2994. Adding X_2 resulted in a reduction of SSE. The question we want to answer is: does adding the variable X_2 lead to a *significant* reduction in SSE?

We use the notation $SSE(x_1)$ to denote the error sum of squares when X_1 is the only independent variable in the model, $SSE(x_1, x_2)$ to denote the error sum of squares when X_1 and X_2 are both in the model, and so on. Hence, the reduction in SSE resulting from adding X_2 to the model involving just X_1 is:

$$SSE(x_1) - SSE(x_1, x_2) = 8.029 - 2.2994 = 5.7296$$

An F test is conducted to determine whether this reduction is significant. The numerator of the F statistic is the reduction in SSE divided by the number of independent variables added to the original model. Here only one variable, X_2, has been added; therefore, the numerator of the F statistic is:

$$\frac{SSE(x_1) - SSE(x_1, x_2)}{1} = 5.7296$$

The result is a measure of the reduction in SSE per independent variable added to the model. The denominator of the F statistic is the mean square error for the model that includes all of the independent variables. For Eurodistributor, this corresponds to the model containing both X_1 and X_2; therefore, $p = 2$ and:

$$MSE = \frac{SSE(x_1, x_2)}{n - p - 1} = \frac{2.2994}{7} = 0.3285$$

The following F statistic provides the basis for testing whether the addition of X_2 is statistically significant.

$$F = \frac{\dfrac{SSE(x_1) - SSE(x_1, x_2)}{1}}{\dfrac{SSE(x_1, x_2)}{n - p - 1}}$$ **(16.10)**

The numerator degrees of freedom for this F test is equal to the number of variables added to the model, and the denominator degrees of freedom is equal to $n - p - 1$.

For the Eurodistributor problem, we obtain:

$$F = \frac{\dfrac{5.7296}{1}}{\dfrac{2.2994}{7}} = \frac{5.7296}{0.3285} = 17.44$$

Because $F = 17.44 > F_{0.05} = 5.59$, we can reject the null hypothesis that X_2 is not statistically significant. In other words, adding X_2 to the model involving only X_1 results in a significant reduction in the error sum of squares – and therefore a significant improvement in fit.

When we wish to test for the significance of adding only one more independent variable to a model, the result found with the F test just described could also be obtained by using the t test for the significance of an individual parameter. Indeed, the F statistic we just computed is the square of the t statistic used to test the significance of an individual parameter.

Because the t test is equivalent to the F test when only one independent variable is being added to the model, we can now further clarify the proper use of the t test for testing the significance of an individual parameter. If an individual parameter is not significant, the corresponding variable can be dropped from the model. However, if the t test shows that two or more parameters are not significant, no more than one independent variable can ever be dropped from a model on the basis of a t test; if one variable is dropped, a second variable that was not significant initially might become significant.

We now turn to a consideration of whether the addition of more than one independent variable – as a set – results in a significant reduction in the error sum of squares.

General case

Consider the following multiple regression model involving q independent variables, where $q < p$.

$$Y = \beta_0 + \beta_1 x_1 + \beta_2 x_2 + \cdots + \beta_q x_q + \varepsilon$$ **(16.11)**

If we add variables $X_{q+1}, X_q, \ldots, X_p$ to this model, we obtain a model involving p independent variables.

$$Y = \beta_0 + \beta_1 x_1 + \beta_2 x_2 + \cdots + \beta_q x_q$$
$$+ \beta_{q+1} x_{q+1} + \beta_{q+2} x_{q+2} + \cdots + \beta_p x_p + \varepsilon$$ **(16.12)**

To test whether the addition of $X_{q+1}, X_q, \ldots, X_p$ is statistically significant, the null and alternative hypotheses can be stated as follows:

$$H_0: \beta_{q+1} = \beta_{q+2} = \cdots = \beta_p = 0$$

H_1: One or more of the parameters is not equal to zero.

The following F statistic provides the basis for testing whether the additional independent variables are statistically significant.

F test statistic for adding or deleting p–q variables

$$F = \frac{\dfrac{\text{SSE}(x_1, x_2, \ldots, x_q) - \text{SSE}(x_1, x_2, .x_q, x_{q+1}, \ldots x_p)}{p - q}}{\dfrac{\text{SSE}(x_1, x_2, \ldots, x_q, x_{q+1}, \ldots x_p)}{n - p - 1}}$$

(16.13)

This computed F value is then compared with F_α, the table value with $p - q$ numerator degrees of freedom and $n - p - 1$ denominator degrees of freedom. If $F > F_\alpha$ we reject H_0 and conclude that the set of additional independent variables is statistically significant.

Many students find equation (16.13) somewhat complex. To provide a simpler description of this F ratio, we can refer to the model with the smaller number of independent variables as the reduced model and the model with the larger number of independent variables as the full model. If we let SSE (reduced) denote the error sum of squares for the reduced model and SSE (full) denote the error sum of squares for the full model, we can write the numerator of (16.13) as:

$$\frac{\text{SSE(reduced)} - \text{SSE(full)}}{\text{number of extra terms}}$$

(16.14)

Note that 'number of extra terms' denotes the difference between the number of independent variables in the full model and the number of independent variables in the reduced model. The denominator of equation (16.13) is the error sum of squares for the full model divided by the corresponding degrees of freedom; in other words, the denominator is the mean square error for the full model. Denoting the mean square error for the full model as MSE(full) enables us to write F as:

$$F = \frac{\dfrac{\text{SSE(reduced)} - \text{SSE(full)}}{\text{number of extra terms}}}{\text{MSE(full)}}$$

(16.15)

EXERCISES

Methods

9. In a regression analysis involving 27 observations, the following estimated regression equation was developed.

$$\hat{y} = 25.2 + 5.5x_1$$

For this estimated regression equation SST = 1,550 and SSE = 520.

   a. At $\alpha = 0.05$, test whether X_1 is significant.
   Suppose that variables X_2 and X_3 are added to the model and the following regression equation is obtained.

$$\hat{y} = 16.3 + 2.3x_1 + 12.1x_2 - 5.8x_3$$

   For this estimated regression equation SST = 1,550 and SSE = 100.

   b. Use an F test and a 0.05 level of significance to determine whether X_2 and X_3 contribute significantly to the model.

10. In a regression analysis involving 30 observations, the following estimated regression equation was obtained.

$$\hat{y} = 17.6 + 3.8x_1 - 2.3x_2 + 7.6x_3 + 2.7x_4$$

For this estimated regression equation SST = 1,805 and SSR = 1,760.

a. At $\alpha = 0.05$, test the significance of the relationship between the variables.
Suppose variables X_1 and X_4 are dropped from the model and the following estimated regression equation is obtained.

$$\hat{y} = 11.1 - 3.6x_2 + 8.1x_3$$

For this model SST = 1,805 and SSR = 1,705.

a. Compute $SSE(x_1, x_2, x_3, x_4)$.
b. Compute $SSE(x_2, x_3)$.
c. Use an F test and a 0.05 level of significance to determine whether X_1 and X_4 contribute significantly to the model.

Applications

11. In an experiment involving measurements of heat production (calories) at various body masses (kg) and work levels (calories/hour) on a stationary bike, the following results were obtained.

Body mass (M)	Work level (W)	Heat production (H)
43.7	19	177
43.7	43	279
43.7	56	346
54.6	13	160
54.6	19	193
54.6	56	335
55.7	13	169
55.7	26	212
55.7	34.5	244
55.7	43	285
58.8	13	181
58.8	43	298
60.5	19	212
60.5	43	317
60.5	56	347
61.9	13	186
61.9	19	216
61.9	34.5	265
61.9	43	306
61.9	56	348
66.7	13	209
66.7	43	324
66.7	56	352

MUSCLE

a. Develop an estimated regression equation that can be used to predict heat production for a given body mass and work level.
b. Consider adding an independent variable to the model developed in (a) for the interaction between body mass and work level. Develop an estimated regression equation using these three independent variables.
c. At a 0.05 level of significance, test to see whether the addition of the interaction term contributes significantly to the estimated regression equation developed in (a).

12. A ten-year study conducted by the AHA provided data on how age, blood pressure and smoking relate to the risk of strokes. Data from a portion of this study follow. Risk is interpreted as the probability (times 100) that a person will have a stroke over the next ten-year period. For the smoker variable, 1 indicates a smoker and 0 indicates a non-smoker.
   a. Develop an estimated regression equation that can be used to predict the risk of stroke given the age and blood-pressure level.
   b. Consider adding two independent variables to the model developed in (a), one for the interaction between age and blood-pressure level and the other for whether the person is a smoker. Develop an estimated regression equation using these four independent variables.
   c. At a 0.05 level of significance, test to see whether the addition of the interaction term and the smoker variable contribute significantly to the estimated regression equation developed in (a).

Risk	Age (years)	Systolic blood pressure (mmHg)	Smoker
12	57	152	0
24	67	163	0
13	58	155	0
56	86	177	1
28	59	196	0
51	76	189	1
18	56	155	1
31	78	120	0
37	80	135	1
15	78	98	0
22	71	152	0
36	70	173	1
15	67	135	1
48	77	209	1
15	60	199	0
36	82	119	1
8	66	166	0
34	80	125	1
3	62	117	0
37	59	207	1

16.3 VARIABLE SELECTION PROCEDURES

In the stepwise regression, forward selection and backward elimination procedures, the criterion for selecting an independent variable to add or delete from the model at each step is based on the F statistic introduced in Section 16.2. Suppose, for instance, that we are considering adding x_2 to a model involving x_1 or deleting x_2 from a model involving x_1 and x_2. To test whether the addition or deletion of x_2 is statistically significant, the null and alternative hypotheses can be stated as follows:

$$H_0: \beta_2 = 0$$
$$H_1: \beta_2 \neq 0$$

In Section 16.2 (refer to equation (16.10)) we showed that

$$F = \frac{\dfrac{SSE(x_1) - SSE(x_1, x_2)}{1}}{\dfrac{SSE(x_1, x_2)}{n - p - 1}}$$

can be used as a criterion for determining whether the presence of x_2 in the model causes a significant reduction in the error sum of squares. The p-value corresponding to this F statistic is the criterion used to determine whether an independent variable should be added or deleted from the regression model. The usual rejection rule applies: Reject H_0 if p-value $\leq \alpha$.

To provide an illustration of variable selection procedures, we introduce the Cravens data set consisting of 25 observations on eight independent variables. (Permission to use these data was provided by Dr David W. Cravens of the Department of Marketing at Texas Christian University.)

Regression analysis of Cravens example

CRAVENS

The Cravens data are for a company that sells products in several sales territories, each of which is assigned to a single sales representative. A regression analysis was conducted to determine whether a variety of predictor (independent) variables could explain sales in each territory. A random sample of 25 sales territories resulted in the data in Table 16.5; the variable definitions are given in Table 16.6.

TABLE 16.5 Cravens data

Sales	Time	Poten	AdvExp	Share	Change	Accounts	Work	Rating
3,669.88	43.10	74,065.1	4,582.9	2.51	0.34	74.86	15.05	4.9
3,473.95	108.13	58,117.3	5,539.8	5.51	0.15	107.32	19.97	5.1
2,295.10	13.82	21,118.5	2,950.4	10.91	−0.72	96.75	17.34	2.9
4,675.56	186.18	68,521.3	2,243.1	8.27	0.17	195.12	13.40	3.4
6,125.96	161.79	57,805.1	7,747.1	9.15	0.50	180.44	17.64	4.6
2,134.94	8.94	37,806.9	402.4	5.51	0.15	104.88	16.22	4.5
5,031.66	365.04	50,935.3	3,140.6	8.54	0.55	256.10	18.80	4.6
3,367.45	220.32	35,602.1	2,086.2	7.07	−0.49	126.83	19.86	2.3
6,519.45	127.64	46,176.8	8,846.2	12.54	1.24	203.25	17.42	4.9
4,876.37	105.69	42,053.2	5,673.1	8.85	0.31	119.51	21.41	2.8
2,468.27	57.72	36,829.7	2,761.8	5.38	0.37	116.26	16.32	3.1
2,533.31	23.58	33,612.7	1,991.8	5.43	−0.65	142.28	14.51	4.2
2,408.11	13.82	21,412.8	1,971.5	8.48	0.64	89.43	19.35	4.3
2,337.38	13.82	20,416.9	1,737.4	7.80	1.01	84.55	20.02	4.2
4,586.95	86.99	36,272.0	10,694.2	10.34	0.11	119.51	15.26	5.5
2,729.24	165.85	23,093.3	8,618.6	5.15	0.04	80.49	15.87	3.6
3,289.40	116.26	26,878.6	7,747.9	6.64	0.68	136.58	7.81	3.4
2,800.78	42.28	39,572.0	4,565.8	5.45	0.66	78.86	16.00	4.2
3,264.20	52.84	51,866.1	6,022.7	6.31	−0.10	136.58	17.44	3.6
3,453.62	165.04	58,749.8	3,721.1	6.35	−0.03	138.21	17.98	3.1
1,741.45	10.57	23,990.8	861.0	7.37	−1.63	75.61	20.99	1.6
2,035.75	13.82	25,694.9	3,571.5	8.39	−0.43	102.44	21.66	3.4
1,578.00	8.13	23,736.3	2,845.5	5.15	0.04	76.42	21.46	2.7
4,167.44	58.44	34,314.3	5,060.1	12.88	0.22	136.58	24.78	2.8
2,799.97	21.14	22,809.5	3,552.0	9.14	−0.74	88.62	24.96	3.9

As a preliminary step, we consider the sample correlation coefficients between each pair of variables. Figure 16.13 is the correlation matrix obtained using statistical software. Note that the sample correlation coefficient between Sales and Time is 0.623, between Sales and Poten is 0.598 and so on.

Looking at the sample correlation coefficients between the independent variables, we see that the correlation between Time and Accounts is 0.758 and significant; hence, if Accounts were used as an independent variable, Time would not add much more explanatory power to the model. Recall that inclusion of highly correlated independent variables, as discussed in Section 15.5 on multicollinearity, can cause problems for the model. If possible, then, we should avoid including both Time and Accounts in the same regression model. The sample correlation coefficient of 0.549 between Change and Rating is also significant (p-value < 0.05) so this may prove similarly problematic.

TABLE 16.6 Variable definitions for the Cravens data

Variable	Definition
Sales	Total sales credited to the sales representative
Time	Length of time employed in months
Poten	Market potential; total industry sales in units for the sales territory*
AdvExp	Advertising expenditure in the sales territory
Share	Market share; weighted average for the past four years
Change	Change in the market share over the previous four years
Accounts	Number of accounts assigned to the sales representative*
Work	Workload; a weighted index based on annual purchases and concentrations of accounts
Rating	Sales representative overall rating on eight performance dimensions; an aggregate rating on a 1–7 scale

* These data were coded to preserve confidentiality.

FIGURE 16.13 Sample correlation coefficients for the Cravens data

	Sales	Time	Poten	AdvExp	Share	Change	Accounts	Work
Time	0.623							
Poten	0.598	0.454						
AdvExp	0.596	0.249	0.174					
Share	0.484	0.106	-0.21	0.264				
Change	0.489	0.251	0.268	0.377	0.085			
Accounts	0.754	0.758	0.479	0.200	0.403	0.327		
Work	-0.117	-0.179	-0.259	-0.272	0.349	-0.288	-0.199	
Rating	0.402	0.101	0.359	0.411	-0.024	0.549	0.229	-0.277

Looking at the sample correlation coefficients between Sales and each of the independent variables can give us a quick indication of which independent variables are, by themselves, good predictors. We see that the single best predictor of Sales is Accounts, because it has the highest sample correlation coefficient (0.754). Recall that for the case of one independent variable, the square of the sample correlation coefficient is the coefficient of determination.

Thus, Accounts can explain $(0.754)^2(100)$, or 56.85 per cent, of the variability in Sales. The next most important independent variables are Time, Poten and AdvExp, each with a sample correlation coefficient of approximately 0.6.

Although there are potential multicollinearity problems, we consider developing an estimated regression equation using all eight independent variables. Computer software was used to provide the results in Figure 16.14. The eight-variable multiple regression model has an adjusted coefficient of determination of 88.3 per cent. Note, however, that the p-values for the t tests of individual parameters show that only Poten, AdvExp and Share are significant at the $\alpha = 0.05$ level, given the effect of all the other variables. Hence, we might be inclined to investigate the results that would be obtained if we used just those three variables. Figure 16.15 shows the results obtained for the estimated regression equation with those three variables. We see that the estimated regression equation has an adjusted coefficient of determination of 82.7 per cent, which, although not quite as good as that for the eight-independent-variable estimated regression equation, is still high.

FIGURE 16.14
Output for the model involving all eight independent variables

Analysis of Variance

Source	DF	Adj SS	Adj MS	F-value	p-value
Regression	8	38,153,712	4,769,214	23.66	0.000
Error	16	3,225,837	201,615		
Total	24	41,379,549			

Model Summary

S	R-sq	R-sq(adj)
499.015	92.20%	88.31%

Coefficients

Term	Coef	SE Coef	t-value	p-value
Constant	-1508	779	-1.94	0.071
Time	2.01	1.93	1.04	0.313
Poten	0.03721	0.00820	4.54	0.000
AdvExp	0.1510	0.0471	3.21	0.006
Share	199.0	67.0	2.97	0.009
Change	291	187	1.56	0.139
Accounts	5.55	4.78	1.16	0.262
Work	19.8	33.7	0.59	0.565
Rating	8	128	0.06	0.950

Regression Equation

Sales = -1508 + 2.01 Time + 0.03721 Poten + 0.1510 AdvExp + 1.99 Share
+ 291 Change + 5.55 Accounts + 19.8 Work + 8 Rating

FIGURE 16.15
Output for the model involving Poten, AdvExp and Share

Analysis of Variance

Source	DF	Adj SS	Adj MS	F-value	p-value
Regression	3	35,130,228	11,710,076	39.35	0.000
Error	21	6,249,321	297,587		
Total	24	41,379,549			

Model Summary

S	R-sq	R-sq(adj)
545.515	84.90%	82.74%

Coefficients

Term	Coef	SE Coef	t-value	p-value
Constant	-1604	506	-3.17	0.005
Poten	0.05429	0.00747	7.26	0.000
AdvExp	0.1675	0.0443	3.78	0.001
Share	282.7	48.8	5.80	0.000

Regression Equation

Sales = -1604 + 0.05429 Poten + 0.1675 AdvExp + 282.7 Share

How can we find an estimated regression equation that will do the best job given the data available? One approach is to compute all possible regressions. That is, we could develop eight one-variable estimated regression equations (each of which corresponds to one of the independent variables), 28 two-variable estimated regression equations (the number of combinations of eight variables taken two at a time) and so on.

In all, for the Cravens data, 255 different estimated regression equations involving one or more independent variables would have to be fitted to the data.

With the powerful computer packages available today, it is possible to compute all possible regressions. But doing so requires the model builder to review a large volume of computer output, much of which is associated with obviously poor models. Statisticians prefer a more systematic approach to selecting the subset of independent variables that provide the best estimated regression results. We now introduce some of the more popular approaches.

Stepwise regression

Based on this statistic, the stepwise regression procedure begins each step by determining whether any of the variables *already in the model* should be removed. If none of the independent variables can be removed from the model, the procedure checks to see whether any of the independent variables that are not currently in the model can be entered.

Due to the nature of the stepwise regression procedure, an independent variable can enter the model at one step, be removed at a subsequent step and then enter the model at a later step. The procedure stops when there are no independent variables to be removed from or entered into the model.

Figure 16.16 shows the results obtained by computer using the stepwise regression procedure for the Cravens data using values of 0.05 for *Alpha to remove* and 0.05 for *Alpha to enter*. (These are the technical settings used by software for deciding whether an independent variable should be removed or entered into the model.) The stepwise procedure terminated after four steps. The estimated regression equation identified by the stepwise regression procedure is:

$$\hat{y} = -1442 + 0.03822 \text{ Poten} + 0.1750 \text{ AdvExp} + 190.1 \text{ Share} + 9.21 \text{ Accounts}$$

Note also in Figure 16.16 that $s = \sqrt{\text{MSE}}$. MSE has been reduced from 881.093 with the best one-variable model (using Accounts) to 453.836 after four steps. The value of *R*-sq has been increased from 56.85 per cent to 90.04 per cent, and the recommended estimated regression equation has an *R*-sq(adj) value of 88.05 per cent.

Forward selection

The forward selection procedure starts with no independent variables. It adds variables one at a time using the same procedure as stepwise regression for determining whether an independent variable should be entered into the model. However, the forward selection procedure does not permit a variable to be removed from the model once it has been entered.

The estimated regression equation obtained using the forward selection procedure is:

$$\hat{y} = -1442 + 0.03822 \text{ Poten} + 0.1750 \text{ AdvExp} + 190.1 \text{ Share} + 9.21 \text{ Accounts}$$

Thus, for the Cravens data, the forward selection procedure leads to the same estimated regression equation as the stepwise procedure.

Backward elimination

The backward elimination procedure begins with a model that includes all the independent variables. It then deletes one independent variable at a time using the same procedure as stepwise regression. However, the backward elimination procedure does not permit an independent variable to be re-entered once it has been removed.

The estimated regression equation obtained using the backward elimination procedure for the Cravens data is:

$$\hat{y} = -1312 + 3.82 \text{ Time} + 0.04440 \text{ Poten} + 0.1525 \text{ AdvExp} + 259.5 \text{ Share}$$

Comparing the estimated regression equation identified using the backward elimination procedure to the estimated regression equation identified using the forward selection procedure, we see that three independent variables – AdvExp, Poten and Share – are common to both. However, the backward elimination procedure has included Time instead of Accounts.

Forward selection and backward elimination are the two extremes of model building; the forward selection procedure starts with no independent variables in the model and adds independent variables one at a time, whereas the backward elimination procedure starts with all independent variables in the model and deletes variables one at a time. The two procedures may lead to the same estimated regression equation. It is possible, however, for them to lead to two different estimated regression equations, as we saw with the Cravens data. Deciding which estimated regression equation to use remains a topic for discussion. Ultimately, the analyst's judgement must be applied. The best-subsets model-building procedure we discuss next provides additional model-building information to be considered before a final decision is made.

FIGURE 16.16   Stepwise regression output for the Cravens data

```
Stepwise Selection of Terms

Candidate terms: Time, Poten, AdvExp, Share, Change, Accounts, Work, Rating

              ----Step 1----    ---Step 2----    ---Step 3----    ---Step 4----

                Coef      P      Coef      P      Coef      P      Coef      P

Constant        709              50              -327             -1442
Accounts       21.72   0.000    19.05   0.000    15.55   0.000     9.21    0.004
AdvExp                          0.2265  0.000    0.2161  0.000    0.1750   0.000
Poten                                            0.02192 0.019    0.03822  0.000
Share                                                             190.1    0.001

S              881.093          650.392         582.636          453.836
R-sq            56.85%           77.51%          82.77%           90.04%
R-sq(adj)       54.97%           75.47%          80.31%           88.05%
R-sq(pred)      43.32%           70.04%          76.41%           85.97%
Mallows'Cp      67.56            27.16           18.36             5.43

α-to-enter = 0.05, α-to-leave = 0.05


Analysis of Variance

Source        DF      Adj SS       Adj MS     F-value    p-value
Regression     4    37,260,200   9,315,050     45.23      0.000
Error         20     4,119,349     205,967
Total         24    41,379,549


Model Summary

      S         R-sq      R-sq(adj)
  453.836      90.04%       88.05%


Coefficients

Term        Coef      SE Coef     t-value     p-value
Constant    -1442       424        -3.40       0.003
Poten      0.03822    0.00798       4.79       0.000
AdvExp     0.1750     0.0369        4.74       0.000
Share      190.1      49.7          3.82       0.000
Accounts    9.21      2.87          3.22       0.004


Regression Equation

Sales = -1442 + 0.03822 Poten + 0.1750 AdvExp + 190.1 Share + 9.21 Accounts
```

Best-subsets regression

Stepwise regression, forward selection and backward elimination are approaches to choosing the regression model by adding or deleting independent variables one at a time. None of them guarantees that the best model for a given number of variables will be found. Hence, these one-variable-at-a-time methods are properly viewed as heuristics for selecting a good regression model.

Some software packages use a procedure called best-subsets regression that enables the user to find, given a specified number of independent variables, the best regression model. Figure 16.17 shows a portion of the computer output obtained by using the best-subsets procedure for the Cravens data set.

This output identifies the two best one-variable estimated regression equations, the two best two-variable equations, the two best three-variable equations and so on. Usually the criterion used in determining which estimated regression equations are best for any number of predictors is the value of the adjusted coefficient of determination (R-sq(adj)). For instance, Accounts, with an R-sq(adj) = 55.0 per cent, provides the best estimated regression equation using only one independent variable; AdvExp and Accounts, with an R-sq(adj) = 75.5 per cent, provides the best estimated regression equation using two independent variables; and Poten, AdvExp and Share, with an R-sq(adj) = 82.7 per cent, provides the best estimated regression equation with three independent variables.

For the Cravens data, the adjusted coefficient of determination, R-sq(adj) = 89.4 per cent, is largest for the model with six independent variables: Time, Poten, AdvExp, Share, Change and Accounts. However, the best model with four independent variables (Poten, AdvExp, Share, Accounts) has an adjusted coefficient of determination almost as high (88.1 per cent). All other things being equal, a simpler model with fewer variables is usually preferred.

FIGURE 16.17 Portion of best-subsets regression output

Model	Number	RSquare	RMSE
Accounts	1	0.5685	881.09
Time	1	0.3880	1049.33
AdvExp, Accounts	2	0.7751	650.39
Poten, Share	2	0.7461	691.11
Poten, AdvExp, Share	3	0.8490	545.52
Poten, AdvExp, Accounts	3	0.8277	582.64
Poten, AdvExp, Share, Accounts	4	0.9004	453.84
Time, Poten, AdvExp, Share	4	0.8960	463.93
Time, Poten, AdvExp, Share, Change	5	0.9150	430.22
Poten, AdvExp, Share, Change, Accounts	5	0.9124	436.75
Time, Poten, AdvExp, Share, Change, Accounts	6	0.9203	427.99
Poten, AdvExp, Share, Change, Accounts, Work	6	0.9165	438.20
Time, Poten, AdvExp, Share, Change, Accounts, Work	7	0.9220	435.66
Time, Poten, AdvExp, Share, Change, Accounts, Rating	7	0.9204	440.29
Time, Poten, AdvExp, Share, Change, Accounts, Work, Rating	8	0.9220	449.02

Making the final choice

The analysis performed on the Cravens data to this point is good preparation for choosing a final model, but more analysis should be conducted before the final choice is made. As we noted in Chapters 14 and 15, a careful analysis of the residuals should be undertaken. We want the residual plot for the chosen model to resemble approximately a horizontal band. Let us assume the residuals are not a problem and that we want to use the results of the best-subsets procedure to help choose the model.

The best-subsets procedure shows us that the best four-variable model contains the independent variables Poten, AdvExp, Share and Accounts. This result also happens to be the four-variable model identified with the stepwise regression procedure. Note also that the S and R-sq(adj) results are virtually identical between the two models. Furthermore, there is very little difference between the corresponding R-sq values.

EXERCISES

Applications

13. Brownlee (1965) presents stack loss data for a chemical plant involving 21 observations on 4 variables, namely:

Airflow: Flow of cooling air

Temp: Cooling water inlet temperature

Acid: Concentration of acid [per 1,000, minus 500]

Loss: Stack loss (the dependent variable) is 10 times the percentage of the ingoing ammonia to the plant that escapes from the absorption column unabsorbed, that is, an (inverse) measure of the overall efficiency of the plant.

Loss	Airflow	Temp	Acid
42	80	27	89
37	80	27	88
37	75	25	90
28	62	24	87
18	62	22	87
18	62	23	87
19	62	24	93
20	62	24	93
15	58	23	87
14	58	18	80
14	58	18	89
13	58	17	88
11	58	18	82
12	58	19	93
8	50	18	89
7	50	18	86
8	50	19	72
8	50	19	79
9	50	20	80
15	56	20	82
15	70	20	91

Develop an estimated regression equation that can be used to predict loss. Briefly discuss the process you used to develop a recommended estimated regression equation for these data.

14. A study investigated the relationship between audit delay (Delay), the length of time from a company's fiscal year-end to the date of the auditor's report, and variables that describe the client and the auditor. Some of the independent variables that were included in this study follow.

Industry   A dummy variable coded 1 if the firm was an industrial company or 0 if the firm was a bank, savings and loan, or insurance company.

Public   A dummy variable coded 1 if the company was traded on an organized exchange or over the counter; otherwise coded 0.

Quality   A measure of overall quality of internal controls, as judged by the auditor, on a five-point scale ranging from 'virtually none' (1) to 'excellent' (5).

Finished   A measure ranging from 1 to 4, as judged by the auditor, where 1 indicates 'all work performed subsequent to year end' and 4 indicates 'most work performed prior to year end'.

A sample of 40 companies provided the following data.

AUDIT
DELAY

Delay	Industry	Public	Quality	Finished
62	0	0	3	1
45	0	1	3	3
54	0	0	2	2
71	0	1	1	2
91	0	0	1	1
62	0	0	4	4
61	0	0	3	2
69	0	1	5	2
80	0	0	1	1
52	0	0	5	3
47	0	0	3	2
65	0	1	2	3
60	0	0	1	3
81	1	0	1	2
73	1	0	2	2
89	1	0	2	1
71	1	0	5	4
76	1	0	2	2
68	1	0	1	2
68	1	0	5	2
86	1	0	2	2
76	1	1	3	1
67	1	0	2	3
57	1	0	4	2
55	1	1	3	2
54	1	0	5	2
69	1	0	3	3
82	1	0	5	1
94	1	0	1	1
74	1	1	5	2
75	1	1	4	3
69	1	0	2	2
71	1	0	4	4
79	1	0	5	2

(Continued)

Delay	Industry	Public	Quality	Finished
80	1	0	1	4
91	1	0	4	1
92	1	0	1	4
46	1	1	4	3
72	1	0	5	2
85	1	0	5	1

a. Develop the estimated regression equation using all of the independent variables.
b. Did the estimated regression equation developed in part (a) provide a good fit? Explain.
c. Develop a scatter diagram showing Delay as a function of Finished. What does this scatter diagram indicate about the relationship between Delay and Finished?
d. On the basis of your observations about the relationship between Delay and Finished, develop an alternative estimated regression equation to the one developed in (a) to explain as much of the variability in Delay as possible.

LAYOFFS

15. A study provided data on variables that may be related to the number of weeks a person has been jobless. The dependent variable in the study (Weeks) was defined as the number of weeks a person has been jobless due to a layoff. The following independent variables were used in the study:

Age	The age of the person
Educ	The number of years of education
Married	A dummy variable; 1 if married, 0 otherwise
Head	A dummy variable; 1 if the head of household, 0 otherwise
Tenure	The number of years on the previous job
Manager	A dummy variable; 1 if management occupation, 0 otherwise
Sales	A dummy variable; 1 if sales occupation, 0 otherwise

The data are available in the file 'LAYOFFS'.
a. Develop the best one-variable estimated regression equation.
b. Use the stepwise procedure to develop the best estimated regression equation. Use values of 0.05 for α-to-enter and α-to-leave.
c. Use the forward selection procedure to develop the best estimated regression equation. Use a value of 0.05 for α-to-enter.
d. Use the backward elimination procedure to develop the best estimated regression equation. Use a value of 0.05 for α-to-leave.
e. Use the best-subsets regression procedure to develop the best estimated regression equation.

16. Refer to the data in Exercise 14.
a. Develop an estimated regression equation that can be used to predict Delay by using Industry and Quality.
b. Plot the residuals obtained from the estimated regression equation developed in (a) as a function of the order in which the data are presented. Does any autocorrelation appear to be present in the data? Explain.
c. At the 0.05 level of significance, test for any positive autocorrelation in the data.

ONLINE RESOURCES

For the data files, additional questions and answers, and the software section for Chapter 16, go to the online platform.

SUMMARY

In this chapter we discussed several concepts used by model builders in identifying the best estimated regression equation. First, we introduced the concept of a general linear model to show how the methods discussed in Chapters 14 and 15 could be extended to handle curvilinear relationships and interaction effects. Then we discussed how transformations involving the dependent variable could be used to account for problems such as non-constant variance in the error term.

In many applications of regression analysis, a large number of independent variables are considered. We presented a general approach based on an F statistic for adding or deleting variables from a regression model. We then introduced a larger problem involving 25 observations and 8 independent variables. We saw that one issue encountered in solving larger problems is finding the best subset of the independent variables. To help in that task, we discussed several variable selection procedures: stepwise regression, forward selection, backward elimination and best-subsets regression.

KEY TERM

General linear model

KEY FORMULAE

General linear model

$$Y = \beta_0 + \beta_1 z_1 + \beta_2 z_2 + \ldots + \beta_p z_p + \varepsilon \tag{16.1}$$

F test statistic for adding or deleting $p-q$ variables

$$F = \frac{\dfrac{\text{SSE}(x_1, x_2, \ldots, x_q) - \text{SSE}(x_1, x_2, . x_q, x_{q+1}, \ldots, x_p)}{p - q}}{\dfrac{\text{SSE}(x_1, x_2, \ldots, x_q, x_{q+1}, \ldots, x_p)}{n - p - 1}} \tag{16.13}$$

CASE PROBLEM 1

House prices

The data relate to bungalow and two-storey homes located in ten selected neighbourhoods of Canada. Each home was listed and sold individually through the Multiple Listing System.

Apart from the dependent variable list price, basic house descriptive variables were categorized into two groups as shown in Table 1, which cover house attributes and lot attributes.

Managerial report

Use the methods presented in this and previous chapters to analyze this data set. Present a

summary of your analysis, including key statistical results, conclusions and recommendations, in a managerial report. Include any appropriate technical material (computer output, residual plots, etc.) in an appendix.

HOMESALES

© MediaProduction/iStock

(Continued)

TABLE 1 Definition of variables

Variable	Definition
House attributes	
STYLE	1 if bungalow, 2 if two-storey
R	Number of rooms
B	Number of bathrooms
BR	Number of bedrooms
S	Living area (square metres)
A	Age (years)
BAS	Basement (from 1 (open) to 3 (finished))
G	Number of garage space
ATT	Dummy variable, 1 if attached, 0 detached
F	Number of fireplaces (woodburning)
C	Number of chattels (appliances e.g. stove, fridge, etc.)
Lot attributes	
LOTS	Lot size (square metres)
CO	1 if corner lot, 0 otherwise
CUL	1 if cul-de-sac, 0 otherwise
LA	1 if lane behind, 0 otherwise
E	Exposure of yard (N, NE, $E = 1$, otherwise 0)
Z0	Dummy variable represents zone 0
Z1	Dummy variable represents zone 1
Z2	Dummy variable represents zone 2
Z3	Dummy variable represents zone 3
Z4	Dummy variable represents zone 4
Z5	Dummy variable represents zone 5
Z6	Dummy variable represents zone 6
Z7	Dummy variable represents zone 7
Z8	Dummy variable represents zone 8
Z9	Dummy variable represents zone 9
TIME	Month of sale

CASE PROBLEM 2

Treating obesity*

Obesity is a major health risk throughout Europe and the USA, leading to a number of possibly life-threatening diseases. Developing a successful treatment for obesity is therefore important, as a reduction in weight can greatly reduce the risk of illness. A sustained weight loss of 5–10 per cent of initial body weight reduces the health risks associated with obesity. Diet and exercise are useful in weight control but may not always be successful in the long term. An integrated programme of diet, exercise and drug treatment may be beneficial for obese patients.

The study

In 1998 Knoll Pharmaceuticals received authorization to market sibutramine for the treatment of obesity

* Source: Freeman, J., Redfern, E. and Bedford, S. (1995) 'Designing statistics courseware for interdisciplinary study', pp. 195–203. In Saunders, D. (ed.) *The Simulation and Gaming Yearbook Volume 3: Games and Simulations for Business.* Kogan Page.

in the USA. One of their suite of studies involved 37 obese patients who followed a treatment regime comprising a combination of diet, exercise and drug treatment. Patients taking part in this study were healthy adults (aged 18 to 65 years) and were between 30 per cent and 80 per cent above their ideal body weight. Rigorous criteria were defined to ensure that only otherwise healthy individuals took part.

Patients received either the new drug or placebo for an eight-week period and body weight was recorded at the start (week 0, also known as baseline) and at week 8. The information recorded for each patient was:

- Age (years)
- Gender (F: female, M: male)
- Height (cm)
- Family history of obesity? (N: no, Y: yes) Missing for patient number 134
- Motivation rating (1: some, 2: moderate, 3: great)
- Number of previous weight loss attempts
- Age of onset of obesity (1: 11 years, 2: 12–17 years, 3: 18–65 years)
- Weight at week 0 (kg)
- Weight at week 8 (kg)
- Treatment group (1 = placebo, 2 = new drug) Results are shown below for a selection of 10 of the 37 patients that took part in the study:

Age	Gender	Height	Family history?	Motivation rating	Previous weight loss attempts	Age of onset	Weight at week 0	Weight at week 8	Treatment group
40	F	170	N	2	1	3	83.4	75.0	2
50	F	164	Y	2	5	2	102.2	96.3	1
39	F	154	Y	2	1	3	84.0	82.6	1
40	F	169	Y	1	7	3	103.7	95.7	2
44	F	169	N	2	1	1	99.2	99.2	2
44	M	177	Y	2	2	2	126.0	123.2	2
38	M	171	Y	1	1	1	103.7	95.5	2
42	M	175	N	2	4	3	117.9	117.0	1
53	M	177	Y	2	3	3	112.4	111.8	1
52	F	166	Y	1	3	3	85.0	80.0	2

Clinical trials

The study is an example of a clinical trial commonly used to assess the effectiveness of a new treatment. Clinical trials are subject to rigorous controls to ensure that individuals are not unnecessarily put at risk and that they are fully informed and give their consent to take part in the study. As giving any patient a treatment may have a psychological effect, many studies compare a new drug with a dummy treatment (placebo) where, to avoid bias, neither the patient nor the doctor recording information knows whether the patient is on the new treatment or placebo as the tablets/capsules look identical; this approach is known as double-blinding. Bias could also occur if the treatment given to a patient was based on their characteristics; for example, if the more overweight patients were given the new treatment rather than the placebo they would have a greater chance of weight loss.

To avoid such bias the decision as to which individuals will receive the new treatment or placebo must be made using a process known as randomization. Using this approach each individual has the same chance of being given either the new treatment or the placebo.

Managerial report

1. Use the methods presented in this and previous chapters to analyze this data set. The priority is to use regression modelling to help determine which variables most influence weight loss. The treatment group variable is a particular concern in this respect.

2. Present a summary of your analysis, including key statistical results, conclusions and recommendations, in a managerial report. Include any appropriate technical material (computer output, residual plots, etc.) in an appendix.

OBESITY

CASE PROBLEM 3

Rating wines from the Piedmont region of Italy

Wine Spectator magazine contains articles and reviews on every aspect of the wine industry, including ratings of wine from around the world. In a recent issue they reviewed and scored 475 wines from the Piedmont region of Italy using a 100-point scale. The following table shows how the *Wine Spectator* score each wine received is used to rate each wine as being classic, outstanding, very good, good, mediocre or not recommended.

Score	Rating
95–100	Classic: a great wine
90–94	Outstanding: a wine of superior character and style
85–89	Very good: a wine with special qualities
80–84	Good: a solid, well-made wine
75–79	Mediocre: a drinkable wine that may have minor flaws
below 75	Not recommended

A key question for most consumers is whether paying more for a bottle of wine will result in a better wine. To investigate this question for wines from the Piedmont region we selected a random sample of 100 wines from the 475 wines that *Wine Spectator* reviewed. The data, contained in the file 'WineRatings', shows the price ($), the *Wine Spectator* score and the rating for each wine.

Managerial report

1. Develop a table that shows the number of wines that were classified as classic, outstanding, very good, good, mediocre and not recommended, and the average price. Does there appear to be any relationship between the price of the wine and the *Wine Spectator* rating? Are there any other aspects of your initial summary of the data that stand out?
2. Develop a scatter diagram with price on the horizontal axis and the *Wine Spectator* score on the vertical axis. Does the relationship between price and score appear to be linear?
3. Using linear regression, develop an estimated regression equation that can be used to predict the score given the price of the wine.
4. Using a second-order model, develop an estimated regression equation that can be used to predict the score given the price of the wine.
5. Compare the results from fitting a linear model and fitting a second-order model.
6. As an alternative to fitting a second-order model, fit a model using the natural logarithm of price as the independent variable. Compare the results with the second-order model.
7. Based upon your analysis, would you say that spending more for a bottle of wine will provide a better wine?
8. Suppose that you want to spend a maximum of $30 for a bottle of wine. In this case, will spending closer to your upper limit for price result in a better wine than a much lower price?

WINERATINGS

© Fani Kurti/iStock

17

Time Series Analysis and Forecasting

CHAPTER CONTENTS

Statistics in Practice Immigration

LEARNING OBJECTIVES After reading this chapter and doing the exercises, you should be able to:

1 Understand that the long-run success of an organization is often closely related to how well management is able to predict future aspects of the operation.

2 Understand the various components of a time series.

3 Use smoothing techniques such as moving averages and exponential smoothing.

4 Use either least squares or Holt's smoothing method to identify the trend component of a time series.

5 Understand how the classical time series model can be used to explain the pattern or behaviour of the data in a time series and to develop a forecast for the time series.

6 Determine and use seasonal indices for a time series.

7 Understand how regression models can be used in forecasting.

8 Provide a definition of the following terms:

  8.1   time series
  8.2   forecast
  8.3   trend component
  8.4   cyclical component
  8.5   seasonal component
  8.6   irregular component
  8.7   mean squared error
  8.8   moving averages
  8.9   weighted moving averages
 8.10   smoothing constants
 8.11   seasonal constant.

STATISTICS IN PRACTICE
Immigration

Large inflows of Albanian migrants now account for a third of total trafficked arrivals. The current situation contrasts somewhat with successive pledges by recent government administrations to tackle illegal immigration.

The numbers of undocumented immigrants crossing the English Channel in small boats from France to the UK has grown exponentially since 2018.

Year	Number of undocumented immigrants crossing the channel in small boats
2018	299
2019	1,843
2020	8,466
2021	28,526
2022	45,756

Source:
Glaze, B. (2023) 'Rishi Sunak can't solve the migrant crisis, says majority of voters', *Daily Mirror*, 5 January. Available at www.bbc.co.uk/news/explainers-63473022.

The purpose of this chapter is to provide an introduction to time series analysis and forecasting. Suppose you are asked to provide quarterly **forecasts** of sales for one of your company's products over the coming one-year period. Production schedules, raw material purchasing, inventory policies and sales quotas will all be affected by the quarterly forecasts you provide. Consequently, poor forecasts may result in poor planning and increased costs for the company. How should you go about providing the quarterly sales forecasts? Good judgement, intuition and an awareness of the state of the economy may give you a rough idea or 'feeling' of what is likely to happen in the future, but converting that feeling into a number that can be used as next year's sales forecast is difficult.

Forecasting methods can be classified as qualitative or quantitative. Qualitative methods generally involve the use of expert judgement to develop forecasts. Such methods are appropriate when historical data on the variable being forecast are either not applicable or unavailable. Quantitative forecasting methods can be used when (1) past information about the variable being forecast is available, (2) the information can be quantified, and (3) it is reasonable to assume that the pattern of the past will continue into the future. In such cases, a forecast can be developed using a time series method or a causal method. We will focus exclusively on quantitative forecasting methods in this chapter.

If the historical data are restricted to past values of the variable to be forecast, the forecasting procedure is called a *time series method* and the historical data are referred to as a time series. The objective of time series analysis is to discover a pattern in the historical data or time series and then extrapolate the pattern into the future; the forecast is based solely on past values of the variable and/or on past forecast errors.

Causal forecasting methods are based on the assumption that the variable we are forecasting has a cause–effect relationship with one or more other variables. In the discussion of regression analysis in Chapters 14, 15 and 16, we showed how one or more independent variables could be used to predict the value of a single dependent variable. Looking at regression analysis as a forecasting tool, we can view the time series value that we want to forecast as the dependent variable. Hence, if we can identify a good set of related independent or explanatory variables, we may be able to develop an estimated regression equation for predicting or forecasting the time series. For instance, the sales for many products are influenced by advertising expenditures, so regression analysis may be used to develop an equation showing how sales and advertising are related. Once the advertising budget for the next period is determined, we could substitute this value into the equation to develop a prediction or forecast of the sales volume for that period. Note that if a time series method were used to develop the forecast, advertising expenditures would not be considered; that is, a time series method would base the forecast solely on past sales.

By treating time as the independent variable and the time series as a dependent variable, regression analysis can also be used as a time series method. To help differentiate the application of regression analysis in these two cases, we use the terms *cross-sectional regression* and *time series regression*. Thus, time series regression refers to the use of regression analysis when the independent variable is time. As our focus in this chapter is on time series methods, we leave the discussion of the application of regression analysis as a causal forecasting method to more advanced texts on forecasting.

17.1 TIME SERIES PATTERNS

A time series is a sequence of observations on a variable measured at successive points in time or over successive periods of time. The measurements may be taken every hour, day, week, month or year, or at any other regular interval.[*] The pattern of the data is an important factor in understanding how the time series has behaved in the past. If such behaviour can be expected to continue in the future, we can use the past pattern to guide us in selecting an appropriate forecasting method.

To identify the underlying pattern in the data, a useful first step is to construct a time series plot. A time series plot is a graphical presentation of the relationship between time and the time series variable; time is on the horizontal axis and the time series values are shown on the vertical axis. Let us review some of the common types of data patterns that can be identified when examining a time series plot.

Horizontal pattern

A horizontal pattern exists when the data fluctuate around a constant mean. To illustrate a time series with a horizontal pattern, consider the 12 weeks of data in Table 17.1. These data show the number of litres of petrol sold by a petrol distributor in Sitges, Spain, over the past 12 weeks. The average value or mean for this time series is 19,250 litres per week. Figure 17.1 shows a time series plot for these data. Note how the data fluctuate around the sample mean of 19,250 litres. Although random variability is present, we would say that these data follow a horizontal pattern.

PETROL

TABLE 17.1 Petrol sales time series

Week	Sales (000 litres)
1	17
2	21
3	19
4	23
5	18
6	16
7	20
8	18
9	22
10	20
11	15
12	22

---

[*]We limit our discussion to time series in which the values of the series are recorded at equal intervals. Cases in which the observations are made at unequal intervals are beyond the scope of this text.

FIGURE 17.1

Petrol sales time series plot

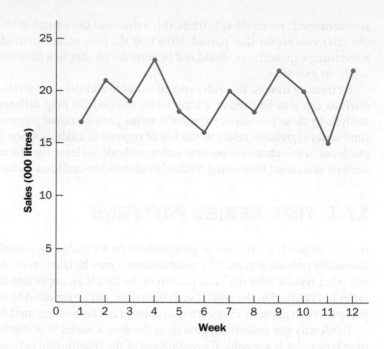

The term **stationary time series**[*] is used to denote a time series whose statistical properties are independent of time. In particular this means that:

1 The process generating the data has a constant mean.
2 The variability of the time series is constant over time.

A time series plot for a stationary time series will always exhibit a horizontal pattern. But simply observing a horizontal pattern is not sufficient evidence to conclude that the time series is stationary. More advanced texts on forecasting discuss procedures for determining if a time series is stationary and provide methods for transforming a time series that is not stationary into a stationary series.

Changes in business conditions can often result in a time series that has a horizontal pattern shifting to a new level. For instance, suppose the petrol distributor signs a contract with the Guardia Civil to provide petrol for police cars located in northern Spain. With this new contract, the distributor expects to see a major increase in weekly sales starting in week 13. Table 17.2 shows the number of litres of petrol sold for the original time series and for the ten weeks after signing the new contract. Figure 17.2 shows the corresponding time series plot. Note the increased level of the time series beginning in week 13. This change in the level of the time series makes it more difficult to choose an appropriate forecasting method. Selecting a forecasting method that adapts well to changes in the level of a time series is an important consideration in many practical applications.

Trend pattern

Although time series data generally exhibit random fluctuations, a time series may also show gradual shifts or movements to relatively higher or lower values over a longer period of time. If a time series plot exhibits this type of behaviour, we say that a **trend pattern** exists. A trend is usually the result of long-term factors such as population increases or decreases, changing demographic characteristics of the population, technology and/or consumer preferences.

---

[*]For a formal definition of stationarity refer to Box, G. E. P., Jenkins, G. M. and Reinsell, G. C. (1994) *Time Series Analysis: Forecasting and Control*, 3rd ed. Prentice Hall, p. 23.

TABLE 17.2 Petrol sales time series after obtaining the contract with the Guardia Civil

Week	Sales (000 litres)
1	17
2	21
3	19
4	23
5	18
6	16
7	20
8	18
9	22
10	20
11	15
12	22
13	31
14	34
15	31
16	33
17	28
18	32
19	30
20	29
21	34
22	33

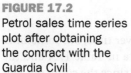

PETROL
REVISED

FIGURE 17.2
Petrol sales time series plot after obtaining the contract with the Guardia Civil

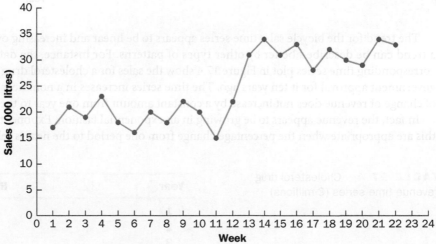

To illustrate a time series with a trend pattern, consider the time series of bicycle sales for a particular manufacturer over the past ten years, as shown in Table 17.3 and Figure 17.3. Note that 21,600 bicycles were sold in year one, 22,900 were sold in year two and so on. In year 10, the most recent year, 31,400 bicycles were sold. Visual inspection of the time series plot shows some up and down movement over the past ten years, but the time series also seems to have a systematically increasing or upward trend.

BICYCLE

TABLE 17.3
Bicycle sales time series

Year	Sales (000)
1	21.6
2	22.9
3	25.5
4	21.9
5	23.9
6	27.5
7	31.5
8	29.7
9	28.6
10	31.4

FIGURE 17.3
Bicycle sales time series plot

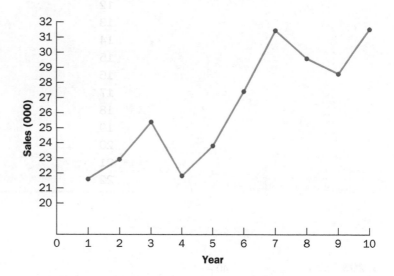

The trend for the bicycle sales time series appears to be linear and increasing over time, but sometimes a trend can be described better by other types of patterns. For instance, the data in Table 17.4 and the corresponding time series plot in Figure 17.4 show the sales for a cholesterol drug since the company won government approval for it ten years ago. The time series increases in a nonlinear fashion; that is, the rate of change of revenue does not increase by a constant amount from one year to the next.

In fact, the revenue appears to be growing in an exponential fashion. Exponential relationships such as this are appropriate when the percentage change from one period to the next is relatively constant.

TABLE 17.4   Cholesterol drug revenue time series (€ millions)

Year	Revenue
1	23.1
2	21.3
3	27.4
4	34.6
5	33.8
6	43.2
7	59.5
8	64.4
9	74.2
10	99.3

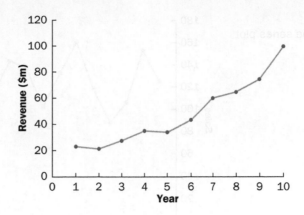

FIGURE 17.4
Cholesterol drug revenue time series plot ($ millions)

Seasonal pattern

The trend of a time series can be identified by analyzing multiyear movements in historical data. Seasonal patterns are recognized by seeing the same repeating patterns over successive periods of time. For example, a manufacturer of swimming pools expects low sales activity in the autumn and winter months, with peak sales in the spring and summer months. Manufacturers of snow removal equipment and heavy clothing, however, expect just the opposite yearly pattern. Not surprisingly, the pattern for a time series plot that exhibits a repeating pattern over a one-year period due to seasonal influences is called a **seasonal pattern**. While we generally think of seasonal movement in a time series as occurring within one year, time series data can also exhibit seasonal patterns of less than one year in duration. For example, daily traffic volume shows within-the-day 'seasonal' behaviour, with peak levels occurring during rush hours, moderate flow during the rest of the day and early evening, and light flow from midnight to early morning.

As an example of a seasonal pattern, consider the number of umbrellas sold at a clothing store over the past five years. Table 17.5 shows the time series and Figure 17.5 shows the corresponding time series plot.

TABLE 17.5 Umbrella sales time series

UMBRELLA

Year	Quarter	Sales
1	1	125
	2	153
	3	106
	4	88
2	1	118
	2	161
	3	133
	4	102
3	1	138
	2	144
	3	113
	4	80
4	1	109
	2	137
	3	125
	4	109
5	1	130
	2	165
	3	128
	4	96

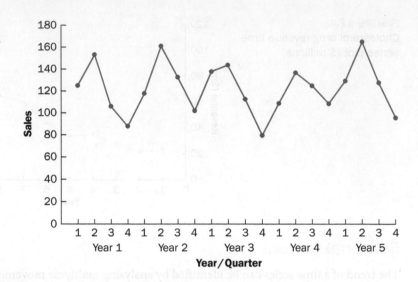

FIGURE 17.5
Umbrella sales time series plot

The time series plot does not indicate any long-term trend in sales. In fact, unless you look carefully at the data, you might conclude that the data follow a horizontal pattern. But closer inspection of the time series plot reveals a regular pattern in the data. That is, the first and third quarters have moderate sales, the second quarter has the highest sales, and the fourth quarter tends to have the lowest sales volume. Thus, we would conclude that a quarterly seasonal pattern is present.

Trend and seasonal pattern

Some time series include a combination of a trend and seasonal pattern. For instance, the data in Table 17.6 and the corresponding time series plot in Figure 17.6 show smartphone sales for a particular manufacturer over the past four years. Clearly, an increasing trend is present. However, Figure 17.6 also indicates that sales are lowest in the second quarter of each year and increase in quarters 3 and 4. Thus, we conclude that a seasonal pattern also exists for smartphone sales. In such cases we need to use a forecasting method that has the capability to deal with both trend and seasonality.

SMART
PHONE

TABLE 17.6   Quarterly
smartphone sales time series

Year	Quarter	Sales (000)
1	1	4.8
	2	4.1
	3	6.0
	4	6.5
2	1	5.8
	2	5.2
	3	6.8
	4	7.4
3	1	6.0
	2	5.6
	3	7.5
	4	7.8
4	1	6.3
	2	5.9
	3	8.0
	4	8.4

FIGURE 17.6
Quarterly smartphone sales
time series plot

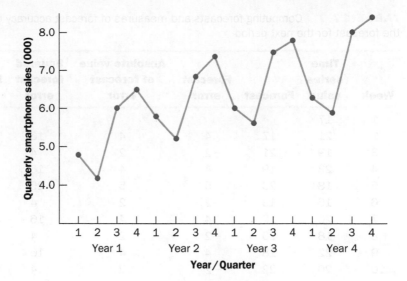

Cyclical pattern

A **cyclical pattern** exists if the time series plot shows an alternating sequence of points below and above the trend line lasting more than one year. Many economic time series exhibit cyclical behaviour with regular runs of observations below and above the trend line. Often, the cyclical component of a time series is due to multiyear business cycles.

For example, periods of moderate inflation followed by periods of rapid inflation can lead to time series that alternate below and above a generally increasing trend line (e.g. a time series for housing costs). Business cycles are extremely difficult, if not impossible, to forecast. As a result, cyclical effects are often combined with long-term trend effects and referred to as trend-cycle effects. In this chapter we do not deal with cyclical effects that may be present in the time series.

Selecting a forecasting method

The underlying pattern in the time series is an important factor in selecting a forecasting method. Thus, a time series plot should be one of the first things developed when trying to determine what forecasting method to use. If we see a horizontal pattern, then we need to select a method appropriate for this type of pattern.

Similarly, if we observe a trend in the data, then we need to use a forecasting method that has the capability to handle trend effectively. The next two sections illustrate methods that can be used in situations where the underlying pattern is horizontal; in other words, no trend or seasonal effects are present. We then consider methods that are appropriate when trend and/or seasonality are present in the data.

17.2 FORECAST ACCURACY

In this section we begin by developing forecasts for the petrol time series shown in Table 17.1 using the simplest of all the forecasting methods: an approach that uses the most recent week's sales volume as the forecast for the next week. For instance, the distributor sold 17,000 litres of petrol in week 1; this value is used as the forecast for week 2. Next, we use 21, the actual value of sales in week 2, as the forecast for week 3 and so on.

The forecasts obtained for the historical data using this method are shown in Table 17.7 in the column labelled Forecast. Because of its simplicity, this method is often referred to as a *naive* forecasting method.

TABLE 17.7 Computing forecasts and measures of forecast accuracy using the most recent value as the forecast for the next period

Week	Time series value	Forecast	Forecast error	Absolute value of forecast error	Squared forecast error	Percentage error	Absolute value of percentage error
1	17						
2	21	17	4	4	16	19.05	19.05
3	19	21	−2	2	4	−10.53	10.53
4	23	19	4	4	16	17.39	17.39
5	18	23	−5	5	25	−27.78	27.78
6	16	18	−2	2	4	− 12.50	12.50
7	20	16	4	4	16	20.00	20.00
8	18	20	−2	2	4	−11.11	11.11
9	22	18	4	4	16	18.18	18.18
10	20	22	−2	2	4	−10.00	10.00
11	15	20	−5	5	25	−33.33	33.33
12	22	15	7	7	49	31.82	31.82
		Totals	**5**	**41**	**179**	**1.19**	**211.69**

How accurate are the forecasts obtained using this *naive* forecasting method? To answer this question we will introduce several measures of forecast accuracy. These measures are used to determine how well a particular forecasting method is able to reproduce the time series data that are already available. By selecting the method that has the best accuracy for the data already known, we hope to increase the likelihood that we will obtain better forecasts for future time periods.

The key concept associated with measuring forecast accuracy is **forecast error**, defined as:

$$\text{Forecast error} = \text{Actual value} - \text{Forecast}$$

For instance, because the distributor actually sold 21,000 litres of petrol in week 2 and the forecast, using the sales volume in week 1, was 17,000 litres, the forecast error in week 2 is:

$$\text{Forecast error in week 2} = 21 - 17 = 4$$

The fact that the forecast error is positive indicates that in week 2 the forecasting method underestimated the actual value of sales. Next, we use 21, the actual value of sales in week 2, as the forecast for week 3. Since the actual value of sales in week 3 is 19, the forecast error for week 3 is $19 - 21 = -2$. In this case, the negative forecast error indicates that in week 3 the forecast overestimated the actual value. Thus, the forecast error may be positive or negative, depending on whether the forecast is too low or too high. A complete summary of the forecast errors for this naive forecasting method is shown in Table 17.7 in the column labelled forecast error.

A simple measure of forecast accuracy is the mean or average of the forecast errors. Table 17.7 shows that the sum of the forecast errors for the petrol sales time series is 5; thus, the mean or average forecast error is 5/11 = 0.45. Note that although the petrol time series consists of 12 values, to compute the mean error we divided the sum of the forecast errors by 11 because there are only 11 forecast errors. Because the mean forecast error is positive, the method is under-forecasting; in other words, the observed values tend to be greater than the forecast values. Because positive and negative forecast errors tend to offset one another, the mean error is likely to be small and thus not a very useful measure of forecast accuracy.

The **mean absolute error**, denoted MAE, is a measure of forecast accuracy that avoids the problem of positive and negative forecast errors offsetting one another. As you might expect given its name, MAE is the average of the absolute values of the forecast errors. Table 17.7 shows that the sum of the absolute values of the forecast errors is 41; therefore:

$$\text{MAE} = \text{average of the absolute value of forecast errors} = \frac{41}{11} = 3.73$$

Another measure that avoids the problem of positive and negative forecast errors offsetting each other is obtained by computing the average of the squared forecast errors. This measure of forecast accuracy, referred to as the mean squared error, is denoted MSE. From Table 17.7, the sum of the squared errors is 179; therefore:

$$\text{MSE} = \text{average of the sum of squared forecast errors} = \frac{179}{11} = 16.27$$

The size of MAE and MSE depends upon the scale of the data. As a result, it is difficult to make comparisons for different time intervals, such as comparing a method of forecasting monthly petrol sales to a method of forecasting weekly sales, or to make comparisons across different time series. To make comparisons like these we need to work with relative or percentage error measures. The **mean absolute percentage error**, denoted MAPE, is such a measure. To compute MAPE we must first compute the percentage error for each forecast. For example, the percentage error corresponding to the forecast of 17 in week 2 is computed by dividing the forecast error in week 2 by the actual value in week 2 and multiplying the result by 100. For week 2 the percentage error is computed as follows:

$$\text{Percentage error for week 2} = \frac{4}{21}(100) = 19.05\%$$

Thus, the forecast error for week 2 is 19.05 per cent of the observed value in week 2. A complete summary of the percentage errors is shown in Table 17.7 in the column labelled percentage error. In the next column, we show the absolute value of the percentage error.

Table 17.7 shows that the sum of the absolute values of the percentage errors is 211.69; therefore:

$$\text{MAPE} = \text{average of the absolute value of percentage forecast errors} = \frac{211.69}{11} = 19.24\%$$

Summarizing, using the naive (most recent observation) forecasting method, we obtained the following measures of forecast accuracy:

$$\text{MAE} = 3.73$$
$$\text{MSE} = 16.27$$
$$\text{MAPE} = 19.24\%$$

These measures of forecast accuracy simply measure how well the forecasting method is able to forecast historical values of the time series. Now, suppose we want to forecast sales for a future time period, such as week 13. In this case the forecast for week 13 is 22, the actual value of the time series in week 12. Is this an accurate estimate of sales for week 13? Unfortunately, there is no way to address the issue of accuracy associated with forecasts for future time periods. But, if we select a forecasting method that works well for the historical data, and we think that the historical pattern will continue into the future, we should obtain results that will ultimately be shown to be good.

Before closing this section, let's consider another method for forecasting the petrol sales time series in Table 17.1. Suppose we use the average of all the historical data available as the forecast for the next period. We begin by developing a forecast for week 2. Since there is only one historical value available prior to week 2, the forecast for week 2 is just the time series value in week 1; thus, the forecast for week 2 is 17,000 litres of petrol. To compute the forecast for week 3, we take the average of the sales values in weeks 1 and 2:

$$\text{Forecast for week 3} = \frac{17 + 21}{2} = 19$$

Similarly, the forecast for week 4 is:

$$\text{Forecast for week 4} = \frac{17 + 21 + 19}{3} = 19$$

The forecasts obtained using this method for the petrol time series are shown in Table 17.8 in the column labelled forecast. Using the results shown in Table 17.8, we obtained the following values of MAE, MSE and MAPE:

$$MAE = \frac{26.81}{11} = 2.44$$

$$MSE = \frac{89.07}{11} = 8.10$$

$$MAPE = \frac{141.34}{11} = 12.85\%$$

We can now compare the accuracy of the two forecasting methods we have considered in this section by comparing the values of MAE, MSE and MAPE for each method.

	Naive method	Average of past values
MAE	3.73	2.44
MSE	16.27	8.10
MAPE	19.24%	12.85%

For every measure, the average of past values provides more accurate forecasts than using the most recent observation as the forecast for the next period. In general, if the underlying time series is stationary, the average of all the historical data will always provide the best results.

TABLE 17.8 Computing forecasts and measures of forecast accuracy using the average of all the historical data as the forecast for the next period

Week	Time series value	Forecast	Forecast error	Absolute value of forecast error	Squared forecast error	Percentage error	Absolute value of percentage error
1	17						
2	21	17.00	4.00	4.00	16.00	19.05	19.05
3	19	19.00	0.00	0.00	0.00	0.00	0.00
4	23	19.00	4.00	4.00	16.00	17.39	17.39
5	18	20.00	−2.00	2.00	4.00	−11.11	11.11
6	16	19.60	−3.60	3.60	12.96	−22.50	22.50
7	20	19.00	1.00	1.00	1.00	5.00	5.00
8	18	19.14	−1.14	1.14	1.31	−6.35	6.35
9	22	19.00	3.00	3.00	9.00	13.64	13.64
10	20	19.33	0.67	0.67	0.44	3.33	3.33
11	15	19.40	−4.40	4.40	19.36	−29.33	29.33
12	22	19.00	3.00	3.00	9.00	13.64	13.64
		Totals	**4.53**	**26.81**	**89.07**	**2.76**	**141.34**

But suppose that the underlying time series is not stationary. In Section 17.1 we mentioned that changes in business conditions can often result in a time series that has a horizontal pattern shifting to a new level. We discussed a situation in which the petrol distributor signed a contract with the Guardia Civil to provide petrol for police cars located in northern Spain. Table 17.2 shows the number of litres of petrol sold for the original time series and the ten weeks after signing the new contract, and Figure 17.2 shows the corresponding time series plot. Note the change in level in week 13 for the resulting time series. When a shift to a new level like this occurs, it takes a long time for the forecasting method that uses the average of all the historical data to adjust to the new level of the time series. But, in this case, the simple naive method adjusts very rapidly to the change in level because it uses the most recent observation available as the forecast.

Measures of forecast accuracy are important factors in comparing different forecasting methods, but we have to be careful not to rely upon them too heavily. Good judgement and knowledge about business conditions that might affect the forecast also have to be carefully considered when selecting a method. And historical forecast accuracy is not the only consideration, especially if the time series is likely to change in the future.

In the next section we will introduce more sophisticated methods for developing forecasts for a time series that exhibits a horizontal pattern. Using the measures of forecast accuracy developed here, we will be able to determine if such methods provide more accurate forecasts than we obtained using the simple approaches illustrated in this section. The methods that we will introduce also have the advantage of adapting well in situations where the time series changes to a new level. The ability of a forecasting method to adapt quickly to changes in level is an important consideration, especially in short-term forecasting situations.

EXERCISES

Methods

1. Consider the following time series data.

Week	1	2	3	4	5	6
Value	18	13	16	11	17	14

Using the naive method (most recent value) as the forecast for the next week, compute the following measures of forecast accuracy.
a. Mean absolute error.
b. What is the forecast for week 7?
c. Mean absolute percentage error.
d. Mean squared error.

2. Refer to the time series data in Exercise 1. Using the average of all the historical data as a forecast for the next period, compute the following measures of forecast accuracy.
a. Mean absolute error.
b. Mean squared error.
c. Mean absolute percentage error.
d. What is the forecast for week 7?

3. Exercises 1 and 2 used different forecasting methods. Which method appears to provide the more accurate forecasts for the historical data? Explain.

4. Consider the following time series data.

Month	1	2	3	4	5	6	7
Value	24	13	20	12	19	23	15

a. Compute MSE using the most recent value as the forecast for the next period. What is the forecast for month 8?
b. Compute MSE using the average of all the data available as the forecast for the next period. What is the forecast for month 8?
c. Which method appears to provide the better forecast?

17.3 MOVING AVERAGES AND EXPONENTIAL SMOOTHING

In this section we discuss three forecasting methods that are appropriate for a time series with a horizontal pattern: moving averages, weighted moving averages and exponential smoothing. These methods also adapt well to changes in the level of a horizontal pattern such as we saw with the extended petrol sales time series (Table 17.2 and Figure 17.2). However, without modification they are not appropriate when significant trend, cyclical or seasonal effects are present. Because the objective of each of these methods is to 'smooth out' the random fluctuations in the time series, they are referred to as smoothing methods. These methods are easy to use and generally provide a high level of accuracy for short-range forecasts, such as a forecast for the next time period.

Moving averages

The **moving averages** method uses the average of the most recent k data values in the time series as the forecast for the next period. Mathematically, a moving average forecast of order k is as follows:

Moving average forecast of order k

$$F_{t+1} = \frac{\Sigma(\text{most recent } k \text{ data values})}{k} = \frac{Y_t + Y_{t-1} + \cdots + Y_{t-k+1}}{k} \qquad (17.1)$$

where:

$$F_{t+1} = \text{forecast of the times series for period } t+1$$
$$Y_t = \text{actual value of the time series in period } t$$

The term *moving* is used because every time a new observation becomes available for the time series, it replaces the oldest observation in the equation and a new average is computed. As a result, the average will change, or move, as new observations become available.

To illustrate the moving averages method, let us return to the petrol sales data in Table 17.1 and Figure 17.1. The time series plot in Figure 17.1 indicates that the petrol sales time series has a horizontal pattern. Thus, the smoothing methods of this section are applicable.

To use moving averages to forecast a time series, we must first select the order, or number of time series values, to be included in the moving average. If only the most recent values of the time series are considered relevant, a small value of k is preferred. If more past values are considered relevant, then a larger value of k is better. As mentioned earlier, a time series with a horizontal pattern can shift to a new level over time. A moving average will adapt to the new level of the series and resume providing good forecasts in k periods. Thus, a smaller value of k will track shifts in a time series more quickly. But larger values of k will be more effective in smoothing out the random fluctuations over time. So managerial judgement based on an understanding of the behaviour of a time series is helpful in choosing a good value for k.

To illustrate how moving averages can be used to forecast petrol sales, we will use a three-week moving average ($k = 3$). We begin by computing the forecast of sales in week 4 using the average of the time series values in weeks 1–3.

$$F_4 = \text{average of weeks } 1 - 3 = \frac{17 + 21 + 19}{3} = 19$$

Thus, the moving average forecast of sales in week 4 is 19 or 19,000 litres of petrol. Because the actual value observed in week 4 is 23, the forecast error in week 4 is $23 - 19 = 4$.

Next, we compute the forecast of sales in week 5 by averaging the time series values in weeks 2–4.

$$F_5 = \text{average of weeks } 2 - 4 = \frac{21 + 19 + 23}{3} = 21$$

Hence, the forecast of sales in week 5 is 21 and the error associated with this forecast is $18 - 21 = -3$. A complete summary of the three-week moving average forecasts for the petrol sales time series is provided in Table 17.9. Figure 17.7 shows the original time series plot and the three-week moving average forecasts. Note how the graph of the moving average forecasts has tended to smooth out the random fluctuations in the time series.

To forecast sales in week 13, the next time period in the future, we simply compute the average of the time series values in weeks 10, 11 and 12.

$$F_{13} = \text{average of weeks } 10 - 12 = \frac{20 + 15 + 22}{3} = 19$$

Thus, the forecast for week 13 is 19 or 19,000 litres of petrol.

FIGURE 17.7

Petrol sales time series plot and three-week moving average forecasts

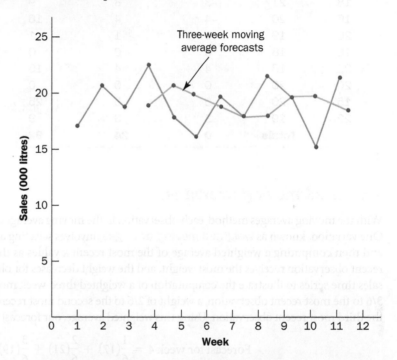

Forecast accuracy

In Section 17.2 we discussed three measures of forecast accuracy: MAE, MSE and MAPE. Using the three-week moving average calculations in Table 17.9, the values for these three measures of forecast accuracy are:

$$\text{MAE} = \frac{24}{9} = 2.67$$

$$\text{MSE} = \frac{92}{9} = 10.22$$

$$\text{MAPE} = \frac{129.21}{9} = 14.36\%$$

In Section 17.2 we also showed that using the most recent observation as the forecast for the next week (a moving average of order $k = 1$) resulted in values of MAE = 3.73, MSE = 16.27 and MAPE = 19.24 per cent. Thus, in each case the three-week moving average approach provided more accurate forecasts than simply using the most recent observation as the forecast.

To determine if a moving average with a different order k can provide more accurate forecasts, we recommend using trial and error to determine the value of k that minimizes MSE. For the petrol sales time series, it can be shown that the minimum value of MSE corresponds to a moving average of order $k = 6$ with MSE = 6.79. If we are willing to assume that the order of the moving average that is best for the historical data will also be best for future values of the time series, the most accurate moving average forecasts of petrol sales can be obtained using a moving average of order $k = 6$.

TABLE 17.9   Summary of three-week moving average calculations

Week	Time series value	Forecast	Forecast error	Absolute value of forecast error	Squared forecast error	Percentage error	Absolute value of percentage error
1	17						
2	21						
3	19						
4	23	19	4	4	16	17.39	17.39
5	18	21	-3	3	9	-16.67	16.67
6	16	20	-4	4	16	-25.00	25.00
7	20	19	1	1	1	5.00	5.00
8	18	18	0	0	0	0.00	0.00
9	22	18	4	4	16	18.18	18.18
10	20	20	0	0	0	0.00	0.00
11	15	20	-5	5	25	-33.33	33.33
12	22	19	3	3	9	13.64	13.64
		Totals	**0**	**24**	**92**	**-20.79**	**129.21**

Weighted moving averages

With the moving averages method, each observation in the moving average calculation receives the same weight. One variation, known as **weighted moving averages**, involves selecting a different weight for each data value and then computing a weighted average of the most recent k values as the forecast. In most cases, the most recent observation receives the most weight, and the weight decreases for older data values. Let us use the petrol sales time series to illustrate the computation of a weighted three-week moving average. We assign a weight of 3/6 to the most recent observation, a weight of 2/6 to the second most recent observation and a weight of 1/6 to the third most recent observation. Using this weighted average, our forecast for week 4 is computed as follows:

$$\text{Forecast for week 4} = \frac{1}{6}(17) + \frac{2}{6}(21) + \frac{3}{6}(19) = 19.33$$

Note that for the weighted moving average method the sum of the weights is equal to 1.

Forecast accuracy

To use the weighted moving averages method, we must first select the number of data values to be included in the weighted moving average and then choose weights for each of the data values. In general, if we believe that the recent past is a better predictor of the future than the distant past, larger weights should be given to the more recent observations. However, when the time series is highly variable, selecting approximately equal weights for the data values may be best. The only requirement in selecting the weights is that their sum must equal 1. To determine whether one particular combination of a number of data values and weights provides a more accurate forecast than another combination, we recommend using MSE as the measure of forecast accuracy. That is, if we assume that the combination that is best for the past will also be best for the future, we would use the combination of number of data values and weights that minimizes MSE for the historical time series to forecast the next value in the time series.

Exponential smoothing

Exponential smoothing also uses a weighted average of past time series values as a forecast; it is a special case of the weighted moving averages method in which we select only one weight – the weight for the most recent observation. The weights for the other data values are computed automatically and become smaller as the observations move further into the past. The exponential smoothing equation is as follows:

Exponential smoothing forecast

$$F_{t+1} = \alpha Y_t + (1 - \alpha)F_t \qquad \textbf{(17.2)}$$

where:

$$F_{t+1} = \text{forecast of the time series for period } t + 1$$
$$Y_t = \text{actual value of the time series in period } t$$
$$F_t = \text{forecast of the time series for period } t$$
$$\alpha = \text{smoothing constant } (0 \le \alpha \le 1)$$

Equation (17.2) shows that the forecast for period $t + 1$ is a weighted average of the actual value in period t and the forecast for period t. The weight given to the actual value in period t is the smoothing constant α and the weight given to the forecast in period t is $1 - \alpha$. It turns out that the exponential smoothing forecast for any period is actually a weighted average of *all the previous actual values* of the time series. Let us illustrate by working with a time series involving only three periods of data: Y_1, Y_2 and Y_3.

To initiate the calculations, we let F_1 equal the actual value of the time series in period 1; that is, $F_1 = Y_1$. Hence, the forecast for period 2 is:

$$\begin{aligned} F_2 &= \alpha Y_1 + (1 - \alpha)F_1 \\ &= \alpha Y_1 + (1 - \alpha)Y_1 \\ &= Y_1 \end{aligned}$$

We see that the exponential smoothing forecast for period 2 is equal to the actual value of the time series in period 1.

The forecast for period 3 is:

$$F_3 = \alpha Y_2 + (1 - \alpha)F_2 = \alpha Y_2 + (1 - \alpha)Y_1$$

Finally, substituting this expression for F_3 in the expression for F_4, we obtain:

$$\begin{aligned} F_4 &= \alpha Y_3 + (1 - \alpha)F_3 \\ &= \alpha Y_3 + (1 - \alpha)[\alpha Y_2 + (1 - \alpha)Y_1] \\ &= \alpha Y_3 + (1 - \alpha)\alpha Y_2 + (1 - \alpha)^2 Y_1 \end{aligned}$$

We now see that F_4 is a weighted average of the first three time series values. The sum of the coefficients, or weights, for Y_1, Y_2 and Y_3 equals 1. A similar argument can be made to show that, in general, any forecast F_{t+1} is a weighted average of all the previous time series values.

Despite the fact that exponential smoothing provides a forecast that is a weighted average of all previous observations, past data do not all need to be saved to compute the forecast for the next period. In fact, equation (17.2) shows that once the value for the smoothing constant α is selected, only two pieces of information are needed to compute the forecast F_{t+1}: Y_t, the actual value of the time series in period t, and F_t, the forecast for period t.

To illustrate the exponential smoothing approach, let us again consider the petrol sales time series in Table 17.1 and Figure 17.1. As indicated previously, to start the calculations we set the exponential smoothing forecast for period 2 equal to the actual value of the time series in period 1. Thus, with $Y_1 = 17$, we set $F_2 = 17$ to initiate the computations. Referring to the time series data in Table 17.1, we find an actual time series value in period 2 of $Y_2 = 21$. Thus, period 2 has a forecast error of $21 - 17 = 4$.

Continuing with the exponential smoothing computations using a smoothing constant of $\alpha = 0.2$, we obtain the following forecast for period 3:

$$F_3 = 0.2Y_2 + 0.8F_2 = 0.2(21) + 0.8(17) = 17.8$$

Once the actual time series value in period 3, $Y_3 = 19$, is known, we can generate a forecast for period 4 as follows:

$$F_4 = 0.2Y_3 + 0.8F_3 = 0.2(19) + 0.8(17.8) = 18.04$$

Continuing the exponential smoothing calculations, we obtain the weekly forecast values shown in Table 17.10. Note that we have not shown an exponential smoothing forecast or a forecast error for week 1 because no forecast was made. For week 12, we have $Y_{12} = 22$ and $F_{12} = 18.48$. We can use this information to generate a forecast for week 13:

$$F_{13} = 0.2Y_{12} + 0.8F_{12} = 0.2(22) + 0.8(18.48) = 19.18$$

Thus, the exponential smoothing forecast of the amount sold in week 13 is 19.18, or 19,180 litres of petrol. With this forecast, the firm can make plans and decisions accordingly.

TABLE 17.10 Summary of the exponential smoothing forecasts and forecast errors for the petrol sales time series with smoothing constant $\alpha = 0.2$

Week	Time series value	Forecast	Forecast error	Squared forecast error
1	17			
2	21	17.00	4.00	16.00
3	19	17.80	1.20	1.44
4	23	18.04	4.96	24.60
5	18	19.03	−1.03	1.06
6	16	18.83	−2.83	8.01
7	20	18.26	1.74	3.03
8	18	18.61	−0.61	0.37
9	22	18.49	3.51	12.32
10	20	18.49	0.81	0.66
11	15	19.19	−4.35	18.92
12	22	18.48	3.52	12.39
		Totals	**10.92**	**98.80**

Figure 17.8 shows the time series plot of the actual and forecast time series values. Note in particular how the forecasts 'smooth out' the irregular or random fluctuations in the time series.

FIGURE 17.8
Actual and forecast petrol sales time series with smoothing constant $\alpha = 0.2$

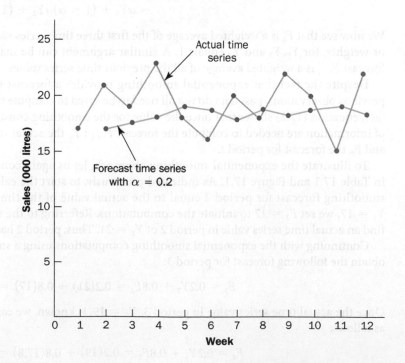

Forecast accuracy

In the preceding exponential smoothing calculations, we used a smoothing constant of $\alpha = 0.2$. Although any value of α between 0 and 1 is acceptable, some values will yield better forecasts than others. Insight into choosing a good value for α can be obtained by rewriting the basic exponential smoothing model as follows:

$$F_{t+1} = \alpha Y_t + (1 - \alpha)F_t \qquad \text{(17.3)}$$
$$F_{t+1} = \alpha Y_t + F_t - \alpha F_t$$
$$F_{t+1} = F_t + \alpha(Y_t - F_t)$$

Thus, the new forecast F_{t+1} is equal to the previous forecast F_t plus an adjustment, which is the smoothing constant α times the most recent forecast error, $Y_t - F_t$. That is, the forecast in period $t + 1$ is obtained by adjusting the forecast in period t by a fraction of the forecast error. If the time series contains substantial random variability, a small value of the smoothing constant is preferred. The reason for this choice is that if much of the forecast error is due to random variability, we do not want to overreact and adjust the forecasts too quickly. For a time series with relatively little random variability, forecast errors are more likely to represent a change in the level of the series. Thus, larger values of the smoothing constant provide the advantage of quickly adjusting the forecasts; this allows the forecasts to react more quickly to changing conditions.

The criterion we will use to determine a desirable value for the smoothing constant α is the same as the criterion we proposed for determining the order or number of periods of data to include in the moving averages calculation. That is, we choose the value of α that minimizes the MSE. A summary of the MSE calculations for the exponential smoothing forecast of petrol sales with $\alpha = 0.2$ is shown in Table 17.10. Note that there is one less squared error term than the number of time periods because we had no past values with which to make a forecast for period 1. The value of the sum of squared forecast errors is 98.80; hence MSE $= 98.80/11 = 8.98$. Would a different value of α provide better results in terms of a lower MSE value? Perhaps the most straightforward way to answer this question is simply to try another value for α. We will then compare its mean squared error with the MSE value of 8.98 obtained by using a smoothing constant of $\alpha = 0.2$.

The exponential smoothing results with $\alpha = 0.3$ are shown in Table 17.11. The value of the sum of squared forecast errors is 102.83; hence MSE $= 102.83/11 = 9.35$. With MSE $= 9.35$, we see that, for the current data set, a smoothing constant of $\alpha = 0.3$ results in less forecast accuracy than a smoothing constant of $\alpha = 0.2$. Thus, we would be inclined to prefer the original smoothing constant of $\alpha = 0.2$. Using a trial-and-error calculation with other values of α, we can find a 'good' value for the smoothing constant. This value can be used in the exponential smoothing model to provide forecasts for the future. At a later date, after new time series observations are obtained, we analyze the newly collected time series data to determine whether the smoothing constant should be revised to provide better forecasting results.

TABLE 17.11 Summary of the exponential smoothing forecasts and forecast errors for the petrol sales time series with smoothing constant $\alpha = 0.3$

Week	Time series value	Forecast	Forecast error	Squared forecast error
1	17			
2	21	17.00	4.00	16.00
3	19	18.20	0.80	0.64
4	23	18.44	4.56	20.79
5	18	19.81	−1.81	3.28
6	16	19.27	−3.27	10.69
7	20	18.29	1.71	2.92
8	18	18.80	−0.80	0.64
9	22	18.56	3.44	11.83
10	20	19.59	0.41	0.17
11	15	19.71	−4.71	22.18
12	22	18.30	3.70	13.69
		Totals	**8.03**	**102.83**

EXERCISES

Methods

5. Consider the following time series data.

Week	1	2	3	4	5	6
Value	18	13	16	11	17	14

   a. Construct a time series plot. What type of pattern exists in the data?
   b. Develop the three-week moving average forecasts for this time series. Compute MSE and a forecast for week 7.
   c. Use $\alpha = 0.2$ to compute the exponential smoothing forecasts for the time series. Compute MSE and a forecast for week 7.
   d. Compare the three-week moving average approach with the exponential smoothing approach using $\alpha = 0.2$. Which appears to provide more accurate forecasts based on MSE? Explain.
   e. Use a smoothing constant of $\alpha = 0.4$ to compute the exponential smoothing forecasts. Does a smoothing constant of 0.2 or 0.4 appear to provide more accurate forecasts based on MSE? Explain.

6. Consider the following time series data.

Month	1	2	3	4	5	6	7
Value	24	13	20	12	19	23	15

   Construct a time series plot. What type of pattern exists in the data?
   a. Develop the three-week moving average forecasts for this time series. Compute MSE and a forecast for week 8.
   b. Use $\alpha = 0.2$ to compute the exponential smoothing forecasts for the time series. Compute MSE and a forecast for week 8.
   c. Compare the three-week moving average approach with the exponential smoothing approach using $\alpha = 0.2$. Which appears to provide more accurate forecasts based on MSE?
   d. Use a smoothing constant of $\alpha = 0.4$ to compute the exponential smoothing forecasts. Does a smoothing constant of 0.2 or 0.4 appear to provide more accurate forecasts based on MSE? Explain.

7. Refer to the petrol sales time series data in Table 17.1.
   a. Compute four-week and five-week moving averages for the time series.
   b. Compute the MSE for the four-week and five-week moving average forecasts.
   c. What appears to be the best number of weeks of past data (three, four or five) to use in the moving average computation? Recall that MSE for the three-week moving average is 10.22.

PETROL

8. Refer again to the petrol sales time series data in Table 17.1.
   a. Using a weight of 1/2 for the most recent observation, 1/3 for the second most recent observation, and 1/6 for the third most recent observation, compute a three-week weighted moving average for the time series.
   b. Compute the MSE for the weighted moving average in part (a). Do you prefer this weighted moving average to the unweighted moving average? Remember that the MSE for the unweighted moving average is 10.22.
   c. Suppose you are allowed to choose any weights as long as they sum to 1. Could you always find a set of weights that would make the MSE at least as small for a weighted moving average than for an unweighted moving average? Why or why not?

9. With the petrol time series data from Table 17.1, show the exponential smoothing forecasts using $\alpha = 0.1$.
   a. Applying the MSE measure of forecast accuracy, would you prefer a smoothing constant of $\alpha = 0.1$ or $\alpha = 0.2$ for the petrol sales time series?
   b. Are the results the same if you apply MAE as the measure of accuracy?
   c. What are the results if MAPE is used?

10. With a smoothing constant of $\alpha = 0.2$, equation (17.2) shows that the forecast for week 13 of the petrol sales data from Table 17.1 is given by $F_{13} = 0.2Y_{12} + 0.8F_{12}$. However, the forecast for week 12 is given by $F_{12} = 0.2Y_{11} + 0.8F_{11}$. Thus, we could combine these two results to show that the forecast for week 13 can be written:

$$F_{13} = 0.2Y_{12} + 0.8(0.2Y_{11} + 0.8F_{11}) = 0.2Y_{12} + 0.16Y_{11} + 0.64F_{11}$$

   a. Making use of the fact that $F_{11} = 0.2Y_{10} + 0.8F_{10}$ (and similarly for F_{10} and F_9), continue to expand the expression for F_{13} until it is written in terms of the past data values Y_{12}, Y_{11}, Y_{10}, Y_9, Y_8, and the forecast for period 8.
   b. Refer to the coefficients or weights for the past values Y_{12}, Y_{11}, Y_{10}, Y_9, Y_8. What observation can you make about how exponential smoothing weights past data values in arriving at new forecasts? Compare this weighting pattern with the weighting pattern of the moving averages method.

Applications

11. For SIS Cargo Services in Dubai, the monthly percentages of all shipments received on time over the past 12 months are 80, 82, 84, 83, 83, 84, 85, 84, 82, 83, 84 and 83.
   a. Construct a time series plot. What type of pattern exists in the data?
   b. Compare the three-month moving average approach with the exponential smoothing approach for $\alpha = 0.2$. Which provides more accurate forecasts using MSE as the measure of forecast accuracy?
   c. What is the forecast for next month?

12. The values of Austrian building contracts (in millions of euros) for a 12-month period follow.

240 350 230 260 280 320 220 310 240 310 240 230

   a. Construct a time series plot. What type of pattern exists in the data?
   b. Compare the three-month moving average approach with the exponential smoothing forecast using $\alpha = 0.2$. Which provides more accurate forecasts based on MSE?
   c. What is the forecast for the next month?

13. The following data represent indices for the merchandise imports by broad economic category for New Zealand from 2013 to 2017:

2013	Jun	1,177	2016	Mar	945
	Sep	1,202		Jun	972
	Dec	1,173		Sep	969
2014	Mar	1,162		Dec	971
	Jun	1,133	2017	Mar	1,013
	Sep	1,149			
	Dec	1,136			
2015	Mar	1,022			
	Jun	1,041			
	Sep	1,118			
	Dec	1,035			

NZ IMPORTS

a. Compute three- and four-quarter moving averages for this time series. Which moving average provides the better forecast for the second quarter of 2017?

b. Plot the data. Do you think the exponential smoothing model would be appropriate for forecasting in this case?

17.4 TREND PROJECTION

We present three forecasting methods in this section that are appropriate for time series exhibiting a trend pattern. First, we show how simple linear regression can be used to forecast a time series with a linear trend. We then illustrate how to develop forecasts using Holt's linear exponential smoothing, an extension of single exponential smoothing that uses two smoothing constants: one to account for the level of the time series and a second to account for the linear trend in the data. Finally, we show how the curve fitting capability of regression analysis can also be used to forecast time series with a curvilinear or nonlinear trend.

Linear trend regression

In Section 17.1 we used the bicycle sales time series in Table 17.3 and Figure 17.3 to illustrate a time series with a trend pattern. We now exploit this time series to illustrate how simple linear regression can be used to forecast a time series with a linear trend. The data for the bicycle time series are repeated in Table 17.12 and Figure 17.9.

TABLE 17.12 Bicycle sales time series

Year	Sales (000)
1	21.6
2	22.9
3	25.5
4	21.9
5	23.9
6	27.5
7	31.5
8	29.7
9	28.6
10	31.4

FIGURE 17.9
Bicycle sales time series plot

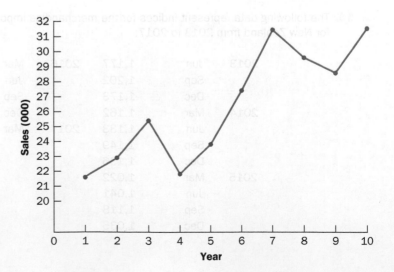

Although the time series plot in Figure 17.9 shows some up and down movement over the past ten years, we might agree that the linear trend line shown in Figure 17.10 provides a reasonable approximation of the long-run movement in the series. We can use the methods of simple linear regression (refer to Chapter 14) to develop such a linear trend line for the bicycle sales time series.

FIGURE 17.10

Trend represented by a linear function for bicycle sales

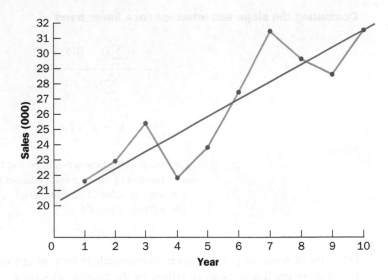

In Chapter 14, the estimated regression equation describing a straight-line relationship between an independent variable x and a dependent variable y is written as:

$$\hat{y} = b_0 + b_1 x$$

where $\hat{y}$ is the estimated or predicted value of y. To emphasize the fact that in forecasting the independent variable is time, we will replace x with t and $\hat{y}$ with T_t to emphasize that we are estimating the trend for a time series. Thus, for estimating the linear trend in a time series we will use the following estimated regression equation:

Linear trend equation

$$T_1 = b_0 + b_1 t \qquad \qquad \textbf{(17.4)}$$

where:

T_1 = linear trend forecast in period t
b_0 = intercept of the linear trend line
b_1 = slope of the linear trend line
t = time period

In equation (17.4), the time variable begins at $t = 1$ corresponding to the first time series observation (year 1 for the bicycle sales time series) and continues until $t = n$ corresponding to the most recent time series observation (year 10 for the bicycle sales time series). Thus the bicycle sales time series $t = 1$ corresponds to the oldest time series value and $t = 10$ corresponds to the most recent year.

Formulae for computing the estimated regression coefficients (b_1 and b_0) in equation (17.4) are as follows:

Computing the slope and intercept for a linear trend*

$$b_1 = \frac{\sum_{t-1}^{n}(t - \bar{t})(Y_t - \bar{Y})}{\sum_{t-1}^{n}(t - \bar{t})^2} \tag{17.5}$$

$$b_0 = \bar{Y} - b_1\bar{t} \tag{17.6}$$

where:

Y_t = value of the time series in period t
n = number of time periods (number of observations)
$\bar{Y}$ = average value of the time series
$\bar{t}$ = average value of t

This form of equation (17.5) is often recommended when using a calculator to compute b_1.

To compute the linear trend equation for the bicycle sales time series, we begin the calculations by computing $\bar{t}$ and $\bar{Y}$ using the information in Table 17.12.

$$\bar{t} = \frac{\sum_{t-1}^{n}t}{n} = \frac{55}{10} = 5.5$$

$$\bar{Y} = \frac{\sum_{t-1}^{n}Y_t}{n} = \frac{264.5}{10} = 26.45$$

Using these values and the information in Table 17.13, we can compute the slope and intercept of the trend line for the bicycle sales time series.

$$b_1 = \frac{\sum_{t-1}^{n}(t - \bar{t})(Y_t - \bar{Y})}{\sum_{t-1}^{n}(t - \bar{t})^2} = \frac{90.75}{82.5} = 1.1$$

$$b_0 = \bar{Y} - b_1\bar{t} = 26.45 - 1.1(5.5) = 20.4$$

Therefore, the linear trend equation is:

$$T_t = 20.4 + 1.1t$$

The slope of 1.1 indicates that over the past ten years the firm experienced an average growth in sales of about 1,100 units per year. If we assume that the past ten-year trend in sales is a good indicator of the future, this trend equation can be used to develop forecasts for future time periods. For example, substituting $t = 11$ into the equation yields next year's trend projection or forecast, T_{11}.

$$T_{11} = 20.4 + 1.1(11) = 32.5$$

---

*An alternate formula for b_1 is:

$$b_i = \frac{\sum_{t=1}^{n}tY_t - \left(\sum_{t-1}^{n}\sum_{t-1}^{n}Y_t\right)/n}{\sum_{t-1}^{n}t^2 - \left(\sum_{t=1}^{n}t\right)^2/n}$$

TABLE 17.13 Summary of linear trend calculations for the bicycle sales time series

t	Y_t	$t - \bar{t}$	$Y_t - \bar{Y}$	$(t - \bar{t})(Y_t - \bar{Y})$	$(t - \bar{t})^2$
1	21.6	−4.5	−4.85	21.825	20.25
2	22.9	−3.5	−3.55	12.425	12.25
3	25.5	−2.5	−0.95	2.375	6.25
4	21.9	−1.5	−4.55	6.825	2.25
5	23.9	−0.5	−2.55	1.275	0.25
6	27.5	0.5	1.05	0.525	0.25
7	31.5	1.5	5.05	7.575	2.25
8	29.7	2.5	3.25	8.125	6.25
9	28.6	3.5	2.15	7.525	12.25
10	31.4	4.5	4.95	22.275	20.25
Totals	**55**	**264.5**			**90.750**	**82.50**

Thus, using trend projection, we would forecast sales of 32,500 bicycles next year.

To compute the accuracy associated with the trend projection forecasting method, we will use the MSE. Table 17.14 shows the computation of the sum of squared errors for the bicycle sales time series. Thus, for the bicycle sales time series:

$$\text{MSE} = \frac{\sum_{t-1}^{n}(Y_t - F_t)^2}{n} = \frac{30.7}{10} = 3.07$$

Because linear trend regression in forecasting is the same as the standard regression analysis procedure applied to time series data, we can use statistical software to perform the calculations. Figure 17.11 shows a portion of the computer output for the bicycle sales time series. Figure 17.12 shows the linear trend and the values of MAPE, MAD and MSE.

TABLE 17.14 Summary of the linear trend forecasts and forecast errors for the bicycle sales time series

Year	Sales (000) Y_t	Forecast T_t	Forecast error	Squared forecast error
1	21.6	21.5	0.1	0.01
2	22.9	22.6	0.3	0.09
3	25.5	23.7	1.8	3.24
4	21.9	24.8	−2.9	8.41
5	23.9	25.9	−2.0	4.00
6	27.5	27.0	0.5	0.25
7	31.5	28.1	3.4	11.56
8	29.7	29.2	0.5	0.25
9	28.6	30.3	−1.7	2.89
10	31.4	31.4	0.0	0.00
			Total	**30.70**

As linear trend regression in forecasting uses the same regression analysis procedure introduced in Chapter 14, we can use the standard regression analysis procedures In Excel to perform the calculations. Figure 17.11 shows the computer output for the bicycle sales time series obtained using a regression analysis module.

In Figure 17.11 the value of MSE in the ANOVA table is:

$$\text{MSE} = \frac{\text{Sum of squares due to error}}{\text{Degrees of freedom}} = \frac{30.7}{8} = 3.837$$

FIGURE 17.11 Regression output for the bicycle sales time series

```
Analysis of variance

Source              DF              SS              MS          F          P
Regression          1            99.82          99.825      26.01      0.001
Residual error      8            30.70           3.837
Total               9           130.52

S = 1.95895          R-sq = 76.5%             R-sq(adj) = 73.5%

Predictor           Coef          SE Coef         T          P
Constant            20.40          1.338        15.24      0.000
Period               1.10          0.2157        5.10      0.001
```

The regression equation is Y = 20.4 + 1.10t

This value of MSE differs from the value of MSE that we computed previously because the sum of squared errors is divided by 8 instead of 10; thus, MSE in the regression output is not the average of the squared forecast errors.

Most forecasting packages, however, compute MSE by taking the average of the squared errors. Thus, when using time series packages to develop a trend equation, the value of MSE that is reported may differ slightly from the value you would obtain using a general regression approach.

Figure 17.12 shows the graphical portion of computer output obtained fitting an exponential trend equation to the bycycle sales time series data.

FIGURE 17.12 Time series linear trend analysis output for the bicycle sales time series

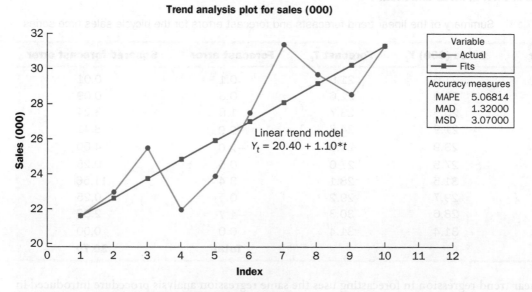

Holt's linear exponential smoothing

Charles Holt developed a version of exponential smoothing that can be used to forecast a time series with a linear trend. Recall that the exponential smoothing procedure discussed in Section 17.3 uses the smoothing constant to 'smooth out' the randomness or irregular fluctuations in a time series; and, forecasts for time period $t + 1$ are obtained using the equation:

$$F_{t+1} = \alpha Y_t + (1 - \alpha)F$$

Forecasts for Holt's linear exponential smoothing method are obtained using two smoothing constants, α and β, and three equations.

> ### Equations for Holt's linear exponential smoothing
>
> $$L_t = \alpha Y_t + (1 - \alpha)(L_{t-1} + b_{t-1}) \qquad \textbf{(17.7)}$$
>
> $$b_t = \beta(L_t - L_{t-1})\,(1 - \beta)b_{t-1} \qquad \textbf{(17.8)}$$
>
> $$F_{t+1} = L_t + b_t k \qquad \textbf{(17.9)}$$
>
> where:
>
> L_t = estimate of the level of the time series in period t
> b_t = estimate of the slope of the time series in period t
> α = smoothing constant for the level of the time series
> β = smoothing constant for the slope of the time series
> F_{t+k} = forecast for k periods ahead
> k = the number of periods ahead to be forecast

Let us apply Holt's method to the bicycle sales time series in Table 17.12 using $\alpha = 0.1$ and $\beta = 0.2$. To start the method off, we need values for L_1, the estimate of the level of the time series in year 1, and b_1, the estimate of the slope of the time series in year 1. A common approach is to set $L_1 = Y_1$ and $b_1 = Y_2 - Y_1$. Using this start-up procedure, we obtain:

$$L_1 = Y_1 = 21.6$$
$$b_1 = Y_2 - Y_1 = 22.9 - 21.6 = 1.3$$

Using equation (17.9) with $k = 1$, the forecast of sales in year 2 is $F_2 = L_1 + b_1 = 21.6 + 1.3(1) = 22.9$. Then we move on using equations (17.7) to (17.9) to compute estimates of the level and trend for year 2 as well as a forecast for year 3.

First we use equation (17.7) and the smoothing constant $\alpha = 0.1$ to compute an estimate of the level of the time series in year 2.

$$L_2 = 0.1(22.9) + 0.9(21.6 + 1.3) = 22.9$$

Note that $21.6 + 1.3$ is the forecast of sales for year 2. Thus, the estimate of the level of the time series in year 2 obtained using equation (17.7) is simply a weighted average of the observed value in year 2 (using a weight of = 0.1) and the forecast for year 2 (using a weight of $1 - \alpha = 1 - 0.1 = 0.9$). In general, large values of α place more weight on the observed value (Y_t), whereas smaller values place more weight on the forecast value $(L_{t-1} + b_{t-1})$.

Next we use equation (17.8) and the smoothing constant $\beta = 0.2$ to compute an estimate of the slope of the time series in year 2.

$$b_2 = 0.2(22.9 - 21.6) + (1 - 0.2)(1.3) = 1.3$$

The estimate of the slope of the time series in year 2 is a weighted average of the difference in the estimated level of the time series between year 2 and year 1 (using a weight of $\beta = 0.2$) and the estimate of the slope in year 1 (using a weight of $1 - \beta = 1 - 0.2 = 0.8$). In general, higher values of β place more weight on the difference between the estimated levels, whereas smaller values place more weight on the estimate of the slope from the last period.

Using the estimates of L_2 and b_2 just obtained, the forecast of sales for year 3 is computed using equation (17.9).

$$F_3 = L_2 + b_2 = 22.9 + 1.3(1) = 24.2$$

The other calculations are made in a similar manner and are shown in Table 17.15. The sum of the squared forecast errors is 39.678; hence MSE = 39.678/9 = 4.41.

TABLE 17.15 Summary calculations for Holt's linear exponential smoothing for the bicycle sales time series using $\alpha = 0.1$ and $\beta = 0.2$

Year	Sales (000) Y_t	Estimated level L_t	Estimated trend b_t	Forecast F_t	Forecast error	Squared forecast error
1	21.6	21.600	1.300			
2	22.9	22.900	1.300	22.900	0.000	0.000
3	25.5	24.330	1.326	24.200	1.300	1.690
4	21.9	25.280	1.251	25.656	−3.756	14.108
5	23.9	26.268	1.198	26.531	−2.631	6.924
6	27.5	27.470	1.199	27.466	0.034	0.001
7	31.5	28.952	1.256	28.669	2.831	8.016
8	29.7	30.157	1.245	30.207	−0.507	0.257
9	28.6	31.122	1.189	31.402	−2.802	7.851
10	31.4	32.220	1.171	32.311	−0.911	0.830
					Total	**39.678**

Will different values for the smoothing constants α and β provide more accurate forecasts? To answer this question we would have to try different combinations of α and β, to determine if a combination can be found that will provide a value of MSE lower than 4.41, the value we obtained using smoothing constants $\alpha = 0.1$ and $\beta = 0.2$. Searching for good values of α and β can be done by trial and error or using more advanced statistical software packages that have an option for selecting the optimal set of smoothing constants.

Note that the estimate of the level of the time series in year 10 is $L_1 = 32.220$, and the estimate of the slope in year 10 is $b_1 = 1.171$. If we assume that the past ten-year trend in sales is a good indicator of the future, equation (17.9) can be used to develop forecasts for future time periods. For example, substituting $t = 11$ into equation (17.9) yields next year's trend projection or forecast, F_{11}.

$$F_{11} = L_{10} + b_{10}(1) = 32.220 + 1.171 = 33.391$$

Thus, using Holt's linear exponential smoothing we would forecast sales of 33,391 bicycles next year.

Linear trend regression is based upon finding the estimated regression equation that minimizes the sum of squared forecast errors and therefore MSE. So, we would expect linear trend regression to outperform Holt's linear exponential smoothing in terms of MSE. For example, for the bicycle sales time series, the value of MSE using linear trend regression is 3.07 as compared to a value of 3.97 using Holt's linear exponential smoothing. Linear trend regression also provides a more accurate forecast using the MAE measure of forecast accuracy; for the bicycle sales time series, linear trend regression results in a value of MAE of 1.32 versus a value of 1.67 using Holt's linear method.

However, based on MAPE, Holt's linear exponential smoothing (MAPE = 5.07 per cent) outperforms linear trend regression (6.42 per cent). Hence, for the bicycle sales time series, deciding which method provides the more accurate forecasts depends upon which measure of forecast accuracy is used.

Nonlinear trend regression

CHOLESTEROL

The use of a linear function to model trend is common. However, as we discussed previously, sometimes time series have a curvilinear or nonlinear trend. As an example, consider the annual revenue in millions of dollars for a cholesterol drug for the first ten years of sales. This example featured earlier in Section 17.1. The data and time series plot are repeated in Table 17.16 and Figure 17.13 respectively. For instance, revenue in year 1 was $23.1 million, revenue in year 2 was $21.3 million and so on. The time series plot indicates an overall increasing or upward trend. But, unlike the bicycle sales time series, a linear trend does not appear to be appropriate. Instead, a curvilinear function appears to be needed to model the long-term trend.

Quadratic trend equation

A variety of nonlinear functions can be used to develop an estimate of the trend for the cholesterol time series. For instance, consider the following quadratic trend equation:

$$T_t = b_0 + b_1 t + b_2 t^2 \qquad\qquad (17.10)$$

TABLE 17.16 Cholesterol revenue time series

Year (t)	Revenue ($ million)
1	23.1
2	21.3
3	27.4
4	34.6
5	33.8
6	43.2
7	59.5
8	64.4
9	74.2
10	99.3

FIGURE 17.13

Cholesterol revenue time series plot ($ million)

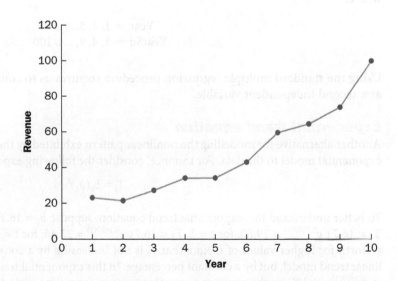

For the cholesterol time series, $t = 1$ corresponds to year 1, $t = 2$ corresponds to year 2 and so on.

The general linear model discussed in Section 16.1 can be used to compute the values of b_0, b_1 and b_2. There are two independent variables, year and year squared, and the dependent variable is the sales revenue in millions of dollars. Thus, the first observation is 1, 1, 23.1; the second observation is 2, 4, 21.3; the third observation is 3, 9, 27.4; and so on. Figure 17.14 shows the computer output for the quadratic trend model.

FIGURE 17.14   Quadratic trend regression output for the bicycle sales time series

```
Analysis of variance

Source                   DF              SS              MS           F           P
Regression                2           5770.1          2885.1       182.52       0.000
Residual Error            7            110.6            15.8
Total                     9           5880.8

S = 3.97578            R-sq = 98.1%              R-sq(adj) = 97.6%

Predictor              Coef          SE Coef           T           P
Constant             24.182          4.676           5.17         0.001
Year                 -2.106          1.953          -1.08         0.317
YearSq               0.9216          0.1730           5.33         0.001
The regression equation is Revenue = 24.2 - 2.11 Year + 0.922 YearSq
```

The estimated regression equation is:

$$\text{Revenue (\$ millions)} = 24.2 - 2.11\,\text{Year} + 0.922\,\text{YearSq}$$

where:

$$\text{Year} = 1, 2, 3, \ldots, 10$$
$$\text{YearSq} = 1, 4, 9, \ldots, 100$$

Using the standard multiple regression procedure requires us to compute the values for year squared as a second independent variable.

Exponential trend equation

Another alternative for modelling the nonlinear pattern exhibited by the cholesterol time series is to fit an exponential model to the data. For instance, consider the following exponential trend equation:

$$T_t = b_0(b_1)^t \tag{17.11}$$

To better understand this exponential trend equation, suppose $b_0 = 16.71$ and $b_1 = 0.1697$. Then, for $t = 1$, $T_1 = 16.71\,e^{0.1697(1)} = 19.80$; for $t = 2$, $T_2 = 16.71\,e^{0.1697(2)} = 23.46$; for $t = 3$, $T_3 = 16.71\,e^{0.1697(3)} = 27.80$, and so forth for higher values of t. Note that T_t is not increasing by a constant amount as in the case of the linear trend model, but by a constant percentage. In this exponential trend model, the multiplicative factor is $e^{0.1697(1)} = 1.185$, so the constant percentage increase from time period to time period is 18.5 per cent.

Many statistical software packages have the capability to compute an exponential trend equation directly. Some software packages only provide linear trend, but by applying a natural log transformation to both sides of the equality in equation (17.8) we can apply the equivalent linear form:

$$\ln T_t = \ln b_0 + b_1 t$$

Figure 17.15 shows the graphical portion of computer output obtained by fitting an exponential trend equation to the cholesterol sales time series data.

FIGURE 17.15

Time series exponential growth trend analysis output for the cholesterol sales time series

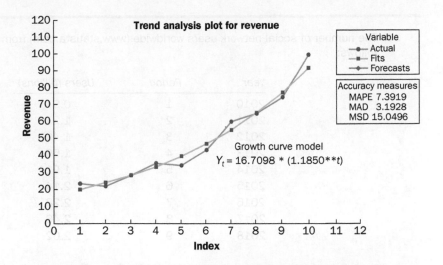

Trend analysis plot for revenue

Variable
Actual
Fits
Forecasts

Accuracy measures
MAPE 7.3919
MAD 3.1928
MSD 15.0496

Growth curve model
$Y_t = 16.7098 * (1.1850**t)$

EXERCISES

Methods

14. Consider the following time series data:

t	1	2	3	4	5
Y_t	6	11	9	14	15

  a. Construct a time series plot. What type of pattern exists in the data?
  b. Develop the linear trend equation for this time series.
  c. What is the forecast for $t = 6$?

15. Refer to the time series in Exercise 14. Use Holt's linear exponential smoothing method with $\alpha = 0.3$ and $\beta = 0.5$ to develop a forecast for $t = 6$.

16. Refer to the number of undocumented immigrants data in the Statistics in Practice vignette at the beginning of the chapter.
  a. Construct a time series plot. What type of pattern exists in the data?
  b. Develop an exponential trend equation for this time series.
  c. Use the equation developed in (b) to forecast numbers for 2023, 2024 and 2025.

Applications

17. Tesla is an electric vehicle manufacturer. The file 'TeslaSales' contains the annual sales of Tesla vehicles from 2013 to 2020.

Year	Period	Sales (units)
2013	1	22,442
2014	2	31,655
2015	3	50,517
2016	4	76,243
2017	5	103,091
2018	6	245,491
2019	7	367,656
2020	8	499,535

TESLA
SALES

  a. Construct a time series plot. What type of pattern exists in the data?
  b. Develop an exponential trend equation for this time series.
  c. Use the equation developed in (b) to forecast sales in 2021 (period 9).

18. The number of social network users worldwide (www.statista.com) from 2010 to 2018 was as follows:

Year	Period	Users (billions)
2010	1	0.97
2011	2	1.22
2012	3	1.40
2013	4	1.59
2014	5	1.91
2015	6	2.14
2016	7	2.28
2017	8	2.46
2018	9	2.62

a. Construct a time series plot. What type of pattern exists?
b. Develop a quadratic trend.

19. Numbers of overseas visitors to Ireland (000) estimated by Failte Ireland for the years 2009–17 are as follows:

2009	2010	2011	2012	2013	2014	2015	2016	2017
6,578	5,945	6,240	6,286	6,686	7,105	8,036	8,742	9,023

a. Graph the data and assess its suitability for linear trend projection.
b. Use a linear trend projection to forecast this time series for 2018–20.

GDP
SINGAPORE

20. GDP (USD billion) for Singapore 2008–17 are tabulated below (tradingeconomics.com/singapore/gdp):

Year	USD (billions)
2008	192.23
2009	192.41
2010	236.42
2011	275.97
2012	290.67
2013	304.45
2014	311.54
2015	304.10
2016	309.76
2017	323.91

a. Graph this time series. Does a linear trend appear to be present?
b. Develop a linear trend equation for this time series.
c. Use the trend equation to estimate the GDP for the years 2018–20.

21. The following data show Google revenue from 2011 (period 1) to 2020 (period 10) in billions of dollars (Alphabet, Inc. annual reports). These data are in the file 'GoogleRevenue'.

GOOGLE
REVENUE

Year	Period	Annual revenue (in $ billion)
2011	1	37.91
2012	2	50.18
2013	3	55.50
2014	4	66.00
2015	5	74.99
2016	6	90.27
2017	7	110.86
2018	8	136.82
2019	9	161.86
2020	10	182.53

a. Construct a time-series plot. What type of pattern exists?
b. Develop a quadratic trend equation.

17.5 SEASONALITY AND TREND

In this section we show how to develop forecasts for a time series that has a seasonal pattern. To the extent that seasonality exists, we need to incorporate it into our forecasting models to ensure accurate forecasts. We begin by considering a seasonal time series with no trend and then discuss how to model seasonality with trend.

Seasonality without trend

As an example, consider the number of umbrellas sold at a clothing store over the past five years. This example featured earlier in Section 17.1. The data and time series plot are repeated in Table 17.17 and Figure 17.16 respectively. The time series plot does not indicate any long-term trend in sales. In fact, unless you look carefully at the data, you might conclude that the data follow a horizontal pattern and that single exponential smoothing could be used to forecast sales. But closer inspection of the time series plot reveals a pattern in the data. That is, the first and third quarters have moderate sales, the second quarter has the highest sales and the fourth quarter tends to be the lowest quarter in terms of sales volume. Therefore, we would conclude that a quarterly seasonal pattern is present.

In Chapter 15 we showed how dummy variables can be used to deal with categorical independent variables in a multiple regression model. We can use the same approach to model a time series with a seasonal pattern by treating the season as a categorical variable. Recall that when a categorical variable has k levels, $k - 1$ dummy variables are required. So, if there are four seasons, we need three dummy variables. For instance, in the umbrella sales time series season there is a categorical variable with four levels: quarter 1, quarter 2, quarter 3 and quarter 4. Thus, to model the seasonal effects in the umbrella time series we need $4 - 1 = 3$ dummy variables. The three dummy variables can be coded as follows:

$$Qtr1 = \begin{cases} 1 & \text{if Quarter 1} \\ 0 & \text{otherwise} \end{cases} \quad Qtr2 = \begin{cases} 1 & \text{if Quarter 2} \\ 0 & \text{otherwise} \end{cases} \quad Qtr3 = \begin{cases} 1 & \text{if Quarter 3} \\ 0 & \text{otherwise} \end{cases}$$

Using $\hat{Y}$ to denote the estimated or forecast value of sales, the general form of the estimated regression equation relating the number of umbrellas sold to the quarter the sales take place is as follows:

$$\hat{y} = b_0 + b_1 \, Qtr1 + b_2 \, Qtr2 + b_3 \, Qtr3$$

UMBRELLA

FIGURE 17.16
Umbrella sales time
series plot

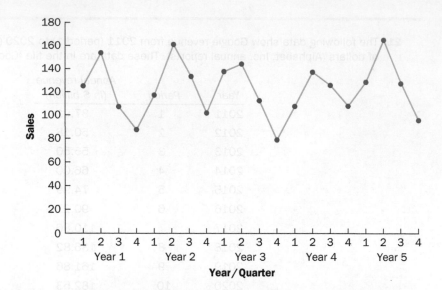

TABLE 17.17   Umbrella sales time series

Year	Quarter	Sales
1	1	125
	2	153
	3	106
	4	88
2	1	118
	2	161
	3	133
	4	102
3	1	138
	2	144
	3	113
	4	80
4	1	109
	2	137
	3	125
	4	109
5	1	130
	2	165
	3	128
	4	96

Table 17.18 is the umbrella sales time series with the coded values of the dummy variables shown. For the data in Table 17.18 we obtained the computer output shown in Figure 17.17. The estimated multiple regression equation obtained is:

$$\text{Sales} = 95.0 + 29.0 \text{ Qtr1} + 57.0 \text{ Qtr2} + 26.0 \text{ Qtr3}$$

We can use this equation to forecast quarterly sales for next year.

Quarter 1 : Sales = 95.0 + 29.0(1) + 57.0(0) + 26.0(0) = 124
Quarter 2 : Sales = 95.0 + 29.0(0) + 57.0(1) + 26.0(0) = 152
Quarter 3 : Sales = 95.0 + 29.0(0) + 57.0(0) + 26.0(1) = 121
Quarter 4 : Sales = 95.0 + 29.0(0) + 57.0(1) + 26.0(0) = 95

TABLE 17.18   Umbrella sales time series with dummy variables

Year	Quarter	Qtr1	Qtr2	Qtr3	Sales
1	1	1	0	0	125
	2	0	1	0	153
	3	0	0	1	106
	4	0	0	0	88
2	1	1	0	0	118
	2	0	1	0	161
	3	0	0	1	133
	4	0	0	0	102
3	1	1	0	0	138
	2	0	1	0	144
	3	0	0	1	113
	4	0	0	0	80
4	1	1	0	0	109
	2	0	1	0	137
	3	0	0	1	125
	4	0	0	0	109
5	1	1	0	0	130
	2	0	1	0	165
	3	0	0	1	128
	4	0	0	0	96

It is interesting to note that we could have obtained the quarterly forecasts for next year simply by computing the average number of umbrellas sold in each quarter, as shown in the following table.

Year	Quarter 1	Quarter 2	Quarter 3	Quarter 4
1	125	153	106	88
2	118	161	133	102
3	138	144	113	80
4	109	137	125	109
5	130	165	128	96
Average	124	152	121	95

Nonetheless, the regression output shown in Figure 17.17 provides additional information that can be used to assess the accuracy of the forecast and determine the significance of the results. And, for more complex types of problem situations, such as dealing with a time series that has both trend and seasonal effects, this simple averaging approach will not work.

FIGURE 17.17
Regression output for the umbrella sales time series

Predictor	Coef	SE Coef	T	P
Constant	95.000	5.065	18.76	0.000
Qtr1	29.000	7.162	4.05	0.001
Qtr2	57.000	7.162	7.96	0.000
Qtr3	26.000	7.162	3.63	0.002

The regression equation is

Sales = 95.0 + 29.0 Qtr1 + 57.0 Qtr2 + 26.0 Qtr3

Seasonality and trend

We now extend the regression approach to include situations where the time series contains both a seasonal effect and a linear trend by showing how to generate forecasts for the quarterly smartphone sales time series introduced in Section 17.1. The data for the smartphone sales time series plot are repeated in Table 17.19 and Figure 17.18 for convenience.

The latter plot in Figure 17.18 indicates that sales are lowest in the second quarter of each year and increase in quarters 3 and 4. Thus, we conclude that a seasonal pattern exists for smartphone sales. But the time series also has an upward linear trend that will need to be accounted for in order to develop accurate forecasts of quarterly sales. This is easily handled by combining the dummy variable approach for seasonality with the time series regression approach discussed in Section 17.3 for handling linear trend.

SMART
PHONE

TABLE 17.19 Smartphone sales time series

Year	Quarter	Sales (000)
1	1	4.8
	2	4.1
	3	6.0
	4	6.5
2	1	5.8
	2	5.2
	3	6.8
	4	7.4
3	1	6.0
	2	5.6
	3	7.5
	4	7.8
4	1	6.3
	2	5.9
	3	8.0
	4	8.4

FIGURE 17.18
Smartphone sales time series plot

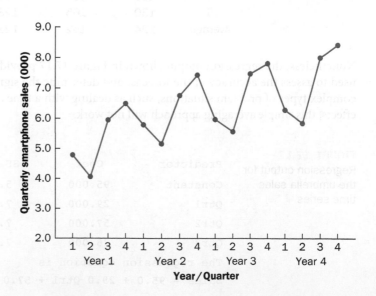

The general form of the estimated multiple regression equation for modelling both the quarterly seasonal effects and the linear trend in the smartphone time series is as follows:

$$\hat{Y}_t = b_0 + b_1\,\text{Qtr1} + b_2\,\text{Qtr2} + b_3\,\text{Qtr3} + b_4\,t$$

where:

$\hat{Y}_t$ = estimate or forecast of sales in period t
Qtr1 = 1 if time period t corresponds to the first quarter of the year; 0 otherwise
Qtr2 = 1 if time period t corresponds to the second quarter of the year; 0 otherwise
Qtr3 = 1 if time period t corresponds to the third quarter of the year; 0 otherwise
t = time period

Table 17.20 provides the revised smartphone sales time series including the coded values of the dummy variables and the time period t. Applying regression procedure to these data, we obtain the computer output shown in Figure 17.19. The estimated multiple regression equation is:

$$\text{Sales} = 6.07 - 1.36\,\text{Qtr1} - 2.03\,\text{Qtr2} - 0.304\,\text{Qtr3} - 0.146t \qquad \textbf{(17.12)}$$

TABLE 17.20 Smartphone sales time series with dummy variables and time period

Year	Quarter	Qtr1	Qtr2	Qtr3	Period	Sales (000)
1	1	1	0	0	1	4.8
	2	0	1	0	2	4.1
	3	0	0	1	3	6.0
	4	0	0	0	4	6.5
2	1	1	0	0	5	5.8
	2	0	1	0	6	5.2
	3	0	0	1	7	6.8
	4	0	0	0	8	7.4
3	1	1	0	0	9	6.0
	2	0	1	0	10	5.6
	3	0	0	1	11	7.5
	4	0	0	0	12	7.8
4	1	1	0	0	13	6.3
	2	0	1	0	14	5.9
	3	0	0	1	15	8.0
	4	0	0	0	16	8.4

FIGURE 17.19
Regression output for the smartphone sales time series

Predictor	Coef	SE Coef	T	P
Constant	6.0688	0.1625	37.35	0.000
Qtr1	−1.3631	0.1575	−8.66	0.000
Qtr2	−2.0337	0.1551	−13.11	0.000
Qtr3	−0.3044	0.1537	−1.98	0.073
Period	0.14562	0.01211	12.02	0.000

The regression equation is

Sales (000) = 6.07 - 1.36 Qtr1 + 2.03 Qtr2 - 0.304
Qtr3 + 0.146 Period

We can now use equation (17.12) to forecast quarterly sales for time periods 17, 18, 19 and 20.

Forecast for Time Period 17 (Quarter 1 in Year 5):

$$\text{Sales} = 6.07 - 1.36(1) - 2.03(0) - 0.304(0) + 0.146(17) = 7.19$$

Forecast for Time Period 18 (Quarter 2 in Year 5):

$$\text{Sales} = 6.07 - 1.36(0) - 2.03(1) - 0.304(0) + 0.146(18) = 6.67$$

Forecast for Time Period 19 (Quarter 3 in Year 5):

$$\text{Sales} = 6.07 - 1.36(0) - 2.03(0) - 0.304(1) + 0.146(19) = 8.54$$

Forecast for Time Period 20 (Quarter 4 in Year 5):

$$\text{Sales} = 6.07 - 1.36(0) - 2.03(0) - 0.304(0) + 0.146(20) = 8.99$$

Thus, accounting for the seasonal effects and the linear trend in smartphone sales, the estimates of quarterly sales in year 5 are 7,190, 6,670, 8,540 and 8,990.

The dummy variables in the estimated multiple regression equation actually provide four estimated multiple regression equations, one for each quarter. For instance, if time period t corresponds to quarter 1, the estimate of quarterly sales is:

$$\text{Quarter 1: Sales} = 6.07 + 1.36(1) - 2.03(1) - 0.304(0) + 0.146t = 4.71 + 0.146t$$

Similarly, if time period t corresponds to quarters 2, 3 and 4, the estimates of quarterly sales are:

$$\text{Quarter 2: Sales} = 6.07 + 1.36(0) - 2.03(1) - 0.304(0) + 0.146t = 4.04 + 0.146t$$
$$\text{Quarter 3: Sales} = 6.07 + 1.36(0) - 2.03(0) - 0.304(1) + 0.146t = 5.77 + 0.146t$$
$$\text{Quarter 4: Sales} = 6.07 + 1.36(0) - 2.03(0) - 0.304(0) + 0.146t = 6.07 + 0.146t$$

The slope of the trend line for each quarterly forecast equation is 0.146, indicating a growth in sales of about 146 smartphones per quarter. The only difference in the four equations is that they have different intercepts. For instance, the intercept for the quarter 1 equation is 4.71 and the intercept for the quarter 4 equation is 6.07. Thus, sales in quarter 1 are $4.71 - 6.07 = -1.36$ or 1,360 smartphones less than in quarter 4. In other words, the estimated regression coefficient for Qtr1 in equation (17.12) provides an estimate of the difference in sales between quarter 1 and quarter 4. Similar interpretations can be provided for -2.03, the estimated regression coefficient for dummy variable Qtr2, and -0.304, the estimated regression coefficient for dummy variable Qtr3.

Models based on monthly data

In the preceding smartphone sales example, we showed how dummy variables can be used to account for the quarterly seasonal effects in the time series. Because there were four levels for the categorical variable season, three dummy variables were required. However, many businesses use monthly rather than quarterly forecasts. For monthly data, season is a categorical variable with 12 levels and thus $12 - 1 = 11$ dummy variables are required. For example, the 11 dummy variables could be coded as follows:

$$\text{Month1} = \begin{cases} 1 \text{ if January} \\ 0 \text{ otherwise} \end{cases}$$

$$\text{Month2} = \begin{cases} 1 \text{ if February} \\ 0 \text{ otherwise} \end{cases}$$

$$\vdots$$

$$\text{Month11} = \begin{cases} 1 \text{ if November} \\ 0 \text{ otherwise} \end{cases}$$

Other than this change, the multiple regression approach for handling seasonality remains the same.

EXERCISES

Methods

22. Consider the following time series.

Quarter	Year 1	Year 2	Year 3
1	71	68	62
2	49	41	51
3	58	60	53
4	78	81	72

a. Construct a time series plot. What type of pattern exists in the data?

b. Use the following dummy variables to develop an estimated regression equation to account for seasonal effects in the data: Qtr1 = 1 if Quarter 1, 0 otherwise; Qtr2 = 1 if Quarter 2, 0 otherwise; Qtr3 = 1 if Quarter 3, 0 otherwise.

c. Compute the quarterly forecasts for next year.

23. Consider the following time series data.

Quarter	Year 1	Year 2	Year 3
1	4	6	7
2	2	3	6
3	3	5	6
4	5	7	8

a. Construct a time series plot. What type of pattern exists in the data?

b. Use the following dummy variables to develop an estimated regression equation to account for any seasonal and linear trend effects in the data: Qtr1 = 1 if Quarter 1, 0 otherwise; Qtr2 = 1 if Quarter 2, 0 otherwise; Qtr3 = 1 if Quarter 3, 0 otherwise.

c. Compute the quarterly forecasts for next year.

Applications

24. Electric power consumption is measured in kilowatt-hours (kWh). The local utility company offers an interrupt programme whereby commercial customers that participate receive favourable rates but must agree to cut back consumption if the utility requests them to do so. Timko Products has agreed to cut back consumption from noon to 8.00 p.m. on Thursday. To determine Timko's savings, the utility must estimate Timko's normal power usage for this period of time. Data on Timko's electric power consumption for the previous 72 hours are as follows.

Time Period	Monday	Tuesday	Wednesday	Thursday
12–4 A.M.	—	19,281	31,209	27,330
4–8 A.M.	—	33,195	37,014	32,715
8–12 noon	—	99,516	119,968	152,465
12–4 P.M.	124,299	123,666	156,033	
4–8 P.M.	113,545	111,717	128,889	
8–12 midnight	41,300	48,112	73,923	

POWER

a. Construct a time series plot. What type of pattern exists in the data?

b. Use the following dummy variables to develop an estimated regression equation to account for any seasonal effects in the data.

Time1 = 1 for time period 12–4 a.m.; 0 otherwise
Time2 = 1 for time period 4–8 a.m.; 0 otherwise
Time3 = 1 for time period 8–12 noon; 0 otherwise
Time4 = 1 for time period 12–4 p.m.; 0 otherwise
Time5 = 1 for time period 4–8 p.m.; 0 otherwise

c. Use the estimated regression equation developed in (b) to estimate Timko's normal usage over the period of interrupted service.

d. Let Period = 1 to refer to the observation for Monday in the time period 12–4 p.m.; Period = 2 to refer to the observation for Monday in the time period 4–8 p.m.; . . . and Period = 18 to refer to the observation for Thursday in the time period 8–12 noon. Using the dummy variables defined in (b) and Time period, develop an estimated regression equation to account for seasonal effects and any linear trend in the time series.

e. Using the estimated regression equation developed in (d), estimate Timko's normal usage over the period of interrupted service.

POLLUTION

25. Air pollution control specialists in northern Poland monitor the amount of ozone, carbon dioxide and nitrogen dioxide in the air on an hourly basis. The hourly time series data exhibit seasonality, with the levels of pollutants showing patterns that vary over the hours in the day. On July 15, 16 and 17, the following levels of nitrogen dioxide were observed for the 12 hours from 6.00 a.m. to 6.00 p.m.

July 15:	25	28	35	50	60	60	40	35	30	25	25	20
July 16:	28	30	35	48	60	65	50	40	35	25	20	20
July 17:	35	42	45	70	72	75	60	45	40	25	25	25

a. Construct a time series plot. What type of pattern exists in the data?

b. Use the following dummy variables to develop an estimated regression equation to account for the seasonal effects in the data.

- Hour 1 = 1 if the reading was made between 6.00 a.m. and 7.00 a.m.; 0 otherwise.
- Hour 2 = 1 if the reading was made between 7.00 a.m. and 8.00 a.m.; 0 otherwise.
   . . .
- Hour 11 = 1 if the reading was made between 4.00 p.m. and 5.00 p.m.; 0 otherwise.

Note that when the values of the 11 dummy variables are equal to 0, the observation corresponds to the 5.00 p.m. to 6.00 p.m. hour.

c. Using the estimated regression equation developed in (a) compute estimates of the levels of nitrogen dioxide for July 18.

d. Let $t = 1$ to refer to the observation in hour 1 on July 15; $t = 2$ to refer to the observation in hour 2 of July 15; 0... and $t = 36$ to refer to the observation in hour 12 of July 17. Using the dummy variables defined in (b) and t, develop an estimated regression equation to account for seasonal effects and any linear trend in the time series. Based on the seasonal effects in the data and linear trend, compute estimates of the levels of nitrogen dioxide for July 18.

17.6 TIME SERIES DECOMPOSITION

In this section we turn our attention to what is called **time series decomposition**. Time series decomposition can be used to separate or decompose a time series into seasonal, trend and **irregular components**. While this method can be used for forecasting, its primary applicability is to obtain a better understanding of the time series. Many business and economic time series are maintained and published by agencies such as Eurostat and the OECD. These agencies use time series decomposition to create deseasonalized time series.

Understanding what is really going on with a time series often depends upon the use of deseasonalized data. For instance, we might be interested in learning whether electrical power consumption is increasing in our area. Suppose we learn that electric power consumption in September is down 3 per cent from the previous month. Care must be exercised in using such information, because whenever a seasonal influence is present, such comparisons may be misleading if the data have not been deseasonalized.

The fact that electric power consumption is down 3 per cent from August to September might be only the seasonal effect associated with a decrease in the use of air conditioning and not because of a long-term decline in the use of electric power. Indeed, after adjusting for the seasonal effect, we might even find that the use of electric power increased. Many other time series, such as unemployment statistics, home sales and retail sales, are subject to strong seasonal influences. It is important to deseasonalize such data before making a judgement about any long-term trend.

Time series decomposition methods assume that Y_t, the actual time series value at period t, is a function of three components: a trend component, a seasonal component and an irregular or error component. How these three components are combined to generate the observed values of the time series depends upon whether we assume the relationship is best described by an additive or a multiplicative model.

An **additive decomposition model** takes the following form:

$$Y_t = \text{Trend}_t + \text{Seasonal}_t + \text{Irregular}_t \qquad (17.13)$$

where:

$$\text{Trend}_t = \text{trend value at time period } t$$
$$\text{Seasonal}_t = \text{seasonal value at time period } t$$
$$\text{Irregular}_t = \text{irregular value at time period } t$$

In an additive model the values for the three components are simply added together to obtain the actual time series value Y_t. The irregular or error component accounts for the variability in the time series that cannot be explained by the trend and seasonal components.

An additive model is appropriate in situations where the seasonal fluctuations do not depend upon the level of the time series. The regression model for incorporating seasonal and trend effects in Section 17.5 is an additive model. If the sizes of the seasonal fluctuations in earlier time periods are about the same as the sizes of the seasonal fluctuations in later time periods, an additive model is appropriate. However, if the seasonal fluctuations change over time, growing larger as the sales volume increases because of a long-term linear trend, then a multiplicative model should be used. Many business and economic time series follow this pattern.

A **multiplicative decomposition model** takes the following form:

$$Y_t = \text{Trend}_t \times \text{Seasonal}_t \times \text{Irregular}_t \qquad (17.14)$$

where:

$$\text{Trend}_t = \text{trend value at time period } t$$
$$\text{Seasonal}_t = \text{seasonal index at time period } t$$
$$\text{Irregular}_t = \text{irregular index at time period } t$$

In this model, the trend and seasonal and irregular components are multiplied to give the value of the time series. Trend is measured in units of the item being forecast. However, the seasonal and irregular components are measured in relative terms, with values above 1.00 indicating effects above the trend and values below 1.00 indicating effects below the trend.

As this is the method most often used in practice, we will restrict our discussion of time series decomposition to showing how to develop estimates of the trend and seasonal components for a multiplicative model. For illustration we will work with the quarterly smartphone sales time series introduced in Section 17.5; the quarterly sales data are shown in Table 17.19 and the corresponding time series plot is presented in Figure 17.18. After demonstrating how to decompose a time series using the multiplicative model, we will then show how the seasonal indices and trend component can be recombined to develop a forecast.

Calculating the seasonal indices

Figure 17.18 indicates that sales are lowest in the second quarter of each year and increase in quarters 3 and 4. Thus, we conclude that a seasonal pattern exists for the smartphone sales time series. The computational procedure used to identify each quarter's seasonal influence begins by computing a moving average to remove the combined seasonal and irregular effects from the data, leaving us with a time series that contains only trend and any remaining random variation not removed by the moving average calculations.

As we are working with a quarterly series, we will use four data values in each moving average. The moving average calculation for the first four quarters of the smartphone sales data is:

$$\text{First moving average} = \frac{4.8 + 4.1 + 6.0 + 6.5}{4} = \frac{21.4}{4} = 5.35$$

Note that the moving average calculation for the first four quarters yields the average quarterly sales over year 1 of the time series. Continuing the moving average calculations, we next add the 5.8 value for the first quarter of year 2 and drop the 4.8 for the first quarter of year 1. Thus, the second moving average is:

$$\text{Second moving average} = \frac{4.1 + 6.0 + 6.5 + 5.8}{4} = \frac{22.4}{4} = 5.60$$

Similarly, the third moving average calculation is $(6.0 + 6.5 + 5.8 + 5.2)/4 = 5.875$.

Before we proceed with the moving average calculations for the entire time series, let us return to the first moving average calculation, which resulted in a value of 5.35. The 5.35 value is the average quarterly sales volume for year 1. As we look back at the calculation of the 5.35 value, associating 5.35 with the 'middle' of the moving average group makes sense. Note, however, that with four quarters in the moving average, there is no middle period. The 5.35 value really corresponds to period 2.5, the last half of quarter 2 and the first half of quarter 3. Similarly, if we go to the next moving average value of 5.60, the middle period corresponds to period 3.5, the last half of quarter 3 and the first half of quarter 4.

The two moving average values we computed do not correspond directly to the original quarters of the time series. We can resolve this difficulty by computing the average of the two moving averages. Since the centre of the first moving average is period 2.5 (half a period or quarter early) and the centre of the second moving average is period 3.5, the average of the two moving averages is centred at quarter 3, exactly where it should be. This moving average is referred to as a *centred moving average*. Thus, the centred moving average for period 3 is $(5.35 + 5.60)/2 = 5.475$. Similarly, the centred moving average value for period 4 is $(5.60 + 5.875)/2 = 5.738$. Table 17.21 shows a complete summary of the moving average and centred moving average calculations for the smartphone sales data.

TABLE 17.21 Centred moving average calculations for the smartphone sales time series

Year	Quarter	Sales (000)	Four-quarter moving average	Centred moving average
1	1	4.8		
1	2	4.1	5.350	
1	3	6.0	5.600	5.475
1	4	6.5	5.875	5.738
2	1	5.8	6.075	5.975
2	2	5.2	6.300	6.188
2	3	6.8	6.350	6.325
2	4	7.4	6.450	6.400
3	1	6.0	6.625	6.538
3	2	5.6	6.725	6.675
3	3	7.5	6.800	6.763
3	4	7.8	6.875	6.838
4	1	6.3	7.000	6.938
4	2	5.9	7.150	7.075
4	3	8.0		
4	4	8.4		

What do the centred moving averages in Table 17.21 tell us about this time series? Figure 17.20 shows a time series plot of the actual time series values and the centred moving average values. Note particularly how the centred moving average values tend to 'smooth out' both the seasonal and irregular fluctuations in the time series. The centred moving averages represent the trend in the data and any random variation that was not removed by using moving averages to smooth the data.

By dividing each side of equation (17.14) by the trend component T_t, we can identify the combined seasonal-irregular effect in the time series.

$$\frac{Y_t}{\text{Trend}_t} = \frac{\text{Trend}_t \times \text{Seasonal}_t \times \text{Irregular}_t}{\text{Trend}_t} = \text{Seasonal}_t \times \text{Irregular}_t$$

FIGURE 17.20

Quarterly smartphone sales time series and centred moving average

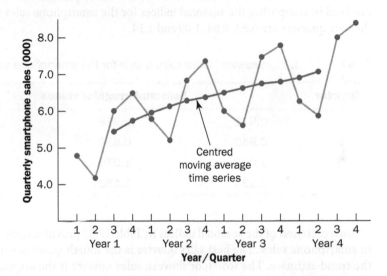

For example, the third quarter of year 1 shows a trend value of 5.475 (the centred moving average). So 6.0/5.475 = 1.096 is the combined seasonal-irregular value. Table 17.22 summarizes the seasonal-irregular ('detrended') values for the entire time series.

TABLE 17.22   Seasonal irregular values for the smartphone sales time series

Year	Quarter	Sales (000)	Centred moving average	Seasonal-irregular value
1	1	4.8		
1	2	4.1		
1	3	6.0	5.475	1.096
1	4	6.5	5.738	1.133
2	1	5.8	5.975	0.971
2	2	5.2	6.188	0.840
2	3	6.8	6.325	1.075
2	4	7.4	6.400	1.156
3	1	6.0	6.538	0.918
3	2	5.6	6.675	0.839
3	3	7.5	6.763	1.109
3	4	7.8	6.838	1.141
4	1	6.3	6.938	0.908
4	2	5.9	7.075	0.834
4	3	8.0		
4	4	8.4		

Consider the seasonal-irregular values for the third quarter: 1.096, 1.075 and 1.109. Seasonal-irregular values greater than 1.00 indicate effects above the trend estimate and values below 1.00 indicate effects below the trend estimate. Thus, the three seasonal-irregular values for quarter 3 show an above-average effect in the third quarter. Since the year-to-year fluctuations in the seasonal-irregular values are primarily due to random error, we can average the computed values to eliminate the irregular influence and obtain an estimate of the third-quarter seasonal influence.

$$\text{Seasonal effect of quarter 3} = \frac{1.096 + 1.075 + 1.109}{3} = 1.09$$

We refer to 1.09 as the *seasonal index* for the third quarter. Table 17.23 summarizes the calculations involved in computing the seasonal indices for the smartphone sales time series. The seasonal indices for the four quarters are 0.93, 0.84, 1.09 and 1.14.

TABLE 17.23   Seasonal index calculations for the smartphone sales time series

Quarter	Seasonal-irregular values			Seasonal index
1	0.971	0.918	0.908	0.93
2	0.840	0.839	0.834	0.84
3	1.096	1.075	1.109	1.09
4	1.133	1.156	1.141	1.14

Interpretation of the seasonal indices in Table 17.23 provides some insight into the seasonal component in smartphone sales. The best sales quarter is the fourth quarter, with sales averaging 14 per cent above the trend estimate. The worst, or slowest, sales quarter is the second quarter; its seasonal index of 0.84 shows that the sales average is 16 per cent below the trend estimate. The seasonal component corresponds clearly to the intuitive expectation that smartphone use and thus smartphone purchase patterns tend to peak in the fourth quarter because of the coming holiday season and an increase in gift purchases. The low second-quarter sales reflect reduced smartphone sales due to the buying habits of potential customers post-holiday season.

One final adjustment is sometimes necessary in obtaining the seasonal indices. As the multiplicative model requires that the average seasonal index equals 1.00, the sum of the four seasonal indices in Table 17.23 must equal 4.00. In other words, the seasonal effects must even out over the year. The average of the seasonal indices in our example is equal to 1.00, and hence this type of adjustment is not necessary. In other cases, a slight adjustment may be necessary. To make the adjustment, multiply each seasonal index by the number of seasons divided by the sum of the unadjusted seasonal indices. For instance, for quarterly data, multiply each seasonal index by 4/(sum of the unadjusted seasonal indices). Some of the exercises later will require this adjustment to obtain the appropriate seasonal indices.

Deseasonalizing the time series

A time series that has had the seasonal effects removed is referred to as a **deseasonalized time series**, and the process of using the seasonal indices to remove the seasonal effects from a time series is referred to as deseasonalizing the time series. Using a multiplicative decomposition model, we deseasonalize a time series by dividing each observation by its corresponding seasonal index.

By dividing each time series observation (Y_t) in equation (17.14) by its corresponding seasonal index, the resulting data show only trend and random variability (the irregular component). The deseasonalized time series for smartphone sales is summarized in Table 17.24. A graph of the deseasonalized time series is shown in Figure 17.21.

FIGURE 17.21
Deseasonalized smartphone sales time series

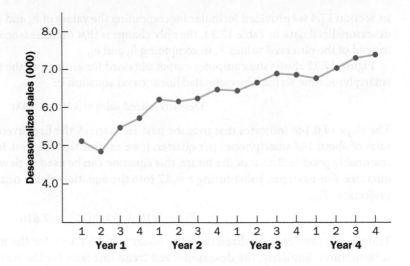

TABLE 17.24   Deseasonalized values for the smartphone sales time series

Year	Quarter	Time period	Sales (000)	Seasonal index	Deseasonalized sales
1	1	1	4.8	0.93	5.16
	2	2	4.1	0.84	4.88
	3	3	6.0	1.09	5.50
	4	4	6.5	1.14	5.70
2	1	5	5.8	0.93	6.24
	2	6	5.2	0.84	6.19
	3	7	6.8	1.09	6.24
	4	8	7.4	1.14	6.49
3	1	9	6.0	0.93	6.45
	2	10	5.6	0.84	6.67
	3	11	7.5	1.09	6.88
	4	12	7.8	1.14	6.84
4	1	13	6.3	0.93	6.77
	2	14	5.9	0.84	7.02
	3	15	8.0	1.09	7.34
	4	16	8.4	1.14	7.37

Using the deseasonalized time series to identify trend

The graph of the deseasonalized smartphone sales time series shown in Figure 17.21 appears to have an upward linear trend. To identify this trend, we will fit a linear trend equation to the deseasonalized time series using the same method shown in Section 17.4. The only difference is that we will be fitting a trend line to the deseasonalized data instead of the original data.

Recall that for a linear trend the estimated regression equation can be written as:

$$T_t = b_0 + b_1 t$$

where:

T_t = linear trend forecast in period t
b_0 = intercept of the linear trend line
b_1 = slope of the trend line
t = time period

In Section 17.4 we provided formulae for computing the values of b_0 and b_1. To fit a linear trend line to the deseasonalized data in Table 17.24, the only change is that the deseasonalized time series values are used instead of the observed values Y_t in computing b_0 and b_1.

Figure 17.22 shows the computer output obtained for estimating the trend line for the deseasonalized smartphone time series. The estimated linear trend equation is:

$$\text{Deseasonalized sales} = 5.10 + 0.148t$$

The slope of 0.148 indicates that over the past 16 quarters, the firm averaged a deseasonalized growth in sales of about 148 smartphones per quarter. If we assume that the past 16-quarter trend in sales data is a reasonably good indicator of the future, this equation can be used to develop a trend projection for future quarters. For example, substituting $t = 17$ into the equation yields next quarter's deseasonalized trend projection, T_{17}.

$$T_{17} = 5.10 + 0.148(17) = 7.616$$

Thus, using the deseasonalized data, the linear trend forecast for the next quarter (period 17) is 7,616 smartphones. Similarly, the deseasonalized trend forecasts for the next three quarters (periods 18, 19 and 20) are 7,764, 7,912 and 8,060 smartphones, respectively.

FIGURE 17.22
Regression output for the deseasonalized smartphone sales time series

```
Analysis of variance
```

Source	DF	SS	MS	F	P
Regression	1	7.4068	7.4068	158.78	0.000
Residual error	14	0.6531	0.0466		
Total	15	8.0599			

```
S = 0.215985        R-sq = 91.9%          R-sq(adj) = 91.3%
```

Predictor	Coef	SE Coef	T	P
Constant	5.1050	0.1133	45.07	0.000
Period	0.14760	0.01171	12.60	0.000

The regression equation is Deseasonalized sales = 5.10 + 0.148 Period

Seasonal adjustments

The final step in developing the forecast when both trend and seasonal components are present is to use the seasonal indices to adjust the deseasonalized trend projections. Returning to the smartphone sales example, we achieved a deseasonalized trend projection for the next four quarters. Now we must adjust the forecast for the seasonal effect. The seasonal index for the first quarter of year 5 ($t = 17$) is 0.93, so to obtain the quarterly forecast we multiply the deseasonalized forecast based on trend ($T_{17} = 7,616$) by the seasonal index (0.93), yielding the result $7,616(0.93) = 7,083$.

Table 17.25 shows the quarterly forecast for quarters 17 through 20. The high-volume fourth quarter has a 9,188-unit forecast, and the low-volume second quarter has a 6,522-unit forecast.

TABLE 17.25 Quarterly forecasts for the smartphone sales time series

Year	Quarter	Deseasonalized trend forecast	Seasonal index	Quarterly forecast
5	1	7,616	0.93	(7,616)(0.93) = 7,083
	2	7,764	0.84	(7,764)(0.84) = 6,522
	3	7,912	1.09	(7,912)(1.09) = 8,624
	4	8,060	1.14	(8,060)(1.14) = 9,188

Models based on monthly data

In the preceding smartphone sales example, we used quarterly data to illustrate the computation of seasonal indices. However, many businesses use monthly rather than quarterly forecasts. In such cases, the procedures introduced in this section can be applied with minor modifications. First, a 12-month moving average replaces the four-quarter moving average; second, 12 monthly seasonal indices, rather than four quarterly seasonal indices, must be computed. Other than these changes, the computational and forecasting procedures are identical.

Cyclical component

Mathematically, the multiplicative model of equation (17.14) can be expanded to include a cyclical component.

$$Y_t = \text{Trend}_t \times \text{Cyclical}_t \times \text{Seasonal}_t \times \text{Irregular}_t \qquad \textbf{(17.15)}$$

The cyclical component, like the seasonal component, is expressed as a percentage of trend. As mentioned in Section 17.1, this component is attributable to multiyear cycles in the time series. It is analogous to the seasonal component, but over a longer period of time. However, because of the length of time involved, obtaining enough relevant data to estimate the cyclical component is often difficult. Another difficulty is that cycles usually vary in length. As it is so difficult to identify and/or separate cyclical effects from long-term trend effects, in practice these effects are often combined and referred to as a combined trend-cycle component. We leave further discussion of the cyclical component to specialized texts on forecasting methods.

EXERCISES

Methods

26. Consider the following time series data:

Quarter	Year 1	Year 2	Year 3
1	4	6	7
2	2	3	6
3	3	5	6
4	5	7	8

   a. Construct a time series plot. What type of pattern exists in the data?
   b. Show the four-quarter and centred moving average values for this time series.
   c. Compute seasonal indices and adjusted seasonal indices for the four quarters.

27. Refer to Exercise 26.
   a. Deseasonalize the time series using the adjusted seasonal indices computed in (c) of Exercise 26.
   b. Compute the linear trend regression equation for the deseasonalized data.
   c. Compute the deseasonalized quarterly trend forecast for Year 4.
   d. Use the seasonal indices to adjust the deseasonalized trend forecasts computed in (c).

Applications

28. The quarterly sales data (number of copies sold) for a college textbook over the past three years are as follows:

Quarter	Year 1	Year 2	Year 3
1	1,690	1,800	1,850
2	940	900	1,100
3	2,625	2,900	2,930
4	2,500	2,360	2,615

a. Show the four-quarter and centred moving average values for this time series.
b. Compute the seasonal and adjusted seasonal indices for the four quarters.
c. When does the publisher have the largest seasonal index? Does this result appear reasonable? Explain.
d. Deseasonalize the time series.
e. Compute the linear trend equation for the deseasonalized data and forecast sales for each of the quarters in year 4 using the linear trend equation.
f. Adjust the linear trend forecasts using the adjusted seasonal indices computed in (b).

29. Quarterly sales data for the number of houses sold over the past four years or so by a national chain are as follows:

Year	Q1	Q2	Q3	Q4
1	200	212	229	207
2	195	204	216	202
3	201	209	221	205
4	208	217	231	213
5	218			

a. Decompose the series into trend, seasonal and random components using a multiplicative model.
b. Hence derive forecasts of the number of houses that will be sold in the next four quarters.
c. Comment on the quality of your modelling results.

30. The following table shows the number of passengers per quarter (in thousands) who flew with MBI Junior for the first quarter of this year and the three years preceding:

Year	Q1	Q2	Q3	Q4
1	44	92	156	68
2	60	112	180	80
3	64	124	200	104
4	76			

a. Decompose the series into trend, seasonal and random components.
b. Hence derive forecasts of the passenger numbers in the next four quarters.
c. Comment on the quality of your modelling.

ONLINE RESOURCES

For the associated date files, additional questions and answers, and the software section for Chapter 17, visit the online platform.

SUMMARY

This chapter provided an introduction to the basic methods of time series analysis and forecasting. First, we showed that the underlying pattern in the time series can often be identified by constructing a time series plot. Several types of data pattern can be distinguished, including a horizontal pattern, a trend pattern and a seasonal pattern. The forecasting methods we have discussed are based on which of these patterns are present in the time series.

For a time series with a horizontal pattern, we showed how moving averages and exponential smoothing can be used to develop a forecast. The moving averages method consists of computing an average of past data values and then using that average as the forecast for the next period. With the exponential smoothing method, a weighted average of past time series values is used to compute a forecast. These methods also adapt well when a horizontal pattern shifts to a different level and resumes a horizontal pattern.

An important factor in determining which forecasting method to use involves the accuracy of the method. We discussed three measures of forecast accuracy: mean absolute error (MAE), mean squared error (MSE) and mean absolute percentage error (MAPE). Each of these measures is designed to determine how well a particular forecasting method is able to reproduce the time series data that are already available. By selecting a method that has the best accuracy for the data already known, we hope to increase the likelihood that we will obtain better forecasts for future time periods.

For time series that have only a long-term linear trend, we showed how simple time series regression can be used to make trend projections. We also discussed how an extension of single exponential smoothing, referred to as Holt's linear exponential smoothing, can be used to forecast a time series with a linear trend. For a time series with a curvilinear or nonlinear trend, we showed how multiple regression can be used to fit a quadratic trend equation or an exponential trend equation to the data.

For a time series with a seasonal pattern, we showed how the use of dummy variables in a multiple regression model can be used to develop an estimated regression equation with seasonal effects. We then extended the regression approach to include situations where the time series contains both a seasonal and a linear trend effect by showing how to combine the dummy variable approach for handling seasonality with the time series regression approach for handling linear trend.

In the last section of the chapter we showed how time series decomposition can be used to separate or decompose a time series into seasonal and trend components and then to deseasonalize the time series. We showed how to compute seasonal indices for a multiplicative model, how to use the seasonal indices to deseasonalize the time series and how to use regression analysis on the deseasonalized data to estimate the trend component. The final step in developing a forecast when both trend and seasonal components are present is to use the seasonal indices to adjust the trend projections.

KEY TERMS

Additive decomposition model
Causal forecasting methods
Cyclical pattern
Deseasonalized time series
Exponential smoothing
Forecasts
Forecast error
Horizontal pattern
Irregular components
Mean absolute error (MAE)
Mean absolute percentage error (MAPE)

Moving averages
Multiplicative decomposition model
Seasonal pattern
Smoothing constant
Stationary time series
Time series
Time series decomposition
Time series plot
Trend pattern
Weighted moving averages

KEY FORMULAE

Moving average forecast of order *k*

$$F_{t+1} = \frac{\Sigma(\text{most recent } k \text{ data values})}{k} = \frac{Y_t + Y_{t-1} + \ldots + Y_{t-k+1}}{k} \tag{17.1}$$

Exponential smoothing forecast

$$F_{t+1} = \alpha Y_t + (1 - \alpha)F_t \tag{17.2}$$

Linear trend equation

$$T_t = b_0 + b_1 t \tag{17.4}$$

where:

$$b_1 = \frac{\sum_{t-1}^{n}(t - \bar{t})(Y_t - \overline{Y})}{\sum_{t-1}^{n}(t - \bar{t})^2} \tag{17.5}$$

$$b_0 = \overline{Y} - b_1\bar{t} \tag{17.6}$$

Holt's linear exponential smoothing

$$L_t = \alpha Y_t + (1 - \alpha)(L_{t-1} + b_{t-1}) \tag{17.7}$$

$$b_t = \beta(L_t - L_{t-1}) + (1 - \beta)b_{t-1} \tag{17.8}$$

$$F_{t+k} = L_t + b_t k \tag{17.9}$$

Quadratic trend equation

$$T_t = b_0 + b_1 t + b_2 t^2 \tag{17.10}$$

Exponential trend equation

$$T_t = b_0(b_1)^t \tag{17.11}$$

Additive decomposition model

$$Y_t = \text{Trend}_t + \text{Seasonal}_t + \text{Irregular}_t \tag{17.13}$$

Multiplicative decomposition model

$$Y_t = \text{Trend}_t \times \text{Seasonal}_t \times \text{Irregular}_t \tag{17.14}$$

CASE PROBLEM 1

Forecasting food and beverage sales

The Vesuvius Restaurant near Naples, Italy, is owned and operated by Luigi Marconi. The restaurant has just completed its third year of operation. During that time, Luigi sought to establish a reputation for the restaurant as a high-quality dining establishment that specializes in fresh seafood. Through the efforts of Luigi and the staff, the restaurant has become one of the best and fastest growing restaurants in the area.

Luigi believes that, to plan for the growth of the restaurant in the future, he needs to develop a system that will enable him to forecast food and beverage sales by month for up to one year in advance. Luigi compiled the following data (in thousands of euros) on total food and beverage sales for the three years of operation.

Month	First year	Second year	Third year
January	242	263	282
February	235	238	255
March	232	247	265
April	178	193	205
May	184	193	210
June	140	149	160
July	145	157	166
August	152	161	174

Month	First year	Second year	Third year
September	110	122	126
October	130	130	148
November	152	167	173
December	206	230	235

Managerial report

Perform an analysis of the sales data for the Vesuvius Restaurant. Prepare a report for Luigi that summarizes your findings, forecasts and recommendations. Include the following:

1. A graph of the time series.
2. An analysis of the seasonality of the data. Indicate the seasonal indices for each month, and comment on the high and low seasonal sales months. Do the seasonal indices make intuitive sense? Discuss.
3. A forecast of sales for January through December of the fourth year.
4. Recommendations as to when the system that you develop should be updated to account for new sales data.
5. Any detailed calculations of your analysis in the appendix of your report.

Assume that January sales for the fourth year turn out to be 295,000. What was your forecast error? If this error is large, Luigi may be puzzled about the difference between your forecast and the actual sales value. What can you do to resolve his uncertainty about the forecasting procedure?

VESUVIUS

CASE PROBLEM 2

RAC

Allocating patrols to meet future demand for vehicle rescue

The data below summarize actual monthly demands for RAC rescue services over a five-year time period. (The Royal Automobile Club is one of the major motoring organizations that offer emergency breakdown cover in the UK.)

To meet the national demand for its services in the coming year, the RAC's human resources planning department forecasts the number of members expected, using historical data and market forecasts. It then predicts the average number of breakdowns and number of rescue calls expected, by referring to the probability of a member's vehicle breaking down each year. In Year 1, an establishment of approximately 1,400 patrols was available to deal with the expected workload. Note that this figure had to be reviewed monthly since it was an average for the year and did not take into account fluctuations in demand 'in different seasons'.

Monthly demand for RAC rescue services years 1–5:

| | Year | | | | |
Month	1	2	3	4	5
January	270,093	248,658	253,702	220,332	241,489
February	216,050	210,591	216,575	189,223	193,794
March	211,154	208,969	220,903	188,950	206,068
April	194,909	191,840	191,415	196,343	191,359
May	200,148	194,654	190,436	189,627	179,592
June	195,608	189,892	175,512	177,653	183,712
July	208,493	203,275	193,900	182,219	193,306
August	215,145	213,357	197,628	190,538	199,947
September	200,477	196,811	183,912	183,481	191,231
October	216,821	225,182	213,909	214,009	198,514
November	222,128	244,498	219,336	239,104	202,219
December	250,866	257,704	246,780	254,041	254,217

Managerial report

1. By undertaking an appropriate statistical analysis of the information provided, describe how you would advise the RAC on its patrol allocation in Year 6.

2. State your assumptions.

3. Comment on the validity of your results or otherwise.

CASE PROBLEM 3

Forecasting lost sales

The Carlson Department Store suffered heavy damage when a hurricane struck on August 31. The store closed for four months (September–December). Carlson is now involved in a dispute with its insurance company about the amount of lost sales during the time the store was closed. Two key issues must be resolved:

1 the amount of sales Carlson would have made if the hurricane had not struck; and

2 whether Carlson is entitled to any compensation for excess sales due to increased business activity after the storm. More than $8 billion in federal disaster relief and insurance money came into the county, resulting in increased sales at department stores and numerous other businesses.

Table 1 gives Carlson's sales data for the 48 months preceding the storm. Table 2 reports

TABLE 1 Sales for Carlson Department store ($m)

Month	Year 1	Year 2	Year 3	Year 4	Year 5
January		1.45	2.31	2.31	2.56
February		1.80	1.89	1.99	2.28
March		2.03	2.02	2.42	2.69
April		1.99	2.23	2.45	2.48
May		2.32	2.39	2.57	2.73
June		2.20	2.14	2.42	2.37
July		2.13	2.27	2.40	2.31
August		2.43	2.21	2.50	2.23
September	1.71	1.90	1.89	2.09	
October	1.90	2.13	2.29	2.54	
November	2.74	2.56	2.83	2.97	
December	4.20	4.16	4.04	4.35	

CARLSON SALES

TABLE 2 Department store sales for the county ($m)

Month	Year 1	Year 2	Year 3	Year 4	Year 5
January		46.80	46.80	43.80	48.00
February		48.00	48.60	45.60	51.60
March		60.00	59.40	57.60	57.60
April		57.60	58.20	53.40	58.20
May		61.80	60.60	56.40	60.00
June		58.20	55.20	52.80	57.00
July		56.40	51.00	54.00	57.60
August		63.00	58.80	60.60	61.80
September	55.80	57.60	49.80	47.40	69.00
October	56.40	53.40	54.60	54.60	75.00
November	71.40	71.40	65.40	67.80	85.20
December	117.60	114.00	102.00	100.20	121.80

COUNTY SALES

(Continued)

total sales for the 48 months preceding the storm for all department stores in the county as well as the total sales in the county for the four months the Carlson Department Store was closed. Carlson's managers asked you to analyze these data and develop estimates of the lost sales at the Carlson Department Store for the months of September through to December. They have also asked you to determine whether a case can be made for excess storm-related sales during the same period. If such a case can be made, Carlson is entitled to compensation for excess sales it would have earned in addition to ordinary sales.

Managerial report

Prepare a report for the managers of the Carlson Department Store that summarizes your findings, forecasts and recommendations. Include the following:

1. An estimate of sales for Carlson Department Store had there been no hurricane.

2. An estimate of countywide department store sales had there been no hurricane.

3. An estimate of lost sales for the Carlson Department Store for September to December.

18

Non-Parametric Methods

CHAPTER CONTENTS

Statistics in Practice   Coffee lovers' preferences: Costa, Starbucks and Caffè Nero

18.1 Sign test
18.2 Wilcoxon signed-rank test
18.3 Mann-Whitney-Wilcoxon test
18.4 Kruskal-Wallis test
18.5 Rank correlation

LEARNING OBJECTIVES   After studying this chapter and doing the exercises, you should be able to:

1 Explain the essential differences between parametric and non-parametric methods of inference.

2 Recognize the circumstances in which it is appropriate to apply the following non-parametric statistical procedures; calculate the appropriate sample statistics; use these statistics to carry out a hypothesis test; interpret the results.

2.1 Sign test.

2.2 Wilcoxon signed-rank test.

2.3 Mann-Whitney-Wilcoxon test.

2.4 Kruskal-Wallis test.

2.5 Spearman rank correlation.

The inferential methods presented previously in this book are generally known as **parametric methods**. These methods begin with an assumption about the population distribution, often that this distribution is normal. Based on this assumption, a sampling distribution is determined, for a relevant sample statistic, that enables inferences to be made about one or more parameters of the population, such as the population mean μ or the population standard deviation σ. For example, in

Chapter 9 we presented a method for making an inference about a population mean based on the assumption that the population had a normal distribution with unknown parameters μ and σ. The test statistic for making an inference about the population mean was identified as having a t distribution; this involved using the sample standard deviation s to estimate the population standard deviation σ. The t distribution was then used to compute confidence intervals and do hypothesis tests about the mean of a normally distributed population.

In this chapter we present non-parametric methods that can be used to make inferences about a population without requiring an assumption about the specific form of the population distribution. For this reason, these non-parametric methods are also called distribution-free methods.

Most parametric methods require quantitative data, whereas non-parametric methods allow inferences based sometimes on categorical (qualitative) data, and sometimes on either ranked or quantitative data. In the first section of the chapter, we show how the binomial distribution can be used, as a sampling distribution, to make an inference about a population median. In Sections 18.2 to 18.4, we show how rank-ordered data are used in non-parametric tests about two or more populations. In the final Section 18.5, we use rank-ordered data to compute the rank correlation for two variables.

STATISTICS IN PRACTICE
Coffee lovers' preferences: Costa, Starbucks and Caffè Nero

Some years ago, Costa Coffee ran a vigorous promotional campaign in the UK under the headline **Sorry Starbucks – the people have voted**. The byline was 'In head-to-head tests, seven out of ten coffee lovers preferred Costa cappuccino to Starbucks.'

The market research behind the claim was carried out by an independent market research organization, Tangible Branding Limited, in three UK towns (High Wycombe, Glasgow and Sheffield). Each participant was asked to undertake a two-way blind tasting test: either Costa versus Starbucks or Costa versus Caffè Nero. 'Runners' transported the coffees to the tasting venue from nearby coffee houses. The order of tasting was rotated. Over the three tasting venues, the total Costa versus Starbucks sample size was 166, and the Costa versus Caffè Nero sample was 168.

In the Costa versus Caffè Nero comparisons, 64 per cent of tasters preferred Costa. In the Costa versus Starbucks tests, 66 per cent preferred Costa. Among self-identified 'coffee lovers', 69 per cent preferred the Costa coffee to Caffè Nero coffee, and 72 per cent preferred Costa to Starbucks. Among Caffè Nero regulars, 72 per cent expressed a preference for Costa's cappuccino, while 67 per cent of Starbucks regulars preferred Costa's cappuccino. The Costa website noted that 'All results are significant at the 95% confidence level.'

The data on which the results are based are qualitative data: a simple expression of preference between two options. The kind of statistical test needed for data such as these is known as a non-parametric test. Non-parametric tests are the subject of the present chapter. The chapter begins with a discussion of the sign test, a test particularly appropriate for the research situation described by Costa in its advertising.

© chpua/iStock

18.1 SIGN TEST

The sign test is a versatile non-parametric method for hypothesis testing that uses the binomial distribution with $\pi = 0.50$ as the sampling distribution. It does not require an assumption about the distribution of the population. In this section we present two applications of the sign test: one involves a hypothesis test about a population median and the other involves a matched-samples test about the difference between two populations.

Hypothesis test about a population median

In Chapter 9 we described hypothesis tests for a population mean. When a population distribution is skewed, the median is often preferred over the mean as a measure of central location for the population. In this section we show how the sign test can be used as a non-parametric procedure for testing a hypothesis about the value of a population median.

To demonstrate the sign test, we consider the weekly sales of MotherEarth Potato Snacks by the Lineker chain of convenience stores. Lineker's management decision to stock the new product is based on the manufacturer's estimate that median sales will be €450 per week per store. After stocking the product for three months, Lineker's management requested the following hypothesis test regarding the population median weekly sales:

$$H_0: \text{Median} = 450$$
$$H_1: \text{Median} \neq 450$$

Data showing one-week sales at 12 randomly selected Lineker's stores are in Table 18.1.

TABLE 18.1 Lineker sample data for the sign test about the population median weekly sales

Store ID	One-week sales (€)	Sign	Store ID	One-week sales (€)	Sign
56	485	+	63	474	+
19	562	+	39	662	+
93	499	+	21	492	+
36	415	−	84	380	−
128	860	+	102	515	+
12	426	−	44	721	+

In the sign test, we compare each sample observation to the hypothesized value of the population median. If the observation is greater than the hypothesized value, we record a plus sign ' + '. If the observation is less than the hypothesized value, we record a minus sign ' − '. If an observation is exactly equal to the hypothesized median value, the observation is eliminated from the sample and the analysis proceeds with the smaller sample size, using only the observations where a plus or a minus sign has been recorded. The conversion of the sample data to either a plus or minus sign gives the non-parametric method its name: the sign test.

Consider the sample data in Table 18.1. The first observation, 485, is greater than the hypothesized median 450: a plus sign is recorded. The second observation, 562, is greater than the hypothesized median 450: a plus sign is recorded. Continuing with the 12 observations in the sample provides the nine plus signs and three minus signs shown in Table 18.1.

Assigning the plus and minus signs has made the situation a binomial distribution application. The sample size $n = 12$ is the number of trials. There are two possible outcomes per trial, a plus sign or a minus sign, and the trials are independent. Let π denote the probability of a plus sign.

If the population median is 450, $\pi = 0.50$, as there should be 50 per cent plus signs and 50 per cent minus signs in the population. So, in terms of the binomial probability π, the sign test hypotheses regarding the population median:

$$H_0: \text{Median} = 450$$
$$H_1: \text{Median} \neq 450$$

are converted to the following hypotheses about the binomial probability π:

$$H_0: \pi = 0.50$$
$$H_1: \pi \neq 0.50$$

If H_0 is rejected, we can conclude that π is not equal to 0.50 and that the population median is not equal to 450. If H_0 is not rejected, we cannot conclude that π is different from 0.50 and neither can we conclude that the population median is different from 450.

With $n = 12$ stores or trials and $\pi = 0.50$, we calculate the binomial probabilities for the number of plus signs under the assumption that H_0 is true (refer to Chapter 5 if you need a reminder about computing binomial probabilities). Figure 18.1 shows a graphical representation of this binomial distribution.

FIGURE 18.1

Binomial sampling distribution for the number of plus signs when $n = 12$ and $\pi = 0.50$

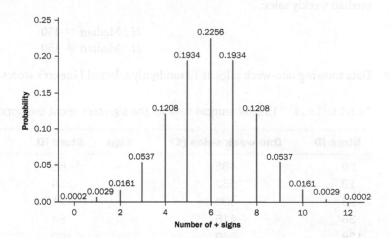

We now use the binomial distribution to test the hypothesis about the population median. We shall use $\alpha = 0.10$ for the test. Since the observed number of plus signs for the sample data (9) is in the upper tail of the binomial distribution, we begin by computing the probability of obtaining nine or more plus signs, that is, the probability of 9, 10, 11 or 12 plus signs. Adding these probabilities, we have $0.0537 + 0.0161 + 0.0029 + 0.0002 = 0.0729$. Since we are using a two-tailed hypothesis test, this upper-tail probability is doubled to obtain the p-value $2(0.0729) = 0.1458$. With p-value $> \alpha$, we cannot reject H_0. In terms of the binomial probability π, we cannot reject $H_0: \pi = 0.50$, and so we cannot reject the hypothesis that the population median is €450.

In this example, the hypothesis test was formulated as a two-tailed test. One-tailed sign tests are also possible. For example, suppose the test had been formulated as an upper-tail test with the following null and alternative hypotheses:

$$H_0: \text{Median} \leq 450$$
$$H_1: \text{Median} > 450$$

The appropriate p-value is the binomial probability that the number of plus signs is greater than or equal to nine found in the sample. As indicated by the calculations above, this one-tailed p-value is 0.0729.

With a computer program such as Excel, R, SPSS or Minitab it is relatively easy to calculate and sum the appropriate binomial probabilities even for large sample sizes. An alternative for large samples is to make use of the normal distribution approximation of the binomial distribution to compute the p-value. A large-sample application of the sign test is illustrated next.

Suppose that one year ago the median price of a new house was €236,000. However, a current downturn in the economy prompts an estate agent to use sample data on recent house sales to determine if the population median price of a new house is less today than it was a year ago. The hypothesis test about the population median price of a new house is as follows:

$$H_0: \text{Median} \geq 236{,}000$$
$$H_1: \text{Median} < 236{,}000$$

We will use $\alpha = 0.05$ to do this test.

A random sample of 61 recent new house sales found 22 houses sold for more than €236,000, 38 houses sold for less than €236,000 and 1 house sold for precisely €236,000. After deleting the house that sold for the hypothesized median price of €236,000, the sign test continues with 22 plus signs, 38 minus signs and a sample of 60 houses.

HOUSESALES

The null hypothesis that the population median $\geq$ €236,000 is expressed by the binomial distribution hypothesis $H_0: \pi \geq 0.50$. If H_0 were true as an equality, we would expect $0.50(60) = 30$ houses to have a plus sign. The sample result showing 22 plus signs is in the lower tail of the binomial distribution. So the p-value is the probability of 22 or fewer plus signs when $\pi = 0.50$. Although it is possible to compute the exact binomial probabilities for 0, 1, 2, …, 22 and sum these probabilities, we will use the normal distribution approximation of the binomial distribution to make this computation easier. For this approximation, the mean and standard deviation of the normal distribution are as follows:

Normal approximation of the sampling distribution of the number of plus signs for $H_0: \pi = 0.50$.

$$\text{Mean: } \mu = 0.50n \qquad \qquad \textbf{(18.1)}$$

$$\text{Standard deviation: } \sigma = \sqrt{0.25n} \qquad \qquad \textbf{(18.2)}$$

Distribution form: approximately normal for $n > 20$.

Using equations (18.1) and (18.2) with $n = 60$ houses and $\pi = 0.50$, the sampling distribution for the number of plus signs can be approximated by a normal distribution with:

$$\mu = 0.50n = 0.50(60) = 30$$
$$\sigma = \sqrt{0.25n} = \sqrt{0.25(60)} = 3.873$$

We now use this distribution to approximate the binomial probability of 22 or fewer plus signs. Remember that the binomial probability distribution is discrete and the normal probability distribution is continuous. To take account of this, the binomial probability of 22 is computed using the normal probability interval 21.5 to 22.5. The 0.5 added to and subtracted from 22 is called the continuity correction. To obtain the p-value for 22 or fewer plus signs we use the normal distribution with $\mu = 30$ and $\sigma = 3.873$ to calculate the probability that the normal random variable, X, has a value less than or equal to 22.5.

$$p\text{-value} = P(X \leq 22.5) = P\left(Z \leq \frac{22.5 - 30}{3.873}\right) = P(Z \leq -1.94)$$

Using the normal probability distribution table, we see that the cumulative probability for $z = -1.94$ provides the p-value $= 0.0262$. With $0.0262 < 0.05$ we reject the null hypothesis and conclude that the median price of a new house is less than the €236,000 median price a year ago.

Hypothesis test with matched samples

In Chapter 10 we introduced a matched-samples experimental design where each of n experimental units provided a pair of observations, one from population 1 and one from population 2. We used the t distribution to make an inference about the difference between the means of the two populations, using quantitative data and assuming that the differences between the pairs of matched observations were normally distributed.

In the following example we use the non-parametric sign test to analyze matched-samples data. Unlike the t distribution procedure, the sign test enables us to analyze categorical as well as quantitative data and requires no assumption about the distribution of the differences. This type of matched-samples design occurs in market research when a sample of n potential customers is asked to compare two brands of a product such as coffee, soft drinks or detergents (refer to Statistics in Practice at the beginning of the chapter). Without obtaining a quantitative measure of each individual's preference for the brands, each individual is asked to state a brand preference. Consider the following example.

Sunny Vale Farms produces Citrus Delight orange juice. A competitor produces Tropical Orange juice. In a study of consumer preferences for the two brands, 14 individuals were given unmarked samples of each product. The brand each individual tasted first was selected randomly. After tasting the two products, the individuals were asked to state a preference for one of the two brands. The purpose of the study was to determine whether consumers in general prefer one product to the other.

If the individual preferred Citrus Delight, a plus sign was recorded. If the individual preferred Tropical Orange, a minus sign was recorded. If the individual was unable to express a difference in preference, no sign was recorded. The data for the 14 individuals in the study are shown in Table 18.2.

TABLE 18.2 Preference data for the Sunny Vale Farms taste test

Individual	Preference	Sign	Individual	Preference	Sign
1	Tropical Orange	−	8	Tropical Orange	−
2	Tropical Orange	−	9	Tropical Orange	−
3	Citrus Valley	+	10	No Preference	
4	Tropical Orange	−	11	Tropical Orange	−
5	Tropical Orange	−	12	Citrus Valley	+
6	No Preference		13	Tropical Orange	−
7	Tropical Orange	−	14	Tropical Orange	−

The data show two plus signs and ten minus signs for the $n = 12$ individuals who expressed a preference for one of the two brands (deleting the two individuals who could not express a preference). Letting π indicate the proportion of the population of customers who prefer Citrus Delight, we want to test the hypotheses that there is no difference between the preferences for the two brands as follows:

$$H_0: \pi = 0.50$$
$$H_1: \pi \neq 0.50$$

If H_0 cannot be rejected, we shall have no evidence indicating a difference in preference for the two brands of orange juice. If H_0 can be rejected, we can conclude that the consumer preferences are different for the two brands. In that case, the brand selected by the greater number of consumers can be considered the preferred brand. We shall use a 0.05 level of significance.

We conduct the sign test exactly as we did earlier in this section. The sampling distribution for the number of plus signs is a binomial distribution with $\pi = 0.50$ and $n = 12$. The binomial probabilities for the number of plus signs are the same ones shown in Figure 18.1. Under the assumption H_0 is true, we would expect $0.50n = 0.50(12) = 6$ plus signs. With only two plus signs in the sample, the results are in the lower tail of the binomial distribution. To compute the p-value for this two-tailed test, we first compute the probability of two or fewer plus signs and then double this value. Using the binomial probabilities of 0,

1 and 2 shown in Figure 18.1, the p-value is $2(0.0002 + 0.0029 + 0.0161) = 0.0384$. With $0.0384 < 0.05$, we reject H_0. The taste test provides evidence that consumer preference differs for the two brands of orange juice. We would advise Sunny Vale Farms of this result and conclude that the competitor's Tropical Orange is the preferred brand. Sunny Vale Farms can then pursue a strategy to address this issue.

As with other uses of the sign test, one-tailed tests may be used depending upon the application. Also, as the sample size becomes large, the normal distribution approximation of the binomial distribution will ease the computations, as shown earlier in this section. While the Sunny Vale Farms sign test for matched samples used categorical preference data, the sign test for matched samples can be used with quantitative data as well. This would be particularly helpful if the paired differences are not normally distributed and are skewed. In this case a positive difference is assigned a plus sign, a negative difference is assigned a negative sign, and a zero difference is removed from the sample. The sign test computations proceed as before.

EXERCISES

Methods

1. The following table lists the preferences indicated by ten individuals in taste tests involving two brands of a product. A plus indicates a preference for Brand A over Brand B.

Individual	Brand A versus Brand B	Individual	Brand A versus Brand B
1	+	6	+
2	+	7	−
3	+	8	+
4	−	9	−
5	+	10	+

   Using $\alpha = 0.05$, test for a difference in the preferences for the two brands.

2. The following hypothesis test is to be conducted:

$$H_0: \text{Median} \leq 150$$
$$H_1: \text{Median} > 150$$

   A sample of size 30 yields 22 cases in which a value greater than 150 is obtained, 3 cases in which a value of exactly 150 is obtained and 5 cases in which a value less than 150 is obtained. Conduct the hypothesis test using $\alpha = 0.01$.

Applications

3. Competition in the desktop computer market is intense. A sample of 450 purchases showed 202 Brand A computers, 175 Brand B computers and 73 other computers. Use a 0.05 level of significance to test the null hypothesis that Brand A and Brand B have the same share of the personal computer market. What is your conclusion?

4. Previous research by SNL Securities suggested that stock splits in the banking industry tended to increase the value of an individual's stock holding. Assume that of a sample of 20 recent stock splits, 14 led to an increase in value, 4 led to a decrease in value and 2 resulted in no change. Suppose a sign test is to be used to determine whether stock splits continue to be beneficial for holders of bank stocks.
   a. What are the null and alternative hypotheses?
   b. With $\alpha = 0.05$, what is your conclusion?

5. In an opinion survey in the UK, respondents were asked to imagine that a brand they liked was prepared to sponsor them. They were then asked whether they would be willing to have the brand logo and the words 'sponsored by ...' on all the photographs they posted on social media. Of the 489 Instagram users surveyed, 199 said they would be willing, 240 said they would not be willing and 50 were unsure. Do the data indicate a statistically significant tendency towards willingness or towards unwillingness? Use a 0.05 level of significance.

6. Suppose a national survey in Sweden has shown that the median annual income adults say would make their dreams come true is €152,000. Suppose further that, of a sample of 225 individuals in Gothenburg, 122 individuals report that the amount of income needed to make their dreams come true is less than €152,000 and 103 report that the amount needed is more than €152,000. Test the null hypothesis that the median amount of annual income needed to make dreams come true in Gothenburg is €152,000. Use $\alpha = 0.05$. What is your conclusion?

7. The median number of part-time employees at fast-food restaurants in a particular city was known to be 15 last year. The city council thinks the use of part-time employees may have increased this year. A sample of nine fast-food restaurants showed that more than 15 part-time employees worked at seven of the restaurants, one restaurant had exactly 15 part-time employees and one had fewer than 15 part-time employees. Test at $\alpha = 0.05$ to see whether the median number of part-time employees has increased.

8. In a recent poll, 600 adults were asked a series of questions about the state of the economy and their children's future. One question was, 'Do you expect your children to have a better life than you have had, a worse life or a life about the same as yours?' The responses showed 242 better, 310 worse and 48 about the same. Use the sign test and $\alpha = 0.05$ to determine whether there is a difference between the proportions of adults who feel their children will have a better life compared to a worse life. What is your conclusion?

18.2 WILCOXON SIGNED-RANK TEST

The Wilcoxon signed-rank test is a procedure for analyzing data from a matched-samples study. The test uses quantitative data but does not require the assumption that the differences between the paired observations are normally distributed. It requires the less stringent assumption that the differences between the paired observations have a symmetrical distribution. The test examines whether the population differences are centred on the value zero (i.e. have a mean or median equal to zero). We demonstrate the Wilcoxon signed-rank test with the following example.

Suppose a manufacturer was attempting to determine whether two production methods differ in task completion time. A sample of 11 workers was selected, and each worker completed a production task using each of the two production methods. The production method that each worker used first was selected randomly. Each worker in the sample therefore provided a pair of observations, as shown in the first three columns of Table 18.3. A positive difference in task completion times (column 4 of Table 18.3) indicates that method 1 required more time, and a negative difference indicates that method 2 required more time. Do the data indicate that the methods are different on average in terms of task completion times?

We are considering two populations of task completion times associated with the two production methods. The following hypotheses will be tested:

$$H_0: \text{The populations are identical}$$
$$H_1: \text{The populations are not identical}$$

TABLE 18.3   Production task completion times (minutes) and ranking of absolute differences

Worker	Method 1	Method 2	Difference	Absolute value of difference	Rank	Signed rank
1	10.2	9.5	0.7	0.7	8	+8
2	9.6	9.8	−0.2	0.2	2	−2
3	9.2	8.8	0.4	0.4	3.5	+3.5
4	10.6	10.1	0.5	0.5	5.5	+5.5
5	9.9	10.3	−0.4	0.4	3.5	−3.5
6	10.2	9.3	0.9	0.9	10	+10
7	10.6	10.5	0.1	0.1	1	+1
8	10.0	10.0	0.0	0.0	–	–
9	11.2	10.6	0.6	0.6	7	+7
10	10.7	10.2	0.5	0.5	5.5	+5.5
11	10.6	9.8	0.8	0.8	9	+9
				Sum of signed ranks		**+44.0**

If H_0 can be rejected, we shall conclude that the two methods differ in task completion time. If H_0 cannot be rejected, we shall not have convincing evidence to conclude that the task completion times differ for the two methods.

The first step of the Wilcoxon signed-rank test requires a ranking of the *absolute values* of the differences between the two methods. We discard any differences of zero and then rank the remaining absolute differences from lowest to highest. Tied differences are assigned the average ranking of their positions. The ranking of the absolute difference values is shown in the sixth column of Table 18.3. Note that the difference of zero for worker 8 is discarded from the rankings. Then the smallest absolute difference of 0.1 is assigned rank 1. The ranking continues until the largest absolute difference of 0.9 is assigned rank 10. The tied absolute differences for workers 3 and 5 are assigned the average rank of 3.5 and the tied absolute differences for workers 4 and 10 are assigned the average rank of 5.5.

When the ranking is complete, the ranks are given the signs of the original differences in the data. For example, the 0.1 difference for worker 7, assigned rank 1, is given the value +1 because the observed difference between the two methods was positive. The 0.2 difference (worker 2), assigned rank 2, is given the value −2 because the observed difference was negative for worker 2. The complete list of signed ranks, as well as their sum, is shown in the last column of Table 18.3.

The null hypothesis is identical population distributions of task completion times for the two methods. In that case, we would expect the positive ranks and the negative ranks to cancel each other, so that the sum of the signed rank values would be approximately zero. Hence, the Wilcoxon signed-rank test involves determining whether the computed sum of signed ranks (+44 in our example) is significantly different from zero.

Let T denote the sum of the signed-rank values. The procedure assumes that the distribution of differences between matched pairs is symmetrical, but not necessarily normal in shape. It can be shown that if the two populations are identical and the number of matched pairs of data is 10 or more, the sampling distribution of T can be approximated by a normal distribution as follows:

Sampling distribution of T for identical populations

$$\text{Mean: } \mu_T = 0 \qquad \qquad \textbf{(18.3)}$$

$$\text{Standard deviation: } \sigma_T = \sqrt{\frac{n(n+1)(2n+1)}{6}} \qquad \qquad \textbf{(18.4)}$$

Distribution form: approximately normal for $n \geq 10$.

For the example, we have $n = 10$ after discarding the observation with the difference of zero (worker 8). Using equation (18.4), we have:

$$\text{Standard deviation: } \sigma_T = \sqrt{\frac{(10)(11)(21)}{6}} = 19.62$$

We shall use a 0.05 level of significance to draw a conclusion. We calculate the following value for the test statistic, using the sum of the signed-rank values $T = 44$.

$$z = \frac{T - \mu_T}{\sigma_T} = \frac{44 - 0}{19.62} = 2.24$$

Using the standard normal distribution table and $z = 2.24$, we find the two-tailed p-value $= 2(1 - 0.9875)$ $= 0.025$. With p-value $< \alpha = 0.05$, we reject H_0. We conclude that the two populations are not identical and that the methods differ in task completion time. Method 2's shorter completion times for eight of the workers lead us to conclude that method 2 is the preferred production method.

EXERCISES

Applications

9. Two fuel additives are tested on family cars to investigate their effect on litres of fuel consumed per 100 kilometres travelled. Test results for 12 cars follow. Each car was tested with both fuel additives. Use $\alpha = 0.05$ and the Wilcoxon signed-rank test to see whether there is a difference between the additives.

	Additive			Additive	
Car	1	2	Car	1	2
1	7.02	7.82	7	8.74	8.21
2	6.00	6.49	8	7.62	9.43
3	6.41	6.26	9	6.46	7.05
4	7.37	8.28	10	5.83	6.68
5	6.65	6.65	11	6.09	6.20
6	5.70	5.93	12	5.65	5.96

10. A study examined the effects of a relaxant on the time required to fall asleep for male adults. Data for ten male participants showing the number of minutes required to fall asleep with and without the relaxant follow. Use a 0.05 level of significance to assess whether the relaxant reduces the time required to fall asleep. What is your conclusion?

Participant	Without relaxant	With relaxant	Participant	Without relaxant	With relaxant
1	15	10	6	7	5
2	12	10	7	8	10
3	22	12	8	10	7
4	8	11	9	14	11
5	10	9	10	9	6

11. A test was conducted of two overnight mail delivery services. Two samples of identical deliveries were set up so that both delivery services were notified of the need for a delivery at the same time. The hours required for each delivery follow. Do the data shown suggest a difference in the delivery times for the two services? Use a 0.05 level of significance for the test.

| | Service | |
Delivery	1	2
1	24.5	18.0
2	26.0	25.5
3	28.0	32.0
4	21.0	20.0
5	18.0	19.5
6	36.0	28.0
7	25.0	29.0
8	21.0	22.0
9	24.0	23.5
10	26.0	29.5
11	31.0	30.0

12. Ten test-market cities in France were selected as part of a market research study designed to evaluate the effectiveness of a particular advertising campaign. The sales in euros for each city were recorded for the week prior to the promotional programme. Then the campaign was conducted for two weeks and new sales data were collected for the week immediately after the campaign. The two sets of sales data (in thousands of euros) follow.

City	Pre-campaign sales	Post-campaign sales
Bordeaux	130	160
Strasbourg	100	105
Nantes	120	140
St Etienne	95	90
Lyon	140	130
Rennes	80	82
Le Havre	65	55
Amiens	90	105
Toulouse	140	152
Marseilles	125	140

What conclusion would you draw about the effect of the advertising programme? Use $\alpha = 0.05$.

18.3 MANN-WHITNEY-WILCOXON TEST

In Chapter 10 we introduced a hypothesis test for the difference between the means of two populations using two independent samples, one from population 1 and one from population 2. This parametric test required quantitative data and the assumption that both populations had normal distributions. In the case where the population standard deviations σ_1 and σ_2 were unknown, the sample standard deviations s_1 and s_2 provided estimates of σ_1 and σ_2 and the t distribution was used to make an inference about the difference between the means of the two populations.

In this section we present a non-parametric method that can be used to determine whether a difference exists between two populations. This test, unlike the signed-rank test, is not based on matched samples. Two independent samples are used, one from each population. The test was developed jointly by Mann and Whitney and by Wilcoxon. It is sometimes called the *Mann-Whitney test* and sometimes the *Wilcoxon rank-sum test*. The Mann-Whitney and Wilcoxon versions of this test are equivalent. We refer to it as the **Mann-Whitney-Wilcoxon (MWW) test**.

The MWW test does not require interval data nor the assumption that the populations are normally distributed. The only requirement of the MWW test is that the measurement scale for the data is at least ordinal. The MWW test examines whether the two populations are identical:

H_0: The two populations are identical
H_1: The two populations are not identical

If H_0 is rejected, we are using the test to conclude that the populations are not identical and that population 1 tends to provide either smaller or larger values than population 2.

We shall first illustrate the MWW test using small samples with rank-ordered data. This will give you an understanding of how the rank-sum statistic is computed and how it is used to determine whether to reject the null hypothesis that the two populations are identical. Later in the section, we shall introduce a large-sample approximation based on the normal distribution that will simplify the calculations required by the MWW test.

Consider on-the-job performance ratings for employees at a CineMax 20-screen multiplex. During an employee performance review, the multiplex manager rated all 35 employees on a scale from 0 (very poor) to 100 (excellent). Knowing that the part-time employees were primarily university and senior school students, the multiplex manager asked if there was evidence of a difference in performance for university students compared to senior school students. In terms of the population of university students and the population of senior school students who could be considered for employment at the multiplex, the hypotheses were stated as follows:

H_0: University and senior school student populations are identical in terms of performance
H_1: University and senior school student populations are not identical in terms of performance

We will use a 0.05 level of significance for this test.

We begin by selecting a random sample of four university students and a random sample of five senior school students working at the CineMax multiplex (these sample numbers are chosen arbitrarily for illustrative purposes). The multiplex manager's performance rating was recorded for each of these employees, as shown in Table 18.4. The first university student selected was given a rating of 81 (out of 100), the second university student selected was given a rating of 92 and so on.

TABLE 18.4   Ranks for the nine students in the CineMax combined samples

University student	Manager's performance rating	Rank	Senior school student	Manager's performance rating	Rank
1	81	6	1	73	5
2	92	9	2	70	4
3	62	3	3	46	1
4	87	8	4	83	7
			5	57	2
	Sum of ranks	**26**		**Sum of ranks**	**19**

The next step in the MWW procedure is to rank the *combined* samples from low to high. Since there is a total of nine students, we rank the performance rating data in Table 18.4 from 1 to 9. The lowest performance rating of 46 for senior school student 3 receives a rank of 1, the second lowest rating of 57 for senior school student 5 receives a rank of 2, … , the highest rating of 92 for university student 2 receives a rank of 9. The ranks for all nine students are shown in Table 18.4.

Next we sum the ranks for each sample as shown in Table 18.4. The MWW procedure may use the sum of the ranks for either sample. We will follow the common practice of using the first sample, which is the sample of four university students. The sum of ranks for the first sample will be the test statistic W for the MWW test. This sum, as shown in Table 18.4, is $W = 6 + 9 + 3 + 8 = 26$.

Let us consider why the sum of the ranks will help us select between the two hypotheses: H_0, the two populations are identical and H_1, the two populations are not identical. Letting U denote a university student and S denote a senior school student, suppose the ranks of the nine students had the following order:

Rank	1	2	3	4	5	6	7	8	9
Student	U	U	U	U	S	S	S	S	S

This permutation or ordering separates the two samples, with the university students all having a lower rank than the senior school students. This is a strong indication that the two populations are not identical. The sum of ranks for the university students in this case is $W = 1 + 2 + 3 + 4 = 10$.

Now consider the following ranking:

Rank	1	2	3	4	5	6	7	8	9
Student	S	S	S	S	S	U	U	U	U

This permutation or ordering separates the two samples again, but this time the university students all have a higher rank than the senior school students. This is another strong indication that the two populations are not identical. The sum of ranks for the university students in this case is $W = 6 + 7 + 8 + 9 = 30$. So we see that the sum of the ranks for the university students must be between 10 and 30. Values of W near 10 imply that university students have lower ranks than the senior school students, whereas values of W near 30 imply that university students have higher ranks than the senior school students. Either of these extremes would signal the two populations are not identical. However, if the two populations are identical, we would expect a mix in the ordering so that the sum of ranks W is closer to the average of the two extremes, or nearer to $(10 + 30)/2 = 20$.

Evaluation of the exact sampling distribution of the W statistic needs a computer program because it is not straightforward. However, there are published tables of critical values, such as those in Table 6 of Appendix B, for cases in which both sample sizes are less than or equal to 10. In that table, n_1 refers to the sample size corresponding to the sample whose rank sum is being used in the test. The null hypothesis of identical populations should be rejected only if W is strictly less than the lower-tail critical value W_L or strictly greater than the upper-tail critical value W_U. The value of W_L is read directly from the table and the value of W_U is computed from equation (18.5).

$$W_U = n_1(n_1 + n_2 + 1) - W_L \tag{18.5}$$

Using Table 6 of Appendix B with a 0.05 level of significance, we see that the lower-tail critical value for the MWW statistic with $n_1 = 4$ (university students) and $n_2 = 5$ (senior school students) is $W_L = 12$. The upper-tail critical value for the MWW statistic computed by using equation (18.5) is:

$$W_U = 4(4 + 5 + 1) - 12 = 28$$

The MWW decision rule indicates that the null hypothesis of identical populations can be rejected if the sum of the ranks for the first sample (university students) is less than 12 or greater than 28. The rejection rule can be written as:

$$\text{Reject } H_0 \text{ if } W < 12 \text{ or if } W > 28$$

Referring to Table 18.4, we see that $W = 26$. The MWW test conclusion is that we cannot reject the null hypothesis that the populations of university and senior school students are identical. The sample of four university students and the sample of five senior school students did not provide statistical evidence to conclude there is a difference between the two populations. Further study with larger samples should be considered before drawing a final conclusion.

As noted above, the exact sampling distribution of the W statistic is not straightforward to evaluate. Some statistical programs do this and give an exact p-value. For example, SPSS includes exact versions of several non-parametric methods. For the CineMax employee ratings, SPSS gives a two-tailed p-value of 0.190 (confirming our conclusion that we do not have sufficient evidence to reject H_0).

Most applications of the MWW test involve larger sample sizes than shown in this first example. For such applications, a large-sample approximation of the sampling distribution of W based on the normal distribution can be used. We will use the same combined-samples ranking procedure that we used in the previous example but will use the normal distribution approximation to compute the p-value rather than using the tables of critical values for W. We illustrate the large sample case by considering a situation at People's Bank.

People's Bank has two branch offices. Data collected from two independent simple random samples, one from each branch, are given in Table 18.5 (the rankings in this table are explained below). What do the data indicate regarding the hypothesis that the populations of current account balances at the two branch banks are identical?

TABLE 18.5   Current account balances for two branches of People's Bank, and combined ranking of the data

Branch 1			Branch 2		
Account	**Balance (€)**	**Rank**	**Account**	**Balance (€)**	**Rank**
1	1,095	20	1	885	7
2	955	14	2	850	4
3	1,200	22	3	915	8
4	1,195	21	4	950	12.5
5	925	9	5	800	2
6	950	12.5	6	750	1
7	805	3	7	865	5
8	945	11	8	1,000	16
9	875	6	9	1,050	18
10	1,055	19	10	935	10
11	1,025	17			
12	975	15			
Sum of ranks		**169.5**	**Sum of ranks**		**83.5**

The first step in the MWW test is to rank the *combined* data from the lowest to the highest values. Using the combined set of 22 observations in Table 18.5, we find the lowest data value of €750 (sixth item of sample 2) and assign to it a rank of 1. Continuing the ranking gives us the following list:

Balance (€)	Item	Assigned rank
750	6th of sample 2	1
800	5th of sample 2	2
805	7th of sample 1	3
850	2nd of sample 2	4
.	.	.
1,195	4th of sample 1	21
1,200	3rd of sample 1	22

In ranking the combined data, we may find that two or more data values are the same. In that case, the tied values are given the *average* ranking of their positions in the combined data set. For example, the balance of €945 (eighth item of sample 1) will be assigned the rank of 11. However, the next two values in the data set are tied with values of €950 (the sixth item of sample 1 and the fourth item of sample 2). These two values would be assigned ranks of 12 and 13 if they were distinct, so they are both assigned the rank of 12.5. The next data value of €955 is then assigned the rank of 14. Table 18.5 shows the assigned rank of each observation.

The next step in the MWW test is to sum the ranks for each sample. The sums are given in Table 18.5. The test procedure can be based on the sum of the ranks for either sample. We use the sum of the ranks for the sample from branch 1. So, for this example, $W = 169.5$.

Given that the sample sizes are $n_1 = 12$ and $n_2 = 10$, we can use the normal approximation to the sampling distribution of the rank sum W. The appropriate sampling distribution is given by the following expressions:

Sampling distribution of W for identical populations

$$\text{Mean: } \mu_W = 0.5 n_1 (n_1 + n_2 + 1) \tag{18.6}$$

$$\text{Standard deviation: } \sigma_T = \sqrt{\frac{n_1 n_2 (n_1 + n_2 + 1)}{12}} \tag{18.7}$$

Distribution form: approximately normal provided $n_1 \geq 10$ and $n_2 \geq 10$

For branch 1, we have:

$$\mu_W = 0.5(12)(12 + 10 + 1) = 138$$

$$\sigma_W = \sqrt{\frac{(12)(10)(12 + 10 + 1)}{12}} = 15.17$$

We shall use a 0.05 level of significance to draw a conclusion. With the sum of the ranks for branch 1, $W = 169.5$, we calculate the following value for the test statistic.

$$z = \frac{W - \mu_W}{\sigma_W} = \frac{169.5 - 138}{15.17} = 2.08$$

Using the standard normal distribution table and $z = 2.08$, we find the two-tailed p-value $= 2(1 - 0.9812)$ $= 0.0376$. With p-value $< \alpha = 0.05$, we reject H_0 and conclude that the two populations are not identical, that is, the populations of current account balances at the branch banks are not the same. The evidence suggests that the balances at branch 1 tend to be higher (and therefore be assigned higher ranks) than the balances at branch 2.

In summary, the MWW rank-sum test consists of the following steps to determine whether two independent random samples are selected from identical populations:

1　Rank the combined sample observations from lowest to highest, with tied values being assigned the average of the tied rankings.
2　Compute W, the sum of the ranks for the first sample.
3　In the large-sample case, make the test for significant differences between the two populations by using the observed value of W and comparing it with the sampling distribution of W for identical populations using equations (18.6) and (18.7). The value of the standardized test statistic z and the p-value provide the basis for deciding whether to reject H_0. In the small-sample case, use Table 6 in Appendix B to find the critical values for the test.

The parametric statistical tests described in Chapter 10 test the equality of two population means. When we reject the hypothesis that the means are equal, we conclude that the populations differ in their means. When we reject the hypothesis that the populations are identical using the MWW test, we cannot state how they differ. The populations could have different means, different medians, different variances or different forms. However, if we believe the populations are the same in every respect except central location – i.e. mean or median – a rejection of H_0 by the non-parametric method implies that the means (or medians) differ.

EXERCISES

Applications

13. A company's price/earnings (P/E) ratio is the company's current stock price divided by the latest 12 months' earnings per share. Listed below are the P/E ratios for a sample of eight Japanese and nine US companies in late 2022 (Yahoo Finance (uk.finance.yahoo.com)). Is there a difference on average between P/E ratios in the two countries? Use the MWW test and $\alpha = 0.10$ to support your conclusion.

	Japan		USA	
Company	P/E ratio	Company		P/E ratio
Fuji Corp	9.46	American Elec Power Co Inc		20.43
Heiwa Corp	22.91	Ford Motor Co		6.02
Kinden Corp	12.76	Ingersoll-Rand plc		40.84
Seibu Elec & Mach Co Ltd	13.79	Motorola Solns Inc		39.09
Shiseido Co Ltd	98.60	Northern Oil & Gas, Inc		3.31
Sumitomo Corporation	4.68	Oracle Corp		38.52
Suzuki Motor Corp	12.94	Schlumberger Ltd		24.36
Toho Gas Co Ltd	14.26	The Gap Inc		92.60
		Winnebago Industries Inc		4.68

14. Two fuel additives are being tested to determine their effect on petrol consumption. Seven cars were tested with additive 1 and nine cars were tested with additive 2. The following data show the litres of fuel used per 100 kilometres. Use $\alpha = 0.05$ and the MWW test to see whether there is a difference on average in petrol consumption for the two additives.

Additive 1	Additive 2
8.20	7.52
7.69	7.94
7.41	6.62
8.47	6.71
7.75	6.41
7.58	7.52
8.06	7.14
	6.80
	6.99

15. Samples of annual salary plus benefits packages for individuals entering the public accounting and financial planning professions follow. The figures are in thousands of euros.

Public accountant	Public accountant	Financial planner	Financial planner
45.2	50.0	44.0	48.6
53.8	45.9	44.2	44.7
51.3	54.5	48.1	48.9
53.2	52.0	50.9	46.8
49.2	46.9	46.9	43.9

a. What are the sample mean annual salaries plus benefits for the two professions?

b. Using $\alpha = 0.05$, test the hypothesis that there is no difference between the annual salaries plus benefits levels of public accountants and financial planners. What is your conclusion?

16. A confederation of house builders provided data on the cost (in £) of the most popular home remodelling projects. Use the MWW test to see whether it can be concluded that the cost of kitchen remodelling differs from the cost of master bedroom remodelling. Use a 0.05 level of significance.

Kitchen	Master bedroom
13,200	6,000
5,400	10,900
10,800	14,400
9,900	12,800
7,700	14,900
11,000	5,800
7,700	12,600
4,900	9,000
9,800	
11,600	

17. The gap between the earnings of men and women with equal education is narrowing in many countries but has not closed. Sample data from the United Arab Emirates for seven men and seven women with Bachelor's degrees are as follows. Data of monthly earnings are shown in thousands of dirham.

Men	12.2	30.2	18.1	24.9	15.3	20.0	22.1
Women	17.8	14.2	11.2	16.2	10.3	19.0	9.9

a. What is the median salary for men? For women?
b. Use $\alpha = 0.05$ and conduct the hypothesis test for identical populations. What is your conclusion?

18.4 KRUSKAL-WALLIS TEST

The MWW test in Section 18.3 can be used to test whether two populations are identical. Kruskal and Wallis extended the test to the case of three or more populations. The Kruskal-Wallis test is based on the analysis of independent random samples from each of the populations. The hypotheses for the Kruskal-Wallis test with $k \geq 3$ populations can be written as follows:

$$H_0: \text{All } k \text{ populations are identical}$$
$$H_1: \text{Not all } k \text{ populations are identical}$$

In Chapter 13 we showed that analysis of variance (ANOVA) can be used to test for the equality of means among three or more populations. The ANOVA procedure requires interval- or ratio-level data and the assumption that the k populations are normally distributed. The non-parametric Kruskal-Wallis test can be used with ordinal data as well as with interval or ratio data. In addition, the Kruskal-Wallis test does not require the assumption of normally distributed populations. We demonstrate the Kruskal-Wallis test by using it in an employee selection example.

Williams Manufacturing hires its management staff from three local universities. Recently the company's personnel department began collecting and reviewing annual performance ratings to determine whether there are differences in performance between the managers hired from these universities. Performance rating data are available from independent samples of seven employees from university A, six employees from university B and seven employees from university C. These data are summarized in Table 18.6. The overall performance rating of each manager is given on a 0–100 scale, with 100 being the highest possible performance rating. The rankings shown in the table are explained below.

TABLE 18.6   Performance evaluation ratings for 20 Williams employees

University A	Rank	University B	Rank	University C	Rank
25	3	60	9	50	7
70	12	20	2	70	12
60	9	30	4	60	9
85	17	15	1	80	15.5
95	20	40	6	90	18.5
90	18.5	35	5	70	12
80	15.5			75	14
Sum of ranks	**95**		**27**		**88**

Suppose we want to test whether the three populations are identical in terms of performance evaluations. We shall use a 0.05 level of significance. The Kruskal-Wallis test statistic, which is based on the sum of ranks for each of the samples, can be computed as follows:

Kruskal-Wallis test statistic

$$W = \left[\frac{12}{n_T(n_T+1)} \sum_{i=1}^{k} \frac{R_i^2}{n_i} \right] - 3(n_T+1) \tag{18.8}$$

where:

k = the number of populations

n_i = the number of items in sample i

$n_T = \Sigma n_i$ = total number of items in all k samples

R_i = sum of the ranks for sample i

Kruskal and Wallis were able to show that, under the null hypothesis that the populations are identical, the sampling distribution of W can be approximated by a chi-squared distribution with $k - 1$ degrees of freedom. This approximation is generally acceptable if each of the sample sizes is greater than or equal to five. The null hypothesis of identical populations will be rejected if the test statistic is large. As a result, the procedure involves an upper-tail test.

To compute the W statistic for our example, we first rank all 20 data items. The lowest data value of 15 from the university B sample receives a rank of 1, whereas the highest data value of 95 from the university A sample receives a rank of 20. The ranks and the sums of the ranks for the three samples are given in Table 18.6. Note that we assign the average rank to tied items;* for example, the data values of 60, 70, 80 and 90 had ties. The sample sizes are:

$$n_1 = 7 \qquad n_2 = 6 \qquad n_3 = 7$$

and:

$$n_T = \Sigma n_i = 7 + 6 + 7 = 20$$

We compute the W statistic by using equation (18.8).

$$W = \left[\frac{12}{(20)(21)} \right] \left[\frac{(95)^2}{7} + \frac{(27)^2}{6} + \frac{(88)^2}{7} \right] - 3(20 + 1) = 8.92$$

---

* If numerous tied ranks are observed, equation (18.8) must be modified. The modified formula is given in W. J. Conover (1999) *Practical Nonparametric Statistics*, 3rd ed. Wiley.

We can now use the χ^2 distribution table (Table 3 of Appendix B) to determine the p-value for the test. Using $k - 1 = 3 - 1 = 2$ df, we find $\chi^2 = 7.378$ has a probability of 0.025 in the upper tail of the distribution and $\chi^2 = 9.21$ has a probability of 0.01 in the upper tail. For $W = 8.92$, between 7.378 and 9.21, the probability in the upper tail is between 0.025 and 0.01. As it is an upper tail test, we can conclude that the p-value is between 0.025 and 0.01. (A calculation in SPSS, Minitab, R or Excel shows p-value $= 0.0116$.)

As p-value $< \alpha = 0.05$, we reject H_0 and conclude that the three populations are not identical. There is a statistically significant difference in manager performance depending on the college attended. Furthermore, because the performance ratings are lowest for college B, it would be reasonable for the company to either cut back recruiting from college B or at least evaluate its graduates more thoroughly.

EXERCISES

Applications

18. Three preparation programmes for university admission tests are being evaluated. The scores obtained by a sample of 20 people who used the test preparation programmes provided the following data. Use the Kruskal-Wallis test to determine whether there is a difference between the scores for the three test preparation programmes. Use $\alpha = 0.01$.

	Programme	
A	B	C
540	450	600
400	540	630
490	400	580
530	410	490
490	480	590
610	370	620
	550	570

19. Forty-minute workouts of one of the following activities three days a week may lead to a loss of weight. The following sample data show the number of calories burned during 40-minute workouts for three different activities. Do these data indicate differences in the amount of calories burned for the three activities? Use a 0.05 level of significance. What is your conclusion?

Swimming	Tennis	Cycling
408	415	385
380	485	250
425	450	295
400	420	402
427	530	268

20. *Condé Nast Traveller* magazine conducts an annual survey of its readers in order to rate the top 80 cruise ships in the world. The overall ratings for a sample of ships from the Holland America, Princess and Royal Caribbean cruise lines are shown here (100 is the highest possible rating). Use the Kruskal-Wallis test with $\alpha = 0.05$ to determine whether the overall ratings differ between the three cruise lines.

Holland America		Princess		Royal Caribbean	
Ship	Rating	Ship	Rating	Ship	Rating
Amsterdam	84.5	Coral	85.1	Adventure	84.8
Maasdam	81.4	Dawn	79.0	Jewel	81.8
Ooterdam	84.0	Island	83.9	Mariner	84.0
Volendam	78.5	Princess	81.1	Navigator	85.9
Westerdam	80.9	Star	83.7	Serenade	87.4

21. Course-evaluation ratings for four instructors follow. Use the Kruskal-Wallis procedure with $\alpha = 0.05$ to test for a difference in the ratings.

Instructor	Course-evaluation rating								
Black	88	80	79	68	96	69			
Jennings	87	78	82	85	99	99	85	94	81
Swanson	88	76	68	82	85	82	84	83	
Wilson	80	85	56	71	89	87			

18.5 RANK CORRELATION

The Pearson product moment correlation coefficient (refer to Chapter 3, Section 3.5 and Chapter 15, Section 15.5) is a measure of the linear association between two variables for which interval or ratio data are available. In this section, we consider the Spearman rank-correlation coefficient r_S, which is a measure of association between two variables applicable when only ordinal data are available.

Spearman rank-correlation coefficient

$$r_S = 1 - \frac{6\sum d_i^2}{n(n^2 - 1)}$$

(18.9)

where:

$n = $ the number of items or individuals being ranked

$x_i = $ the rank of item i with respect to one variable

$y_i = $ the rank of item i with respect to the second variable

$d_i = x_i - y_i$

Suppose a company wants to determine whether individuals who were expected at the time of employment to be better sales executives actually turn out to have better sales records. To investigate this question, the personnel manager carefully reviewed the original job interview summaries, academic records and letters of recommendation for ten current members of the firm's sales force. After the review, the personnel manager ranked the ten individuals in terms of their potential for success, basing the assessment solely on the information available at the time of employment. Then a list was obtained of the number of units sold by each sales executive over the first two years.

On the basis of actual sales performance, a second ranking of the ten sales executives was carried out. Table 18.7 gives the relevant data and the two rankings. In the ranking of potential, rank 1 means lowest potential, rank 2 next lowest and so on. The statistical question is whether there is agreement between the ranking of potential at the time of employment and the ranking based on the actual sales performance over the first two years.

TABLE 18.7   Sales potential and actual two-year sales data for ten sales executives, and computation of the Spearman rank-correlation coefficient

Sales executive	x_i = Ranking of potential	Two-year sales (units)	y_i = Ranking of sales performance	$d_i = x_i - y_i$	d_i^2
A	9	400	10	−1	1
B	7	360	8	−1	1
C	4	300	6	−2	4
D	10	295	5	5	25
E	5	280	4	1	1
F	8	350	7	1	1
G	1	200	1	0	0
H	2	260	3	−1	1
I	3	220	2	1	1
J	6	385	9	−3	9
					$\Sigma d_i^2 = 44$

$$r_S = 1 - \frac{6\Sigma d_i^2}{n(n^2 - 1)} = 1 - \frac{(6)(44)}{(10)(100 - 1)} = 0.73$$

The computations for the Spearman rank-correlation coefficient are summarized in Table 18.7. The rank-correlation coefficient r_S is (positive) 0.73. The Spearman rank-correlation coefficient ranges from -1.0 to $+1.0$ and its interpretation is similar to that of the Pearson correlation coefficient, in that positive values near 1.0 indicate a strong positive association between the rankings: as one rank increases, the other rank increases. Rank correlations near -1.0 indicate a strong negative association between the rankings: as one rank increases, the other rank decreases. The value $r_S = 0.73$ indicates a positive correlation between potential and actual performance. Individuals ranked high on potential tend to rank high on performance.

At this point, we may want to use the sample results to make an inference about the population rank correlation ρ_S. To do this, we test the following hypotheses:

$$H_0: \rho_S = 0$$
$$H_1: \rho_S \neq 0$$

Under the null hypothesis of no population rank correlation ($\rho_S = 0$), the sampling distribution of r_S is as follows:

Sampling distribution of r_S

$$\text{Mean: } \mu_{r_s} = 0 \qquad \qquad \text{(18.10)}$$

$$\text{Standard deviation: } \sigma_{r_s} = \sqrt{\frac{1}{n-1}} \qquad \qquad \text{(18.11)}$$

Distribution form: approximately normal provided $n \geq 10$

The sample rank-correlation coefficient for sales potential and sales performance is $r_S = 0.73$. From equation (18.10) we have $\mu_{r_s} = 0$ and from (18.11) we have:

$$\sigma_{r_s} = \sqrt{1/(10 - 1)} = 0.33$$

Using the test statistic, we have:

$$z = \frac{r_S - \mu_{r_S}}{\sigma_{r_S}} = \frac{0.73 - 0}{0.33} = 2.20$$

Using the standard normal distribution table and $z = 2.20$, we find the p-value $= 2(1 - 0.9861) = 0.0278$. With a 0.05 level of significance, p-value $< \alpha = 0.05$ leads to the rejection of the hypothesis that the population rank correlation is zero. We can conclude that there is a positive rank correlation between sales potential and sales performance.

EXERCISES

Methods

22. Consider the following set of rankings for a sample of ten elements:

Element	x_i	y_i	Element	x_i	y_i
1	10	8	6	2	7
2	6	4	7	8	6
3	7	10	8	5	3
4	3	2	9	1	1
5	4	5	10	9	9

   a. Compute the Spearman rank-correlation coefficient for the data.
   b. Use $\alpha = 0.05$ and test for statistically significant rank correlation. What is your conclusion?

23. Consider the following two sets of rankings for six items:

	Case One			Case Two	
Item	First ranking	Second ranking	Item	First ranking	Second ranking
A	1	1	A	1	6
B	2	2	B	2	5
C	3	3	C	3	4
D	4	4	D	4	3
E	5	5	E	5	2
F	6	6	F	6	1

   Note that in the first case the rankings are identical, whereas in the second case the rankings are exactly opposite. What value should you expect for the Spearman rank-correlation coefficient for each of these cases? Explain. Calculate the rank-correlation coefficient for each case.

Applications

24. The following two lists show how ten IT companies ranked in a national survey in terms of reputation and the percentage of respondents who said they would purchase the company's shares. A positive rank correlation is anticipated because it seems reasonable to expect that a company with a higher reputation would be a more desirable purchase.

Company	Reputation	Probable purchase
Microsoft	1	3
Intel	2	4
Dell	3	1
Lucent	4	2
Texas Instruments	5	9
Cisco Systems	6	5
Hewlett-Packard	7	10
IBM	8	6
Motorola	9	7
Yahoo	10	8

a. Compute the rank correlation between reputation and probable purchase.
b. Test for a statistically significant positive rank correlation. What is the p-value?
c. At $\alpha = 0.05$, what is your conclusion?

25. The file 'UNIV TABLES' on the online platform contains data for a sample of 12 UK universities, taken from two of the university 'league tables' published in 2022. The variables in the file are as follows: (1) university name, (2) a 'student satisfaction' score taken from the *Complete University Guide* (CUG_SS), (3) a 'research quality' score taken from the CUG (CUG_RQ), (4) the total points score for the university in the CUG (CUG_Tot), and (5) the total points score for the university in a different league table, published by *The Guardian* newspaper (GUG_Tot). The first three lines of the table are shown below.

UNIV
TABLES

University	CUG_SS	CUG_RQ	CUG_Tot	GUG_Tot
London School of Economics & Politics	3.87	3.53	960	95.2
University of Southampton	3.89	3.41	793	76.2
University of Glasgow	3.93	3.43	726	82.1

a. Calculate the Spearman rank-correlation coefficient between the CUG total points score and the *Guardian* total points score. Test the coefficient for statistical significance and report your conclusions.
b. Calculate the Spearman rank-correlation coefficient between the SS and RQ scores from the CUG. Test the coefficient for statistical significance and report your conclusions.

26. A sample of 15 students received the following rankings on mid-term and final examinations in a statistics course.

Rank		Rank		Rank	
Mid-term	Final	Mid-term	Final	Mid-term	Final
1	4	6	2	11	14
2	7	7	5	12	15
3	1	8	12	13	11
4	3	9	6	14	10
5	8	10	9	15	13

Compute the Spearman rank-correlation coefficient for the data and test for a statistically significant correlation, with $\alpha = 0.10$.

ONLINE RESOURCES

For the data sets, additional questions and answers, and the software section for Chapter 18, visit the online platform.

SUMMARY

In this chapter we presented several statistical procedures that are classified as non-parametric methods. Because non-parametric methods can be applied to ordinal and in some cases nominal data, as well as interval and ratio data, and because they require less restrictive population distribution assumptions, they expand the class of problems that can be subjected to statistical analysis.

The sign test is a non-parametric procedure for identifying differences between two populations when the data available are nominal data. In the small-sample case, the binomial probability distribution can be used to determine the p-value for the sign test; in the large-sample case, a normal approximation can be used. The Wilcoxon signed-rank test is a procedure for analyzing matched-samples data whenever interval- or ratio-scaled data are available for each matched pair. The procedure tests the hypothesis that the two populations being considered are identical. The procedure assumes that the distribution of differences between matched pairs is symmetrical, but not necessarily normal in shape.

The Mann-Whitney-Wilcoxon test is a non-parametric method for testing for a difference between two populations based on two independent random samples. Tables were presented for the small-sample case, and a normal approximation was provided for the large-sample case. The Kruskal-Wallis test extends the Mann-Whitney-Wilcoxon test to the case of three or more populations. The Kruskal-Wallis test is the non-parametric analogue of the parametric ANOVA for differences between population means.

We introduced the Spearman rank-correlation coefficient as a measure of association for two ordinal or rank-ordered sets of items.

KEY TERMS

Distribution-free methods
Kruskal-Wallis test
Mann-Whitney-Wilcoxon (MWW) test
Non-parametric methods

Parametric methods
Sign test
Spearman rank-correlation coefficient
Wilcoxon signed-rank test

KEY FORMULAE

Sign test (large-sample case)

$$\text{Mean: } \mu = 0.50n \tag{18.1}$$

$$\text{Standard deviation: } \sigma = \sqrt{0.25n} \tag{18.2}$$

Wilcoxon signed-rank test

$$\text{Mean: } \mu_T = 0 \tag{18.3}$$

Standard deviation: $\sigma_T = \sqrt{\dfrac{n(n + 1)(2n + 1)}{6}}$ (18.4)

Mann-Whitney-Wilcoxon test (large-sample)

Mean: $\mu_W = 0.5n_1(n_1 + n_2 + 1)$ (18.6)

Standard deviation: $\sigma_W = \sqrt{\dfrac{n_1 n_2(n_1 + n_2 + 1)}{12}}$ (18.7)

Kruskal-Wallis test statistic

$$W = \left[\frac{12}{n_T(n_T + 1)} \sum_{i=1}^{k} \frac{R_i^2}{n_i} \right] - 3(n_T + 1)$$ (18.8)

Spearman rank-correlation coefficient

$$r_S = 1 - \frac{6\Sigma d_i^2}{n(n^2 - 1)}$$ (18.9)

Sampling distribution of r_S in test for significant rank correlation

Mean: $\mu_{r_S} = 0$ (18.10)

Standard deviation: $\sigma_{r_S} = \sqrt{\dfrac{1}{n - 1}}$ (18.11)

CASE PROBLEM

Brand origin recognition accuracy and consumer cosmopolitanism

At the end of Chapter 10 we presented a case problem about consumers' awareness of the country of origin of the products and brands they buy. Marketing researchers refer to this as brand origin recognition accuracy (BORA). Typically, consumers' BORA is far from perfect.

The file 'COSMO' on the online platform contains results from a survey of residents in England, who were asked to identify the country of origin of 16 well-known brands. These included Tesco, Aldi, Dyson, Samsung

© tumsasedgars/iStock

and Superdry. Eight of the brands were of UK origin, and eight of non-UK origin. The data in the file are for 150 survey respondents. They comprise demographic characteristics of the respondents, as well as figures for the number of UK brands correctly identified as being of UK origin, the number of non-UK

(Continued)

brands correctly identified as being of non-UK origin, and the number of non-UK brands for which the specific country of origin was correctly identified. In the data file these latter three variables are labelled respectively as BORA_UK, BORA_non_UK and BORA_Country. The file also contains values for a consumer 'cosmopolitanism' score. This is derived from responses to a five-item scale including items such as 'I enjoy exchanging ideas with people from other cultures and countries'. The first few lines of data have been reproduced below.

	Gender	Age	Education	Cosmopolitanism	BORA_UK	BORA_non_UK	BORA_Country
1	Female	35–44 years old	Master's degree or...	7	3	6	3
2	Female	25–34 years old	Bachelor's degree	7	8	7	5
3	Female	35–44 years old	Master's degree or...	6	7	8	5
4	Female	25–34 years old	Bachelor's degree	7	5	9	2
5	Male	25–34 years old	Bachelor's degree	7	7	8	3
6	Female	18–24 years old	Bachelor's degree	5	4	9	5
7	Female	45–54 years old	GCSE	3	6	8	4
8	Female	25–34 years old	Bachelor's degree	4	7	8	6
9	Female	55–64 years old	Bachelor's degree	7	8	9	8
10	Female	25–34 years old	Bachelor's degree	6	6	8	8

COSMO

Analyst's report

Prepare a marketing background report that addresses the following.

1. Summarize the distribution of the 'BORA_UK', 'BORA_non_UK', 'BORA_Country' and 'Cosmopolitanism' figures.
2. Is there evidence that, in the population from which the respondents were selected, the average level of recognition of the UK brands was different from that for the non-UK brands? Use a non-parametric method of testing.
3. Is there evidence that, in the population from which the respondents were selected, there were differences in average brand origin recognition accuracy between men and women, or between older people and younger people? Use non-parametric methods of testing.
4. Is there evidence of a correlation between 'Cosmopolitanism' on the one hand, and 'BORA_UK' and 'BORA_non_UK' on the other hand? Use a non-parametric method of testing.

APPENDIX A
References and Bibliography

GENERAL

Azevedo, V. and Robles. M. (2013) 'Multidimensional targeting: Identifying beneficiaries of conditional cash transfer programs'. *Social Indicators Research* 112(2): 447–475.

Barlow, J. F. (2005) *Excel Models for Business & Operations Management*, 2nd ed. John Wiley & Sons.

Crawley, M. J. (2012) *The R Book*, 2nd ed. Wiley-Blackwell.

Crawley, M. J. (2014) *Statistics: an Introduction Using R*, 2nd ed. Wiley-Blackwell.

Davies, T. M. (2016) *The Book of R*. No Starch Press.

Feng, W., Wang, T. and Rui, G. (2019) 'Influence of number magnitude in luxury brand names on consumer preference'. *Social Behaviour and Personality*, 47(5). Available at doi.org.10.2224/sbp.7486 (accessed September 2022).

Field, A. (2017) *Discovering Statistics Using IBM SPSS Statistics*, 5th ed. Sage.

Freedman, D., Pisani, R. and Purves, R. (2013) *Statistics*, 4th revised ed. W. W. Norton.

Green, S. B. and Salkind, N. J. (2021) *Using SPSS for Windows and Macintosh: Analyzing and Understanding Data*, 8th ed. Pearson.

Hand, D. J. (ed.), Lunn, D., Ostrowski, E., Daly, F. and McConway, K. (1994) *A Handbook of Small Data Sets*. Chapman and Hall.

Hare, C. T. and Bradow, R. L. (1977) Light duty diesel emissions correction factors for ambient conditions. SAE Paper 770717.

Hogg, R. V., McKean, J. and Craig, A. T. (2019) *Introduction to Mathematical Statistics: Pearson New International Edition*, 8th ed. Pearson.

Joiner, B., Cryer, J. and Ryan, B. F. (2012) *Minitab Handbook: Update for Release 16*, 6th ed. Wadsworth.

Kabacoff, R. L. (2022) *R in Action*: Data analysis and graphics with R and Tidyverse, 3rd ed. Manning.

Kinnear, P. R. and Gray, C. D. (2011) *SPSS 19 Made Simple*. Psychology Press.

MacPherson, G. (2010) *Applying and Interpreting Statistics: A Comprehensive Guide*, 2nd ed. Springer.

Mazess, R. B., Peppler, W. W. and Gibbons, M. (1984) 'Total body composition by dual-photon (^{153}Gd) absorptiometry'. *American Journal of Clinical Nutrition* 40: 834–839.

Meng, X.-L. (2018) 'Statistical paradises and paradoxes in big data (i): Law of large populations, big data paradox, and the 2016 US presidential election'. *The Annals of Applied Statistics* 12(2): 685–726.

Miller, I. and Miller, M. (2014) *John E. Freund's Mathematical Statistics*. 7th ed. Prentice Hall.

Moore, D. S., McCabe, G. P. and Craig, B. (2017) *Introduction to the Practice of Statistics*, 9th ed. Freeman.

Nappo, N. (2019) 'Is there an association between working conditions and health? An analysis of the Sixth European Working Conditions Survey data'. *PLoS ONE* 14(2): e0211294. doi.org/10.1371/journal.pone.0211294.

OECD (1982) *Forecasting car ownership and use: A report by the Road Research Group*. OECD.

Rossman, A. J. (1994) 'Televisions, physicians and life expectancy'. *Journal of Statistics Education* 2(2).

Royal Statistical Society (RSS) and the Institute and Faculty of Actuaries (IFoA) (2019) *A Guide for Ethical Data Science*. RSS and IFoA.

Spiegelhalter, D. (2020) *The Art of Statistics: Learning from Data*. Penguin.

Szumilas, M. (2010) 'Explaining odds ratios'. *J. Can. Acad. Child Adolesc. Psychiatry* 19(3): 227–229.

Tanur, J. M. (2002) *Statistics: A Guide to the Unknown*, 4th ed. Brooks/Cole.

Tomkinson, B. and Freeman, J. (2011) 'Problems of assessment'. International Conference on Engineering Education (ICEE-2011) Belfast, Northern Ireland. 21–26 August 2011.

Tukey, J. W. (1977) *Exploratory Data Analysis*. Addison-Wesley.

UK Statistics Authority, National Statistician's Data Ethics Advisory Committee (2022) *Guidelines on Using the Ethics Self-Assessment Process*, 30 March.

Wisniewski, M. (2016) *Quantitative Methods for Decision-makers*, 6th ed. Financial Times/Prentice Hall.

EXPERIMENTAL DESIGN

Cochran, W. G. and Cox, G. M. (1992) *Experimental Designs*, 2nd ed. Wiley.

Hicks, C. R. and Turner, K. V. (1999) *Fundamental Concepts in the Design of Experiments*, 5th ed. Oxford University Press.

Montgomery, D. C. (2017) *Design and Analysis of Experiments*, 9th ed. Wiley.

Wu, C. F. J. and Hamada, M. (2009) *Experiments: Planning, Analysis and Parameter Optimization*, 2nd ed. Wiley.

TIME SERIES AND FORECASTING

Bowerman, B. L. and O'Connell, R. T. (2000) *Forecasting and Time Series: An Applied Approach*, 3rd ed. Brooks/Cole.

Box, G. E. P., Jenkins, G. and Reinsel, G. C. (1994) *Time Series Analysis: Forecasting and Control*, 3rd ed. Prentice Hall.

Makridakis, S., Wheelwright, S. C. and Hyndman, R. J. (1998) *Forecasting: Methods and Applications*, 3rd ed. Wiley.

INDEX NUMBERS

Allen, R. G. D. (2014) *Index Numbers in Theory and Practice*. (1975 1st edition) Palgrave MacMillan.

ONS (2014) *Consumer Price Indices Technical Manual*.

Richardson, I. (1999) *Producer Price Indices: Principles and Procedures*. UK Office for National Statistics.

NON-PARAMETRIC METHODS

Conover, W. J. (1999) *Practical Nonparametric Statistics*, 3rd ed. John Wiley & Sons.

Gibbons, J. D. and Chakraborti, S. (2010) *Nonparametric Statistical Inference*, 5th ed. Chapman & Hall/CRC.

Nussbaum, E. M. (2015) *Categorical and Nonparametric Data Analysis*. Routledge.

Siegel, S. and Castellan, N. J. (1988) *Nonparametric Statistics for the Behavioral Sciences*, 2nd ed. McGraw-Hill.

Sprent, P. and Smeeton, N. C. (2007) *Applied Nonparametric Statistical Methods*, 4th ed. Chapman & Hall/CRC.

PROBABILITY

Hogg, R. V., Tanis, E. A. and Zimmerman, D. (2014) *Probability and Statistical Inference*, 9th (global) ed. Pearson.

Ross, S. M. (2014) *Introduction to Probability Models*, 11th ed. Academic Press.

Wackerly, D. D., Mendenhall, W. and Scheaffer, R. L. (2007) *Mathematical Statistics with Applications*, 7th ed. Duxbury Press.

QUALITY CONTROL

Deming, W. E. (1982) *Quality, Productivity and Competitive Position*. MIT.

Evans, J. R. and Lindsay, W. M. (2010) *The Management and Control of Quality*, 8th ed. South-Western.

Gryna, F. M. and Juran, I. M. (1993) *Quality Planning and Analysis: From Product Development Through Use*, 3rd ed. McGraw-Hill.

Ishikawa, K. (1991) *Introduction to Quality Control*. Kluwer Academic.

Montgomery, D. C. (2013) *Introduction to Statistical Quality Control*, 7th ed. Wiley.

REGRESSION ANALYSIS

Belsley, D. A. (1991) *Conditioning Diagnostics: Collinearity and Weak Data in Regression*. Wiley.

Brownlee, K. A. (1965) *Statistical Theory and Methodology in Science and Engineering*, 2nd ed. Wiley.

Chatterjee, S., Hadi, A. S. and Price, B. (2012) *Regression Analysis by Example*, 5th ed. Wiley.

Draper, N. R. and Smith, H. (1998) *Applied Regression Analysis*, 3rd ed. Wiley.

Hosmer, D. W. and Lemeshow, S. (2013) *Applied Logistic Regression*, 3rd ed. Wiley.

Kadiyala, K. R. (1970) 'Testing for the independence of regression disturbances'. *Econometrics* 38(1): 97–117.

Lunn, D., Hand, D. J., Ostrowski, E., Daly, F. and McConway, K. A. (1994) *Handbook of Small Data Sets*. Chapman and Hall.

Mendenhall, M. and Sincich, T. (2011) *A Second Course in Statistics: Regression Analysis*, 7th ed. Prentice Hall.

DECISION ANALYSIS

Clemen, R. T. and Reilly, T. (2001) *Making Hard Decisions with Decision Tools*. Duxbury Press.

Goodwin, P. and Wright, G. (2004) *Decision Analysis for Management Judgment*, 3rd ed. Wiley.

Skinner, D. C. (2009) *Introduction to Decision Analysis*, 3rd ed. Wiley.

SAMPLING

Cochran, W. G. (1977) *Sampling Techniques*, 3rd ed. Wiley.

Deming, W. E. (1984) *Some Theory of Sampling*. Dover.

Hansen, M. H., Hurwitz, W. N., Madow, W. G. and Hanson, M. N. (1993) *Sample Survey Methods and Theory*. Wiley.

Kish, L. (1995) *Survey Sampling*. Wiley.

Levy, P. S. and Lemeshow, S. (2008) *Sampling of Populations: Methods and Applications*, 4th ed. Wiley-Blackwell.

Scheaffer, R. L., Mendenhall, W., Ott, R. L. and Gerow, K. G. (2011) *Elementary Survey Sampling*, 7th ed. Duxbury.

DATA VISUALIZATION

Cleveland, W. S. (1993) *Visualizing Data*. Hobart Press.

Cleveland, W. S. (1994) *The Elements of Graphing Data*, 2nd ed. Hobart Press.

Few, S. (2009) *Now You See It: Simple Visualization Techniques for Quantitative Analysis*. Analytics Press.

Few, S. (2012) *Show Me the Numbers: Designing Tables and Graphs to Enlighten*, 2nd ed. Analytics Press.

Few, S. (2012) *Information Dashboard Design: The Effective Visual Communication of Data*, 2nd ed. O'Reilly Media.

Fry, B. (2008) *Visualizing Data: Exploring and Explaining Data with the Processing Environment*. O'Reilly Media.

Robbins, N. B. (2013) *Creating More Effective Graphs*. Chart House.

Telea, A. C. (2008) *Data Visualization Principles and Practice*. A. K. Peters Ltd.

Tufte, E. R. (1990) *Envisioning Information*. Graphics Press.

Tufte, E. R. (1990) *The Visual Display of Quantitative Information*, 2nd ed. Graphics Press.

Tufte, E. R. (1997) *Visual Explanations: Images and Quantities, Evidence and Narrative*. Graphics Press.

Tufte, E. R. (1997) *Visual and Statistical Thinking: Displays of Evidence for Making Decisions*. Graphics Press.

Tufte, E. R. (2006) *Beautiful Evidence*. Graphics Press.

Wong, D. M. (2010) *The Wall Street Journal Guide to Information Graphics*. W. W. Norton & Company.

Young, F. W., Valero-Mora, P. M. and Friendly M. (2006) *Visual Statistics: Seeing Data with Dynamic Interactive Graphics*. Wiley.

BUSINESS ANALYTICS

Camm, J. D., Cochran, J. J., Fry, M. J., Ohlmann, J. W., Anderson, D. R., Sweeney, D. J. and Williams, T. A. (2022) *Business Analytics*, 3rd ed. Cengage Learning.

APPENDIX B
Tables

TABLE 1 Cumulative probabilities for the standard normal distribution

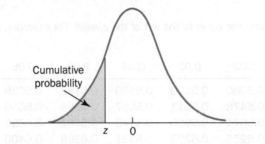

Cumulative
probability

Entries in the table give the area under the curve to the left of the z value. For example, for $z = -0.85$, the cumulative probability is 0.1977.

z	0.00	0.01	0.02	0.03	0.04	0.05	0.06	0.07	0.08	0.09
−3.0	0.0013	0.0013	0.0013	0.0012	0.0012	0.0011	0.0011	0.0011	0.0010	0.0010
−2.9	0.0019	0.0018	0.0018	0.0017	0.0016	0.0016	0.0015	0.0015	0.0014	0.0014
−2.8	0.0026	0.0025	0.0024	0.0023	0.0023	0.0022	0.0021	0.0021	0.0020	0.0019
−2.7	0.0035	0.0034	0.0033	0.0032	0.0031	0.0030	0.0029	0.0028	0.0027	0.0026
−2.6	0.0047	0.0045	0.0044	0.0043	0.0041	0.0040	0.0039	0.0038	0.0037	0.0036
−2.5	0.0062	0.0060	0.0059	0.0057	0.0055	0.0054	0.0052	0.0051	0.0049	0.0048
−2.4	0.0082	0.0080	0.0078	0.0075	0.0073	0.0071	0.0069	0.0068	0.0066	0.0064
−2.3	0.0107	0.0104	0.0102	0.0099	0.0096	0.0094	0.0091	0.0089	0.0087	0.0084
−2.2	0.0139	0.0136	0.0132	0.0129	0.0125	0.0122	0.0119	0.0116	0.0113	0.0110
−2.1	0.0179	0.0174	0.0170	0.0166	0.0162	0.0158	0.0154	0.0150	0.0146	0.0143
−2.0	0.0228	0.0222	0.0217	0.0212	0.0207	0.0202	0.0197	0.0192	0.0188	0.0183
−1.9	0.0287	0.0281	0.0274	0.0268	0.0262	0.0256	0.0250	0.0244	0.0239	0.0233
−1.8	0.0359	0.0351	0.0344	0.0336	0.0329	0.0322	0.0314	0.0307	0.0301	0.0294
−1.7	0.0446	0.0436	0.0427	0.0418	0.0409	0.0401	0.0392	0.0384	0.0375	0.0367
−1.6	0.0548	0.0537	0.0526	0.0516	0.0505	0.0495	0.0485	0.0475	0.0465	0.0455
−1.5	0.0668	0.0655	0.0643	0.0630	0.0618	0.0606	0.0594	0.0582	0.0571	0.0559
−1.4	0.0808	0.0793	0.0778	0.0764	0.0749	0.0735	0.0721	0.0708	0.0694	0.0681
−1.3	0.0968	0.0951	0.0934	0.0918	0.0901	0.0885	0.0869	0.0853	0.0838	0.0823
−1.2	0.1151	0.1131	0.1112	0.1093	0.1075	0.1056	0.1038	0.1020	0.1003	0.0985
−1.1	0.1357	0.1335	0.1314	0.1292	0.1271	0.1251	0.1230	0.1210	0.1190	0.1170
−1.0	0.1587	0.1562	0.1539	0.1515	0.1492	0.1469	0.1446	0.1423	0.1401	0.1379
−0.9	0.1841	0.1814	0.1788	0.1762	0.1736	0.1711	0.1685	0.1660	0.1635	0.1611
−0.8	0.2119	0.2090	0.2061	0.2033	0.2005	0.1977	0.1949	0.1922	0.1894	0.1867
−0.7	0.2420	0.2389	0.2358	0.2327	0.2296	0.2266	0.2236	0.2206	0.2177	0.2148
−0.6	0.2743	0.2709	0.2676	0.2643	0.2611	0.2578	0.2546	0.2514	0.2483	0.2451
−0.5	0.3085	0.3050	0.3015	0.2981	0.2946	0.2912	0.2877	0.2843	0.2810	0.2776
−0.4	0.3446	0.3409	0.3372	0.3336	0.3300	0.3264	0.3228	0.3192	0.3156	0.3121
−0.3	0.3821	0.3783	0.3745	0.3707	0.3669	0.3632	0.3594	0.3557	0.3520	0.3483
−0.2	0.4207	0.4168	0.4129	0.4090	0.4052	0.4013	0.3974	0.3936	0.3897	0.3859
−0.1	0.4602	0.4562	0.4522	0.4483	0.4443	0.4404	0.4364	0.4325	0.4286	0.4247
−0.0	0.5000	0.4960	0.4920	0.4880	0.4840	0.4801	0.4761	0.4721	0.4681	0.4641

TABLE 1 (Continued)

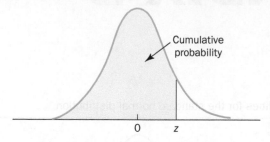

Cumulative probability

Entries in the table give the area under the curve to the left of the z value. For example, for z = 1.25, the cumulative probability is 0.8944.

z	0.00	0.01	0.02	0.03	0.04	0.05	0.06	0.07	0.08	0.09
0.0	0.5000	0.5040	0.5080	0.5120	0.5160	0.5199	0.5239	0.5279	0.5319	0.5359
0.1	0.5398	0.5438	0.5478	0.5517	0.5557	0.5596	0.5636	0.5675	0.5714	0.5753
0.2	0.5793	0.5832	0.5871	0.5910	0.5948	0.5987	0.6026	0.6064	0.6103	0.6141
0.3	0.6179	0.6217	0.6255	0.6293	0.6331	0.6368	0.6406	0.6443	0.6480	0.6517
0.4	0.6554	0.6591	0.6628	0.6664	0.6700	0.6736	0.6772	0.6808	0.6844	0.6879
0.5	0.6915	0.6950	0.6985	0.7019	0.7054	0.7088	0.7123	0.7157	0.7190	0.7224
0.6	0.7257	0.7291	0.7324	0.7357	0.7389	0.7422	0.7454	0.7486	0.7517	0.7549
0.7	0.7580	0.7611	0.7642	0.7673	0.7704	0.7734	0.7764	0.7794	0.7823	0.7852
0.8	0.7881	0.7910	0.7939	0.7967	0.7995	0.8023	0.8051	0.8078	0.8106	0.8133
0.9	0.8159	0.8186	0.8212	0.8238	0.8264	0.8289	0.8315	0.8340	0.8365	0.8389
1.0	0.8413	0.8438	0.8461	0.8485	0.8508	0.8531	0.8554	0.8577	0.8599	0.8621
1.1	0.8643	0.8665	0.8686	0.8708	0.8729	0.8749	0.8770	0.8790	0.8810	0.8830
1.2	0.8849	0.8869	0.8888	0.8907	0.8925	0.8944	0.8962	0.8980	0.8997	0.9015
1.3	0.9032	0.9049	0.9066	0.9082	0.9099	0.9115	0.9131	0.9147	0.9162	0.9177
1.4	0.9192	0.9207	0.9222	0.9236	0.9251	0.9265	0.9279	0.9292	0.9306	0.9319
1.5	0.9332	0.9345	0.9357	0.9370	0.9382	0.9394	0.9406	0.9418	0.9429	0.9441
1.6	0.9452	0.9463	0.9474	0.9484	0.9495	0.9505	0.9515	0.9525	0.9535	0.9545
1.7	0.9554	0.9564	0.9573	0.9582	0.9591	0.9599	0.9608	0.9616	0.9625	0.9633
1.8	0.9641	0.9649	0.9656	0.9664	0.9671	0.9678	0.9686	0.9693	0.9699	0.9706
1.9	0.9713	0.9719	0.9726	0.9732	0.9738	0.9744	0.9750	0.9756	0.9761	0.9767
2.0	0.9772	0.9778	0.9783	0.9788	0.9793	0.9798	0.9803	0.9808	0.9812	0.9817
2.1	0.9821	0.9826	0.9830	0.9834	0.9838	0.9842	0.9846	0.9850	0.9854	0.9857
2.2	0.9861	0.9864	0.9868	0.9871	0.9875	0.9878	0.9881	0.9884	0.9887	0.9890
2.3	0.9893	0.9896	0.9898	0.9901	0.9904	0.9906	0.9909	0.9911	0.9913	0.9913
2.4	0.9918	0.9920	0.9922	0.9925	0.9927	0.9929	0.9931	0.9932	0.9934	0.9936
2.5	0.9938	0.9940	0.9941	0.9943	0.9945	0.9946	0.9948	0.9949	0.9951	0.9952
2.6	0.9953	0.9955	0.9956	0.9957	0.9959	0.9960	0.9961	0.9962	0.9963	0.9964
2.7	0.9965	0.9966	0.9967	0.9968	0.9969	0.9970	0.9971	0.9972	0.9973	0.9974
2.8	0.9974	0.9975	0.9976	0.9977	0.9977	0.9978	0.9979	0.9979	0.9980	0.9981
2.9	0.9981	0.9982	0.9982	0.9983	0.9984	0.9984	0.9985	0.9985	0.9986	0.9986
3.0	0.9986	0.9987	0.9987	0.9988	0.9988	0.9989	0.9989	0.9989	0.9990	0.9990

TABLE 2 *t* Distribution

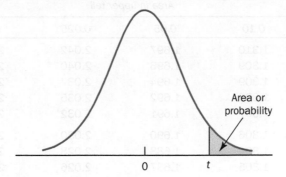

Entries in the table give *t*-values for an area or probability in the upper tail of the *t* distribution. For example, with ten degrees of freedom and 0.05 area in the upper tail, $t_{0.05} = 1.812$.

Degrees of freedom	Area in upper tail					
	0.20	0.10	0.05	0.025	0.01	0.005
1	1.376	3.078	6.314	12.706	31.821	63.656
2	1.061	1.886	2.920	4.303	6.965	9.925
3	0.978	1.638	2.353	3.182	4.541	5.841
4	0.941	1.533	2.132	2.776	3.747	4.604
5	0.920	1.476	2.015	2.571	3.365	4.032
6	0.906	1.440	1.943	2.447	3.143	3.707
7	0.896	1.415	1.895	2.365	2.998	3.499
8	0.889	1.397	1.860	2.306	2.896	3.355
9	0.883	1.383	1.833	2.262	2.821	3.250
10	0.879	1.372	1.812	2.228	2.764	3.169
11	0.876	1.363	1.796	2.201	2.718	3.106
12	0.873	1.356	1.782	2.179	2.681	3.055
13	0.870	1.350	1.771	2.160	2.650	3.012
14	0.868	1.345	1.761	2.145	2.624	2.977
15	0.866	1.341	1.753	2.131	2.602	2.947
16	0.865	1.337	1.746	2.120	2.583	2.921
17	0.863	1.333	1.740	2.110	2.567	2.898
18	0.862	1.330	1.734	2.101	2.552	2.878
19	0.861	1.328	1.729	2.093	2.539	2.861
20	0.860	1.325	1.725	2.086	2.528	2.845
21	0.859	1.323	1.721	2.080	2.518	2.831
22	0.858	1.321	1.717	2.074	2.508	2.819
23	0.858	1.319	1.714	2.069	2.500	2.807
24	0.857	1.318	1.711	2.064	2.492	2.797
25	0.856	1.316	1.708	2.060	2.485	2.787
26	0.856	1.315	1.706	2.056	2.479	2.779
27	0.855	1.314	1.703	2.052	2.473	2.771
28	0.855	1.313	1.701	2.048	2.467	2.703
29	0.854	1.311	1.699	2.045	2.462	2.756

(Continued)

TABLE 2   *(Continued)*

Degrees of freedom	Area in upper tail					
	0.20	0.10	0.05	0.025	0.01	0.005
30	0.854	1.310	1.697	2.042	2.457	2.750
31	0.853	1.309	1.696	2.040	2.453	2.744
32	0.853	1.309	1.694	2.037	2.449	2.738
33	0.853	1.308	1.692	2.035	2.445	2.733
34	0.852	1.307	1.691	2.032	2.441	2.728
35	0.852	1.306	1.690	2.030	2.438	2.724
36	0.852	1.306	1.688	2.028	2.434	2.719
37	0.851	1.305	1.687	2.026	2.431	2.715
38	0.851	1.304	1.686	2.024	2.429	2.712
39	0.851	1.304	1.685	2.023	2.426	2.708
40	0.851	1.303	1.684	2.021	2.423	2.704
41	0.850	1.303	1.683	2.020	2.421	2.701
42	0.850	1.302	1.682	2.018	2.418	2.698
43	0.850	1.302	1.681	2.017	2.416	2.695
44	0.850	1.301	1.680	2.015	2.414	2.692
45	0.850	1.301	1.679	2.014	2.412	2.690
46	0.850	1.300	1.679	2.013	2.410	2.687
47	0.849	1.300	1.678	2.012	2.408	2.685
48	0.849	1.299	1.677	2.011	2.407	2.682
49	0.849	1.299	1.677	2.010	2.405	2.680
50	0.849	1.299	1.676	2.009	2.403	2.678
51	0.849	1.298	1.675	2.008	2.402	2.676
52	0.849	1.298	1.675	2.007	2.400	2.674
53	0.848	1.298	1.674	2.006	2.399	2.672
54	0.848	1.297	1.674	2.005	2.397	2.670
55	0.848	1.297	1.673	2.004	2.396	2.668
56	0.848	1.297	1.673	2.003	2.395	2.667
57	0.848	1.297	1.672	2.002	2.394	2.665
58	0.848	1.296	1.672	2.002	2.392	2.663
59	0.848	1.296	1.671	2.001	2.391	2.662
60	0.848	1.296	1.671	2.000	2.390	2.660
61	0.848	1.296	1.670	2.000	2.389	2.659
62	0.847	1.295	1.670	1.999	2.388	2.657
63	0.847	1.295	1.669	1.998	2.387	2.656
64	0.847	1.295	1.669	1.998	2.386	2.655
65	0.847	1.295	1.669	1.997	2.385	2.654
66	0.847	1.295	1.668	1.997	2.384	2.652
67	0.847	1.294	1.668	1.996	2.383	2.651
68	0.847	1.294	1.668	1.995	2.382	2.650
69	0.847	1.294	1.667	1.995	2.382	2.649

TABLE 2   *(Continued)*

Degrees of freedom	Area in upper tail					
	0.20	0.10	0.05	0.025	0.01	0.005
70	0.847	1.294	1.667	1.994	2.381	2.648
71	0.847	1.294	1.667	1.994	2.380	2.647
72	0.847	1.293	1.666	1.993	2.379	2.646
73	0.847	1.293	1.666	1.993	2.379	2.645
74	0.847	1.293	1.666	1.993	2.378	2.644
75	0.846	1.293	1.665	1.992	2.377	2.643
76	0.846	1.293	1.665	1.992	2.376	2.642
77	0.846	1.293	1.665	1.991	2.376	2.641
78	0.846	1.292	1.665	1.991	2.375	2.640
79	0.846	1.292	1.664	1.990	2.374	2.639
80	0.846	1.292	1.664	1.990	2.374	2.639
81	0.846	1.292	1.664	1.990	2.373	2.638
82	0.846	1.292	1.664	1.989	2.373	2.637
83	0.846	1.292	1.663	1.989	2.372	2.636
84	0.846	1.292	1.663	1.989	2.372	2.636
85	0.846	1.292	1.663	1.988	2.371	2.635
86	0.846	1.291	1.663	1.988	2.370	2.634
87	0.846	1.291	1.663	1.988	2.370	2.634
88	0.846	1.291	1.662	1.987	2.369	2.633
89	0.846	1.291	1.662	1.987	2.369	2.632
90	0.846	1.291	1.662	1.987	2.368	2.632
91	0.846	1.291	1.662	1.986	2.368	2.631
92	0.846	1.291	1.662	1.986	2.368	2.630
93	0.846	1.291	1.661	1.986	2.367	2.630
94	0.845	1.291	1.661	1.986	2.367	2.629
95	0.845	1.291	1.661	1.985	2.366	2.629
96	0.845	1.290	1.661	1.985	2.366	2.628
97	0.845	1.290	1.661	1.985	2.365	2.627
98	0.845	1.290	1.661	1.984	2.365	2.627
99	0.845	1.290	1.660	1.984	2.364	2.626
100	0.845	1.290	1.660	1.984	2.364	2.626
∞	0.842	1.282	1.645	1.960	2.326	2.576

TABLE 3 Chi-squared distribution

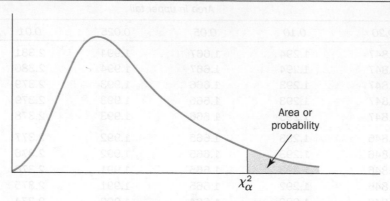

Area or probability

χ_α^2

Entries in the table give χ_α^2 values, where α is the area or probability in the upper tail of the chi-squared distribution. For example, with ten degrees of freedom and 0.01 area in the upper tail, $\chi_{0.01}^2 = 23.209$

Degrees of freedom	Area in upper tail									
	0.995	0.99	0.975	0.95	0.90	0.10	0.05	0.025	0.01	0.005
1	0.000	0.000	0.001	0.004	0.016	2.706	3.841	5.024	6.635	7.879
2	0.010	0.020	0.051	0.103	0.211	4.605	5.991	7.378	9.210	10.597
3	0.072	0.115	0.216	0.352	0.584	6.251	7.815	9.348	11.345	12.838
4	0.207	0.297	0.484	0.711	1.064	7.779	9.488	11.143	13.277	14.860
5	0.412	0.554	0.831	1.145	1.610	9.236	11.070	12.832	15.086	16.750
6	0.676	0.872	1.237	1.635	2.204	10.645	12.592	14.449	16.812	18.548
7	0.989	1.239	1.690	2.167	2.833	12.017	14.067	16.013	18.475	20.278
8	1.344	1.647	2.180	2.733	3.490	13.362	15.507	17.535	20.090	21.955
9	1.735	2.088	2.700	3.325	4.168	14.684	16.919	19.023	21.666	23.589
10	2.156	2.558	3.247	3.940	4.865	15.987	18.307	20.483	23.209	25.188
11	2.603	3.053	3.816	4.575	5.578	17.275	19.675	21.920	24.725	26.757
12	3.074	3.571	4.404	5.226	6.304	18.549	21.026	23.337	26.217	28.300
13	3.565	4.107	5.009	5.892	7.041	19.812	22.362	24.736	27.688	29.819
14	4.075	4.660	5.629	6.571	7.790	21.064	23.685	26.119	29.141	31.319
15	4.601	5.229	6.262	7.261	8.547	22.307	24.996	27.488	30.578	32.801
16	5.142	5.812	6.908	7.962	9.312	23.542	26.296	28.845	32.000	34.267
17	5.697	6.408	7.564	8.672	10.085	24.769	27.587	30.191	33.409	35.718
18	6.265	7.015	8.231	9.390	10.865	25.989	28.869	31.526	34.805	37.156
19	6.844	7.633	8.907	10.117	11.651	27.204	30.144	32.852	36.191	38.582
20	7.434	8.260	9.591	10.851	12.443	28.412	31.410	34.170	37.566	39.997
21	8.034	8.897	10.283	11.591	13.240	29.615	32.671	35.479	38.932	41.401
22	8.643	9.542	10.982	12.338	14.041	30.813	33.924	36.781	40.289	42.796
23	9.260	10.196	11.689	13.091	14.848	32.007	35.172	38.076	41.638	44.181
24	9.886	10.856	12.401	13.848	15.659	33.196	36.415	39.364	42.980	45.558
25	10.520	11.524	13.120	14.611	16.473	34.382	37.652	40.646	44.314	46.928
26	11.160	12.198	13.844	15.379	17.292	35.563	38.885	41.923	45.642	48.290
27	11.808	12.878	14.573	16.151	18.114	36.741	40.113	43.195	46.963	49.645
28	12.461	13.565	15.308	16.928	18.939	37.916	41.337	44.461	48.278	50.994
29	13.121	14.256	16.047	17.708	19.768	39.087	42.557	45.722	49.588	52.335

TABLE 3 *(Continued)*

Degrees of freedom	Area in upper tail									
	0.995	*0.99*	*0.975*	*0.95*	*0.90*	*0.10*	*0.05*	*0.025*	*0.01*	*0.005*
30	13.787	14.953	16.791	18.493	20.599	40.256	43.773	46.979	50.892	53.672
35	17.192	18.509	20.569	22.465	24.797	46.059	49.802	53.203	57.342	60.275
40	20.707	22.164	24.433	26.509	29.051	51.805	55.758	59.342	63.691	66.766
45	24.311	25.901	28.366	30.612	33.350	57.505	61.656	65.410	69.957	73.166
50	27.991	29.707	32.357	34.764	37.689	63.167	67.505	71.420	76.154	79.490
55	31.735	33.571	36.398	38.958	42.060	68.796	73.311	77.380	82.292	85.749
60	35.534	37.485	40.482	43.188	46.459	74.397	79.082	83.298	88.379	91.952
65	39.383	41.444	44.603	47.450	50.883	79.973	84.821	89.177	94.422	98.105
70	43.275	45.442	48.758	51.739	55.329	85.527	90.531	95.023	100.425	104.215
75	47.206	49.475	52.942	56.054	59.795	91.061	96.217	100.839	106.393	110.285
80	51.172	53.540	57.153	60.391	64.278	96.578	101.879	106.629	112.329	116.321
85	55.170	57.634	61.389	64.749	68.777	102.079	107.522	112.393	118.236	122.324
90	59.196	61.754	65.647	69.126	73.291	107.565	113.145	118.136	124.116	128.299
95	63.250	65.898	69.925	73.520	77.818	113.038	118.752	123.858	129.973	134.247
100	67.328	70.065	74.222	77.929	82.358	118.498	124.342	129.561	135.807	140.170

TABLE 4 F distribution

Entries in the table give F_α values, where α is the area or probability in the upper tail of the F distribution. For example, with four numerator degrees of freedom, eight denominator degrees of freedom, and 0.05 area in the upper tail, $F_{0.05} = 3.84$.

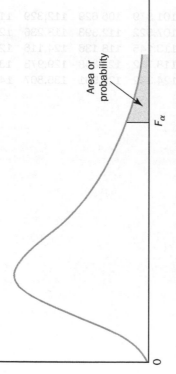

Denominator degrees of freedom	Area in upper tail	Numerator degrees of freedom																	
		1	2	3	4	5	6	7	8	9	10	15	20	25	30	40	60	100	1000
1	0.10	39.86	49.50	53.59	55.83	57.24	58.20	58.91	59.44	59.86	60.19	61.22	61.74	62.05	62.26	62.53	62.79	63.01	63.30
	0.05	161.45	199.50	215.71	224.58	230.16	233.99	236.77	238.88	240.54	241.88	245.95	248.02	249.26	250.10	251.14	252.20	253.04	254.19
	0.025	647.79	799.48	864.15	899.60	921.83	937.11	948.20	956.64	963.28	968.63	984.87	993.08	998.09	1001.40	1005.60	1009.79	1013.16	1017.76
	0.01	4052.18	4999.34	5403.53	5624.26	5763.96	5858.95	5928.33	5980.95	6022.40	6055.93	6156.97	6208.66	6239.86	6260.35	6286.43	6312.97	6333.92	6362.80
2	0.10	8.53	9.00	9.16	9.24	9.29	9.33	9.35	9.37	9.38	9.39	9.42	9.44	9.45	9.46	9.47	9.47	9.48	9.49
	0.05	18.51	19.00	19.16	19.25	19.30	19.33	19.35	19.37	19.38	19.40	19.43	19.45	19.46	19.46	19.47	19.48	19.49	19.49
	0.025	38.51	39.00	39.17	39.25	39.30	39.33	39.36	39.37	39.39	39.40	39.43	39.45	39.46	39.46	39.47	39.48	39.49	39.50
	0.01	98.50	99.00	99.16	99.25	99.30	99.33	99.36	99.38	99.39	99.40	99.43	99.45	99.46	99.47	99.48	99.48	99.49	99.50
3	0.10	5.54	5.46	5.39	5.34	5.31	5.28	5.27	5.25	5.24	5.23	5.20	5.18	5.17	5.17	5.16	5.15	5.14	5.13
	0.05	10.13	9.55	9.28	9.12	9.01	8.94	8.89	8.85	8.81	8.79	8.70	8.66	8.63	8.62	8.59	8.57	8.55	8.53
	0.025	17.44	16.04	15.44	15.10	14.88	14.73	14.62	14.54	1447	14.42	14.25	14.17	14.12	14.08	14.04	13.99	13.96	13.91
	0.01	34.12	30.82	29.46	28.71	28.24	27.91	27.67	27.49	27.34	27.23	26.87	26.69	26.58	26.50	26.41	26.32	26.24	26.14
4	0.10	4.54	4.32	4.19	4.11	4.05	4.01	3.98	3.95	3.94	3.92	3.87	3.84	3.83	3.82	3.80	3.79	3.78	3.76
	0.05	7.71	6.94	6.59	6.39	6.26	6.16	6.09	6.04	6.00	5.96	5.86	5.80	5.77	5.75	5.72	5.69	5.66	5.63
	0.025	12.22	10.65	9.98	9.60	9.36	9.20	9.07	8.98	8.90	8.84	8.66	8.56	8.50	8.46	8.41	8.36	8.32	8.26
	0.01	21.20	18.00	16.69	15.98	15.52	15.21	14.98	14.80	14.66	14.55	14.20	14.02	13.91	13.84	13.75	13.65	13.58	13.47

TABLE 4 (Continued)

Denominator Area in

Denominator degrees of freedom	upper tail	Numerator degrees of freedom																	
		1	2	3	4	5	6	7	8	9	10	15	20	25	30	40	60	100	1000
5	0.10	4.06	3.78	3.62	3.52	3.45	3.40	3.37	3.34	3.32	3.30	3.324	3.21	3.19	3.17	3.16	3.14	3.13	3.11
	0.05	6.61	5.79	5.41	5.19	5.05	4.95	4.88	4.82	4.77	4.74	4.62	4.56	4.52	4.50	4.46	4.43	4.41	4.37
	0.025	10.01	8.43	7.76	7.39	7.15	6.98	6.85	6.76	6.68	6.62	6.43	6.33	6.27	6.23	6.18	6.12	6.08	6.02
	0.01	16.26	13.27	12.06	11.39	10.97	10.67	10.46	10.29	10.16	10.05	9.72	9.55	9.45	9.38	9.29	9.20	9.13	9.03
6	0.10	3.78	3.46	3.29	3.18	3.11	3.05	3.01	2.98	2.96	2.94	2.87	2.84	2.81	2.80	2.78	2.76	2.75	2.72
	0.05	5.99	5.14	4.76	4.53	4.39	4.28	4.21	4.15	4.10	4.06	3.94	3.87	3.83	3.81	3.77	3.74	3.71	3.67
	0.025	8.81	7.26	6.60	6.23	5.99	5.82	5.70	5.60	5.52	5.46	5.27	5.17	5.11	5.07	5.01	4.96	4.92	4.86
	0.01	13.75	10.92	9.78	9.15	8.75	8.47	8.26	8.10	7.98	7.87	7.56	7.40	7.30	7.23	7.14	7.06	6.99	6.89
7	0.10	3.59	3.26	3.07	2.96	2.88	2.83	2.78	2.75	2.72	2.70	2.63	2.59	2.57	2.56	2.54	2.51	2.50	2.47
	0.05	5.59	4.74	4.35	4.12	3.97	3.87	3.79	3.73	3.68	3.64	3.51	3.44	3.40	3.38	3.34	3.30	3.27	3.23
	0.025	8.07	6.54	5.89	5.52	5.29	5.12	4.99	4.90	4.82	4.76	4.57	4.47	4.40	4.36	4.31	4.25	4.21	4.15
	0.01	12.25	9.55	8.45	7.85	7.46	7.19	6.99	6.84	6.72	6.62	6.31	6.16	6.06	5.99	5.91	5.82	5.75	5.66
8	0.10	3.46	3.11	2.92	2.81	2.73	2.67	2.62	2.59	2.56	2.54	2.46	2.42	2.40	2.38	2.36	2.34	2.32	2.30
	0.05	5.32	4.46	4.07	3.84	3.69	3.58	3.50	3.44	3.39	3.35	3.22	3.15	3.11	3.08	3.04	3.01	2.97	2.93
	0.025	7.57	6.06	5.42	5.05	4.82	4.65	4.53	4.43	4.36	4.30	4.10	4.00	3.94	3.89	3.84	3.78	3.74	3.68
	0.01	11.26	8.65	7.59	7.01	6.63	6.37	6.18	6.03	5.91	5.81	5.52	5.36	5.26	5.20	5.12	5.03	4.96	4.87
9	0.10	3.36	3.01	2.81	2.69	2.61	2.55	2.51	2.47	2.44	2.42	2.34	2.30	2.27	2.25	2.23	2.21	2.19	2.16
	0.05	5.12	4.26	3.86	3.63	3.48	3.37	3.29	3.23	3.18	3.14	3.01	2.94	2.89	2.86	2.83	2.79	2.76	2.71
	0.025	7.21	5.71	5.08	4.72	4.48	4.32	4.20	4.10	4.03	3.96	3.77	3.67	3.60	3.56	3.51	3.45	3.40	3.34
	0.01	10.56	8.02	6.99	6.42	6.06	5.80	5.61	5.47	5.35	5.26	4.96	4.81	4.71	4.65	4.57	4.48	4.41	4.32
10	0.10	3.29	2.92	2.73	2.61	2.52	2.46	2.41	2.38	2.35	2.32	2.24	2.20	2.17	2.16	2.13	2.11	2.09	2.06
	0.05	4.96	4.10	3.71	3.48	3.33	3.22	3.14	3.07	3.02	2.98	2.85	2.77	2.73	2.70	2.66	2.62	2.59	2.54
	0.025	6.94	5.46	4.83	4.47	4.24	4.07	3.95	3.85	3.78	3.72	3.52	3.42	3.35	3.31	3.26	3.20	3.15	3.09
	0.01	10.04	7.56	6.55	5.99	5.64	5.39	5.20	5.06	4.94	4.85	4.56	4.41	4.31	4.25	4.17	4.08	4.01	3.92
11	0.10	3.23	2.86	2.66	2.54	2.45	2.39	2.34	2.30	2.27	2.25	2.17	2.12	2.10	2.08	2.05	2.03	2.01	1.98
	0.05	4.84	3.98	3.59	3.36	3.20	3.09	3.01	2.95	2.90	2.85	2.72	2.65	2.60	2.57	2.53	2.49	2.46	2.41
	0.025	6.72	5.26	4.63	4.28	4.04	3.88	3.76	3.66	3.59	3.53	3.33	3.23	3.16	3.12	3.06	3.00	2.96	2.89
	0.01	9.65	7.21	6.22	5.67	5.32	5.07	4.89	4.74	4.63	4.54	4.25	4.10	4.01	3.94	3.86	3.78	3.71	3.61

(Continued)

TABLE 4 (Continued)

												Numerator degrees of freedom								
Denominator degrees of freedom	Area in upper tail	1	2	3	4	5	6	7	8	9	10	15	20	25	30	40	60	100	1000
12	0.10	3.18	2.81	2.61	2.48	2.39	2.33	2.28	2.24	2.21	2.19	2.10	2.06	2.03	2.01	1.99	1.96	1.94	1.91
	0.05	4.75	3.89	3.49	3.26	3.11	3.00	2.91	2.85	2.80	2.75	2.62	2.54	2.50	2.47	2.43	2.38	2.35	2.30
	0.025	6.55	5.10	4.47	4.12	3.89	3.73	3.61	3.51	3.44	3.37	3.18	3.07	3.01	2.96	2.91	2.85	2.80	2.73
	0.01	9.33	6.93	5.95	5.41	5.06	4.82	4.64	4.50	4.39	4.30	4.01	3.86	3.76	3.70	3.62	3.54	3.47	3.37
13	0.10	3.14	2.76	2.56	2.43	2.35	2.28	2.23	2.20	2.16	2.14	2.05	2.01	1.98	1.96	1.93	1.90	1.88	1.85
	0.05	4.67	3.81	3.41	3.18	3.03	2.92	2.83	2.77	2.71	2.67	2.53	2.46	2.41	2.38	2.34	2.30	2.26	2.21
	0.025	6.41	4.97	4.35	4.00	3.77	3.60	3.48	3.39	3.31	3.25	3.05	2.95	2.88	2.84	2.78	2.72	2.67	2.60
	0.01	9.07	6.70	5.74	5.21	4.86	4.62	4.44	4.30	4.19	4.10	3.82	3.66	3.57	3.51	3.43	3.34	3.27	3.18
14	0.10	3.10	2.73	2.52	2.39	2.31	2.24	2.19	2.15	2.12	2.10	2.01	1.96	1.93	1.91	1.89	1.86	1.83	1.80
	0.05	4.60	3.74	3.34	3.11	2.96	2.85	2.76	2.70	2.65	2.60	2.46	2.39	2.34	2.31	2.27	2.22	2.19	2.14
	0.025	6.30	4.86	4.24	3.89	3.66	3.50	3.38	3.29	3.21	3.15	2.95	2.84	2.78	2.73	2.67	2.61	2.56	2.50
	0.01	8.86	6.51	5.56	5.04	4.69	4.46	4.28	4.14	4.03	3.94	3.66	3.51	3.41	3.35	3.27	3.18	3.11	3.02
15	0.10	3.07	2.70	2.49	2.36	2.27	2.21	2.16	2.12	2.09	2.06	1.97	1.92	1.89	1.87	1.85	1.82	1.79	1.76
	0.05	4.54	3.68	3.29	3.06	2.90	2.79	2.71	2.64	2.59	2.54	2.40	2.33	2.28	2.25	2.20	2.16	2.12	2.07
	0.025	6.20	4.77	4.15	3.80	3.58	3.41	3.29	3.20	3.12	3.06	2.86	2.76	2.69	2.64	2.59	2.52	2.47	2.40
	0.01	8.68	6.36	5.42	4.89	4.56	4.32	4.14	4.00	3.89	3.80	3.52	3.37	3.28	3.21	3.13	3.05	2.98	2.88
16	0.10	3.05	2.67	2.46	2.33	2.24	2.18	2.13	2.09	2.06	2.03	1.94	1.89	1.86	1.84	1.81	1.78	1.76	1.72
	0.05	4.49	3.63	3.24	3.01	2.85	2.74	2.66	2.59	2.54	2.49	2.35	2.28	2.23	2.19	2.15	2.11	2.07	2.02
	0.025	6.12	4.69	4.08	3.73	3.50	3.34	3.22	3.12	3.05	2.99	2.79	2.68	2.61	2.57	2.51	2.45	2.40	2.32
	0.01	8.53	6.23	5.29	4.77	4.44	4.20	4.03	3.89	3.78	3.69	3.41	3.26	3.16	3.10	3.02	2.93	2.86	2.76
17	0.10	3.03	2.64	2.44	2.31	2.22	2.15	2.10	2.06	2.03	2.00	1.91	1.86	1.83	1.81	1.78	1.75	1.73	1.69
	0.05	4.45	3.59	3.20	2.96	2.81	2.70	2.61	2.55	2.49	2.45	2.31	2.23	2.18	2.15	2.10	2.06	2.02	1.97
	0.025	6.04	4.62	4.01	3.66	3.44	3.28	3.16	3.06	2.98	2.92	2.72	2.62	2.55	2.50	2.44	2.38	2.33	2.26
	0.01	8.40	6.11	5.19	4.67	4.34	4.10	3.93	3.79	3.68	3.59	3.31	3.16	3.07	3.00	2.92	2.83	2.76	2.66
18	0.10	3.01	2.62	2.42	2.29	2.20	2.13	2.08	2.04	2.00	1.98	1.89	1.84	1.80	1.78	1.75	1.72	1.70	1.66
	0.05	4.41	3.55	3.16	2.93	2.77	2.66	2.58	2.51	2.46	2.41	2.27	2.19	2.14	2.11	2.06	2.02	1.98	1.92
	0.025	5.98	4.56	3.95	3.61	3.38	3.22	3.10	3.01	2.93	2.87	2.67	2.56	2.49	2.44	2.38	2.32	2.27	2.20
	0.01	8.29	6.01	5.09	4.58	4.25	4.01	3.84	3.71	3.60	3.51	3.23	3.08	2.98	2.92	2.84	2.75	2.68	2.58
19	0.10	2.99	2.61	2.40	2.27	2.18	2.11	2.06	2.02	1.98	1.96	1.86	1.81	1.78	1.76	1.73	1.70	1.67	1.64
	0.05	4.38	3.52	3.13	2.90	2.74	2.63	2.54	2.48	2.42	2.38	2.23	2.16	2.11	2.07	2.03	1.98	1.94	1.88
	0.025	5.92	4.51	3.90	3.56	3.33	3.17	3.05	2.96	2.88	2.82	2.62	2.51	2.44	2.39	2.33	2.27	2.22	2.14
	0.01	8.18	5.93	5.01	4.50	4.17	3.94	3.77	3.63	3.52	3.43	3.15	3.00	2.91	2.84	2.76	2.67	2.60	2.50

(Continued)

TABLE 4 *(Continued)*

Denominator degrees of freedom	Area in upper tail	Numerator degrees of freedom																	
		1	2	3	4	5	6	7	8	9	10	15	20	25	30	40	60	100	1000
20	0.10	2.97	2.59	2.38	2.25	2.16	2.09	2.04	2.00	1.96	1.94	1.84	1.79	1.76	1.74	1.71	1.68	1.65	1.61
	0.05	4.35	3.49	3.10	2.87	2.71	2.60	2.51	2.45	2.39	2.35	2.20	2.12	2.07	2.04	1.99	1.95	1.91	1.85
	0.025	5.87	4.46	3.86	3.51	3.29	3.13	3.01	2.91	2.84	2.77	2.57	2.46	2.40	2.35	2.29	2.22	2.17	2.09
	0.01	8.10	5.85	4.94	4.43	4.10	3.87	3.70	3.56	3.46	3.37	3.09	2.94	2.84	2.78	2.69	2.61	2.54	2.43
21	0.10	2.96	2.57	2.36	2.23	2.14	2.08	2.02	1.98	1.95	1.92	1.83	1.78	1.74	1.72	1.69	1.66	1.63	1.59
	0.05	4.32	3.47	3.07	2.84	2.68	2.57	2.49	2.42	2.37	2.32	2.18	2.10	2.05	2.01	1.96	1.92	1.88	1.82
	0.025	5.83	4.42	3.82	3.48	3.25	3.09	2.97	2.87	2.80	2.73	2.53	2.42	2.36	2.31	2.25	2.18	2.13	2.05
	0.01	8.02	5.78	4.87	4.37	4.04	3.81	3.64	3.51	3.40	3.31	3.03	2.88	2.79	2.72	2.64	2.55	2.48	2.37
22	0.10	2.95	2.56	2.35	2.22	2.13	2.06	2.01	1.97	1.93	1.90	1.81	1.76	1.73	1.70	1.67	1.64	1.61	1.57
	0.05	4.30	3.44	3.05	2.82	2.66	2.55	2.46	2.40	2.34	2.30	2.15	2.07	2.02	1.98	1.94	1.89	1.85	1.79
	0.025	5.79	4.38	3.78	3.44	3.22	3.05	2.93	2.84	2.76	2.70	2.50	2.39	2.32	2.27	2.21	2.14	2.09	2.01
	0.01	7.95	5.72	4.82	4.31	3.99	3.76	3.59	3.45	3.35	3.26	2.98	2.83	2.73	2.67	2.58	2.50	2.42	2.32
23	0.10	2.94	2.55	2.34	2.21	2.11	2.05	1.99	1.95	1.92	1.89	1.80	1.74	1.71	1.69	1.66	1.62	1.59	1.55
	0.05	4.28	3.42	3.03	2.80	2.64	2.53	2.44	2.37	2.32	2.27	2.13	2.05	2.00	1.96	1.91	1.86	1.82	1.76
	0.025	5.75	4.35	3.75	3.41	3.18	3.02	2.90	2.81	2.73	2.67	2.47	2.36	2.29	2.24	2.18	2.11	2.06	1.98
	0.01	7.88	5.66	4.76	4.26	3.94	3.71	3.54	3.41	3.30	3.21	2.93	2.78	2.69	2.62	2.54	2.45	2.37	2.27
24	0.10	2.93	2.54	2.33	2.19	2.10	2.04	1.98	1.94	1.91	1.88	1.78	1.73	1.70	1.67	1.64	1.61	1.58	1.54
	0.05	4.26	3.40	3.01	2.78	2.62	2.51	2.42	2.36	2.30	2.25	2.11	2.03	1.97	1.94	1.89	1.84	1.80	1.74
	0.025	5.72	4.32	3.72	3.38	3.15	2.99	2.87	2.78	2.70	2.64	2.44	2.33	2.26	2.21	2.15	2.08	2.02	1.94
	0.01	7.82	5.61	4.72	4.22	3.90	3.67	3.50	3.36	3.26	3.17	2.89	2.74	2.64	2.58	2.49	2.40	2.33	2.22
25	0.10	2.92	2.53	2.32	2.18	2.09	2.02	1.97	1.93	1.89	1.87	1.77	1.72	1.68	1.66	1.63	1.59	1.56	1.52
	0.05	4.24	3.39	2.99	2.76	2.60	2.49	2.40	2.34	2.28	2.24	2.09	2.01	1.96	1.92	1.87	1.82	1.78	1.72
	0.025	5.69	4.29	3.69	3.35	3.13	2.97	2.85	2.75	2.68	2.61	2.41	2.30	2.23	2.18	2.12	2.05	2.00	1.91
	0.01	7.77	5.57	4.68	4.18	3.85	3.63	3.46	3.32	3.22	3.13	2.85	2.70	2.60	2.54	2.45	2.36	2.29	2.18
26	0.10	2.91	2.52	2.31	2.17	2.08	2.01	1.96	1.92	1.88	1.86	1.76	1.71	1.67	1.65	1.61	1.58	1.55	1.51
	0.05	4.23	3.37	2.98	2.74	2.59	2.47	2.39	2.32	2.27	2.22	2.07	1.99	1.94	1.90	1.85	1.80	1.76	1.70
	0.025	5.66	4.27	3.67	3.33	3.10	2.94	2.82	2.73	2.65	2.59	2.39	2.28	2.21	2.16	2.09	2.03	1.97	1.89
	0.01	7.72	5.53	4.64	4.14	3.82	3.59	3.42	3.29	3.18	3.09	2.81	2.66	2.57	2.50	2.42	2.33	2.25	2.14

(Continued)

TABLE 4 (Continued)

								Numerator degrees of freedom											
Denominator degrees of freedom	Area in upper tail	1	2	3	4	5	6	7	8	9	10	15	20	25	30	40	60	100	1000
27	0.10	2.90	2.51	2.30	2.17	2.07	2.00	1.95	1.91	1.87	1.85	1.75	1.70	1.66	1.64	1.60	1.57	1.54	1.50
	0.05	4.21	3.35	2.96	2.73	2.57	2.46	2.37	2.31	2.25	2.20	2.06	1.97	1.92	1.88	1.84	1.79	1.74	1.68
	0.025	5.63	4.24	3.65	3.31	3.08	2.92	2.80	2.71	2.63	2.57	2.36	2.25	2.18	2.13	2.07	2.00	1.94	1.86
	0.01	7.68	5.49	4.60	4.11	3.78	3.56	3.39	3.26	3.15	3.06	2.78	2.63	2.54	2.47	2.38	2.29	2.22	2.11
28	0.10	2.89	2.50	2.29	2.16	2.06	2.00	1.94	1.90	1.87	1.84	1.74	1.69	1.65	1.63	1.59	1.56	1.53	1.48
	0.05	4.20	3.34	2.95	2.71	2.56	2.45	2.36	2.29	2.24	2.19	2.04	1.96	1.91	1.87	1.82	1.77	1.73	1.66
	0.025	5.61	4.22	3.63	3.29	3.06	2.90	2.78	2.69	2.61	2.55	2.34	2.23	2.16	2.11	2.05	1.98	1.92	1.84
	0.01	7.64	5.45	4.57	4.07	3.75	3.53	3.36	3.23	3.12	3.03	2.75	2.60	2.51	2.44	2.35	2.26	2.19	2.08
29	0.10	2.89	2.50	2.28	2.15	2.06	1.99	1.93	1.89	1.86	1.83	1.73	1.68	1.64	1.62	1.58	1.55	1.52	1.47
	0.05	4.18	3.33	2.93	2.70	2.55	2.43	2.35	2.28	2.22	2.18	2.03	1.94	1.89	1.85	1.81	1.75	1.71	1.65
	0.025	5.59	4.20	3.61	3.27	3.04	2.88	2.76	2.67	2.59	2.53	2.32	2.21	2.14	2.09	2.03	1.96	1.90	1.82
	0.01	7.60	5.42	4.54	4.04	3.73	3.50	3.33	3.20	3.09	3.00	2.73	2.57	2.48	2.41	2.33	2.23	2.16	2.05
30	0.10	2.88	2.49	2.28	2.14	2.05	1.98	1.93	1.88	1.85	1.82	1.72	1.67	1.63	1.61	1.57	1.54	1.51	1.46
	0.05	4.17	3.32	2.92	2.69	2.53	2.42	2.33	2.27	2.21	2.16	2.01	1.93	1.88	1.84	1.79	1.74	1.70	1.63
	0.025	5.57	4.18	3.59	3.25	3.03	2.87	2.75	2.65	2.57	2.51	2.31	2.20	2.12	2.07	2.01	1.94	1.88	1.80
	0.01	7.56	5.39	4.51	4.02	3.70	3.47	3.30	3.17	3.07	2.98	2.70	2.55	2.45	2.39	2.30	2.21	2.13	2.02
40	0.10	2.84	2.44	2.23	2.09	2.00	1.93	1.87	1.83	1.79	1.76	1.66	1.61	1.57	1.54	1.51	1.47	1.43	1.38
	0.05	4.08	3.23	2.84	2.61	2.45	2.34	2.25	2.18	2.12	2.08	1.92	1.84	1.78	1.74	1.69	1.64	1.59	1.52
	0.025	5.42	4.05	3.46	3.13	2.90	2.74	2.62	2.53	2.45	2.39	2.18	2.07	1.99	1.94	1.88	1.80	1.74	1.65
	0.01	7.31	5.18	4.31	3.83	3.51	3.29	3.12	2.99	2.89	2.80	2.52	2.37	2.27	2.20	2.11	2.02	1.94	1.82
60	0.10	2.79	2.39	2.18	2.04	1.95	1.87	1.82	1.77	1.74	1.71	1.60	1.54	1.50	1.48	1.44	1.40	1.36	1.30
	0.05	4.00	3.15	2.76	2.53	2.37	2.25	2.17	2.10	2.04	1.99	1.84	1.75	1.69	1.65	1.59	1.53	1.48	1.40
	0.025	5.29	3.93	3.34	3.01	2.79	2.63	2.51	2.41	2.33	2.27	2.06	1.94	1.87	1.82	1.74	1.67	1.60	1.49
	0.01	7.08	4.98	4.13	3.65	3.34	3.12	2.95	2.82	2.72	2.63	2.35	2.20	2.10	2.03	1.94	1.84	1.75	1.62
100	0.10	2.76	2.36	2.14	2.00	1.91	1.83	1.78	1.73	1.69	1.66	1.56	1.49	1.45	1.42	1.38	1.34	1.29	1.22
	0.05	3.94	3.09	2.70	2.46	2.31	2.19	2.10	2.03	1.97	1.93	1.77	1.68	1.62	1.57	1.52	1.45	1.39	1.30
	0.025	5.18	3.83	3.25	2.92	2.70	2.54	2.42	2.32	2.24	2.18	1.97	1.85	1.77	1.71	1.64	1.56	1.48	1.36
	0.01	6.90	4.82	3.98	3.51	3.21	2.99	2.82	2.69	2.59	2.50	2.22	2.07	1.97	1.89	1.80	1.69	1.60	1.45
1000	0.10	2.71	2.31	2.09	1.95	1.85	1.78	1.72	1.68	1.64	1.61	1.49	1.43	1.38	1.35	1.30	1.25	1.20	1.08
	0.05	3.85	3.00	2.61	2.38	2.22	2.11	2.02	1.95	1.89	1.84	1.68	1.58	1.52	1.47	1.41	1.33	1.26	1.11
	0.025	5.04	3.70	3.13	2.80	2.58	2.42	2.30	2.20	2.13	2.06	1.85	1.72	1.64	1.58	1.50	1.41	1.32	1.13
	0.01	6.66	4.63	3.80	3.34	3.04	2.82	2.66	2.53	2.43	2.34	2.06	1.90	1.79	1.72	1.61	1.50	1.38	1.16

TABLE 5   Critical values for the Durbin-Watson test for autocorrelation

Entries in the table give the critical values for a one-tailed Durbin-Watson test for autocorrelation. For a two-tailed test, the level of significance is doubled.

Significance points of d_L and d_U: $\alpha = 0.05$
Number of independent variables

n	d_L (1)	d_U (1)	d_L (2)	d_U (2)	d_L (3)	d_U (3)	d_L (4)	d_U (4)	d_L (5)	d_U (5)
15	1.08	1.36	0.95	1.54	0.82	1.75	0.69	1.97	0.56	2.21
16	1.10	1.37	0.98	1.54	0.86	1.73	0.74	1.93	0.62	2.15
17	1.13	1.38	1.02	1.54	0.90	1.71	0.78	1.90	0.67	2.10
18	1.16	1.39	1.05	1.53	0.93	1.69	0.82	1.87	0.71	2.06
19	1.18	1.40	1.08	1.53	0.97	1.68	0.86	1.85	0.75	2.02
20	1.20	1.41	1.10	1.54	1.00	1.68	0.90	1.83	0.79	1.99
21	1.22	1.42	1.13	1.54	1.03	1.67	0.93	1.81	0.83	1.96
22	1.24	1.43	1.15	1.54	1.05	1.66	1.96	1.80	0.86	1.94
23	1.26	1.44	1.17	1.54	1.08	1.66	0.99	1.79	0.90	1.92
24	1.27	1.45	1.19	1.55	1.10	1.66	1.01	1.78	0.93	1.90
25	1.29	1.45	1.21	1.55	1.12	1.66	1.04	1.77	0.95	1.89
26	1.30	1.46	1.22	1.55	1.14	1.65	1.06	1.76	0.98	1.88
27	1.32	1.47	1.24	1.56	1.16	1.65	1.08	1.76	1.01	1.86
28	1.33	1.48	1.26	1.56	1.18	1.65	1.10	1.75	1.03	1.85
29	1.34	1.48	1.27	1.56	1.20	1.65	1.12	1.74	1.05	1.84
30	1.35	1.49	1.28	1.57	1.21	1.65	1.14	1.74	1.07	1.83
31	1.36	1.50	1.30	1.57	1.23	1.65	1.16	1.74	1.09	1.83
32	1.37	1.50	1.31	1.57	1.24	1.65	1.18	1.73	1.11	1.82
33	1.38	1.51	1.32	1.58	1.26	1.65	1.19	1.73	1.13	1.81
34	1.39	1.51	1.33	1.58	1.27	1.65	1.21	1.73	1.15	1.81
35	1.40	1.52	1.34	1.58	1.28	1.65	1.22	1.73	1.16	1.80
36	1.41	1.52	1.35	1.59	1.29	1.65	1.24	1.73	1.18	1.80
37	1.42	1.53	1.36	1.59	1.31	1.66	1.25	1.72	1.19	1.80
38	1.43	1.54	1.37	1.59	1.32	1.66	1.26	1.72	1.21	1.79
39	1.43	1.54	1.38	1.60	1.33	1.66	1.27	1.72	1.22	1.79
40	1.44	1.54	1.39	1.60	1.34	1.66	1.29	1.72	1.23	1.79
45	1.48	1.57	1.43	1.62	1.38	1.67	1.34	1.72	1.29	1.78
50	1.50	1.59	1.46	1.63	1.42	1.67	1.38	1.72	1.34	1.77
55	1.53	1.60	1.49	1.64	1.45	1.68	1.41	1.72	1.38	1.77
60	1.55	1.62	1.51	1.65	1.48	1.69	1.44	1.73	1.41	1.77
65	1.57	1.63	1.54	1.66	1.50	1.70	1.47	1.73	1.44	1.77
70	1.58	1.64	1.55	1.67	1.52	1.70	1.49	1.74	1.46	1.77
75	1.60	1.65	1.57	1.68	1.54	1.71	1.51	1.74	1.49	1.77
80	1.61	1.66	1.59	1.69	1.56	1.72	1.53	1.74	1.51	1.77
85	1.62	1.67	1.60	1.70	1.57	1.72	1.55	1.75	1.52	1.77
90	1.63	1.68	1.61	1.70	1.59	1.73	1.57	1.75	1.54	1.78
95	1.64	1.69	1.62	1.71	1.60	1.73	1.58	1.75	1.56	1.78
100	1.65	1.69	1.63	1.72	1.61	1.74	1.59	1.76	1.57	1.78

(Continued)

TABLE 5   (Continued)

Significance points of d_L and d_U: $\alpha = 0.025$
Number of independent variables

	k	1		2		3		4		5	
n		d_L	d_U	d_L	d_U	d_L	d_U	d_L	d_U	d_L	d_U
15		0.95	1.23	0.83	1.40	0.71	1.61	0.59	1.84	0.48	2.09
16		0.98	1.24	0.86	1.40	0.75	1.59	0.64	1.80	0.53	2.03
17		1.01	1.25	0.90	1.40	0.79	1.58	0.68	1.77	0.57	1.98
18		1.03	1.26	0.93	1.40	0.82	1.56	0.72	1.74	0.62	1.93
19		1.06	1.28	0.96	1.41	0.86	1.55	0.76	1.72	0.66	1.90
20		1.08	1.28	0.99	1.41	0.89	1.55	0.79	1.70	0.70	1.87
21		1.10	1.30	1.01	1.41	0.92	1.54	0.83	1.69	0.73	1.84
22		1.12	1.31	1.04	1.42	0.95	1.54	0.86	1.68	0.77	1.82
23		1.14	1.32	1.06	1.42	0.97	1.54	0.89	1.67	0.80	1.80
24		1.16	1.33	1.08	1.43	1.00	1.54	0.91	1.66	0.83	1.79
25		1.18	1.34	1.10	1.43	1.02	1.54	0.94	1.65	0.86	1.77
26		1.19	1.35	1.12	1.44	1.04	1.54	0.96	1.65	0.88	1.76
27		1.21	1.36	1.13	1.44	1.06	1.54	0.99	1.64	0.91	1.75
28		1.22	1.37	1.15	1.45	1.08	1.54	1.01	1.64	0.93	1.74
29		1.24	1.38	1.17	1.45	1.10	1.54	1.03	1.63	0.96	1.73
30		1.25	1.38	1.18	1.46	1.12	1.54	1.05	1.63	0.98	1.73
31		1.26	1.39	1.20	1.47	1.13	1.55	1.07	1.63	1.00	1.72
32		1.27	1.40	1.21	1.47	1.15	1.55	1.08	1.63	1.02	1.71
33		1.28	1.41	1.22	1.48	1.16	1.55	1.10	1.63	1.04	1.71
34		1.29	1.41	1.24	1.48	1.17	1.55	1.12	1.63	1.06	1.70
35		1.30	1.42	1.25	1.48	1.19	1.55	1.13	1.63	1.07	1.70
36		1.31	1.43	1.26	1.49	1.20	1.56	1.15	1.63	1.09	1.70
37		1.32	1.43	1.27	1.49	1.21	1.56	1.16	1.62	1.10	1.70
38		1.33	1.44	1.28	1.50	1.23	1.56	1.17	1.62	1.12	1.70
39		1.34	1.44	1.29	1.50	1.24	1.56	1.19	1.63	1.13	1.69
40		1.35	1.45	1.30	1.51	1.25	1.57	1.20	1.63	1.15	1.69
45		1.39	1.48	1.34	1.53	1.30	1.58	1.25	1.63	1.21	1.69
50		1.42	1.50	1.38	1.54	1.34	1.59	1.30	1.64	1.26	1.69
55		1.45	1.52	1.41	1.56	1.37	1.60	1.33	1.64	1.30	1.69
60		1.47	1.54	1.44	1.57	1.40	1.61	1.37	1.65	1.33	1.69
65		1.49	1.55	1.46	1.59	1.43	1.62	1.40	1.66	1.36	1.69
70		1.51	1.57	1.48	1.60	1.45	1.63	1.42	1.66	1.39	1.70
75		1.53	1.58	1.50	1.61	1.47	1.64	1.45	1.67	1.42	1.70
80		1.54	1.59	1.52	1.62	1.49	1.65	1.47	1.67	1.44	1.70
85		1.56	1.60	1.53	1.63	1.51	1.65	1.49	1.68	1.46	1.71
90		1.57	1.61	1.55	1.64	1.53	1.66	1.50	1.69	1.48	1.71
95		1.58	1.62	1.56	1.65	1.54	1.67	1.52	1.69	1.50	1.71
100		1.59	1.63	1.57	1.65	1.55	1.67	1.53	1.70	1.51	1.72

TABLE 5   *(Continued)*

Significance points of d_L and d_U: $\alpha = 0.01$
Number of independent variables

n	k 1 d_L	d_U	2 d_L	d_U	3 d_L	d_U	4 d_L	d_U	5 d_L	d_U
15	0.81	1.07	0.70	1.25	0.59	1.46	0.49	1.70	0.39	1.96
16	0.84	1.09	0.74	1.25	0.63	1.44	0.53	1.66	0.44	1.90
17	0.87	1.10	0.77	1.25	0.67	1.43	0.57	1.63	0.48	1.85
18	0.90	1.12	0.80	1.26	0.71	1.42	0.61	1.60	0.52	1.80
19	0.93	1.13	0.83	1.26	0.74	1.41	0.65	1.58	0.56	1.77
20	0.95	1.15	0.86	1.27	0.77	1.41	0.68	1.57	0.60	1.74
21	0.97	1.16	0.89	1.27	0.80	1.41	0.72	1.55	0.63	1.71
22	1.00	1.17	0.91	1.28	0.83	1.40	0.75	1.54	0.66	1.69
23	1.02	1.19	0.94	1.29	0.86	1.40	0.77	1.53	0.70	1.67
24	1.04	1.20	0.96	1.30	0.88	1.41	0.80	1.53	0.72	1.66
25	1.05	1.21	0.98	1.30	0.90	1.41	0.83	1.52	0.75	1.65
26	1.07	1.22	1.00	1.31	0.93	1.41	0.85	1.52	0.78	1.64
27	1.09	1.23	1.02	1.32	0.95	1.41	0.88	1.51	0.81	1.63
28	1.10	1.24	1.04	1.32	0.97	1.41	0.90	1.51	0.83	1.62
29	1.12	1.25	1.05	1.33	0.99	1.42	0.92	1.51	0.85	1.61
30	1.13	1.26	1.07	1.34	1.01	1.42	0.94	1.51	0.88	1.61
31	1.15	1.27	1.08	1.34	1.02	1.42	0.96	1.51	0.90	1.60
32	1.16	1.28	1.10	1.35	1.04	1.43	0.98	1.51	0.92	1.60
33	1.17	1.29	1.11	1.36	1.05	1.43	1.00	1.51	0.94	1.59
34	1.18	1.30	1.13	1.36	1.07	1.43	1.01	1.51	0.95	1.59
35	1.19	1.31	1.14	1.37	1.08	1.44	1.03	1.51	0.97	1.59
36	1.21	1.32	1.15	1.38	1.10	1.44	1.04	1.51	0.99	1.59
37	1.22	1.32	1.16	1.38	1.11	1.45	1.06	1.51	1.00	1.59
38	1.23	1.33	1.18	1.39	1.12	1.45	1.07	1.52	1.02	1.58
39	1.24	1.34	1.19	1.39	1.14	1.45	1.09	1.52	1.03	1.58
40	1.25	1.34	1.20	1.40	1.15	1.46	1.10	1.52	1.05	1.58
45	1.29	1.38	1.24	1.42	1.20	1.48	1.16	1.53	1.11	1.58
50	1.32	1.40	1.28	1.45	1.24	1.49	1.20	1.54	1.16	1.59
55	1.36	1.43	1.32	1.47	1.28	1.51	1.25	1.55	1.21	1.59
60	1.38	1.45	1.35	1.48	1.32	1.52	1.28	1.56	1.25	1.60
65	1.41	1.47	1.38	1.50	1.35	1.53	1.31	1.57	1.28	1.61
70	1.43	1.49	1.40	1.52	1.37	1.55	1.34	1.58	1.31	1.61
75	1.45	1.50	1.42	1.53	1.39	1.56	1.37	1.59	1.34	1.62
80	1.47	1.52	1.44	1.54	1.42	1.57	1.39	1.60	1.36	1.62
85	1.48	1.53	1.46	1.55	1.43	1.58	1.41	1.60	1.39	1.63
90	1.50	1.54	1.47	1.56	1.45	1.59	1.43	1.61	1.41	1.64
95	1.51	1.55	1.49	1.57	1.47	1.60	1.45	1.62	1.42	1.64
100	1.52	1.56	1.50	1.58	1.48	1.60	1.46	1.63	1.44	1.65

This table is reprinted by permission of Oxford University Press on behalf of The Biometrika Trustees from J. Durbin and G. S. Watson, 'Testing for serial correlation in least square regression II', *Biometrika* 38 (1951): 159–178.

TABLE 6 T_L Values for the Mann-Whitney-Wilcoxon test

Reject the hypothesis of identical populations if the sum of the ranks for the n_1 items is *less* than the value T_L shown in the following table or if the sum of the ranks for the n_1 items is *greater* than the value T_U where:

$$T_U = n_1(n_1 + n_2) - T_L$$

$\alpha = 0.10$					n_2				
n_1	2	3	4	5	6	7	8	9	10
2	3	3	3	4	4	4	5	5	5
3	6	7	7	8	9	9	10	11	11
4	10	11	12	13	14	15	16	17	18
5	16	17	18	20	21	22	24	25	27
6	22	24	25	27	29	30	32	34	36
7	29	31	33	35	37	40	42	44	46
8	38	40	42	45	47	50	52	55	57
9	47	50	52	55	58	61	64	67	70
10	57	60	63	67	70	73	76	80	83

$\alpha = 0.05$					n_2				
n_1	2	3	4	5	6	7	8	9	10
2	3	3	3	3	3	3	4	4	4
3	6	6	6	7	8	8	9	9	10
4	10	10	11	12	13	14	15	15	16
5	15	16	17	18	19	21	22	23	24
6	21	23	24	25	27	28	30	32	33
7	28	30	32	34	35	37	39	41	43
8	37	39	41	43	45	47	50	52	54
9	46	48	50	53	56	58	61	63	66
10	56	59	61	64	67	70	73	76	79

Addition law A probability law used to compute the probability of the union of two events. It is $P(A \cup B) = P(A) + P(B) - P(A \cap B)$. For mutually exclusive events, $P(A \cap B) = 0$; in this case the addition law reduces to $P(A \cup B) = P(A) + P(B)$.

Additive decomposition model In an additive model the values for the Trend, Seasonal and Irregular components are simply added together to obtain the actual time series value, Y_t.

Adjusted multiple coefficient of determination A measure of the goodness of fit of the estimated multiple regression equation that adjusts for the number of independent variables in the model and thus avoids overestimating the impact of adding more independent variables.

Alternative hypothesis The hypothesis concluded to be true if the null hypothesis is rejected.

Analytics The scientific process of transforming data into insight for making better decisions.

ANOVA table A table used to summarize the analysis of variance computations and results. It contains columns showing the source of variation, the sum of squares, the degrees of freedom, the mean square and the F value(s).

Autocorrelation Correlation in the errors that arises when the error terms at successive points in time are related.

Bar chart A graphical device for depicting categorical data that have been summarized in a frequency, relative frequency or percentage frequency distribution. Also known as a bar graph.

Bar graph A graphical device for depicting categorical data that have been summarized in a frequency, relative frequency or percentage frequency distribution. Also known as a bar chart.

Basic requirements for assigning probabilities Two requirements that restrict the manner in which probability assignments can be made: (1) for each experimental outcome E_i we must have $0 \le P(E_i) \le 1$; (2) considering all experimental outcomes, we must have $P(E_1) + P(E_2) + \ldots + P(E_n) = 1.0$.

Bayes' theorem A theorem that enables the use of sample information to revise prior probabilities.

Big data A set of data that cannot be managed, processed or analyzed with commonly available software in a reasonable amount of time. Big data are characterized by great volume (a large amount of data), high velocity (fast collection and processing), or wide variety (could include non-traditional data such as video, audio and text).

Binomial experiment An experiment having the four properties stated at the beginning of Chapter 5, Section 5.5.

Binomial probability distribution A probability distribution showing the probability of x successes in n trials of binomial experiments.

Binomial probability function The function used to compute binomial probabilities.

Bivariate probability distribution A probability distribution involving two random variables. A discrete bivariate probability distribution provides a probability for each pair of values that may occur for the two random variables.

Blocking The process of using the same or similar experimental units for all treatments. The purpose of blocking is to remove a source of variation from the error term and hence provide a more powerful test for a difference in population or treatment means.

Box plot A graphical summary of data based on a five-number summary.

Categorical data Non-numeric data which include labels or names used to identify an attribute of each element of a data set.

Categorical independent variables Independent variables based on categorical data.

Categorical variable A variable based on categorical data.

Causal forecasting methods Forecasting methods that relate a time series to other variables that are believed to explain or cause its behaviour.

Census An enumeration or survey to collect data on the entire population.

Central limit theorem A theorem that enables one to use the normal probability distribution to approximate the sampling distribution of $\bar{X}$ when the sample size is large.

Chebyshev's theorem A theorem that can be used to make statements about the proportion of data values that must be within a specified number of standard deviations of the mean.

Class midpoint The value halfway between the lower and upper class limits in a frequency distribution.

Classical method A method of assigning probabilities that is appropriate when all the experimental outcomes are equally likely.

Clustered bar charts Graphical displays showing the joint distribution of two categorical variables. Clusters of bars are placed side by side, with each cluster showing the distribution of variable 2 for one of the categories of variable 1.

Coefficient of determination A measure of the goodness of fit of the estimated regression equation. It can be interpreted as the proportion of the variability in the dependent variable Y that is explained by the estimated regression equation.

Coefficient of variation A measure of relative variability computed by dividing the standard deviation by the mean and multiplying by 100.

Combination In an experiment we may be interested in determining the number of ways n objects may be selected from among N objects without regard to the order in which the n objects are selected. Each selection of n objects is called a combination and the total number of combinations of N objects taken n at a time is CN:

$$^N C_n = \binom{N}{n} = \frac{N!}{n!(N-n)!}$$

Comparisonwise Type I error rate The probability of a Type I error associated with a single pairwise comparison.

Complement of A The event consisting of all sample points that are not in A.

Completely randomized design An experimental design in which the treatments are randomly assigned to the experimental units.

Conditional probability The probability of an event given that another event already occurred. The conditional probability of A given B is $P(A \mid B) = P(A \cap B)/P(B)$.

Confidence coefficient The confidence level expressed as a decimal value. For example, 0.95 is the confidence coefficient for a 95 per cent confidence level.

Confidence interval A confidence interval is an interval estimate of a parameter produced by a procedure that has a defined probability of including the population parameter within the interval. For example, a 95 per cent confidence interval is produced by a procedure that provides intervals such that 95 per cent of the intervals formed using the procedure will include the population parameter.

Confidence level The confidence associated with an interval estimate. For example, if an interval estimation procedure provides intervals such that 95 per cent of the intervals formed using the procedure will include the population parameter, the interval estimate is said to be constructed at the 95 per cent confidence level.

Contingency table A frequency table resulting from the cross-classification of two or more categorical variables.

Continuity correction factor A value of 0.5 that is added to or subtracted from a value of X when the continuous normal distribution is used to approximate the discrete binomial distribution.

Continuous random variable A random variable that may assume any numerical value in an interval or collection of intervals.

Cook's distance measure A measure of the influence of an observation based on both the leverage of observation i and the residual for observation i.

Correlation coefficient A measure of association between two variables. Usually a correlation coefficient takes on values between -1 and $+1$. Values near $+1$ indicate a strong positive relationship, values near -1 indicate a strong negative relationship. Values near zero indicate the lack of a relationship. Pearson's product-moment correlation coefficient measures linear association between two variables.

Covariance A measure of linear association between two variables. Positive values indicate a positive relationship; negative values indicate a negative relationship.

Coverage error A type of non-sampling error which means that, because the research objective and the population from which the sample is to be drawn are not aligned, the data collected will not help accomplish its research objective.

Critical value A value that is compared with the test statistic to determine whether H_0 should be rejected.

Cross-sectional data Data collected at the same or approximately the same point in time.

Cross tabulation A tabular summary of data for two variables. The classes for one variable are represented by the rows; the classes for the other variable are represented by the columns.

Cumulative frequency distribution A tabular summary of quantitative data showing the number of items with values less than or equal to the upper class limit of each class.

Cumulative percentage frequency distribution A tabular summary of quantitative data showing the percentage of items with values less than or equal to the upper class limit of each class.

Cumulative relative frequency distribution A tabular summary of quantitative data showing the fraction or proportion of items with values less than or equal to the upper class limit of each class.

Cyclical pattern One that shows an alternating sequence of points below and above a trend line lasting more than one year.

Data The facts and figures collected, analyzed and summarized for presentation and interpretation.

Data dashboard A set of visual displays that organizes and presents information that is used to monitor the performance of a company or organization in a manner that is easy to read, understand and interpret.

Data mining The process of converting data in a warehouse into useful information using a combination of procedures from statistics, mathematics and computer science.

Data set All the data collected in a particular study.

Data visualization A term used to describe the use of graphical displays to summarize and present information about a data set. The goal of data visualization is to communicate as effectively and clearly as possible, the key information about the data.

Degrees of freedom The term 'degrees of freedom' was introduced in Chapter 8, as a parameter of the t distribution. When the t distribution is used in the computation of an interval estimate of a population mean, the appropriate t distribution has $n - 1$ degrees of freedom, where n is the size of the simple random sample. The number of degrees of freedom is governed by the number of independent terms used in the calculation of the variance estimate used in the interval estimate. (Also a parameter of the χ^2 distribution and the F distribution.)

Dependent variable The variable that is being predicted or explained. It is denoted by Y.

Descriptive analytics The set of analytical techniques that describe what has happened in the past.

Descriptive statistics Tabular, graphical and numerical summaries of data.

Deseasonalized time series A time series from which the effect of season has been removed by dividing each original time series observation by the corresponding seasonal index.

Discrete random variable A random variable that may assume either a finite number of values or an infinite sequence of values.

Discrete uniform probability distribution A probability distribution for which each possible value of the random variable has the same probability.

Distribution-free methods Statistical methods that make no assumption about the distributional form of the population.

Dot plot A graphical device that summarizes data by the number of dots above each data value on the horizontal axis.

Dummy variable A variable used to model the effect of categorical independent variables. A dummy variable may take only the value zero or one.

Durbin-Watson test A test to determine whether first-order correlation is present.

Elements The entity on which data are collected.

Empirical discrete distribution A discrete probability distribution for which the relative frequency method is used to assign the probabilities.

Empirical rule A rule that can be used to approximate the percentage of data values that are within one, two and three standard deviations of the mean for data that exhibit a bell-shaped distribution.

Estimated logistic regression equation The estimate of the logistic regression equation-based on sample data; that is:

$$\hat{y} = \text{estimate of } P(Y = 1 | x_1, x_2, \ldots x_p)$$
$$= \frac{e^{b_0 + b_1 x_1 + b_2 x_2 + \ldots + b_p x_p}}{1 + e^{b_0 + b_1 x_1 + b_2 x_2 + \ldots + b_p x_p}}$$

Estimated logit An estimate of the logit based on sample data; that is:

$$\hat{g}(x_1, x_2, \ldots, x_p) = b_0 + b_1 x_1 + b_2 x_2 + \cdots + b_p x_p$$

Estimated multiple regression equation The estimate of the multiple regression equation based on sample data and the least squares method; it is:

$$\hat{y} = b_0 + b_1 x_1 + b_2 x_2 + \cdots + b_p x_p$$

Estimated regression equation The estimate of the regression equation developed from sample data by using the least squares method. For simple linear regression, the estimated regression equation is:

$$\hat{y} = b_0 + b_1 x$$

Ethical behaviour Characterized by honesty, fairness and equity in interpersonal, professional and academic relationships and in research and scholarly activities. Ethical behaviour respects the dignity, diversity and rights of individuals and groups of people.

Event A collection of sample points.

Expected value A measure of the central location of a random variable. For a chance node, it is the weighted average of the payoffs. The weights are the state-of-nature probabilities.

Experiment A process that generates well-defined outcomes.

Experimental units The objects of interest in the experiment.

Experimentwise Type I error rate The probability of making a Type I error on at least one of several pairwise comparisons.

Exponential probability distribution A continuous probability distribution that is useful in computing probabilities for the time it takes to complete a task.

Exponential smoothing A forecasting technique that uses a weighted average of past time series values as the forecast. It is also referred to as simple exponential smoothing.

Factor Another word for a categorical independent variable.

Factorial experiment An experimental design that allows statistical conclusions about two or more factors.

Finite population correction factor The term $\sqrt{(N-n)/(N-1)}$ that is used in the formulae for $\sigma_{\bar{x}}$ and σ_P when a finite population, rather than an infinite population, is being sampled. The generally accepted rule of thumb is to ignore the finite population correction factor whenever $n/N \leq 0.05$.

Five-number summary An exploratory data analysis technique that uses five numbers to summarize the data: smallest value, first quartile, median, third quartile and largest value.

Forecasts Predictions of future values of a time series.

Forecast error The forecast error is the difference between the actual value of a time series and its forecast.

Four Vs The processes that generate big data: volume, variety, veracity and velocity.

Frequency distribution A tabular summary of data showing the number (frequency) of items in each of several non-overlapping classes.

General linear model A model of the form $Y = \beta_0 + \beta_1 z_1 + \beta_2 z_2 + \cdots + \beta_p z_p + \varepsilon$, where each of the independent variables z_j ($j = 1, 2, \ldots, p$) is a function of $x_1, x_2, \ldots, x_k$, the variables for which data have been collected.

Goodness of fit test A statistical test conducted to determine whether to reject a hypothesized probability distribution for a population.

High leverage points Observations with extreme values for the independent variables.

Histogram A graphical presentation of a frequency distribution, relative frequency distribution or percentage frequency distribution of quantitative data, constructed by placing the class intervals on the horizontal axis and the frequencies, relative frequencies or percentage frequencies on the vertical axis.

Horizontal pattern A horizontal pattern exists when the data fluctuate around a constant mean.

Hypergeometric probability distribution A probability distribution showing the probability of x successes in n trials from a population with r successes and $N - r$ failures.

Hypergeometric probability function The function used to compute hypergeometric probabilities.

Independent events Two events A and B where $P(A | B) = P(A)$ or $P(B | A) = P(B)$; that is, the events have no influence on each other.

Independent samples Where, e.g. two groups of workers are selected and each group uses a different method to collect production time data.

Independent variable The variable that is doing the predicting or explaining. It is denoted by X.

Influential observation An observation that has a strong influence or effect on the regression results.

Interaction The effect of two independent variables acting together.

Interquartile range (IQR) A measure of variability, defined to be the difference between the third and first quartiles.

Intersection of A and B The event containing the sample points belonging to both A and B. The intersection is denoted $A \cap B$.

Interval estimate An estimate of a population parameter that provides an interval believed to contain the value of the parameter.

Interval scale The scale of measurement for a variable if the data demonstrate the properties of ordinal data and the interval between values is expressed in terms of a fixed unit of measure. Interval data are always numeric.

Irregular components Components of the time series that reflect the random variation of the time series values beyond what can be explained by the trend, cyclical and seasonal components.

ith residual The difference between the observed value of the dependent variable and the value predicted using the estimated regression equation; for the ith observation the ith residual is $y_i - \hat{y}_i$.

Joint probabilities The probabilities of two events both occurring; that is, the probability of the intersection of two events.

Kruskal-Wallis test A non-parametric test for identifying differences among three or more populations on the basis of independent samples.

Least squares method The method used to develop the estimated regression equation. It minimizes the sum of squared residuals (the deviations between the observed values of the dependent variable, y, and the estimated values of the dependent variable, $\hat{y}_i$).

Level of significance The probability of making a Type I error when the null hypothesis is true as an equality.

Leverage A measure of how far the values of the independent variables are from their mean values.

Logistic regression equation The mathematical equation relating $E(Y)$, the probability that $Y = 1$, to the values of the independent variables; that is,

$$E(Y) = P(Y = 1 | x_1, x_2, \ldots x_p) = \frac{e^{\beta_0 + \beta_1 x_1 + \ldots + \beta_p x_p}}{1 + e^{\beta_0 + \beta_1 x_1 + \ldots + \beta_p x_p}}$$

Logit The natural logarithm of the odds in favour of $Y = 1$; that is, $g(x_1, x_2, \ldots, x_p) = \beta_0 + \beta_1 x_1 + \beta_2 x_2 + \cdots + \beta_p x_p$

Mann-Whitney-Wilcoxon (MWW) test A non-parametric statistical test for identifying differences between two populations based on the analysis of two independent samples.

Margin of error The value added to and subtracted from a point estimate in order to construct an interval estimate of a population parameter.

Marginal probabilities Values in the margins of a joint probability table that provide the probabilities of each event separately.

Matched samples Where, e.g. only a sample of workers is selected and each worker uses first one and then the other method, with each worker providing a pair of data values.

Mean A measure of central location computed by summing the data values and dividing by the number of observations.

Mean absolute error (MAE) The average of the absolute forecast errors.

Mean absolute percentage error (MAPE) The average of the ratios of absolute forecast errors to actual values expressed as a percentage.

Mean square error (MSE) A measure of the accuracy of a forecasting method. This measure is the average of the sum of the squared differences between the forecast values and the actual time series values.

Measurement error A type of non-sampling error that involves incorrect measurement of the characteristic of interest.

Median A measure of central location provided by the value in the middle when the data are arranged in ascending order.

Mode A measure of location, defined as the value that occurs with greatest frequency.

Moving averages A method of forecasting or smoothing a time series that uses the average of the most recent n data values in the time series as the forecast for the next period.

Multicollinearity The term used to describe the correlation between the independent variables.

Multinomial population A population in which each element is assigned to one and only one of several categories. The multinomial distribution extends the binomial distribution from two to three or more outcomes.

Multiple coefficient of determination A measure of the goodness of fit of the estimated multiple regression equation. It can be interpreted as the proportion of the variability in the dependent variable that is explained by the estimated regression equation.

Multiple comparison procedures Statistical procedures that can be used to conduct statistical comparisons between pairs of population means.

Multiple regression analysis Regression analysis involving two or more independent variables.

Multiple regression equation The mathematical equation relating the expected value or mean value of the dependent variable to the values of the independent variables; that is:

$$E(Y) = \beta_0 + \beta_1 x_1 + \beta_2 x_2 + \cdots + \beta_p x_p$$

Multiple regression model The mathematical equation that describes how the dependent variable Y is related to the independent variables $x_1, x_2, \ldots x_p$ and an error term ε.

Multiplication law A probability law used to compute the probability of the intersection of two events. It is $P(A \cap B) = P(B)P(A \mid B)$ or $P(A \cap B) = P(A)P(B \mid A)$. For independent events it reduces to $P(A \cap B) = P(A)P(B)$.

Multiplicative decomposition model In a multiplicative model the values for the trend, seasonal and irregular components are simply multiplied together to obtain the actual time series value.

Mutually exclusive events Events that have no sample points in common; that is, $A \cap B$ is empty and $P(A \cap B) = 0$.

Nominal scale The scale of measurement for a variable when the data use labels or names to identify an attribute of an element. Nominal data may be non-numeric or numeric.

Non-response error A type of non-sampling error that occurs when segments of the target population are systematically underrepresented or overrepresented in the sample.

Non-parametric methods Statistical methods that require relatively few assumptions about the population probability distributions and the level of measurement. These methods can be applied when nominal or ordinal data are available.

Normal probability distribution A continuous probability distribution. Its probability density function is bell shaped and determined by its mean μ and standard deviation σ.

Normal probability plot A graph of the standardized residuals plotted against values of the normal scores. This plot helps determine whether the assumption that the error term has a normal probability distribution appears to be valid.

Null hypothesis The hypothesis tentatively assumed true in the hypothesis testing procedure.

Observation The set of measurements obtained for a particular element.

Odds in favour of an event occurring The probability the event will occur divided by the probability the event will not occur.

Odds ratio The odds that $Y = 1$ given that one of the independent variables increased by one unit ($odds_1$) divided by the odds that $Y = 1$ given no change in the values for the independent variables ($odds_0$); that is: Odds ratio = $odds_1/odds_0$.

One-tailed test A hypothesis test in which rejection of the null hypothesis occurs for values of the test statistic in one tail of its sampling distribution.

Ordinal scale The scale of measurement for a variable if the data exhibit the properties of nominal data and the order or rank of the data is meaningful. Ordinal data may be non-numeric or numeric.

Outlier A data point or observation that does not fit the pattern shown by the remaining data, often unusually small or unusually large.

p-value A probability computed using the test statistic. It is the probability of getting a value for the test statistic as extreme or more extreme than the value served in the sample, assuming the null hypothesis to be true as an equality. For a lower tail test, the p-value is the probability of obtaining a value for the test statistic at least as small as that provided by the sample. For an upper tail test, the p-value is the probability of obtaining a value for the test statistic at least as large as that provided by the sample. For a two-tailed test, the p-value is the probability of obtaining a value for the test statistic at least as unlikely as that provided by the sample.

Parameters Numerical characteristics of a population, such as a population mean μ, a population standard deviation σ, a population proportion π.

Parametric methods Statistical methods that begin with an assumption about the distributional shape of the population. This is often that the population follows a normal distribution.

Partitioning The process of allocating the total sum of squares and degrees of freedom to the various components.

Percentage frequency distribution A tabular summary of data showing the percentage of items in each of several non-overlapping classes.

Percentile A value such that at least p per cent of the observations are less than or equal to this value and at least $(100 - p)$ per cent of the observations are greater than or equal to this value. The 50th percentile is the median.

Permutation In an experiment we may be interested in determining the number of ways n objects may be selected from among N objects when the order in which the n objects are selected is important. Each ordering of n objects is called a permutation and the total number of permutations of N objects taken n at a time is:

$$^{N}P_n = n!\binom{N}{n} = \frac{N!}{(N - n)!}$$

Pie chart A graphical presentation of categorical data based on the subdivision of a circle into sectors that correspond to the relative frequency for each category.

Point estimate The value of a point estimator used in a particular instance as an estimate of a population parameter.

Point estimator The sample statistic, such as $\overline{X}$, S or P, that provides the point estimate of the population parameter.

Poisson probability distribution A probability distribution showing the probability of x occurrences of an event over a specified interval of time or space.

Poisson probability function The function used to compute Poisson probabilities.

Pooled estimate of π A weighted average of P_1 and P_2.

Population The set of all elements of interest in a particular study.

Population parameters Numerical values used as a summary measure for a population (e.g. the population mean μ, the population variance σ^2 and the population standard deviation σ).

Posterior probabilities Revised probabilities of events based on additional information.

Power The probability of correctly rejecting H_0 when it is false.

Power curve A graph of the probability of rejecting H_0 for all possible values of the population parameter not satisfying the null hypothesis. The power curve provides the probability of correctly rejecting the null hypothesis.

Practical significance The real-world impact that result of statistical inference will have on business decisions.

Prediction interval The interval estimate of an individual value of Y for a given value of X.

Predictive analytics The set of analytical techniques that use models constructed from past data to predict the future or assess the impact of one variable on another.

Prescriptive analytics The set of analytical techniques that yield a best course of action.

Prior probability The probability of the state of nature prior to obtaining sample information.

Probability A numerical measure of the likelihood that an event will occur.

Probability density function A function used to compute probabilities for a continuous random variable. The area under the graph of a probability density function over an interval represents probability.

Probability distribution A description of how the probabilities are distributed over the values of the random variable.

Probability function A function, denoted by $p(x)$, that provides the probability that X assumes a particular value, x for a discrete random variable.

Qualitative data Labels or names used to identify an attribute of each element. Qualitative data use either the nominal or ordinal scale of measurement and may be non-numeric or numeric.

Quantitative data Numerical values that indicate how much or how many of something.

Quantitative variable A variable based on quantitative data.

Quartiles The 25th, 50th and 75th percentiles, referred to as the first quartile, the second quartile (median) and third quartile, respectively. The quartiles can be used to divide a data set into four parts, with each part containing approximately 25 per cent of the data.

Random variable A numerical description of the outcome of an experiment.

Randomized block design An experimental design incorporating blocking.

Range A measure of variability, defined to be the largest value minus the smallest value.

Ratio scale The scale of measurement for a variable if the data demonstrate all the properties of interval data and the ratio of two values is meaningful. Ratio data are always numeric.

Regression equation The equation that describes how the mean or expected value of the dependent variable is related to the independent variable; in simple linear regression, $E(Y) = \beta_0 + \beta_1 x$.

Regression model The equation describing how Y is related to X and an error term; in simple linear regression, the regression model is $Y = \beta_0 + \beta_1 x + \varepsilon$.

Relative frequency distribution A tabular summary of data showing the fraction or proportion of data items in each of several non-overlapping classes.

Relative frequency method A method of assigning probabilities that is appropriate when data are available to estimate the proportion of the time the experimental outcome will occur if the experiment is repeated a large number of times.

Replications The number of times each experimental condition is repeated in an experiment.

Residual analysis The analysis of the residuals used to determine whether the assumptions made about the regression model appear to be valid. Residual analysis is also used to identify outliers and influential observations.

Residual plot Graphical representation of the residuals that can be used to determine whether the assumptions made about the regression model appear to be valid.

Response variable Another term for dependent variable.

Sample A subset of the population.

Sample point An element of the sample space. A sample point represents an experimental outcome.

Sample space The set of all experimental outcomes.

Sample statistics Numerical values used as a summary measure for a sample (e.g. the sample mean $\bar{X}$, the sample variance S^2 and the sample standard deviation S).

Sample survey A survey to collect data on a sample.

Sampled population The population from which the sample is taken.

Sampling distribution A probability distribution consisting of all possible values of a sample statistic.

Sampling error The difference between the value of a sample statistic (such as the sample mean, sample standard deviation or sample proportion) and the value of the corresponding population parameter (population mean, population standard deviation or population proportion) that occurs because a random sample is used to estimate the population parameter.

Sampling frame A list of the sampling units for a study. The sample is drawn by selecting units from the sampling frame.

Sampling with replacement Once an element has been included in the sample, it is returned to the population. A previously selected element can be selected again and therefore may appear in the sample more than once.

Sampling without replacement Once an element has been included in the sample, it is removed from the population and cannot be selected a second time.

Scatter diagram A graphical presentation of the relationship between two quantitative variables. One variable is shown on the horizontal axis and the other variable is shown on the vertical axis.

Seasonal pattern The same repeating pattern in observations over successive periods of time.

Serial correlation Same as autocorrelation.

σ (sigma) known The condition existing when historical data or other information provide a good estimate or value for the population standard deviation prior to taking a sample. The interval estimation procedure uses this known value of σ in computing the margin of error.

σ (sigma) unknown The condition existing when no good basis exists for estimating the population standard deviation prior to taking the sample. The interval estimation procedure uses the sample standard deviation s in computing the margin of error.

Sign test A non-parametric statistical test for identifying differences between two populations based on the analysis of nominal data.

Simple linear regression Regression analysis involving one independent variable and one dependent variable in which the relationship between the variables is approximated by a straight line.

Simple random sampling Finite population: a sample selected such that each possible sample of size n has the same probability of being selected. Infinite population: a sample selected such that each element comes from the same population and the elements are selected independently.

Simpson's paradox Conclusions drawn from two or more separate cross-tabulations that can be reversed when the data are aggregated into a single cross-tabulation.

Single-factor experiment An experiment involving only one factor with k populations or treatments.

Skewness A measure of the shape of a data distribution. Data skewed to the left result in negative skewness; a symmetrical data distribution results in zero skewness; and data skewed to the right result in positive skewness.

Smoothing constant A parameter of the exponential smoothing model that provides the weight given to the most recent time series value in the calculation of the forecast value.

Spearman rank-correlation coefficient A correlation measure based on rank-ordered data for two variables.

Stacked bar chart A graphical display showing the joint distribution of two qualitative (categorical) variables. Each bar in a bar chart for variable 1 is divided into segments to show the distribution of variable 2 for that category of variable 1.

Standard deviation A measure of variability computed by taking the positive square root of the variance.

Standard error The standard deviation of a point estimator.

Standard error of the estimate The square root of the mean square error, denoted by s. It is the estimate of σ, the standard deviation of the error term ε.

Standard normal probability distribution A normal distribution with a mean of zero and a standard deviation of one.

Standardized residual The value obtained by dividing a residual by its standard deviation.

Stationary time series One whose statistical properties are independent of time.

Statistical inference The process of using data obtained from a sample to make estimates or test hypotheses about the characteristics of a population.

Statistics The art and science of collecting, analyzing, presenting and interpreting data.

Stem-and-leaf display An exploratory data analysis technique that simultaneously rank orders quantitative data and provides insight about the shape of the distribution.

Studentized deleted residuals Standardized residuals that are based on a revised standard error of the estimate obtained by deleting observation i from the data set and then performing the regression analysis and computations.

Subjective method A method of assigning probabilities on the basis of judgement.

t distribution A family of probability distributions that can be used to develop an interval estimate of a population mean when the population standard deviation σ is unknown and is estimated by the sample standard deviation s.

Tall data A big data set that has so many observations that traditional statistical inference has little meaning.

Target population The population about which inferences are made.

Test statistic A statistic whose value helps determine whether a null hypothesis can be rejected.

Time series A set of observations on a variable measured at successive points in time or over successive periods of time.

Time series data Data collected over several time periods.

Time series decomposition This technique can be used to separate or decompose a time series into seasonal, trend and irregular components.

Time series plot A graphical presentation of the relationship between time and the time series variable; time is on the horizontal axis and the time series values are shown on the vertical axis.

Treatment Different levels of a factor.

Tree diagram A graphical representation that helps in visualizing a multiple-step experiment.

Trend line A line that provides an approximation of the relationship between two variables.

Trend pattern Gradual shifts or movements to relatively higher or lower values over a longer period of time.

Two-tailed test A hypothesis test in which rejection of the null hypothesis occurs for values of the test statistic in either tail of its sampling distribution.

Type I error The error of rejecting H_0 when it is true.

Type II error The error of not rejecting H_0 when it is false.

Unbiasedness A property of a point estimator that is present when the expected value of the point estimator is equal to the population parameter it estimates.

Uniform probability distribution A continuous probability distribution for which the probability that the random variable will assume a value in any interval is the same for each interval of equal length.

Union of A and B The event containing all sample points belonging to A or B or both. The union is denoted $A \cup B$.

Variable A characteristic of interest for the elements.

Variance A measure of variability based on the squared deviations of the data values about the mean.

Variance inflation factor (VIF) A measure of how correlated an independent variable is with all other independent predictors in a multiple regression model.

Venn diagram A graphical representation for showing symbolically the sample space and operations involving events in which the sample space is represented by a rectangle and events are represented as circles within the sample space.

Weighted mean The mean obtained by assigning each observation a weight that reflects its importance.

Weighted moving averages A method of forecasting or smoothing a time series by computing a weighted average of past data values. The sum of the weights must equal one.

Wide data A big data set that has so many variables that simultaneous consideration of all variables is infeasible.

Wilcoxon signed-rank test A non-parametric statistical test for identifying differences between two populations based on the analysis of two matched or paired samples.

z-score A value computed by dividing the deviation about the mean $(x_i - \bar{x})$ by the standard deviation s. A z-score is referred to as a standardized value and denotes the number of standard deviations x_i is from the mean.

INDEX